an alphabetical guide to motion picture, television and videotape production

other reference works of interest

Barton HANDBOOK OF ADVERTISING MANAGEMENT

Buell HANDBOOK OF MODERN MARKETING

Dichter HANDBOOK OF CONSUMER MOTIVATIONS

Focal Press THE FOCAL ENCYCLOPEDIA OF PHOTOGRAPHY

Markus ELECTRONICS AND NUCLEONICS DICTIONARY

Maynard HANDBOOK OF BUSINESS ADMINISTRATION

Melcher and Larrick PRINTING AND PROMOTION HANDBOOK

Quick ARTISTS' AND ILLUSTRATORS' ENCYCLOPEDIA

Stephenson HANDBOOK OF PUBLIC RELATIONS

Stevenson GRAPHIC ARTS ENCYCLOPEDIA

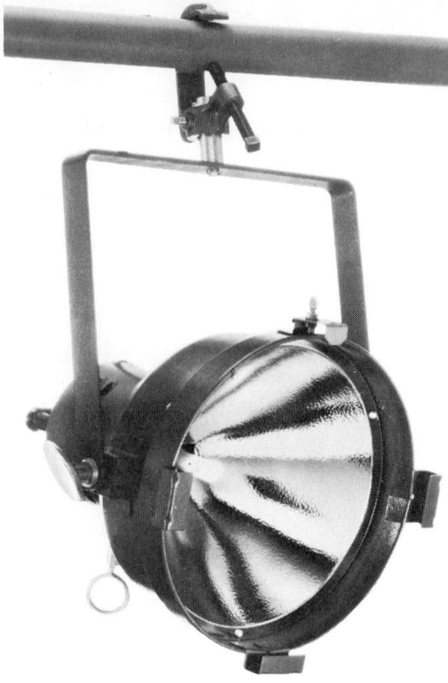

A O

Eli L. Levitan

Society of Motion Picture and Television Engineers
Society of Photographic Scientists and Engineers
Academy of Television Arts and Sciences
National Visual Presentation Association
International Alliance of Theatrical Stage Employees
Florida Education Commission; Miami-Dade Junior College

an alphabetical guide to motion picture, television and videotape production

new york st. louis McGRAW-HILL BOOK COMPANY
san francisco düsseldorf johannesburg kuala lumpur
london mexico montreal new delhi panama paris
são paulo singapore sydney tokyo toronto

To those talented and dedicated people with whom I have had the pleasure of working, or whose paths have crossed mine during my 36 years in the industry, my sincere thanks.

an alphabetical guide to motion picture, television, and videotape production

Library of Congress Catalog Card Number 69-13610

37384

4567890 VHVH 7865432109

SPONSORING EDITOR Daniel N. Fischel
DIRECTOR OF PRODUCTION Stephen J. Boldish
EDITING SUPERVISORS D. A. Douglas/Winifred Eisler
DESIGNER Naomi Auerbach
EDITING AND PRODUCTION STAFF Gretlyn Blau, Carol Ann Ferrari, Teresa F. Leaden, George E. Oechsner

editorial advisory board

PROCESSING	**Burton 'Bud' Stone** Mecca Film Industries, Inc. *Society of Motion Picture and Television Engineers* *Association of Cinema Laboratories*
PRODUCTION	**Gerald Hirschfeld,** MPO Videotronics *Society of Motion Picture and Television Engineers* *American Society of Cinematographers* *International Photographers of the Motion Picture Industry*
SOUND AND RECORDING	**Chet Stewart** Reeves Sound Studios, Inc.
SPECIAL EFFECTS	**Maurice Levy** Em-Ell Production Service, Inc. *Fellow, Society of Motion Picture and Television Engineers* *Academy of Television Arts and Sciences*
VIDEOTAPE	**George K. Gould** President, Teletronics International, Inc.
TELEVISION	*Our special thanks to the engineers and technicians of the Radio Corporation of America and the National Broadcasting Company for their cooperation and the graphic material they provided.*

preface

How is the invisible man photographed? And how are the various props in that same motion picture series made to move without visible help? How are artificial fog, mist, snow, and rain effects produced? And cobwebs or dust? How are night scenes filmed in broad daylight?

On the commercial side, the provocative effects are intended to be less subtle. How, for example, do you pour a cup and a half of coffee? Who taught margarine to twist? And cigarettes to march?

How can a driver get from one place to another without a car? How can a car get from one place to another without a driver? How is a car driven over water? And, how do they get the man into the driver's seat, and what was he doing up in the air in the first place?

The answers to these questions are to be found in the carefully guarded books in which most motion picture technicians make periodic entries and notations. The accumulated information covering every phase of production, in time, becomes a personal and permanent record of processes and techniques, film types and exposure tables, editing notes and optical data. Some of these jottings are referred to constantly. Others, by comparison, are entered and forgotten. Many notations are made possible only after long periods of painstaking experimentation and research.

The editors and author, after much discussion, agreed that the reader would best be served by arranging this information in alphabetical order. This type of arrangement, it was felt, would enable a researcher to locate information pertaining to a specific subject quickly.

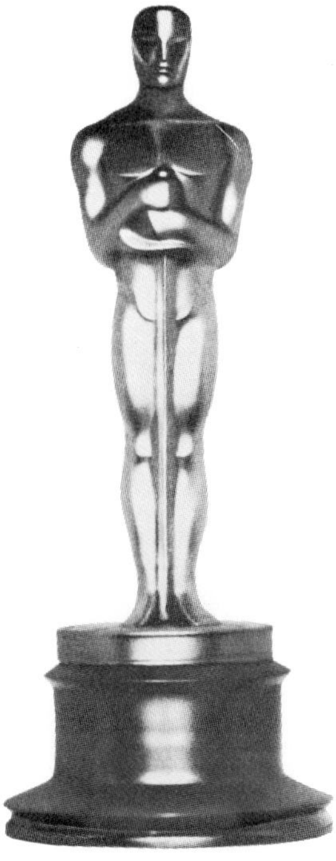

The coveted Oscar, an award presented annually by the Academy of Motion Picture Arts and Sciences to deserving performers and technicians on both sides of the camera.

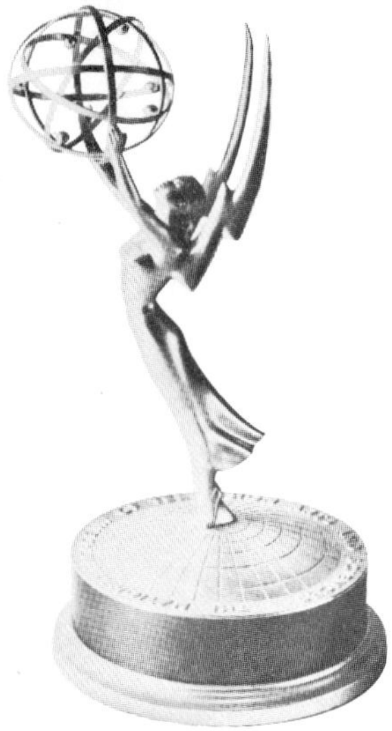

The Emmy, presented by the Academy of Television Arts and Sciences, to those whose contributions merit similar recognition.

As a ready reference, the information and the subject matter in the pages that follow provide the professional, as well as the amateur, with an immediate answer to almost any question on the subject of motion picture, television, and videotape production and the equipment used in each of these mediums. With technical terminology held to a minimum, the descriptions and explanations of processes and techniques should also serve as a guide for the achievement of known effects as well as for the creation of new ones.

Similarly, the text and illustrative material can be expected to influence the future efforts of commercial film producers and advertising agency creative groups. Used as a guide, the material within these covers should make their search for and subsequent achievement of new and eye-arresting effects a great deal easier. Generally, the description of the more frequently used processes and techniques should provide the interested reader with an informative behind-the-scenes look at motion picture, television, and videotape production.

The descriptions and explanations of the processes and techniques in the pages that follow place the emphasis on the essential information needed for their practical application rather than on the superficial and more technical aspects of the processes and techniques themselves. Tables and other statistical information, where they are deemed necessary, accompany the text and eliminate the need for additional research. In many instances, background information relative to some aspect of production or the development of a technique is included in order to provide a more complete picture of that specific area of production.

The film clips and photographs that are used to supplement the text illustrate the practical applications of the various processes and techniques and are taken from theatrical as well as commercial film and videotape productions. They serve as graphic examples that illustrate the imagination and creative effort that today, as in the future, remain an important part of all motion picture, television, and videotape production.

Eli L. Levitan

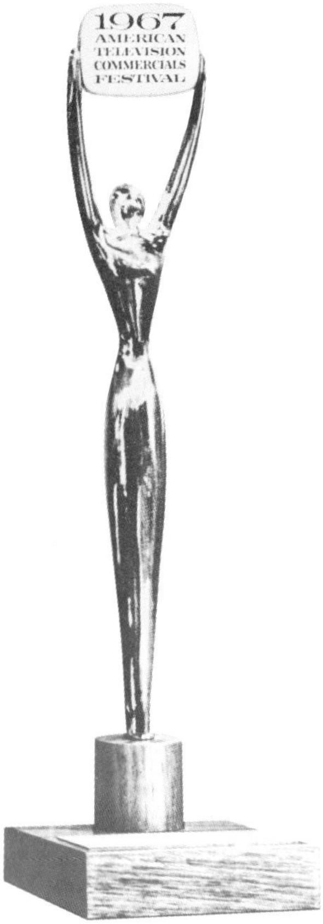

The Clio, awarded by the Directors of the American Television Commercials Festival to those creative personnel whose efforts follow the words ". . . and now, an important message. . . ."

introduction

The word photography is formed from a combination of two Greek words, meaning "light" and "write." It is defined as the process or art of producing images on sensitized surfaces by the chemical action of light or of other forms of radiant energy.

A camera is the photographic apparatus in which sensitive plates or film are exposed, the image being formed by means of a lens. The lens, in turn, is a transparent substance, usually glass, that makes possible the convergence of light rays and their subsequent exposure on the sensitized surface of the film in the camera.

Chronologically, the principles of the camera had been described, rather indefinitely, by a gentleman named Alhazen as early as 1100 A.D. Roger Bacon, about a century and a half later, was a little more definite in his description. A short time after Leonardo da Vinci's death in 1519, the first detailed explanations and drawings of a camera were discovered in one of his many unpublished manuscripts. Approximately thirty-five years later, another Italian, Giovanni Battista della Porta, built the first crude working model. During the next three hundred years, many other models were designed, developed and, in time, forgotten. In 1834, Fox Talbot, an Englishman, began a long series of experiments and developed the Calotype process, the first practical method for obtaining a picture and rendering it permanent.

A scientific study analyzing the movements of people and objects was made by Peter Mark Roget of England in 1824. His article, "The Persistence of Vision with Regard to Moving Objects," was directly responsible for the development of the Zoetrope and other devices for viewing a series of pictures. It also helped disprove the theory that the hand was, in fact, quicker than the eye.

The eye, that amazing organ that makes visual perception possible, needs but a mere quarter of a second to record an image and transmit it to the brain for registration and interpretation. This information led to the reasoning that if two images were recorded within the same quarter of a second, they would be superimposed, thereby creating the illusion of motion. The inevitable search for a method or process that could make "still" pictures move followed. Scores of patents were issued for cameras, projection devices, and other mechanisms designed for that specific purpose.

In early viewing devices, two disks were mounted on a common rotating shaft. The pictures or drawings, arranged in sequence, were mounted on one disk. The viewer looked through slots in the second disk. By 1853, the magic lantern was added to the disk thereby making it possible to project the pictures upon a screen or any other available flat surface.

The combination of picture and light source influenced the development of other projection devices such as the Phenakistoscope, Daedaleum, Phasmatrope, Pantoptikon, Eidoloscope, Vitascope, Theatrograph, Cinematograph, and Kinetoscope. The Kinetograph, a camera that weighed almost a ton, was one of Thomas Edison's early contributions and complemented the Kinetoscope, a peep show arrangement that consisted of a magnifying lens and a light source. Between those two items, lengths of film ran continuously. The Cinematograph that preceeded Edison's Kinetoscope was a rather ingenious and versatile device that served as a camera, a film printing machine and, with a few minor adjustments, as a projector.

Edison had been able to solve all photographic and projection problems as they arose. In order to demonstrate the effectiveness of his new product, motion pictures, Edison also discovered that he would first have to produce them. The Wizard of Menlo Park accepted the challenge reluctantly. A nearby barn, with the roof and doors removed, became the first semipermanent motion picture studio.

Oddly, Edison's experience as a producer led him, at one time, to make the rather pessimistic observation—"exhibitions to large audiences would too rapidly exhaust the novelty of the pictures." Fortunately the inventor's pessimism was not shared by other film producers.

An Edison cameraman, Edwin S. Porter, introduced the "story" to the motion picture industry in 1903 with "The Life of an American Fireman." A more ambitious production followed. The first full reeler, "The Great Train Robbery," was shown at a nickelodeon in Pittsburgh in November of 1905.

Some film producers, during the industry's formative stages, were interested in the motion picture as a means of expression and as another form of artistic endeavor. The more practical among the pioneers were completely optimistic. The motion picture, these enthusiasts felt, would become *the* means of mass communication, hurdle all language barriers, and, in time, provide entertainment for millions. The more curious contented themselves with researching. All, however, were interested in winning audience acceptance for their collective product.

The search for new and better filming techniques began early and gathered momentum quickly. One of the searchers was George Melies, a Frenchman with insatiable curiosity—and a camera. Melies was also a magician, a fact that undoubtedly influenced his efforts as a cinematographer and, in turn, the future of the entire industry.

Fades, dissolves, double-exposure effects, and matte shots were used by him for the first time. The year? 1896. And, although the quality of the print could be described as amateurish, his film presentations were completely professional, even when measured by today's standards. The fact that Melies was a magician did not (fortunately) set the pattern or dictate the requirements for future motion picture technicians and production personnel.

Oddly enough, before 1927, the motion picture industry was self-sufficient and could function without the aid of electrical science. Its equipment and processes were either mechanical, optical, or chemical. In 1927, however, the industry found its voice, made itself heard and became dependent, for the first time, upon electronics. No one, at that time, could have predicted that the voice, or *sound track,* would some day be used to extol the virtues of a headache remedy or urge listeners to "leave the driving to us." This development, however, was only a part of the continuing search for the new technique and unusual illusory effect that paralleled the growth of the motion picture industry.

Today, as an art form or means of expression, the motion picture enjoys a rather secure and enviable position in the world of entertainment. Similarly, the educational, scientific, documentary, industrial, or commercial film presentation, each designed to fill a specific need in its own particular field, is highly regarded as educator, goodwill ambassador, and supersalesman. In reaching this lofty status, producers, directors, and technicians have used every available motion picture process and technique to create visual effects that capture and hold the interest of the audience.

With that goal foremost in mind, the mere telling of a story on film has been just about abandoned by professional filmmakers. Even scientific films, whose technical subject matter is usually interesting in itself and needs little if any dramatizing, merit the same amount of consideration as does the high-budget production designed specifically for entertainment purposes.

In the hands or under the guidance of a creative and imaginative director or film editor, the uneventful and commonplace scene, as well as the simple demonstration staged for commercial purposes, is transformed into a dramatic, eye-arresting film offering.

Normal actions are either speeded up, slowed down, or reversed while specially prepared musical scores and mood lighting help ease the viewer into a desired mental attitude. Or visual interest is heightened by the introduction of exciting illusory effects through high-speed and time-lapse photography, stop-motion effects, or a combination of techniques, including animation. If necessary, day can literally be turned into night while the unbelievable is given an air of authenticity. All these processes and techniques, collectively, have helped put all types of film productions into the entertainment category.

The mysterious driving force that prevents individuals from sitting back contentedly and resting on previously gained laurels is, perhaps, more in evidence in the motion picture industry than in almost any other field of endeavor. Personal recognition, public acceptance and the general awareness that the originator of a new and different technique could only hope to keep production processes secret for a relatively short period of time are the major motivating forces behind the continuing scramble for the novel and unusual effect.

The fact that the industry hires a hall each year for the purpose of honoring its deserving personnel is incentive enough for the most sophisticated. The Oscars, Emmys, and Clios awarded at these affairs, for group as well as for individual effort and achievement on both sides of the camera, supply the necessary initiative for future endeavor. Congratulatory messages fade, and the gold statuettes lose their luster quickly. Uneasy lies the head that wears the crown. The search for the new technique and unusual illusory effect goes on—and on.

acknowledgments

If one picture is, in fact, worth a thousand words, the nice people listed below have certainly helped make our task easier, the book thinner, and the editors happier. We are extremely grateful for the photographs and illustrative material made available by them.

Agfa-Gevaert, Inc.
American Oil Co.
American Tobacco Co.
Ampex Corp.
Animation Equipment Corp.
Arco Productions
Arriflex Corp. of America
Bach Auricon, Inc.
P. Ballantine & Sons
Bausch & Lomb, Inc.
Beckman & Whitley
Behrends, Inc.
Bell & Howell Co.
Benton & Bowles, Inc.
Birns & Sawyer Cine Equipment Co.
Brillo Manufacturing Co.
D. P. Brothers & Co.
Camera Mart
Campbell-Ewald Co.
Canon Cameras
Capital Film Laboratories, Inc.
Cinema Beaulieu
Clairol
Color Service Co. Inc.
ColorTran Industries, Inc.
Compton Advertising Co.
Creative Opticals, Inc.
Dancer, Fitzgerald, Sample, Inc.
Andre Debrie Manufacturing Co.
Walt Disney Productions
Doiflex Cameras
Doyle, Dane, Bernbach, Inc.
DuKane Corp.
E. I. duPont de Nemours & Co. Inc.
Eastman Kodak Co.
Eclair Corp. of America
Wm. Esty Co.
Max Factor & Co.
Fairchild Cameras,
Max Fleischer Studios
Flight Research, Inc.
Foote, Cone & Belding, Adv't'g.
Gamma Scientific, Inc.
General Aniline & Film Corp.
General Electric Corp.
General Film Laboratories
General Foods Corp.

General Motors
Giannini Scientific Corp.
Gordon Enterprises
Green Giant Co.
Grey Advertising Co.
Harrison & Harrison
Karl Heitz, Inc.
Ilford, Ltd.
Kimberly-Clark
Lipsner-Smith Corp.
Macbeth Instrument Corp.
Magnasync/Moviola Corp.
J. M. Mathes, Advertising
J. A. Maurer, Inc.
McCann-Erickson, Inc.
Metro/Kalvar, Inc.
Minnesota Mining & Manufacturing Co.
Mitchell Camera Corp.
Mole-Richardson Co.
Movielab, Inc.
MPO Videotronics, Inc.
Nagra Magnetic Recorders, Inc.
National Broadcasting Co., Inc.
Needham, Louis & Brorby, Inc.
Nikon, Inc.
Norman, Craig & Kummel, Inc.
Ogilvy, Benson & Mather Adv't'g.
Paillard Corp.
Paramount Pictures Corp.
Photo Research Corp.
Proctor & Gamble
Producers Sales Corp.
Radio Corp. of America
Reeves Sound Studios, Inc.
Reinhold Publishing Corp.
Research Products, Inc.
Ryder Magnetic Sales Corp.
S.O.S. Photo-Cine-Optics, Inc.
Sylvania Electric Products, Inc.
Technicolor Corp.
Todd-AO Corp.
J. Walter Thompson Co.
Toho Co., Ltd.
Twentieth Century-Fox
Underwater Sports, Inc.
United Artists
Videotape Productions of N. Y.
Wall Cameras
Westinghouse Electric Corp.
Weston Instruments, Inc.
Xerox Corp.

Our special thanks to Gordon Enterprises for allowing us to include descriptions of some of the motion picture and sound equipment listed in their all-inclusive catalog.

contents

A

A-B composite system (videotape) Also called **linelock** and **intersync**. An electronic process that is used for combining the picture information recorded on two different reels of tape in order to produce a composite effect on a third reel.

In principle, the process is similar to the checkerboard method of preparing A & B rolls used by a film editor. Odd-numbered scenes are spliced together on one roll, with corresponding lengths of blank leader between them that match the length of the even-numbered scenes on the B roll. During the mixing process, the two reels are fed through a switching console and blended electronically, each scene in turn, to produce the final composite tape. See Figure A-1 and A & B ROLLS.

A & B editing—film A method of editing and subsequently printing the picture information from two separate rolls of 16mm film to produce a composite print.

Arranged and spliced together by the film editor in checkerboard fashion, the picture information in each roll, the A and the B, is reexposed by the laboratory on raw stock, thus producing a composite on one length of film.

Assuming that each scene in a motion picture is numbered consecutively, all the odd-numbered scenes are spliced into roll A and the even-numbered scenes into roll B. Each scene on the roll is separated from the next scene on the same roll by a length of black, opaque leader film equal to the number of frames in the scenes on the B

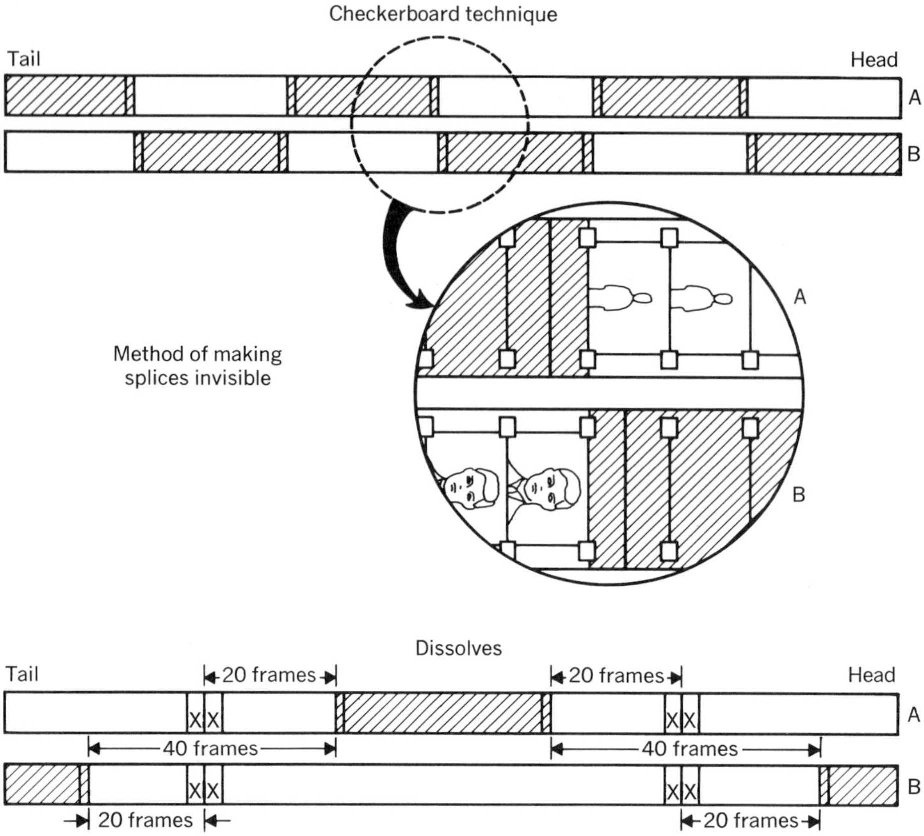

Fig. A-1 Standards for preparation of original 16mm A & B rolls. (*Color Service Co., Inc.*)

Invisible splices: In order to make 16mm splices invisible, it is recommended that the A & B rolls be prepared using the "checkerboard" technique. This technique is accomplished by alternating the scenes back and forth on the A & B rolls. The splices should be made in a splicer that makes one of the cuts on the frame line and the other cut in the picture area. When splicing the scene to black leader, the scene should be placed in the splicer in such a manner that the scraped portion and visible cut of the splice occur only in the black leader. After splicing the head of the scene in this manner, you will note that in order to correctly make the splice at the tail of the scene, it will be necessary to turn the scene around in the splicer.

Double- or single-perforated leaders: It is recommended that only double-perforated leader be used when all the original is double-perforated. When some or all of the original is single-perforated, the use of single-perforated black leader is recommended to prevent possible future damage to the original. The perforations in the black leader should, of course, be on the same edge as the perforations in the original.

roll. During the laboratory printing process, scene 1 on the A roll is printed first. Scene 2, on the B roll, is exposed in turn on the raw stock used for the purpose.

Transitional effects such as fades and cross-dissolves can be added during the printing process by the laboratory. The standard method used by the film editor to indicate effects, generally, is shown in Figure A-2.

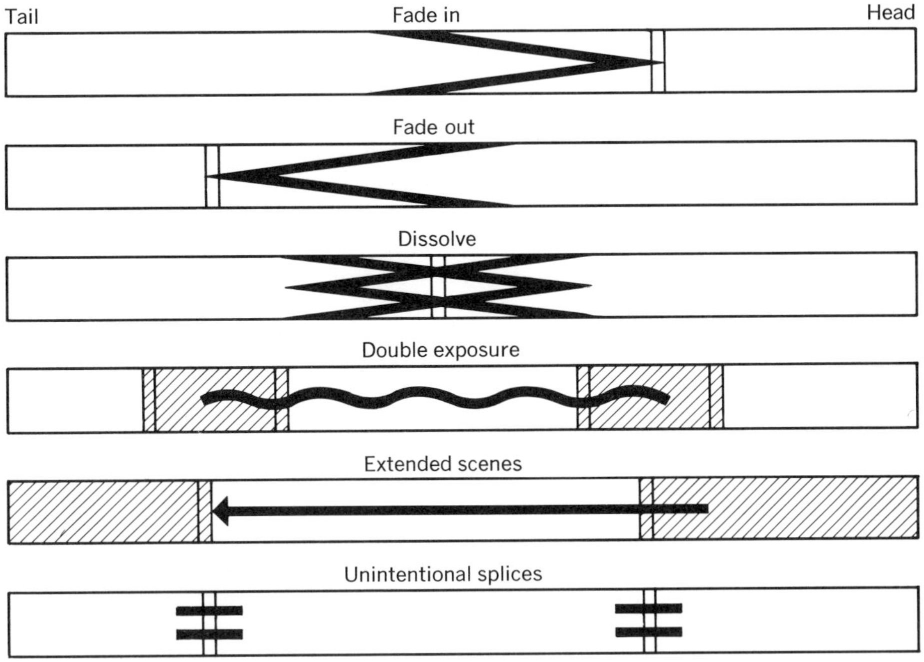

Fig. A-2 Method of marking work prints to indicate effects. (*Color Service Co., Inc.*)

Recommendations of the Committee on Preprint Preparation of the Association of Cinema Laboratories, Inc.:

Fade-in: A fade-in should be marked by two straight lines forming a long V. These two lines run diagonally for the length of the effect from the edges of the film and converge at the center of the frame at the scene change.

Fade-out: The fade-out should be the reverse of the fade-in, the lines extending for the length of the effect and meeting at the center of the frame at the scene change to indicate where the screen will be totally dark.

Dissolve: A dissolve is in actual practice a fade-in of one scene superimposed over a fade-out of another scene. To indicate a dissolve on the work print, the markings should be the same as those used for a fade-in and a fade-out, but they should in this case overlap each other. The midpoint of both markings should occur at the splice in the work print, and the markings should extend an equal distance into both scenes.

Double exposures: Double exposures, superimposed titles, etc., should be indicated by cutting into the background scene of the work print a few inches of edge-numbered work print of the new superimposed scene, to indicate where the double exposure begins. Several inches of the superimposed scene should then be cut in to indicate the end of the double exposure. These two pieces should be connected together by a wavy line.

To ensure proper matching, it is essential that one of these small pieces of work print be edge-numbered or identified in some way.

Extended scenes: Scenes to be extended in matching that are indicated in the work print by leader should be marked with a long arrow, the shaft of which is drawn through the leader, and the head of which comes up to the last frame of the leader to indicate the length of the extension.

Unintentional splices: Splices occurring in the work print that are to be disregarded in matching should be indicated by two short parallel straight lines drawn through the splice and at right angles to it. This would mean that the scene is not to be cut at that point.

Note: It is suggested that markings on the work print be made with a yellow grease pencil. This color is recommended because it shows up well on both color and black and white. If any cues are made on the work print for recording or mixing, it is suggested that a red grease pencil be used.

In some instances, titles are spliced together on C & D rolls and combined with the picture information during the printing process to produce the composite effect. See EDITING.

A & B editing—splicing leaders The film editor, when splicing leaders to the A & B rolls, as well as to the sound-track roll, should first place the matched rolls emulsion up in the synchronizer with the accompanying sound track in editorial sync. The next step is to turn the rolls in the synchronizer to a point a few frames in advance of the first picture or sound, whichever comes first. Marks are made on corresponding frames on all rolls at this point, and the head leaders are spliced at the indicated marks. The head leader identified as "A roll" should be spliced onto the roll that has the first frame of picture. A few frames beyond the last picture or sound, whichever of the two occurs later, marks are made on all the rolls, and the tail leaders are spliced at that point. Preparing standard leaders in advance and then splicing them to the rolls when they are aligned in editorial synchronism practically eliminates the possibility of error, either by the editor or by the laboratory at the time of printing.

If the film is to be used on television, or if the client wishes to have leaders with cue numbers, then it is recommended that the SMPTE Society leader be used. The Society leader would then, of course, precede the picture and the two feet of black leader would not be necessary. See Figure A-3.

A & B editing—videotape The success of A & B roll editing in videotape production depends on the skill of the editor preparing the rolls. His sense of timing, accuracy, and awareness of the final processing procedures all contribute to a successful production.

A & B rolls for videotape production are prepared in the same way as for motion picture production. And, as in motion picture production, the objective of A & B roll editing is to assemble the scenes in the most accurate manner possible, faithfully carrying out the creative concept. To accomplish this, transitional and special effects should be carefully planned and the timing must produce a "pace" or sense of "flow."

After the editing process has been completed, the A & B rolls are set up for the video mix. At that time, the transitional effects are added between scenes, and special effects and titles are inserted where they are needed.

When preparing A & B roll material for transfer to videotape, the editor should use 35mm footage computations for timing. Although standard SMPTE (Society of Motion Picture and Television Engineers) leader is preferable, a 12-foot start mark may be substituted for synchronization purposes.

Overlap footage should be included after the last frame of an

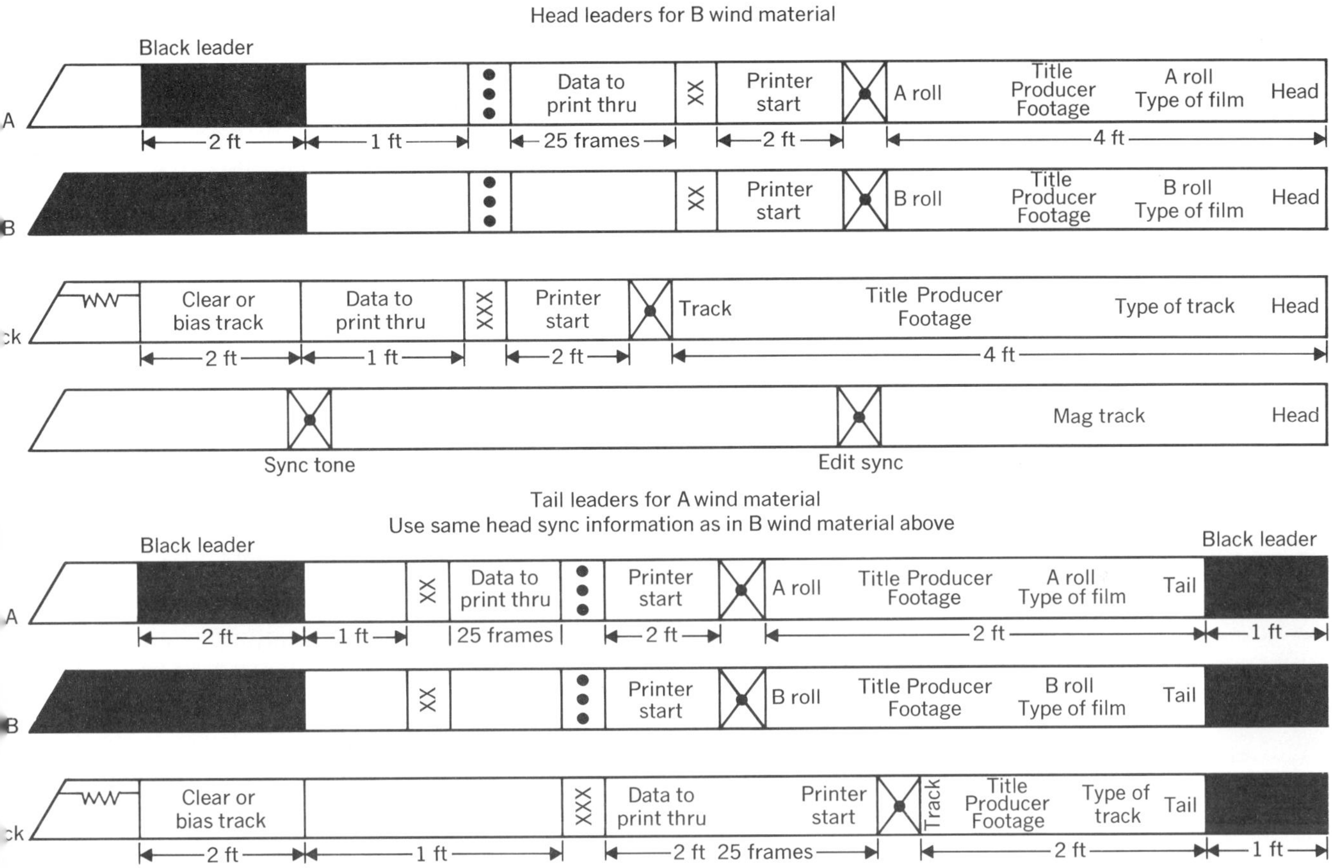

Fig. A-3 **Preparation of 16mm printing leaders. *(General Film Laboratories.)***

outgoing scene, or before the first frame of an incoming scene. When transitional effects are required, the overlap footage must be as long as the effect. If, for example, a 2-foot cross-dissolve is required, the editor must include at least that amount of footage at the head of the incoming scene, as well as at the end of the outgoing scene. The generous editor will allow more overlap if he can, thereby permitting more control during the production of the effect.

Special matteing effects may be incorporated during the transfer, the introduction of a product over the final scene of a television commercial, for example. This type of effect can be achieved by either of two methods. If the subject is not in motion, a slide, art card, or length of high-contrast film on which the shape of the product is accurately depicted in white on a black background, can be used to key out the product's shape from the background scene. The product is then inserted in the previously matted area.

When there is motion in the product shot, the chroma-key technique is used. This technique is comparable to blue matteing in film production. (See CHROMA-KEY and MATTES.)

All footage should be color-corrected before the mix. Running the footage several times, to correct for faults in density or color balance, during the mix, can be a costly process.

Accurate frame and footage counts also help speed the mixing process. The editor should know, for example, exactly how much reading time is required for credits so that they can be properly integrated with the program material.

The picture information and magnetic sound track should be on separate reels. Each reel should be wound head out and have proper start marks inscribed on the leaders.

The videotape transfer is actually a dubbing process since the master recording (like the original film negative) remains intact. The tape usage is, therefore, reduced to the footage that is actually required for the composite.

Because the editing and mixing procedure is a sequential process (each scene being added in building-block fashion), it is very difficult to go back to earlier parts of the program to make changes. The first decision with regard to effects and scene length should also be the last one because after the tape is mixed, returning to one scene for the purpose of making changes means reediting the entire production.

Finally, unlike film where a loss in quality is expected in duplicate material, the high-band recorders in use since 1965 produce second- and third-generation duplicates of such high quality that they are virtually indistinguishable from the original. See EDITING and ELECTRONIC EDITING.

A & B printing materials See Table A-1.

A & B rolls—emulsion positions of picture and track

A-wind Picture. In single-perforated film, the image will read correctly when viewed through the emulsion surface of the film with the left side of a title, for example, toward the perforations.

A-wind Track. With the start mark of the roll of film to the right of the viewer, the track is toward him when seen through the emulsion surface.

B-wind Picture. In single-perforated film, the image will be correctly oriented when viewed through the base side with the left side of a title, for example, toward the perforations.

B-wind Track. With the start mark of the roll of film to the right of the viewer, the track is away from the viewer when seen through the emulsion surface.

TABLE A-1 A & B Printing Materials

Printing material	Sound track	Optical or laboratory print
Black and white		
Original negative	B wind, negative	A wind, positive
Dupe negative from master positive	B wind, negative	A wind, positive
Original reversal	B wind, positive	A wind, reversal
Internegative from original reversal	A wind, negative	B wind, positive
Color		
Original negative	B wind, negative	A wind, positive
Original reversal	B wind, negative	A wind, reversal
Dupe negative from master positive	B wind, negative	A wind, positive
Reversal master from original reversal	A wind, negative	B wind, reversal
Internegative from original reversal	A wind, negative	B wind, positive

A simple rule to remember is that when the picture reads through the emulsion side of the film, an A-wind track is required. If the picture reads through the base side of the film, a B-wind track is needed.

On positive prints, the perforated area on the sound-track side of the film should be fogged or blacked out when 16mm black-and-white or positive color prints are intercut with reversal color or black-and-white originals. This procedure prevents undesirable exposure on the sound-track edge that could influence the fidelity of the sound on the composite print.

A & B windings 16mm film is available in two different types of winding, A & B, each designation indicating the position of the sprocket holes in relation to the film frame. With the emulsion on the inside of the roll in both windings, the A-wind sprocket holes are on the left side of the film and closest to the viewer when the roll is unwound from the top and the leading end of the film is extended clockwise to the right. With the roll in the same position, the sprocket holes in the B winding are located along the right edge of the film—away from the viewer.

The A-type windings are used as the duplicating film for making contact prints. The B-type windings are generally used for optical work, projection printing, and in-camera raw stock. Reversal film is available in both A & B windings.

The 16mm rolls of A & B raw stock can be rewound and used interchangeably during the photographic process. The negative, after developing, may have the sound-track area on either side of the film frame, depending on the film's position in the camera at the time of exposure—an important consideration in the preparation of 16mm negative sound tracks for composite printing purposes.

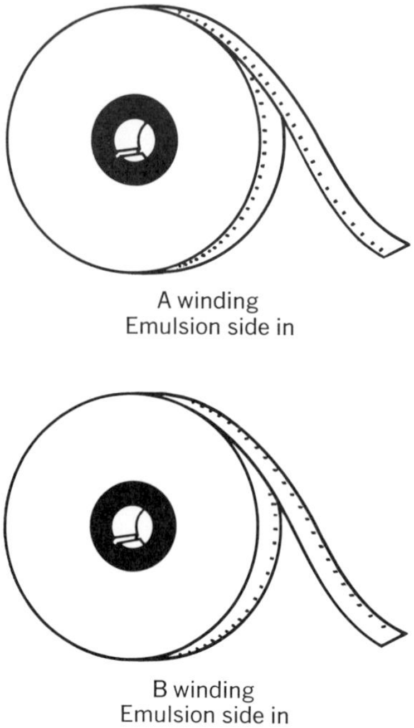

Fig. A-4a

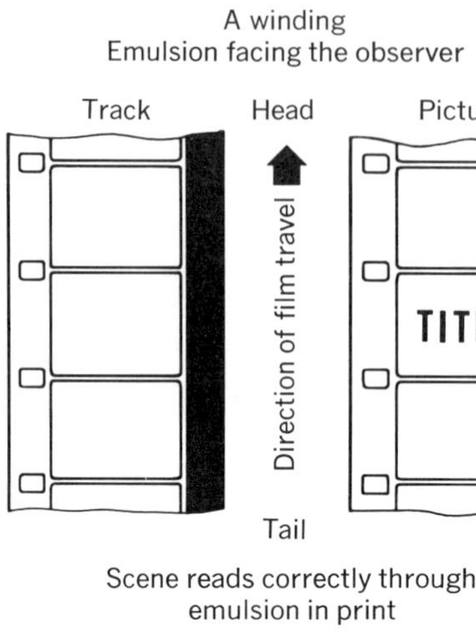

Fig. A-4b

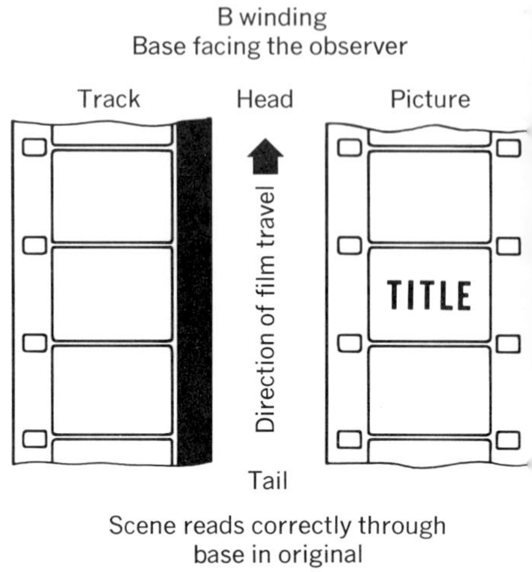

Fig. A-4c

In A windings, the developed negative will have the sound-track area on the right side of the film frame when the emulsion faces the viewer. The sound-track area, in B-wind negatives, is located on the left side of the film frame when the emulsion side of the film faces the viewer. See Figures A-4*a* to *c*.

abaxial Those parallel rays of light that are farthest from the optical axis of a lens. Compare PARAXIAL.

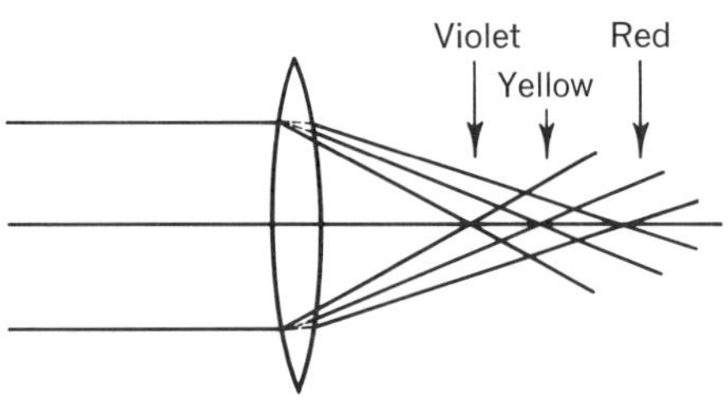

Fig. A-5 Chromatic aberration.

aberration, chromatic A fault or imperfection in a double-convex lens made from a single piece of glass. Because each color has its own characteristic wavelength and, therefore, comes to a focus at a different point, filming through lenses with this type of aberration produces color fringing or an overlapping of color in the resultant print. See Figure A-5. Compare ABERRATION, SPHERICAL.

aberration, spherical Imperfections in a lens that cause distortion in the shapes of photographed objects. Rays striking the outer portions of these lenses are focused at a point nearer to the lens than the light rays falling on the center of the lens. Correction is possible through the use of parabolic reflectors or anastigmat lenses. Parabolic reflectors are shaped like flattened eggs and designed so that light striking their geometrical focus is reflected as a parallel beam. Anastigmat lenses are corrected to eliminate the possibility of spherical aberration. Although aberrations, in general, make geometrically perfect focus practically impossible, they do not affect the visual results too seriously. See Figure A-6. Compare ABERRATION, CHROMATIC. See ANASTIGMAT LENS.

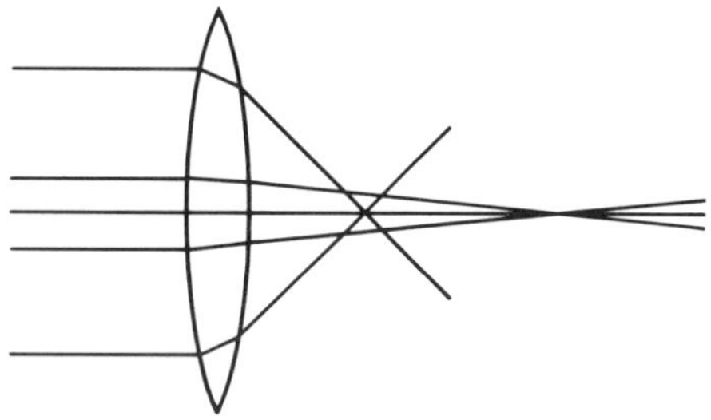

Fig. A-6 Spherical aberration.

abrasion marks Visible spots or fine black streaks that become evident in a negative after developing. These markings are caused by friction, pressure, rubs, or, in general, mishandling.

absorption The process wherein a portion of the light striking a surface is absorbed by that surface. Materials such as colored glass or filters utilize the principle of absorption to alter a specific color or change the color balance of an entire scene. A green filter, for example, will allow that particular color to be transmitted through it while absorbing all other colors.

abstract background A background relatively free of any detail that might distract from the subject matter being filmed or taped.

accelerator The alkali in a developing solution that speeds development by making the gelatin (in the emulsified mixture of gelatin and silver salts) porous.

acetate base The transparent, nonflammable cellulose acetate base that supports the film's layers of light-sensitive emulsion. Strength and dimensional stability in a film base are both quite important since unstable or inferior film supports tend to shrink or stretch during laboratory processing. The resultant physical changes in the film make critical matteing jobs difficult if not impossible. Stronger and dimensionally more stable than the highly flammable nitrate bases that were used during the motion picture industry's infancy, acetate film supports vary in thickness from approximately 0.0003 to 0.0009 inch.

acetone A highly flammable solvent with several uses in photography. The colorless liquid is used as a thinner in processing solutions that contain substances which do not dissolve readily when mixed with water. Acetone is also used for thinning the film cement that is used by an editor for splicing film.

achromatic lens A lens designed to eliminate chromatic aberration and its resultant color fringing. Actually a combination of two lenses, one made of crown glass and the other of flint glass, an achromatic lens brings light rays of two colors to the same focus. Optical printers use achromatic light sources as the printing light for rephotographing previously processed film on the raw stock in the camera. See ABERRATION, CHROMATIC and APOCHROMATIC LENS.

acid, acetic A chemical compound used by laboratories during the fixing process. Also used for bleaching and toning, acetic acid is colorless and smells like vinegar. Mixing readily with water, the corrosive liquid in its pure form becomes a solid at temperatures below 62°F. See ACID, BORACIC.

acid, boracic A chemical compound, soluble in water, that is used by laboratories in fine-grain developers and in fixing baths. Also known as boric acid, the crystalline powder is white and can be used to extend the life of fixers which contain potash alum. Compare ACID, ACETIC and ACID, CITRIC.

acid, citric A chemical compound used by laboratories during the fixing process in place of acetic acid. The colorless crystals dissolve readily in both water and alcohol. Citric acid is also used in clearing solutions. Compare ACID, ACETIC and ACID, BORACIC.

acoustics The science dealing with sound, generally. Specifically, the process of recording sound and its subsequent reproduction and transmission. Also, the design and inherent characteristics of areas such as recording studios, stages, and auditoriums, in which sound is recorded and reproduced.

The term *acoustic response,* used in connection with studios, stages, and auditoriums, refers to the measurable frequency and general characteristics of the rooms used for recording and reproducing sound. These characteristics can be altered by employing any of the available types of acoustic material designed for that purpose. Most of these materials have sound-absorbent properties.

Too much absorbent material can "deaden" a room by absorbing a large amount of the high-frequency sounds. Other rooms used for recording and reproducing sound are referred to as "brilliant" because of the dominance of high-frequency sounds. Studios with reflective ceilings, walls, and floors also allow high frequencies to predominate and are known in the industry as "hot" or "live." Walls that are either convex-shaped or not parallel are used to diffuse sound in recording studios and theaters. Nonparallel ceilings have a similar effect on the recording. The greater the amount of diffusion, the better the acoustical response. The standard coefficient of absorption with regard to materials used for acoustic purposes is one square foot of air with little or no reflection. The absorption coefficients of acoustic materials vary considerably. Poured concrete, for example, with one or more layers of paint covering its surface has a coefficient of .014. A person seated in a theater, by comparison, is at the other extreme, with an absorption coefficient of 3.800 per square foot. Velour draperies and carpets resting on a felt base are rated at .350 and .370, respectively. In addition to the standard tiles that are used for ceilings, the more frequently used acoustical materials are building paper, rock wool, and fiber glass.

actinic light Any light capable of affecting a film's emulsion.

action (1) The general movement of performers during the filming process. (2) The director's signal to bring cameras up to proper operating speed and for the performers to begin acting.

acutance In optics, the sharpness of an image.

adaptation A script treatment designed for a particular medium. A book, for example, may be rewritten for motion picture use. Similarly, a drama or musical comedy, in its original form, may need revising in order to make it more suitable for television programming or motion picture production.

adapter ring A device for holding filters in front of the camera lens. It is threaded so it can be screwed into position in front of the lens, a threaded tube or threaded barrel.

additive color-printing system Designed around the light valve, an electromechanical light control, program decoding and memory storage unit, the additive color-printing system as used with the Bell & Howell model C printer, provides flexibility in laboratory film printing. The additive color printer requires perforated tape for automatic operation. High-speed scene-to-scene color compensation is controlled by this tape system. The preselected color-timing information is programmed so that as one scene is being printed, the color values for the next scene are in the memory unit awaiting the next cue. Light changes may be controlled by a standard film notch or with the transistorized radio-frequency patch cue system.

Color stability and control are obtained through the use of dichroic mirrors which separate a single 1,000-watt light source of the proximity reflector type into three primary color beams. The dichroic mirror system filters and refines the three color beams into narrow-band wavelengths specifically selected in the red, green, and blue portions of the spectrum for the greatest printing efficiency on photographic color emulsions. See Figures A-7*a* and *b*.

additive correction The laboratory process in which a single printing light is separated into three primary color beams, each controlling the intensity of one of the primary colors. The printing light is used to expose the duplicating stock used during the printing process. See ADDITIVE COLOR-PRINTING SYSTEM.

additive synthesis The process of reproducing a specific color by mixing combinations of the three primary colors in their required proportions.

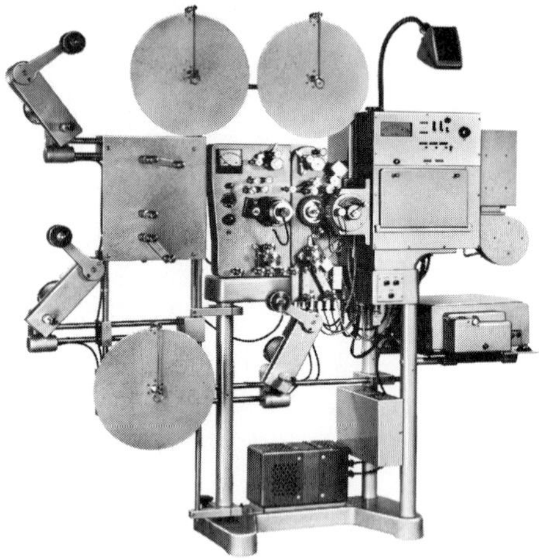

Fig. A-7*a* **Bell & Howell's additive color printer.**

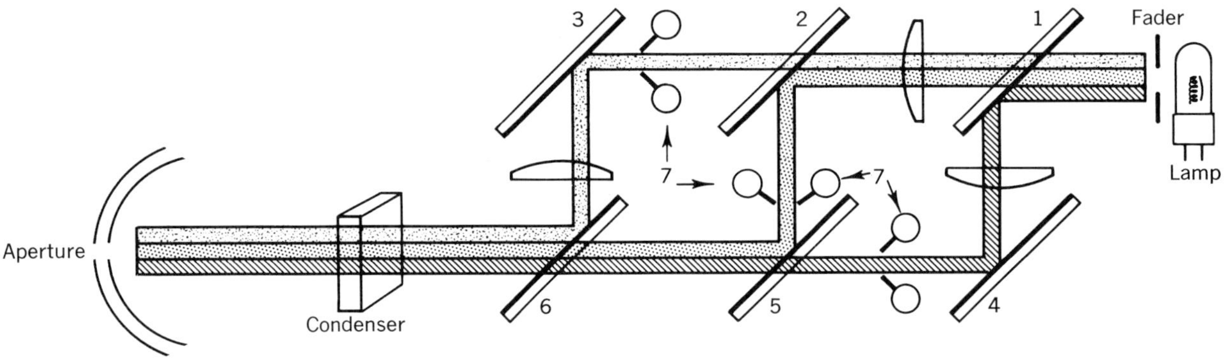

Fig. A-7b **Additive color system.**

Dichroic mirror	*Reflects*	*Transmits*
No. 1	Red	Green and blue
No. 2	Red and green (Since red has already been removed from this beam, it reflects green only.)	Blue
No. 3	Blue only, further reducing bandwidth	Unusable wavelengths
No. 4	Red only, reducing bandwidth	Unusable wavelengths
No. 5	Green	Red
No. 6	Blue	Red and green
No. 7	Light valves	

advance, sound track to picture The spacing, in terms of frames of film, between any specific point in the sound track and the corresponding picture during projection. When the sound track is printed on the same length of film that contains the picture information, as in the composite print, for example, the track's dialogue and music precede the corresponding picture by 20 frames in the 35mm print and by 26 frames in the 16mm print. The advance is made necessary because of the location of the gates in projection equipment. The gate through which the picture is projected is situated above or ahead of the gate through which the film's sound track is scanned by the exciter lamp. Picture and sound, with this arrangement, are each in their respective gates at the same instant. See PROJECTION and EXCITER LAMP.

aerial image photography The aerial image unit is a self-contained background projector that is electrically synchronized with the stop-motion motor that drives the animation camera. With the unit, a cameraman can combine animation or other artwork with previously photographed live-action scenes without using mattes. In operation, the live-action film is projected onto the ground glass in the center of the compound table of the animation stand. Top-lighted transparent sheets of Celluloid, containing the animation or artwork, are positioned on the registration pegs and photographed simultaneously with the projected live-action film. The aerial image process makes possible a wide variety of effects at great savings in both time and optical costs. See Figure A-8.

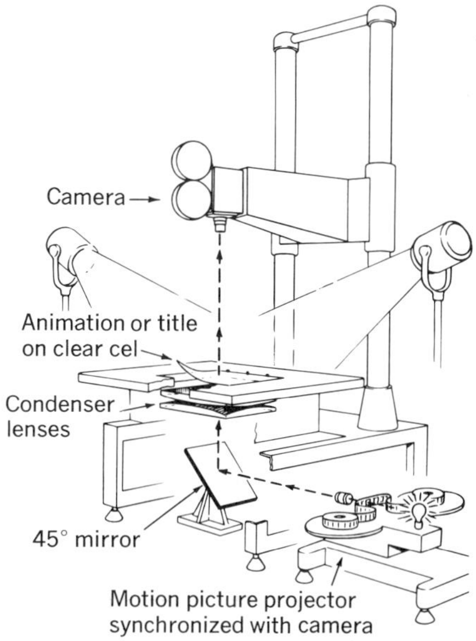

Fig. A-8 The aerial image projection unit attached to the Oxberry animation stand.

aerial photography The first photographs taken from above ground level preceded the invention of the airplane by approximately a half-century. Taken in 1858 from the swaying gondola of a captive balloon, the wet collodion plates were, literally, out of this world.

Ironically, the filming technique was developed out of necessity during World War I. The plate cameras of that period were quickly modified and fitted with magazines capable of exposing the plates in rapid succession. The thousands of photographs taken each day with the relatively crude equipment kept military personnel well-informed with regard to troop movements, supply lines, and defensive positions.

The large number of photographs taken during that period seems insignificant in comparison with the millions of photographs, both still and motion picture, taken each day during World War II and the Korean war. The number of photographs taken during the Vietnam action surpassed even that figure. Actually, our military intelligence received almost 80 percent of its information from these pictures.

In each instance, however, the experience contributed greatly to the development of new equipment and techniques—for peacetime purposes. Today's specially designed cameras use a wide variety of lenses for recording picture information on high-speed panchromatic film. Color, as well as infrared film, is also used for specialized purposes in aerial photography.

For haze penetration, in long-distance or high-altitude filming, panchromatic film is used with filters that absorb blue light from the atmosphere. Infrared film, especially useful for detecting camouflaged areas, is also extremely effective for penetrating dense haze. Deep-red filters, when used for this specific purpose, provide amazing results. For identifying objects on land as well as those just below the surface of the water, color film produces the needed contrasts in color values.

Because the camera moves at the same speed as the airplane on which it is mounted, minimum exposure time is required in order to prevent blurring. The combination of high-speed shutters and extrasensitive film eliminates this possibility.

Aerial cameras are designed to provide an overlapping of picture information. Each photograph overlaps the preceding one by approximately 60 percent. The superimposition of the processed film permits stereoscopic study of the images. This is especially useful for plotting maps and analyzing ground detail.

Aside from their original use for military purposes, aerial photographs have many important peacetime uses. High-altitude surveys, for example, have provided much needed information for forestry-conservation and soil-erosion programs. For scientific exploration, as well as for geological and archeological surveys, the filming technique is extremely useful. Large-scale surveys, in which many aerial photographs are joined and analyzed as a single unit, are used for many commercial purposes. It is entirely possible that the property on which the reader's home was built, if it is a part of a fairly large real estate development, was originally laid out from this type of survey.

aerial photography filters Filters used for the purpose of reducing the effect of atmospheric haze. In spite of the apparent clearness of the area selected for filming, varying amounts of atmospheric haze are normally present in almost all location filming and outdoor photography. See Table A-2.

Aero-Vision Also called **helicopter mount**. An antivibration mount designed for use in helicopters, airplanes, and other moving vehicles. Its patented features provide almost 100 percent picture steadiness and freedom from vibration, even when using long lenses. See Figure A-9.

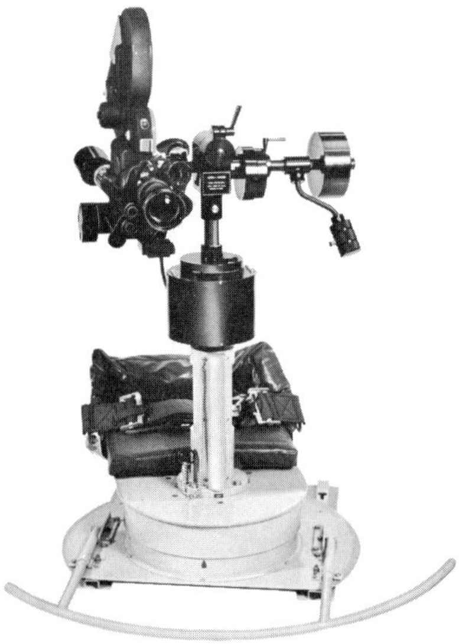

Fig. A-9 The Aero-Vision helicopter mount.

TABLE A-2 Filters for Black-and-White Aerial Motion Picture Photography*

Altitude, feet	Relatively no haze (very low humidity)	Light haze (moderate humidity)	Heavy haze, may appear bluish (high humidity)
Low obliques and verticals			
Below 1,000	No filter	K-2 (2×) light yellow	G (3×) deep yellow
1,000–2,000	No filter	G (3×)	A-25 (8×) light red
2,660–5,000	No filter	G (3×)	A-25 (8×)
5,000–10,000	K-2 (2×)	A-25 (8×)	Deep red or infrared†
Above 10,000	G (3×)	A-25 (8×)	Deep red or infrared†
High obliques (including horizon)			
Below 2,000	K-2 (2×)	G (3×)	A-25 (8×)
Above 2,000	G (3×)	A-25 (8×)	Deep red or infrared†

Courtesy of Eastman Kodak.
*Filter factors appear in parentheses.
Filter-factor allowance: 2× Open up 1 *f*-stop
3× Open up 1½ *f*-stops
8× Open up 3 *f*-stops
†When used with infrared film.

afocal lens Lenses that increase or reduce the focal length of the camera lens, without affecting either the *f*-stop or back focus, by shifting the plane closer to or further away from the film.

afterimage Also called **tails**. A secondary or ghost image that trails the main or more distinct image. The undesirable effect is caused by cameras or other photographic equipment in which the shutter and transport mechanism are not in phase (improperly synchronized). Compare PERSISTENCE OF VISION.

agglomeration The granular formations in negative and print material after processing. These formations, actually silver in the film's emulsion, are groupings of the silver deposits which vary in size and thickness depending on the density of the image. Increases in granularity are proportionate to the density of the negative.

In relatively large areas, especially where the tonal values are on the light side, graininess is even more apparent. Fine-grain film should be used for making prints from dupe negatives in order to improve the quality of the print with respect to graininess. Similarly, prints made on reversal material are less grainy than prints produced from a negative material, assuming that each has more or less the same degree of sensitivity.

Although the information in these paragraphs deals with graininess in black-and-white film, the basic effects are qualitatively the same for color film.

aging and damage effects The aging process, in performers, is characterized by changes in hair coloring and an ever-increasing number of lines and wrinkles. If a major portion of the performer's life has been exposed to the elements, that fact will also be reflected by a weather-beaten complexion.

The ravaging effects of time upon less animate subject matter also leave telltale scars that reflect wear and deterioration. In place of gray hair or balding pates, the house of ancient vintage may be discolored, or have loose shingles, broken windows, or a dilapidated appearance, generally. Inside, bits of fallen plaster, loose wallpaper, torn curtains, and discolored furniture stir memories along with great quantities of dust—actually fine powder or fuller's earth.

These conditions are, of course, extreme and can easily be duplicated by turning the selected filming "location" into a temporary playground for energetic children. Most studio sets, fortunately, must either present a lived-in appearance or, at worst, show only a minimum amount of wear.

Rugs, in the lived-in home constructed for the filming, should show worn spots near doors, chairs, and other areas where the traffic would normally be heaviest. Upholstery, in keeping with the theme, should be more faded in some spots than in others. Furniture finishes, similarly, may have lost some of their gloss, and the protective doilies under accessories may need minor repairs or laundering. Although these signs of natural aging may have escaped the notice of the "occupants" themselves, the overall effect is immediately apparent to the moviegoer. For studio prop departments, the assignment is not a difficult one. In abundant supply, all these materials can be made available in an amazingly short time.

Simulating damage caused by fire, flood, and earthquakes, however, requires planning and, in many cases, time-consuming preparation. If the destruction is to be shown as a gradual process, in a number of scenes, the results of the studio technician's planning and preparation should be carefully concealed in the opening shot of the sequence. In subsequent scenes, the "precut" walls will fall on cue, and wallpaper and picture frames will drop as the camouflaged threads and wires that held them in place earlier are cut.

Sections of wall, furniture, and other objects that either fall or are thrown at the performers during the filming of scenes calling for such action inflict little, if any, damage. The ketchupy blood and glycerin tears that spoil the performer's makeup have a greater effect on the viewer than the harmless cork, balsa, and papier-mâché props had on the actor or actress.

agitation The process of keeping either the solution or exposed film in constant motion during the developing, fixing, and washing processes. This constant movement allows different parts of the solution to come in contact with the film, thus assuring an even application of fresh solution to all parts of the film throughout the entire processing procedure.

Airmobile video system The Airmobile video* system is a packaged system of color television cameras, videotape recorders, and control equipment specifically designed for, and installed in, standardized jet airfreight shipping containers. Each package of equipment is complete for its particular function and thus becomes a self-contained module.

Each modulus has all necessary monitoring, mixing, and effects for teleproduction requirements, and appropriate test and signal generation equipment for all maintenance. This system makes worldwide location videotaping—for news and feature programming—economically practical.

alcohol A colorless liquid used as a rapid drier in laboratory processing. Mixing readily with water, it is especially useful in solutions containing materials that do not dissolve easily. In its denatured form, it is suitable for processing but *not* for drinking.

Alekan-Gerard process See FRONT-SCREEN PROJECTION.

alkali Metallic hydroxides used in processing solutions. The degree of alkalinity of a solution determines the speed of development. The higher the degree of alkalinity, the more rapid the developing process.

alum, ammonium A white crystalline salt used by laboratories in fixing baths for the purpose of hardening and clearing the film emulsion. It dissolves rapidly in water and can be used in place of potash alum in fixing baths. Compare ALUM, CHROME and ALUM, POTASH.

alum, chrome An astringent compound used by laboratories in fixing baths. Dissolving readily in water, the violet crystals form a purple solution which turns green when heated. Compare ALUM, AMMONIUM and ALUM, POTASH.

alum, potash Colorless crystals used by laboratories as a hardener in fixing baths. Also available as a crystalline powder (either form

*Reeves Sound Studios, New York.

is soluble in water), potash alum is known as the common alum. Compare ALUM, AMMONIUM and ALUM, CHROME.

aluminum screen A highly reflective screen originally designed for 3-D projection. This type of screen intensifies colors and provides relatively brilliant images in rooms that cannot be sufficiently darkened during projection. The viewing angle of these screens is not as wide as the viewing angles made possible by screens with beaded surfaces. Aluminum screens need tension devices for best performance. See PROJECTION SCREENS.

Amici prism A 90-degree roof angle prism whose coated surfaces allow a minimum of light loss. These prisms can be used in elbow-type telescopes, in erecting systems to revert an image, or as shown in Figure A-10, in conjunction with the animation stand for insert shots or tabletop photography. See TABLETOP PHOTOGRAPHY.

Amidol A whitish powder used by laboratories as a developing agent. Soluble in water, the powder should be stored in airtight containers as it turns brown when exposed to air. The solution is unstable and should be mixed just before it is used.

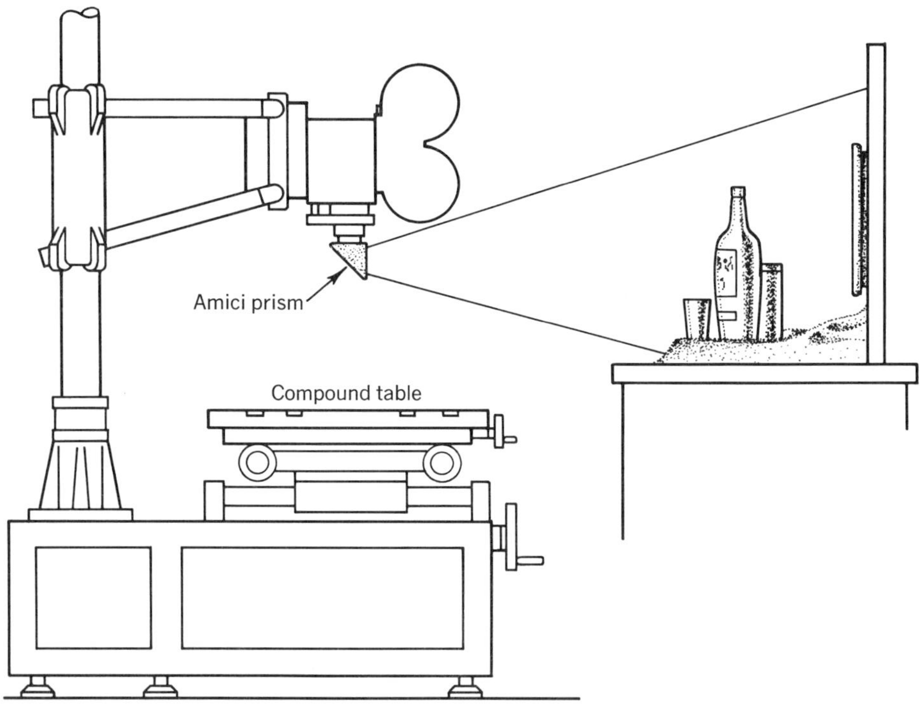

Fig. A-10 **The Amici prism positioned on an animation stand for tabletop photography.**

ammonium bromide A white crystalline powder that is used as a warm tone–developer restrainer. Readily soluble in water, the hygroscopic chemical can also be used as a restrainer for paper emulsions. Compare AMMONIUM CHLORIDE.

ammonium chloride A chemical used with or as a replacement for hypo in rapid fixing baths. The white crystalline powder dissolves quickly in water and can also be used as a bleaching agent. Compare AMMONIUM BROMIDE.

amplifier A device used for the purpose of increasing or reinforcing the strength of a signal or an electric impulse through power supplied from an auxiliary source other than the input. In electronic circuits, the active element is either a vacuum tube or a transistor.

amplitude The relative loudness of sounds. During recording, the degree of loudness can be adjusted by the volume controls in the recording equipment. See DECIBEL.

Amtec An accessory trademarked by Ampex which eliminates skewing and quadrature error in playback. The letters stand for automatic time element compensator.

anaglyphic projection The projection process used for three-dimensional motion pictures. To get the desired effect, the audience wears eyeglasses with a red filter in front of one eye and a green filter in front of the other. See STEREOSCOPIC PHOTOGRAPHY.

anamorphic image A squeezed image produced by photographing the subject matter through a lens that compresses the width of the image but not the height. The image is restored to its proper proportions, during projection, by a complementary deanamorphic lens. Anamorphic lenses provide the means for obtaining wide-screen effects with conventional camera and projection equipment and standard film.

anamorphic lens A type of lens used for compressing images, on standard-width film, for subsequent wide-screen projection. See ANAMORPHIC IMAGE.

anamorphic print A print in which the picture information has been squeezed or compressed on standard-width film. During projection, the images are restored to their normal proportions by a complementary deanamorphic lens. See ANAMORPHIC IMAGE and ANAMORPHIC LENS.

anastigmat lens A photographic lens that is used to correct the effects of astigmatism. The front surface of this type of lens corrects for spherical aberration while the rear surface of the lens eliminates astigmatism in the outer areas of the field. See ABERRATION, SPHERICAL and ASTIGMATISM.

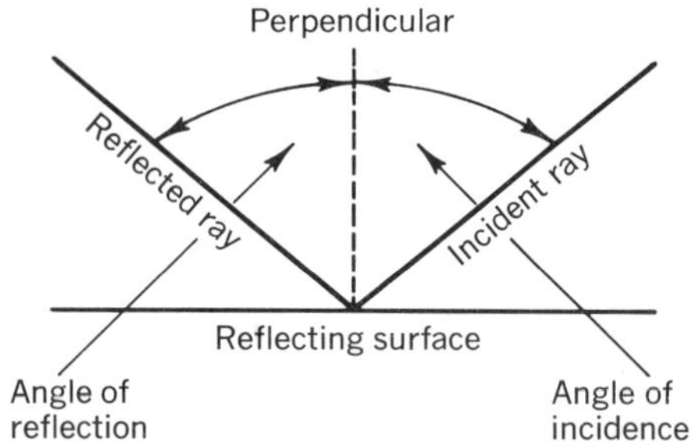

Fig. A-11 Angle of incidence and angle of reflection.

angle, camera The position of the camera in relation to the subject matter.

angle of incidence The angle at which light waves strike a reflective surface. During the filming process, gobos and similar shielding devices are used to prevent undesirable light from reaching or reflecting into the camera lens. The angle of incidence is equal to the angle of reflection. See Figure A-11. Compare ANGLE OF REFLECTION.

angle of reflection The angle at which a light wave leaves a reflective surface. See Figure A-11. Compare ANGLE OF INCIDENCE.

angle of view An area within two imaginary lines that extend from the center of a camera lens to the horizontal limits of the field, as seen on the ground-glass reticle in the camera's viewfinder. Some typical angles of view for different types of lenses are shown in Figure A-12.

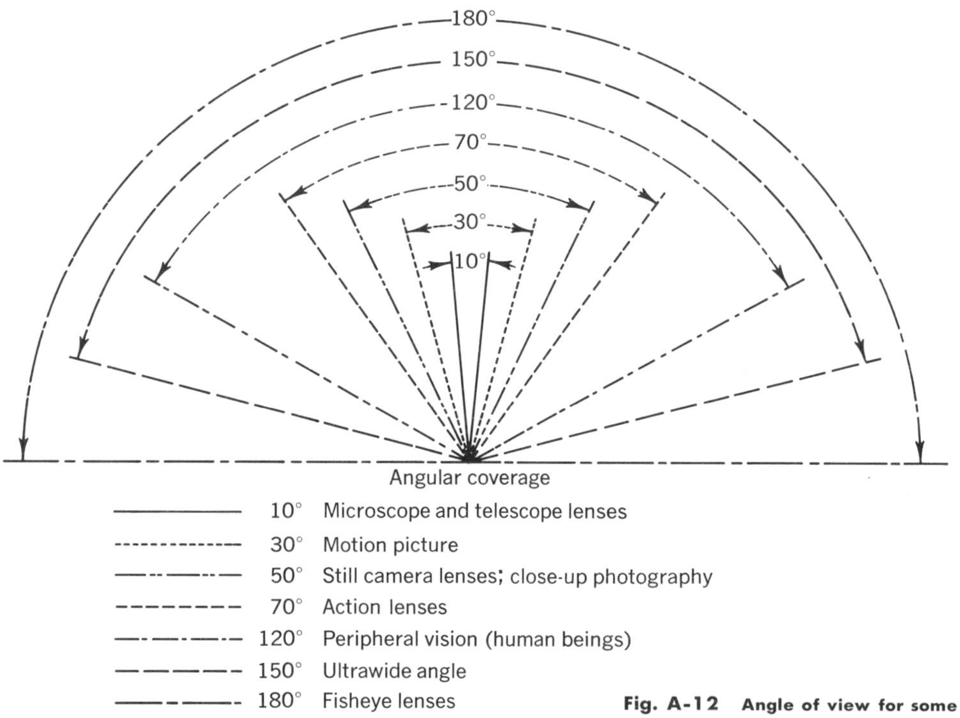

Fig. A-12 Angle of view for some typical lenses.

angle-of-view formula for lenses

$$\tfrac{1}{2} \text{ angle of view} = \frac{\tfrac{1}{2} \text{ aperture in inches (width or height)}}{\text{focal length of lens in inches}}$$

angstrom Equal to $\frac{1}{10}$ millimicron, the angstrom is used to measure light waves. Approximately 254 million angstroms equal one inch. A millimicron is about a billionth of a meter. Light waves in the visible portion of the spectrum vary from 400 to 700 millimicrons in length.

Color	*Wavelength millimicrons*
Red	700
Red-orange	615
Orange	590
Yellow	578
Yellow-green	565
Green	508
Green-blue	492
Blue	484
Blue-purple	475
Violet	400

Spectrophotometers are measuring devices for determining the component wavelengths of a color.

animation Visitors, after being shown through the various departments of an animation studio, will, inevitably, turn to their guide and repeat the question asked at the beginning of the tour. "Yes, but what makes them move?"

Basically, the answer lies in the mountainous pile of animated drawings that are photographed in sequence on successive frames of motion picture film. The illusion of motion is achieved with stop-motion photographic processes which transform the many individual drawings into smooth-flowing, continuous action.

From a nontechnical point of view, the animator is the prime mover. Assisted by a small army of dedicated artists and technicians, with each contributing his or her talents to the composite production effort, the process of instilling lifelike movement in the animated cartoon characters is a complicated, time-consuming procedure.

One of the most versatile art forms known, animation is also considerably older than any other motion picture technique. The first patent for showing "moving drawings" was issued to William Lincoln two years after the Civil War ended. Thirty-three years earlier, in 1834, an Englishman named William George Horner invented the *Daedaleum,* or wheel of life, for the same purpose. Continuing backward chronologically, in 1831 a Frenchman named Joseph Antoine Plateau developed a device for showing "motion pictorially."

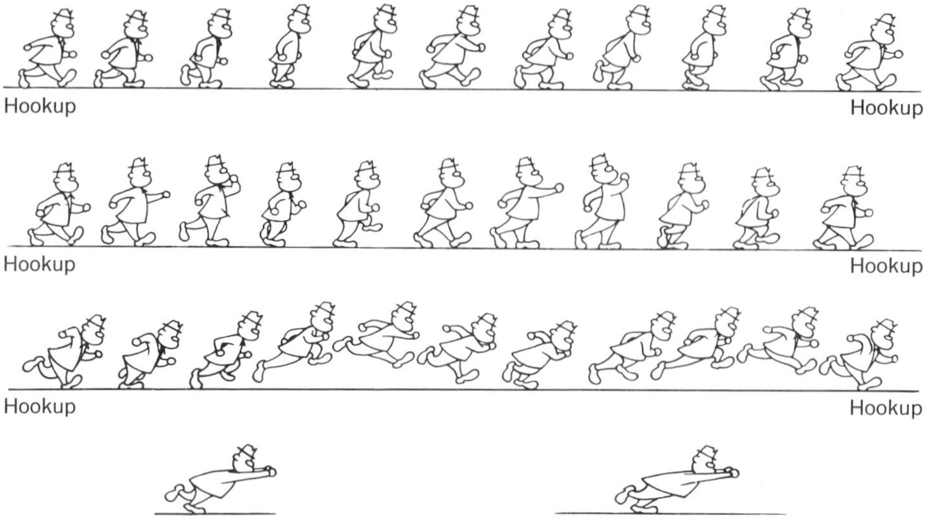

Fig. A-13*a* **Animation cycles.** A number of animated drawings that are used to repeat an action. In a walk cycle, for example, the same drawings can be used over and over again for as much film footage as is needed. The drawings in the cycle are prepared in such a manner that the first and last drawings of the action are similar and, therefore, interchangeable.

Fig. A-13*b* **Live action combined with animation.** Live action scenes can be combined with animated sequences, with any of several matteing processes, to provide an infinite variety of unusual effects. In this frame from a commercial production, animation has been combined with both live action and stop motion.

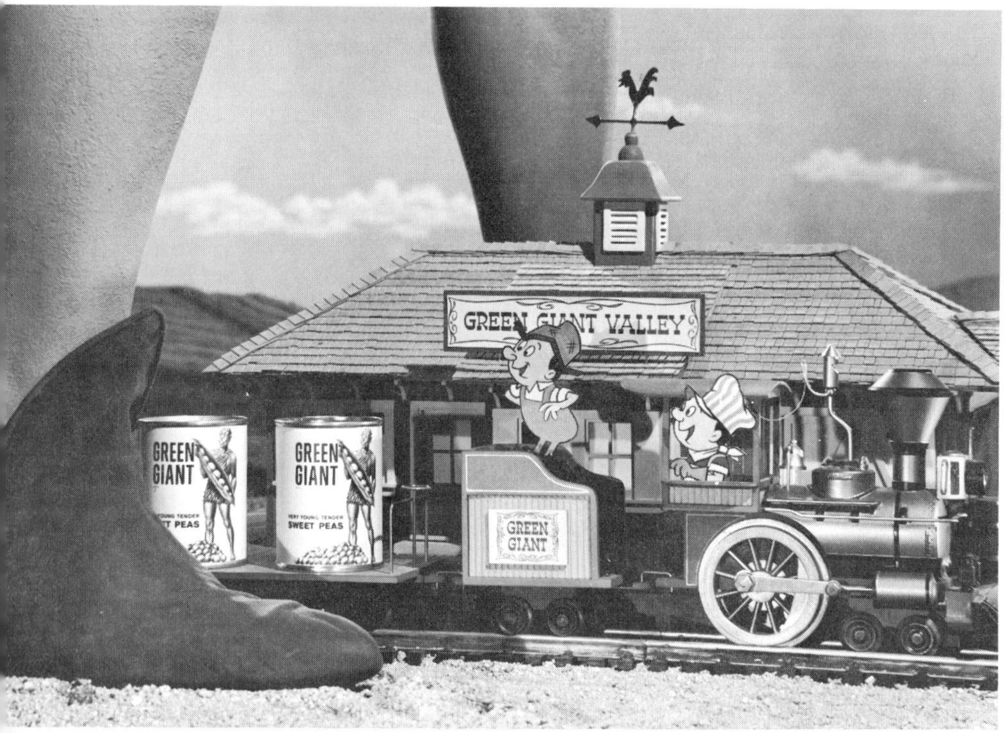

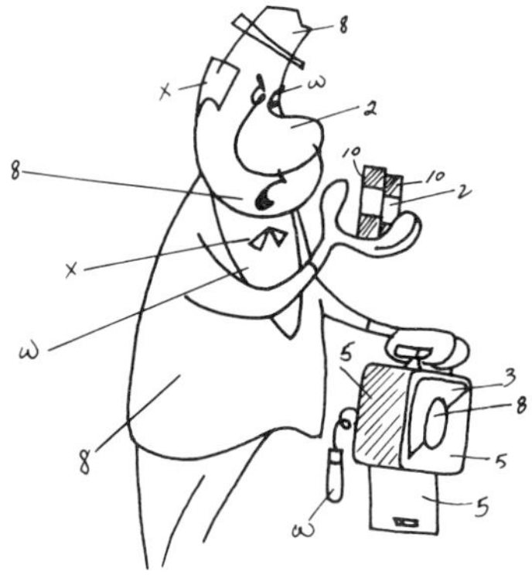

Fig. A-13c Model drawing.

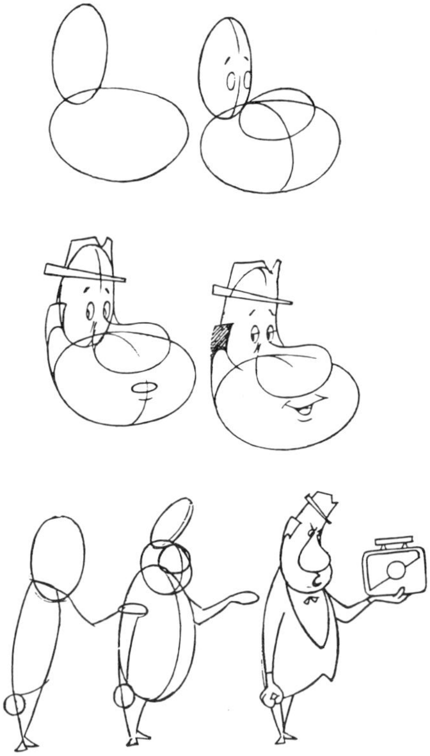

Fig. A-13*d* Animated cartoon figure. Two small circles for the head and a larger one for the body. The proportions of an animated cartoon figure, in the first drawing of a scene or the last, must be consistent.

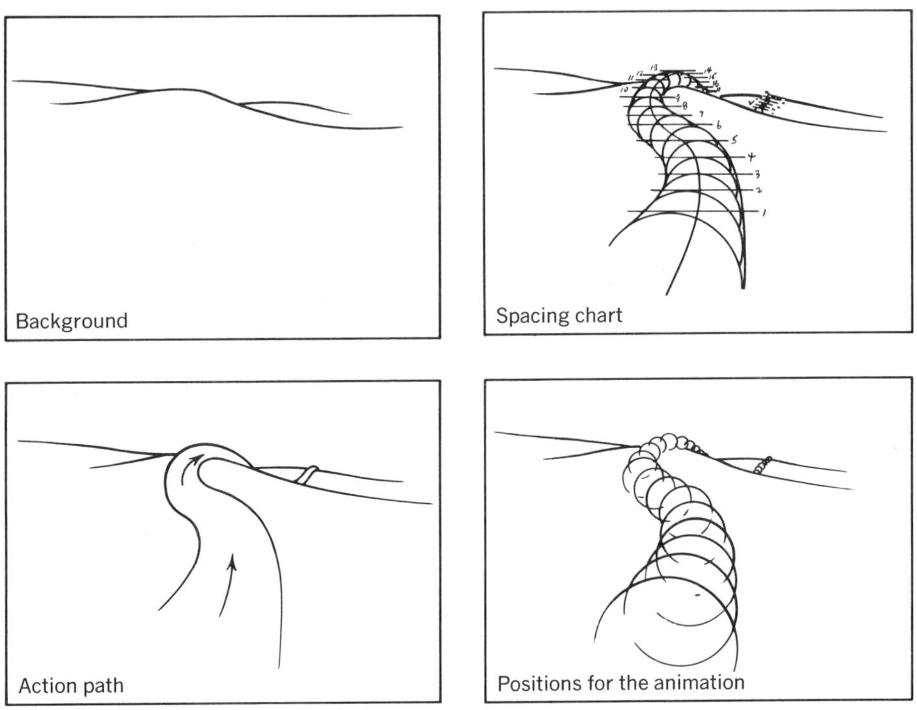

Fig. A-13e Spacing and timing for animation. The spacing for the drawings and the timing of animated actions are governed by the frame counts indicated on the bar or lead sheets prepared by the animation director.

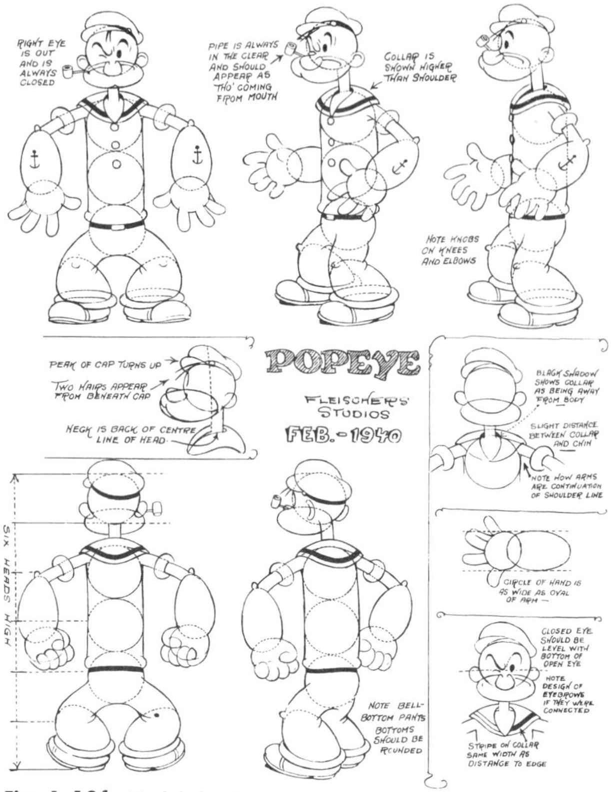

Fig. A-13f Model sheet.

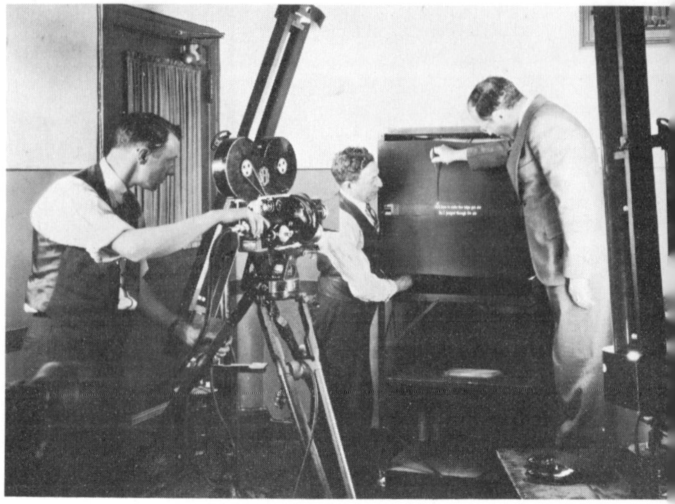

Fig. A-13g If you're old enough to remember the "Sing-Along" animated cartoons, you'll probably find this old photograph quite interesting.

As each line of the song was brought into position, a ping pong ball glued to a stick held by the music director beat out the rhythm that, hopefully, kept audiences in sync.

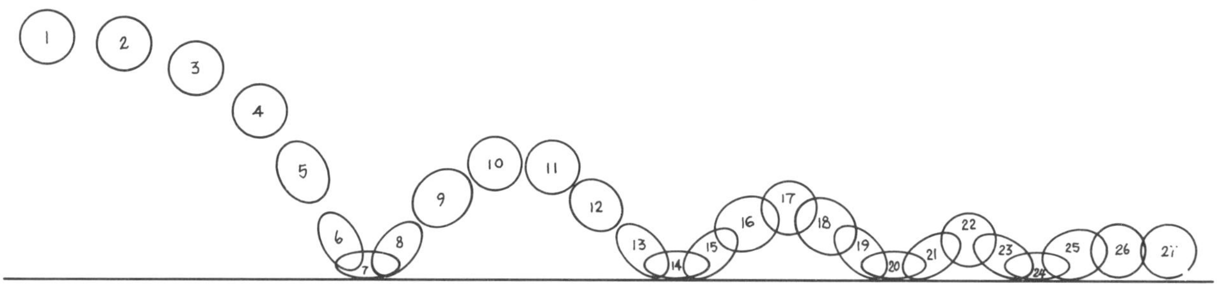

Fig. A-13*h* Principles used in animation. The action of a bouncing ball illustrates almost all of the principles used in animation.

Squash and stretch, for example, are shown by the flattening of an object as it makes contact with a solid surface and the apparent elongation of an object as it accelerates or decelerates. The return of an object to its normal shape after it has reached a more or less constant speed and the subsequent reduction in the speed of a moving object when momentum is lost can also be seen in the diagram.

All these natural physical laws are followed by the animator in order to achieve a feeling of realism in the animated cartoon characters that come to life on the motion picture screen.

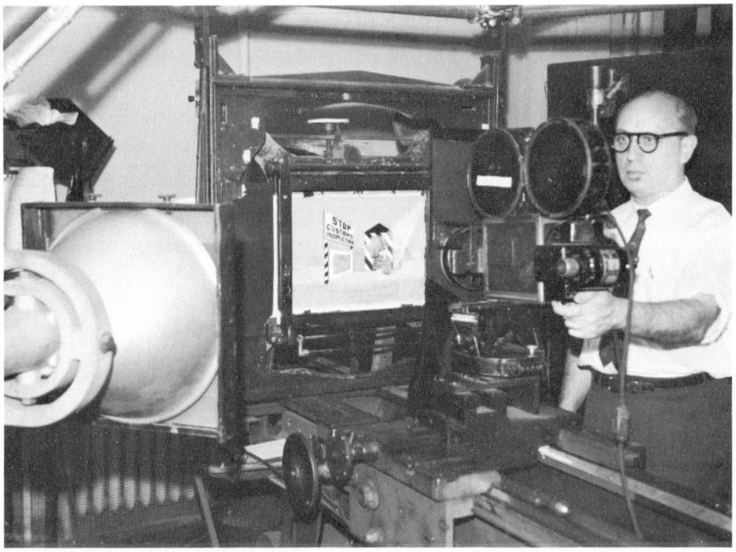

Fig. A-13*i* Animation camera. Until the mid-'40s, animation cameras, with their stop-motion motors, were mounted on lathe beds. The background and cels were registered on pegs mounted on a large metal frame. A large glass platen was used to lock the background and the cels in position, and the frame was then lifted into a vertical position to face the camera mounted on the lathe bed. This procedure was followed for each frame of film in the production. Zoom effects were achieved by moving the camera, for each frame of the zoom, to a predetermined position on the lathe bed.

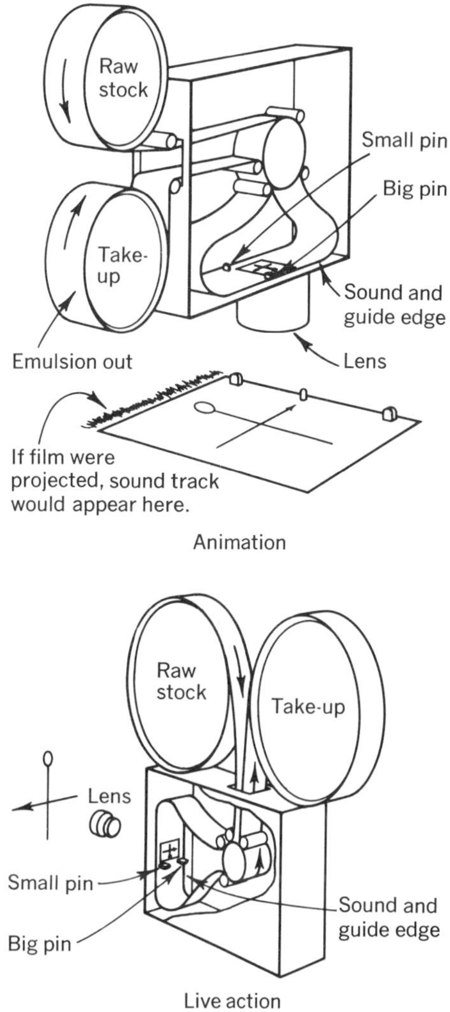

Fig. A-13*j* How camera works for animation and for live action.

Known as a *Phenakistoscope,* the device crudely incorporated some of the features that make modern photography and projection possible. The first of the two disks in the device, ingeniously mounted on a simple shaft, carried the drawings and served as the projector. The second disk, through which the drawings were viewed, proved the accuracy of the "persistence of vision" theory and led to the development of the shuttle.

As an all-purpose technique, animation is capable of producing a variety of effects and visual representations that are unobtainable with other motion picture techniques. The impossible or implausible which has no real-life counterpart and cannot, therefore, be photographed by other processes makes ideal subject matter for the animator. Basically an art medium, animation can be produced solely for entertainment purposes or combined with other film techniques to provide a variety of unusual illusory effects. Equally valuable for scientific, industrial, and commercial purposes, the animated production not only sustains interest over longer periods of time but also can boast of giving an audience better subject retention than any other film technique.

Many of the technical production processes that help make the animated cartoon the highlight of the program for moviegoing audiences are quite similar to those used for producing the "live" motion picture. Although the majority of the artists work with pencils, some use pens and brushes that are dipped into ink or opaque water colors. Production tools not directly associated with the artistic endeavor include specially designed cameras with stop-motion motors, moviolas, sound readers, and miscellaneous recording equipment.

The first of the many steps in the production process, as with other film-making techniques, is the preparation of the script. Illustrated by drawings that are arranged in comic-strip fashion, the script, in this visual form, is used as a guide for the production processes that follow. Usually drawn by members of the layout department, the cartoon characters are shown in relation to the backgrounds that are subsequently rendered by the artists who specialize in that area of production.

Though the sound track is an integral part of any motion picture, be it a feature film or a one-minute television commercial, it is especially important to the animated production. Unlike live film presentations, many of the visual effects in animation depend on the accompanying music and sound for the touches of realism that make an audience forget that the animated performers are not actually alive.

Some sound tracks are recorded after the drawings have been completed and subsequently traced on transparent sheets of Celluloid and photographed. This is post-sync recording. In the absence of a sound track, the editor must prepare a work print according to the educated guesses of the animation director and his rough timings of action and lip sync. Fortunately, a majority of the studios engaged in the production of animated films for entertainment purposes and almost all those busily producing animated television commercials work with prerecorded tracks. With either system, post-sync or pre-sync, the tonal qualities of the recording artists must be identified with the logical personality of the animated cartoon characters that they represent. Association and recognition must be immediate if the cartoon figure is to be believable.

After the recording has been completed, the tape is transferred to either 35mm magnastripe or 35mm magnetic full-base film. In this form, the track can be analyzed and used with editing equipment for achieving the necessary synchronization between picture and sound track. The sound reader, a device capable of reproducing recorded sound, and the synchronizer, for measuring lengths of film, are both used for analyzing sound tracks.

In making the analysis, the film editor reduces words to syllables and notes the phonetic spelling of the dialogue in the blank areas of the film. Accents and beats in the music track are also converted into frame and footage counts. These precise film measurements, entered on *bar* or *lead* sheets by the animation director, account for every frame of film in a scene or sequence and provide a visual synopsis that shows the film length of each action and the timing for each word in the track. Musical beats and the position of camera effects are also indicated. Used as a guide by the animator for timing the cartoon character's movements, the bar sheets show the exact number of frames allotted to each specific action in a scene.

Correct timing, taken for granted in live-action filming, helps make an action believable and is, therefore, of the utmost importance in animation. Unless some special effect is required, timings for animated actions should be consistent with the timings of similar actions performed live. As a general rule, timings are governed by the frame count indicated on the bar sheets and by the beats in the musical track. In the absence of a musical background, a suitable beat should be established by the animator and used as the rhythmic basis for the action.

Since the attention of the audience is usually directed to the cartoon character's face during exchanges of dialogue, lip sync or mouth actions are as important as accompanying movement. Although production charts indicate a different position of the mouth for each of the many sounds used in everyday conversation, the animator should keep lip-sync action as simple as possible. It is not necessary, for example, to represent each of the vowels and consonants in the sound track with a separate drawing. Such attempts usually result in a meaningless pile of drawings and much wasted effort. For the most effective lip-sync animation, the animator usually studies his own expressions in a mirror and selects the ones that emphasize and accent the more important sounds in a phrase. With the phonetic spelling in the film editor's analysis as an additional guide, properly drawn mouth actions lend authenticity to the cartoon character's speech and add immeasurably to the overall effect.

The animator, after plotting and timing an action, makes key drawings, or *extremes*. These drawings, registered on pegs in the animator's underlit drawing board, show the essential phases of a complete action and its continuation with respect to the cartoon character's subsequent movements. If, for example, 24 drawings are needed to illustrate a particular movement, the animator might make drawings 1, 6, 11, 16, 20, and 24. References to timing and other pertinent information included on these drawings serve as a spacing guide for the assistant animator and in-betweener, the artists responsible for completing the sequence.

Although flaws in an animated action will become apparent by flipping the papers on which the drawings are made, a process similar to those used in viewing devices at amusement parks, the drawings are usually photographed on the animation stand. Referred to as a pencil test, the developed negative is reviewed by the animation director and animator. If the flow of movement does not meet with approval, the action is changed, rephotographed, and again reviewed. If the action is satisfactory, and meets story board requirements, the drawings are traced on thin, transparent sheets of Celluloid (cels). In the last of the production processes prior to filming, opaque water colors are applied to the reverse side of the cels according to a predetermined color scheme. Checked for numerical continuity, the painted cels are placed over the background, and the cartoon characters are matched to doors, windows, and other props. After this "dry run" has been completed, the cels and background are ready for the camera.

During the photographic process, the animation cameraman is guided by exposure sheets prepared by the animator. Like the bar sheets, they account for every frame in the production and contain pertinent information for all departments along with camera instructions. For each exposure, the cameraman places each of the numbered cels over the rendered backgrounds in the sequence indicated on the exposure sheets. Since each of the cartoon characters in a scene is usually traced and opaqued on a separate cel, the cameraman may have to expose as many as four or five cel levels over the background. Panning instructions for backgrounds, zooms, fades, and cross-dissolves are also indicated on the exposure sheets. Since these effects are added at the time of filming with frame-by-frame stop-motion processes, the start, stop, and intermediate camera positions for zooms must be plotted before filming begins.

Animation, as a motion picture technique and art form, might never have emerged from its crawling stage without the mechanical contributions of the more technical minded pioneers who helped develop the highly specialized filming equipment. The animation stand with its camera, stop-motion motor, and compound table, the movable working area on which the artwork is positioned, assure maximum control with respect to registration and a minimum of effort by the cameraman.

The imaginative use of animation, in combination with other motion picture techniques, places a wide variety of illusory effects at the disposal of the film producer. In television commercial production, for example, matteing processes provide the means for optically combining animation with live scenes. Sharing the spotlight with live performers, in the same frame of film, cartoon characters and animated titles help demonstrate products in a most convincing manner.

Because of the insatiable demands of the television industry, animation has been used as a basis for the development of other techniques. A relative newcomer, the visual-squeeze technique permits an advertiser to "squeeze" more audio-visual information into the 1,440 frames of film, in a one-minute commercial, than could be included with other techniques. Originating with the animated "flip" books, an idea as old as animation itself, photographs are cut out, positioned on cels, and exposed with frame-by-frame stop-motion processes. Zooms and other camera movements help create the illusion of movement in the "still" photographs. In a more recent variation of this technique, hundreds of photographs are combined with other visual material and exposed on the animation stand to create a collective impression or subliminal thought. Unlike the visual-squeeze technique, each of the still pictures used in this quick-cut technique remains on screen for a mere 3 or 4 frames.

Each of the drawings used in the limited animation type of production, by comparison, is designed to produce the maximum in effects with a minimum of animation effort. In most cases, the eyes and mouths of the cartoon characters animate while the heads remain in a static pose. Similarly, legs and arms move, during walking and running cycles, while the bodies, traced and opaqued on separate cels, remain motionless.

Technical animation, as the name implies, is an adaptation that has been used successfully to fill many needs. Originally designed for educational purposes, technical animation has not only speeded training programs but has been used for commercial demonstrations as well. Statistics showing that more than 70 percent of the men in uniform and those working in war plants during World War II were trained with films that included technical animation sequences offer convincing evidence of the effectiveness of the technique. Unlike cartoon animation, the subject matter usually consists of charts, cutout figures, titles, directional lines, and similar graphic material.

As a technique that is sufficient unto itself or capable of being combined with others, animation's position in the motion picture industry is secure and unchallenged. In addition to its value as an entertainment medium, its versatility as an all-purpose technique has made animation one of the most powerful means of communication known. The effectiveness of this stop-motion technique is reflected through its steadily increasing use by advertising agencies, industrial organizations, and educational institutions. As an educator, the medium has earned its reputation as a teacher par excellence. On the commercial level, animation needs no references to substantiate its abilities as a supersalesman.

Although animation has its serious moments and the screen is no longer an exclusive playground for mice with big ears; ducks with rasping voices; and one-eyed, pipe-smoking cartoon characters in sailor suits; Mickey Mouse, Donald Duck, and Popeye the Sailor will always be welcome sights to children of all ages, from 9 to 90. See Figures A-13*a* to *j*.

animation—computerized One of the most exciting new motion picture production techniques is the computer-controlled method for the production of animated effects. Developed by Al Stahl, one of the country's outstanding animation directors, the system permits the director to plot actions and effects on a teletypewriter which feeds a preprogrammed digital computer. This programmer directs the animation stand through all the motions needed to put the effect on film.

The system consists of a teleprinter, an automatic tape punch, an 8,000-word digital computer, an elaborate camera-control console, and a highly complex animation stand. The stand features ultra-precise indexing, 1/1,000-inch movements, automatic lens iris, automatic platen lift, and special pulse-action motors for all movements.

The computer-controlled stand can superimpose actions, make complicated zoom and pan movements, skip-frame, make fades and cross-dissolves, etc. In fact, once the director has punched his control tape, an assistant can operate the stand since there are any number of controls built in to ensure against error. Even undesirable strobe, caused by improper panning speeds, is a thing of the past with the computer-controlled animation stand. The automated animator is programmed to reject—with a prompt warning—any information ordering improper panning speeds. See Figure A-14.

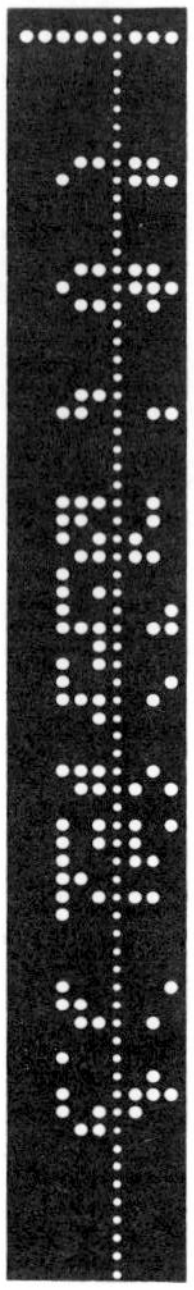

Fig. A-14 Tape for computerized animation.

animation—multiplane effects Although the dimensional backgrounds used for multiplane shots are built to scale, the animated characters are inked and opaqued on conventional cels.

During the photographic process, the cels are registered on pegs and enclosed in a glass frame. The background and the frame containing the cels are lit separately. Each of the several background planes can be moved independently at different speeds. The distance between the inked and opaqued cartoon characters and the multiplane background makes it possible to achieve dimension and perspective that cannot be duplicated on the animation stand. See Figures A-15*a* and *b*.

animation board A drawing board or table designed to fill the specific needs of the animator. A rectangular, translucent glass over a light box under the center of the table allows the animator to see the lines on a number of drawings that are registered to pegs. In

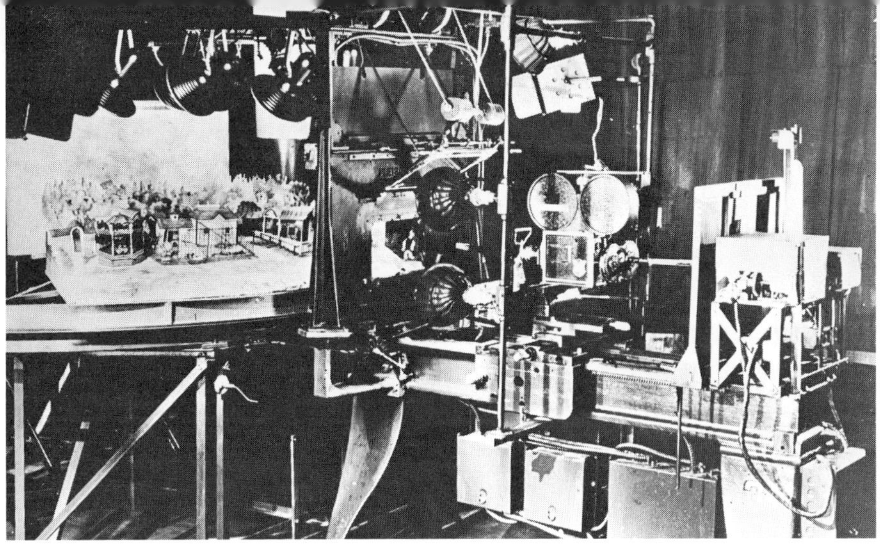

Fig. A-15*a* **Multiplane animation photography.**

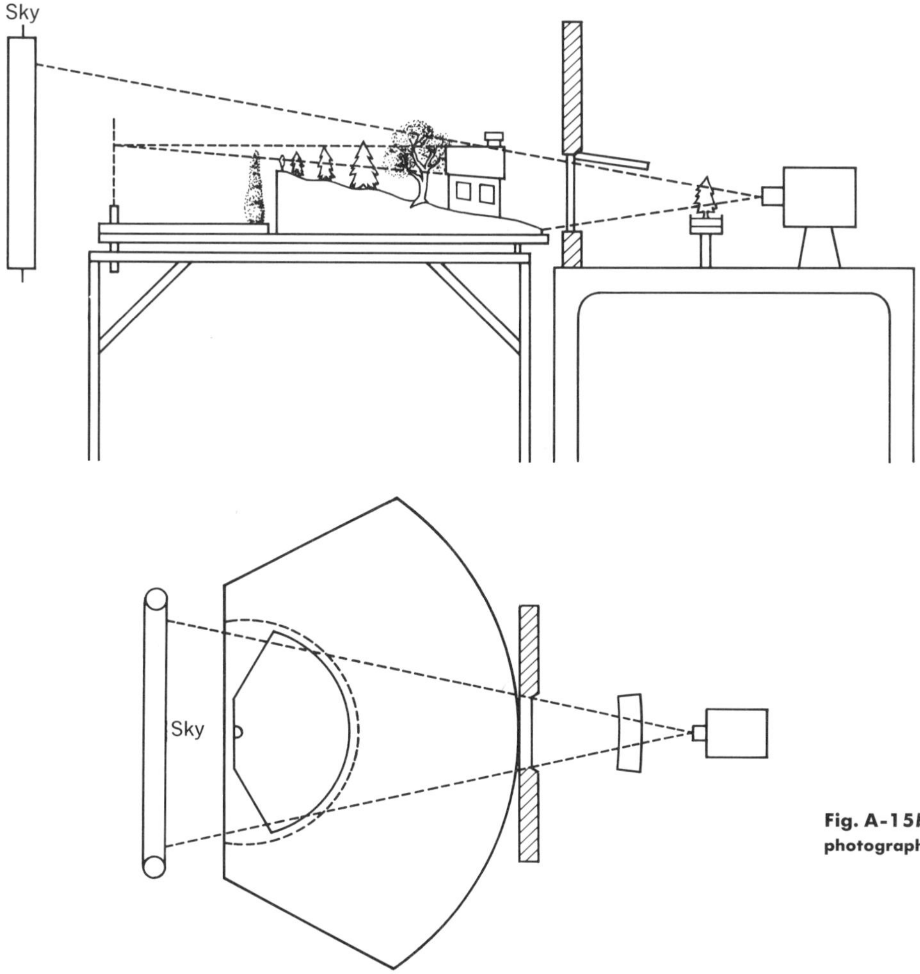

Fig. A-15*b* **Diagram of multiplane animation photography.**

some cases, the cutout in the center of the table is circular to accommodate the specially made disks that have become standard equipment in the animation studio. These disks have traveling peg bars that are used for positioning and registering pan backgrounds and sliding cels that are frequently used in animation production. See Figure A-16.

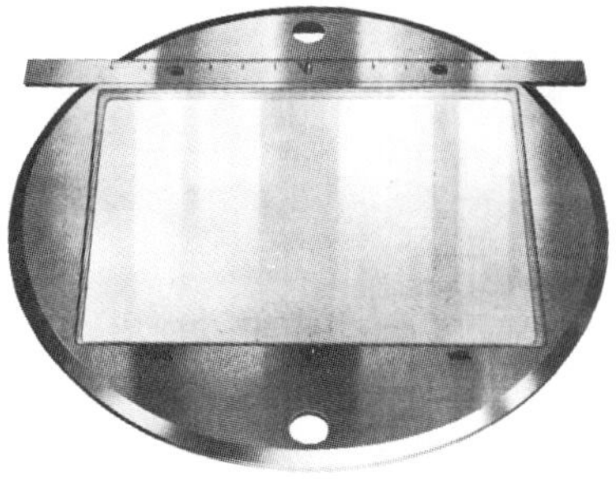

Fig. A-16 **Acme animation board. (*Producers Sales Corp., Burbank, Calif.*)** The top pegs used for registering the animated drawings and backgrounds on the animator's drawing board are mounted on metal slides with calibrations graduated in 20ths of an inch. Some bottom peg bars are built to slide; others are permanently fixed to the drawing board. The board is recessed on its under side so that it can be rotated over the cutout area in the supporting table.

The light source underneath the transparent glass in the center of the drawing board enables the animator to follow the action in a number of drawings simultaneously.

animation camera A motion picture camera driven by a stop-motion motor with provision for frame-by-frame photography. Mounted on columns, the camera can travel vertically, in relation to the artwork, to create zoom effects. If necessary, frames of processed film can be projected through the camera onto the flat surface of the compound table on which the artwork is registered. This method of projection makes it possible to match graphic material to previously exposed (and processed) picture information to produce a wide variety of unusual illusory effects. See Figures A-17*a* to *c*. Compare AERIAL IMAGE PHOTOGRAPHY.

animation stand An all-inclusive term for the camera, its mounting, and the compound table on which the artwork is registered.

Fig. A-17*a* **The animation stand, compound table, camera, peg bars, platen, and N/S and E/W controls.**

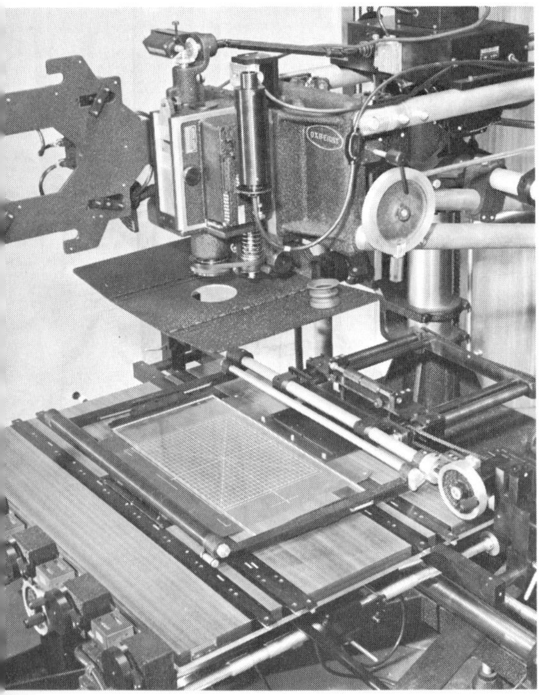

Fig. A-17*b* Each cel is placed over the background and exposed according to the instructions on the exposure sheets.

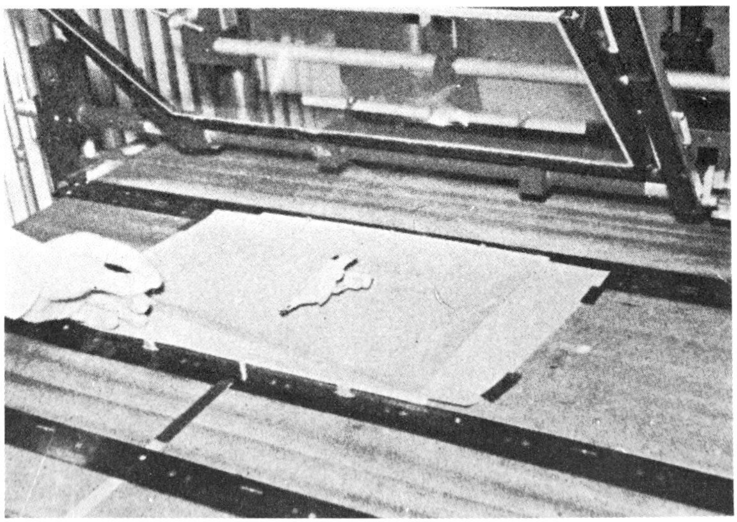

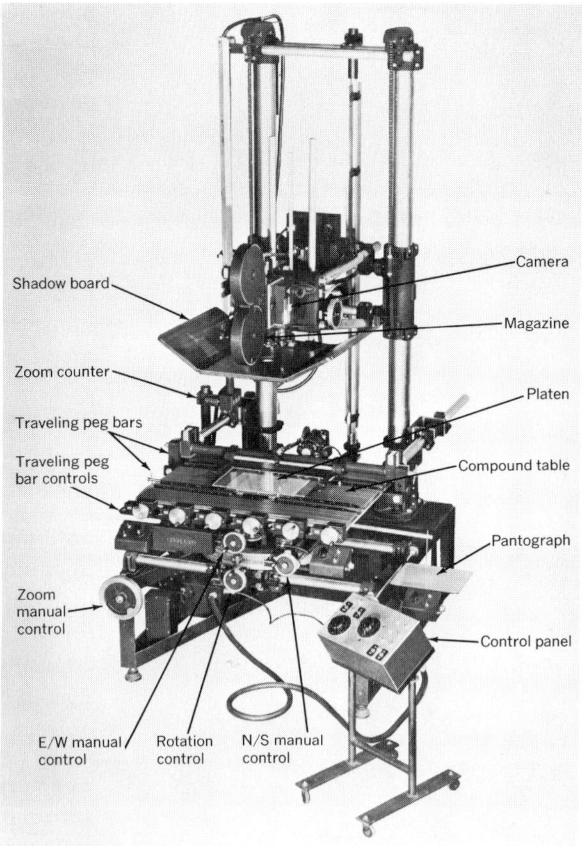

Fig. A-17c The animation stand, camera, and control panel. (Oxberry *Animation Equipment Corp.*)

animator The artist or cartoonist who draws the key positions of an action and is generally responsible for the illusion of motion created from the mountainous pile of drawings produced under his guidance. See Figure A-18.

In some cases, gimmicks such as cutouts, scratch-offs and camera movement serve as effective substitutes for the many drawings that are usually needed to depict an action.

Fig. A-18 Each line on the animator's drawing is carefully traced on cels or acetates by the inker.

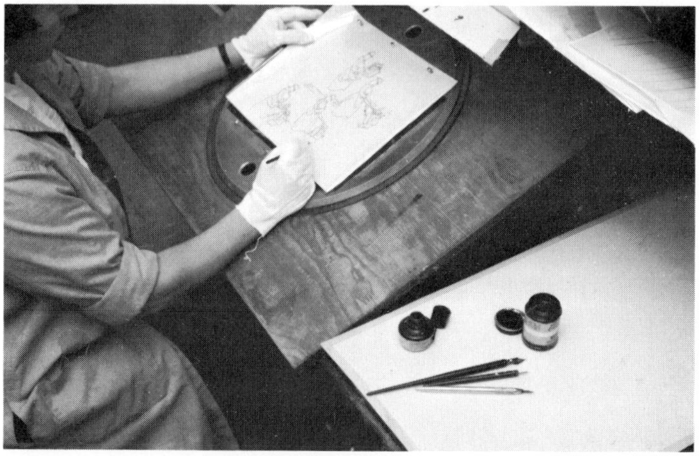

anodized reflector A coating used for reflective purposes in lamp housings.

answer print Also called **optical print**. The first composite print. If approved, it becomes the guide for the subsequent production of release prints.

antenna A tall structure used for transmitting or receiving electromagnetic waves or other signals. Positioned in open areas or atop high structures where the waves or signals meet with little or no interference, an antenna may be in the form of a system of wires, metal rods, or reflecting disks.

anticipation A preliminary action that precedes an accented pose. In animation or live action, its purpose is to give added emphasis to the main action that follows.

antihalation backing Gray dye or jet-black coatings applied to in-camera film bases that prevent undesirable halation. These coatings have no effect on the quality of the print. They are only present for the purpose of preventing light from passing through the film's emulsion layer, striking the acetate base, and reflecting back to the sensitive emulsion to cause halation or light flares around light objects.

aperture The opening in a lens, usually referred to as the *f*-value, through which light passes to expose the film. *F*-stops are settings that indicate the amount of light that is transmitted through the lens. Each *f*-stop increases or decreases the amount of transmitted light by one-half of the total amount. See *f*-STOP. Compare APERTURE, EFFECTIVE, EQUIVALENT, RELATIVE, and WORKING.

aperture, camera The opening in the base or aperture plate of a camera shuttle through which light passes and exposes the film during the interval of exposure, while the film is in a stationary position. The size of the aperture is what determines the framing and field size of the film area being exposed. See Table A-3 and Figures A-19*a* and *b*.

aperture, effective Because even the finest lenses absorb or reflect a percentage of the light passing through them, the *f*-setting used for filming is rarely a true indication of the amount of light reaching the film in the camera. To find the effective aperture, a cameraman would first have to know the transmission characteristics of the lens as well as the percentage of light that is lost during exposure. That information would then be used to make *f*-stop compensations that would offset the known light loss. The effective aperture is also referred to as the **working aperture**.

TABLE A-3 Camera Apertures

Film width	Inches: Width × Height	Millimeters: Width × Height
8mm	0.173 × 0.130	4.39 × 3.30
Super 8mm	0.211 × 0.158	5.39 × 4.03
16mm	0.410 × 0.294	10.41 × 7.47
35mm, nonanamorphic	0.868 × 0.631	22.05 × 16.03
35mm, anamorphic	0.870 × 0.735	22.10 × 18.67
35mm full screen (no sound track)	0.980 × 0.735	24.89 × 18.67
35mm, instrumentation photography	0.980 × 0.735	24.89 × 18.67
35mm, 1.85:1	0.825 × 0.446	20.96 × 11.38
35mm, 2.35:1, CinemaScope	0.870 × 0.735	22.10 × 18.67
65mm	2.072 × 0.906	52.628 × 23.012

Fig. A-19a Camera apertures for different size film. *A* = width; *B* = height.

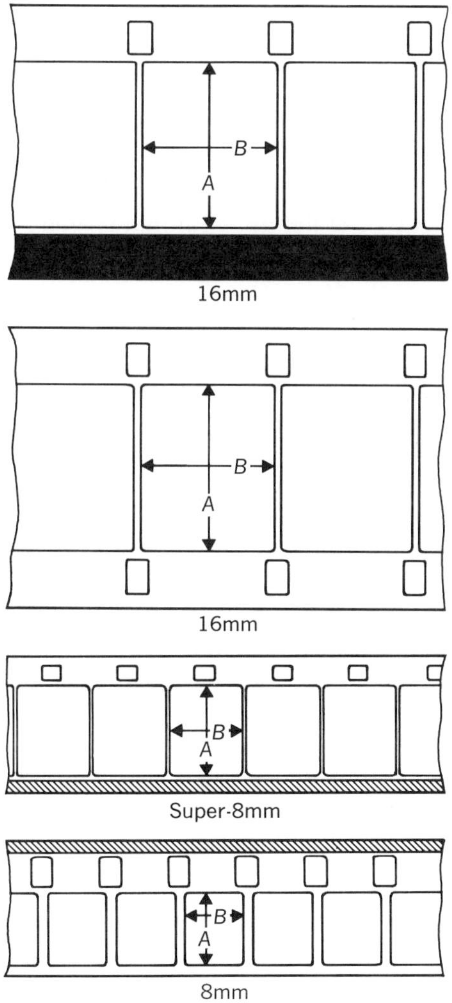

Fig. A-19b Relative sizes of the proposed frame and the standard Academy aperture in current use. The new 35mm dimensions proposed by George A. Mitchell, for increasing the picture area, would still preserve the 1.33:1 aspect ratio. Mr. Mitchell, the designer of the Mitchell camera, is an Honorary Academy Award winner and one of those dedicated individuals who has devoted a lifetime of effort toward the development of new equipment and methods for use by the industry.

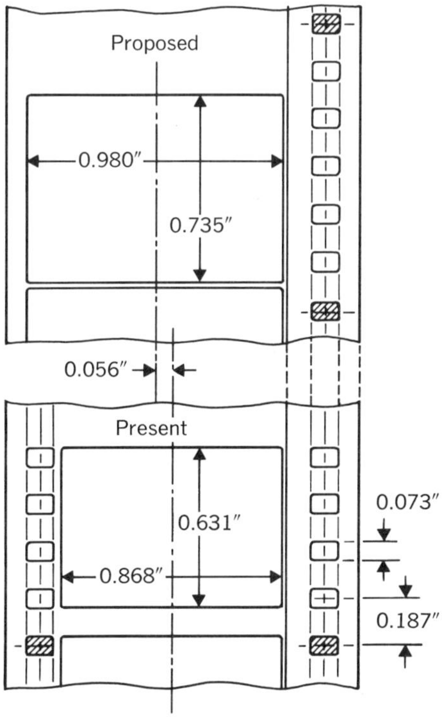

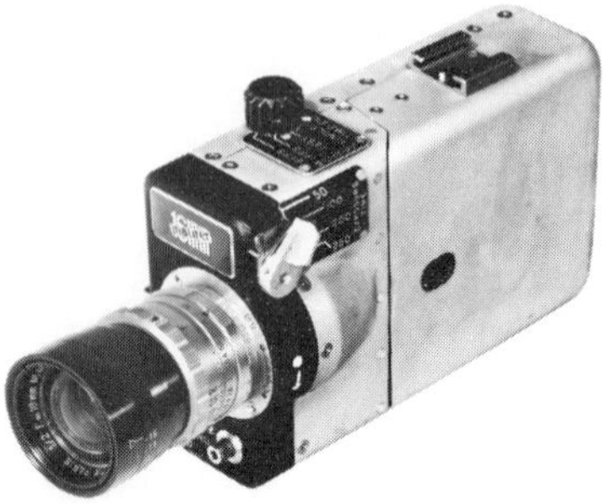

Fig. A-20 Apochromatic lens. The Gemini 10 and 11 space flights were the proving ground for Kinoptik apochromatic lenses. Two 16mm Maurer sequence cameras on board Gemini 10 and 11 were equipped with Kinoptic 18mm, *f*/1.8 apochromatic lenses and filmed essential rendezvous portions of the space venture. Released to the TV networks, the film was admired by millions of viewers for their clarity and detailed reproduction of America's space achievements.

aperture, equivalent The efficiency of a lens, based on its optical characteristics (light transmission capabilities, for example), rather than its size. Compare APERTURE, EFFECTIVE and APERTURE, RELATIVE.

aperture, relative The relationship between the aperture of the diaphragm and the focal length of a lens. The measure of that relationship is used to determine the effective aperture for other lenses—regardless of their size or focal length. In practice, different lenses set at relative apertures produce images that are similar in brightness and quality. Compare APERTURE, EFFECTIVE and APERTURE, EQUIVALENT.

aperture, working See APERTURE, EFFECTIVE.

aplanat lens A corrected lens that shows practically no spherical aberration when set at full aperture.

apochromatic lens A lens that is corrected for all three primary colors. Achromatic lenses, by comparison, are corrected for only two colors. See Figure A-20. Compare ACHROMATIC LENS.

arc light A type of lighting unit that is used for simulating sunlight. The illumination is produced by the flow of current between pieces of carbon positioned within the lamp housing.

arctic and tropical filming The tongue-in-cheek suggestion that places the blame on temperature for the problems associated with motion picture production in hot or cold climates is completely in order. The filming process, generally, is more complicated and hazardous in arctic and tropical areas than in comparatively temperate climates. Precautionary measures for the handling of film and equipment are required. Photographic procedures, on the other hand, are relatively routine.

Motion picture production in arctic regions is, at best, a chilling experience for the production personnel as well as for the equipment they must handle. The insinuation that production problems, under tropical skies, are equally too hot to handle would not be amusing to uncomfortable camera crews in the arctic. In either case, sympathy from these unsung heroes would not be wasted on studio technicians who complain about hot lights and temperamental air-conditioning systems.

In subzero temperatures, motion picture film, even when stored properly, becomes quite brittle. In that state, the edges are sharp enough to cause serious damage to fingers. The obvious advantage of loading film indoors is offset by the potential dangers created by humid conditions. Friendly warmth, in this case, can become an immediate enemy in the form of condensation. The evaporating moisture, when the camera is taken outdoors, will freeze quickly on lens, housing, and moving parts, making the photographic process difficult, if not impossible. Since numb fingers cannot be expected to thread roll film in cameras with any degree of efficiency, the apparent solution to the problem lies in magazine loading.

Cameras scheduled for filming under arctic conditions must be checked and prepared for the freezing temperatures in which they are to be used. Part of this winterizing process includes the removal of oil from the camera and its replacement with lubricants that are more efficient at low temperatures. In some cases, camera parts are machined and readjusted. As a final check, the camera is placed in a freezer or refrigerator and made to run for several hours.

The choice of equipment is limited by the unusual circumstances and conditions that go hand in glove with frigid filming. The advantages of lightweight equipment makes the 16mm camera an immediate favorite. Because of the unreliability of power sources in subzero temperatures, spring-driven motors are used more frequently than either electric power or batteries. Miniature heaters, installed in cameras, help keep the movable parts from freezing and the film from becoming brittle and cracking. The handling of unpainted motion picture equipment also presents certain problems and an element of real risk, since unprotected hands will stick to exceedingly cold metal surfaces. When not in use, cameras should be kept indoors in sealed rubber bags or other protective covering that will eliminate the danger of condensation. If the containers cannot be brought indoors, they should be made airtight in order to keep windblown snow from entering the camera and interfering with moving parts. Cameras, even in cold climates, should never be exposed to direct sunlight.

Photographic procedures, in comparison with the care and handling of the equipment, are basically routine. Slow or medium-speed, fine-grained film usually provides better results than high-speed film types. Because of the unusual brilliance in normal lighting sources, sun and reflective snow, accurate exposure-meter readings are difficult. When one is in doubt, it is safer to underexpose than to overexpose. In extremely cold temperatures, it is usually necessary to make an adjustment in exposure to compensate for the reduction in the speed of the film. (One lens opening slower at subzero temperatures is equal to normal exposure in moderate climates.)

Normal film loops may become affected by the cold and make rewinding a risky process. Because of the brittle condition of the film, great care must be taken to keep it from tearing or breaking during rewinding. Finally, film should not be kept in the camera any longer than the minimum time needed for the actual photography. Cameras should be loaded just before filming is to begin and then unloaded immediately after the roll has been exposed. If the film cannot be processed within a reasonable amount of time, it should be kept in containers at outdoor temperatures.

Under tropical skies, humidity again plays the part of the villain in motion picture production. In this climate, however, "heat" is cast in a supporting role. The pair, acting together, create problems that make filming a memorable experience and adventure. Favorable conditions for the growth of fungus and the attraction of insects, both capable of damaging expensive equipment (and personnel), are added hazards associated with tropical motion picture production.

As in the winterizing process for arctic equipment, cameras must be readied for the high temperatures in which they will be used. Precautions must be taken against damage during transportation. Sealed cans and waterproof containers that contain sufficient amounts of desiccating material for absorbing moisture provide adequate protection for long periods.

To prevent corrosion and the growth of fungus, cameras and equipment must be thoroughly cleaned at regular intervals. Camera lenses, also, need protection from excessive heat and fine, windblown sand. Temperature can soften the cement that holds the glass elements together at 120°, and cause it to melt completely at the 140° mark. Fine grains of sand, insignificant in themselves, can cause irreparable damage to the surface of a lens. An inexpensive piece of clear optical glass, in front of the lens, will provide adequate protection without seriously affecting the quality of the picture or necessitating any more than a fraction of an *f*-stop increase in exposure.

As a precautionary measure, film should be stored in dry containers to guard against the ever-present danger of mildew. Camera loading should take place immediately before filming is to begin. Small rolls, enough for the immediate filming assignment, are preferable to larger rolls which must be left in the camera for much longer periods. Because of the humid conditions, film should be unloaded from cameras as soon as possible after filming ends, and sealed in airtight, moistureproof containers.

Basic photographic procedures that are followed in moderate climates also apply to filming in tropical areas. Cameramen should not take for granted the fact that sky is sky and water is water; exposure that is used for one can be used successfully when filming another. Beaches and open stretches of water around the equator, for example, are much brighter than similar locations in moderate climates and so make adjustments in exposure necessary. In photographing this type of scene, the required decrease in exposure may be as much as one full stop. Filming when the sun is directly overhead should be avoided, if possible. In close-ups, the relative position of the sun puts the actor's eyes in deep shadow and creates effects similar to those achieved with backlighting.

Haze from smoke or water can sometimes be cut with a knife but hardly ever with filters. Exposure increases, in such cases, do not help as much as the choice of a new and more opportune shooting date. Foliage, as well as other background detail, can influence the overall color balance of a scene because of its reflective qualities.

Two words complete the list of suggestions regarding motion picture production in arctic and tropical areas—"Dress properly!"

artificial light Light produced either by electrical or chemical action as opposed to natural light reaching us from the sun, moon, or stars.

ASA American Standards Association. Changed first to the United States of America Standards Institute, but now the American National Standards Institute (ANSI). ASA (ANSI) standards, accompanied by a number, indicate the speed of sensitized materials used by the motion picture industry. The body also publishes recommendations and specifications with respect to equipment and procedures.

ASA exposure indexes Veteran cameramen who have been relying on ASA ratings for years may be shocked by the news that ASA exposure indexes are not intended for use in connection with motion picture films. Stated exposure values and effective film speeds, rated for still photographic procedures, are nevertheless used as practical guides by most cinematographers, who also take into consideration differences in equipment and lighting conditions along with photographic and laboratory procedures.

ash can A studio term for a multiple-arc light source that is suspended over a motion picture set.

ASI United States of America Standards Institute. Now changed to ANSI (American National Standards Institute).

aspect ratio Basically, the width of the screen image in relation to its height. The standard ratio, width to height, is 1.33:1. In wide-screen presentations, the relative width is considerably greater.

Since 1896, the year that Raoul Grimoin-Sanson projected 10 images simultaneously onto a circular screen, motion picture producers have experimented with and produced films using every type of screen format imaginable. The 10-image 10-projector process was called *Cineorama* and should not be confused with *Cinerama,* a more recent wide-screen process which first saw the light of a projection lamp in 1952, more than a half-century later.

Some of the early wide-screen films were produced with short-focal-length lenses. Others were projected on deeply curved screens to create the illusion of depth. In one case, a producer used two projectors and two separate lengths of film to show pictures with a potential aspect ratio of 2.66:1. These pictures were projected so that the two images were seen side by side, on an oversized screen. Appropriately enough, the process was known as *Widescope.*

Magnascope, a system in which the image was magnified up to several times its normal size by using a special lens during projection, and *Natural Vision,* a 70mm film process in which the images were projected on two screens, one positioned in back of the other, to create a dimensional effect, both made brief appearances during the 1920s. *Magnifilm,* a Paramount process and *Realife,* by MGM, were other wide-screen processes that made unspectacular appearances at approximately the same time.

There have been wide-screen processes in which the film was exposed as well as projected horizontally and other processes in which the members of the audience sat in chairs that could be made to revolve so that they could view any portion of the screen that surrounded them.

Whether or not wide-screen processes would have reached their present state of development in a financially healthy industry is a matter of opinion. And so, although necessity is usually referred to as the mother of invention, adversity, in the form of sagging box office receipts, must share the responsibilities for parenthood in this case.

The first modern major attempt to change the viewing habits of motion picture audiences reached its peak shortly after 1950. Although the spectacular, stereoscopic effects in 3-D presentations made audiences blink in amazement, they were also responsible for many cases of eyestrain. With the composite picture divided into separate lengths of film, one reel for each eye, dual projectors were too often incapable of providing the precise registration needed for proper and comfortable viewing. Although the polarized spectacles worn by the viewers helped create the impression of depth and dimension, the superimposed images, projected simultaneously, remained unsynchronized. In time the disenchanted audiences grew tired of the repetitious subject matter and turned in their goggles. Even from the comparative safety of comfortable theater seats, dizzying rides and flying objects had lost their novelty. Dimensional pictures fell "flat" on their respective effects.

The mild display of enthusiasm that greeted the 3-D presentations seemed like wild acclaim compared with the cooler receptions that were given to the newer wide-screen motion pictures by the theater owners. Realizing that their theater equipment would have to be redesigned or scrapped, the owners had little alternative but to accept the trend and dig deeper into already dented bank accounts. Optimistically, the screens were widened. From the standard 1.33:1 aspect ratio of the 35mm film frame to the 360-degree Circarama, wide-screen images have been produced, printed, and projected in just about every possible format imaginable. In some wide-screen systems, the images are recorded by several cameras and projected on deeply curved screens by a synchronous projection system. Other wide-screen processes use a single camera with an anamorphic lens that compresses or squeezes the image during the photographic process and restores it to its normal proportions during projection. Although standard-width 35mm film is used as the negative or in-camera stock for many wide-screen systems, others use 55, 65, and 70mm film for recording and printing purposes. In some cases, the images are exposed on 35mm film that travels horizontally through the camera. With this latter arrangement, the double-frame negative is approximately twice the size of the standard 35mm frame. To complement some wide-screen photographic processes, specially designed lenses, "bug-eyes," widen the camera range to as much as 150 degrees. In almost all cases, however, the wide-angle images can be reproduced on standard-width 35mm film for distribution to theaters that are not equipped with anamorphic lenses and curved screens.

The optical or photographic sound track that accompanies the 35mm film has kept pace with the constantly changing dimensions of the growing screens. With the picture stretching from one side of the theater to the other, stereophonic sound, from as many as seven separate sound tracks, has been substituted for the single optical track.

Because some theaters are equipped for magnetic stereo and others for optical sound, the magoptical sound system provides a compatible compromise. Differences in sound-reproducing systems are resolved by coating the anamorphic release prints with magnetic stripes. The subsequent combination, standard optical track and magnetic tracks, can be used with all sound-reproducing systems. A more recent development, *Perspecta* sound, is an ingenious method for producing stereo effects from a single optical sound track. With this sound-reproducing system, cueing control tones direct the sound to speakers located in different parts of the theater, or in the conventional manner, to the speaker system behind the motion picture screen.

The fact that the standards set by Thomas Edison have outlasted all the systems and processes mentioned previously attests to this inventor's genius and foresight. Although a square picture might have been the more practical shape, the great American inventor chose the 4-to-3 proportions for the film frame because he felt that those near-classic dimensions would offer the best possibilities for picture composition. That Edison was correct in his judgment is evidenced by the fact that the original aspect ratio, or the relationship between picture width and height, has withstood the test of time nobly.

Today, the numbers 1.33:1 are as familiar as the shape of a projected frame of film. To motion picture technicians, the two are synonymous—the number 1.33 indicating the horizontal dimension and the number 1 identifying the vertical. Mathematically, those proportions, or aspect ratio, remain constant, regardless of the size of the screen used for projection. A standard 35mm frame of film, even when projected on a wide screen, for example, still has the 1.33:1 proportions. *Wide screen,* a term used interchangeably, if not always with complete accuracy, for aspect ratio, is generally accepted as an all-inclusive designation for projected pictures that are wider than the 1.33:1 standard of the 35mm film frame.

Since the industry's infancy, many wide-screen processes have been exhibited at fairs and expositions throughout the world. The only consistency, with respect to the many processes that have been presented for the public's approval throughout the years, is the fact that there has been an inconsistency insofar as any attempts at standardization are concerned. See ANAMORPHIC IMAGE, ANAMORPHIC LENS, ANAMORPHIC PRINT, CINEMASCOPE, CINEMIRACLE, CINERAMA, DIMENSION 150, LENS SQUEEZE RATIO, MGM CAMERA 65, NATURAMA, PANAVISION, SCANOSCOPE, SQUEEZE LENS, SUPER PANAVISION 70, SUPERSCOPE, TECHNIRAMA, TECHNISCOPE, TODD-AO, ULTRA PANAVISION 70, ULTRA SEMISCOPE, VISTAVISION, WIDE-ANGLE LENS, WIDE SCREEN, and WIDE-SCREEN PRINT.

aspherical lens A lens whose surface does not follow the shape of a sphere. (Most lenses have spherical surfaces.) Aspherical lenses are used in some types of projection systems to correct for aberrations in other lenses in the same optical system.

assistant animator The artist or cartoonist who draws the intermediate positions of an action laid out by the animator. See ANIMATOR.

astigmatism An off-axis aberration in a lens that causes light rays to converge unequally, thus resulting in imperfect images. The resultant print, because of this type of aberration, may be sharp in some areas and out of focus in others. See Figure A-21.

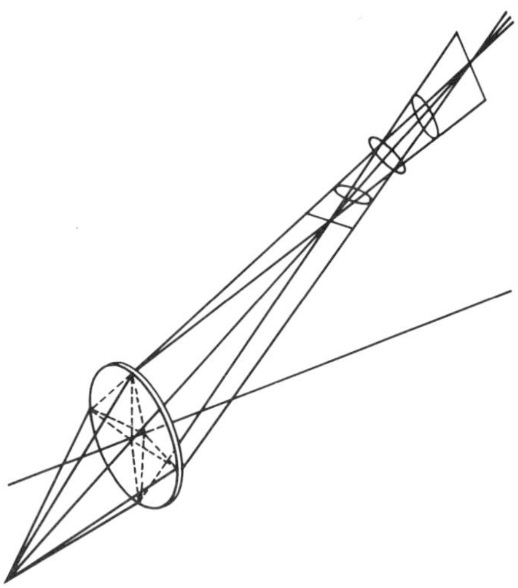

Fig. A-21 Astigmatism.

astronomical photography The photography of subject matter beyond the earth's immediate atmosphere. The process makes possible the observation and study of bodies that do not reflect sufficient amounts of light to permit adequate exposure. Highly sensitive film and time-lapse photographic equipment attached to powerful telescopes are used for recording and analyzing the movements and composition of heavenly bodies. Tracking devices keep the camera lens fixed on specific bodies for long periods of time. Astronomical studies have shown that, in some cases, the photographed subject matter was more than a billion light-years from the earth at the time of filming. Spectrograms are used for determining the direction in which a star is moving as well as its temperature and distance from earth.

asymmetrical lens A lens in which the elements are not arranged symmetrically. See Figure A-22.

audio From the Latin—"I hear." An all-inclusive term used to describe the recording, transmission, reproduction, and reception of sound.

audio cue channel The audio cue record/playback unit in videotape recorders permits the recording of cueing information along the edge of the tape. This information can be in the form of voice, tone, or digital cues. The audio cue unit permits independent recording of the audio signals without disturbing the prerecorded video.

audio head The device in the tape-transport mechanism that records the sound. On videotape, the sound signal precedes the corresponding picture information by $9\frac{1}{4}$ inches.

audio signal, television Although the television picture information and sound are recorded simultaneously, and transmitted over the same channel, the video signal is amplitude-modulated whereas the sound signal is frequency-modulated. The sound is transmitted over a frequency that is 4.5 megacycles higher than the frequency used for the video signal. The difference between FM broadcast bands ($\pm$75 kilocycles) and TV FM broadcast bands ($\pm$25 kilocycles) is the width of the sidebands. The frequencies used for FM broadcasts range from 88 to 108 megacycles. The frequencies used for TV FM broadcasts range from 59.75 to 889.75 megacycles. See VIDEO BAND.

autofocal A device, such as a cam or mechanical attachment, that automatically changes the focus to correspond with changes in the distance between the camera and the subject matter being photographed.

auxiliary lens A supplementary lens which, when added to the lens combination in use, causes changes in the focus of the lens system.

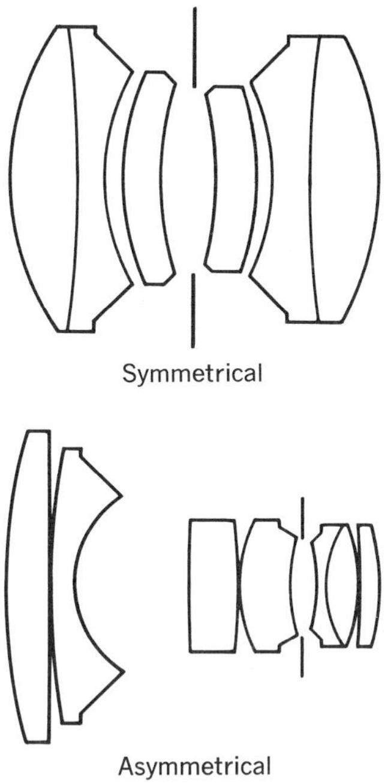

Fig. A-22 Diagram—Schneider lenses.

available light The filming of subject matter, usually in well-lit areas, where no supplementary lighting is required.

axis, optical An imaginary horizontal line passing through the center of every component part of an optical system.

azimuth In magnetic recorders, the degree of alignment between the operative gap in a recording or reproducing magnetic head and the lines of magnetism formed in the recording tape. Unless zero azimuth exists, the reproduced signal will be of low volume and contain distortion.

azimuth loss The signal loss due to misalignment of the playback head gap and the recorded signal.

B

baby Solarspot An all-purpose spotlight with flood to spot control. Its 6-inch Fresnel lens is capable of providing a circle of light approximately 10 feet in diameter from a distance of 10 feet when set at maximum flood. The lamps used with this unit range from 500 to 750 watts.

backdrop Draping or similar material that is arranged to serve as an appropriate background for the subject matter during the photographic process.

back focus The distance between the rear surface of a lens and the film in position for exposure.

background The general setting or location in which an action takes place. Also, the "flats" or semipermanent sets that are constructed and painted by scenic artists.

background, animation An opaque rendering or card that serves as the background for the inked and opaqued cels, titles, etc., that are photographed on the animation stand. See Figures B-1*a* and *b*.

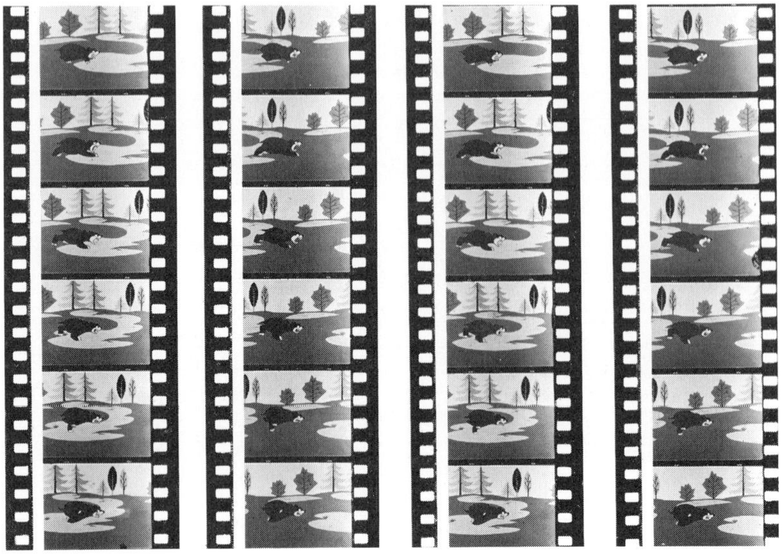

Fig. B-1*a* Animation background. A pan background for an animated scene. The animation is photographed over the background, via stop-motion processes, and the background is moved for each exposure according to the instructions on the exposure sheet prepared by the animator.

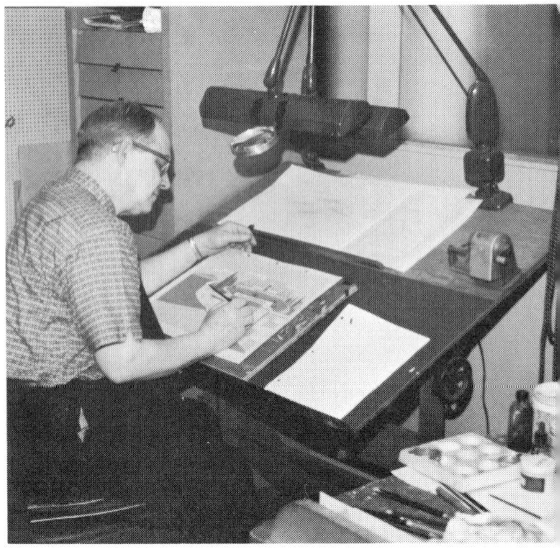

Fig. B-1*b* The background artist.

background, sound and recording Recorded sound (music, crowd noises, sound effects, etc.) that is used as a background for the dialogue in the sound track.

background plate The projected print or slide that serves as a background during the filming of process shots.

background projector The projector used for process cinematography. This type of projector must be capable of projecting rock-steady, evenly illuminated backgrounds and produce a perfectly flat field through its optical system. The newer models offer precision movement with positive registration pins, remote-focusing mobile base, rotating projector head, rotary distributor, water-cooled with high-intensity arc housing, automatic fire shutter and footage counters.

Hansard process background projector, 35mm. The Hansard projector's movement has two registration and two pulldown pins similar to those found in Bell & Howell and Mitchell cameras. The lenses used for projection are high-speed Bausch & Lomb or Kolmorgan in focal lengths from 4 to 9 inches.

The projector's optical system (lamphouse condenser, relay, and projection lens) is complemented by a cooling system. High-intensity carbon-arc lamps which accommodate carbons with a 170- to 250-ampere output provide the illumination.

A 220-volt three-phase synchronous motor assures synchronization between the camera and projector. A strobe light provides the means for checking that synchronization—even while the camera is running. The projector base features a built-in pan and tilt mechanism, and an intercom system which makes possible communication between the cameraman and the projectionist.

Mitchell background projector, 35mm. The Mitchell projector has dual registration pins and a movement that is similar to those found in the Mitchell cameras. Its optical system (condenser, relay, and lens) throws an evenly illuminated image over the entire surface of the translucent screens that are used for process shots. The lens supplied with the projector is an *f*/2.0 Bausch & Lomb Super Cinephor (5-, 6-, and 7-inch focal lengths) which can be focused either at the projector or, remotely, from a position near the camera.

The projector uses high-amperage arcs (170 to 250 amperes output) and is equipped with an iris control for regulating the amount of light at the aperture. An interlocking motor drive (220 volts, three phase) is used to synchronize the transport mechanisms in both the projector and camera. Frame and footage counters make it possible to locate specific points on the 1,000-foot spools or reels that the magazine accommodates.

The projector is mounted on a wheeled base that offers a solid support while providing mobility. The projector can be panned 180 degrees and tilted 12 degrees in either direction from its normal center position. The projector head can also be moved in a horizontal plane and locked in any desired position. A built-in intercom system makes communication possible between the projectionist and the cameraman.

backing That portion of a motion picture background which is visible through an opening in the set, such as window or door, for example. Actually, an auxiliary background. The backing for a window could contain clouds or distant buildings; and for a door, a hallway, landing, steps, or a wall with another door.

backlight—animation A source of illumination under the animation stand. The light is focused on the underside of the translucent glass positioned in the center of the compound table. This type of lighting is used for filming transparencies, the animator's drawings for test purposes, creating silhouette effects, and for photographing mattes that will subsequently be used in conjunction with other lengths of film during the optical process.

backlight—lighting Illumination aimed in the general direction of the camera, from the rear portion of a motion picture set. This type of lighting is used for the purpose of silhouetting or highlighting the subject matter.

baffle Basically, a shield or partition. In loudspeakers, a baffle consists of one or more sound-absorbing screens positioned within the speaker's enclosure. These screens prevent sound waves from bouncing back and forth within the unit, thus improving the fidelity of the sound being reproduced. A directional baffle, positioned in front of a loudspeaker, controls the direction of the sound emanating from it.

balance, lighting The proper proportioning of key and fill lighting in the illumination of a set.

balance, sound The proper relationship among the several sound elements such as music, dialogue, or sound effects in the recorded or composite sound track.

balancing stripe The narrow stripe of magnetic tape running lengthwise along one side of the film. Unlike the wider magnetic stripe that is used for recording, the narrow stripe is merely used to equalize the thickness of the roll so that it can be wound evenly.

ball-and-socket head A camera mount that is attached to a tripod by means of a threaded screw.

Balop Also **Telop**. Both are acronyms—**Balop**ticon projector* and **tel**evision **op**tical. The trade name for the graphic material that is positioned on easels and photographed by the television camera during the telecast. The cards containing the advertising messages or announcements are prepared with an aspect ratio of 1.33:1 in keeping with the standard proportions of the 35mm film frame. The upper right quadrant, in many cases, is reserved for the television station's call letters.

band, spectral Subdivisions of the spectrum. In color photography, the visible portion of the spectrum is generally thought of in terms of red, green, and blue bands of color.

banding, color In videotape recording, a disturbance characterized by horizontal bands (variation of hue in each band) caused when a video recording is played back on a videotape recorder with a small

*A Balopticon projector is a device that projects pictures by reflection.

difference in head speed. This can be corrected with the addition of an automatic velocity compensator to the playback machine.

banding, noise A disturbance characterized by grainy playback in one headband. Compare BANDING, COLOR.

bandwidth The spread between the high and low limiting frequencies (for sound or color) as measured in cycles per second.

bank A number of incandescent lamps mounted in a single reflector housing.

bar See COLOR BARS.

barn door A metal frame with two or four adjustable leaves or flaps that control the intensity and direct the path of light emanating from a light source.

A barn door is also a wipe effect in which the vertical, center portion of the scene is seen first. The remainder of the scene is revealed progressively on successive frames of film for the length of the wipe effect. The name is derived from the similarity of the effect to the opening of a pair of doors.

barney An electrically heated, insulated covering specifically designed to keep film in the camera pliable even at subzero temperatures. The same type of covering is also used to protect the camera and film from summer heat, moisture, and humidity.

barrel distortion Image distortion caused by uncorrected lenses or because of the relationship between the diaphragm and lens elements. With the diaphragm positioned in front of the lens, the lines forming a rectangle bulge outward. Positioned behind the lens, the same lines lean toward the center of the rectangle. Correction is possible by the addition of lens elements with complementary errors. See Figure B-2.

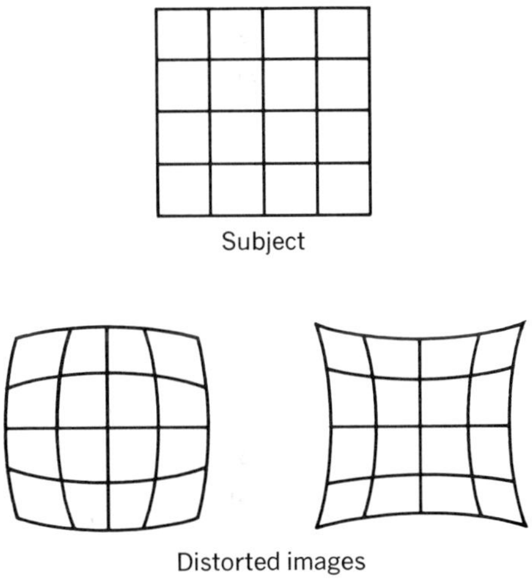

Fig. B-2 Barrel distortion.

barrel markings Imprints on lens mountings that indicate the focal length of the lens, the serial number, and the name of the manufacturer.

bar sheets, animation Also called **lead sheets**. Cue sheets, in some ways resembling music staffs, that provide a visual synopsis of an animated sequence. Prepared by the animation director, bar sheets show the exact number of film frames allotted to an action, the location of each musical beat in the accompanying sound track in

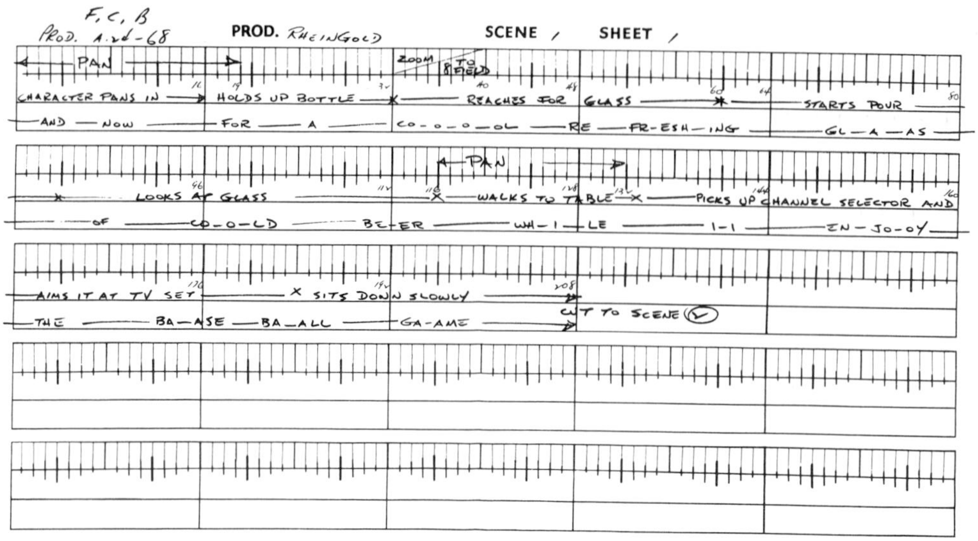

Fig. B-3 **Bar sheet.**

relation to specific frames of film, and the frames in which lip sync occurs. See Figure B-3.

base, film The material, usually acetate or polyester, that supports the photographic emulsion.

base, lenses Any of the triangular sides of a prism.

base light Diffused illumination used for the purpose of reducing shadows to negligible proportions.

batch A quantity of film, manufactured at the same time and stamped with the same serial number to indicate that the emulsion in each roll of the batch has the same characteristics and degree of sensitivity.

batten A length of pipe or wood used to support a light unit. A connecting cable or receptacle is usually attached to the pipe or wood so that the light unit can be connected to the power supply.

bayonet lock A fitting that makes it possible to attach or detach a lens in its housing quickly and easily. The lens is pressed into position against spring tension and twisted to lock it into position.

bazooka A device used to support lighting units positioned over the edge of an overhead catwalk. The bazooka's plate is nailed to the catwalk and fits over the handrail at a 45-degree angle. See Figure L-9 under LIGHTING ACCESSORIES.

beaded screen A white plastic or canvas screen used for projection. Its surface has tiny glass beads or globes embedded in it. Highly directional, this type of screen provides a relatively brilliant image for the viewer seated at an angle of approximately 15 degrees from the screen's center. For this reason, beaded screens are particularly effective in narrow rooms. Beaded screens, unprotected against daylight for extended periods, may turn yellowish and seriously affect the quality of the projected image. Compare ALUMINUM SCREEN and MATTE WHITE SCREEN.

beam A ray, or group of parallel rays of light, or other radiation.

beam coverage See COVERAGE, BEAM.

beam lumens See LUMENS, BEAM.

beam splitter The arrangement of prisms in cameras, optical printers, and projection systems that make it possible to divide or split a beam of light. It is used in self-matteing photographic processes, where the subject matter is photographed on two lengths of film, through a common lens, with the aid of the beam splitter. The image for the matte is exposed on high-contrast film. The same image is photographed simultaneously on black-and-white or color film and is subsequently used as the insert. See SELF-MATTEING PROCESSES.

bear trap A spring-clamp device used on motion picture and television sets for supporting such items as small light units, flags, or nets. The bear trap can be clamped to posts, rails, or pipes.

beat The musical tempo used for timing or synchronizing sound.

bel Abbreviated **b**. A unit of energy for measuring the relative volume of sound. It is equal to 10 decibels. The unit of measurement was named after Alexander Graham Bell. See DECIBEL.

bell lamp A bowl-shaped flood unit with a reflecting surface. A number of these units can be used in overlapping combinations to serve as the fill light for a large area. The unit accommodates bulbs ranging from 500 to 1,500 watts.

bellows Accordionlike extensions used to move the lens further away from the film plane in order to produce extreme close-ups with good definition. Their function is similar to that of an extension tube.

B-H curve A plotting for showing the characteristic curve of magnetic materials. The characteristics of a magnetic tape are plotted on a graph by the application of a magnetizing force to the tape. This force, applied at the minimum level, is increased until it reaches the point of saturation. At that point, the magnetizing force is brought back to zero.

The characteristics of magnetic tape, like those of motion picture film, vary with the type of tape produced by the various manufacturers. Tape varies considerably with respect to output levels, degree of noise, and distortion levels. For quality recording, tape should have a maximum variation of ½ decibel to a specific reference signal.

bias A high-frequency current which is applied to the recording head, during the recording process, for the purpose of increasing the signal-to-noise ratio and the strength of relatively weak signals while reducing harmonic distortion. Less technically, the introduction of a high-frequency signal during recording in order to minimize distortion normally present.

bias, peak An optimum setting or level, in relation to the maximum capacity of the equipment, for the recording of sound. Although it is possible to record at higher settings, the chances of losing quality in the recording are proportionately greater. Overbiasing degrades the frequency response. For best sound performance and maximum output, the bias control should be adjusted until the oscillator indicates a constant input level. Compare BIAS.

big eyes A 10,000-watt light source.

billboard An industry term for credits.

binary exposure system Also known as the **Norwood binary system**, this method of exposure determination measures reflected light from the camera location. Aimed in the general direction of the scene, the meter measures the spot brightness and the field brightness in a single operation. The meter evaluates both, favoring the foreground subject matter, thus assuring scene-to-scene consistency with respect to the skin tones of the performers.

binaural The process of producing two sync recordings on a single ¼-inch tape by means of a dual magnetic recording head. During the playback, each of the previously recorded signals is reproduced through separate speakers, thereby creating a feeling of depth to the sound. The process is also known as **stereophonic recording**.

bipack (opticals) A process for combining the picture information on processed film with graphics positioned on the animation stand, or with a second length of processed film, in order to obtain a direct

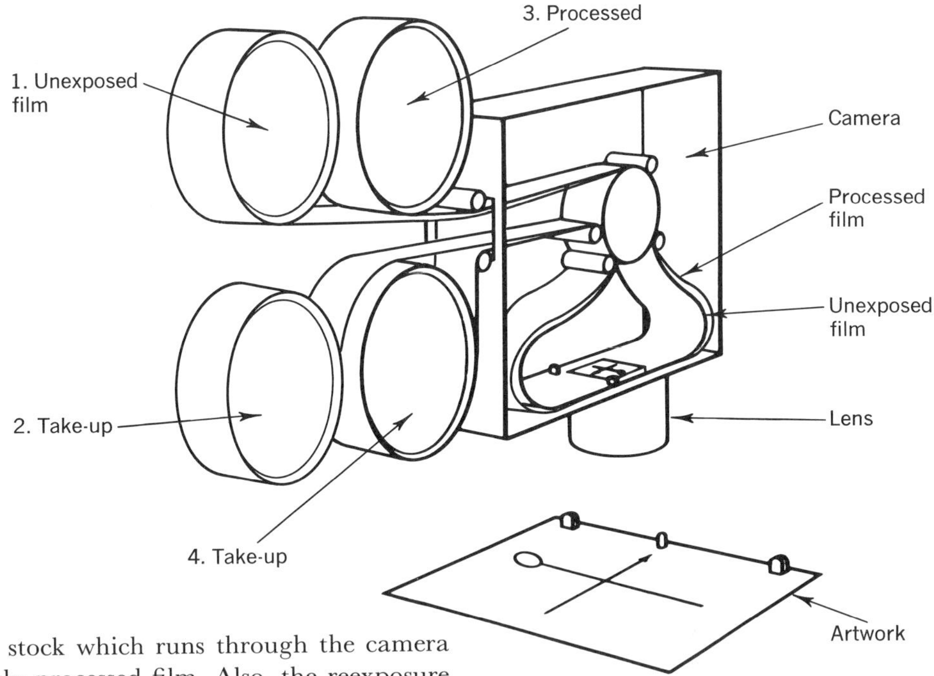

Fig. B-4*a* The camera and bipack magazine.

composite effect on the raw stock which runs through the camera in contact with the previously processed film. Also, the reexposure and combination of picture information from several lengths of processed film on the raw stock in the camera. The bipack is sometimes referred to as "the poor man's optical printer."

bipack magazine A four-sectioned film container. Mounted on the camera, the processed film that is to be reexposed is loaded into one of the feed compartments. The raw stock is loaded into the second feed compartment. The two sections on the other side of the magazine are used for take-up purposes. See Figures B-4*a* to *c*.

Fig. B-4*b* The bipack magazine.

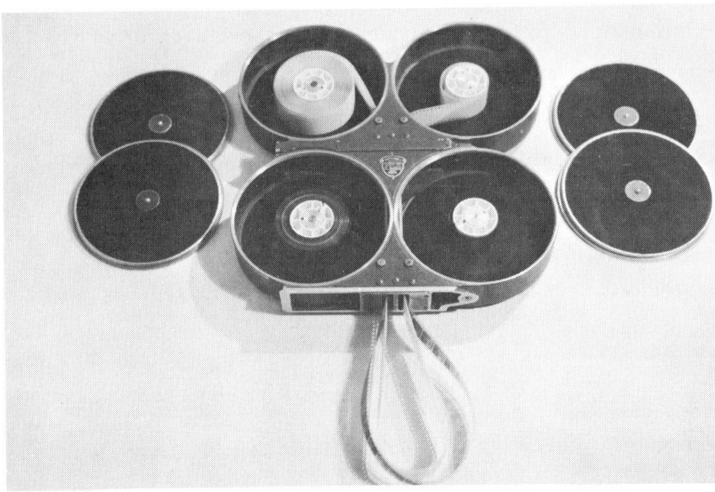

Fig. B-4*c* The camera and bipack magazine shown in relation to artwork registered on the pegs of the animation-stand compound table.

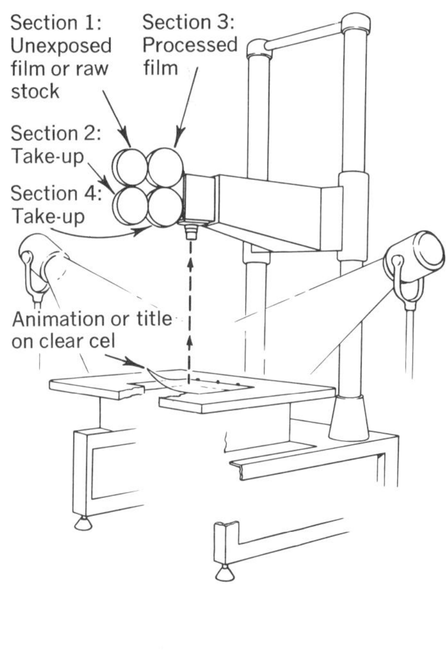

bipost base Lamp bases with twin posts that serve as the contacts. Capable of carrying heavy currents, up to 10 kilowatts, the metal posts support the filaments within the lamp.

bit part A minor role or walk-on part in which the performer may have few or no speaking lines.

black-and-white film Film with an emulsion layer that is incapable of reproducing color values in print material. Brightness values in the print are reproduced as whites, blacks, and intermediate shades of gray.

black-backing traveling matte See MATTES and SELF-MATTEING PROCESSES.

black balance A television camera control that varies the black level of color channels individually.

black body Black-body radiator is a concept in physics used to state the spectral composition of light and, in photography, to measure color temperature. A black body is a perfect radiator that reflects no light. It emits light only when subjected to heat. Thus the color temperature of a lamp, for example, is the temperature in degrees Kelvin to which a black body would have to be heated in order to emit similar light.

Pieces of burning coal, for example, glow with a deep red color that gradually changes to red, orange, yellow and, in turn, white as the temperature increases. If heated to greater temperatures, the white would, in time, turn to blue.

black level A television camera control that selects manual or automatic black-level control. Black level indicates the minimum signal needed for transmitting the darkest portion of the picture.

black light Neither black nor light, black light refers to invisible radiation such as ultraviolet. The name is applicable since it is possible to see objects in the dark by coating them with dyes which fluoresce when exposed to this invisible radiation.

blackout switch A master switch that controls and can, therefore, be used to turn off all the light sources on a stage or studio set.

blanker An auxiliary module in vidicon film cameras that adds the final blanking to the video signal. Blanking is that portion of the composite video signal whose instantaneous amplitude makes the vertical and horizontal retrace invisible. More simply, it is the interval during which the scanning beam, after completing the tracing of one picture, returns to the starting point to begin tracing the next picture. The process is repeated 30 times during the scanning of each field, or a total of 60 times each second.

bleaching A laboratory process in which a developed silver image is converted into a silver salt. Bleaching lightens an image and prepares it for the laboratory processes that follow. It initiates many of the toning processes, in those film types where metal compounds are needed to bring out the color values of the image. The process is also used to produce matrices that are used for making color prints.

bleed An undesirable effect caused by the seepage of dyes to areas beyond those intended for them. Where bleeding occurs, separation lines between colors are usually soft, untrue with respect to color, and, in other ways, distracting and objectionable.

blimp The camera housing that prevents undesirable noises, some from the camera itself, from interfering with the recording of the sound image.

bloom Glare caused by undesirable reflections striking the television camera lens. See CRITICAL ANGLE.

bloop (editing) An opaque triangle, usually of tape or plastic, that is placed over the sound-track area at splicing points. Without these masks, each splice would be accented by an undesirable sound during projection.

bloop light and blooping signal A cueing light attached to the camera that serves the same purpose as the clapstick that is used to indicate the start of a scene. The blooping signal is conveyed from the camera to the recorder by the sync cable and recorded on the sound track by the sync head. See SYNC GENERATOR.

blow-up (editing) The process of duplicating the picture information in 16mm film to the 35mm film size. The results, in some cases, are unsatisfactory because the grain in the 16mm film used for duplicating is magnified proportionately.

The term is also used to indicate the magnification of a portion of a film frame so that the specified area fills the entire frame. The desired blow-up, or magnification, is accomplished optically.

blue backing A background used for producing the mattes needed for special-effects photography. Used with color-blind or nonsensitive film, a product may be photographed in front of this type of background for subsequent superimposition over live-action scenes.

blue-backing self-matteing process See MATTES and SELF-MATTEING PROCESSES.

blue glass A blue viewing filter that eliminates much of the color contrast from a scene, thus indicating to the cameraman and director just how a scene will photograph in black and white.

blur An indistinct image caused by vibration, poor focus, or by any movement of the camera during exposure.

blurring Normally, a certain amount of blur can be expected when filming subjects that are moving at constant speeds during the interval of exposure. Blurring occurs with slow-moving objects as well as with those moving at greater rates of speed—and photographed at proportionately faster shutter speeds. Since the amount of blur, in "still" work as well as in motion picture filming, must be wider than that amount the eye can normally detect (approximately $^1/_{100}$ inch in normal-sized prints), most image blur is not objectionable.

An object that is moving from one side of the film frame to the other side of the frame will have longer "tails" and a greater degree of blur than the same object would have when moving toward or directly away from the camera. Objects that move closer to the camera or farther away from it merely show proportionate increases or decreases in size with a minimum of blur.

In motion picture photography, the cameraman should merely select a filming speed that is compatible with the lighting conditions, the speed of the film being used, and the most suitable lens aperture. Of the several influencing factors that must be considered, the shutter speed, with respect to blurring, is the least important.

Blurring, it must be remembered, may be considered an effect, depending upon the scene's content. As such, it is not objectionable but highly desirable. Blurring, in itself, is objectionable and disturbing when it occurs in subject matter that is intended for scientific study or analysis. See SHUTTER SPEED, MINIMUM.

boom A mobile, cranelike device to which the microphone is attached during the filming process. Mounted on wheels, the boom can be moved from one part of the set to another in order to follow the sound and action of the scene.

boom light A lighting unit that is attached to the end of an adjustable boom. The boom is used to follow the scene's action.

booster signal Also called **relay** and **microwave link**. An intermediate point at which the original telecast is received, amplified, and retransmitted to more distant areas.

booth A room in which projection equipment is housed on a more or less permanent basis. Small windows in the front of the booth are lined up with the several projectors so that each of their lenses faces the screen and projects an image that has a minimum of keystoning. In some theaters, the projectionist also controls the house lights and stage drapes from within the booth. Also, the control room at a sound recording studio.

borax A crystalline powder used in fine-grain developers. This comparatively weak alkali dissolves readily when mixed in hot water.

borderlight A striplight mounted above the acting area of a motion picture set or television stage. Compare CONCERT BORDER.

breakdown In animation, a notation on an animator's extreme drawings indicating that the assistant animator or in-betweener must make the intermediate drawings that are needed to complete the scene's action. See EXTREMES.

In motion picture production, the detailed analysis of the shooting script. Such pertinent facts as the scenes having the same general setting or background and the properties, cameras, and lighting equipment needed during the filming process are all indicated in the breakdown or production sheets.

breakers The switches that control the flow of current to studio lights.

breakup In television and videotape recording, a momentary image distortion.

bridge A transitional scene that is inserted in a motion picture for the purpose of improving the general continuity.

In musical backgrounds, those bars of music that act as transitions between the major portions of the score. These bridges are used to change the mood, key, or tempo.

brightness A term indicating the relationship between a source of illumination and the area being illuminated. The light emitting power of a source is expressed in candles per unit area.

Since brightness is a comparative term, it is quite possible to have two sources of illumination that differ in size but not in candlepower readings. Under those circumstances, the smaller of the two light sources would be considered brighter.

brightness range The relationship or ratio between the lightest and darkest areas in a frame of processed film.

brightness separation overlay Also called **montage amplifier**, **combined** and **moving electronic matte**. See MATTES and SELF-MATTEING PROCESSES.

brilliance (1) Also called **luminosity**. The intensity of the illumination that is reflected from a surface. (2) The efficiency of a lens as indicated by the ratio of transmitted light to incident light. Because of differences in the brilliance of lenses, two lenses of identical aperture may need completely different exposures in order to produce similar results.

brilliant A processed print, rich in quality. In prints referred to as brilliant, the colors in the original subject matter are faithfully reproduced, and detail is evident in both dark and light areas.

broad A source of illumination (2,000 watts) that provides a uniform light over relatively large areas. This all-purpose floodlight is used for illuminating backgrounds and minimizing shadows.

broadcast band Also called **channel**. The frequency range assigned to each commercial station by the Federal Communications Commission.

bromide Any of the salts used for photographic purposes—potassium bromide or silver bromide, for example. Bromides are made from bromine, a brownish red, nonmetallic element.

Potassium bromide is a photographic chemical that is used in processing solutions. Silver bromide is extremely light-sensitive and one of the principal components of photographic emulsions.

brute A large arc spotlight generally used for key lighting.

BSI speeds The standards set by the British Standards Institute for rating the speed of an emulsion. The British standards are numerically similar to the ASA standards.

bubble A studio term for an incandescent electric bulb.

buckling Because film expands in warm weather, contracts in cold, swells when wet, and shrinks when dry, it can, when affected by any of these conditions, "buckle" or ride off the sprockets in the transport mechanism of a camera or projector. Buckling will also occur when the film in the camera or projector is not threaded properly. Under any of these conditions, the film will jam in the camera.

When this happens, a mechanism called the buckle trip breaks the electric circuit to the motors and stops the camera, in order to prevent damage to the movement and other vital parts, as well as to inform the operator of camera malfunction. Shuttles must be checked and cleaned thoroughly after a buckle occurs and its cause is corrected.

buffer A salt, sodium carbonate, for example, which is used to keep a developing solution at its proper degree of alkalinity.

buildup A laboratory term indicating increases in density caused by the release of metallic silver in negative or print material during processing. See AGGLOMERATION and GRAIN.

bulb An all-inclusive lighting term for glass lighting elements, whether incandescent or gaseous. Bulbs used for lighting motion picture and television sets must be capable of producing an intense light for illuminating large areas. See Figure B-5.

bumper Extra footage that precedes the opening scene or follows the closing scene of a television commercial. Also referred to as **safety footage**, the practice of including additional frames at the beginning and end of a time spot is a precautionary measure that ensures against a blank screen if and when the television station's projectionist misses a cue.

burn-in A photographic process wherein white titles, usually on high-contrast film, are superimposed over a scene. The term is taken from the fact that the titles are given additional exposure during the optical combination of title and scene.

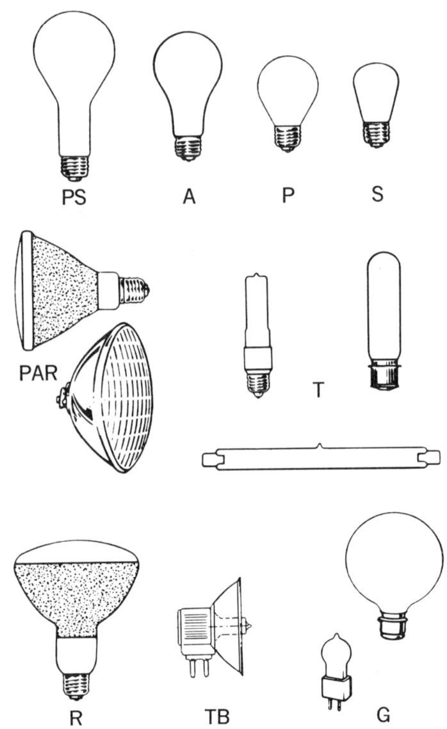

Fig. B-5 Common incandescent bulb shapes—not to scale. (*Sylvania Electric Products Inc.*) Bulb shapes normally are identified by a letter(s) describing the basic shape or some salient feature. T, for example, stands for tubular, R for reflector lamp, S for straight side.

The descriptive letter(s) is followed by a number expressing the approximate maximum diameter in eighths of an inch; a G-48 bulb is spherical in form with a maximum diameter of approximately 6 inches (48/8). Sometimes a suffix is used with the lamp designation to indicate the bulb finish; common ones include IF for inside frost, CL for clear, and various letters for colored bulbs. Suffixes are also used at times to designate special features such as design for rough service.

burst A signal or pulse of varying frequencies which is used for testing and monitoring the performance of a television transmitter. The frequency burst is inserted once, during the vertical blanking interval, for each field of scan—a total of 60 times each second. The 18th scanning line of the 525 lines that make up the complete picture is the one that is used for the burst signal.

business The on-stage actions and movements of performers, or specific aspects of these, during a scene.

butterfly A square-shaped aluminum reflector that is used for softening shadows in strong sunlight.

butt splicer This type of splicer is designed for butt-welding motion picture and magnetic film. Alignment is quite accurate with no overlap to throw frames out of focus, and no frame loss to throw double-system sound tracks out of sync.

buzz track A sound track that is used for testing the alignment of the sound head in projection and recording equipment. The sound track contains two frequencies—300 cycles per second on one side of the track and 1,000 cycles per second on the other side. If the scanning beam is lined up properly, no sound is heard. If the sound head is not in its precise position, one of the frequencies will sound off and produce a buzzing sound.

C

C mounts Lens mounts with standard threading that accept many types of lenses produced by the various manufacturers. Unlike bayonet locks, in which the lenses are snapped into position in the mount, lenses designed for use in C mounts are screwed into position.

cabinet, drying A dust-free, metal cabinet (usually heated) that is used by the laboratory for drying film.

cable release A trip device, usually a plunger and connecting wire in a flexible cable, that is attached to a camera and used to activate the shutter.

calibrations In animation, the regularly spaced markings on pan backgrounds that show the cameraman where the background must be positioned for each frame of the panning action. Also, the mathematical computations, based on the zoom, north/south and east/west counter numbers on the animation stand, that serve as a guide for making zoom and pan movements.

In lenses, the markings on a lens housing. Also, the calculation and subsequent selection of lens settings based on the sensitivity of the film and lighting being used.

call sheet A posted studio schedule announcing which scenes are to be photographed on a specific day and the time the performers are expected to be made up and ready for filming, along with other pertinent information.

cam In animation, a stationary bar on the animation stand to which the lens mount is mechanically linked. This arrangement makes it possible for the camera lens to remain in precise focus throughout zooms by automatically changing the focal plane.

In motion picture cameras, the cam is the motivating device that controls the movement of the claws that engage the film's sprockets and advance each frame into position for the interval of exposure.

camera The device used for obtaining a photographic image is known as a camera. Also, a device in which sensitized film is exposed by light transmitted through a lens.

Although cameras differ greatly in design and construction, the basic components are, of necessity, similar. Early cameras were not dependent on lenses for the formation of images. Pinholes, in light-tight wooden boxes, permitted light to enter and expose the film. Lens caps acted as shutters, and images were recorded by simply removing the caps for the length of time needed for the exposure. The mechanism that held the film in position was equally crude by comparison with modern camera movements and registration devices. Composition, framing, and the determination of picture area were based on educated guesses and the accuracy of the cameraman's aim.

Although photographic manuals are constantly describing new equipment, the basic camera components are practically the same as they were in the early days of picture making. Refinements in movements, lens mounts, and focusing and finder arrangements have made the cameraman's lot considerably easier. The development of frame and footage counters, motor drives and antibuckle switches, as well as the addition of built-in filter and matte holders, are other features that have contributed toward making modern cinematography a science rather than an art.

In all motion picture cameras, movements are designed to ensure precise positioning of the film at the time of exposure. In some cameras, fixed pilot pins engage one of the sprocket holes along the edges of the film and guide each frame into position in the shuttle, the ingenious mechanism that holds the film in place during the interval of exposure. Some cameras have special pressure plates that can accommodate several thicknesses of film without any adjustment.

Shutters, the mechanical devices that allow light to enter the camera and expose the film for a predetermined amount of time, are positioned in one of two planes, at the lens or at the film. Theoretically, the perfect camera shutter does not exist. Mechanical perfection would permit the shutter to open and close instantaneously. The perfect shutter would, therefore, remain in its open position for the entire interval of the preset exposure time.

With settings to 180 degrees, shutters allow the cameraman to control the amount of light that reaches and exposes the film. They may be locked and preset at any degree for achieving ghost images and other special effects. In-camera fades and cross-dissolves can also be made by unlocking the shutter and setting the indicator for the length of the desired effect.

Lens mounts, too, have been redesigned and modernized to include provision for rapid changeover from one lens to another. Many have micrometer focusing mounts and turrets that revolve in either direction.

Some cameras permit visual focusing, through the taking lens, even when the camera is in operation. In others, critical focusing is achieved by means of calibrated lens scales. To simplify the cameraman's task further, the images seen on the ground-glass finder are not only read correctly, as to right and left, but they can be magnified up to 20 times the normal viewing size as well.

Today's cameramen frown upon the ancient practice of hand-cranking the film, and find motors, spring drives, and batteries far more practical and definitely more accurate and less tiring. The film in today's cameras can be made to run forward or backward at a predetermined number of frames per second. Gearboxes with adjustments for different speeds make it possible for the photographer to take not only the standard 24, but from 1 to as many as 128 pictures per second. Even the latter speed seems slow when compared with the several thousands of frames per second that can be filmed with high-speed photographic equipment.

During the photographic process, unexposed film unwinds from a spool in the feed compartment of the lightproof magazines that fit snugly on top of the camera. The film passes through the camera shuttle in a stop-and-go intermittent movement. Registration pins in the shuttle engage sprocket holes along the edges of the film and hold each frame in position while the shutter opens and allows the subject matter to be exposed for a predetermined amount of time.

After the exposure has been made, the film is advanced, and the next frame is brought into position for the interval of exposure. The exposed film passes over the sprockets and continues on its way to the take-up compartment of the magazine. This process is repeated 24 times each second when the camera is operating at normal sound speed and 18 times per second at silent speed. The camera and magazine, when sound and picture are being recorded simultaneously, are covered by a "blimp," a housing designed to hold camera motor and other noises at a minimum level.

Some cameras are hand-held. Others are supported by tripods. Some are moved about on dollies known as *crabs* while others are set on top of cranes that are referred to as *cherry pickers.* For following and filming action at ground level, cameras are often attached to automobiles. For air-to-air or air-to-ground shots, cameras are taken aloft in airplanes and helicopters where they are either hand-held or mounted on devices that keep the cameras steady during the filming process. When necessary, cameras are taken to the very depths of the ocean in watertight and pressureproof housings.

In cold weather, they are protected from the elements by coverings known as *barneys.* In tropical countries, they are shielded from the hot sun by umbrellas and other protective coverings. In recent years, cameras have been used quite efficiently by orbiting astronauts.

The images on the processed motion picture film are used for entertainment as well as for scientific and commercial purposes. Such graphic records provide a permanent and convenient means for viewing, analysis, and study of subject matter. Indeed, photographic processes surpass, in some ways, the capabilities of the human eye as an optical instrument. The eye, for example, cannot record information in the invisible range of the spectrum. Neither is it capable of accurately following or recording images that travel at significant rates of speed. See Figures C-1*a* to *d* and Table C-1.

Arriflex model 2C-T, 35mm. The 2C-T has a special film movement and matching film gate for the two-perforation-high Techniscope frame.* The design takes advantage of the shorter pulldown and provides a 200-degree shutter opening.

The 2C-T does not require converted magazines. Regular Arriflex 200- and 400-foot magazines can be used and interchanged among all Arriflex 35 cameras. The camera is equipped with mirror-shutter reflex viewing, interchangeable ground glass, larger viewfinder field, a divisible eyepiece which permits use of the periscope-finder attachment, an automatic-closure eyepiece (optional extra), cardioid film movement, three lenses, and a ground glass marked for the Techniscope format. The 2C-T uses all regular accessories. A ready-to-shoot Arriflex with a 400-foot magazine and handgrip motor weighs 13½ pounds.

Arriflex model 2C-V, 35mm. The Arriflex model 2C-V is equipped with a variable shutter. The model 2CGS has a 60-cycle generator and an automatic clapstick. Basically similar to the other cameras in the 2C series, the 2CS is equipped for operation to 80 frames per second.

*The Techniscope process saves on the cost of original film and processing and utilizes inexpensive lenses. From the two-perforation-high frame original, it yields optical prints in most of the release formats, conventional or anamorphic, 35 or 16mm.

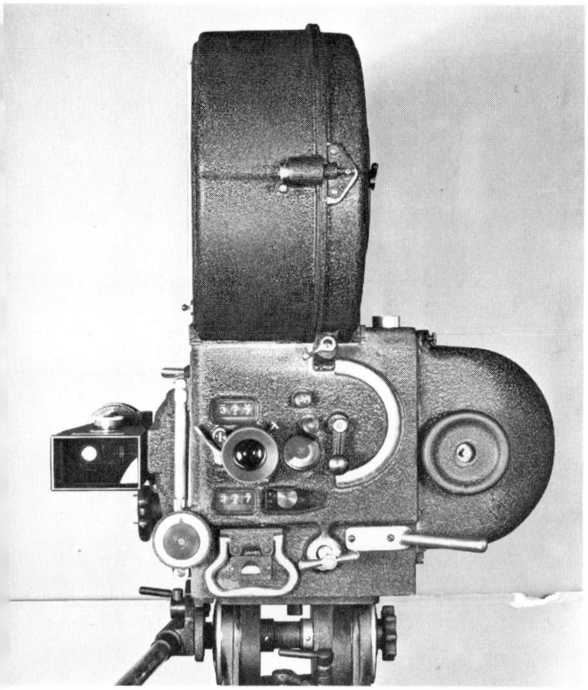

Fig. C-1*a* The camera from the cameraman's position. Shown in the photograph are the viewfinder, counters, and the blimp covering the magazine.

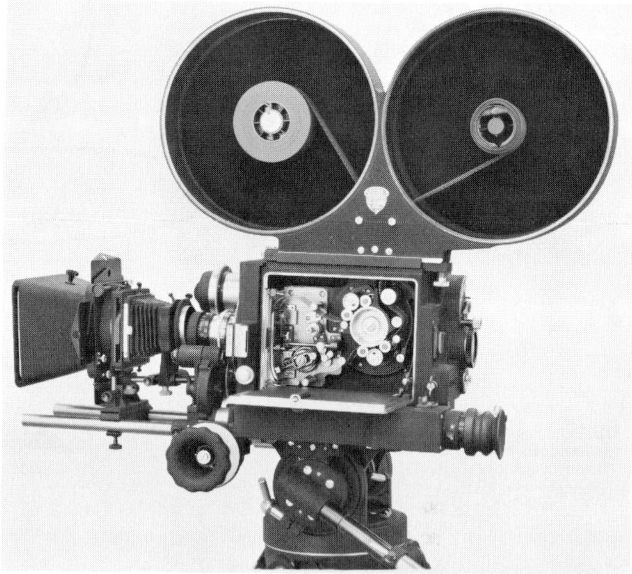

Fig. C-1*b* The loaded camera with the magazine covers removed and the camera box open. (*Meriman-Renish, Los Angeles.*) Also shown are the lens mount, eyepiece, and matte box.

Arriflex model 16S, 16mm. The 16S model is compact and lightweight. The ready-to-film camera weighs just over 8 pounds with a 100-foot daylight spool, three lenses, motor, and matte box. The camera features a mirror-shutter reflex viewfinder, precise registration-pin film movement, a three-lens divergent turret, a frames-per-second tachometer, and a film capacity from 100-foot internal spools to 400-foot magazines.

Arriflex model 16M, 16mm. Standard equipment for the 16M model includes a 60-cycle signal generator and automatic clapstick for location sound filming. The Arriflex 16M features three quick-change magazines (M-200-foot, M-400-foot, and M-1,200-foot), a mirror-shutter brilliant reflex finder, precision registration-pin movement, three-lens divergent turret, tachometer, frame and footage counter, and a complete system of professional accessories.

Arriflex model 16BL, 16mm. The Arriflex 16BL is a compact location sound camera. For double-system or single-system sound, the camera is convertible without the use of special tools. The Arri recording module provides for this capability. Residual noise level of the 16BL is approximately 31 decibels. The camera weighs 18 pounds, including the 12 to 120mm Angenieux zoom lens, 12-volt

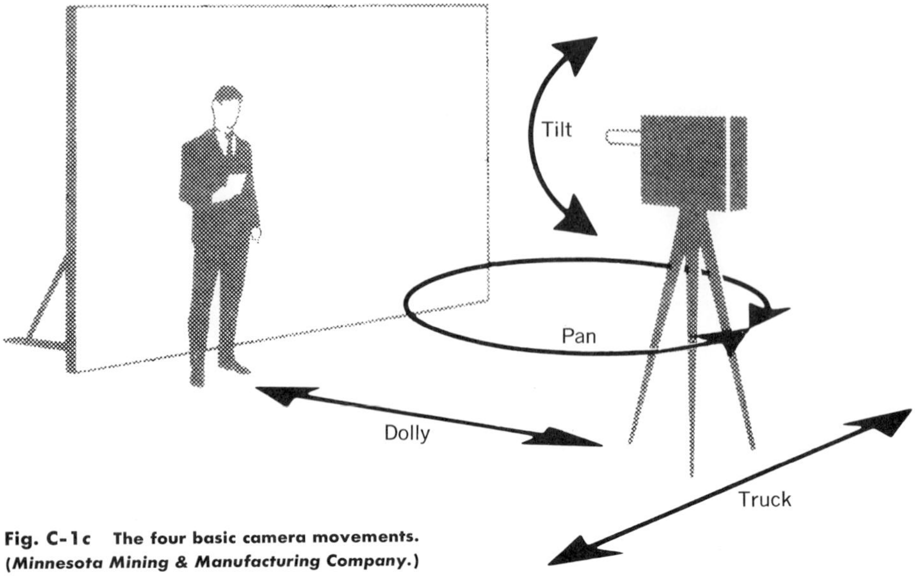

Fig. C-1c The four basic camera movements. *(Minnesota Mining & Manufacturing Company.)*

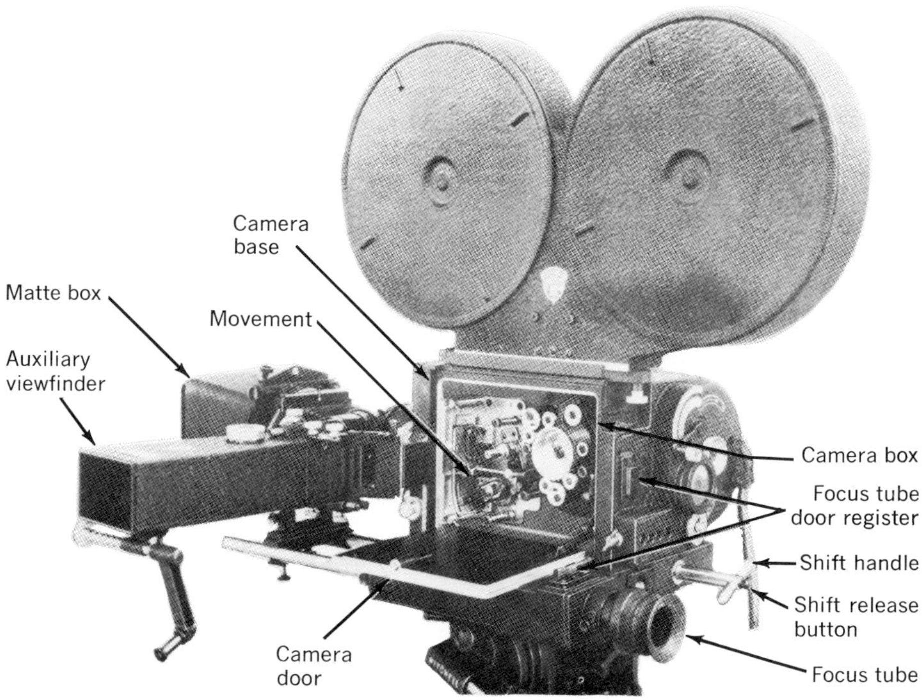

Fig. C-1*d* Access to movement (Mitchell NC).

TABLE C-1 Camera Troubleshooting Chart (NC-BNC)

Trouble	Probable cause	Remedy
	Troubles encountered during operation	
Film will not enter race in threading.	Registration pins not retracted.	Retract registration pin from film race.
Camera door will not close.	Pulldown claw not retracted.	Rotate drive mechanism until pulldown claw retracts from film race.
	Registration pin not engaged.	Move registration pin knob forward until in position.
	Sprocket film guides not closed.	Turn sprocket film guide knobs to position rollers on sprocket.
Blurred image or no image in focusing tube.	Magnifying knob in middle position.	Move knob entire distance in direction desired.
Motor will not run.	Buckle trip switch not closed.	Pull out buckle trip reset knob.
	Insufficient power at source, poor plug connections or defective cable.	Check power input line and power source. Look for damaged or corroded pins on plugs.
	Temperature too low for proper operation.	Test motor off camera. If motor runs but will not drive camera, it may indicate that camera parts have tightened owing to cold. Pull off magazine drive belt and let camera run without film until warmed.
Film does not take up.	Take-up belt broken or slipping.	Replace or tighten belt.
	Film not entering take-up core properly.	Check film in take-up drum. If scenes have been shot, open drums in darkroom.
Film being scratched.	Dirty aperture plate.	Clean aperture plate.
	Pressure plate rollers jammed.	Clean and lubricate pressure plate.
	Loop too large or too small.	Correct loop size.
	Magazine or magazine rollers are dirty.	Clean magazine.

TABLE C-1 Camera Troubleshooting Chart (NC-BNC) (cont.)

Trouble	Probable cause	Remedy
	Troubles encountered during operation	
Film being torn.	Improper threading.	Check upper and lower loops.
Film being punched or perforations being torn.	Improper threading.	Rethread properly.
	Troubles apparent on examining film	
No image on film.	Shutter entirely closed.	Check shutter before taking.
	Lens cap not removed.	Remove lens cap before taking.
	Camera not racked over to photographing position.	Shift camera into shooting position.
Picture obstructed.	Matte box improperly adjusted or bellow not fully collapsed and flush against 35mm lens. Sunshade not removed when using 25mm lens.	Remove sunshade.
	Matte left in matte box.	Remove matte when taking straight shot.
Picture incorrectly framed.	Viewfinder not adjusted correctly for parallax.	Check parallax adjustment.
	Viewfinder matte adjusting knobs not set to match lens in use.	Set knobs properly.
	Viewfinder matte adjusting knobs incorrectly calibrated.	Correct calibration.
Faulty focus.	Lens not correctly calibrated when focusing with measuring tape.	Check lens calibration. Readjust if necessary.
	Lens elements loose.	Check and tighten lens elements.
Film fogged.	Magazine covers not securely screwed in.	Check covers and tighten.

TABLE C-1 Camera Troubleshooting Chart (NC-BNC) (cont.)

Trouble	Probable cause	Remedy
	Troubles apparent on examining film	
Film fogged.	Camera door not completely closed.	Check door latch and door edges for light leak. If door is sprung, camera must be overhauled.
	Loose or worn magazine light trap rollers.	Replace magazine and send worn magazine to factory for repair.
Picture fuzzy or blurred.	Lenses or filters dirty or covered with moisture.	Clean lenses and filters.
Pictures flicker.	Excessive variations in line voltage.	Check with voltmeter.
	Loose electrical connections.	Check connections.
	Film loops too small.	Rethread and adjust loops properly.
Pictures unsteady.	Movement worn or register pins out of register.	Send camera to factory for repair.
	Tripod not steady.	Be sure tripod is securely positioned before taking.
Frames improperly exposed, unevenly spaced, or vertically blurred (double image).	Camera out of synchronization.	Time (synchronize) camera.
Harmonic beat (picket-fence effect in pan shots)	Shutter closed down too far.	Keep shutter open 90 degrees or more for pan shots.

Courtesy of Mitchell Camera Corp.

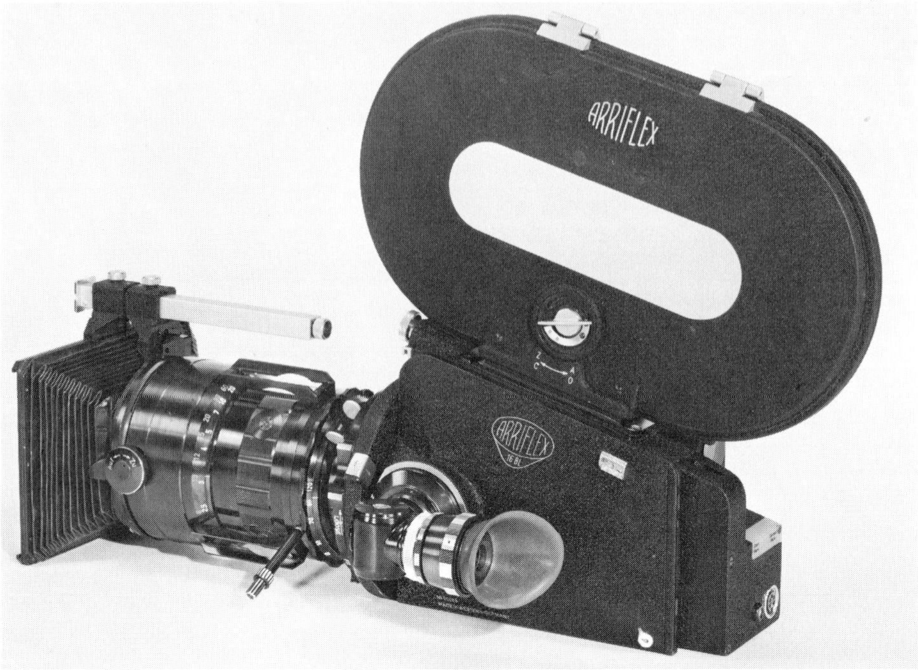

Fig. C-2 **Arriflex 16BL.**

d-c governor-controlled motor and BL-400-foot magazine. See Figure C-2.

Auricon cameras. Auricon cameras are self-blimped sound-on-film cameras. They can be equipped with either optical (RCA variable-area type or Western Electric variable-density type) or Auricon Filmagnetic sound heads for recording lip-sync 16mm sound motion pictures. All models have a sinusoidal, intermittent movement, utilizing a self-engaging single-claw film pulldown of hardened tool steel. All models have 175-degree fixed shutters ($^1/_{50}$ second at 24 frames per second).

Auricon Cinevoice II, 16mm. This lightweight 16mm camera comes equipped with either a three-lens turret (C mount) or single-lens mount for zoom-lens use. Although the camera has a 115-volt a-c constant-speed motor, synchronous motors are also available. The finder, with parallax adjustment, covers various lens fields from 13 to 75mm. Available with variable-area or variable-density optical recording galvanometer, or with Filmagnetic recording head, the camera can be supplied without sound equipment for silent use, or complete with amplifier and other recording accessories. In addition to the 100-foot internal-daylight-loading spools, Gordon Enterprises offers custom modifications for various external magazines to increase the camera's film capacity.

Auricon Pro-600, 16mm. This 16mm camera is equipped for

600-foot magazines and has a built-in heater with thermostat control, electromatic film take-up, and synchronous motor drive. The camera may be used for single- or double-system sound recording. Variable-density, variable-area, or Filmagnetic recording heads are available. The camera can be supplied with single-lens C mount for zoom lenses, or with a three-lens turret. The turret is equipped with a telefinder optical system which provides a 10× magnified viewfinder image for Telephoto lenses. Three matching finder objective lenses revolve with the turret to match the lens in use. A critical ground-glass focusing eyepiece is provided on the turret model. The lens and turret "blimping" hood prevents camera noise from reaching the microphone even when filming extreme close-ups. The camera's electromatic take-up operates without clutches or belts. A small amount of tension is applied to the take-up spool even when the camera drive motor is turned off. The Pro-600 is available in various camera and sound combinations.

Auricon Pro-600 special, 16mm. This 16mm camera is a lightweight, portable version of the Pro-600 model. The 400-foot magazine takes either a lab-pack film or daylight-loading spools. The camera is driven by a synchronous Auricon sound-drive motor. The camera is available in single-lens C mount for zoom lenses or with a three-lens turret. The camera is prepared for the Filmagnetic sound unit. Film take-up and other features are the same as in the Pro-600. The camera, with 400-foot magazine and zoom lens, weighs 24 pounds.

Auricon Super 1200, 16mm. This 16mm camera has a stainless steel pressure plate that is removable for cleaning. An external push button retracts the claw for threading and opens the shutter for optical, 10 × magnified, full-frame, ground-glass focusing. The three-lens C-mount turret has matching telefinder lenses. The variable shutter permits fade-in and fade-out. It may also be locked for exposures from $^1/_{50}$ to $^1/_{200}$ second. Variable-density, variable-area, or Filmagnetic recording heads are available. The self-blimped Super 1200 requires no external blimp for studio operation. Indicator lights are provided at front and rear to signal that the camera is running. A safety switch arm signals a warning when the film is not functioning normally. The camera's 1,200-foot magazines record 33 minutes of continuous sound film. The lens and turret blimping hood prevents camera noise from reaching the microphone even when filming extreme close-ups. The Super 1200 is equipped with lock-in type, heavy-duty electric connectors for sound recording and power inputs and footage and frame counters.

Beaulieu Super 8, model 2008S. This Super 8 camera has many advanced features including a transistorized electronic system powered by rechargeable batteries. This electronic element drives the camera at precisely regulated speeds from 2 to 50 frames per second for single-frame and radio-controlled filming. Monitoring facilities include a frame counter, footage counter, photocell pointer, film drive, and a battery-charge monitor.

The camera features a two-way push-button trip (lockable for continuous filming or threaded for the connection of a flexible release), provision for interchangeable lenses, a three-position master switch (battery check, manual, auto), a variable shutter for fade-in/fade-out transitions, a wide range of filming speeds (controlled by transistorized electronic regulator), an adjustable viewfinder eyepiece with 20× magnification, and a battery circuit plunger switch. (Light pressure cuts the batteries in; release cuts them out.) See Figure C-3*a*.

Beaulieu model R16 sync, 16mm. This Beaulieu 16mm camera features professional-sound electronic regulation—a transistorized system covered by the manufacturer's patents. The Synchro-pilot, an alternator driven by the camera motor, is connected by cable to the tape recorder's sync head. When the camera is running, it generates a current which, once recorded on the magnetic tape, ensures sound-image synchronization. The R16 sync features a three-lens rotating turret, or fixed mount for zoom-type lenses, a high-luminosity reflex finder which provides 10× image magnification, a high-precision CdS cell built into the viewfinder system, eight calibrated filming speeds from 2 to 64 frames per second, footage counters graduated in feet and meters, and frame counters graduated from 0 to 100. The camera accepts 50- and 100-foot reels of film.

Fig. C-3*a* **Cinema Beaulieu 2008S.**

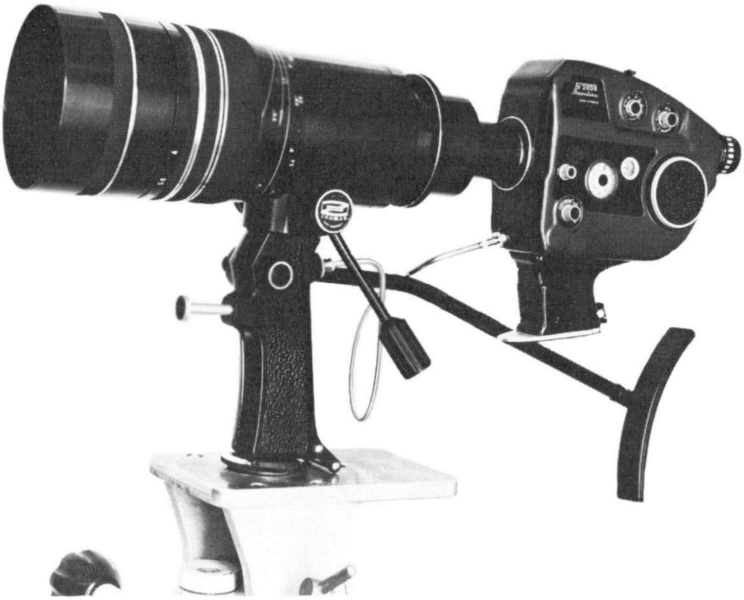

Beaulieu model R16ES sync, 16mm. This Beaulieu 16mm camera features an elliptical movement which utilizes a self-engaging single-claw pulldown, film speeds from 2 to 64 frames per second, a guillotine (up and down) sliding mirror-surfaced shutter, and reflex-focusing on a divided-grain ground glass with 10× magnification.

The camera's three-lens turret, fitted with three actuating knobs, accepts all standard C-mount lenses. Light entering the taking lens is metered by a CdS photocell. The camera employs a tachometer-generator system that reads the oscillator's beat against its own rotation rate. The camera accepts 100-foot internal-daylight-loading spools. The R16ES is provided with a socket for mounting a 50- or 60-cycle sync generator, which automatically couples to the camera shutter, for providing sync pulse to ¼-inch magnetic tape recorder for synchronous double-system sound filming. The camera's master switch controls the operation of film in either forward or reverse shooting. See Figure C-3*b*.

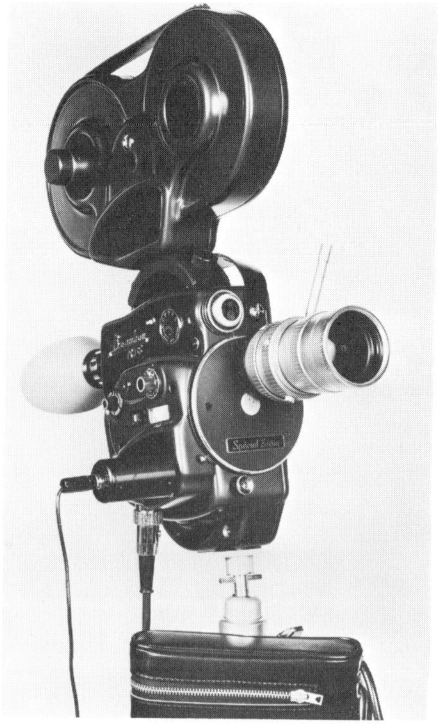

Fig. C-3*b* Cinema Beaulieu R16ES sync.

Beckman & Whitley model CM16, 16mm. This model weighs 15 pounds with a zoom lens and 400 feet of film and measures 3 by 8 by 16⅜ inches. The slim, rear-mounted magazine places the center of gravity on the cameraman's shoulder and can be controlled with one hand, leaving the other hand free to follow focus or adjust controls. The reflex viewer is positioned so that the cameraman can see the full 180 degrees in front of the camera. The on/off switch is at the index fingertip on the grip. All the other controls—footage counter, tachometer, aperture dial, lens setting—are visible on the operator side.

The CM16 is designed to record single-system sound and provide double-system sound-recording capability. During operation, the film is engaged by a tracking register pin. The continuous film movement provides for easy threading with no pins or claws to retract, engage, or prealign. In addition to the objective lens iris, the camera has its own internal iris to control film exposure. This permits full-stop viewing at all times and provides more control over focus than is possible with other internal viewers. The 12× magnified, erect image seen in the viewer is horizontally 20 percent larger than the field so that the cameraman can anticipate action entering from left or right. The eyepiece can be adjusted ±3 diopters to accommodate the cameraman's vision. All viewer optics are coated and color-corrected.

Because the entire movement is sound-isolated from the camera body, no blimp is needed. The 200-, 400-, and 1,200-foot magazines place supply and take-up reels side by side so that there is no perceptible change in camera balance as the film flows from one part of the magazine to the other.

The stroboscopic tachometer which detects drifts in film speed is positioned on the operator side where the cameraman can see it at all times. The internal filter slot is positioned in parallel light to eliminate image shift due to filter thickness. The camera lens mount accepts all C-mount fixed-focal-length lenses. Some lens designs are not, however, optically compatible with the CM16. Adapters are available for Arriflex-type mounts. Both camera and magazines are equipped with thermostatically controlled heaters. These permit camera operation over a wide temperature range from −30 to +120°F.

Built for military applications, the synchronous motor is customized for sound-film camera requirements. It is designed so that speed is always constant and does not change when the load changes. It operates on 15 volts a-c at 60 cycles and needs less than 20 watts of power to drive the camera and 400-foot magazine.

In addition to the CM16, Beckman & Whitley manufactures some 40 different high-speed cameras that record at rates from 200 to 200 million pictures per second.

Bell & Howell model 70, 16mm. Used by the armed forces, Bell & Howell's 16mm model 70 camera includes the following features: a gear-coupled lens and viewfinder turret which automatically positions the viewfinder and matching lens. Long-footage film run is provided by a spring-driven motor with ratchet-type winding key. The motor is governor-controlled for constant speed and has an automatic run-down stop. The camera's film-footage indicator registers up to 100 feet of film.

The seven operating speeds, from 8 to 64 frames per second are governor-controlled and can be changed without stopping the camera. The 100-foot capacity of single- or double-perforated film can be increased to 400 feet with a magazine accessory. A removable hand crank, with frame counter, backs up the film for double exposures.

The starting button can be operated only when the proper lens is in position, or it can be locked for continuous run. The camera housing itself is made of die-cast aluminum alloy for light weight and strength. A positive viewfinder shows the exact field covered by the lens. The eyepiece focuses to compensate for normal eyesight variations. The parallax adjustment dial is calibrated from 3 feet to infinity. The central area of the image can be magnified by 15 diameters for precise focusing during close-up work. The camera's rotary-type shutter automatically closes if the camera is stopped during normal operation, thus eliminating flash frames. All model 70 cameras can be factory-adapted for motor drive and 400-foot magazine.

Bell & Howell model 70TMR, 16mm. This camera is a modified version of the 70 camera and is designed for time-motion studies. It converts split-second movement into extreme slow motion, permitting careful analysis of motions, methods, or entire procedures. High-speed operations can be measured down to $\frac{1}{4,000}$ of a minute to determine actual time values or to record difficult operation procedures for rapid on-the-job training.

Changes from normal field to wide-angle or telephoto fields are provided by one of the three lenses. The gear-coupled turret head automatically positions each lens and matching viewfinder. Camera speeds conform to the standard decimal system used in time-motion studies. The camera is also calibrated for 16 and 24 frames per second filming. See Figure C-4*a*.

Bell & Howell model 70DR, 16mm. The DR model is spring-driven for use as a hand-held camera or with a tripod and offers seven speeds from 8 to 64 frames per second. The three-lens turret is gear-coupled to matching, positive-viewfinder turret. The camera provides magnified, through-the-lens focusing; a forward-reverse hand crank; and spool loading with a 100-foot film capacity.

Bell & Howell model 70HR, 16mm. The 70HR camera has all of the features found in the 70 DR. It is equipped with a shutter stabilizer and a Veeder footage counter and comes adapted for the addition of an electric motor and external film magazine. See Figure C-4*b*.

Bell & Howell model 70SR, 16mm. This camera is basically the same as the 70HR but is a high-speed camera for making slow-motion films. The camera operates at a speed of 128 frames per second only.

Bell & Howell Eyemo, 35mm. The Eyemo is a versatile camera with a detachable, calibrated crank which counts individual frames and allows backup for dissolves. The camera's spring-driven motor provides filming speeds from 8 to 48 frames per second, with instantaneous starting, stopping, and friction-clutch take-up. The positive viewfinder gives full-size erect image with matched finder and taking lenses. A handle and wrist strap permit one-hand operation. All the Eyemo models take a 100-foot daylight load.

Bell & Howell model 71KM, 35mm. The 71KM is a single-lens camera with speeds from 8 to 48 frames per second. It has a gelatin filter carrier, hand crank, and viewfinder with a 2-inch objective lens. The motor and magazine adapters are extras.

Bell & Howell model 71RM, 35mm. The single-lens model 71RM offers the same features as the 71KM, but includes custom-fitted motor and magazine adapters. It does not include the built-in gelatin filter carrier.

Bell & Howell model 71QM, 35mm. The three-lens spider turret camera model 71QM has six speeds, from 8 to 48 frames per second, a hand crank, and a built-in gelatin filter carrier. The camera also features a parallax-correcting viewfinder with a three-objective turret and a 2-inch objective lens, a ratchet winding hand crank, prismatic focusing magnifier, and adapter for motor and magazine. The 71QM accepts 200- or 400-foot magazines and has a 110-volt a-c motor.

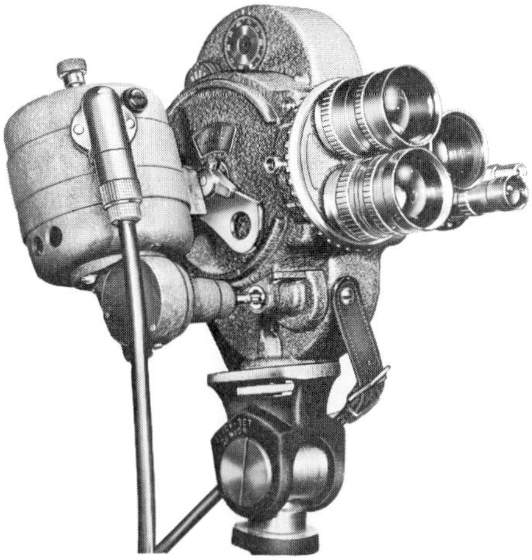

Fig. C-4*a* **Bell & Howell 70TMR time and motion study camera.**

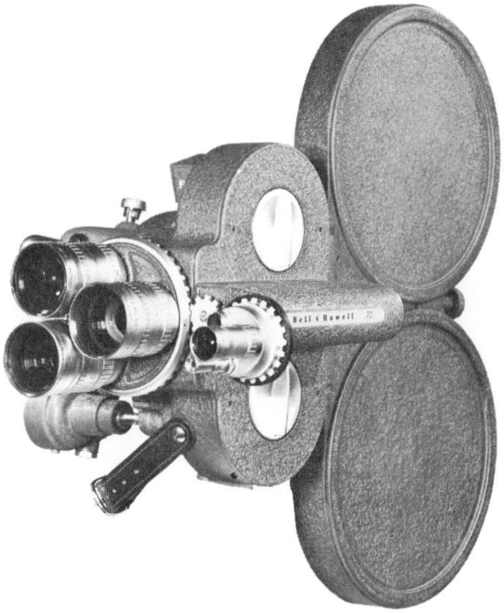

Fig. C-4*b* **Bell & Howell 70HR 16mm spool-loading movie camera with external film magazine.**

Bolex models H-16 REX-5 and H-16 M-5, 16mm. Bolex H-16 cameras are rugged, dependable, and lightweight, making them suitable for one-man filming operations. Faulted only because of the 100-foot film capacity magazines, the Bolex 400-foot magazine was designed to fit both the H-16 REX-5 and the H-16 M-5. The constant-speed motor (24 frames per second) with sync output for lip-sync on $\frac{1}{4}$-inch tape and automatic built-in clapstick provides for easy synchronization.

The H-16 REX-5 camera has a 24-frame-per-second constant-speed motor and a detachable take-up motor on the magazine which eliminates the use of take-up belts. The camera also features reflex viewing and focusing on ground glass, a variable shutter, filter slot, automatic dual frame counters, and a registrator claw for picture steadiness.

The H-16 M-5 is a single-lens camera with viewing through a removable optical finder mounted on the side of the camera. The M-5 takes all standard C-mount lenses or Pan Cinor and Angenieux zoom lenses, both equipped with reflex viewing and focusing. See Figure C-5.

Bolex models H-16 REX-4 and H-16 M-4, 16mm. These cameras offer the same features found in the REX-5 and M-5 but are not equipped with the saddle to accept the 400-foot film magazine. The cameras are suitable for all applications where the 100-foot film capacity is sufficient.

There are 10 fixed-focal-length lenses from 10 to 150mm in the H-16 system, and 7 zoom lenses giving a wide range of zooming ratios from 5 to 1 to 10 to 1, including the Vario Switar 86EE. Accessories available in the H-16 system include motors, close-up attachments, grips, matte box, titler, light meter, carrying cases, and both optical sound and optical/magnetic sound projectors.

Canon Super 8, 518 Zoom. The 518 is an 8mm cine camera that accepts Super 8 film cartridges. The camera has an $f/1.8$ lens with a zoom ratio of 1 : 5. The zoom range can be extended to super-Telephoto with an optional teleconverter.

The viewfinder is a parallax-free, single-lens, reflex type. Also featured is a built-in prism screen focusing system and exposure meter indicator, an eyepiece that can be adjusted to the eyesight of the photographer, and a through-the-lens metering system with a highly sensitive CdS meter. Aperture warning marks for under- or over-exposure are shown in the viewfinder. The CCA filter, built into the camera, automatically corrects for exposure of tungsten film for daylight use.

An electrically powered micrometer provides continuous automatic filming of the full 50-foot length of the film at 18 frames per second. Power for the drive is taken from three 1.5-volt batteries. The footage counter automatically resets when a new cartridge is inserted. A push-button-actuated indicator shows the power level of

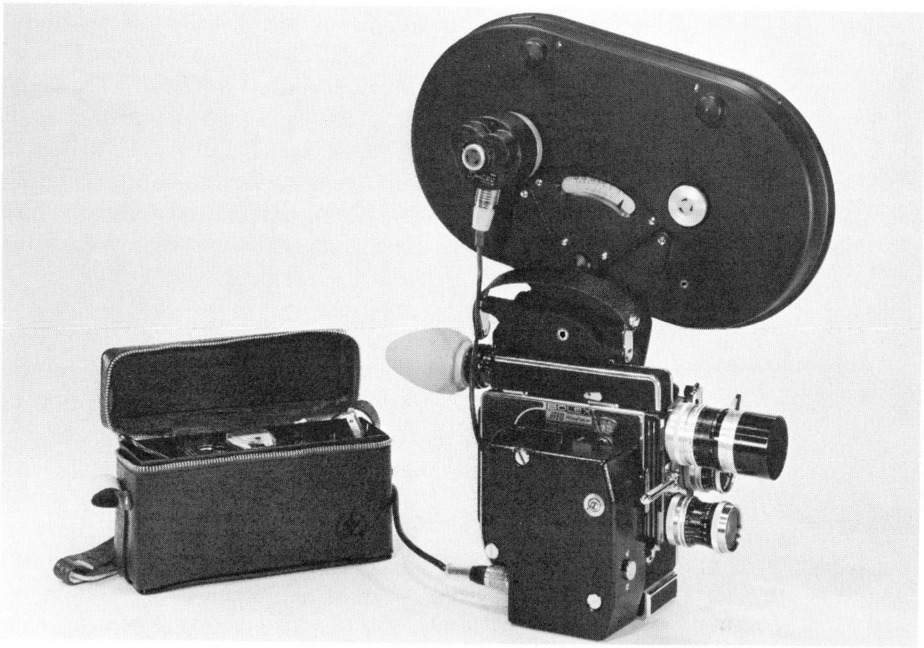

Fig. C-5 Bolex H-16 REX-5 with 400-foot magazine and battery pack.

micrometer batteries. The $8\frac{1}{8}$- by 4- by $1\frac{3}{4}$-inch camera weighs only 2 pounds 1 ounce.

Canon Super 8, 318 Zoom. This camera is similar to the 518 model with the following exceptions. The *f*/1.8 lens has a zoom ratio of 1 : 3. The zone focus system is set by rotation of the front lens. The camera's dimensions are 8- by 4 by $1\frac{3}{4}$ inches. It weighs 2 pounds.

Note: No teleconverter is available for this model.

Canon Super 8, 814 Auto Zoom. The camera is designed for Super 8 cartridge loading. Loaded with a 50-foot length of film (approximately 3 minutes and 20 seconds at 18 frames per second), the film speed is automatically set when the cartridge is loaded into the camera. The camera is equipped with an *f*/1.4 lens. The viewfinder is of the single-lens reflex type.

Also featured is an aperture scale, improper-exposure indicator, a film-ending indicator, and an adjustable eyepiece. A through-the-lens system CdS meter measures the light passing through the taking lens. The built-in CCA filter corrects colors of tungsten-type film for daylight use. Filming speeds are 12, 18, 24 frames per second, and single-frame exposure. Fade-ins and fade-outs are possible with the aperture control ring. Film drive and power zooming are operated by two micromotors. Four 1.5-volt batteries loaded inside the handgrip provide sufficient power for 10 cartridges of film under normal temperature.

The footage counter automatically resets when the side cover is opened. The handgrip is of the folding type, containing film-drive battery chamber. The camera's dimensions are $9\frac{3}{8}$ by $4\frac{1}{4}$ by $2\frac{1}{2}$ inches. It weighs 3 pounds 7 ounces.

Canon Scoopic-16, 16mm. The Canon Scoopic-16 combines automatic CdS exposure control and an integral 13–76mm zoom lens with reflex viewing, electric drive, and automatic loading. The camera has a contoured handgrip with thumb-action shutter release, and is planned for convenience in hand-held action shooting.

The camera uses 16mm film, single- or double-perforated on standard 100-foot spools. The *f*/1.6 coated lens has a zoom ratio of 5.84 : 1, focusing to 5 feet. The CdS exposure control system cross-couples to all running speeds and *f*-stops. The selected aperture shows on a scale in the viewfinder. Running speeds are 16, 24, 32, and 48 frames per second. The camera is self-threading; allows through-the-lens viewing; and has a corrective, adjustable eyepiece and a self-resetting film counter. It is motor-driven by one 12.5-volt rechargeable battery which allows approximately 8 rolls per charge.

Debrie Super-Parvo color camera, 35mm studio model. The two models of the Super-Parvo color camera are available either in type N (normal model), type A (anamorphic model), or type AN combined. (The shutter units N or A and the corresponding aperture plates must necessarily be preset in the Debrie factories.)

The camera features reflex viewing and a mirror shutter which is adjustable from 180 to 0 degrees by exterior control. The film gate has fixed guide rails and interchangeable register pins. A soundproof camera housing, an illuminated footage counter, tachometer and focusing scale, and interior magazines with a 1,000-foot capacity are other features. The camera also has an automatic cutout switch, a safety trip device with automatic camera stop, and an interior interchangeable motor with provision for forward or reverse filming.

The exterior model has the same characteristics as the studio model with less elaborate soundproofing, but with the advantage of halving the weight.

Debrie model CX-16, 16mm. Ready to operate, loaded with 400 feet of film and equipped with a 12 to 120mm Angenieux zoom lens, the CX-16 weighs 16.5 pounds. Most of its weight rests on the operator's shoulder.

The positive-locking, three-lens turret accepts a wide variety of variable- and fixed-focus lenses. Its peripheral bearing design allows transport of the camera with a zoom lens in place. A single-lens version of the camera is also available.

Two interchangeable 400-foot capacity magazines are available—the Mono-bloc and the double or separable type. The CX-16 forms a loop around two rollers during threading. Upon insertion in the camera, the rollers retract, and a correctly sized loop remains. Continuous viewing of the scene, without parallax, allows continuous follow-focus and framing.

The 12-volt d-c motor is supplied in two models; the standard which is centrifugally regulated, and the precision which is controlled by a precise time base such as a quartz crystal or tuning fork. With the latter motor and a similarly controlled tape recorder, no interconnecting cable from camera to recorder is needed. Power is battery-supplied. The CX-16 supplies a 60-cycle sync pulse by cable to the tape recorder when the centrifugally regulated motor is used.

Doiflex 16mm camera. This compact, lightweight camera features a through-the-lens full-reflex viewing system, precision registration-pin pulldown claw movement, a positive-closing disk shutter with reciprocating reflex mirror, and a three-lens revolving turret which accepts standard C-mount lenses.

The camera has a built-in tachometer, an adjustable footage counter, and an interchangeable motor mount which accepts transistorized variable-speed, constant, synchronous, or stop-motion motors. The camera accepts daylight-load roll film, has a built-in viewer light trap, and fits $\frac{1}{4}$- or $\frac{3}{8}$-inch tripod threads.

Accessories include sunshade, matte box, pistol grip, transistorized variable-speed motor, stop-motion motor, rewind device, zoom-lens support bracket, camera carrying case, portable battery, and 400-foot magazines.

Eclair Model NPR, 16mm. The **N**oiseless, **P**ortable, **R**eflex camera features a reflex optical system, an electric drive, governor-controlled motor incorporating a sync-pulse generator for use with a $\frac{1}{4}$-inch tape recorder with sync-pulse system, a 400-foot film magazine, and a two-lens offset turret which permits the use of a wide-angle lens and a long lens at the same time.

Accessories include a 1,200-foot magazine, extra-long-charge featherweight batteries, a choice of silenced motors for various applications, and lens-mount adapters to permit the use of present lenses. A silent, governor-controlled motor includes a sync-pulse generator. The Eclair, when used with a Perfectone or similar-type recorder, does not require a blimp.

Eclair Camerette, 35mm. The camera features true reflex focusing and viewing through the taking lens via an angled mirror shutter. The precision film movement produces perfect registration. Divergent turrets accommodate three lenses. The shutter is adjustable: 40 to 200 degrees. Film magazines of 100, 200, or 400 feet are attached by snapping them into position.

Eclair Camerette, 16mm/35mm. This camera offers the same features found in the 35mm Camerette but will accept both 16 and 35mm film. Changeover is accomplished by inserting or removing the film aperture mask and changing the film-gate magazines. The reflex viewer shows both 16 and 35mm fields.

Fairchild cinephonic camera, 8mm. This single-system sound camera records the sound on magnetic sound-striped black-and-white or color film. The miniaturized sound system is completely inside the camera. The rechargeable nickel-cadmium battery powers both the film drive and sound-recording unit. (A fully charged battery allows 50 minutes running time—enough for 8 rolls of double-8mm film.)

A sensitive, omnidirectional microphone is connected to the camera by a 15-foot cable. A double earpiece is provided for monitoring sound. Picture and sound controls are grouped for one-hand operation. The three-lens turret is equipped with 13mm Cinphar *f*/1.8 fixed-focus lens. 38mm *f*/1.8 Telephoto and 8.5 mm *f*/1.8 wide-angle lenses are optional. The viewfinder has parallax correction for close-ups or distance. An accessory exposure meter screws into the lens turret. Sound is recorded 52 frames (7.8 inches) ahead of picture. The camera is supplied with microphone, headset, battery, charging cable, 13mm lens, and 85B filter.

Kodak reflex special, 16mm. The reflex special has a reflex focusing system incorporating a 170-degree mirror shutter angled at 45 degrees to the film plane. Images entering the taking lens are alternately exposed to the film by an open segment of the shutter and to the ground glass of the viewer system by a mirror mounted on the back of the shutter segment. There is no light loss at the film, and the scene is constantly viewed through the taking lens, thus permitting continuous framing and focusing.

A variable shutter operates behind the mirror shutter. It can be locked at set positions, or varied from open to closed for making fades and dissolves. Fully closed, the camera may be reversed and the film backed up for double exposures. The image is viewed on a ground glass with 10× magnification. The center of the image may be critically focused at 20× magnification. The eyepiece can be individually focused to the operator's eye. A lever closes off the viewing system when not in use. The basic camera permits use of 100-foot internal-daylight-loading spools. 400-foot and 1,200-foot external film magazines are gear-driven.

Three electric motors are available: synchronous, variable, and animation. The three-lens turret is spring-loaded and mounted on ball bearings. The lenses employ the type-R bayonet mount with clamp fasteners. A single large sprocket controls upper and lower film loops. Either single- or double-perforated film, with or without full-width magnetic sound track, may be threaded. The film is transported by a single claw and is positively guided on the sprocket side, spring-loaded from the other side.

Kodak Ekton lenses, ranging in focal length from 10 to 150mm, have been developed for this camera. Also available is an *f*/2.2 zoom lens with variable focus from 17.5 to 70mm. A built-in buckle switch automatically stops the camera in the event of film jam or take-up failure. A bracket on the outside of the viewer permits attaching a Mitchell-type viewfinder.

Cine-Kodak special II, 16mm. The camera features interchangeable 100- or 200-foot film chambers which are prethreaded. The two-lens divergent turret accepts Kodak Cine Ektar lenses in type

C mount. The governor-controlled spring motor exposes 38 feet of film at a single winding. The camera operates from 8 to 64 frames per second as well as single frame. An adjustable 165-degree shutter is provided with open, ½ open, ¼ open, and closed (notched) positions. Automatic fades and dissolves can be produced. A single frame release makes time-lapse photography possible. Six special-effect masks are provided. The camera is available equipped with either 25mm Kodak Cine Ektar *f*/1.4 or 25mm Kodak Cine Ektar *f*/1.9 lens.

Cine-Kodak K-100, 16mm. The camera features a governor-controlled spring motor which exposes 40 feet of film at a single winding. The camera operates from 16 to 64 frames per second and has a fixed 165-degree shutter. The telescopic-type finder is designed so that the 25mm lens shows the subject actual size. The finder focus is adjustable for individual eyesight.

The camera can be hand-cranked either forward or backward for special effects. The camera is adaptable to electric motor drive by using motor-drive shaft assembly. A single-lens model is available equipped with either Cine Ektar 25mm *f*/1.6 or Cine Ektar 25mm *f*/1.9 lens.

Maurer camera model 150. The intermittent movement of this camera has a single pulldown claw with a rectangular rather than arced path. The pilot pins are tapered and can find an out-of-line perforation, enter it, and accurately register the film before the claw is withdrawn. Two spring buttons press the film against the guide strip to position it laterally.

The buckle trip automatically stops the camera if a jam occurs owing to improper threading or film tear. The 235-degree dissolving shutter is adjustable and can make automatic 40 and 60 frame fades and lap dissolves. The shutter indicator at the rear of the camera is calibrated to ½ lens stops and permits changes in exposure while filming.

The camera features a rack-over focusing system with aerial image instead of ground glass. A bright image is possible even at *f*/22. The center of the image can be magnified 140× for hypercritical focusing. A microswitch prevents operation while the camera is in the focusing position. The cameraman sees a 2¼- by 3-inch erect image. A single knob focuses the finder and automatically sets parallax. A second knob moves all four framing wires to change field size. The three-lens turret takes any C-mount lenses as well as a 10:1 zoom ratio lens.

The motor drive is applied directly to the intermittent movement through the one-frame-per-turn shaft. The camera is supplied with a 110-volt 60-cycle single-phase synchronous a-c motor which drives the camera at a constant speed—24 frames per second. A 110-volt a-c/d-c variable-speed motor with rheostat and tachometer—speed range from 2 to 72 frames per second and a 12-volt d-c governor-controlled (24 frames per second) motor are also available.

The sunshade is adjustable for normal, wide-angle, and telephoto lenses. The filter holder takes 2- or 3-inch-square glass or gelatin filters and is also threaded to take series VII round filters. The 400- and 1,200-foot double-compartment-type magazines are gear-driven and feature automatic feed as well as forward and reverse take-up. The magazines accept both darkroom loads and daylight-loading spools. The camera is equipped with counters featuring large numerals for easy reading. The camera has a hand crank (8 frames per turn) for manual operation. See Figure C-6*a*.

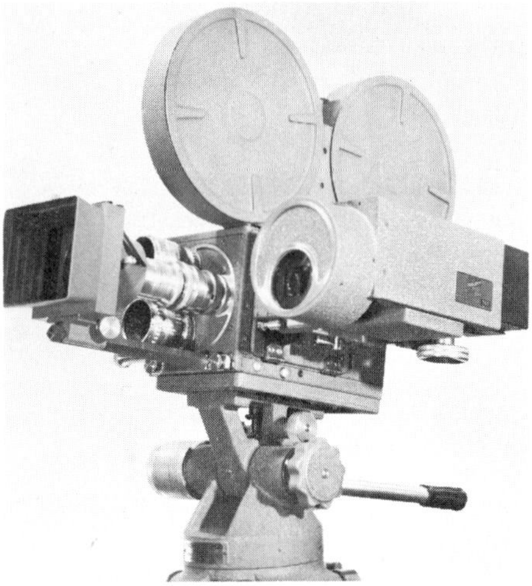

Fig. C-6*a* **Maurer 150 16mm professional camera.**

Mitchell cameras. All 35mm Mitchell cameras, with the exception of the Mitchell R-35 reflex, employ a rack-over focusing system with 5 and 10× magnification. Both focusing finder and film aperture have matching slots for inserting precut mattes employed in trick shots. The focusing telescope has two contrast-viewing filters and an adjustable eyepiece. The viewfinder is attached to the camera with a dovetail bracket and may be adjusted for parallax. Cameras have built-in four-way adjustable mattes. Various types of wild, synchronous, interlock, and stop-motion motors, operating at various alternating and direct voltages, are available.

Each model has a buckle trip that stops the camera in case of film take-up failure. Veeder footage and frame counters operate either forward or in reverse. Shutter openings and lens mountings differ slightly on various models. Camera movements differ in standard, high-speed, and sound cameras but all have eccentric-type film movements employing dual-registration pins and a double-claw pulldown arm engaging four perforations simultaneously. Movements have timing marks to permit error-free insertion without loss of timing.

Mitchell NC sound camera, 35mm. The Mitchell NC magazines are insulated with sound-absorbing material. An insulating plate on the camera prevents contact between the camera box and the magazines. The movement delivers fast pulldown with long dwell on the registration pins. A miniature shutter dial on the back of the camera shows the position of the shutter blades in relation to the aperture. This is an extremely useful feature in synchronizing rear-projected process shots. A hand-dissolve or shutter-adjusting lever at the rear of the camera moves in a calibrated segment, graduated from 0 to 175 degrees. The lever may be locked at fixed shutter openings in increments of 10 degrees.

Mitchell BNC studio sound camera, 35mm. The BNC is a self-blimped version of the NC model. It is smaller, lighter, and quieter than conventionally blimped cameras, and operating controls are never in contact with the inner mechanism while the camera is running. Focusing controls are geared to the camera lens and viewfinder. Both camera taking lens and viewfinder image are in identical focus, and the viewfinder is corrected for parallax at all times. A remote-control selsyn focusing control is available as an accessory. Three focusing positions are provided: two at the front of the camera (on the left and right sides) and one at the back. The bayonet-type single-lens mount employs a locating pin to assure the proper positioning of the lens. A focusing-mechanism release lever disengages the lens for quick lens change.

Mitchell standard camera, 35mm. The standard 35mm camera operates from 1 to 32 frames per second. The camera mechanism has alternate gears made of fiber. Equipped with a 170-degree shutter, the camera cannot be used for high-speed photography because of sleeve-type bearings.

Mitchell high-speed camera, 35mm. This camera will operate from 1 to 128 frames per second. The camera is similar to the standard, but the mechanism has all steel gears and ball bearings. The camera operates at high speed without strain.

Mitchell single-system camera, 35mm. The camera features the same precision construction and features as the NC model. It is equipped with RCA variable-area galvanometer for sound-on-film recording simultaneously with the picture on the same length of film.

Mitchell portable reflex studio camera R-35. The design of the R-35 replaces the rack-over with full-aperture true reflex viewing and focusing when required. The Mitchell dual-registration pins and dual pulldown claws, in an eccentric movement, operate to 120 frames per second. The dual-shutter system retains the best features of both mirror reflex and standard variable shutter. The mirror shutter, set at 45 degrees to the film plane, is used for reflex focusing and viewing. The focal-plane variable-disk 180-degree shutter provides positive exposure control over the entire area of the aperture with maximum exposure efficiency.

The inverted 400-foot film magazine serves as a balanced support when placed on the cameraman's shoulder for hand-held filming. Standard 400-, 1,000-, and 2,000-foot magazines may be used in conventional top position by attaching an adapter to the rear of the camera. The R-35's design permits a complete range of television, wide-screen, and scope-type filming, and photo instrumentation work at high speeds.

Mitchell Monitor, high speed, 16mm. The Monitor features frame rates up to 600 frames per second, variable in 1-frame increments during standstill or in operation, even by remote control or programming with ± 1 percent regulation.

Dual-pin registration keeps the film perfectly aligned horizontally and vertically. The universal motor operates on either 28 volts d-c or 120 volts a-c/d-c, 50 to 1,000 cycles standard. An integral reflex boresight system permits through-the-lens viewing without removal of film. The shutter also is integral and adjustable from 6 to 120 degrees from outside the camera. The camera has a 400-foot internal capacity. The camera also takes 1,200-foot external magazines with breakaway take-up chamber for removing exposed footage only.

A Conex automatic iris system that controls exposure at all frame speeds and makes approximately six *f*-stop changes in a quarter-second is another option.

Mitchell professional camera, 16mm. This camera is a scaled-down version of the 35mm Mitchells, with special features incorporated for 16mm users. The camera has a 230-degree shutter delivering an additional half-stop exposure at the same lens opening as cameras possessing 180-degree shutters. The film movement operates from 1 to 128 frames per second. Sound blimps equipped with follow-focus attachment handle either 400- or 1,200-foot magazines.

Mitchell System 35. The System 35 reflex can be converted to an inverted magazine configuration for portable and close-quarter filming, fitted with a slant-mounted magazine and equipped for remote control and vidicon viewing, or installed in a sound blimp with reflex-viewing side door.

With the reflex, through-the-lens viewing/focusing system, the operator views a brilliant, larger-than-full-aperture field, even with the camera lens stopped down. The camera also features high- and low-power magnification, color and monochrome contrast-viewing filters, matte slot for use in special effects, a nonfogging adjustable eyepiece, and interchangeable aspect-ratio ground glasses.

The camera's built-in pulse generator makes it compatible (regardless of motor type) with all professional audio systems. The basic studio version takes all standard magazines (200 to 1,200 feet), and a complete line of bayonet-mount lenses.

The camera's rotating-mirror, through-the-lens reflex-viewing setup is quite valuable in photoinstrumentation. Connected to a vidicon camera, it makes possible an integrated cinematographic system with simultaneous, parallax-free, closed-circuit TV viewing of the exact camera/film image.

The unit can be completely controlled and monitored as far as 2,000 feet away. For remote operation, the camera is fitted with a vidicon camera, mounted so that it views the on-line image relayed through the camera focus tube. The vidicon unit uses a standard 16mm format, 50mm lens, and a special negative lens cemented to the tube face to correct for field curvature. The vidicon camera produces 700-line horizontal resolution. For extremely critical applications, other vidicon cameras with horizontal resolution as high as 1,250 lines can be substituted.

The vidicon camera scans the reflex image viewed through the camera lens. That image is simultaneously displayed on a blimp-mounted electronic viewfinder as well as on an unlimited number of monitors. In a multiple-camera setup, the director and the director of photography can monitor and coordinate a series of such cameras from a television-type console.

Control of iris adjustment, focal length, and focus—as well as camera orientation—is handled by a single operator at a remote console. Cameras can also be set up and controlled manually for such range applications as multicamera missile tracking. See Figure C-6*b*.

Todd-AO cameras. The AP-65 is an all-purpose 65mm lightweight hand-held reflex camera developed by Mitchell in cooperation with Todd-AO. Dual-registration pins assure steadiness, through all speeds of photography. Dual-forked pulldown claws engage four perforations simultaneously. The adjustable pulldown stroke eliminates perforation picking due to film shrinkage. The removable aperture plate has a built-in matte slot, and can be removed in its entirety for cleaning.

The AP-65 camera features continuous through-the-lens viewing (erect image). Since there is no shutter in the viewing path, this contributes to flickerless viewing. The design of the camera allows viewing more than the full aperture area. The reflex finder also provides for higher magnification for critical focusing and contrast-viewing filters for color photography.

The camera has a quick-release flange for mounting of all Todd-AO focal-length lenses. The combination matte box and sunshade swings out of position for the insertion of lenses. All lenses are geared for manual follow-focus control.

A lightweight internal 28-volt d-c motor is operated by a solid-state speed control. Speed range is 12 to 32 frames per second. A 350-foot displacement-type magazine is provided. The magazine

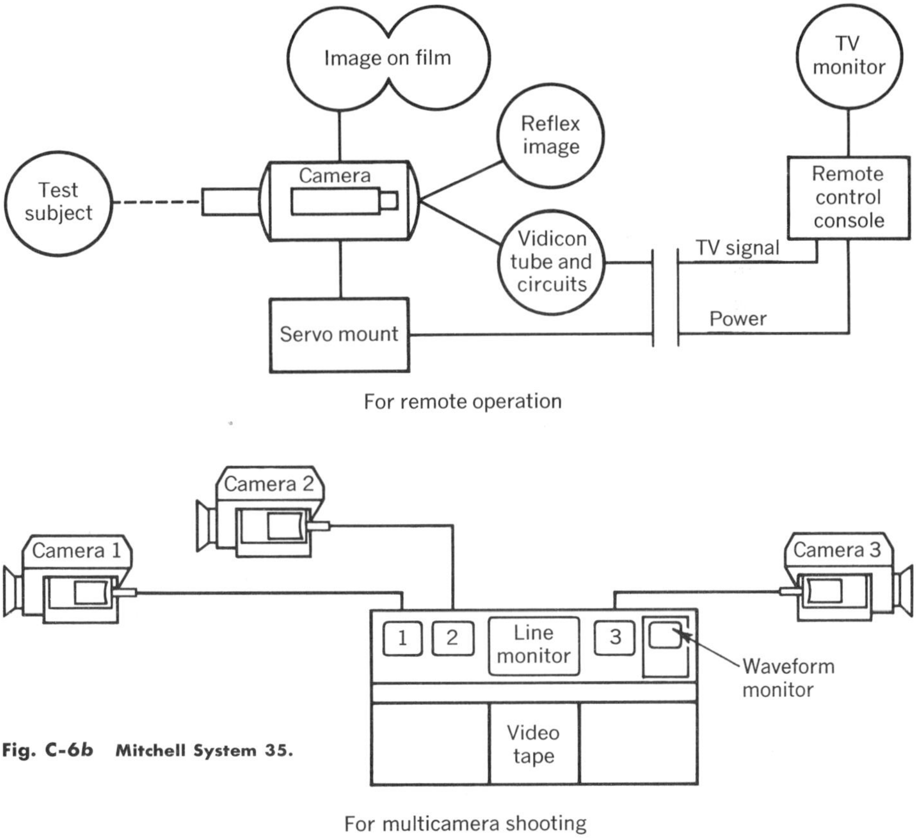

Fig. C-6*b* Mitchell System 35.

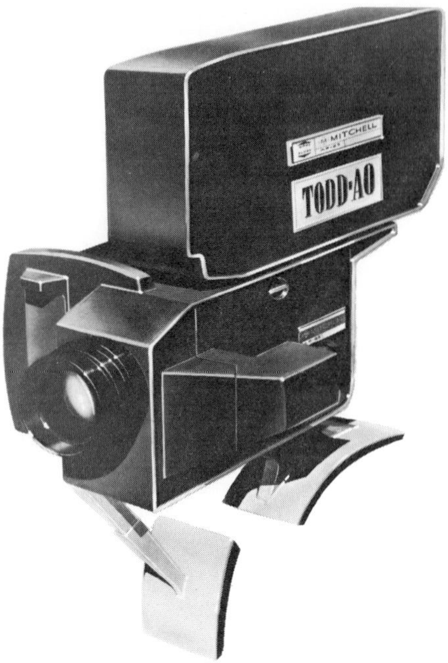

Fig. C-7 Todd-AO AP-65.

take-up is a positive driver through a disk clutch, thus eliminating the spring- or leather-type belts. Magazines with a 1,000-foot capacity are also available.

The large, erect-image viewfinder is calibrated for different focal-length lenses. Dual calibrations for any two aspect ratios are also featured. The camera has a fixed shutter—175 degrees.

Including film, the weight of the camera is 24 pounds. A shoulder support and handgrip are supplied for hand-held operation. The camera can also be mounted on a tripod or boom. A filter slot is built into the camera head. Two thicknesses of gelatin filters may be placed in front of the aperture. A 50- and 60-cycle sync-pulse generator with external plug is built in on the rear of the camera. The AP-65 can be remotely controlled and has a thermostat-controlled heating system with an on/off switch and indicator. In addition to the fixed-focal-length lenses, two lightweight zoom lenses are available. Also available is an underwater blimp.

Batteries are internally mounted; and filming speeds, stops, and focus are externally controlled. The underwater blimp is designed to work in depths of 50 feet or less. A light meter is built into the case. See Figure C-7.

Wall single-system Mark I, 35mm. The Mark I is capable of recording a variable-area or variable-density optical sound track simultaneously with the picture on the same length of film. The camera may be used unblimped. It is equipped with Mitchell-type rack-over, through-the-taking-lens focusing, movement with dual pulldown claws, and dual-registration pins, four-lens turret, Bell & Howell lens mounts, adjustable 170-degree shutter, automatically controlled take-up tension, 24-volt d-c rheostat-controlled motor, provision for standard Mitchell viewfinder, and 400- or 1,000-foot film magazines.

The camera may be purchased with or without sound equipment and may be equipped with many Bell & Howell and Mitchell accessories.

Wall single-system Mark II, 35mm. The Mark II is a modified version of the Mark I by National Cine Equipment Co., Inc. The electric motor and tachometer have been eliminated from the rear of the camera and replaced by a side motor mount that allows mounting any standard Mitchell motor, of any voltage, with attached tachometer. Camera controls are similar to the Mitchell NC.

camera, box As the name suggests, a camera in the shape of a box. This type of nonfolding "still" camera has no bellows and, usually, no provision for focusing the inexpensive lens built into the front of the box.

camera, cirkut A camera capable of taking pictures that are approximately 5 to 10 times as long as they are wide. During the period of exposure, the film travels continually from the feed reel to the take-up reel while the camera moves in an arc. Perspective, in the print material, is usually distorted.

camera, copying This type of camera is used mainly for rephotographing still photographs and other flat graphic material. Copying cameras are usually equipped with extra-long bellows.

camera, direct positive This camera, at one time, was a rather popular attraction at fairs and carnivals. The camera used ferrotype plates or positive paper and produced a "while-you-wait" positive print of the subject in minutes.

camera, enlarging Also known as an enlarger and projection printer, this type of camera is capable of projecting negatives onto sensitized photographic paper in order to produce relatively large prints.

camera, focal plane A still camera with a built-in focal-plane shutter. The shutter is usually made of metal or fabric and has a variable slit through which light reaches the film. Mounted close to the focal plane, the shutter moves across the film and exposes it progressively when the trip is released.

camera, hand The term was originally used to describe still cameras that could be hand-held during the period of exposure rather than cameras that had to be supported by a tripod. Since many motion picture cameras can now be hand-held, the original term is no longer applicable.

camera, identification Actually, a smaller version of the portrait cameras that are used for studio work. This camera produces flat-lit photographs of relatively poor quality—but adequate for passports, licenses, and other forms of identification. In some cases, a number is photographed simultaneously and appears in the print under the subject.

camera, mapping A camera designed to produce a series of overlapping photographs of relatively large areas. Usually taken from an airplane at regular intervals, the photographs are subsequently arranged to form a mosaic.

camera, miniature The classification does not apply to the size of the camera but rather to the size of the negative produced by it—usually $2\frac{1}{4}$ by approximately 3 inches, or smaller.

camera, one-shot A camera designed around beam-splitting devices that are used to expose as many as four films simultaneously to produce color-separation negatives. During the interval of exposure, the divided light is transmitted through color filters positioned in front of each of the negatives in the camera.

camera, photomacro Designed for extreme close-up photography, this camera produces negatives that magnify the subject matter to approximately 5 to 10 times its natural size. Compare CAMERA, PHOTOMICRO.

camera, photomicro In photomicrography, a camera used with a compound microscope to produce images that are magnified to thousands of times the size of the original subject matter.

camera, portrait A relatively large camera designed for studio work. Usually mounted on heavy, stationary supports, the camera is adapted to a number of plate sizes and has built-in adjustments that make it especially suitable for portrait photography. Compare CAMERA, IDENTIFICATION.

camera, press A term usually applied to the 4- by 5-inch plate or cut-film cameras used by newspaper photographers.

camera, reflex A camera with a built-in mirror which reflects an erect image to the ground-glass viewfinder. In some cameras, the mirror flips up during the interval of exposure. In other cameras, the mirror is stationary and transmits light directly to the film—and reflects light to the viewfinder for composing and focusing.

camera, roll film A camera designed to accept rolls of film which are wound around lightproof paper. These rolls usually have a capacity of approximately eight individual pictures, although some cameras accept 35mm film and have a capacity of 24 to 36 pictures.

camera, speed A camera designed around focal-plane shutters and fast lenses for photographing objects moving at relatively fast speeds.

camera, stereo This type of camera takes two photographs, simultaneously, of the same subject. Identical except for slightly differing angles, the two pictures are mounted in holders and viewed through special devices to produce three-dimensional effects.

camera, view An all-purpose, versatile camera which accepts a wide variety of lenses. Light in weight, its many adjustments make it adaptable for professional use in many areas of photography. Also known as a **field camera** because it was originally used for photographing outdoor scenes or, as the name implies, views, the camera uses half plates ($4\frac{3}{4}$ by $6\frac{1}{2}$ inches).

camera angle The position and angle of the camera, or cameras, on a motion picture or television stage in relation to the performers and background. The term is also used in connection with the area covered by the camera as well as its perspective in relation to the set.

camera aperture See APERTURE, CAMERA.

camera base The upright section of the camera base holds the lens turret and contains the four-way matte mechanism. A slot for inserting gelatin filters is located at the rear of the viewfinder dovetail support. The horizontal section of the base incorporates the gibs and racking mechanism for shifting the camera box from side to side. The camera box may be shifted from the photographing position (left) to the focusing position (right) by operating a release button and handle on the rear of the base. A locating plunger in the base fits into either of two recesses in a locating plate beneath the camera box to position the box accurately in the photographing or focusing position. The viewfinder is mounted in a dovetail bracket on the upright section of the base. A bracket on the horizontal side of the base serves to support the viewfinder. This bracket has provisions for adjusting the finder to compensate for parallax. A spirit level is installed on the rear edge of the camera base. The Mitchell camera, model NC, is used here to describe camera bases generally. See Figure C-8.

Fig. C-8 Camera box and base (Mitchell NC).

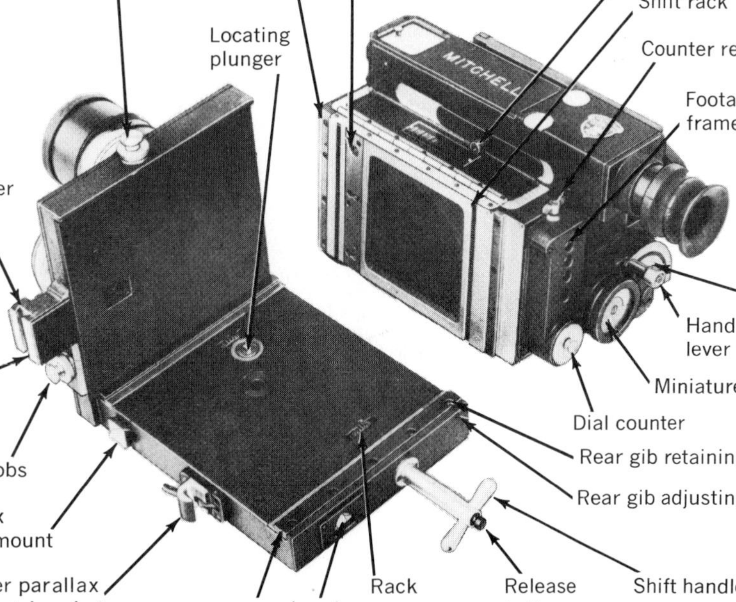

camera box The camera box is divided longitudinally into two compartments. The left-hand or operating side* contains the film-moving mechanism. The right-hand side contains movement drive shaft, two grease plugs, and film buckle-trip plunger. On the Mitchell BNC it also contains the dissolve-rack-lock tension-adjusting knob and electric connector. The operating side has a door hinged at the bottom. The focusing tube is an integral part of this door. The motor is attached to the drive side by means of two thumbscrews and a sliding latch mechanism. On top of the camera is a recessed plate for attaching the film magazine and an opening through which the film is drawn from the magazine to the film compartment of the camera. A dust plate is provided to cover this opening when the magazine is not in place. The Mitchell camera box, model BNC, is used here to describe camera boxes generally. See Figure C-8.

camera cam head† A camera mounting for television and motion picture cameras. Designed to fit on any standard pedestal or tripod, the camera is mounted so that its center of gravity is over the vertical center line of the head, distributing the weight of the camera equally. Two identical cams, on either side of the head, are profiled so that as the camera is tilted up or down, the angular displacement of the cams counterbalances the weight of the camera in any position.

camera case The camera's outer case usually consists of three main sections: the magazine housing, the motor housing, and the main housing. Each section is lined with sound-insulation material.

Mounted on the main housing case are the parallax and focusing control, the matte-box bracket, the carrying handles, the power-input plug, the viewfinder dovetail bracket, the spirit-level tube, and the buckle-trip reset plunger.

The camera case controls consist of the focus-tube magnification lever, filter-operating plungers, eyepiece, hand-dissolve lever, automatic dissolve push buttons, shift-rack handle, and manual threading wheel. These outer-case controls are spring-loaded for instantaneous disconnect when released. This feature is provided to avoid contact with the inner controls during operation of the camera so that no noise is transmitted from the camera box to the outside through these controls.

In the outer case, windows are provided for viewing the miniature shutter, the calibrated dissolve indicator, the footage totalizer, footage counter, and the frame counter. The Mitchell camera case is used here to describe camera cases generally.

*The cameraman's left when he faces the set from a position behind the camera.

†Designed by Houston Fearless.

camera chain (television) The camera and auxiliary electronic equipment needed for producing picture information and preparing it for subsequent transmission.

camera crane Also called **cherry picker**. A mobile camera support that accommodates both camera and cameraman. Adjustable in height, its platform can be raised to considerable heights for overhead or special-angle shots.

cameraman Also called **cinematographer**. The person generally responsible for the lighting arrangement and the success of the photographic process. This person may or may not be the camera operator. In many cases, his assistant is behind the camera during the actual filming.

camera mount A support, such as a helicopter mount, on which a motion picture camera may be placed in order to keep it in a fixed position during the photographic process. (A tripod is not generally considered a mount.)

camera projection The process of projecting frames of processed film on the surface of the animation compound table so that titles and other graphic material may be matched to the film and photographed in that position for subsequent optical superimposition.

camera report A report that contains all the information pertinent to each day's filming during the production of a motion picture. The report lists such matters as the number of takes for each scene, the footage involved for each take, the takes selected for printing, and miscellaneous comments.

camera synchronism A spacing, in terms of frames, between any specific point in the sound track and the corresponding picture, such that the two will project simultaneously.

camera tube In television, electron tubes capable of translating light rays reflected from the subject matter being televised into electric impulses.

can A 1,000-watt lamp in a housing mounted on a stand or pedestal.

Also, a studio term for the headphones worn by some technicians. When worn during the filming process or while telecasting, the headphones serve as an intercom system. On-the-spot instructions are relayed from the director to the control room to the technicians on the studio stage.

candela Abbreviated **cd**, this international unit of measurement for determining the strength of a light source replaced the candle as the standard in 1936. By definition, the candela is the luminous intensity of $\frac{1}{60}$ of one square centimeter of projected area of a black-body radiator operating at 1273.5°C, the temperature of the solidification point of platinum. The unit of measurement, the lumen, is based on this type of measurement—1 candela being equal to 1 lumen per square foot or 10.764 lux. There is less than a 2 percent difference between the candle and the candela, a candela being equal to 98.1 percent of a candle. See CANDLE.

candle The original unit of measurement was based on the light-radiating capacity of a wax candle burning under controlled conditions. Compare CANDELA. See CANDLEPOWER.

candle meter Also known as a **lux**, a candle meter is a unit of illumination equal to the amount of light falling on a surface one meter away from a one-candlepower light source.

candlepower Candlepower is defined as maximum footcandles times the fixture distance squared. It should be noted that the candlepower law is valid only for distances greater than 20 times the diameter of the fixture reflector or lens. There are a few exceptions to this law, but they are not pertinent when one is dealing with flood-type fixtures. With the knowledge of the candlepower of any particular fixture, one can determine the maximum footcandles that same fixture will produce at any given distance. By simply dividing the candlepower figure by the distance squared, one will obtain maximum footcandles. For example, if a fixture is rated at one million candlepower, and it is necessary to know the maximum intensity of that particular fixture at 50 feet, divide one million by 50 squared (or 2,500), which is equal to 400. Therefore, that particular fixture will produce 400 maximum footcandles at a 50-foot distance.

candles per square foot The brightness of an illuminated area is measured in terms of light intensity per unit area—normally candles per square foot. In countries using the metric system, the normal unit is candles per square meter. One candle per square foot is equal to 0.09290304 candle per square meter.

capstan An extremely accurate rotating spindle against which magnetic recording material is held by a pressure or pinch roller. The capstan is designed to transport the tape at exactly the required speed without slippage or variation.

capstan oscillator In videotape recorders, the capstan oscillator is a phase detector which compares the incoming pulse to the local frame pulse and produces a direct voltage proportional to the magnitude of the phase error. This d-c error voltage controls the frequency of the oscillator, which in turn controls the speed of the capstan motor. Tape speed is thereby synchronized to local reference, and the original signals are correctly reproduced.

carbon-arc lamps Light from carbon-arc lamps have color characteristics ranging between floodlight and sunlight. The light originates in the electric arc generated between two carbon electrodes positioned in the lighting unit. In addition to their function as light sources for set lighting, carbon arcs are also used as a source of illumination in projectors. Carbon arcs produce a steadier light when direct current is used rather than alternating current.

carbons Pieces of compressed carbon that serve as the electrodes in carbon-arc lamps.

carbon tetrachloride A nonflammable solvent that is used for removing surface dirt and markings from film—finger smudges, for example. Carbon tet, as it is generally called, is a colorless liquid that does not mix with water. Because its vapor has toxic effects, the solvent should be used only in open areas where there is free circulation of air.

carbro process An early color printing process that is still in use today. Its name is derived from the combination of **car**bon print and **bro**mide print.

The separation negatives needed for the carbro process are usually produced from a transparency. In the first stage of the printing process, bromide prints of each of the primary colors are made from the separation negatives. These are used to prepare the matrices or color-relief images which are registered and superimposed, one over the other, to produce the composite color print. Production costs are generally lower than the comparative cost of prints produced with the dye-transfer process. Compare DYE-TRANSFER PROCESS.

Cardioid film movement A heart-shaped cam used in the film-transport mechanism of cameras (the Arriflex 35, for example), to yield an essentially straight-line motion to the film pulldown claw. This design differs from the more conventional cams which impart a more circular motion to the pulldown claw.

carrier wave The r-f (radio frequency) wave that is used to transmit sound or video signals.

cartop platform A cartop platform* provides an operating surface for a two-man camera crew and equipment. Rubber support cushions distribute the weight over the entire platform area. Steel tie-down clamps and turnbuckle fastenings hold the platform in place. Stair-type ladders allow the cameraman or technician to reach the platform.

cartridge A plastic or metal film container. Also known as a **cassette**, a cartridge holds a relatively short length of film and can be loaded into cameras in daylight. After the last exposure has been made, the film is usually rewound in the cassette so it can be removed from the camera and sent to a laboratory for processing.

cassette A relatively small container or cartridge for roll film that can be loaded into the camera in ordinary light. The cassette itself must, of course, be loaded and unloaded in a darkroom.

cast (1) The process of selecting the performers for specific dramatic roles in a production. (2) The group of performers themselves.

casting director The person responsible for the selection of the cast.

catadioptric system A compact, reflective optical system that uses reflective and refractive lenses as well as mirrored surfaces in order to fold long focal lengths into relatively short light paths.

cathode That part of an electron tube that emits electrons.

cathode-ray tube An electron tube. Within it electrons from the cathode are directed at a fluorescent screen (target) at the end of the tube. Wherever the electrons strike, the fluorescent screen glows in direct proportion to the intensity of the beam.

catoptric system An all-reflecting lens system free of chromatic aberration. Compare DIOPTRIC and CATADIOPTRIC SYSTEMS.

CdS These letters, usually found in descriptions of exposure meters, stand for **Cad**mium **S**ulfide, a photo-resistive chemical compound. The cadmium sulfide cell acts as a valve in light-measuring circuits and generates a current which activates the meter's registering indicator in proportion to the amount of light reaching the cell.

cel A sheet of transparent Celluloid on which the inker traces the lines in an animated drawing. Cels are approximately 0.005 inch

*Carter Carpac.

thick and are punched in the same way as the paper used by the animator. These cels fit the registration pegs on the animation board or similar pegs in the camera compound table of the animation stand.

cel level The position of each cel in relation to the background. A *hold* cel on which only the cartoon character's body has been inked and opaqued would be placed on the pegs first, thus becoming the cel level nearest to the background. This cel would be designated as the *bottom* cel level.

The cartoon character's head, inked and opaqued on another cel, would be placed over the hold cell and be designated as the *middle* cel level. The *top* cel level, the animating mouth or eyes, for example, would complete the figure. The component parts, each traced and opaqued from the animator's drawings on a different cel level, would show the composite figure in action over the scene's background.

Cello-Glass Material placed around a drumlike device that revolves around a light source. Irregularly shaped holes cut into the Cello-Glass produce a variety of moving shadow effects which, in turn, lend a feeling of movement to actions taking place on a stationary set—as in a process shot. Branches, for example, can be fastened to the drum and made to revolve past the light source in order to cast leafy shadow patterns on the performers who would then seem to be walking along a tree-lined street.

Celluloid A flexible, transparent material that serves as the base or support for the layers of film emulsion. Before manufacturers turned to the cellulose acetate safety base as a film support, the base used was the highly flammable nitrate of cellulose. See CELLULOSE.

Cellulose Cotton wool (pure cellulose) is the main ingredient in the manufacture of the acetate bases that are used as supports for the film's sensitive layers of emulsion. See CELLULOID.

Celotex A fibrous material. Because of its excellent sound-absorption qualities, it is ideally suited for acoustic treatment wherever sound is recorded or reproduced.

centrigrade thermometer One of several types of instruments used for measuring temperature. On the centigrade scale, the freezing point of water is absolute zero. The boiling point is 100°. The centigrade scale differs from the Fahrenheit scale in that the freezing point for the latter is 32°, and the boiling point 212°.

Temperature readings on the centigrade scale can be converted to equivalent Fahrenheit readings by multiplying the number of degrees by $^9/_5$ and then adding 32.

changeover The process of switching from the reel of film being projected to the reel that follows without interruption in picture continuity. See PROJECTION.

changing bag A poor man's darkroom. Actually a black cloth bag that is large enough to allow a cameraman to load and unload film in places where no darkroom is available. Two holes in the bag for the cameraman's hands and arms, with elastic bands around the holes to keep the bag lightproof, give the cameraman freedom of movement while handling the film and magazine within the bag.

channel Also called **broadcast band**. The frequency range assigned to each commercial station by the Federal Communications Commission.

character In animation, the drawings or caricatures of persons or animals used in the production of animated cartoons.

In motion pictures, specific types ideally suited for the portrayal of certain parts. Some performers, because of their facial expressions or physical characteristics, are "typed" as successful businessmen, gangsters, or butlers, for example. These performers are almost always cast in the same type of role.

characteristic curve (processing) Also called **H & D curve**. A method of plotting and determining, by graphs, the sensitivity and development characteristics of film.

Characteristic curves are plotted by using the known exposure values as the horizontal scale of the graph. The density readings obtained from the processed film are used as the vertical scale. The resultant arc provides the sensitometric information and image-recording characteristics of the emulsion.

The portion of the arc from *A* to *B* in Figure C-9*a* is known as the toe and shows the continuous increase in density change. *B* to *C,* the straight-line portion, indicates the constant increase, and the shoulder, *C* to *D,* shows the decrease in density change. See also Figure C-9*b*.

H & D stands for Hurter and Driffield, the scientists who developed the system for measuring the sensitivity of a film's emulsion.

cheat The regrouping or rearrangement of performers or props in relation to the background and the position of the background for the purpose of changing the angle of view. In two-shots where only one of the performers is photographed in a position facing the camera, the second or less favored performer is "cheated."

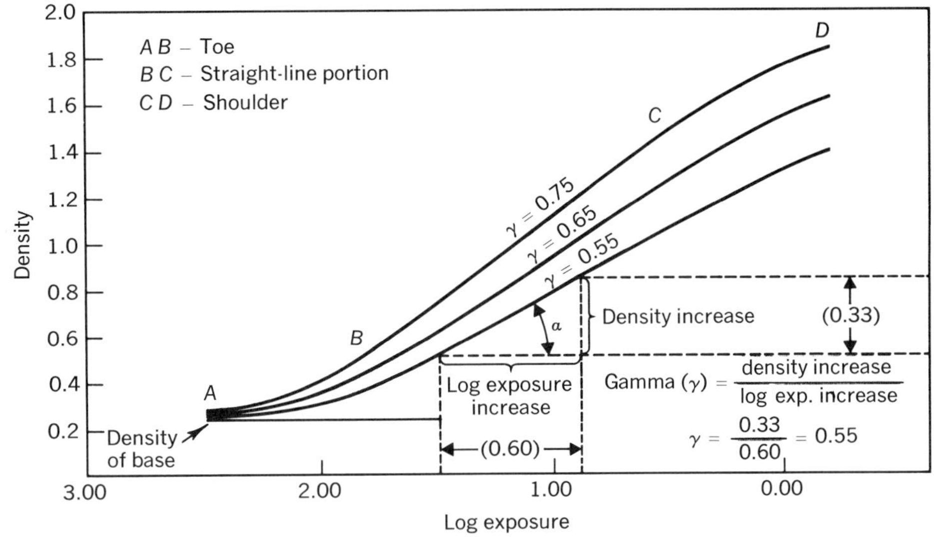

Fig. C-9a **H & D characteristic curves. (*Eastman Kodak.*)**

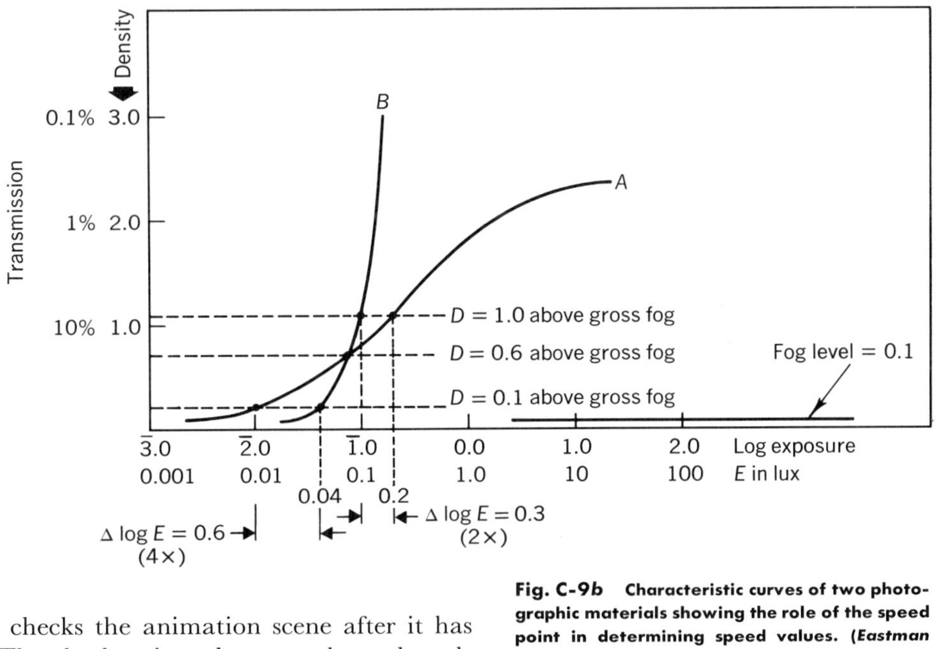

Fig. C-9b **Characteristic curves of two photographic materials showing the role of the speed point in determining speed values. (*Eastman Kodak.*)**

checker The person who checks the animation scene after it has been inked and opaqued. The checker, in a dry run, places the cels over the background and, literally, "photographs" the scene before it is given to the cameraman. During the checking process, the cartoon characters are matched to the props painted on the background, and the opaque colors applied to the cels are checked for consistency.

checkerboard editing A method of preparing A & B roll material for printing. With this method, all odd-numbered scenes are spliced together into one roll (the A roll). The even-numbered scenes are also spliced together on a second roll (the B roll).

Between each two scenes on the A roll, the film editor inserts blank leader that corresponds in length to the even-numbered scenes on the B roll. The same procedure is followed throughout the entire length of each roll.

If, for example, scene 2 on the B roll is exactly 5 feet long, the film editor splices a 5-foot length of blank leader between scenes 1 and 3 on the A roll. The sound track, on a separate roll, is not broken up into 2 rolls as are the even-and odd-numbered scenes. See Figure C-10 and A & B EDITING.

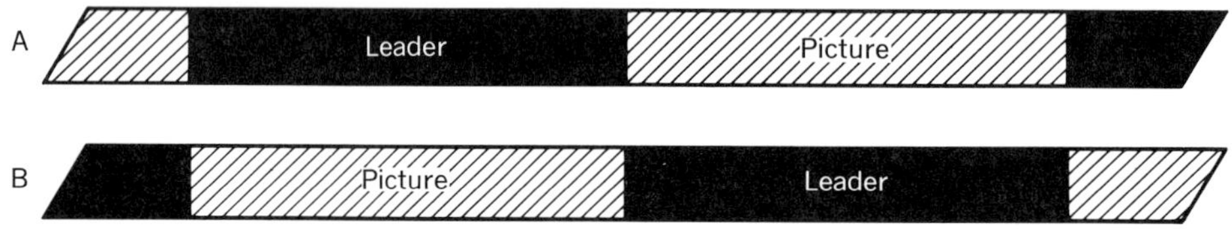

Fig. C-10 A & B Checkerboard splicing.

cheesecake The photography of an attractive young lady, usually for publicity purposes, in which her "characteristic curves" are featured. Compare CHARACTERISTIC CURVE.

chemical fade The process of bleaching the last part of a scene, frame by frame, by immersing the film in solutions in which hypo and ferricyanide (Farmer's reducer) are the principal ingredients. Each frame is immersed for a progressively longer period of time than the one preceding it to give the effect of a shutter fade.

chemical fog An undesirable negative fog caused by overdevelopment.

china girl A slang trade term for a series of frames of film which show a medium close-up of an actress or model. The skin tones, dress, and background contain all of the representative colors. Along the bottom or side of each frame are color patches which serve as a standard or guide for the timer. The optical or laboratory timer, after evaluating the color balance of a scene by comparing values with the colors in the china girl, indicates the amount of correction and the type of filter needed to produce a print that duplicates the colors found in the original subject matter. See Figure C-11 (in color insert).

chippy A studio term for a carpenter.

chroma A control in television cameras that varies the white level in all color channels simultaneously. It does not affect the monochrome white level. See WHITE LEVEL.

chroma amplitude- and velocity-error corrector This videotape recorder accessory measures and corrects two types of errors. The slight changes in tape-to-head contact, during record and playback, result in a variation in the chroma content of the signal. The chroma amplitude-correction section of the accessory measures and equalizes these variations. The velocity-error section measures and corrects for changes in head velocity that occur during record and playback which result in banding within the picture.

chroma-key See MATTES and SELF-MATTEING PROCESSES.

chromatic Pertaining to color.

chromaticity Pertaining to the hue and degree of saturation of colors.

chronograph camera This type of camera is designed for use where an accurate time record of the subject is desired. Such cameras are generally used in pairs so that the subject matter may be tracked from two predetermined positions simultaneously, thus providing references for checking the accuracy of the information that is to be used in subsequent studies.

The optical systems of the two cameras are constructed so that two separate images are registered on each frame of film—the subject and the face and hands of an illuminated chronometer. The unit consists of a 35mm Mitchell high-speed camera mounted on a modified friction tilthead with shim plate and handle. The bracket secured to the tripod head provides for mounting a chronometer in a position, from which it may be photographed through the back of the film by a secondary lens and shutter system incorporated in the camera box.

chronophotography A photographic process in which an action is shown by a series of related photographs—each taken at regular, predetermined intervals. The shutter, for this photographic process, is usually tripped or activated by regularly spaced, automatic electronic devices.

CIE color system Commission Internationale d'Éclairage (International Commission of Illumination). A method of color notation which specifies the exact amounts of each of the primary colors that must be mixed in order to match any specific color or wavelength in the visible spectrum.

cinch marks Undesirable markings or abrasions caused by tightening a roll of film incorrectly. Pulling the free or end leader of a roll of film while holding or applying pressure to the center of the roll will, almost always, produce the undesirable markings. Film should be handled carefully, using reels and cores for winding whenever possible.

CinemaScope See ASPECT RATIO and WIDE-SCREEN PROCESSES.

CinemaScope 55 See ASPECT RATIO and WIDE-SCREEN PROCESSES.

cinematographer A motion picture cameraman. In many cases, the cinematographer also assumes the responsibility for the lighting of the motion picture set.

CineMiracle See ASPECT RATIO and WIDE-SCREEN PROCESSES.

cineradiography A photographic process in which an image is recorded by using x-rays or gamma rays. Unlike the film used for other photographic processes, both sides of the negative are coated with a special emulsion that provides a high degree of contrast.

In cineradiography, the beams of x-rays and gamma rays pass directly through an object and exposes the film, thereby producing a shadow photograph that varies in density. These variations reveal the absorption qualities as well as the composition and thickness of the subject matter.

The photographic technique is as useful for recording physiological changes as they occur, via motion picture processes, as it is for examining the inner structure of a tooth and the extent of decay with "stills." In either case, whether by projection or by examining the negative directly, the technician is not exposed to the dangers of radiation normally associated with these rays. Where necessary, two radiographs are taken from as many positions for stereoscopic viewing. The illusion of depth thus created gives dimension to the subject matter.

Aside from its value for medical studies, radiography is extremely useful for industrial purposes. Practically applied, the information enables physicists and metallurgists to examine the interior composition of the various materials used in construction. Although advertisers have not yet taken full advantage of the technique for commercial demonstrations, it is safe to predict that before long the TV viewer will be taken behind the scenes for a closer look at the details of construction in consumer products that lend themselves to such treatment.

Cinerama See ASPECT RATIO and WIDE-SCREEN PROCESSES.

CineVoice conversion A method developed by F & B/Ceco for converting Auricon CineVoice cameras so that the need for body braces is eliminated. The combination of a handgrip with built-in switch and short-reflex zoom-lens finder offers the cameraman mobility while operating the camera from the shoulder with no other means of support.

Other accessories available in the conversion include Veeder-Root footage counters, sync motors, nonslip drive, adjustable clutch for 400- or 1,200-foot magazines, filter slot, and buckle trip. See Figure C-12.

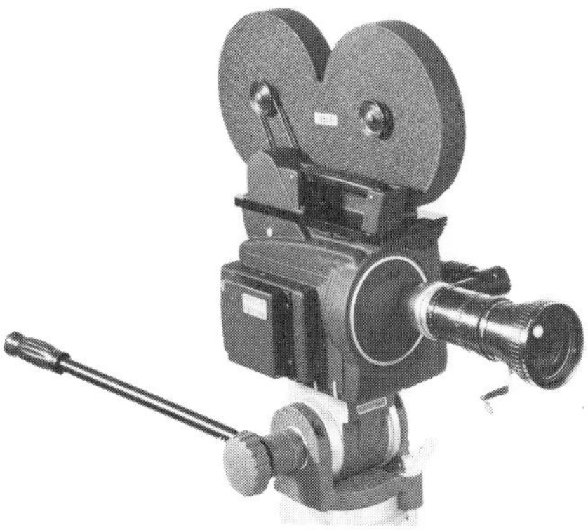

Fig. C-12 CineVoice conversion (Auricon camera).

Cinex A series of frames of film, each containing the same picture information, with varying exposures. These test strips may be made by varying exposure either during the photographic process, by optical printing, or at the laboratory in order to indicate the possible range of color and tonal correction that is possible in the print.

Cinex, color wheel A series of color filters arranged in a circular pattern and positioned in windows cut in a disk. The color wheel is used for exposing representative frames of film from a scene (or scenes) for subsequent color evaluation.

Photographed on the optical printer, each of the filters is brought into position in front of the camera lens, and the selected frame is exposed through each of the filters, in turn, at different lens stops.

The 15-frame Cinex produced by a color scene tester uses the same color filters that are used in the color wheel. The scene tester, however, provides only one frame of the filter color at one exposure. With the color wheel, a number of frames are exposed through each of the filters, thus providing a basis for comparison and more latitude with regard to timing and color correction—the basic reason for the Cinex. The color-wheel Cinex is more expensive than the Cinex produced with the color scene tester. See Figure C-13 (in color insert) and COLOR SCENE TESTER.

Circarama See ASPECT RATIO and WIDE-SCREEN PROCESSES.

circles of confusion The blurred circles around lights that represent objects outside the field of acceptable focus. Finer lenses and more critical focusing tend to reduce the size of the circles of confusion and thus increase the sharpness of the otherwise blurry images.

Measurements intended to define a circle of confusion are based on a viewing distance of 10 inches. The assumption is that any light areas less than $^1/_{100}$ of an inch in diameter would, at that distance, be visible as a sharp point rather than as a blurred spot of light.

The diameter of the circle of confusion, at different viewing distances, varies in direct proportion to the focal length of the lens that is used for exposing. If, for example, a 2-inch lens is used, the image obtained in a contact print would have to be viewed from a distance of 2 inches. As an alternative, the same image would have to be enlarged five times to provide an image that could be viewed at the correct distance—10 inches.

Generally, the diameter of the circle of confusion is consistent, on negatives of similar size, regardless of the focal length of the lens used for exposure. The acceptable circle of confusion for folding cameras is $\frac{1}{200}$ inch; for $2\frac{1}{4}$- by $2\frac{1}{4}$-inch cameras, $\frac{1}{350}$ inch; for miniature cameras and 35mm motion picture cameras, $\frac{1}{500}$ inch; for 16mm motion picture cameras, 1/1,000 inch; and for 8mm motion picture cameras, 1/2,000 inch.

circle of least confusion The lens position at which the subject matter is in acceptable focus and the diameter of the circle of confusion is at a minimum.

clapstick A hinged arm attached to the slate that is photographed at the start of each scene to identify it. The sound made by the arm when it is brought down on the upper edge of the slate is used as a synchronizing guide by the film editor.

Clarke process A process for combining cloud effects with other subject matter, during the actual filming process, to produce a direct, in-camera composite. Developed by Charles Clarke of Twentieth Century-Fox during the 1940s, the patented process earned an Academy Award for its inventor. See CLOUD EFFECTS.

claw The pin in the camera's transport mechanism that engages the film's sprocket holes and brings each frame into position in the shuttle for the interval of exposure.

clearing bath Any of the baths used at various stages during laboratory processing for the purpose of cleaning the film and washing away any of the residual solution used in previous baths during the processing procedure.

click stops A method of setting controls in photographic equipment so that an audible click accompanies the engagement of a control in a preset position. Click stops are used in lens diaphragms, in those parts of an animation stand that are used to control panning, and in other equipment where it is easier or more convenient to listen for a sound than it is to try to locate a control on a darkened set.

click track A cueing track that is used as a guide during the recording process or, in the absence of a sound track, by the film editor for establishing a beat or rhythm for the corresponding picture.

clipping Literally, the shearing off of the amplitude peaks of a videotape signal. Electronic circuits built into videotape recorders are designed to clip signals above or below certain preset frequencies in order to keep the white and black levels of the picture within those limits. The process reduces the intensity of the highlights at one extreme and dense, black areas at the other extreme to provide a better scene balance, thus enhancing the overall picture or signal.

clogged head In videotape recording, a buildup of oxide on the video head during recording, causing noise or loss of image in that band.

closed circuit (television) A system for transmitting picture information to specific audiences at designated points over coaxial cables designed for that purpose. The signal is accessible only to stations physically connected to the transmission cable.

close-up Also **C.U.** A scene or action in which only a small part of the performer or subject matter is visible in the film frame. A close-up of an actor, for example, might show only his head or, in an extreme close-up, only the eyes or mouth.

close-up lenses Close-up or meniscus lenses are designed to fit over a camera's regular lens like a filter. These lenses permit focusing on subject matter only inches from the camera. In addition to their uses in photography, they can also be used in conjunction with enlargers in order to produce higher and lower magnifications.

Zoom and regular lenses can be used for close-up filming in combination with close-up lenses which attach to the front of a lens. Close-up lenses are distinguished by either their diopter power or focal length.

To convert the focal length into diopters, divide 1 by the focal length. For example, a lens engraved with a focal length of 1 meter is a 1-diopter lens. ($1/1 = 1$.) An 0.5-meter close-up lens equals 2 diopters. ($1/0.5 = 2$.)

The focal length of a lens in meters is obtained by dividing 1 by the diopter power of the lens. For example, $1/4 = 0.25$, meaning that a 4-diopter lens has a focal length of 0.25 meter (25 centimeters or about 10 inches). Note that the focal length must always be expressed in meters to make these conversions.

With the basic lens set to infinity, the distance from the close-up lens to the subject, measured from the close-up lens, is always equal to the focal length of the close-up lens, regardless of the focal length of the basic lens.

The power of close-up lenses is rather weak, and the optical quality of the basic lens is hardly affected. If stronger lenses, or two of the weaker ones are used together, it is recommended that the diaphragm be closed one or two stops. A falloff in sharpness may result when the lenses are used wide open.

cloud effects The simulation of cloud effects, like most inventions, was born out of necessity. On many occasions, motion picture studios would select a suitable location for filming and then transport large quantities of personnel and equipment to the selected site. Waiting for a photogenic cloud formation to appear gave the actors and crew plenty of time for relaxation—and the studio accountants, headaches.

The optical combination of cloud effects with cloudless location scenes was the apparent answer. The optical process filled the immediate need but, because of the second-generation duping, at a reduction in print quality.

During the 1940s, Charles Clarke of the Twentieth Century–Fox studios, submitted an alternative method. Mr. Clarke took photographic transparencies that contained cloud formations of every description, mounted them in frames, and simply positioned them in front of the camera.

For scenes in which the camera had to be panned, larger transparencies were fitted into curved frames that were positioned in front of the camera. The transparency thus remained in a stationary position while the camera lens panned across it. Today almost every major studio uses adaptations of the original arrangement.

In order to ensure proper focus between the clouds in the transparency and the scene being filmed, wide-angle lenses are used and set at the smallest lens stop possible. Adjusting the transparency to create a favorable relationship between subject matter and clouds is easily accomplished by the cameraman. Composition, therefore, becomes a matter of choice since the director and cameraman can see the composite arrangement before filming begins. Perspective with regard to cloud formations is of negligible importance.

The cloudless on-location skies, the very reason for the use of these transparencies, provide the printing light during the photographic process. Where necessary, graduated neutral-density filters are used for the purpose of altering the tonal values of the clouds, thereby creating a more favorable balance in the composite scene. Cloud effects, on a smaller scale, can be used in conjunction with miniature sets.

The large variety of realistic cloud formations available to the cameraman quite adequately fill the most specific requirement. Some cloud formations are fleecy, some billowing. Others are dark and threatening. Since each of these formations is also available in front-, back-, and sidelighting as well as in arrangements that provide a suitable composition for any location, cloudless skies are no longer a problem for the itinerant cameraman. See CLARKE PROCESS.

clump Silver salts in the emulsion which, during the developing process, are converted into visible silver, black in color. The density of these silver deposit groupings or clumps, throughout the film frame, increases in direct proportion to the density of the negative image. See AGGLOMERATION and GRAIN.

coated lens A thin, transparent coating of magnesium fluoride, for example, is applied to the surfaces of lenses to reduce the amount of light lost during transmission as the light passes through and is reflected from each of the surfaces of the several elements making up the lens.

coaxial cable A metallic cable used for transmitting television broadcasts at high frequencies with a minimum of power loss. It usually consists of a conductor centered in a tubular jacket by a series of insulating surfaces. The metallic jacket acts as the return, or ground, conductor.

cobwebs Spiders had best look to their laurels. Motion picture technicians, in recent years, have succeeded in duplicating the best efforts of the predacious insects and reducing their position in the film industry to a point of relative insecurity.

Even the most expert eight-legged arachnid capable of spinning a web may find it difficult to distinguish between the real thing and the artificial networks produced by the latest cobweb-spinning unit. Distributed by Mole-Richardson, the device is quite important to the producer of mystery and horror thrillers.

With the cobweb-spinning unit, the gossamer effect can be produced by simply aiming the device at the area needing the webbing. In operation, the rubber cement compound that is used to produce the effect is forced through a tiny valve. A multibladed fan attached to the unit transforms the compound into thin threads of fluid that could fool the most discerning spider.

After the web has been spun, the technician can give the simulated web an added touch of realism by dusting it with any type of powder. See Figures C-14*a* to *c*.

coherent light An ordered succession of light waves of the same frequency. See HOLOGRAPHY.

collage Actually a montage effect created by combining subject matter from several sources (usually photographs or other graphic material) to form a composite effect in the frame of film.

collimator (lenses) A device used for aligning, correcting the line of sight, or obtaining parallel rays of light in optical viewing systems such as telescopes and viewfinders.

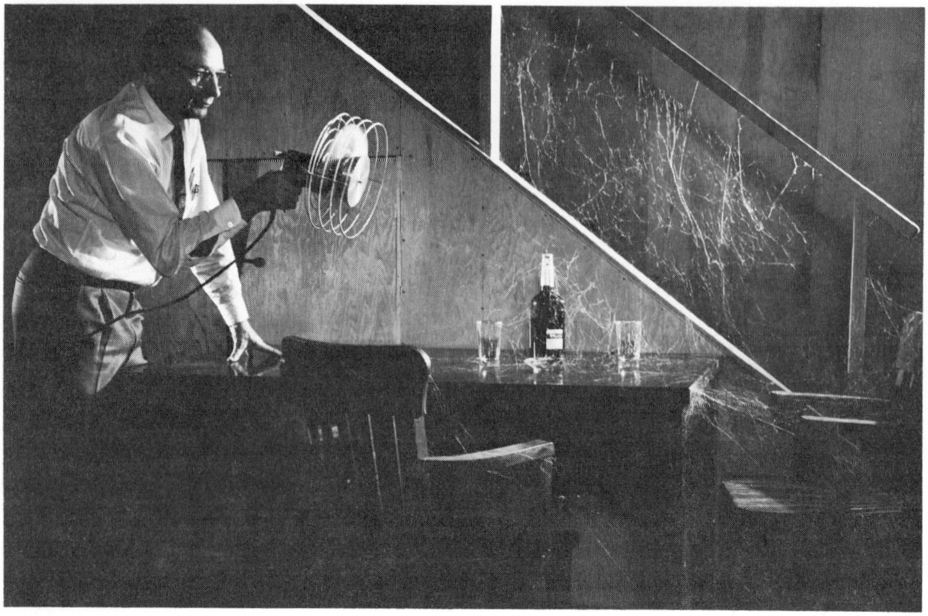

Fig. C-14*a* Producing cobwebs with Mole-Richardson type 1911 cobweb spinner Moleffect.

Fig. C-14*b* Cobwebs produced with Mole-Richardson type 1911 cobweb spinner Moleffect.

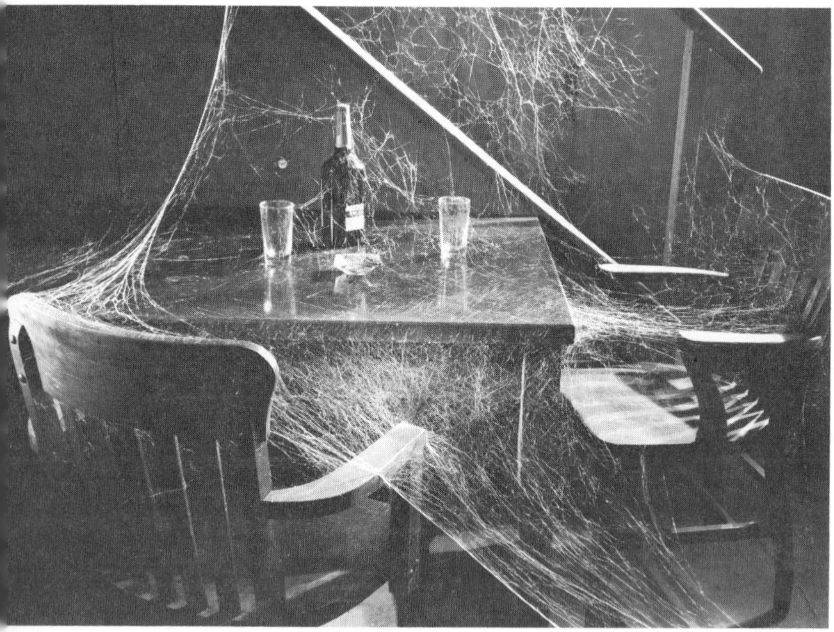

Fig. C-14*c* Mole-Richardson type 1911 cobweb spinner Moleffect.

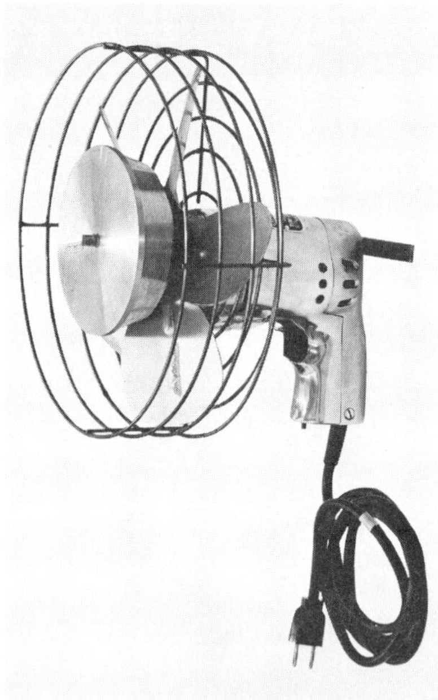

color Red, green, and blue (actually red-orange, green, and blue-violet), are the primary optical colors. A different order governs paints and inks because of the way in which the physical pigments behave when mixed.

All colors are the result of selective absorption and reflection of wavelengths of light. The color (red, for example) is visible in an object because it reflects only that color while absorbing the blue and green wavelengths of light. Objects that absorb all the approximately 3,800 wavelengths that are visible to the eye are black. Conversely, objects that reflect all the colors are white. Colors produced from sunlight dispersed by a prism are the only really pure colors.

Optical illusions, in connection with color perception, are quite common. One side of a color patch, as in the case of step wedges, appears darker than the opposite side of the same patch—depending on the density of the adjoining color. If the adjoining color is darker than the color of the patch next to it, that side of the patch which is butted against the darker patch will appear to be lighter in color. Similarly, an object of one color that is placed in front of a background of a complementary color will seem to be brighter than it actually is.

Color, generally, creates impressions that influence the viewer in more ways than are realized. Red, for example, is associated with warmth. Blue, by comparison, has the opposite effect. In a photograph or painting, red will appear to be closer, in relation to the viewer, than a blue area of the same size. Red, therefore, is regarded as an advancing color, and blue a retreating or receding one. Similar principles relate to black and white values. White is for good guys—black is for villains.

color, complementary When all the visible colors are arranged in a circle, in the order in which they are dispersed by a prism, the color directly opposite any other in the circle would be its complement. Green, for example, would be the complement of red. Any two colors at opposite ends of the circle, added together, produce white light.

The colors in a negative film image are complementary to the values in the original subject matter. See Figure C-15 and COLOR.

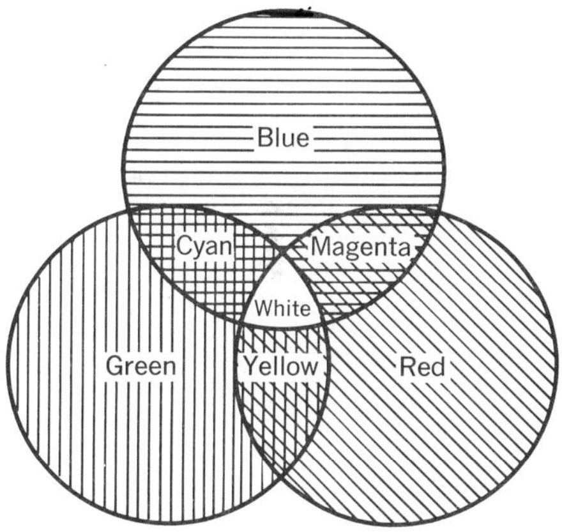

Fig. C-15 **Primary and complementary colors.**

color, local Colors are not always what they appear to be. Only when they are viewed in daylight, under reasonably sunny skies, do they present their true colors. Objects are influenced by the color of the light that illuminates them. A white card against a green background will absorb some of the background color and seem to be tinted green.

color, minus A color made up of all the colors in the spectrum with the exception of one restricted band is known as a minus color. Blue, for example, is not found in yellow. Yellow, therefore, is said to be minus blue.

color-arc lamp (lighting) Lamps developed by Sylvania Electric Products, Inc., and designed to produce a greater amount of illumination with less power requirements than conventional incandescent studio lamps of similar color temperature. Used for color television, motion picture production, and commercial photography floodlighting, the lamps have an average rated life of 7,500 hours in movie studio applications compared with a 25- to 150-hour average rated life of incandescent lamps operating in the same color-temperature range.

color automatic timing corrector The color automatic timing corrector in videotape recorders reduces the residual time-base errors of the system to a figure required for playing back color recordings. Stabilization is achieved by measuring the residual jitter, after prestabilization by the monochrome automatic timing corrector unit, and then applying a further correction through an electronically variable relay line. See MONOCHROME AUTOMATIC TIMING CORRECTOR.

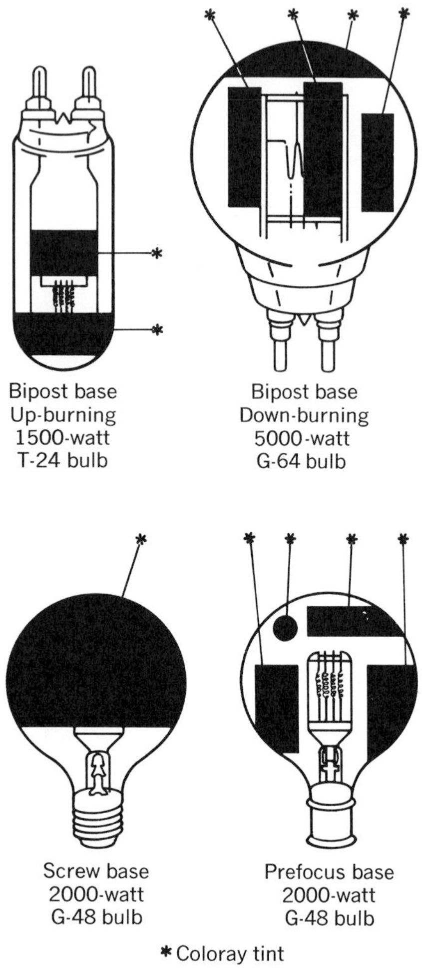

Fig. C-16 Diagrams of Coloray lamps. (*Lee-muiron Rousseau, New York.*)

Coloray lamps Specially designed* color lamps which are used during the photographic process to change, modify, or intensify specific colors. With these lamps, the color of an entire face, or a portion thereof, can be altered and the color of makeup nullified or accented at the discretion of the director. Colors in a scene can be sharply delineated within their spectrum or replaced entirely. Daylight scenes can be turned into night scenes without the use of neutral-density filters or combinations of filters. See Figure C-16.

color bars An established electrical reference signal consisting of bars of color—the photographic primaries and their complements. The color standards, set by the SMPTE, appear at the head of a taped program or commercial. Playback videotape recorders are adjusted to these standards.

color calibrator A device that makes possible the visual comparison (via television monitors) of the peak-white video level with a reference white produced by a lamp within the calibrator. The chromaticity of the peak white is adjusted after comparisons with the standard white produced by the calibrator, thereby assuring a uniform signal between studios and relay points that are transmitting similar program material.

*The lamps and process are patented by Lee-muiron Rousseau, New York.

color chart and gray scale A chart* with a 10-step gray scale (black and white are not included) and 32 color patches. It is used as a guide for checking the manner in which different colors will reproduce as shades of gray. Tests, made by exposing the patches on film or by comparison over closed-circuit television monitors, are used to evaluate and correct the tonal values, assuming that corrective measures are necessary. The gray scale and color patches are quite useful when used as guides, before filming begins, in order to determine how a scene photographed in color will subsequently be reproduced in black-and-white prints.

color-compensating filter As the name implies, the filters in this group make it possible to alter the color balance in a scene and are used to compensate for deficiencies in the quality of the lighting. In the Kodak series, for example, CC-05-R indicates that the filter is *c*olor *c*ompensating (CC), that its intensity and absorption qualities are rated at *0.05* and that, being *r*ed (R), it is capable of absorbing green and blue light. See Table C-2.

*Sharp's color chart and gray scale.

TABLE C-2 Kodak Color-compensating Filters

Peak density	Yellow (absorbs blue)	Exposure increase in stops*	Magenta (absorbs green)	Exposure increase in stops*	Cyan (absorbs red)	Cyan-2 (absorbs red)†	Exposure increase in stops*
0.025	CC 025Y	—	CC 025M	—	CC 025C	CC 025C-2	—
0.05	CC 05Y	—	CC 05M	1/3	CC 05C	CC 05C-2	1/3
0.10	CC 10Y	1/3	CC 10M	1/3	CC 10C	CC 10C-2	1/3
0.20	CC 20Y	1/3	CC 20M	1/3	CC 20C	CC 20C-2	1/3
0.30	CC 30Y	1/3	CC 30M	2/3	CC 30C	CC 30C-2	2/3
0.40	CC 40Y	1/3	CC 40M	2/3	CC 40C	CC 40C-2	2/3
0.50	CC 50Y	2/3	CC 50M	2/3	CC 50C	CC 50C-2	1

Peak density	Red (absorbs blue and green)	Exposure increase in stops*	Green (absorbs blue and red)	Exposure increase in stops*	Blue (absorbs red and green)	Exposure increase in stops*
0.025	CC 025R	—	—	—	—	—
0.05	CC 05R	1/3	CC 05G	1/3	CC 05B	1/3
0.10	CC 10R	1/3	CC 10G	1/3	CC 10B	1/3
0.20	CC 20R	1/3	CC 20G	1/3	CC 20B	2/3
0.30	CC 30R	2/3	CC 30G	2/3	CC 30B	2/3
0.40	CC 40R	2/3	CC 40G	2/3	CC 40B	1
0.50	CC 50R	1	CC 50G	1	CC 50B	1 1/3

*These values are approximate. For critical work, they should be checked by practical test, especially if more than one filter is used.

†The filters in the C-2 series should be used in preference to other cyan filters when cyan filters are required in filter packs used in printing on Kodak Ektacolor and Kodak Ektachrome papers.

color-corrected lens A lens free of chromatic aberration.

color correction The deliberate alteration of specific color values, within a scene, through laboratory processes. See Figure C-17 and TIMING AND COLOR CORRECTION.

color difference self-matteing process See MATTES and SELF-MATTEING PROCESSES.

color dropout compensator The color dropout compensator in videotape recorders is an electronic control which detects a lack of signal from the tape and reinserts the missing information by making needed compensations in signal strength.

color film Film with one or more layers of color-sensitive emulsion. Color values in the original subject matter are reproduced in prints after processing. See FILM and MULTILAYER FILM.

color-film conversion filter A filter that is positioned in front of a camera lens and used to convert daylight balanced film to tungsten—or tungsten to daylight.

Fig. C-17 **Scene-to-scene printing color correction is accomplished at the Technicolor Television Division by operator of Technicolor-designed delineoscope.**

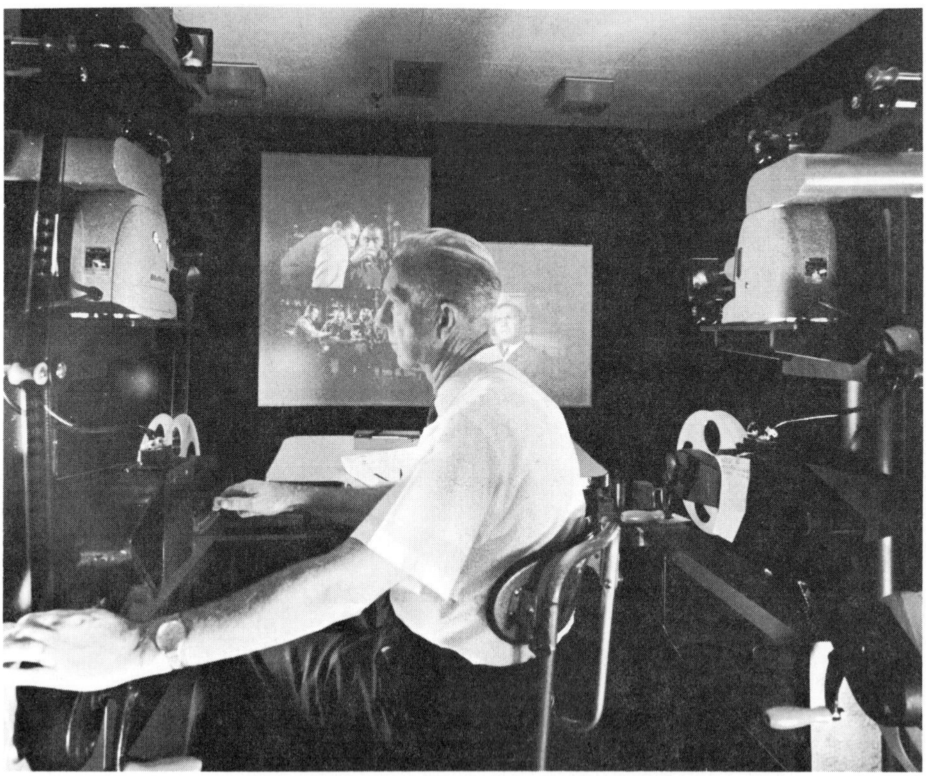

TABLE C-3 The Effects of Filters on Colors

Color	Type of filter needed to lighten the color	Type of filter needed to darken the color
Red-orange	Red or deep red	Blue or green
Yellow	Yellow, orange, or red	Blue
Green	Green or yellow-green	Blue or red
Cyan	Blue or green	Red
Blue	Blue or green	Red or orange
Magenta	Red	Green

color filter A material, either colored gelatin or glass, that is placed in front of a camera lens. The filter allows certain colors or wavelengths to pass through it to expose the film in the camera while absorbing other colors. Filters permit the passage of light rays of their own color but hold back or absorb light rays of complementary colors. See Table C-3.

Colorgard A metering device developed by Gardner Laboratories for measuring the brightness of the three phosphors on a color monitor. Eliminating the problems normally associated with visual comparison, the meters are used to adjust the color balance and assure color uniformity between transmission points.

color image The image produced on color film after processing. In its negative form, the colors are complementary to those in the original subject matter.

colorless filter Also called **ultraviolet filter**. A filter that absorbs visible light. It is sensitive only to light in the invisible spectrum.

color media Any colored transparent material that can be placed in front of a light source for the purpose of adding color to the light being emitted by it. These are often referred to as *gels* (for gelatin), although glass and plastic materials are also used.

color model Also called **model drawing**. In animation, one of several drawings in a scene that serves as a guide for the opaquers. Each of the colors that is to be used for opaquing the cartoon characters, after the drawings have been traced on cels, is indicated by number, thus assuring consistent coloring throughout a scene, even though the many cels in the scene may be opaqued by several people.

color scene tester A unit* designed to make scene-to-scene tests for color and light intensity in motion picture release work. It uses three

*Herrnfeld.

separate light sources; one for the red; one for the green; and one for the blue-sensitive layer of the printing film. The scene tester exposes 15 frames of film per cycle of operation. Each frame receives a different combination of exposures. The film editor selects one frame, representative of the scene in question, in which the colors are closest to those he would like to see duplicated in the print. See Figure C-18 (in color insert).

color sensitivity The response of a film's emulsion to colors in the visible range of the spectrum.

color separation As the name implies, the separation of colors from previously processed film into three panchromatic films that are exposed through three primary filters. The separation prints allow greater latitude in the subsequent correction of individual colors.

color-separation electronic matteing system See MATTES and SELF-MATTEING PROCESSES.

Colortec An accessory trademarked by Ampex that reduces time errors in tape playback long enough for the color circuits in the receiver to decode color properly.

color temperature Measured in degrees Kelvin (273° higher than the readings on the centigrade scale), the color temperature of a light source is equal to the temperature at which a theoretically perfect black-body radiator would have to be heated in order to radiate light of a color similar to that of a light source.

Temperature changes affect the color of the source. Black coal, for example, produces a red flame at its lowest burning temperature. As the coal becomes hotter, the color of the flame changes to orange and, in time, to yellow. At extremely high temperatures, coal would emit light which could subsequently turn to blue. Light-balancing or photometric filters, designed to absorb red or blue, are used to change the color temperature of a light source. These filters are especially useful for balancing the color temperature of the light with the sensitivity of the film. See Table C-4 and MIRED VALUES.

color-temperature corrective filter A type of filter used for balancing the color temperature of the light source with the sensitivity of the film.

Color films are produced for use with specific types of illumination. When the prescribed lighting is not available or cannot be used, conversion filters are used to adjust the color quality of the illumination. The "adjustment" actually consists of bringing the color temperature of the light source to a point at which it is consistent with the color quality of the lighting for which the film was originally designed.

color-temperature meter Also called **three-color meter**. Color-temperature meters are designed to measure the color temperature of a light source in degrees Kelvin as well as the amount of color in that light source in relation to the other colors in the source. Blue/red and green/red scales provide information with respect to degrees Kelvin. Capable of readings ranging from 2000°K to approximately 30,000°K, some meters also indicate the corrective filters that are needed to ensure proper exposure.

ColorTran lighting system The ColorTran lighting system, first introduced in 1939, has become a widely accepted method of controlled illumination for the amateur as well as the professional photographer. Light-intensity and color-temperature control is achieved by increasing the voltage fed to ordinary tungsten lamps, thus changing their intensity and spectral characteristics.

Besides providing dependable illumination for all photographic applications, the low-budget system uses fewer light sources for obtaining equivalent light levels and requires proportionately less setting-up time. The system is based on the fact that an increase in voltage raises a lamp's filament temperature, thereby making it possible for the lamp filament to emit up to four times more light than it was originally designed to produce. Voltages below a lamp's

TABLE C-4 Color-temperature Chart

Light source	Color	Degrees Kelvin
Outdoors (daylight)		
Skylight—clear	Deep blue	30,000
Skylight—clear	Strong blue	25,000
Skylight—clear	Medium blue	16,000
Skylight—clear	Light blue	10,000–13,000
Average shade	Blue	8,000
Light shade	Light blue	7,100
Overcast or hazy sky	White, light blue	6,500–8,000
Direct sunlight—summer	White	5,400–6,500
Sun—noon	White	5,400
Sun—afternoon	White, orange	4,800
Sun—late afternoon	Light orange	4,300
Direct sunlight—winter	White, blue	4,700–5,800
Indoors (artificial)		
Flash	Light orange	3,600–3,800
Photoflood	Light orange	3,400
Tungsten	Slightly reddish	3,200
Mazda	Slightly reddish	2,800–3,000

normal rated value cause the light to appear yellow. With voltages above normal, the light will be bluish white.

The ColorTran converter is the control center that converts the line voltage input to precisely calibrated output voltages. The degree of control, in steps ranging from 2900 to 3450°K, ensures the proper matching of light to the film being used.

These lighting systems are as popular for television production, animation, and still photography as they are for motion picture illumination. See Figures C-19*a* to *f* and Tables C-5 to C-10.

Fig. C-19*a* The effect of the ColorTran converter in spectral response.

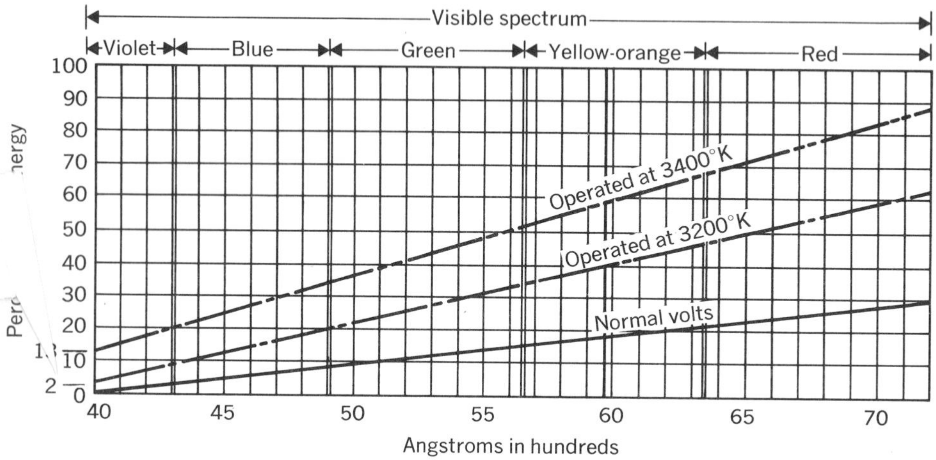

Fig. C-19*b* A typical ColorTran lighting arrangement for a relatively large television stage.

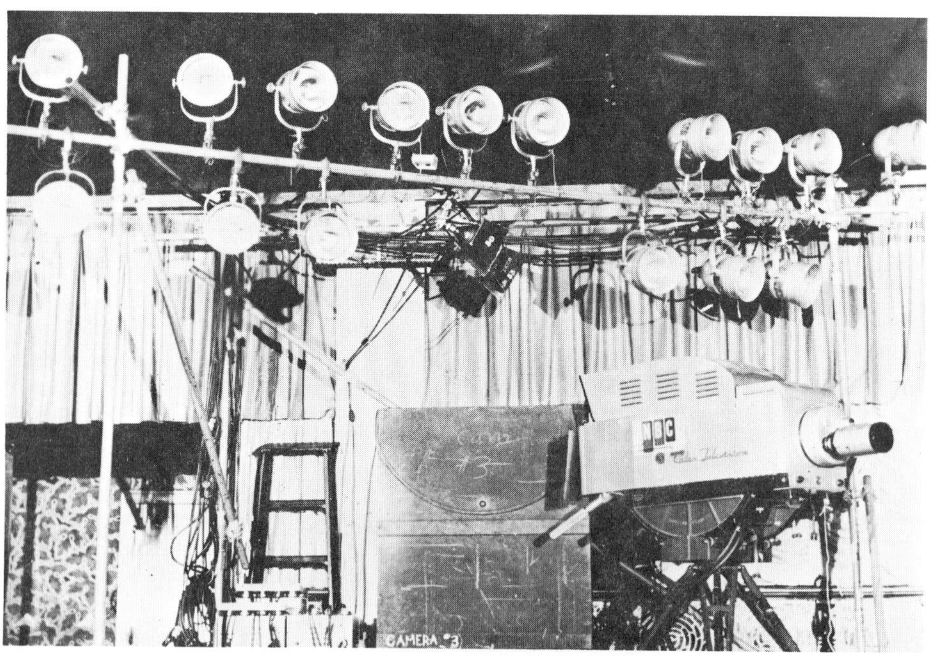

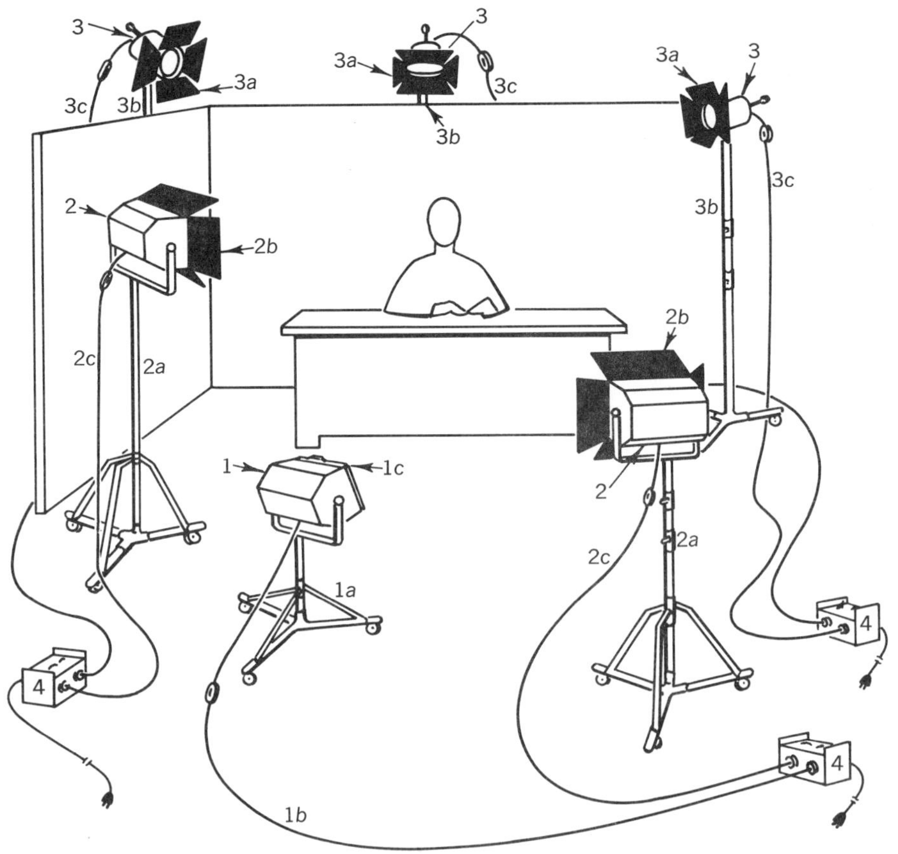

Fig. C-19c ColorTran for lighting television and motion picture sets. Color or black and white: 15 feet wide by 10 feet deep by 10 feet high.

Item	Quantity	Description	Code number
1	1	*Fill/base light—multibroad*	LQBM-10F
1*a*	1	Adjustable stand (low), with casters	SSR-12
1*b*	1	Extension cord with M/F connectors	LEC25RG
1*c*	1	Diffusion lens and holder	DBMF/1
1*d*	1	Q-I lamp, 500-watt, 3200°K, 120-volt	B55-32*
2	2	*Fill/base light—single-broad*	LQBM-10
2*a*	2	Adjustable stands (extra high), with casters	SSRH
2*b*	2	Four leaf barndoors	B4BM-10/E
2*c*	2	Extension cord with M/F connectors	LEC25RG
2*d*	2	Q-I lamps, 1000-watt, 3200°K, 120-volt	B5-32*
3	3	*Back/key light—multibeam 650, including accessory holder*	LQF-6/DQK
3*a*	3	Four leaf barndoors	B4QK-10
3*b*	3	Adjustable stand (extra high), with casters	SSRH
3*c*	3	Extension cord with M/F connectors	LEC25RG
3*d*	3	Q-I lamps, 650-watt, 3200°K, 120-volt	B6-32*
4	3	*Light outlet and control box*—with two 15-ampere receptacles and two switches, 25-foot cable (no. 12/3 conductor) parallel blade U-ground connectors	LDOC2RG/25RG

*1*d*, 2*d*, and 3*d* are the lamps recommended for use with the multibroad, single-broad, and multibeam 650 respectively.

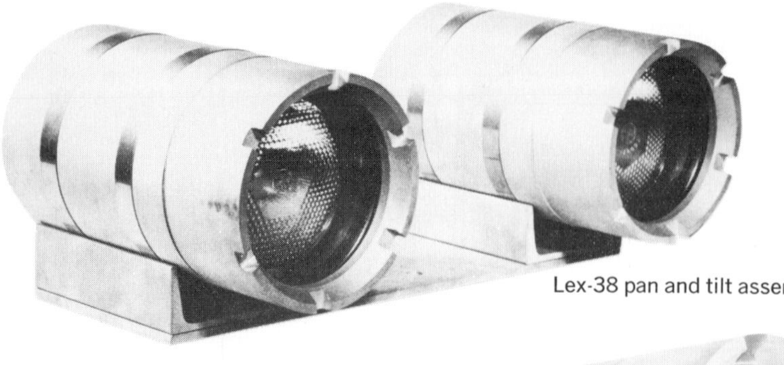

Lex-38 pan and tilt assembly

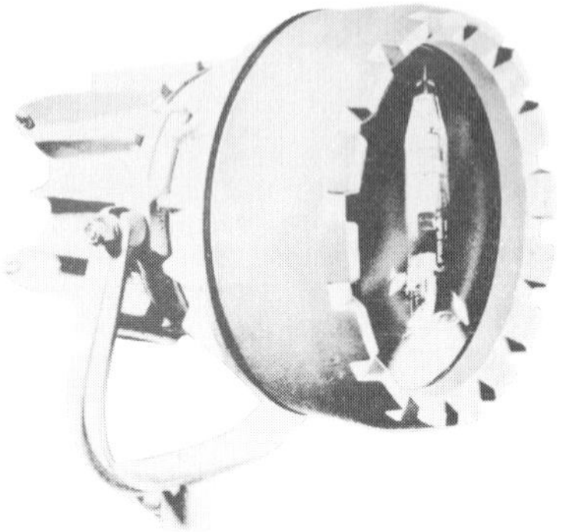

Lex-38
for stand mounting

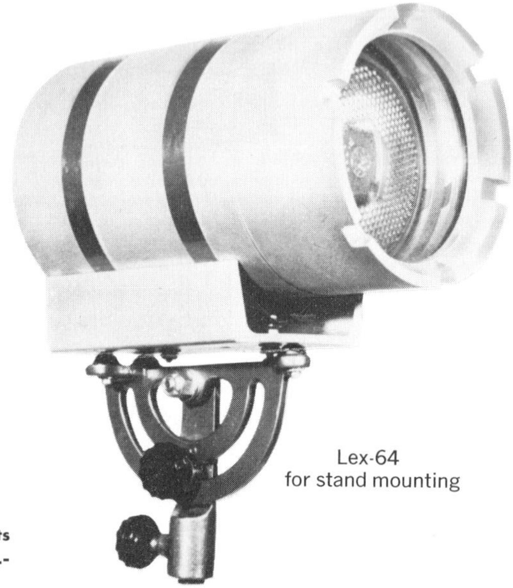

Lex-64
for stand mounting

Fig. C-19*d* **ColorTran explosion-proof lights for PAR-38 and PAR-64 lamps. Certified MIL-5272C procedures III and IV.**

Fig. C-19e **ColorTran converter, Cinemaster Chief, Mark II. (*Natural Lighting Corp., Burbank, Calif.*)**

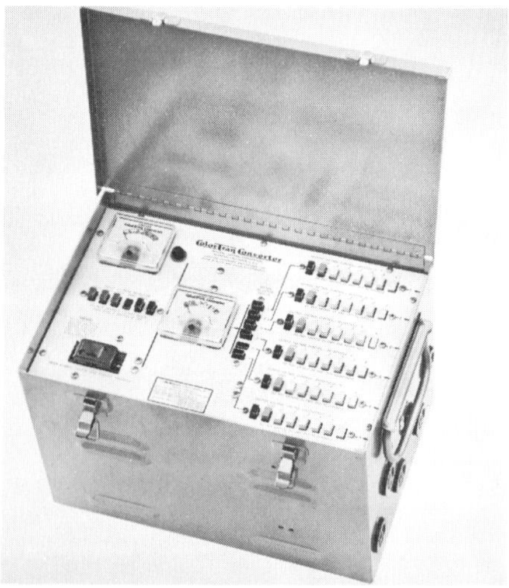

Fig. C-19*f* **ColorTran gaffer grip type A, manufactured by Natural Lighting Corp., Burbank, Calif.**

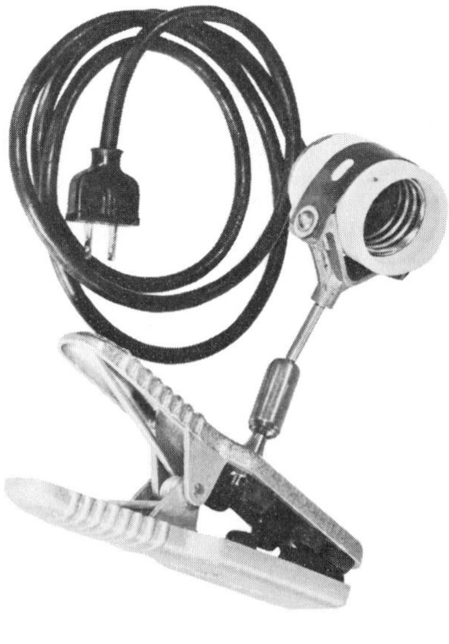

TABLE C-5 Performance of Lamps Boosted with ColorTran

	Input current, amperes			Intensity at 10 feet	
Lamp	Voltage setting	220-volt input	120-volt input	Foot-candles	Coverage
PAR-64 500-W Wide flood	120 160 185	2.5 4.0 4.9	4.7 7.4 9.1	115 300 450	4 ft × 5 ft
PAR-64 500-W Medium flood	120 160 185	2.5 4.0 4.9	4.7 7.4 9.1	370 870 1,400	2 ft × 4½ ft
PAR-64 500-W Narrow spot	120 160 185	2.5 4.0 4.9	4.7 7.4 9.1	1,200 2,720 3,750	2½ ft × 2 ft
PAR-56 300-W Medium flood	120 160 185	1.4 2.2 2.8	2.6 4.3 5.4	242 620 870	1½ ft × 3½ ft
PAR-56 300-W Wide flood	120 160 185	1.4 2.2 2.8	2.6 4.3 5.4	90 230 350	3 ft × 7 ft
PAR-56 300-W Narrow spot	120 160 185	1.4 2.2 2.8	2.6 4.3 5.4	880 2,050 2,950	1½ ft × 1½ ft
R-80 1,500-W Flood	120 160 185	7.0 10.9 13.7	12.8 20.0 25.1	230 610 940	7 ft diameter
R-60 1,000-W Flood	120 160 185	4.7 7.4 9.1	8.7 13.7 16.7	166 345 490	7 ft diameter
R-40 500-W Flood	120 160 185	2.5 4.0 4.9	4.7 7.4 9.1	34 82 125	10 ft diameter
1,500-W PAR-64 (LCQ only) Wide flood 3350° K	120 volts only *Caution:* Do not boost with converter.		12.0	1,500	2½ ft × 4 ft

TABLE C-6 ColorTran Long-life Quartz-Iodine Lamps

Lamp code*	Voltage (a-c or d-c)	Wattage	Color temper-ature, °K	Rated life, hours	Lamp lumens	ColorTran light fixtures (Notes)
B2-34	30	250	3400	6	8,000	1
B10-32 or B10-32F	120	1,000	3200	150	28,000	2
B10-34 or B10-34F	120	1,000	3400	35	33,000	2
B10-32E or B10-32EF	120	1,000	3200	150	28,000	3
B5-32 or B5-32F	120	1,000	3200	500	27,000	4, 5, 6, 7
BS10-32F/1	120	1,000	3200	150	26,000	8
B75-32 or B75-32F	120	750	3200	500	19,800	4, 5, 6, 7
B6-32 or B6-32F	120	650	3200	125	16,500	9
B6-34 or B6-34F	120	650	3400	30	20,000	9
B55-32 or B55-32F	120	500	3200	500	13,000	4, 5, 6, 7
B5-29 or B5-29F	120	500	2900	10,000	8,600	4, 5, 6, 7
B4-30 or B4-30F	120	400	3000	2,000	7,400	9
B25-32 or B25-32F	230	800	3200	250	21,600	4, 5, 6, 7
B8-32E or B8-32EF	230	800	3200	80	20,600	2
B8-32 or B8-32F	230	800	3200	80	20,600	7, 9
B8-34 or B8-34F	230	800	3400	20	25,000	9
BPNS10-32	120	1,000	3200	150	28,000	10, 11
BPNS10-34	120	1,000	3400	35	33,000	10, 11

TABLE C-6 ColorTran Long-life Quartz-Iodine Lamps (*cont.*)

Lamp code*	Voltage (a-c or d-c)	Wattage	Color temperature, °K	Rated life, hours	Lamp lumens	ColorTran light fixtures (Notes)
BPS10-32	120	1,000	3200	150	28,000	10, 11, 12
BPS10-34	120	1,000	3400	35	33,000	10, 11, 12
BPM10-32	120	1,000	3200	150	28,000	12, 13
BPM10-34	120	1,000	3400	35	33,000	12, 13
BPW10-32	120	1,000	3200	150	28,000	12, 13
BPW10-34	120	1,000	3400	35	33,000	12, 13
BQ-101	120	1,300	3200	150	33,000	14
BQ-201	240	1,300	3200	150	33,000	14

*Order code symbol F indicates frosted lamp.

Notes:

1. Multibeam 30 (LQF-3).
   Quartz-King 30 (LQK-3).
2. Quartz-King dual 1000 (LQK-10/DY and /DM).
3. Multibeam 1000 (LQF-10).
4. Quartz-King 500 (LQK/5WYA).
   Mini-lite 10 (LQM10/WYB).
   Back-Lite (LQB/B).
5. Variable broad (LQBF-10).
   Single broad (LQBM-10).
   Multibroad (LQBM-10F).
   Double broad (LQBM-20).
6. Soft-Lites (LQS-7, LQS-10, LQS-20, LQBS-40).
   LQS-7 does not accept B5-32 lamp.
7. Cyc-Strips (LQCY10-1 and LQCY10-4).
8. Focusing 1000 (LQF-10-30 MKII).
9. Multibeam 650 (LQF-6).
   Quartz-King dual 650 (LQK-6/DY and /DS).
   Mini-Lite 6 (LQM6/WYB).
   Quartz adapter (AQB-6).
10. Hi-Spot, with specular intensifier (LQH10-0).
11. Cine-Queen (LCQ).
    LEX 64.
12. Follow spot (LQFS-1).
13. Hi-Spot, with diffuse intensifier (LQH10-1).
14. Super-80 (LQSE).

TABLE C-7 Performance of Colortran Quartz-King Lights

Quartz-King models	Focus position	Maximum intensity, footcandles at 10 feet	Cove
			Beam* diamet feet
Dual 650 LQK6 / DY (using B6-32 lamp)	Medium Flood	270 140	6 10
Dual 1000 LQK10 / DY (using B10-32 lamp)	Medium Flood	480 210	5½ 10
500 LQK / 5WYA (using B5-32 lamp)	Flood (fixed focus)	260	(W × H) 10½ × 6½

*Beam is measured to 50 percent of maximum intensity.
†Field is measured to 10 percent of maximum intensity.
W Width.
H Height.

TABLE C-8 Colortran Long-life Quartz-Iodine Lamps Available for Quartz-King Lights

Quartz-King models	Lamp code*	Voltage, a-c/d-c	Watts	Color temperature, °K	Rated life, hours	Maximum intensity, footcandles at 10 feet — Focus position: Medium	Flood
Dual 650 (LQK6/DY)	B6-32 or B6-32F	120	650	3200	125	270	140
	B6-34 or B6-34F	120	650	3400	30	320	170
	B4-30 or B4-30F	120	400	3000	2,000	120	60
	B8-32 or B8-32F	230	800	3200	80	330	170
	B8-34 or B8-34F	230	800	3400	20	400	210
Dual 1000 (LQK10/DY)	B10-32 or B10-32F	120	1,000	3200	150	480	210
	B10-34 or B10-34F	120	1,000	3400	35	570	250
	B8-32E or B8-32EF	230	800	3200	80	360	150
500 (LQK/5WYA)	B5-32 or B5-32F	120	1,000	3200	500	*Note:* Not applicable: wide-flood focus only. ⟶	260
	B75-32 or B75-32F	120	750	3200	500		190
	B55-32 or B55-32F	120	500	3200	500		130
	B5-29 or B5-29F	120	500	2900	10,000		80
	B25-32 or B-25-32F	230	800	3200	250		210

*F indicates frosted lamp.

feet	Coverage, degrees		Lumen output	
ld† diameter, feet	Beam* diameter	Field† diameter	Beam*	Field†
13	35	66	6,100	11,700
18	52	83	7,800	13,300
12	32	61	8,800	17,200
18½	52	85	12,200	20,600
W × H) 18 × 17	(W × H) 55 × 36	(W × H) 83 × 81	9,900	20,600

TABLE C-9 Performance of ColorTran Broads; Models LQBM-10, LQBM-10F, LQBM-20

Model	Focus position	Maximum intensity, footcandles at 10 feet*	Coverage at 10 feet (width × height)		Maximum candlepower*	Coverage, degrees (width × height)		Lumen output*	
			Beam,† feet	Field,‡ feet		Beam†	Field‡	Beam†	Field‡
LQBM-10 (single-broad)	Fixed	200	12 × 8	19 × 15	27,000	57 × 40	81 × 68	11,000	18,800
LQBM-10F (multibroad)	Medium	230	12 × 7	20 × 14	32,000	55 × 35	82 × 62	11,300	19,200
	Wide	61	13 × 20	24 × 27	10,000	59 × 81	91 × 99	11,500	16,300
LQBM-20 (double-broad)	Fixed	350	12 × 9	20 × 15	48,000	55 × 30	81 × 67	20,400	34,200

*Using B5-32 lamps.
†Beam is measured to 50 percent of maximum intensity.
‡Field is measured to 10 percent of maximum intensity.
Note: Performance data and lamp specifications also apply to equivalent TV models of ColorTran broads.

TABLE C-10 Long-life Quartz-Iodine Lamps Available for ColorTran Broads; Models LQBM-10, LQBM-10F, LQBM-20

Lamp code*	Voltage, a-c/d-c	Watts	Color temperature, °K	Rated life, hours	Maximum intensity, footcandles at 10 feet			
					LQBM-10 (single-broad)	LQBM-10F (multibroad: single) Medium focus	LQBM-10F (multibroad: single) Wide focus	LQBM-20† (double-broad)
B5-32 or B5-32F	120	1,000	3200	500	200	230	61	350
B75-32 or B75-32F	120	750	3200	500	150	170	45	260
B55-32 or B55-32F	120	500	3200	500	96	110	30	170
B25-32 or B25-32F	230	800	3200	250	160	190	49	280

*F indicates frosted lamp.
†Two lamps required.

color transparency A color photograph that must be viewed or projected by light transmitted through it.

coma A defect in a lens that produces tail-like markings in small rounded objects. Actually, an off-axis aberration that causes distortion of the oblique rays passing through the lens. See Figure C-20 and ABERRATION, SPHERICAL.

commentary The off-stage narration that accompanies the on-screen picture in documentary, industrial, training, and educational films.

commercials Lengths of motion picture film or videotape, with a running time from 10 seconds to one minute, designed for the specific purpose of extolling the virtues of the manufacturer or his product. Institutional commercials run a poor second in relation to the number of hard- and soft-sell advertising messages that are seen regularly.

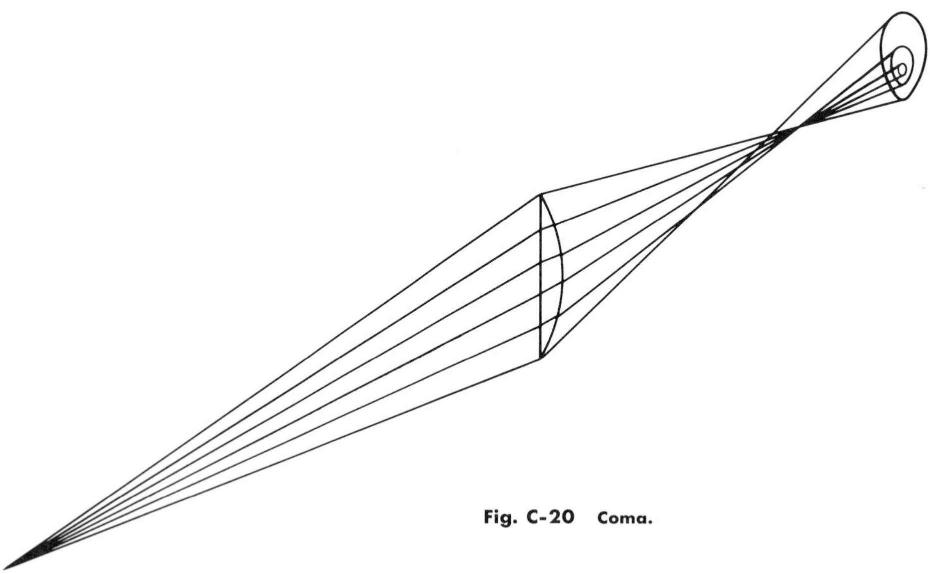

Fig. C-20 Coma.

The frequency and number of commercials that may be included in a specific time segment varies with the geographical location of the commercial station and the time selected for their presentation.

Although most sponsors are careful to avoid misleading claims, monitoring boards such as the NAB "suggest" ground rules in the form of codes that leave little to the imagination.

Most viewers agree that the sound levels for commercials are deliberately higher than the program material surrounding them. This opinion is not shared by studio technicians who have no control over the volume. Spliced together, in their proper program sequence, they may seem to be louder only because they were originally recorded at higher sound levels.

Good, bad, or indifferent, and entertaining or not, commercials and their sponsors make commercial programming possible. (Or would you rather watch a full-hour show explaining the various techniques that can be used for pruning rosebushes without interruption or an occasional announcement that begins with—"And now, a word from . . . .") See Figure C-21.

Fig. C-21 **Commercial film standards. (*National Broadcasting Company, Inc.*)** Diagrams not drawn to scale.

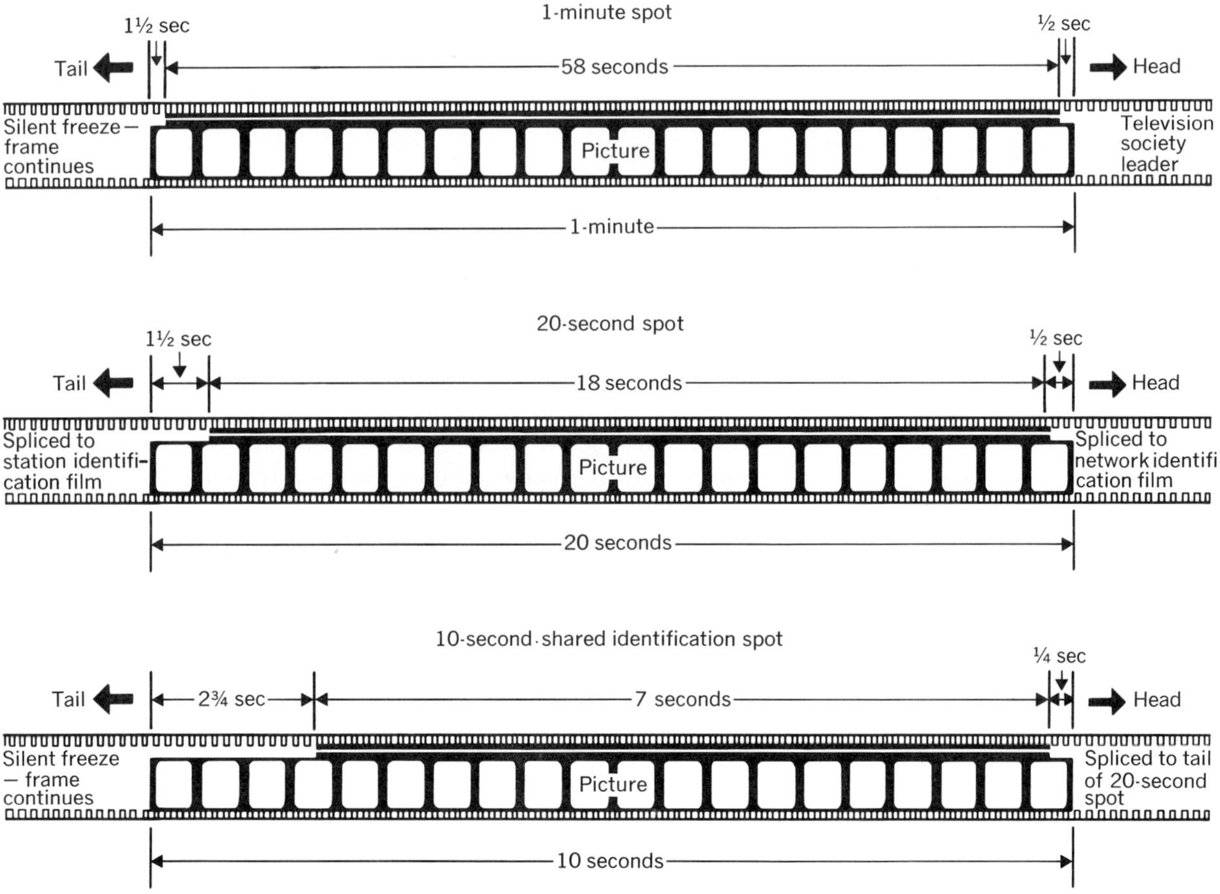

compensating filter The group of filters generally used for restoring the balance and duplicating the color values in the print so that they appear similar to the colors in the original subject matter.

complete drawing Abbreviated **C.D.** The instructions placed on a drawing by the animator when only a part of the cartoon character has been drawn. If the cartoon character is speaking, for example, the animator may draw the mouth action and mark "C.D." on the drawing to indicate that the remainder of the head must be drawn by the assistant animator or in-betweener.

composite daily print A print, in projection synchronism, that is produced from an original in-camera composite negative or from original sound and picture negatives. During the screening, it is checked for photographic and sound quality, action, and general effect.

composite master positive A fine-grain composite print, either black and white or color, that is used for making composite dupe negatives. See COMPOSITE PRINT.

composite print A print, either black and white or color, in which the picture and sound track are combined on the same length of film and printed in projection synchronism.

composite reversal original An original print, produced by reversal processing, in which the sound track and the picture information appear on the same length of film.

composites, electronic Similar in principle, but not in performance, are the matteing systems used in motion picture production and the electronic *inset* and *chroma-key* processes that make composite effects possible in television and videotape production.

In motion picture production, mattes blank out specific portions of the picture in a frame of film—temporarily. In television, the same type of masking is accomplished electronically—but, also temporarily. In both cases, new picture information is subsequently inserted in the unexposed areas of the picture that serves as the background.

In order to produce a composite effect for the motion picture production, previously photographed negative and positive mattes are combined with the various lengths of film that contain the component picture information needed to achieve the desired effect. The optical cameraman, in the first phase of the combining process, exposes the positive matte with the film that contains the background. The exposed film, if developed, would show the background along with an unexposed area that matches the size, shape, and position of the picture information that is to be inserted later. The exposed film, with its latent image, remains in the optical camera—unprocessed.

During the second run or exposure, the negative matte and the insert picture information are photographed together on the previously exposed film. Only that part of the picture which fits within the clear area of the negative matte will be visible in the processed film. With each matte canceling the other, the composite picture shows the insert and the background in the desired relationship. The possibilities of a double exposure, or superimposed images, are thus completely eliminated.

Electronic composites, in both monochrome and color television, are produced by insetting picture information from one television camera into the picture produced by a second camera. The first electronic process for achieving composite effects, patented in 1932, anticipated a need for combining picture information from several sources.

In its simplest application, the video signal from one television camera is used to key out or delete a portion of the picture produced by the second camera. Backgrounds for the composite effect may be miniature representations of full-size sets, artwork, photographs, or, in fact, the picture information in a frame of motion picture film. During the production of a black-and-white composite, one camera remains focused on the background selected for the scene. The second television camera, on another part of the same set or at some remote location, photographs the inset subject matter that has been positioned in front of a black background. If desirable, a third camera may be used to record foreground elements or action appropriate to the composite scene.

In either case, proper perspective and a constant relationship between the several planes are necessary to create the desirable illusion of depth in the composite picture. Since the output from one camera is combined with the video signal from other cameras, each of the component sets used in the multiple filming must be lit separately. The lighting needed for the black backing is considerably less than the amount of illumination needed for either the subject matter or the background for the scene itself. Since the success of the keying system is based on variations in the brightness of the video signal, blacks similar to those in the background must be avoided in both clothing and props. Unless other colors are substituted, the composite picture will contain undesirable double exposures with the scene's background visible in each of the black areas.

The chroma-key process for producing color composites, basically, is similar to the method used for achieving black-and-white composite effects. Photographed separately, the setting for the combined picture information again becomes the background into which the component elements are subsequently inset. Keying action, however, depends more on color values than on brightness variations in the video signal.

The backgrounds for the insert material, lit separately as in the monochrome process, are rendered in colors that complement the values in the subject matter. Although any of the primary colors can be used for these backings, blue is preferable since that color most closely complements skin tones. (Blue is also the favored color in matte preparation for the motion picture production.)

In long shots, or scenes in which a part of the studio floor is visible, that surface must be painted and treated as though it was a part of the backing. Although flat lighting must be used for such backgrounds, the subject matter may be illuminated to produce a wide variety of effects. Obviously, all colors similar to those used for keying purposes must be avoided in clothing, accessories, and product shots.

To the television viewer, the composite effects are electronic magic. To the studio technician, it is merely a matter of combining the video signal from one camera with the output from other cameras. See SELF-MATTEING PROCESSES and ZOOM KEYER.

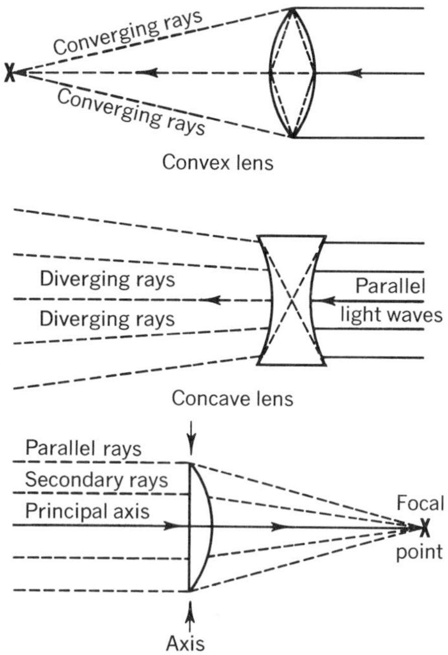

Fig. C-22 **Concave and convex lenses.**

composition The arrangement of subject matter within the film frame, in relation to the background, in order to present both in the most pleasing or desirable manner possible.

compound lens A lens with more than one element.

compound shutter The symmetrical arrangement of the thin, metal leaves that open outward from a central point in front of the optical axis of the camera lens.

compound table That part of the animation stand on which the artwork is positioned during the photographic process. The compound table is fitted with control wheels and counters and can be moved north/south and east/west in order to achieve panning movement. Zoom effects are achieved by moving the camera, which is mounted above the compound table.

concave lens A lens that spreads or disperses light waves. See Figure C-22.

concert border A borderlight mounted immediately behind the proscenium arch. Compare BORDERLIGHT.

condenser An adjustable light-collecting lens (either a planoconvex or two such lenses arranged so that their plane surfaces face outward) which is used to provide even illumination over a field. In lighting units, Fresnel lenses act as condensers and concentrate the beam of light on the area where it is needed. In projection equipment, the condensing lenses are used to focus the light source that is subsequently passed through the film. Electrically, condensers are electrostatic devices capable of storing electric charges.

cone light A floodlight unit designed to cover large areas with soft, diffused light. The inner surface of the housing is painted white and serves as a reflector. Excellent as fill lights for television production, cone lights accept lamps ranging from 750 to 5,000 watts.

constant-density ratio law The theory advanced by Hurter and Driffield which states that density is determined by exposure and not by development. Changes in development time will only affect the contrast—but not the density of the image.

contact precision step printer The contact printer is a daylight operating machine that functions on the step-printing principle. The method of printing, with two films held in contact during the exposure, is used for such operations as printing dupe negatives, color separations, background plates, and master positives.

The film movement incorporated in the printer has fixed registration pins. A single piece of film may be operated through the film movement a number of times and yet maintain perfect registration, regardless of the direction of film travel.

The optical system is designed to achieve a variable balance between spectral and diffuse light and provides a flat, even field of illumination at the aperture. The camera assembly is equipped with both a manually variable and an automatic dissolve shutter.

The dissolve or fade lengths available are from 16 to 128 frames. Other features include the facility for printing edge numbers on either side of the film, buckle switch, a resettable frame and footage counter, light trap rollers in removable carriers, and film notch switches for the actuation of automatic light and shutter-controlling devices. The printer is equipped with a variable-speed synchronous motor that produces synchronous speeds to 24 frames per second.

Manual operation provisions include a Variac and voltmeter for lamp intensity control and switches for the automatic-dissolve mechanism. The shutter may also be used for fully manual operation. Film take-up is provided by individual torque motors for both negative and raw stock. See Figure C-23 and CONTACT PRINTING.

contact printing A printing process in which a negative or positive length of film is in direct contact with the raw stock. Both lengths of film move past the printing light (where the raw stock is exposed) to produce a negative or positive image, depending on the polarity of the original film used for printing.

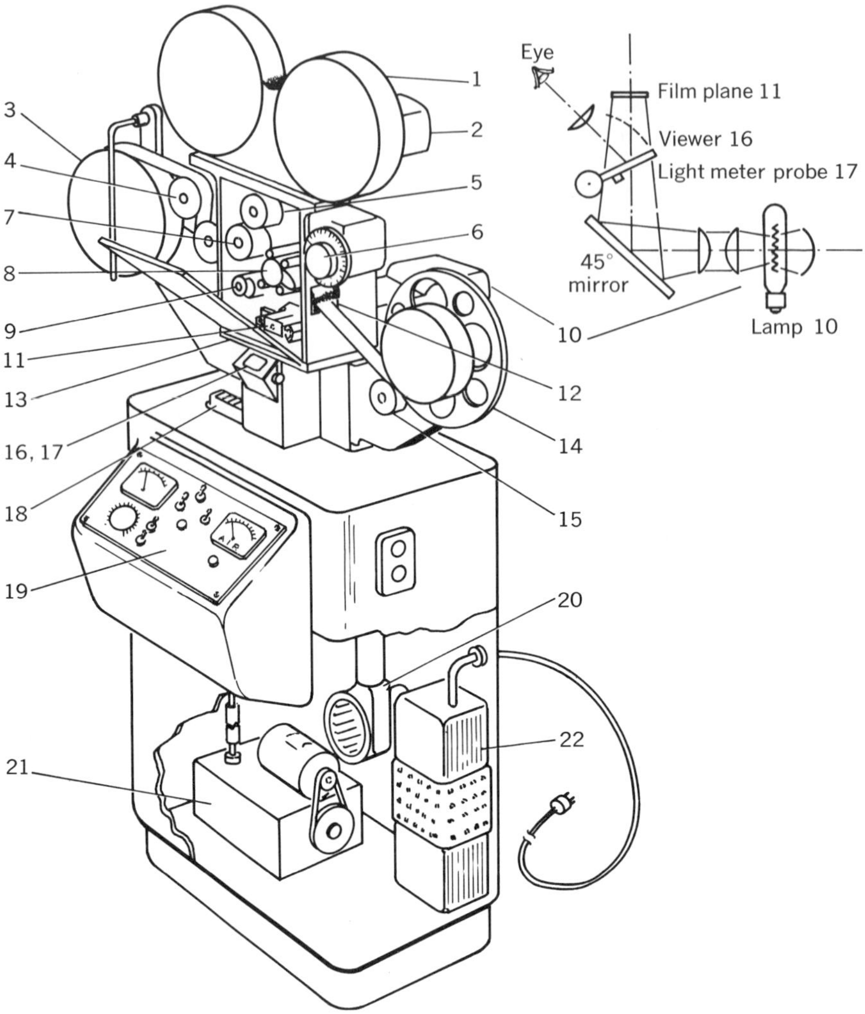

Fig. C-23 **Contact precision step printer. (*Animation Equipment Corp., New Rochelle, N.Y.*)**

1. Magazine, Mitchell type, 1,000-foot capacity, 35 mm.
2. Take-up, automatic, for 400-, 1,000-, and 1,200-foot magazines, variable tension control, take-up motor, variable friction clutch feed.
3. Feed reel (for master negative) with variable friction clutch feed.
4. Cleaning rollers, dry type, for 35mm negative.
5. Buckle switch.
6. Dissolving shutter control, calibrated in degrees, manual type.
7. Cleaning rollers, dry type, for 35mm raw stock.
8. Sprocket assembly, 35mm, with edge numbering light (exposes numbers on either or both edges of film).
9. Film notch actuator for 35mm. Stops printer when notched film passes edge roller switch.
10. Lamphouse and optical system, blower-cooled. Light path reaches film by 45-degree mirror. Two-inch-square filter slot.

11. Shuttle, 35mm, with enlarged aperture for complete frame coverage. Dimension 1.000 × 0.740. Apertures interchangable without stripping shuttle.
12. Light traps with rollers, velvet-covered, removable.
13. Printer housing with built-in facilities for 35mm and 16mm components. Variable 170-degree rotating shutter with direct control for precision repeatability. Printer operates in normal room light.
14. Take-up flange, removable. (May be used in place of tight wind.)
15. Tight wind, 35mm, with variable tension control.
16. Sync mark viewer. May be flipped into light path to check sync marks when loading.
17. Light-meter probe. May be flipped into light path to check light intensity.
18. Counter, frame and footage. 35mm. Resettable.
19. Control panel, hinged for service accessibility.
20. Blower and filter for lamphouse.
21. Drive, 2-speed, synchronous motor, 180 rpm stop motion and continuous, 360 rpm continuous only.
22. Constant-voltage transformer for lamp. Electrical requirements: 115 volts, 60 cycles, 10 amperes.

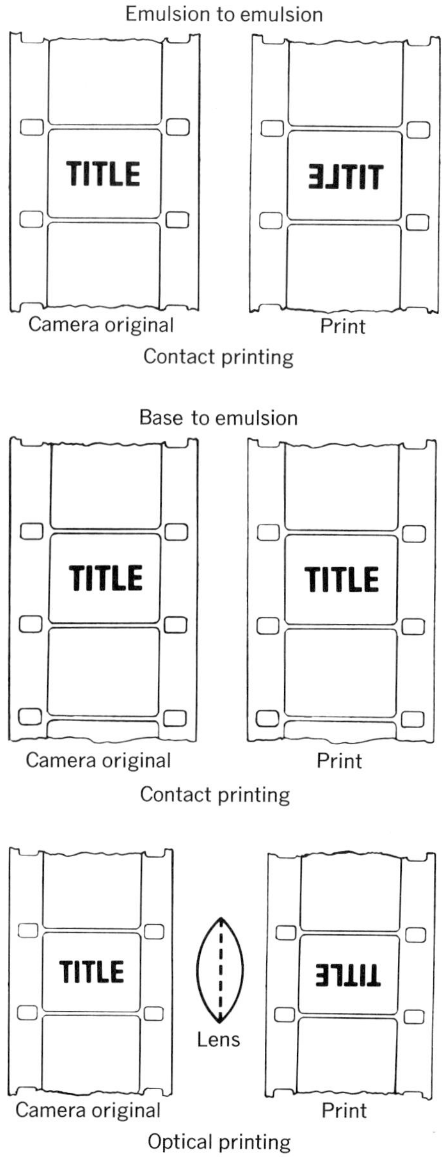

Fig. C-24 **Contact printing.**

Contact printing of a negative will produce a positive image on the raw stock because the clear portion of the film will transmit the most light, thus creating the densest "black" in the raw stock. Thus polarity is reversed, and a positive will also produce a negative. See Figure C-24 and CONTACT PRECISION STEP PRINTER.

continuing zoom A zoom effect, either animated or live, during which part of the subject matter extends beyond the field covered by the camera.

continuity The numerical order of scenes in a shooting script. In the normal course of production, scenes are not necessarily photographed in the order in which they will ultimately be projected.

Scenes that use similar backgrounds, either in the studio or on location, are photographed at the same time—while the performers are still available and the backgrounds, lights, and props are still set up. This type of scheduling not only saves time but also, more important, cuts production costs considerably. Since each scene, regardless of the order in which it is photographed, is identified by its own slate, the editor has no trouble cutting the scenes together in their proper sequence.

continuous-contact printing In continuous-contact printing, the light-modulating film and the raw stock both move continuously past the printing aperture at a constant speed. Compare CONTACT PRINTING. See CONTINUOUS PRINTER.

continuous printer A printing machine used by film laboratories. Unlike step printers which have an intermittent movement similar to those found in cameras and projectors, the film in a continuous printer travels past the light aperture in a continuous motion. Sound tracks especially need continuous printing since the stop-and-go, intermittent action of a step printer would completely ruin the sound by altering the patterns in the variable-area and variable-density sound tracks. See VARIABLE-AREA and VARIABLE-DENSITY SOUND TRACK.

continuous projector A type of projector capable of projecting motion picture film for television scanning. The continuously moving film is projected onto the target of a vidicon tube at the rate of 30 frames per second. Light from xenon flash lamps, directed onto the target area, is used to illuminate the picture being projected. The short duration of the flash, triggered by a transparent fiducial mark on the film, stops the motion of the projected picture to permit scanning.

The corresponding sound is conventionally reproduced by a sound-on-film system. Belt-driven by two 1,725-rpm induction motors, the projectors handle approximately 2,000 feet of film, representing a viewing time of about 20 minutes. The projector can be started or stopped in less than 5 seconds at any time during operation. See FIDUCIAL MARK and VIDICON TUBE.

continuous tone The range of values that black-and-white photography is capable of reproducing—from black through all the intermediate shades of gray to white—in gradual steps too fine to be perceived.

contrast In film, the spread between dark and light areas in the frame or scene.

In lighting, the ratio between the lightest and darkest areas of the motion picture or television stage.

contrast, simultaneous color The effect produced by the proximity of two colors. Complementary colors placed side by side (red and green, for example), will appear more intense than if viewed separately. By comparison, colors that are not complementary, viewed under similar conditions, will appear dull.

contrast filter The group of filters used for lightening or darkening a specific color without affecting the other colors in the subject matter.

contrast ratio In television or videotape transmission, the brightness range between blacks and whites necessary to achieve acceptable transfer characteristics or to ensure that full tonal range will be reproduced.

control panel The monitoring devices used by directors and technicians for controlling sound and picture during the recording and transmission of broadcast material.

converging lens A lens designed to refract light rays toward the optical axis. Generally classified as a positive lens, this type has one convex surface and is thinner at the edges of the lens than it is in the center section.

conversion filter Gelatin or glass filters positioned in front of a light source and used to convert tungsten and quartz lamps to a daylight color balance. Macbeth Whiterlite blue glass and dichroic filters are used for this purpose. The color temperature lost because of their use varies from 30° to 60°, approximately. See Table C-11.

TABLE C-11 Conversion Filters*

To convert				
From	To	Number	Filter color	Exposure increase in stops
3200	5500	80A	Blue	2
3400	5500	80B		$1\frac{2}{3}$
3800	5500	80C		1
4200	5500	80D		$\frac{1}{3}$
5500	3400	85	Orange	$\frac{2}{3}$
5500	3200	85B		$\frac{2}{3}$

Courtesy of Eastman Kodak.

*For critical work, check these values by practical test, especially if more than one filter is used.

converter An electrical device used for converting direct current to alternating current. It is usually a combination of motor and generator. The reverse is generally called a *rectifier.*

convertible lens Either half of a symmetrical lens. The theory is that each half of the lens is precisely corrected and can, therefore, be used by itself. See SYMMETRICAL LENS.

convex lens A lens that causes light rays to converge to a central point. See Figure C-22.

cookie Taken from the Greek word *kukaloris,* meaning the breaking up of light, a cookie is a flat piece of metal with irregularly shaped cutout areas. When placed in front of a light source, soft-focus shadow patterns are created on flat-toned backgrounds. See Figure C-25.

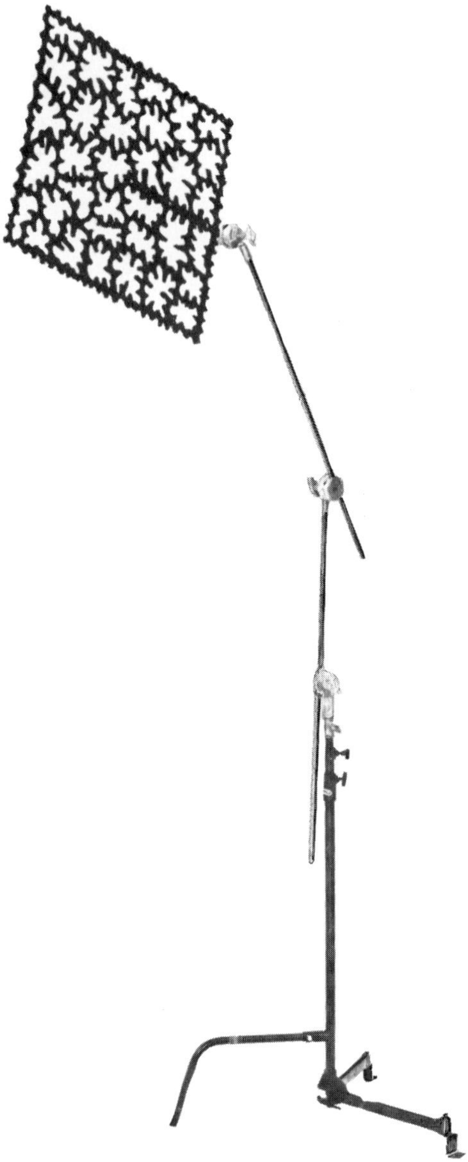

Fig. C-25 A cookie mounted on a century head and pedestal.

Cooper-Hewitt lights Early-vintage light sources. Unsuitable for panchromatic film, these mercury-vapor lights are now little more than a memory.

copper chloride Green hygroscopic crystals that are used as bleachers, intensifiers, and reducers during laboratory processing. They mix readily with water.

copper sulfate Also called **blue vitriol**. Blue crystals used as bleachers, intensifiers, and reducers during laboratory processing. The crystals dissolve readily in water.

cording A means of identifying a portion of the film wound on a roll. Short lengths of string or tape attached to perforations at the beginning and end of a specific portion in the roll indicate the footage needed by the film editor. Fine grains or dupes are then made of these specified sections.

core Round spools, usually plastic, used for winding or rewinding film. Raw stock, in earlier years, was supplied on cores made of wood. See Figure C-26.

corrective filter Also called **compensating filter**. The group of filters generally used for restoring the color balance and, subsequently, duplicating the original color values in the subject matter.

Fig. C-26 Core and spool characteristics. (*DuPont.*)

Size	No.	American Standard	Characteristics
16mm	32	PH22.38	O.D. 1.970 in., I.D. 1.020 in.; keyway slotted
	44	PH1.33	0.320-in. square hole; keyway; 2–0.218-in. round holes, both flanges
	49	PH1.34	Same as no. 44
	60	—	0.320-in. square hole; keyway both flanges
35mm	4	—	O.D. 1.000 in., I.D. 0.235 in.; no slot or keyway
	16	PH1.35	O.D. 3.605 in., I.D. 1.000 in.; 0.320-in. square hole; keyway both flanges
	24	PH22.37	O.D. 1.970 in., I.D. 1.020 in.; keyway through entire width of core
70mm	11	—	O.D. 2.000 in., I.D. 1.045 in.; no slot or keyway; wooden
	58	PH1.36	0.320-in. square holes; keyway both flanges
	59	—	O.D. 3.111 in., I.D. 1.020 in.; keyway, slotted
5.481 in.	68	—	O.D. 2.000 in., I.D. 0.625 in.; round hole; no slot or key-way; wooden

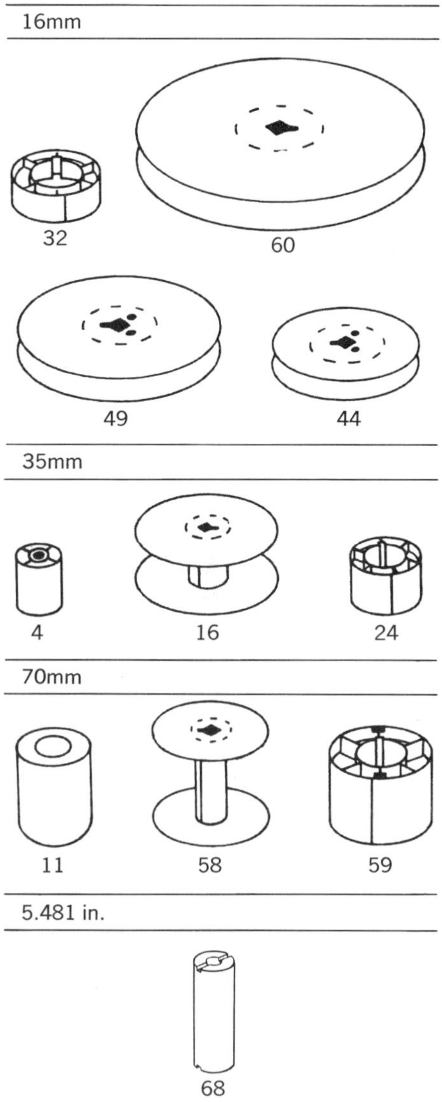

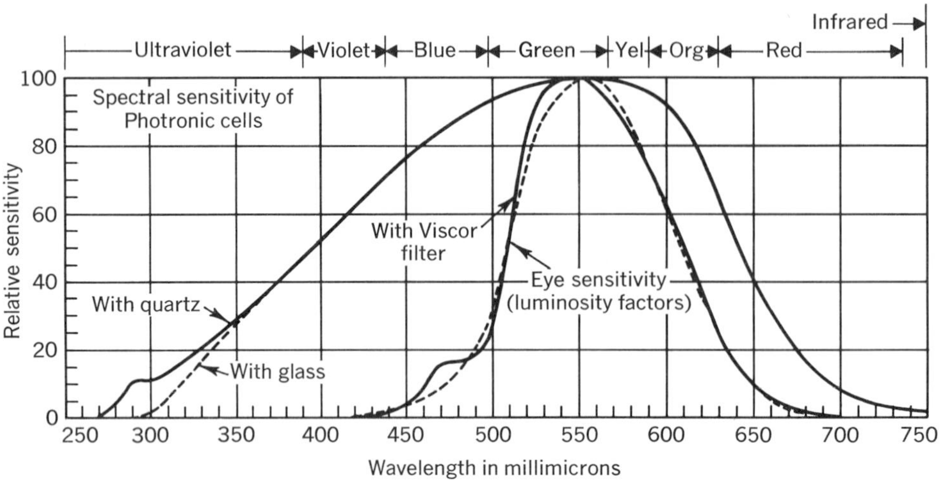

Fig. C-27 Viscor and cosine correction filters. (*Weston Instruments, Inc.*) By means of the Viscor and cosine correction filters over the Photronic cell, it is possible to measure such illuminants as sodium, neon, mercury vapor, or any other type of light source, and the measurements are correct, irrespective of the angle between the Photronic cell and the light source or sources.

cosine correction filter The cosine correction filter corrects the Photronic cell response in accordance with Lambert's cosine law of illumination. See Figure C-27. Compare VISCOR FILTER.

counters Frame and footage counters are built into most camera boxes. These counters record the exact amount of film exposed during the photographic process.

They are especially useful when making in-camera transitional effects such as fades and cross-dissolves. For the latter effect, the film must be rewound, after the outgoing scene has been exposed, before the new or incoming scene can be faded in to complete the effect. See CROSS-DISSOLVE.

coupler development The production of dye images in color photography by chemical reaction. During the developing process, the exposed silver halide in the emulsion is reduced to form metallic silver. Oxidized by the reaction, the developer combines with another chemical known as a coupler to produce a dye in proportion to the amount of metallic silver developed earlier. Being insoluble, the dye remains in the film to form the color image.

In some cases, the Kodachrome process, for example, the coupler components of the dyes are added to the film from solutions. The complex processing procedure, in this case, requires accurate chemical control. In the Ektachrome process, by comparison, the coupler components are added to the layers of emulsion during manufacture.

coverage, beam Beam coverage is defined as coverage to 50 percent of maximum intensity. Photographically speaking, 50 percent of maximum intensity represents one stop difference from the center intensity of the light. Another use of beam coverage is when the light patterns of similar fixtures are arranged for *overlapping.* The overlap point between any two or more similar fixtures is at the border point of 50 percent of maximum intensity (or beam coverage). In this way, coverage of uniform intensity can be provided. Beam is also called **one-half maximum**. Compare COVERAGE, FIELD.

coverage, field Field coverage is defined as coverage to 10 percent of maximum intensity. Field coverage indicates, for all practical purposes, the total size of available light from a particular fixture. In addition, the field measure is useful in indicating the visual limits for use in theater or display lighting. Field is also called **one-tenth maximum**. Compare COVERAGE, BEAM.

coverage, lighting In lighting, coverage is defined as the area or angle included to a defined percentage of maximum intensity. Coverage is measured on a flat plane perpendicular to the light source. When coverage is defined to a certain percentage of maximum intensity, no point within the coverage area is less than that stated percentage of maximum intensity.

Coverage is measured in two ways. First, the area illuminated (width and height or diameter), measured in feet, with the distance of the light source to the subject specified. Second, the angle emanating from the light fixture, measured in degrees. Coverage, when measured in degrees, is independent of distance, and offers the user a visual feeling of the cone of light that is produced. Further, it provides sufficient information to develop coverage data (in feet) for any distance. There are basically two percentages of maximum intensity to which coverage is measured—beam coverage and field coverage.

Also, the positioning of light sources on a motion picture or television stage so that the illumination covers not only the entire camera field but also a marginal area beyond those limits in which the action of the scene is to take place during the filming process.

covering power The ability of a lens to produce sharp images in every part of the film frame when set at the largest *f*-stop opening. Covering power also applies to that area on the plane of the negative in which the image is evenly illuminated and acceptably defined. Beyond that area, the image begins to lose definition, and the illumination falls off gradually as it reaches the edge of the film frame.

cover shot An extra "take" of a scene that is photographed after the other takes of the same scene have already been filmed. This procedure is followed when a reasonable doubt exists in the director's mind as to the effectiveness of the previous takes. This covering shot is also known as an **insurance take**.

cowcatcher A short commercial announcement that precedes the actual program material. Compare HITCHHIKE.

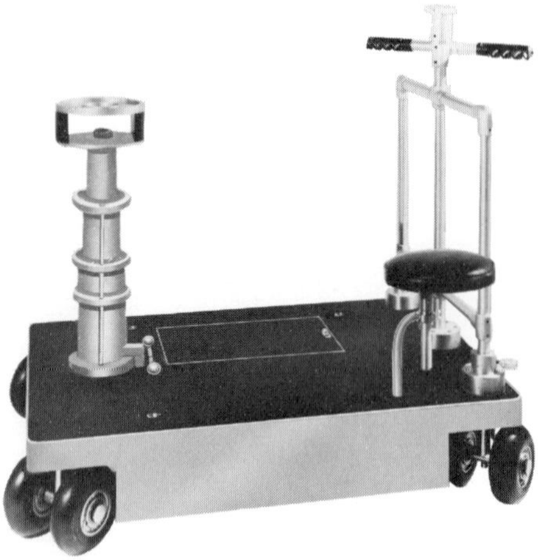

Fig. C-28 ColorTran crab dolly.

crab dolly A mobile platform on which the camera is mounted during the filming process. Most crab dollies feature dual steering of both front and rear wheels, either individually or together. Any combination of camera movements—straight, curved, angled, and right-angle turns—can be easily made, smoothly and silently. One man can quickly position and align a crab dolly for either moving or static shots. Shooting time is drastically cut because the camera can be precisely positioned for static shots or moved for continuous filming in which several setups are combined in one take. The crab dolly is indispensable in modern production shooting because of the moving effects possible and the great savings in setup time. See Figure C-28.

crane A relatively large, mobile device capable of supporting a camera and its operator. Excellent for overhead and bird's-eye shots, a crane is also referred to as a **cherry picker**.

crawl title Also called **running title**. Credits or acknowledgments that pan from the bottom to the top of the screen. During the panning movement, the lines of the title appear at the bottom of the film frame, move up gradually and finally disappear as they pan beyond the top limits of the film frame.

credits The listing of the names of the performers at the beginning of a motion picture. This listing is followed by secondary credits that name the technicians whose herculean efforts helped make the film production possible. See CRAWL TITLE.

critical angle The minimum angle at which a ray of light is refracted after passing from one transparent medium to another. Rays of light change direction, bending toward or away from their original paths, depending on the density of the medium they are entering. The angle of incidence is equal to the angle of reflection. The angle of refraction, however, is the angle at which a ray of light is bent away from its normal path when passing from one medium to another.

The refractive index is the ratio between the sine of the angle of incidence and the sine of the angle of refraction.

$$\text{Refractive index} = \frac{\text{sine of angle of incidence}}{\text{sine of angle of refraction}}$$

Cronar polyethylene terephthalate film The name is almost longer than its definition: a tough, flexible film base or support for photographic emulsions.

crop The camera framing of a picture which excludes a portion of the subject matter near each of the edges of the frame.

cross-dissolve An optical or camera effect in which the outgoing scene fades out and the incoming scene fades in over the same length of film. See Figures C-29*a* to *c*.

cross-fade An optical effect in which the outgoing scene, as well as the incoming scene, fades to a selected color or shade of gray rather than to black as in the usual fade effect. The effect is actually achieved with two cross-dissolves. In the first, the outgoing scene dissolves to the selected color. In the second, the selected color dissolves out as the incoming scene dissolves in to complete the effect. See Figures C-29*a* to *c*.

cross hairs The fine reticle markings engraved on the ground glass in the camera's viewfinder. These lines are used as a guide for centering the subject matter as well as for composition purposes. In some cases, the television field is also engraved along with the cross-hair centering lines.

crosslight Also called **modeling light**. Illumination directed at the performers or subject matter from opposite sides of the set. The main function of crosslighting is to balance the effects of key light on a performer's face. When the key light is used to light the left side of a performer's face, for example, the less intense crosslight is used to model or light the right side of the face.

crosstalk Extraneous sound, from one channel that is being used for recording, overlapping into a second channel.

cue A signal or spoken line preceding an action, or the dialogue used to alert the performers involved in the subsequent action.

cue card Also referred to as an **idiot card**, a cue card serves a purpose similar to that of the teleprompter. As the name implies, cue cards contain cues, portions of speeches, lines of dialogue, or the accompanying lyrics for a song.

Cue cards are positioned, or held, in the vicinity of the camera so that the performer does not give the impression that the information on them is actually being read.

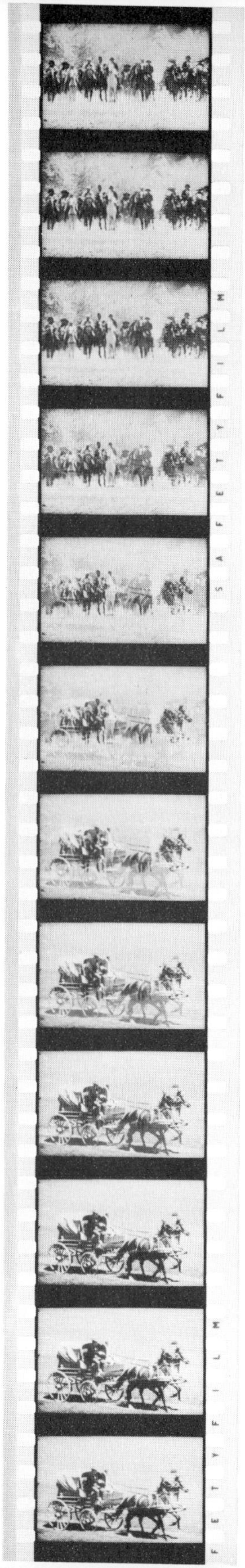

Fig. C-29a **The cross-dissolve.**

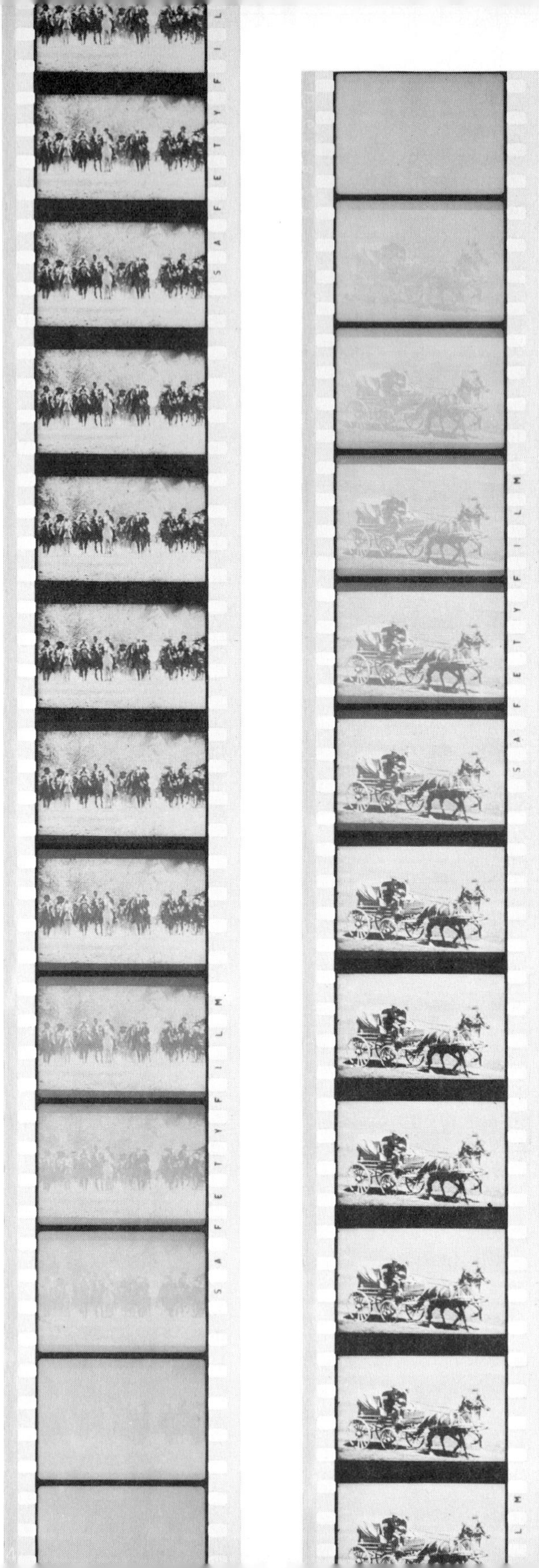

Fig. C-29*b* **The cross-fade.** The scene in the column at the left fades to gray, (actually a cross-dissolve), rather than to black as in a conventional fade. The incoming scene in the column at the right fades in from gray to complete the transitional effect.

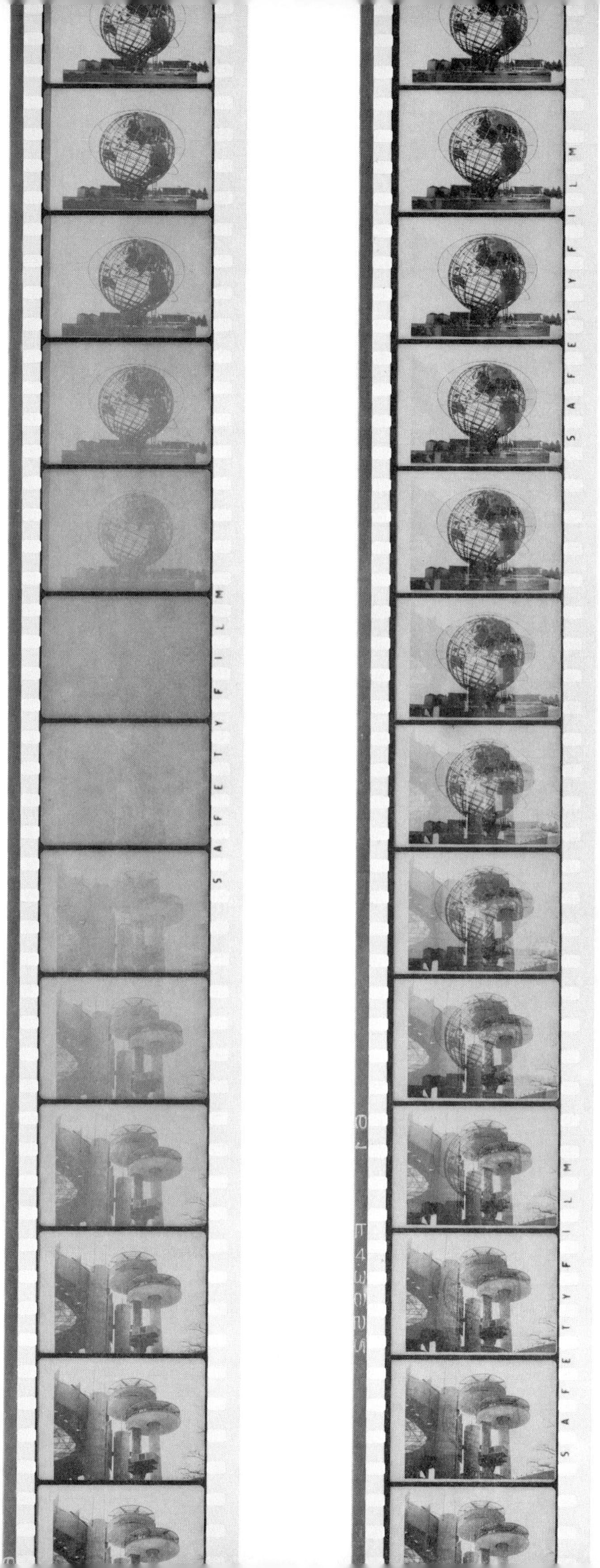

Fig. C-29c **Left: a cross-fade. Right: a cross-dissolve.**

cue sheet The optical layout sheets, containing information that serves as a guide for the optical cameraman—frame numbers, filters needed for color correction, lens stops, and the order of effects, for example. See Figure C-30.

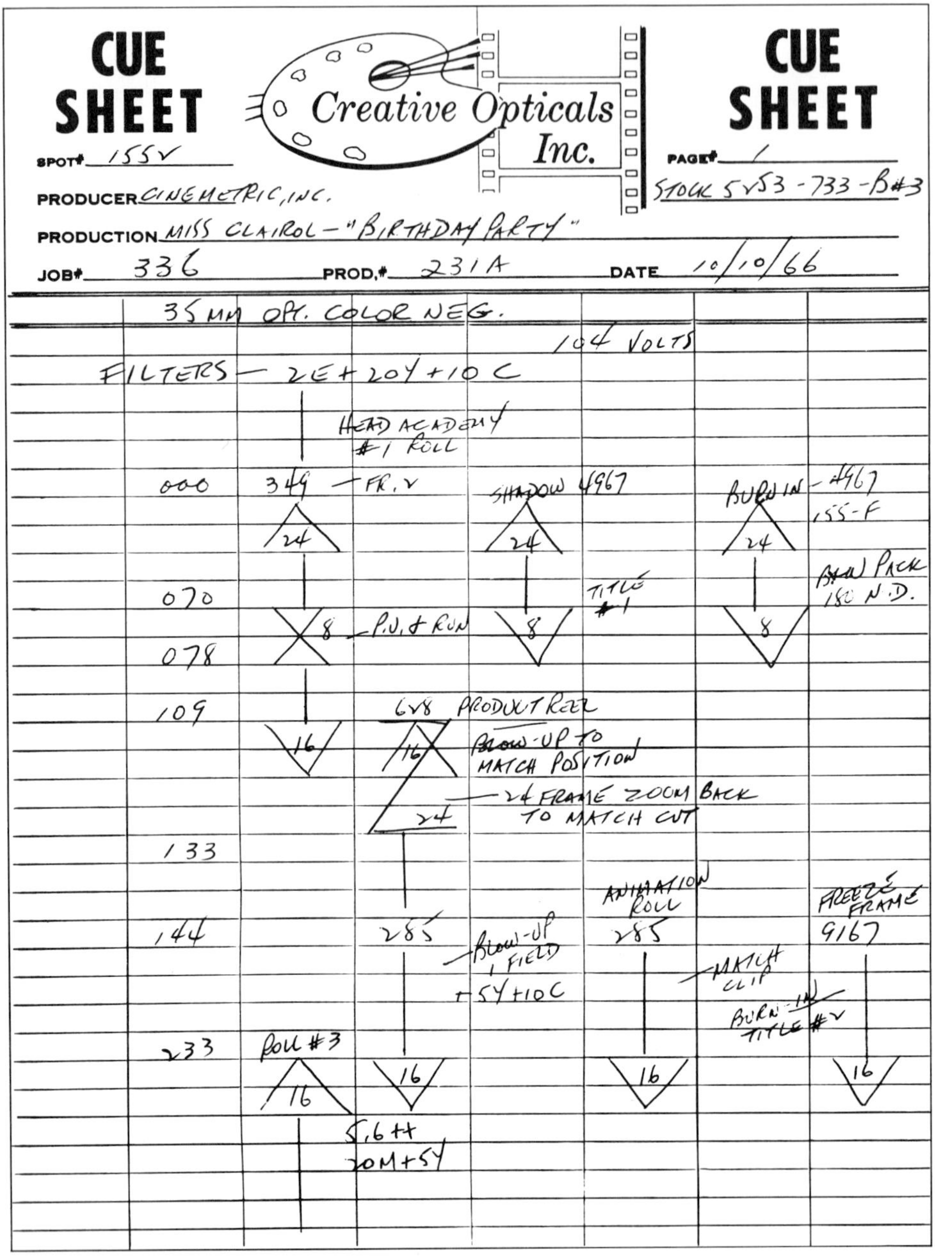

Fig. C-30 Cue sheet.

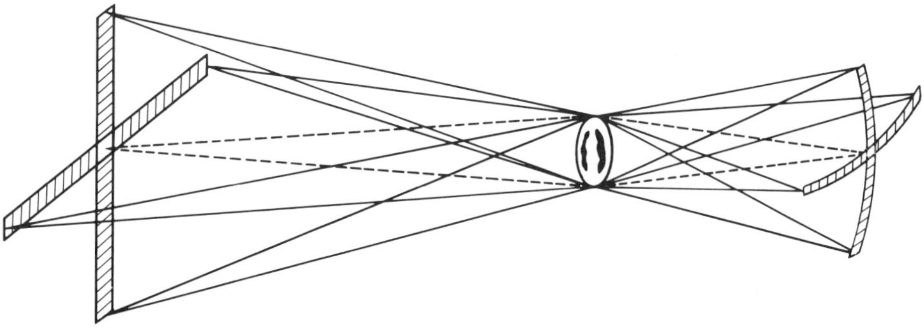

Fig. C-31*a* **Curvature of field.**

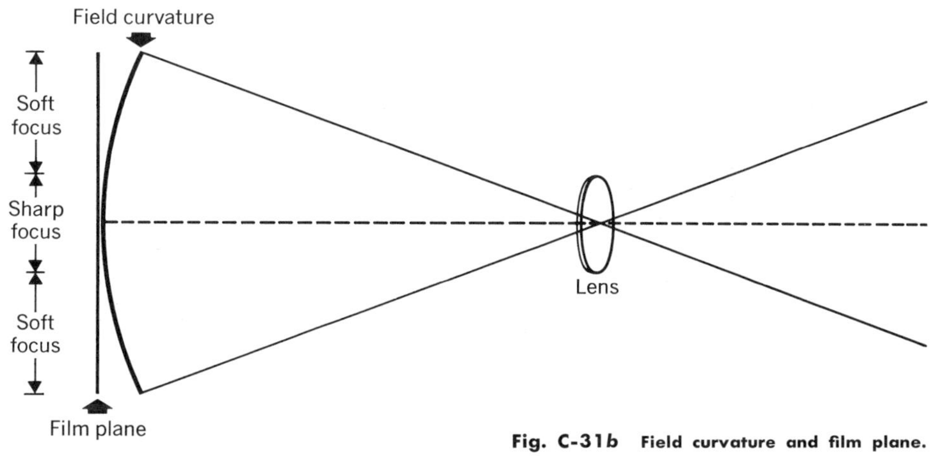

Fig. C-31*b* **Field curvature and film plane.**

curvature of field An off-axis lens aberration that causes light rays to come to a focus at different points, resulting in a curved focal plane. Because of this type of aberration, the image loses sharpness near the outer edges of the film frame. See Figures C-31*a* and *b*.

cut The direct, abrupt end of a scene or action. The last frame of a scene is spliced to the first frame of the scene that follows.

cutaway An insert scene designed to add suspense to a sequence or emphasize the action. During an exciting moment in a football game, for example, a scene may be inserted to show the coach on the sidelines or the people cheering in the stands.

cutback The continuation of a sequence after several insert scenes or cutaway shots.

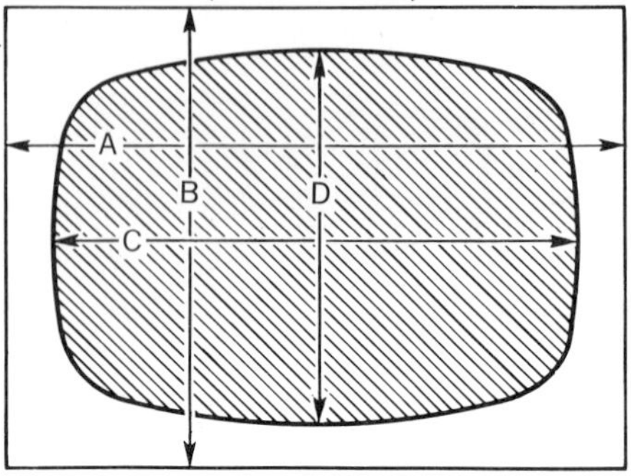

Fig. C-32 Television transmission and safety areas.

Film	Scanned area	Safety area	
		Inches	Percent
16mm	A	0.368	100
	B	0.276	75
	C	0.331	90
	D	0.248	67.5
35mm	A	0.792	100
	B	0.594	75
	C	0.713	90
	D	0.535	67.5
2 x 2-inch slides	A	1.125	100
	B	0.844	75
	C	1.013	90
	D	0.759	67.5

cutoff Because the face of a television tube tapers slightly, only the flat portion of the tube face is used for transmission to avoid image distortion along the outer edges. The term cutoff is used to describe that part of the transmitted picture or motion picture frame that is not visible when seen on television receivers. Titles and other important graphic material must be positioned within the *safety field,* an area that leaves a comfortable margin at the top, bottom, and sides of the transmitted picture. See Figure C-32 and table.

cutout A drawing or part of a drawing that is cut out and then inked and opaqued on thin illustration board instead of on a cel. Placed over the background, or moved according to guides supplied by the animator, cutouts are used to reduce the number of cel levels. See CEL LEVEL.

cutting The process of editing film or tape. The term is used rather loosely to encompass more than the actual removal of footage.

cutting, negative The process of cutting a negative in order to match the "final" version of the work print prepared by the film editor. The negative, in its fine-cut form, is the guide for the preparation of dupe negatives, separation negatives, and other material from which the composite prints are produced. Release prints are almost never made from the original negative. Negative material must be handled carefully—literally, with white gloves.

cyan Technically, one of the primary colors. From a laboratory viewpoint, the subtractive printing color, blue-green.

cycle A number of animated drawings or cels that are used to repeat an action. In a walk cycle, for example, the same drawings may be used over and over again and exposed for as much film footage as is needed. See HOOKUP.

In sound recording, the repetition of sound signals within specific time periods. The rate of repetition is referred to as frequency. See FREQUENCY.

cyclorama A relatively large background for a motion picture set. When the cyclorama is constructed of wood, fiberboard, and similar materials, it is referred to as a *hard cyc*. Usually, however, the background is rendered on heavy cloth that is stretched taut and supported by a wooden frame. See Figure C-33.

cyclorama strip light Also called **cyc strip**. A high-intensity strip light for illuminating relatively large areas. Used individually or in four-light multiples, the units are mounted horizontally at the top or positioned along the bottom of the cyclorama to produce a wide spread of uniform light.

cylindrical lens Cylindrical lenses are not actually lenses but textured surfaces on film bases used in lenticular systems. The surfaces of these film bases have vertical lenticular ribs or ridges, forming a series of cylindrical lenses. This type of film is exposed through the base. The right and left sides of an image are on different sides of each ridge and are projected on screens so that each part of the image is seen on a corresponding surface of the lenticular screen used for stereoscopic projection. See LENTICULAR FILM and LENTICULAR SCREEN.

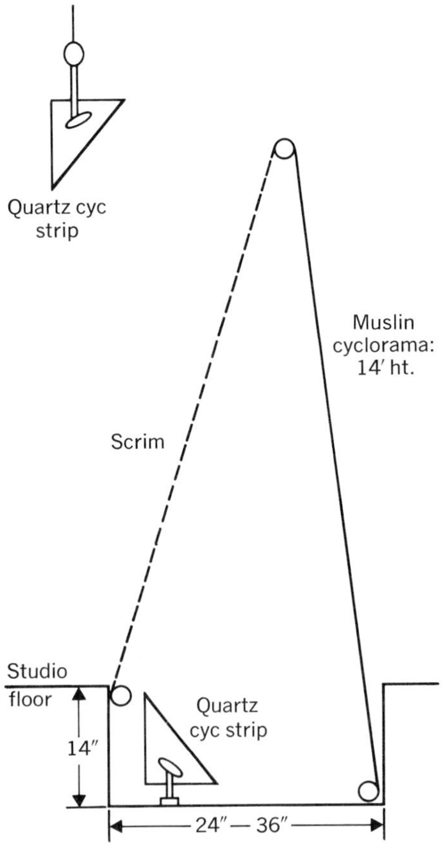

Fig. C-33 Cyclorama lighting. A cyclorama light pit places the horizon at ground level and eliminates the need for ground rows. The opaque white muslin drop is tilted forward to permit more even illumination from the bottom quartz cyc strip. A translucent scrim provides a sense of depth and is lit with a top quartz cyc strip.

D

D The optical symbol for distance.

daguerreotype A nineteenth century photographic process named after Louis Jacques Mandé Daguerre. The subject matter was exposed on a silver plate and was seen as a positive image when viewed under reflected light.

Daguerre-Talbot Necessity has been referred to as the mother of invention. Photography, however, was born out of frustration.

Louis Jacques Mandé Daguerre and William Henry Fox Talbot were both aspiring artists whose proficiency with paint and brush failed to excite even the kindest of critics. The fact that both turned their attentions to mechanical processes for duplicating picture information, during the same quarter of a century, was only coincidental since neither knew of the other's efforts.

Daguerre first announced the success of his daguerreotype photographic process to the French Academy in August, 1839. Previously, Daguerre had collaborated with another Frenchman, Joseph Niepce, who had produced a negative and positive image by the action of sunlight on a silvered copper plate. Mr. Daguerre's announcement prompted another noted inventor, Samuel F. B. Morse, to travel to Paris for the purpose of obtaining additional information about the process before its introduction to the United States.

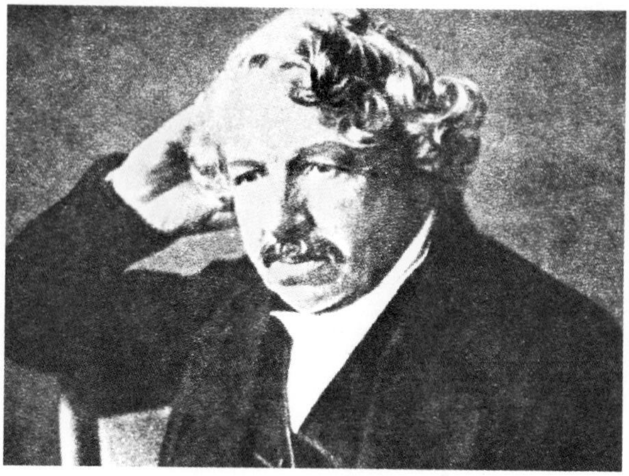

Fig. D-1*a* Louis Jacques Mandé Daguerre.

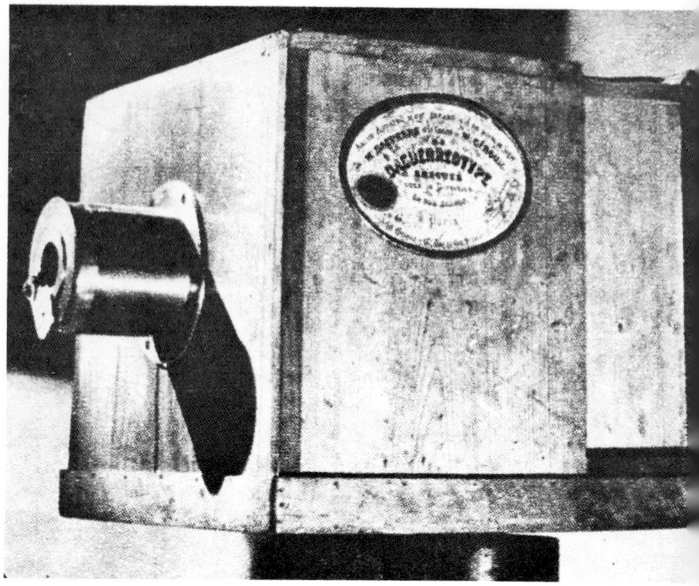

Fig. D-1*b* Daguerre's camera.

Henry Fox Talbot, after a similarly frustrating career as an artist, found a more-than-adequate substitute for reproducing images in his calotype process. Although both Daguerre and Talbot made use of the same photographic principles, Talbot's negatives and prints were both made on sensitized paper rather than on the copper plates used by Louis Daguerre. Since many prints could be made from the paper negative, Talbot's process was the more practical—economically.

Although the daguerreotype process was greeted enthusiastically and received worldwide attention, Talbot's foresight and insistence on protecting his process by patent delayed its exploitation and prompt acceptance for commercial purposes. Both the Daguerre and Talbot processes were perfected during the same decade, the 1830s. See Figures D-1*a* and *b*.

dailies Also called **rushes**. The processed results of the previous day's photographic efforts. These dailies are usually viewed by the director, cameraman, editor, and others for the purpose of checking the action, lighting, the smoothness of zoom and pan movements, and the general content of the scene. If necessary, arrangements are made for reshooting. The rephotographed scenes are referred to as *retakes*.

darkroom A lightproof area that is used for loading and unloading film from magazines. Also, a room used for developing, printing, enlarging, and copying still photographs.

Dawn process A process developed around 1905 in which performers were photographed either in front of or behind backgrounds that were painted on glass for the purpose of producing direct, in-camera composite effects. Norman Dawn, an artist, cameraman, and director, is credited with the development of the process. See GLASS SHOTS.

day-for-night photography, In order to satisfy the insatiable demands of television, many motion picture producers have been forced into round-the-clock production schedules. This increase in production has been directly responsible for the development of new filming techniques.

One of these techniques has helped ease producers' schedules considerably by making available a practical method for simulating night effects with daylight photography. This day-for-night filming technique has been developed to its present state of predictable results by the basic formula of overcorrection and underexposure.

Overcorrection is made possible through the use of filters. The degree of underexposure, $1\frac{1}{2}$ to 2 stops from normal, depends on conditions at the time of shooting. As a general rule, however, the brighter the scene, the greater the amount of underexposure needed to ensure the proper reproduction of the nighttime effect.

Opinions of professional film makers and laboratory technicians on the best filming methods, choice of filters, degree of underexposure, and negative quality, vary considerably. Quite obviously, there are at least as many filming methods as there are opinions.

Some directors and cameramen prefer filming day-for-night scenes when the sun is directly overhead. Others agree that the best results are obtained from the combination of blue sky and strong sunlight. Some cameramen consider dusk as the ideal time for filming and have applied the name "magic hour" to the period.

Laboratory technicians have their own opinions regarding negative quality, density preference, and color correction. Since the results obtained with the various filming methods are all comparable with regard to quality, it must be assumed that each of the theories has merit, is sound in application, and has a place in the production of day-for-night effects.

Obviously, the choice of auxiliary illumination is influenced by natural light sources and sky conditions. In most cases, the angle of the sun is the determining factor in the selection of lighting and camera setups. Shadows directed toward the camera, in combination with auxiliary lighting that follows and accentuates shadow patterns, are preferable. Back crosslighting, from low angles, serves a dual purpose and is best suited for color as well as black-and-white filming. This type of lighting adds dimension to a scene by creating a natural separation between actors and background. Since the background is not illuminated directly with this lighting method, undesirable detail adds little to the overall realism of the effect and is held to a minimum. Frontlighting, by comparison, is usually avoided because the resulting shadow patterns are often objectionable. Backlighting produces acceptable results in long shots but is quite unsatisfactory for close-up and medium shots.

Light sources that add realism to a scene must be accentuated by using lights of higher intensity. Concealed photofloods and reflector floods in streetlights and auto headlights, and behind storefronts and windows, serve adequately for this purpose. In order to create the proper nighttime atmosphere when color film is used, yellow gelatins are positioned in front of these light sources. Because of the realistic appearance of these background light sources during the magic hour, or twilight zone, gelatins and corrective filters are not needed during that time period.

Neutral wedges, or density filters, are used for graying down clear, hot skies. Directors and cameramen, when using these graduated neutral-density filters, should avoid filming actions that would overlap from the filter area to the clear-glass area. Pola screens, by comparison, are used only when it is necessary to darken a portion of the sky. Panning actions should be carefully planned in order to avoid polarization and the resultant changes in density.

Pola screens, like neutral-density filters, can be used in both black-and-white and color filming without influencing or upsetting the overall color balance of the scene. Neither do they have any effect on other filters that are used in day-for-night filming.

Red and green filters, used in combination, correct sky colors and darken flesh tones which normally appear lighter under the influence of red filters. Filters such as the 23A are corrective and cause sharp contrasts in shading. On the light side, complexions are almost white in comparison to the unnatural blacks on the shadow side. Green filters such as the 56 restore the normal balance and present actors and actresses in their true coloring.

The amount of underexposure needed for realistic night effects is governed by the type of filters used during the photographic process. Two methods are generally used for determining the proper lens settings.

In the first method, the increase in exposure made necessary by the use of corrective filters, is disregarded if the increase approximates the lens setting needed for the underexposure. The lens stop remains where it would be for normal exposure if the filters were not present.

In the second method, a neutral-density filter is used in combination with the effect filters. The lens stop, for this method, is set to compensate for the exposure increase necessitated by the use of effect filters, as in normal photography. The addition of the proper neutral-density filter provides the underexposure needed for the night effect. (A 30 percent neutral-density filter is equal to a full lens-stop increase in exposure; a 60 percent neutral-density filter provides for a two-stop increase in exposure.)

For location shooting, this method is preferable since night scenes, for insert shots, can be made quickly and easily by merely adding the proper neutral-density filter without changing exposure. Immediately after the filming of the night shot, the filter is removed, and daylight filming can continue without changing lens settings.

For best effects, in both black-and-white and color filming, there should be good contrast between actors and background. Clothing should be considerably lighter than the background. Dark clothing, in front of comparatively light backgrounds, creates an unrealistic silhouette effect.

The producer is far from limited in his choice of film types for nighttime effects. Black-and-white, color (negative and reversal types), and infrared film are all readily available and capable of producing excellent results. Variations, in choice of filters and amount of underexposure, are only minor considerations and do not affect the methods used for determining proper exposure.

Since it is almost impossible for a cameraman to compensate for one color without influencing or upsetting the color balance in the original negative, color corrections should be left in the hands of optical or laboratory technicians. Their latitude for controlling and balancing color is considerably greater than the compensations that can be made by the cameraman on location.

daylight The difference between daylight and sunlight, in terms of color temperature, can be as great as 20,000°K or more. Sunlight is that amount of light reaching us directly from the sun. Daylight, insofar as color temperature is concerned, is the total amount of measurable light, sunlight and skylight combined. Daylight has a color-temperature range from 6000 to approximately 30,000°K. For most photographic applications, sunlight is simulated at about 5600°K.

daylight loading Film is available on special spools or in containers known as *cassettes* that make it possible for a cameraman to load and unload cameras in subdued light.

daylight loading tank A small tank used for developing relatively short lengths of film. The name is somewhat misleading since the film must be loaded into the tank in a darkroom or other lighttight area. After loading, however, the development process can be continued in daylight.

decamired filters The effect that filters have on color temperature varies with the level of the temperature. For this reason, certain types of filters cannot be used in combination (light-balancing and conversion types, for example). Decamired filters provide the degree of difference that ensures proper exposure at all temperature levels, regardless of the filter or combination of filters being used during the filming process.

The scale used for determining the decamired values is based on the reciprocal of the degree of color temperature. Daylight Kodachrome, with a rating of 5500°K, is used to explain the formula needed to determine the value of a film type in terms of decamireds.

1,000,000 ÷ rating of film, °K = mireds
1,000,000 ÷ 5500 °K = 181 mireds
181 mireds ÷ 10 = 18 decamireds (numbers are rounded off to the nearest tenth)

To find the value of the required decamired filter, the following formula is used:

DM value of the film − DM value of the light source = DM filter value

18 DM (rating of daylight Kodachrome)
−30 DM (light source at 3300°K)
−12 DM

A minus answer indicates the need for cooling filters (blue). A plus answer indicates the need for warming filters (red). The answer to the above example indicates the need for a blue decamired filter with a 12 correction. See Tables D-1*a* and D-1*b*.

decibel One-tenth of a bel, abbreviated **db**. This unit of electric power has no numerical value but is used as a means of establishing a relationship between voltages, currents, impedance, and other elements relating to sound recording and reproduction. It is defined by dividing the power output obtained from a circuit by the power put into it. This quotient is then converted to a logarithm which is the loss or the gain in bels. It is customary to multiply this figure by 10 in order to convert to decibel units.

db = decibel units

P_o = power output $\qquad$ $$\text{db} = 10 \log \frac{P_o}{P_i}$$

P_i = power input

The sound of leaves rustling is equal to 20 decibels. The volume of sound produced by an automobile motor is 70 decibels. Surprisingly, an automobile horn at 120 decibels is 15 decibels higher than the noise made by a riveter. See BEL.

decoder That part of television receiving equipment which takes the encoded signal and sorts out the red, green, and blue components. These signals, after being processed, modulate the intensity of the three electron beams in color television receivers in order to duplicate the colors in the original subject matter. See ENCODER.

TABLE D-1a Decamired Filter Selector Chart*

	Lighting							
Film	Bluish daylight, open shade, light overcast DM 14	Electronic flash and blue flashbulbs DM 16	Average sunlight 9 A.M. to 3 P.M. DM 18	Late afternoon sunlight, reddish DM 20	Clear flashbulbs (not SM, SF or M2) DM 26	Photofloods 3400°K DM 29	SM and SF flashbulbs DM 30	3200°K studio-type floods DM 31
Daylight type DM 18	R3	R1½	None	B1½	B6 and B1½	B9 and B1½	B12	B12
Type F and negative color DM 26	R12	R3 and R6	R6 and R1½	R6	None	B3	B3	B3 and B1½
Type A and 3400°K tungsten DM 29	R12 and R3	R12	R6, R3 and R1½	R6 and R3	R3	None	B1½	B1½
Type B and 3200°K tungsten DM 31	R12, R3 and R1½	R12 and R3	R12	R3, R6 and R1½	R3 and R1½	R1½	R1½	None

Courtesy of Tiffen Manufacturing Corp.

*This chart lets you choose the proper decamired filter or combination of filters, without having to resort to addition or subtraction. Values are figured to the nearest 1 or ½ DM (decamired) value, and are correct for average pictorial use.

TABLE D-1b Decamired Exposure Increase Chart

R series	B series	*f*-stop
R1½	B1½	¼
R3	B3	½
R6	B6	1
R3 + R6 = R9	B3 + B6 = B9	1½
R12	B12	2
R12 + R3 = R15	B12 + B3 = B15	2½
R12 + R6 = R18	B12 + B6 = B18	3

Courtesy of Tiffen Manufacturing Corp.

definition Also called **defining power**. In optics, the ability of a lens to form sharp images by concentrating or dispersing the rays of light from an object. See Figure D-2.

In processing, the sharpness of detail in the processed film.

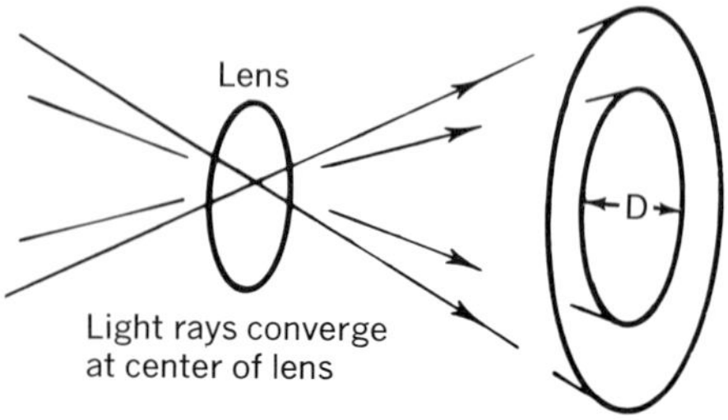

Fig. D-2 The circle of definition increases as the lens is stopped down and a smaller aperture is used. *D*, the diameter of the circle of definition, should be greater than the diagonal of the film.

deflection A module in vidicon film cameras that provides horizontal and vertical deflection signals. The control also provides vidicon protection against scan failure. See YOKE.

Delineoscope A Technicolor designed system for making scene-to-scene printing color corrections.

demodulator output The demodulator output in videotape recorders separates tape sync from tape signal and provides line drivers to feed unprocessed video to monitoring circuits and to the processing amplifier. This unit also contains the postemphasis circuit. See PROCESSOR.

densitometer A densitometer is used to measure the densities of a length of developed film in terms of logarithmic units. The relationship between transmission and density is given by the following formula.

$$D = \log 10 \frac{1}{T}$$

where T is the transmission of the material.

In operation, a sensitometer creates a series of precise exposures on a length of film by means of a step wedge so that the developed result is a number of bands of uniform density. Each exposure band in the series is greater by a calculated amount than the preceding step, and the exposure range covers all practical conditions. These densities are plotted against the logarithm of the exposures which produced them, that is, against the steps on the film. The curve drawn from these plots is the characteristic curve of the film. In plotting such a curve, the processing conditions must be related to the actual conditions of use. See SENSITOMETRY and CHARACTERISTIC CURVE.

Westrex model RA-1100 G (color, black-and-white, infrared). The RA-1100 G densitometer is intended primarily to provide integral-density measurements of color negatives and positives and essentially diffuse visual-density and print-density measurements of black-and-white picture film.

The RA-1100 G is a direct-reading type of electric densitometer, essentially a sensitive light meter in which a photocell is used to measure the ratio between the light transmission with and without the film in the light beam. It utilizes narrow-bandpass Eastman status A and status M color filters, a visual filter, and a printing-density filter. Light from an exciter lamp, held constant by a voltage regulator, is interrupted by a light chopper and directed through filters, through the film and limiting aperture by suitable lenses. After passing through the film, the light produces a signal in the photocell proportional to the amount of transmitted light.

The amplified and rectified output of the photocell is indicated on a special meter calibrated to read in terms of density. The principal components of the instrument are an optical system, a photocell, an amplifier and rectifier system, and a special output meter. See Figure D-3*a*.

Westrex model RA-1100 H. The RA-1100 H is a direct-reading integrating-sphere type of electric densitometer. It is essentially a sensitive light meter in which one of two photocells is used to measure the ratio between the light transmission with and without the film in the light beam. See Figure D-3*b*.

density The degree of opacity in a photographic image. The degree of opacity is determined by the amount of silver deposited throughout the film frame and, consequently, limits the amount of light that will pass through it.

density, analytical The determination of the composition of an image by measurements which indicate the amounts of the individual dyes (yellow, magenta, and cyan) that form the image.

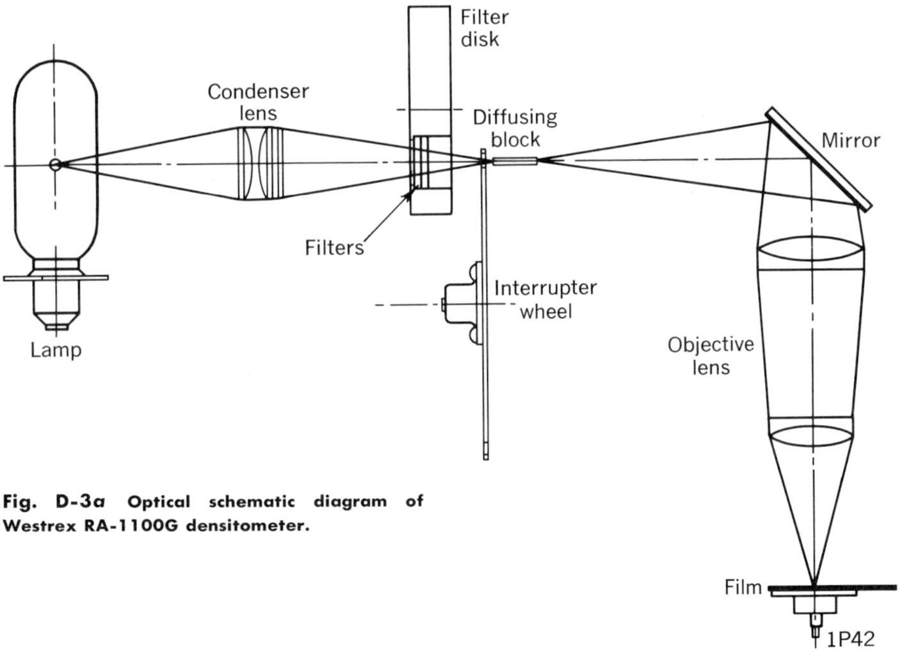

Fig. D-3a Optical schematic diagram of Westrex RA-1100G densitometer.

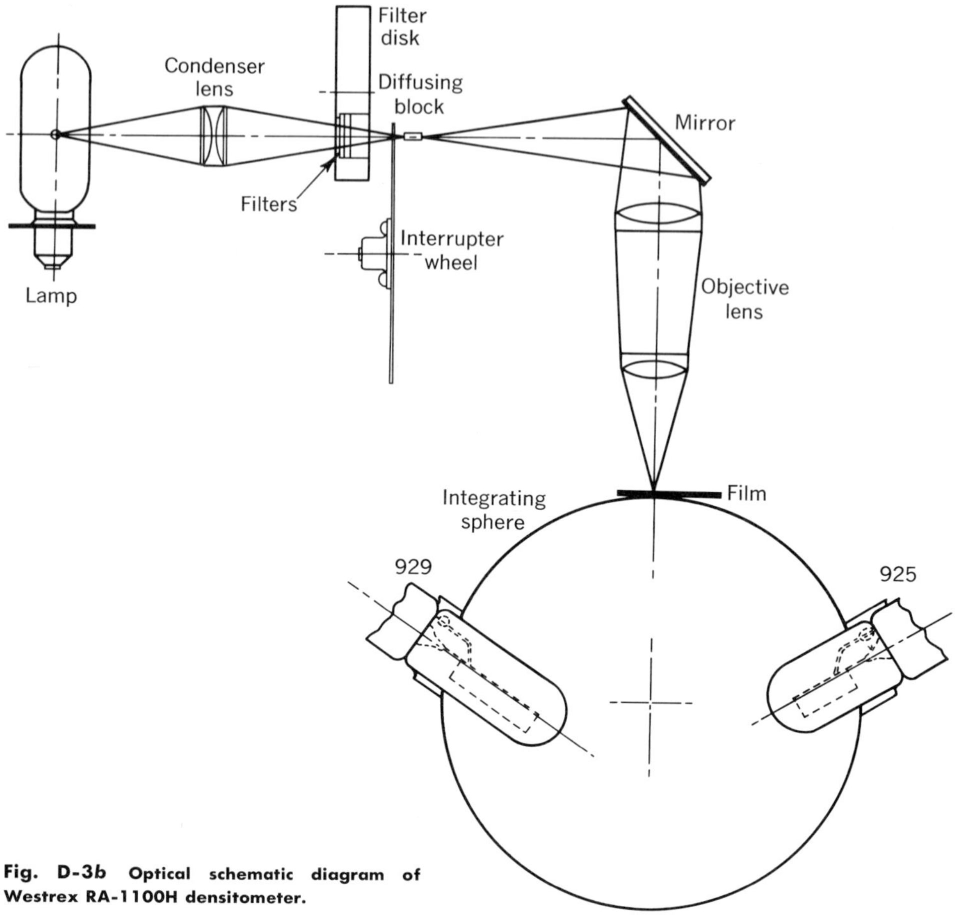

Fig. D-3*b* Optical schematic diagram of Westrex RA-1100H densitometer.

density, arbitrary three-filter The measure of the response of individual colors to red, green, and blue filters in a densitometer. These measurements indicate any dissimilarities in two identical color images produced by the same dyes. The comparison indicates the need for changes in printing densities. The term *arbitrary* is used because the readings vary with the type of filters or photocells that are used at the time the readings are made.

density, colorimetric The measure of the color stimulus of an image, or the response evoked in the visual mechanism, based on the spectral distribution of the individual colors forming the image.

density, equivalent neutral The luminous density of a gray image that could be obtained by adding specific amounts of any two of the dye deposits of the color process to the third dye. Visual comparisons based on this system indicate the relative chromaticities of light and dark images. Measurements of the brightness of the scene (brightness ratio) are used as a guide for subsequent printing corrections.

density, equivalent neutral printing The printing density of an image that could be produced by adding specific amounts of any two of the three dye deposits of the color process to the third dye in order to form an image in which the densities of the component dyes are equal.

density, integral The measure of the effects of absorption by an image on the light beam which is used for printing during the production of positive materials.

density, luminous The measure of the brightness or luminosity response of an image.

density, printing The density of the image that is formed as a result of the development process and its effect on the action of the light beam which it attenuates during the printing process.

density, spectral The density of an image as determined by measurements of light of a single wavelength. Spectral densities, or spectral transmittances, are measured with spectrophotometers.

depth of field The distance in front of and behind the subject in which detail remains in acceptable focus.

depth-of-field formula

(1) Lens to near limit

$$= \frac{\text{hyperfocal distance} \times \text{distance between camera and subject}}{\text{hyperfocal distance} + \text{distance between camera and subject} - \text{focal length of lens}}$$

(2) Lens to far limit

$$= \frac{\text{hyperfocal distance} \times \text{distance between camera and subject}}{\text{hyperfocal distance} - \text{distance between camera and subject} - \text{focal length of lens}}$$

Depth of field is the difference between the solution derived from formula (1) and the solution derived from formula (2).

depth of focus The distance between the lens and film, during exposure, in which the subject matter being photographed will be sharply defined. Depth of field and depth of focus are often used interchangeably—but incorrectly.

depth-of-focus formula

$$\frac{\text{Focal length} \times f\text{-stop}}{1{,}000}$$

desaturation process Also called **inhibition process**. See TECHNICOLOR.

dessicator Chemicals that in their dry state are used to keep film dry. As the chemicals absorb moisture from the air, they slowly change from the powder to the crystallized form. Cans of exposed film, in arctic and tropical climates, are stored in boxes that are lined with dessicated materials before their shipment to laboratories for processing.

detail The sharpness of minute areas in the photographic print.

deuce A 2,000-watt incandescent floodlight.

developer A laboratory term for the chemical agent or solution that is used to bring out or develop the latent image that exposure has left in the emulsion.

development Motion picture films are developed in standard developers using standard laboratory processing techniques. The sensitometric curves for each film indicate the processing conditions employed by processing laboratories, and so describe accurately the characteristics and performance of the film in normal practice.

Sensitometric control of development technique involves accurate monitoring of the composition, temperature, replenishment, and replacement of processing solutions. This is essential in continuous processing machines, and equally important are the calculation and control of the transport speed of the film in the machine. There are variations in equipment and technique between processing laboratories, but the general characteristics of motion picture films are such that these variations do not affect the ultimate image quality of the negative or positive print.

Rating film at higher than average speed ratings will not produce a satisfactory negative if the film is processed by techniques related to average exposure conditions. Similarly, it is not possible to make appreciable improvements in image quality by special processing techniques unless minimum exposure is given.

Basically, development can be simply described as a laboratory process in which the latent image on the exposed emulsion is made visible. This is accomplished by removing the bromine from the exposed portions of silver bromide by chemical reduction.

development, fine-grain A laboratory developing process using solutions which have borax as the accelerator. In fine-grain development, the clumping of silver in the negative and the resultant coarse grain are avoided by the use of these solutions. See AGGLOMERATION and FINE GRAIN.

development fog An undesirable fog effect which is caused by overdevelopment.

dialogue The words spoken by performers during the filming process.

diaphragm Also called **iris** or **stop**. The movable metal blades that adjust the size of the lens opening and thus control the amount of light passing through the lens to expose the film in the camera. ASA and SMPTE standards provide that, to set a diaphragm to the required aperture, it should first be opened to a larger *f*-stop and then closed down to the setting selected for filming. The procedure is recommended to eliminate backlash. See Figure D-4.

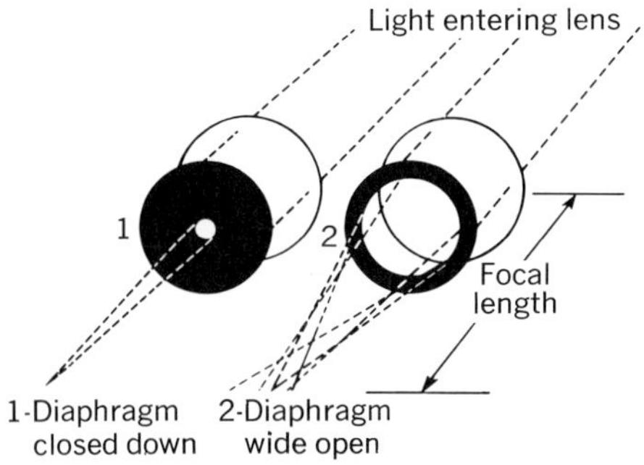

Fig. D-4 The lens diaphragm and light transmission. The diaphragm controls the amount of light that is transmitted by a lens. With a wide-open aperture, light passing through the center of the lens is not focused at the same point as light entering and passing through the edges of the lens, resulting in blurred or unsharp images, depending on the spherical aberration of the lens. With the aperture closed down, only the light passing through the center of the lens is used for the formation of the image. In this way, the parallel rays of light focus at about the same point. Stopping down thus improves the sharpness of the lens image.

diaphragm markings The calibrations, in terms of *f*-stops on the diaphragm, that indicate the size of the aperture.

diapositive A "still" positive image or a transparency. Diapositives are viewed by transmitted light.

dichroism The capability of a crystallized body to transmit or reflect different colors, depending on the direction of the light being transmitted through it.

Also, the property of showing different colors when viewed in different directions.

Dichroic filters are capable of transmitting and reflecting the primary colors of artificial light sources. These filters convert the standard for interior lighting, 3200°K, to 5600°K, a temperature reading that simulates sunlight. Unlike conventional glass filters that add blue to the light source, the coatings on dichroic daylight conversion filters reflect the red portion of the spectrum while allowing the blue to be transmitted.

differential focusing A method of focusing on a scene and the subject matter within it so that the critical area, that part of the set in which the action will occur, is in sharp focus. The scene's background is usually in soft focus. See ZONE FOCUSING.

differential gain The change in amplitude of the electric signal, usually of the color subcarrier, introduced by the overall circuit and measured in decibels or percent.

diffraction The bending, or change in the direction of a light ray, caused by contact with an opaque body. Also, changes in the direction of light waves as they pass from one medium to another.

Diffracted light rays striking the edge of a lens diaphragm reduce image sharpness. A certain amount of correction, however, is possible by stopping down or using higher *f*-stop settings—*f*/8 to *f*/11, for example. See REFRACTIVE INDEX.

diffraction grating A series of fine, parallel lines, approximately one thousand to the inch, etched on a glass surface. Some light rays striking this type of surface pass through the minute spaces between the lines. Other light rays are diffracted.

Light rays diffract differently, depending on their wavelengths. Blue rays, for example, are diffracted to a greater degree than red rays. These gratings, for separating a light ray into its component colors, are especially useful in spectrography.

diffuser (1) In lighting, a translucent material that is used for softening the lighting on a motion picture or television stage. (2) Also called **splay**. Uneven surfaces, usually curved or warped sheets of thin plywood, positioned near the sound source during the recording process. These surfaces improve the acoustical response by adding "liveness" to a recording through increased reverberation, thus providing better definition to narration and music.

diffusion The spreading or scattering of light waves.

diffusion of focus Lack of image sharpness caused by defects in a lens or improper focusing. See SOFT FOCUS.

diffusion and distortion The terms diffusion and distortion, in connection with photography, are often used interchangeably, if not with complete accuracy. Diffusion, in its broadest sense, means spreading or scattering widely. Photographically, the spreading or scattering action refers to light.

Applied practically, the materials used for achieving diffusion effects permit some light from the subject matter to pass through the lens unobstructed while scattering the remaining light in many directions. Unlike out-of-focus effects, in which no part of the image is sharp, this uneven scattering of light exposes the film producing areas of softness and sharpness in the same picture. With highlighted areas influencing the darker portions of an image, contrast is reduced, and the print presents an overall diffused appearance.

In addition to their value for transitional purposes, diffusion effects can add dramatic, suspense-filled moments to a film by creating an air of mystery, an eerie atmosphere, or a sense of vagueness. While some of the materials that help make these effects possible are commercially available, others are custom-made to fill specific needs. Since many of these devices are quite transparent, the light loss is negligible, and no exposure compensation is necessary. In extreme cases, a half-stop more exposure is adequate. Obviously, cinematographers who prefer to make their own diffusing materials must also be prepared to make exposure tests, under different filming conditions, in order to determine the proper lens setting for each condition.

Beside the standard glass disks that have a variety of patterns molded into their surfaces, such fabrics as gauze and white netting are excellent for diffusion. Fabrics with loose weaves that permit more light to pass through them produce sharper images—and a minimum of diffusion. With tighter weaves, and smaller openings between the individual strands, the amount of diffusion is proportionately greater. Mounted in cardboard holders or positioned in the rear frame of the matte box, several thicknesses of fabric, with the threads crossing each other at angles, produce a variety of interesting effects and increase the amount of diffusion. Since white materials have better reflective qualities than either black or colored fabrics, the lighter-colored material is preferable.

Metal window screening and light-colored women's stockings can both be used to achieve diffusion effects. Clear glass slides and sheets of Celluloid are other readily available, ad-lib materials that can be mounted in holders and positioned in the front or rear frames of the matte box. The latter materials, when scored with a glass cutter or smeared with transparent or translucent substances, such as Vaseline or water glass, provide controllable degrees of diffusion. Obviously, the degree of diffusion increases in direct proportion to the number of scored lines and varies with the density of the applied substances.

For certain effects such as softening features in close-ups, the diffusing material remains in a stationary position for the entire shot. If desirable, the amount of diffusion can be increased or decreased gradually by either of two methods. The simplest way of introducing diffusion is by means of a cross-dissolve. With this method, the subject matter is photographed normally, for the required footage, and then faded out. The overlapping fade-in, on the same subject matter, is made with the diffusing material in place.

The second method for increasing or decreasing the amount of diffusion usually involves special preparation and, possibly, preproduction experimentation. If glass slides are to be used for achieving the effect, for example, portions of the slide are left uncoated so that when the clear area of the slide is in front of the lens, the subject matter will be seen normally. As the coated surface of the slide passes in front of the lens, the images are influenced by the applied substances and become diffused. Frame-by-frame changes in diffusion are achieved by moving the diffusing material across the front of the lens for the entire length of the scene or effect.

Under no circumstances should the camera be positioned so that any of the light sources on the set throws its light on the reflective surface of the diffusing materials. The resultant increase in the amount of light scatter produces an effect that is, in many ways, similar to an undesirable film fog. Lacking an alternative, the cameraman should substitute black gauze, screening, or other materials with less reflective surfaces for the conventional light-colored materials.

Although many of the illusions produced by diffusion are used for the same purposes as those achieved by distortion, the processed images and projected results have little in common. Influenced by the various diffusing materials that are used to achieve those effects, dark and light tones blend, and the original color values are altered considerably. In spite of the resultant reduction in contrast, the images retain their original shapes. Distorted images, by comparison, are changed in one or more directions, while the color balance changes little, if at all.

For transitional purposes, as well as for suggesting such moods or conditions as madness or rage, distortion effects can be very effective.

To the cinematographer, any material that can be used to distort an image, be it a broken bottle, mirror, or discarded piece of metal, becomes a valuable accessory. Improvisation is the rule rather than the exception. While certain materials, such as sheets of glass with irregular surfaces, are used regularly and produce near-predictable results, cameramen have been known to use the bases from broken bottles and crumpled sheets of cellophane to achieve the more unusual distortion effects.

Lenses that stretch or elongate images in any direction, depending on the position of the lens at the time of exposure, are commercially available and provide controllable degrees of distortion. Because the amount of distortion decreases proportionately as the auxiliary lens is moved closer to the camera lens, it should be mounted as far as possible from the camera lens, without cutting into the film frame. Although some areas of the frame may not be in completely sharp focus, the shortness of these effects and the fact that the images are already distorted combine to make the loss of focus unobjectionable.

Mirrors and flexible sheets of polished metal, both ideal for producing distortion effects, can make the fat lady of the circus look like a candidate in a Miss America contest or turn a midget into a giant. Although their optical properties are far from excellent, the results they produce are usually adequate.

The position of the diffusion and distortion materials during the photographic process, in relation to the camera, varies considerably. Coated glass slides, diffusion disks, and gauzes may be mounted in either the front or rear frames of the matte box. Larger objects, such as panes of glass, mirrors, and metal sheets, are supported by specially built frames that are usually positioned between the camera and the subject matter.

Frame-by-frame changes in the degree of distortion can be achieved with stop-motion photography. Used frequently in television commercial production, the various materials are placed in front of the camera lens in animation and titling stands and moved slightly for each exposure. Since practically all the materials either are transparent or have highly reflective surfaces, very little exposure compensation is needed.

The viewfinder, in addition to its value for composing purposes, is especially important in the production of diffusion and distortion effects. Originally filmed on a hit-or-miss, trial-and-error basis, the scene can be viewed, in its composite form, with the diffusing or distorting materials in place. Any irregularities or undesirable effects, immediately apparent during this dry run, can be corrected before a single frame of film is exposed. Paradoxically and unfortunately, diffusion and distortion effects sometimes just happen—without planning or preparation. To the cameraman, these unanticipated special effects are synonymous with retakes.

diffusion-effect filters The diffusion effect in Harrison filters is created by grinding a predetermined pattern into a glass support and then laminating the pattern glass, with a transparent adhesive, to a second cover glass. This gives the filter two plain outer surfaces with the optical effect safely sandwiched between the two glass supports.

The optical pattern is exceedingly fine in order to prevent undesirable reflections from reaching the film. Similarly, long streamers of light and circular halos that sometimes follow moving objects across the screen are either minimized or eliminated by the fineness of the diffusion pattern.

Image size, exposure, and lighting contrast—factors that can influence the degree of diffusion during the filming process—must be considered. The relative size of the performers in extreme long shots and close-ups are also influencing factors. All or any of these variables make it almost impossible to duplicate a specific degree of diffusion—if and when a retake is necessary.

The normal range of Harrison diffusion filters is from 1 to 5, with each number optically doubling the degree of diffusion found in the lower number. Two number 1 diffusion filters, for example, are equal to the degree of diffusion in one number 2 filter. Two number 2 filters, in turn, are equal to one number 3 filter.

As with all effects, diffusion should be used only where necessary. The basic purpose of diffusion filters is to soften hard lines in the subject matter. Excessive diffusion can produce results that are, at best, artificial looking and obvious.

Contrast-viewing glasses enable the cameraman to visualize the diffusion effect that will subsequently be produced on film.

diffusion screens, half-and-half Half-and-half diffusion screens, with half of the filter containing an optical pattern and the other half clear screen, enable the photographer to create atmosphere or diffusion effects in half of the picture while the other half remains unaffected.

The filters are used for scenics and indoor atmosphere effects, and for matching foreground areas with background projection plates during the filming of process shots.

Half-and-half diffusion screens are available as solid-diffusion screens, varying in density through seven steps. Contrast-viewing glasses help the cameraman "see" the photographic effect that will be produced by the diffusion screens before the shot is actually made.

dimensional stability (film) Dimensional changes in negative or positive material may be permanent or reversible. Many of the dimensional changes that occur as a result of changes in temperature and relative humidity are, fortunately, reversible, although these changes, if they are not uniform in length and width, may induce some permanent distortions in the material.

The changes that occur as the film ages, however, and most of those changes that occur during processing, are nonreversible. Nonreversible processing changes are of four kinds:

1. Reorientation of the gelatin structure of the emulsion caused by the absorption of processing solutions and subsequent washing and drying
2. Loss of volume in the emulsion layer because of the fixing out of unexposed silver halides
3. Hysteresis effects that occur in drying
4. Changes in the humidity coefficient because of the loss of the stiffening effect of the silver halides in the emulsion layer

The effects of aging can be minimized by careful control of the storage conditions. This obviously applies mainly to negative material or prints that are required for archival purposes.

The processing changes are, in normal circumstances, kept within acceptable limits by virtue of the inherent dimensional stability of the acetate-base material of motion picture films.

dimension 150 See ASPECT RATIO and WIDE-SCREEN PROCESSES.

dimmer An electric voltage-reducing unit that adjusts the intensity of the lights on a motion picture set.

DIN speed Deutsche Industrie Norm (German Standards Organization). The German system of rating the speed of an emulsion. See EXPOSURE INDEX RATINGS.

diopter lens Also called **supplementary lens**. Extremely useful for close-up work, this type of positive lens is positioned in front of the normal camera lens in order to increase the size of the image. With a +1 diopter lens that has a focal length of 1 meter or approximately 40 inches, the camera can be positioned that close to the subject matter for filming.

The focal length of a +2 diopter lens is about 20 inches. The +3 diopter, ⅓ meter, indicates a focal length of approximately 13 inches. When more than one diopter lens is being used, the highest-powered diopter lens is positioned closest to the normal camera lens. See Table D-2.

dioptric system A lens system with refracting elements. This is the type of lens system found in most cameras. Compare CATOPTRIC and CATADIOPTRIC SYSTEMS.

diorama A dimensional-scale model or miniature representation of a setting designed specifically for use in a motion picture or television production.

diplexer A coupling unit capable of combining picture and sound signals so that they can be radiated from the same antenna at the same time.

director The supervisor of production and the person responsible for bringing the script to life on motion picture film.

Also, the person responsible for the animation and the related processes necessary to the completion of an animated film.

direct-reading exposure meter An exposure meter designed to provide reflected-light readings. Setting the exposure index and shutter speed on the meter scale, and then aiming the meter in the direction of the subject matter, gives direct *f*-stop readings.

TABLE D-2 Diopter Close-up Lenses

Lens	Camera settings, feet	Distance from subject, inches*
+1	Infinity	39
	50	37
	25	35
	Fixed	34
	15	32½
	10	29¾
	8	28
	6	25½
	5	23¾
	4	21¾
	3½	20½
+2	Infinity	19½
	50	19
	25	18½
	Fixed	18
	15	17¾
	10	17
	8	16½
	6	15½
	5	15
	4	14
	3½	13½
+3	Infinity	13
	50	12¾
	25	12½
	Fixed	12¼
	15	12
	10	11¾
	8	11½
	6	11¼
	5	10¾
	4	10½
	3½	10

*Round numbers to the nearest quarter-inch.

disappearing objects As a manufacturer of illusion, modern cinematographers make the great Houdini, at the height of his career, seem amateurish. The more routine effects, making objects appear or disappear, for example, are child's play for any cameraman.

With no need for props or special equipment, and with the camera anchored securely, the subject matter is photographed normally in order to establish the effect. Pausing only long enough to remove the previously exposed object, the cameraman completes the magical feat by exposing additional footage without it. Imaginative motion picture technicians can focus their cameras on a meddling mother-in-law and achieve the same amazing results—on film.

Similarly, a bull elephant can be made to disappear completely or allowed to shrink down to the size of a newborn pup, on successive frames of film, by merely substituting a smaller-sized animal for the original subject. The possibilities are, indeed, unlimited.

A Frenchman named George Melies, also a magician by trade, created effects during the industry's infancy that set the pattern for the special-effects technicians who followed. One of his films, "Optical Illusions," produced before 1910, featured multiple exposures and a variety of eye-arresting effects designed to illustrate the magical qualities of the new medium. Achieved without matteing processes or optical printers, the quality of M. Melies' special effects and the techniques he used in order to achieve them were far ahead of the primitive quality of the print on which his efforts were reproduced.

dispersion The separation of a beam of light into its constituent colors, as by a prism.

display A term generally used in motion pictures and television to describe picture information (either on film or as seen on a monitor screen) that is used for rerecording, projection, transmission, or merely for checking equipment.

Holograms, for example, are referred to as *visual displays.* In motion pictures, the term is used in connection with projected images (see PINHOLE OPTICS) that are used to simulate conditions needed for scientific studies. In television, monochrome test patterns, horizontal-rate waveforms, and operational-alignment test charts are all included in the general category.

distance The measurable interval between any two points. In optics, distance refers to either the interval between the lens and the subject matter or the interval between the lens and the film's emulsion in the camera.

distance meter A meter designed to measure specific distances between a camera and the subject matter without leaving the vicinity of the camera. In operation, two separated viewfinders are moved to converge on the subject, and the extent of adjustment is indicated on a scale calibrated in units of distance. See RANGE FINDER.

distance scale The calibrations or markings on a camera or lens mount indicating focus settings for given distances. These settings eliminate the need for focusing visually if the distance between camera and subject matter is known.

distar A supplementary lens that can be added to a camera lens to increase the focal length. This lens makes it possible for the cameraman to change the relationship between the camera and the subject matter without moving the camera from its original position.

distortion Imperfections, sometimes deliberate, in the surface of a lens that cause changes in the horizontal or vertical proportions of an image.

In videotape, the departure, during transmission or amplification, of the received-signal waveform from that of the original transmitted waveform. See WAVEFORM.

divergence The spreading or dispersing of light waves.

documentary films Factual films that record some historic event, the life of a person, the way of life of a group or society, or the like.

dolly A wheeled mount for moving the camera and its operator from one position to another. See CRAB DOLLY.

Doppler effect Increases or decreases in the volume and pitch of sounds caused by changes in the distance between the source of the sound and the position of the observer. These changes are the result of the compression of sound waves. Compression takes place when the movement of a sound source is added to the velocity of the sound wave.

The intensity and pitch of a siren on a police car, for example, seems to increase as it gets closer to the observer's position and decreases after it passes that point.

dot Flat, round disks, usually metal, that are positioned in front of a light source to diffuse it.

double An actor, similar in build and coloring, who takes the place of a star during the filming of a scene when an element of danger exists or when special athletic ability is required. Compare STAND-IN.

double broad A lighting unit used for fill purposes. Like the single broad, the double broad uses 1,000- to 1,500-watt lamps and produces a more intense light over a wider area.

double-coated film Film with layers of emulsion on each side of its acetate base. The image produced on one side of the emulsion, in two-color processes, is red. The image on the reverse side is blue or green, depending on the predominant color and the dye in the emulsion.

In black-and-white double-coated film, the upper layer of emulsion is faster than the under layer in order to widen the range of tonal values and increase the latitude of the film. Double-coated film is used for gamma- and x-ray photography.

double exposure An optical effect produced by reexposing different picture material over film frames that have already been exposed. Ghost images are the usual result. See Figure D-5. Compare SUPERIMPOSURE.

double fog-effect filters The double fog-effect filters made by Harrison are designed to provide a combination of the effects produced by two other filters. The first of these is the regular fog filter that produces slight halos around the highlight areas with increasing loss of definition as the fog effect is increased. The second type creates a flattening effect without loss of definition. The two optical principles, combined into the one filter, produces fog effects ranging from light to extremely heavy without loss of definition.

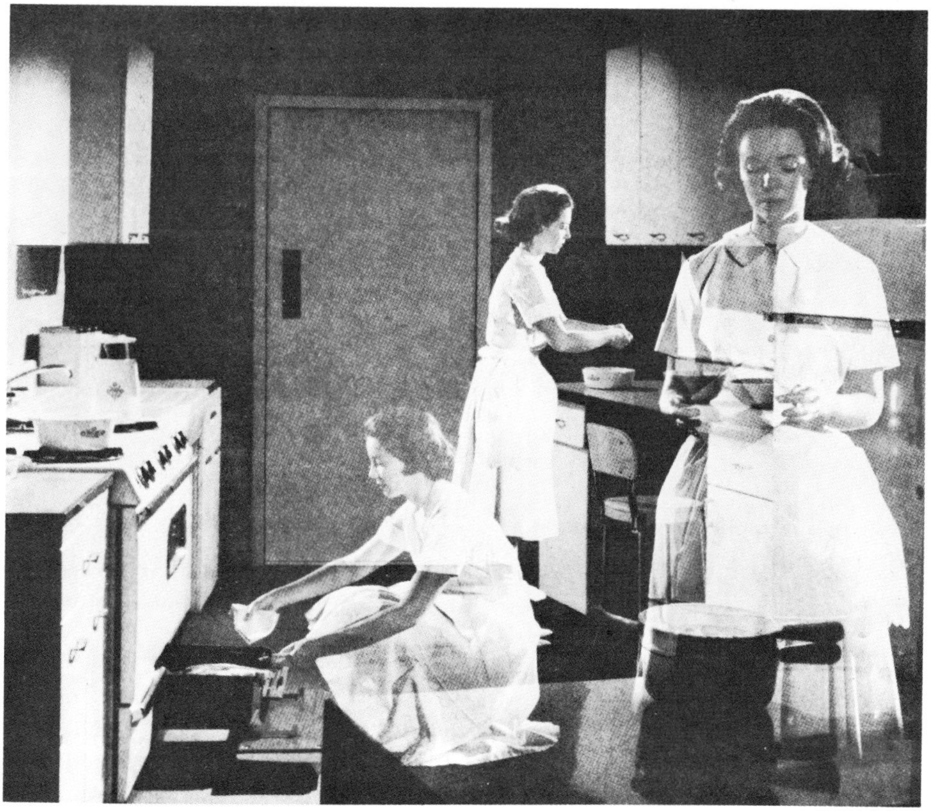

Fig. D-5 Double exposure. Produced for the Corning Glass Works commercial, the effect was achieved by photographing the background on one length of film and, subsequently, the housewife in front of a black background. Each of the housewives was, in turn, superimposed over the background to produce the composite effect shown here.

Since these double fog filters do not produce noticeable halos around highlights or flare in shadow areas, the effect is created almost entirely by lowering contrast—either by graying the scenes's dark areas or by darkening the highlights. In either case, image size must be considered with respect to the choice of filters and the degree of fog in each. Low-contrast scenes, in which the subject matter is relatively small, tolerate only a small degree of fog. Large images, by comparison, in scenes with high lighting contrast, require a greater degree of fog in the filter selected for use. Available in five steps, ranging from light to extremely heavy, the filters give a feeling of increased exposure although no increase in exposure is necessary because of the use of these filters.

double printing The process of combining the picture information from two negatives to produce a single, composite positive.

double system The simultaneous recording of picture and sound as separate elements—the sound being recorded on a magnetic sound recorder in synchronization with the picture information which is recorded on motion picture film. This system provides a greater degree of flexibility for the film editor while improving the overall quality of the recorded sound.

A synchronous motor system drives both the camera and sound-recording mechanism. The sound is recorded on a positive emulsion (to produce a negative image) and is developed separately in a special solution. A direct-positive recording (white image) can be produced by rearranging the elements in the recorder's optical system. An orthochromatic film stock, sensitive to blue, violet, and ultraviolet light, is used for recording.

In single-system filming, where the quality of the sound is of secondary importance (as in newsreel work), the picture and sound are both recorded on the same length of film during the actual photographic process.

double-take An action or piece of "business" by an actor that follows an initial reaction—usually for comic effect.

drapes Cloth hangings that are used as a background or to improve the acoustical properties of a set or recording studio.

dress The process of decorating a set with appropriate props.

dropout In videotape or sound recording, a temporary loss or interruption of the picture or sound signal that is caused by irregularities or insufficient coating, at various points, on a length of magnetic tape. The dropout appears on the television screen as a distracting horizontal streak.

dropout compensator An accessory for videotape recorders which replaces the missing video information with video previously presented.

dropped shadow An optical effect used frequently in order to make superimposed titles easier to read. A white title may have its black counterpart offset below and to the right of the title, for example, in order to make the title seem to stand away from the background, thus improving its readability. See Figure D-6.

Fig. D-6 The dropped shadow. (LEFT TO RIGHT) The title in column 1 was simply burned in over the gray background. The title in the middle column has a slight dropped shadow. The larger dropped shadow in the column at the right causes the title to "separate" from the background making it easier to read.

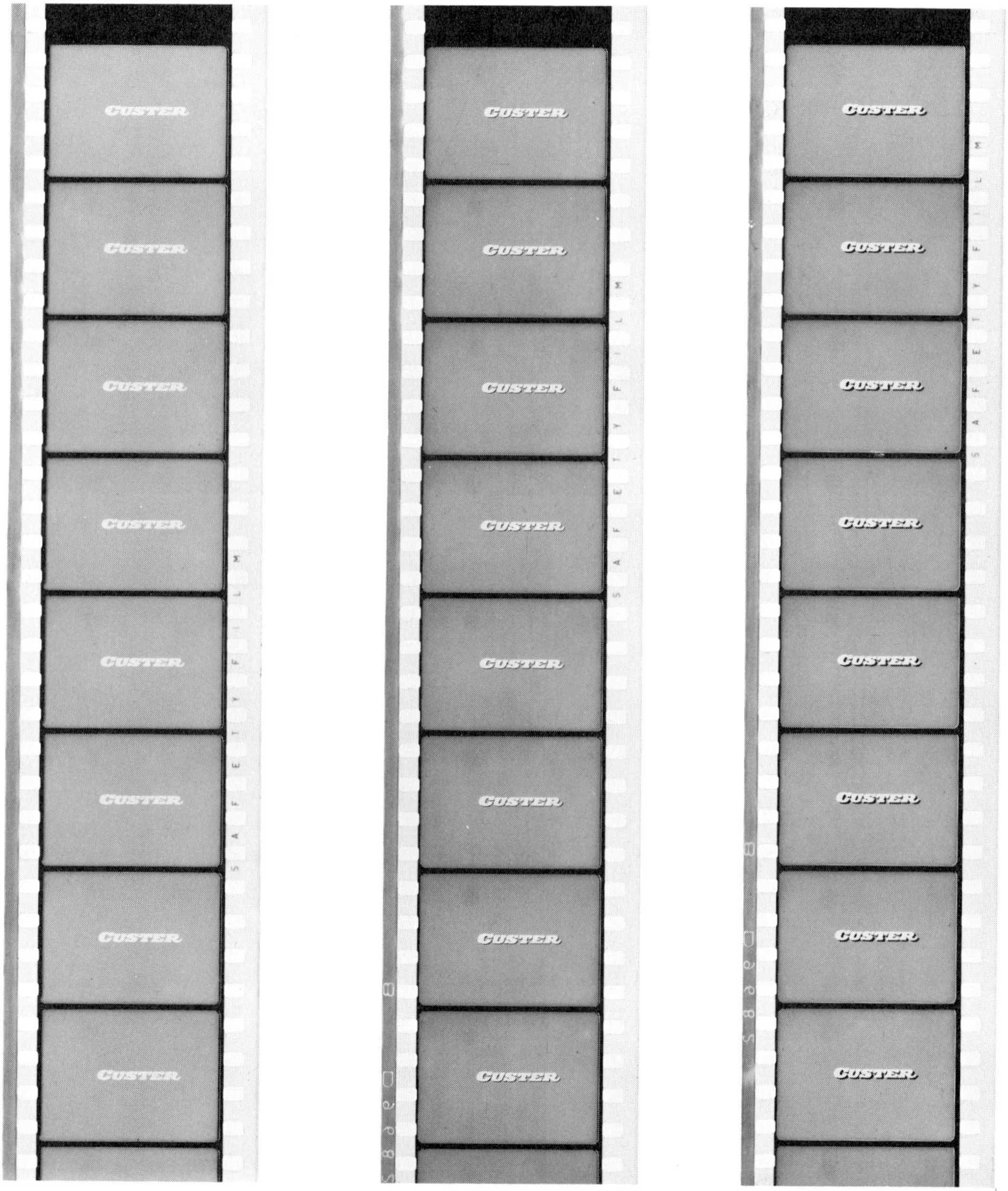

drum A cylindrical device used for drying film when a laboratory drying chamber is not readily available.

drunken screw A curved cam responsible for film advancement in the camera, given this name because of its eccentric motion. Its action is transmitted through the pin that activates the claws. These, in turn, engage the film's sprocket holes and advance the film into position in the camera shuttle. The claws are disengaged during the interval of exposure.

dry run A rehearsal during which the actors speak their lines and go through their actions as though the cameras were actually filming. Cameras, during these dry runs, also move to their assigned positions just as they would during an actual "take."

dual role The portrayal of two or more characterizations by one performer in the same motion picture.

dual track A recording area on $\frac{1}{4}$-inch magnetic tape that occupies slightly less than half the tape's width. Recordings are made on only one-half of the tape and then, by flipping the tape roll, a second recording can be made on the unused track area. The tape travels through the recorder's transport mechanism, during the second recording, in the same direction that it traveled during the initial recording.

Duarc An unfocused double-arc lighting source used for fill purposes in color photography.

dubbing 35mm magnetic sound recording can be used either to record a single track, for example, during the actual filming process, or to record several tracks side by side as in the final dubbing or mixing operation. A feature film normally uses at least four sound tracks: the main dialogue track, the music track, the effects track, and the postsynchronized dialogue track. This last track consists of dialogue recorded in an acoustically perfect area if the original dialogue recorded on the set or on location was not of sufficiently high quality. Each of these sound recordings must eventually be combined into one master recording in an operation known as *dubbing* or *sound mixing*.

Also, a recording process in which new dialogue is substituted for the original dialogue, as is done when foreign pictures are shown to an audience that would not understand the original language. See Figure D-7.

dunning process See MATTES and SELF-MATTEING PROCESSES.

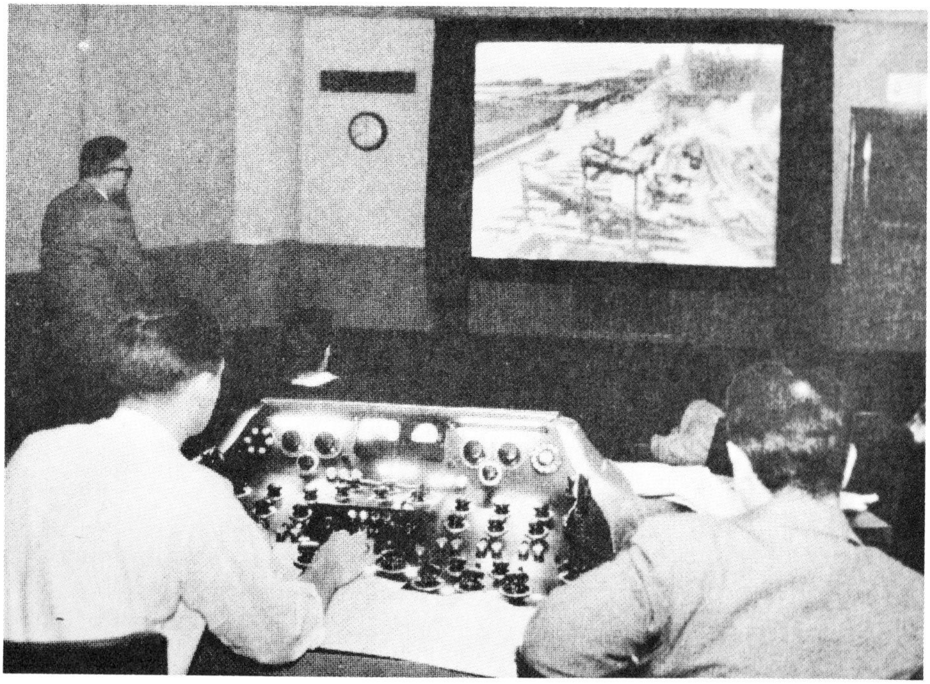

Fig. D-7 Dubbing.

duping The process of making a negative from an existing positive. The term is derived, of course, from *duplicating*.

duplex circuit Also called **full duplex**. Circuits designed so that electric signals can be transmitted between two points, in either direction, simultaneously.

duplicating stock A relatively slow film in which the granular structure of the silver halide in the emulsion is finer than the granular composition of faster films, thus providing higher resolving power in the duplicated image. This film stock is used mainly for optical or laboratory duping or printing.

Duvetyn Black, opaque material that is used to cover gobos, flags, and cutters in order to avoid undesirable reflections from striking the lens during the photographic process.

dye-transfer process A process in which three separation negatives are printed, by either projection or contact, and then processed to produce relief images in the gelatin. The thickness of these relief images is proportional to the density of the image.

These relief films (matrices) absorb dye colors in varying amounts, depending on the height of the relief image. With the gelatin side in contact with the unexposed film, or transfer paper, the dyes from each of the matrices are transferred, in turn, to produce a color composite print in which the tones and values in the original subject matter are duplicated. Since the process is mechanical, registration is automatic.

Dynalens An optical device designed by Juan de la Cierva for the purpose of neutralizing image motion via lens compensations. Equipped with stabilizing gyros, the Dynalens system uses electrical energy to bring a correcting lens into a position that counters an undesirable movement of the camera during the photographic process. The device is especially valuable for eliminating unsteadiness and minimizing vibration in shots made from moving automobiles, helicopters, etc.

dynamic range The audible spectrum, measured in decibels and ranging from the loudest to the softest sounds that can be reproduced through a recording device without distortion.

The dynamic range of a 35mm sound track averages between 40 and 50 decibels. The range of a 16mm sound track is between 35 and 40 decibels.

E

eccentric movement The eccentric movement of the camera advances film intermittently and holds it in position for the exposure of each successive frame. A pulldown claw moves the film into place behind the aperture, and registration pins on a horizontal shaft actuated by a linkage hold the film stationary during exposure. A pressure plate, held in place by a retainer arm, keeps the film flat at the aperture. The aperture plate is usually made of stainless steel, hard-chrome-plated, and has raised surfaces at the sides of the film path which allow the film to touch the plate only along its edges. In some cases, a matte slot in the aperture plate allows the insertion of mattes for multiple exposure work. A space of 0.015 inch in the film race is sufficient for two films or spliced film. See Figure E-1.

echelon sync system The echelon system of sync recording uses two narrow standard-type magnetic recording heads. The slits are in the same orientation as the slit on the audio head. They effectively record two sound tracks in push-pull on the respective edges of the tape.

The echelon synchronizing magnetic head for any $\frac{1}{4}$-inch tape recorder is designed to convert the recorder for synchronous sound recording and for in-sync reproduction with a projector or sprocket

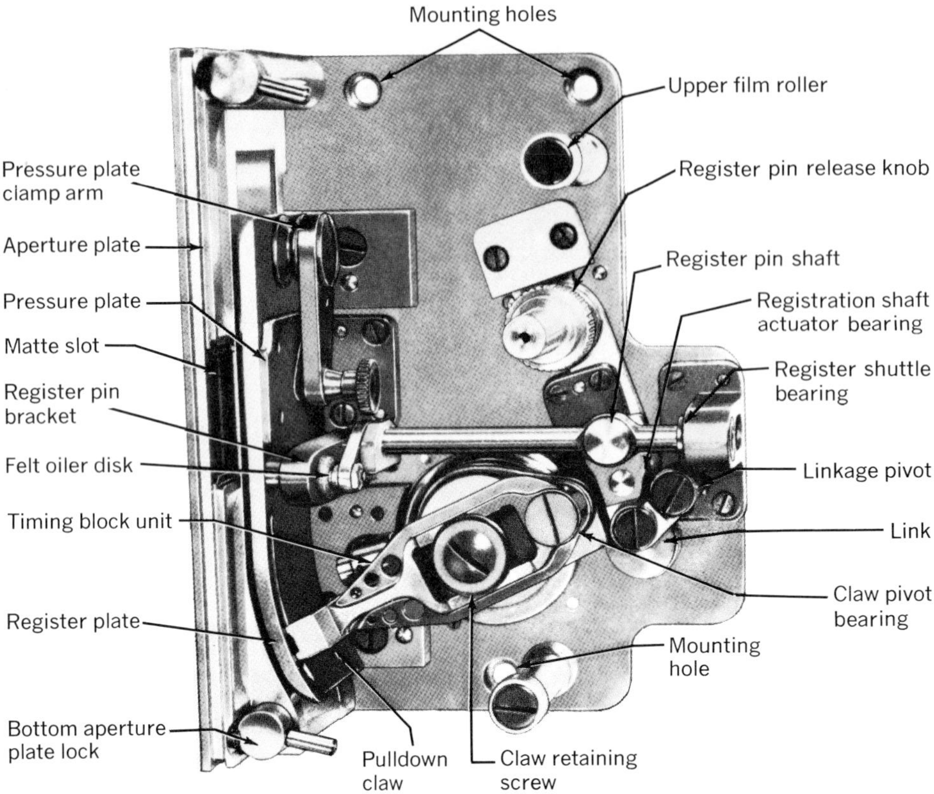

Fig. E-1 Eccentric movement (Mitchell NC and BNC).

recorder. When mounted on a tape recorder, it provides a synchronizing signal on the edges of the tape. During reproduction, this signal is used as an index of synchronization between picture and sound, controlling the relative position of the tape with respect to the sprocket-driven film. The unit operates directly from any Ryder sync generator without the need for auxiliary power, batteries, or amplification. The magnetic head records a fully modulated signal on the edges of the tape. The recording head slits operate in phase and place an inphase but slightly displaced echelon recording on the two edges of the tape. This arrangement eliminates the dirt pile-up problem prevalent on narrow heads and provides a continuity of recording and reproduction over splices and horizontal tape defects. The head is normally mounted between the capstan and take-up reel.

echo A reverberating sound produced by the reflection of sound waves. The time lapse between the original sound source and the point at which that sound reaches a reflective surface cannot be less than $\frac{1}{20}$ second.

echo chamber A long, narrow room with a microphone at one end and a loudspeaker at the other end. Reverberation results when an amplified sound is made to travel back and forth between the loudspeaker and the microphone. The walls and floor of the room, for best results, should have a hard surface (cement or plaster), and the ceiling should not be parallel to the floor.

Also, an acoustical device designed to reflect sound waves or repeat sounds for the purpose of producing echo effects. Reverberation generators such as the Fairchild Reverbatron, for example, are used to produce this type of effect during the recording or mixing process.

Reverberation, generally, enriches a recording by giving the sound a "live" quality. See DIFFUSER (2).

echo chamber, magnetic A recorder in which one head is used for recording while several reproducing heads, positioned at regular intervals in the recording circuit, create a time delay as the tape travels through the recorder. The delayed sound signal is combined with the original recorded sound by the circuit's head selector switch to produce the echo effect.

edge light A halo effect produced by placing a light source directly behind the subject matter.

edge numbers Numbers printed on the edges of the negative film to indicate footages. Duplicated on the print during processing, these numbers are used by the film editor for matching negative to work-print material in the completion stages of production.

Edicomp See ELECTRONIC EDITING.

Editec See ELECTRONIC EDITING.

editing Although the film editor assumes a major share of the responsibility for the completed production, he rarely receives a proportionate amount of the credit for its success and is often referred to as the "unsung hero" of the motion picture industry. The editor's good judgment is, quite often, the single determining factor that separates the "hit" from the "bomb."

With the emphasis on creativity, the editor is expected to do much more than splice scenes in their proper order and match them to previously recorded sound tracks. Today, imagination is as important an ingredient as skill or experience. The infinite variety of photographic possibilities, because of the director's script interpretation and choice of technique, makes it impossible to outline prescribed rules or established editing patterns. In the absence of recognizable standards, the subject matter in each production must be edited and presented in the most interesting manner possible. The editor's ability and the talents of the director are thus blended and judged collectively.

Under the supervision of different editors, similar material will, almost always, reflect individual tastes and treatments. Properly paced and timed, one effort may be cut in award-winning fashion. An improper flow of picture information, by comparison, may find the second editorial effort consigned to the number-two spot on a double-feature program. During the projection of the award-winning film, the audience, quite likely, will sit on the edges of their collective seats, from opening credits to closing titles, and forget the bags of popcorn that had been purchased earlier. The same popcorn vending machines, while the second feature is being projected, become the main attraction and the "scene" of activity.

Optical and magnetic sound readers for synchronizing and analyzing sound tracks, rewinds capable of carrying several feed and take-up reels, synchronizers for measuring film, splicers for joining separate lengths, scissors for cutting them apart, and the all-important moviola for projecting the picture reel and matching it to the track are the tools with which the film editor works. The white cotton gloves he wears while handling original negatives, dupes, printing material, and fine-grain film are not merely wearing apparel but standard equipment. Similarly, the wax pencil for marking film, the conversion chart that converts film footages into frames and projection time, and the wipe charts that visually suggest a variety of transitional effects are all within easy reach on the film editor's underlit table.

Synchronizing the picture reel with the sound track is a relatively small part of the editor's job. Even this chore, however, is complicated by the many tracks that are subsequently mixed to form the single composite. Including the dialogue tracks, the musical backgrounds, and sound effects, a one-minute film sequence may contain as many as 40 individual sounds—each of which must first be synchronized with the corresponding picture and then used for "mixing" purposes.

When necessary, the editor may borrow footage from previous productions for use as inserts. The wide variety of subject matter in these catalogued "stock" shots include crowded city streets as well as dense jungle scenes, locomotives hurtling through the night, airplanes of all types taking off, flying, and landing, ships at sea, battle scenes, and, in fact, almost every conceivable type of action in as many settings. Music libraries, similarly, include classical as well as popular selections—for every possible mood or theme. Sound effects, ranging from a lion's roar to the gentle pitter-patter of raindrops on a window pane are also catalogued and readily available to the film editor. The libraries at some of the larger studios, contain thousands of stock shots, musical backgrounds, and sound effects on both tape and film.

Assembling, sorting, and splicing the many miles of film normally used in the average motion picture is a herculean job, indeed. From the time the first can of "rushes" arrives, until the release print is ready for projection, the editor can be expected to handle approximately 10 to 20 times the amount of film footage seen by motion picture and television audiences. "Out" takes, trims, negatives, dupes, title footage, and negative and positive mattes used for optical combinations account for approximately 90 percent of the film footage that is never seen by anyone but the editor and the optical cameraman.

Obviously, the *rushes* or *dailies* as they are sometimes called, are awaited anxiously. Since the results of the previous day's photographic efforts are almost always delivered directly to the editorial department, each roll of processed film should be reviewed and preedited before it is screened. For reasons politic, only those technically concerned should be allowed to see the processed print in its uncut form. Others can and should wait until after the mistakes or faults have been spliced out of the roll. Since "out" takes sometimes contain usable portions, the editor should review all questionable material with the director and cameraman before relegating the film to the bottom of the barrel. Quite often, the odd lengths contain the exact footage needed for inserts or transitional effects.

From the preparation of the synchronizing leader strips, when the first can of rushes arrives from the lab, until the last scene has been synchronized with the corresponding sound track, the editor must assume complete responsibility for the completion of the picture—in word as well as in deed. His opinions and judgments are, understandably, much more objective than those of the director and cameraman. The latter, in spite of their technical proficiency, should never attempt to edit their own film productions.

Assuming that each of the scenes has been properly slated during the photographic process, the editor, before actual cutting begins, should catalog the contents of the many film cans that contain picture and sound as they arrive from the lab. Numbers, imprinted along the edges of each length of film serve a twofold purpose. At any stage of production, they provide the editor with an immediate means of identification as well as a basis for coding. Most laboratories have facilities for duplicating original edge numbers or for renumbering work prints and other material in accordance with the editor's individual requirements. In the final stages of the editing process, the edge numbers in the work print are matched against those in the original film, thereby assuring an exact duplication of picture information in the composite print.

Obviously, original film should be handled only when necessary and, even then, only when wearing the white gloves mentioned earlier. Since torn or pulled sprocket holes are certain to affect the registration in the final print, original material should never be projected. Running original film is not a calculated risk but a foolhardy gamble. Similarly, any scratches, abrasions, and flaws in the original film, later used for printing purposes, are certain to be reproduced and duplicated in the print.

The preparation of the "rough" cut is the first step in the completion process. Selected takes are spliced together in their proper order, but not necessarily to their actual length. Attempts at synchronization are, at best, temporary and approximate. In the absence of a sound track, or where the editor must time the footage to a script, rough synchronization can be achieved by reading the text several times at a normal pace—about 130 words per minute. Stopwatches that indicate time in fractions of seconds and corresponding film footages provide temporary timings that can be used for sync purposes until the actual sound track has been recorded.

The editor is not expected to use every frame of film exposed by the cameraman. The theory—better too much than not enough—applies only during the photographic process. Since retakes are time-consuming and costly, the professional cameraman overshoots at the beginning and end of each scene, makes "insurance" takes when doubt exists, and covers himself further by reshooting portions of a scene from other camera angles. Editing, often referred to as cutting, should be just that. Excess footage should not be used for lengthening the picture but for adding interest by imaginative intercutting of establishing long shots with close-ups and angle shots.

Since each scene becomes an integral part of the production, it should be judged in relation to other scenes that precede or follow the newly joined length of film. Although the moviola is used for frequent viewing and synchronization purposes, sequences, after the interlock, should be projected on a large screen. Quite often, overlooked and undesirable detail, unnoticed on the small projection head of the moviola, will be immediately apparent on the larger screens in studio projection rooms. Provisional footage, available through the cameraman's foresight and generosity during the photographic process, can readily be spliced into the reel to replace unusable portions or, possibly, entire scenes.

The *work print,* as the name implies, contains all the visual material that will subsequently be duplicated in the final print. By the time the editorial effort will have been finalized, scenes in the work print will have been shortened, lengthened, removed, changed in relation to other scenes, or replaced by similar scenes with different camera angles. The original negative, for printing purposes, will be cut and matched from the completed work print.

Work-print material, from original color negative, may be reproduced in black and white. Since true color values and balance between scenes cannot be accurately determined until the color print is finally made, the editor is forced to rely on judgment alone with respect to scene-to-scene color balance. At best, estimates based on educated guesses can easily negate the economic advantages of the black-and-white work print when the original material is in color.

Frequent screenings, while the picture is in the work-print stage, serve many useful purposes. Audience reaction, for example, may be used as a yardstick for measuring the picture's timing and pacing. In rarer instances, unjaded viewpoints and frank comments may indicate a need for a fresh reappraisal of the edited material. More often, however, because of the editor's familiarity with the subject matter, previously unnoticed and undesirable detail may suddenly be called to the embarrassed editor's attention.

The same basic practices that are used in editing 35mm film also apply to the 16mm production. Variations, such as the number of frames in a foot of film, running time of standard reels, and track advance, are mathematical differences that do not affect the creative effort. Each foot of 35mm film, for example, contains 16 frames. The standard 35mm reel contains 900 feet of film, and the projection time for the entire reel is exactly 10 minutes. The 35mm sound track precedes the corresponding picture by 20 frames.

Each foot of 16mm film, by comparison, contains 40 frames of film. The 360 feet of film, in the standard reel, is projected at the rate of 36 feet per minute. Since 1,440 frames of film are projected in one minute, regardless of film size, the projection time for the 16mm reel is the same as for the 35mm reel—10 minutes. The 16mm optical track is advanced by 26 frames. Magnetic tracks precede the corresponding picture by 28 frames during projection.

The method used for preparing 16mm material for printing is quite different from the comparatively simple process of matching original to work print in the 35mm film production. Instead of preparing cue sheets for optical combinations and transitional effects, the editor must separate the work print into two reels of equal size. Scenes in these A and B rolls are arranged to form a checkerboard pattern. With this method, splices between scenes are rendered invisible; cross-dissolves, fades, and wipes can be achieved without opticals; and titles and other effects can be added or superimposed during the printing process.

Assuming that each scene in the picture is numbered consecutively, all the odd-numbered scenes are spliced into roll A and the even-numbered scenes into roll B. Scene 1 and scene 3 are separated by a length of black leader equal to the number of frames in scene 2 of roll B. Similarly, the black leader spliced between scene 2 and scene 4 of roll B will match the corresponding footage of scene 3 in roll A. Printed together on duplicating stock, the composite, splice-free processed film contains all the scenes from both reels in their proper order, along with transitional effects and titles.

Animation presents additional burdens. The editor, for the simplest of sequences, must analyze the prerecorded sound track and provide the animation director with frame and footage counts. These counts, used as a guide by the animator, determine the screen length of animated actions as well as lip sync. In scenes that contain dialogue, each word is broken into syllables, and the phonetic spelling is written in the clear area of the film. The length of each word and the frame counts used for drawing each of the mouth actions are measured on a synchronizer. Open- and closed-mouth positions, for lip-sync animation, are amazingly realistic when the drawings follow the editor's frame counts.

After the track has been analyzed, the animation director transcribes the frame counts on *bar* or *lead* sheets. With each frame properly entered, the information provides a complete visual synopsis of the entire animated sequence, including the exact number of frames alloted to an action and the position of each musical beat in relation to the animated antic.

In some cases, dialogue, music, and effects tracks are recorded after the animated drawings have been made, and later photographed. With this post-sync method, the editor must match action and lip sync to the recorded track after the actual production processes have been completed. Although the pre-sync tracks must be analyzed before production can begin, the editorial completion process is far less complicated. Synchronization becomes almost mechanical, and, in practically all cases, more precise results can be achieved with prerecorded tracks than with post-sync methods.

The relationship between sound track and picture is especially important in the animated production. Many visual effects in animation depend almost entirely on the accompanying sound and music for added realism. As with live productions, musical backgrounds and segues help establish moods and transitional sounds that help accent the animated antics of the cartoon characters.

In animation, as with the 35mm feature production, the 16mm industrial film or the one-minute television commercial, the approved and accepted answer print that is used as the standard for subsequent prints does not complete the editorial effort. The editor's responsibility does not actually end until after the composite prints, in which the picture and sound track are finally combined on the same length of film, have been approved and the release prints distributed. Only then can the many previously used rolls of film be packed away in cans and stored in film vaults—for posterity. The editor, from rushes to mix, must combine knowledge with imagination, experience with judgment, and patience with diplomacy. For each of the many miles of picture and track that will pass through the synchronizer, from rough cut to release print, there will be an equivalent number of problems.

To the dedicated, hardworking, unsung hero of the motion picture industry, these "routine" problems are accepted philosophically as a regular part of the film editor's job. See Figures E-2*a* and *b*.

editor The person responsible for the postproduction or completion processes of the film or videotape production—mainly, the proper arrangement of scenes and the synchronization of all sound and picture elements that will appear in the release prints that are distributed to theaters and exhibitors.

The editor is responsible for the sorting and cataloging of dailies and sound tracks, the preparation of work prints that are subsequently used as guides for opticals, and the supervision over the mixing process during which all the sound elements are combined to produce a composite sound track that complements the corresponding picture material.

Fig. E-2*a* **Film editing. (*Transfilm Inc., New York.*)**

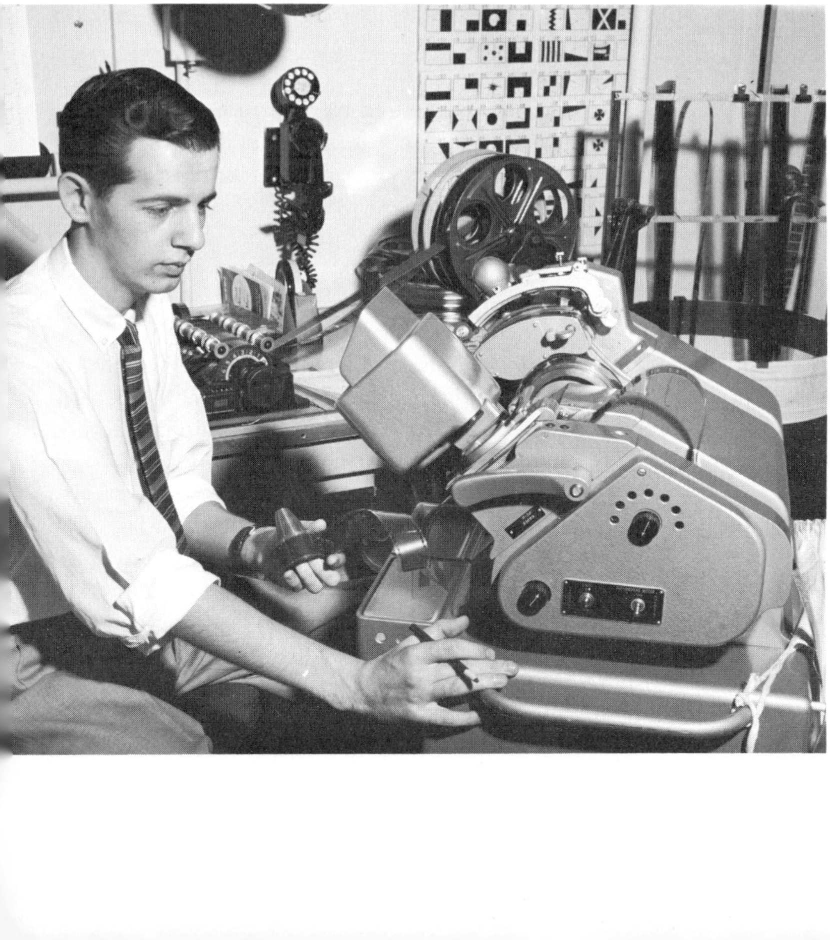

Fig. E-2*b* **The viewer sees 1,440 pictures each minute. (*Transfilm, Inc., New York.*)**

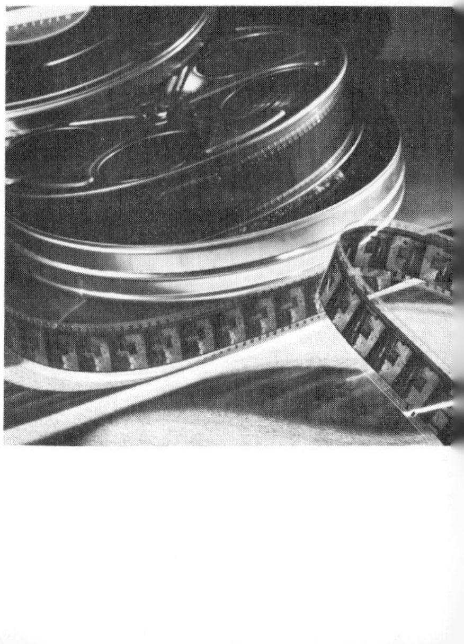

editorial synchronism The spacing, in terms of frames, between any specific point in the sound track and the corresponding picture during the editing process. Unlike projection synchronism, in which the sound track precedes the corresponding picture by 20 frames in the 35mm print and by 26 frames in the 16mm print, the sound track and picture remain aligned during the editing process.

edit sync Audible beeps on the videotape cue track at one-second intervals. These beeps help the editor locate specific cut points. See TALKING CLOCK.

educational films Motion pictures or filmstrips designed for instructional use.

effective aperture Also called **working aperture**. In most cases, the *f*-values of a lens do not take into consideration the light loss caused by absorption and reflection. The effective aperture is the difference between the stated *f*-value of the lens and the actual amount of light the lens is capable of transmitting. See RELATIVE APERTURE.

effects An all-inclusive term for illusory effects. The general term includes fades, cross-dissolves, and wipes, as well as trick shots, montages, split screens, and multiple-image shots. See SPECIAL EFFECTS and TRANSITIONAL EFFECTS.

effects filter Unlike the purposes served by gelatin and glass filters, effects filters are used to produce an overall effect in a scene. The filters are actually pieces of gauze, cheesecloth, or similar semitransparent material that can be positioned in front of a lens to soften the image or produce subtle diffusion effects. See DIFFUSION AND DISTORTION.

Ektachrome A multilayer color-reversal film produced by the Eastman Kodak Company. Unlike Kodachrome film, in which the color-forming dyes are added during processing, the coupler components of the dyes are added to the emulsion during manufacture. See COUPLER DEVELOPMENT. Compare KODACHROME.

Ektacolor A nonreversal, multilayer color film produced by the Eastman Kodak Company. The coupler components of the dyes are added to the emulsion during manufacture. See COUPLER DEVELOPMENT. Compare KODACOLOR.

electromagnet A metallic substance that can be temporarily magnetized by passing an electric current through the wires coiled around it. During the recording process, audible sound is converted into electric current. When that current passes through the recording head (which is itself an electromagnet), it creates a magnetic field. As the magnetic tape passes over the recording head, the iron oxide particles on the tape are magnetized and oriented so that the tape, when played back, will reproduce the identical sounds picked up by the microphone at the time the recording was made.

electron A negative electric charge. The unit of electricity that is released by the cathode in an electron tube.

electron beam The focused stream of electrons that strike the chemical coating or fluorescent screen as in a cathode-ray tube.

electron-beam recorder The primary application of the electron-beam recorder (3M-EBR-100) is to replace kinescope methods in use since the earliest days of television broadcasting. Its capabilities, in addition to kinescope recording, show promise for other scientific and communication areas, among them satellite and spacecraft transmissions, electronic image generation, and aerospace image recording.

Conventional kinescope recording necessitates a phosphor screen, a glass faceplate, and a light optical system as an interface between the electron beam carrying the television signal and the photographic film that will store it. The electron-beam recorder eliminates this interface, using electrons instead of light to record television films of quality comparable to the original signal.

Elimination of the phosphor screen, glass faceplate, and camera lens greatly improves contrast at high resolution and gives noise-free images. The total electron energy is delivered to the film by eliminating light-scattering, light-absorption, electron-to-light conversion losses in the phosphor screen, and by avoiding the inefficient geometry of the kinescope lens system.

In operation, 16mm film, single-perforated, fine-grain positive stock, is held in preloaded 1,200-foot film cassettes in a coaxial configuration. Threading may be carried out under daylight conditions. When running, the film is pulled down by a stable pulldown claw during a blanking cycle of the electron gun. There is no mechanical shutter mechanism.

The drive motor of the camera is locked to the vertical synchronizing pulse of the incoming video signal. The image on the film is formed by the electron beam from the gun. The beam is deflected horizontally and vertically to coincide with the synchronizing pulses from the incoming video signal. A slow-speed, fine-grain film can be used for even further improvement of resolution, as opposed to the kinescope's need for a relatively fast and grainy film. See Figures E-3*a* to *c*.

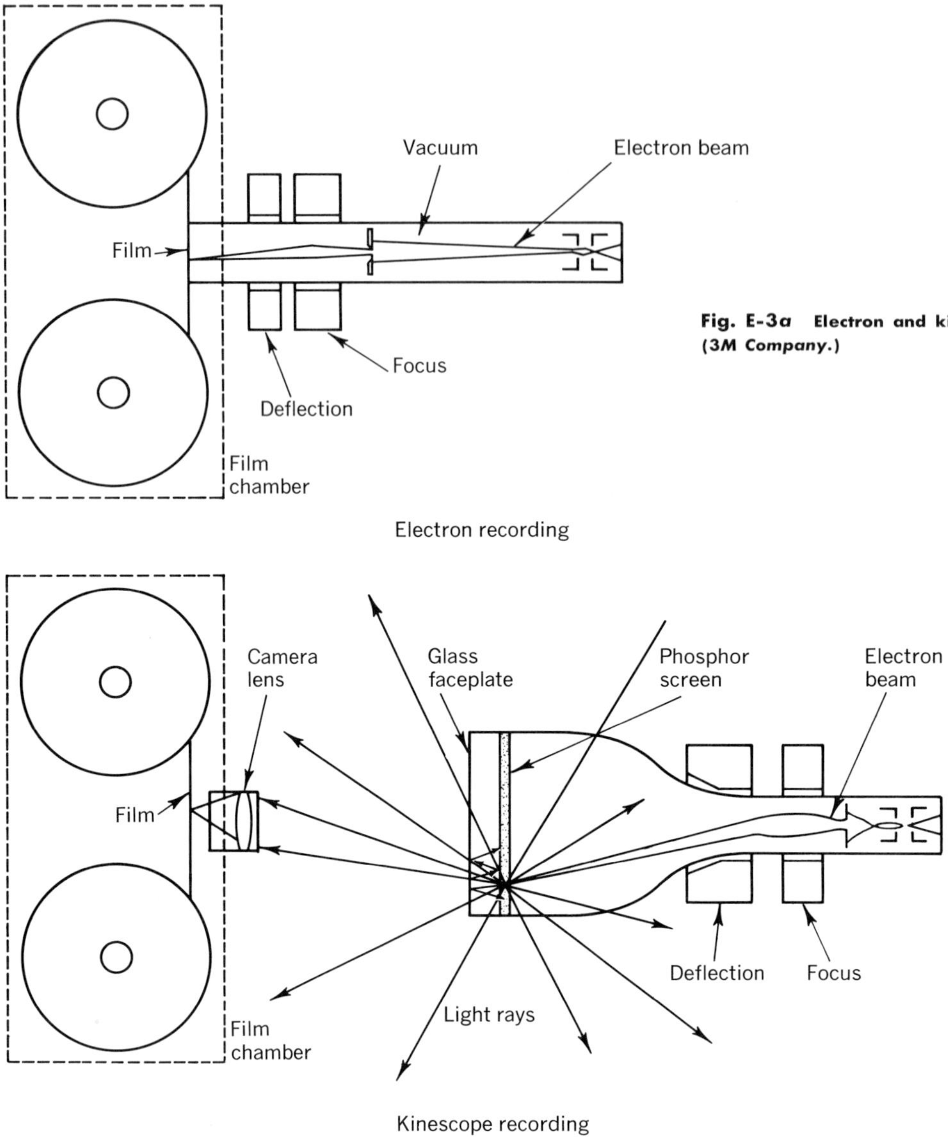

Fig. E-3*a* Electron and kinescope recording. (3*M Company.*)

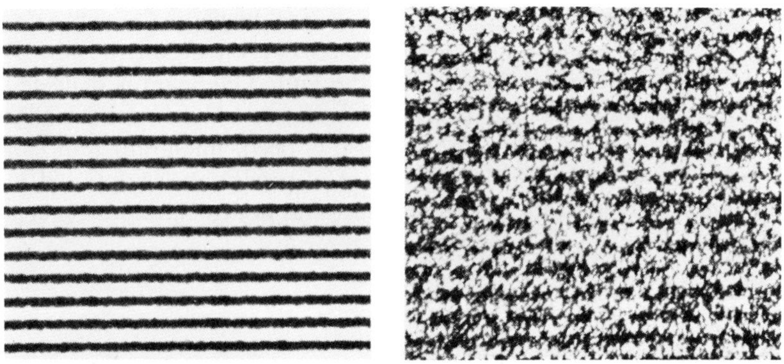

Fig. E-3*b* Television raster lines enlarged from two 16mm film frames. (3*M Company.*) LEFT: electron recording on special 3M fine-grain (less than 0.1 micron electron recording film). RIGHT: kinescope recording on television recording film. Line-to-line spacing in both pictures is approximately 0.00058 inch or 14.7 microns.

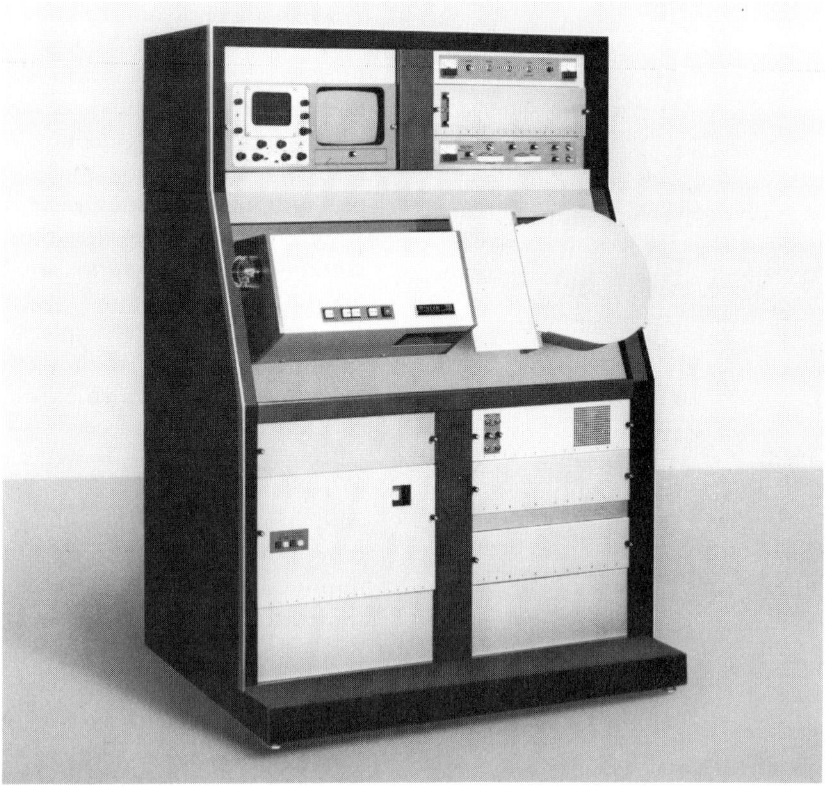

Fig. E-3c Electron-beam recorder model EBR 100. (*3M Company.*)

electron gun A part of the cathode-ray tube where the electrons are formed into a beam and deflected to the fluorescent screen on the inside of the cathode-ray tube.

electronic editing Electronic methods for editing videotape utilize film-editing techniques. Basically, electronic computers accumulate the editing data provided by a kinescope recording which is used as a film-editing master. Interlocked 16mm film transports are used to translate the accumulated data, with respect to scene location, to corresponding videotape scene-location data.

The composite print is produced in one of two ways, either by A & B roll mixing or by the Edicomp or Editec electronic editing processes. In the first (the Edicomp process), an electronic programmer accumulates the editing data which is subsequently used for transferring the original sound and videotape picture to an edited second-generation copy. In the Editec process, the edited picture and sound are prepared in a manner similar to the work prints prepared by a film editor before the optical process. The edited work print is used as a guide for the re-recording process and the production of the composite videotape.

With the Edicomp method, a programmed electronic editor is triggered (after the slate and standard leader have been "exposed"), to begin the frame count for the transfer of the original tape picture and sound to an edited second-generation copy.

In operation, the videotape machine has the ability to switch from playback to record within 4 milliseconds. The end result is the assembly of scene after scene, in building block fashion.

In the VTC film-tape translator electronic editing system, the computer section controls the precise frame that is to be recorded on the final videotape. Thus, a new frame can be recorded each $\frac{1}{30}$ second. The new material may be recorded as video/audio simultaneously, video only, audio only, or in any combination delaying audio or video by any required amount.

Electronic editing has several distinct advantages over other editing systems. The original videotape master, for example, is never physically spliced and is, therefore, always available for future use. Also, since the precise editing is computer-controlled, it can be reprogrammed as often as is necessary for the purpose of making minute adjustments within $\frac{1}{30}$ second. The fact that each step of the editing procedure is logged on a flow sheet so that the entire process, or any portion of it, can be reprogrammed by referring to the log sheets is an additional advantage. Finally, the editor is fed through a switching process which permits the addition of such transitional effects as dissolves or fades and the recording of mattes during the editing process. See Figure E-4.

electronic shutter An electronic control used for transmitting motion pictures over television circuits. Although the film travels at a continuous speed through the projector, the electronic shutter operates during the vertical blanking interval in synchronization with the photocathode of the image-orthicon tube.

ellipsoidal spotlight A light unit housing an ellipsoidal-shaped reflector. The unit accommodates a pattern holder, or similar device, for projecting soft-focus shadow patterns on flat-toned backgrounds. See Figure E-5.

emulsion Practically all motion picture film is supplied on cores and wound with the emulsion facing the center of the roll. Unlike the single layer of emulsion on which black-and-white images are exposed, color film has its silver halide grains spread over three light-sensitive layers of emulsion.

Although the first experiments with color motion pictures were conducted in 1912 by Dr. Herbert Kalmus, the founder of the Technicolor Corporation, the first multilayer film was developed by the Eastman Kodak Company and introduced to the industry in

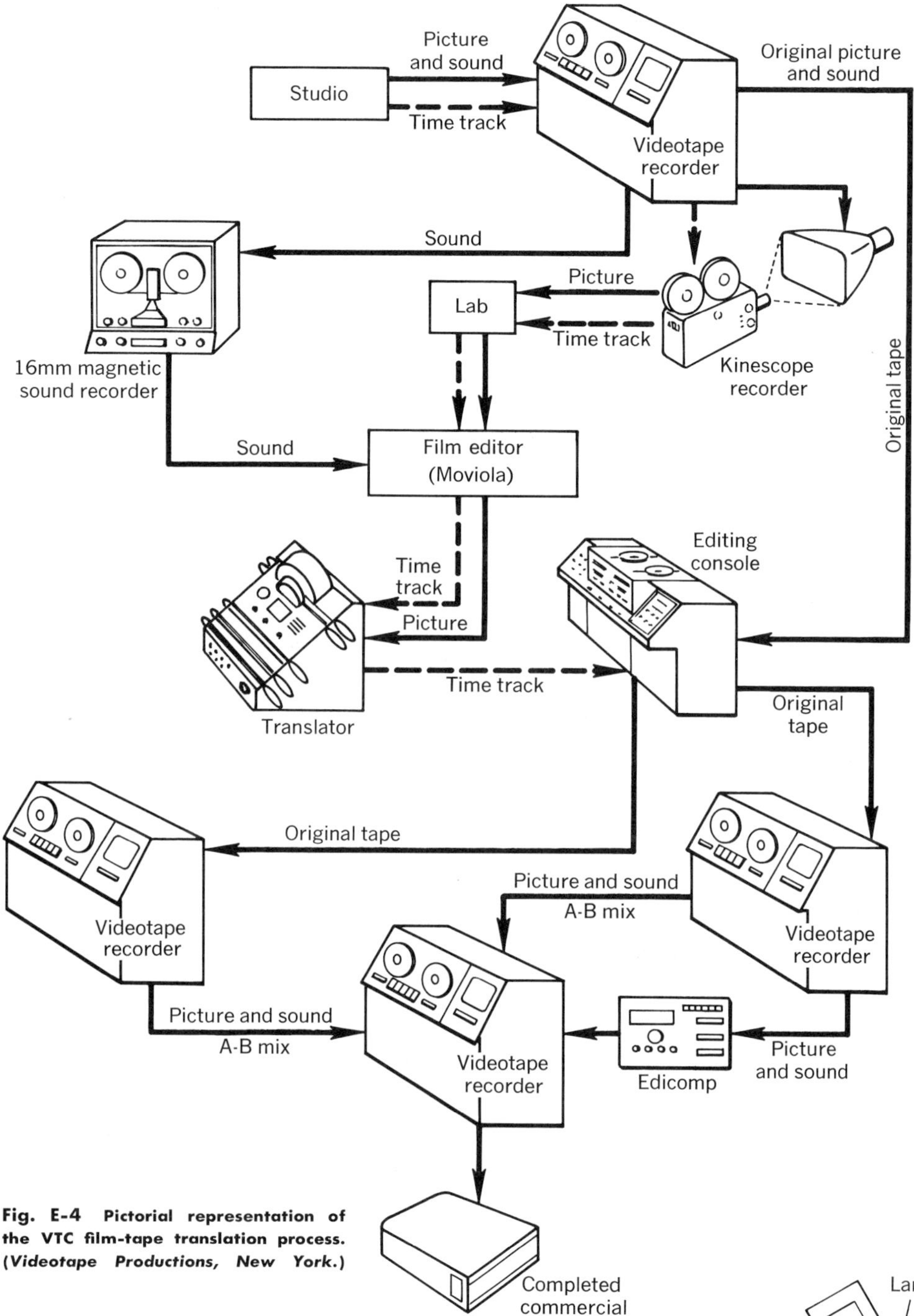

Fig. E-4 Pictorial representation of the VTC film-tape translation process. (*Videotape Productions, New York.*)

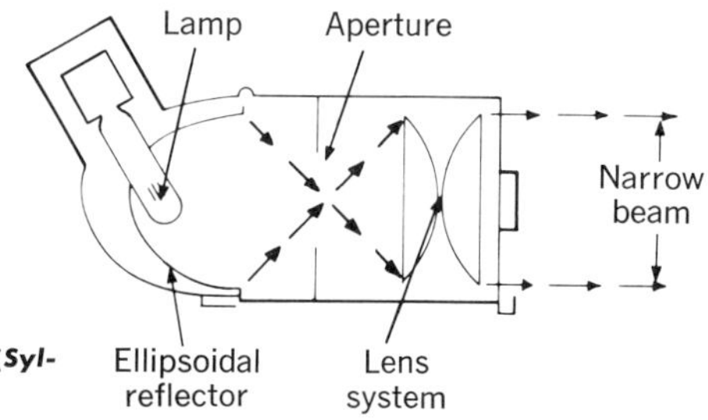

Fig. E-5 Ellipsoidal spotlight system. (*Sylvania Electric Products, Inc.*)

1935. Each of the layers of emulsion is sensitive to a different color of the spectrum.

Sandwiched between these three layers are gelatin filters that allow only one primary color to be exposed on each of the separate emulsions. Accepting the "correct" color while preventing other colors from seeping through to the remaining layers and upsetting the true color balance, these gelatin filters serve their purpose and are subsequently eliminated during the developing process. Oddly enough, the multilayer film, including the gelatin layers, is only slightly thicker than the single layer of emulsion that rests on the black-and-white film base.

The top layer of emulsion is sensitive to blue light. Red and green light are not affected as they pass through this upper layer. The yellow filter directly underneath prevents any blue light from reaching the middle and bottom layers of emulsion. The middle emulsion layer, sensitive to both green and blue, allows only green light to expose the layer—the yellow filter blocking the passage of blue light. The third layer of emulsion (the one nearest to the acetate base) is sensitive to blue and green as well as to red light. The blue, however, is blocked out by the yellow filter, and the sensitivity to green is low enough to be ruled out as inconsequential.

emulsion, standard position—negative and positive The layers of emulsion, in both the 35 and 16mm rolls of raw stock, face the lens at the time of exposure. After processing, the 35mm black-and-white print, as well as the print made from a color negative, reads correctly for direct viewing when the sound-track area appears on the left side of the film frame and the emulsion side of the film (dull side) faces the viewer.

Since 35mm prints made from color-reversal film may have the emulsion in either the standard or nonstandard position, each print should be examined separately and rewound, if necessary, before it is projected. During projection, the emulsion side of the 35mm print faces the light source while the base or shiny side of the film faces in the direction of the projector lens.

As a general rule, the emulsion position of duplicate negatives or master positives in color, black-and-white, and reversal-type films is the same as the emulsion position in the original negative and positive lengths of film. Prints made from an original 35mm negative or from a 35mm duplicate negative, for example, will both be in the standard position, with the sound track on the left side of the frame when the emulsion side of the print faces the viewer. During projection, the emulsion side of the print faces the light source while the base side faces the projector lens.

In both the 16mm A- and B-wind pictures, the sound track is on the right side of the film frame in its head-up position. The perforations are thus on the left side of the film. B-wind film (the film normally exposed in the camera) is referred to as the "original" and reads correctly through the base. Contact prints produced from this type of material are A-wind and read correctly when the emulsion faces the viewer. Since one-to-one prints or reduction prints that are produced optically can be printed either emulsion-to-emulsion or base-to-emulsion, the desired emulsion position of the print must be specified.

During projection, the emulsion in the B-wind print is toward the lens. The emulsion in the A-wind print is away from the lens. See A & B EDITING—FILM and A & B ROLLS—EMULSION POSITIONS OF PICTURE AND TRACK.

emulsion curve A graph of the sensitivity of an emulsion at different color wavelengths. Comparisons that need not be as precise as those obtained by plotting the emulsion curve can be obtained by a spectrogram. See SPECTROGRAMS.

encoder An electronic device used for processing the television color signal. Its function is to combine the three color signals into one signal so that, after transmission, the color components making up the scene can be restored to their proper relationship and duplicate the colors in the original subject matter.

enlarger A darkroom projection printing device used by still photographers for producing prints—usually larger than the size of the negative that is used for copying.

enlarger lens A lens used for copy work. This type of lens is designed to produce a critically sharp image with near-perfect field flatness. Normal camera lenses focus light from varying distances at the film plane. Enlarger lenses, by comparison, must be capable of projecting negatives on flat printing paper to produce pictures that are in critically sharp focus, from one edge of the printing paper to the other.

episode A part of a motion picture. During the industry's infancy, the weekly episode or "cliffhanger" was an extremely popular attraction that brought audiences to the theater on a regular basis.

equalizer Also known as compensators, equalizers are devices that are connected to electric circuits in order to change the circuit's amplitude-frequency characteristics.

Equalization is the process of modifying the amplitude-frequency response characteristics in a recording and reproducing system, for one or both of the following purposes:

1. To produce a flat overall frequency response
2. To match the signal handling capabilities of the recording system to the frequency distribution of the signal to be recorded or to minimize the audible noise of the reproducer

In videotape recorders, the equalizer provides additional aperture compensation for the overall playback equalization characteristic. The equalizer also includes control circuits for the head resonance test mode.

equivalent aperture The efficiency of a lens, based on its optical characteristics (light-transmission capabilities, for example) rather than its size. See RELATIVE APERTURE. Compare EFFECTIVE APERTURE.

erase head The device in the tape-transport mechanism over which the magnetic tape passes on its way to the recording head. As the tape passes over the erase head, it is cleaned to reduce the noise level. Previously recorded signals are erased by the high-frequency current passing through the erase head. Sound can also be removed by bulk erasers or degaussers which demagnetize the entire tape roll.

establishing shot Also called **long shot**. Photographing a relatively large area from some distant point to establish a relationship between the general setting and the scenes or actions within it.

Estar base The Eastman Kodak trade name for its polyester material.

exciter lamp An intense incandescent light source used for recording optical sound tracks or reproducing the sound during the projection process. In projection, the lamp is focused on the film's sound-track area and the patterns normally found in variable-area and variable-density tracks. In recording as in projection, it is called an exciter lamp because of its effect on the photoelectric cell which converts the light waves into sound waves and, subsequently, into audible sound. See Fig. E-6.

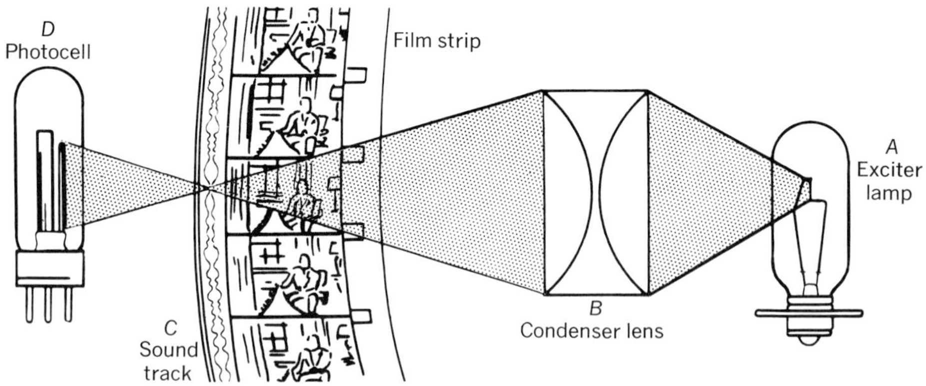

Fig. E-6 Exciter lamp. (*Sylvania Electric Products, Inc.*) The diagram shows how sound is transferred from film to the amplifier. Light from the exciter lamp (*A*)—light transmitter—passes through the lens (*B*). The lens focuses light on the sound track of the film (*C*). This light passes through the sound track and into a photocell (*D*)—light receiver. The photocell receives sound in the form of light impulses and converts it to electric impulses or, in other words, actual sound. These impulses, leaving the photocell, are too faint to be audible. Therefore, they pass from the cell into an amplifier which magnifies the sound to any desired audibility.

These lamps have a rated life of from 50 to 200 hours and a voltage range between 4 and 10. Amperage covers a range from 0.2 to approximately 7.5. See Table E-1.

exhaustion The inability of a developing solution, after excessive use, of reacting on a film's emulsion to produce acceptable images.

exhibitor The owner or operator of a theatre or auditorium where motion pictures are shown.

explosives The effect of bullets striking and ricocheting off land and water surfaces can be made as realistic as the real thing by using plastic bullets, pellets, and glass marbles for ammunition and rapid-fire air guns for propulsion purposes.

The bullets used for achieving the effect are actually cast plastic shells that have been filled with dry colors or a combination of Vaseline and grease. When assembled, the two rounded half shells are pressed together and coated with lacquer. The "ammunition," unaffected by the compressed air used as a propellant, shatters on contact to produce amazingly realistic effects.

Perforated pipes, buried just below the surface of the soil, can also be used to send up spattering puffs of loose earth to simulate the effect of bullets striking the ground. Compressed air, blowing the earth away from each hole in turn, produces the harmless illusion.

On studio sets, bullet holes are predrilled into walls or furniture, filled with a camouflaged dowel, and packed with fine sawdust or similar material of the same coloring as the background. At the end of the "bullet's" flight, the dowel is sprung on cue to reveal the hole. The accompanying spray of dust makes the effect quite realistic.

Artillery fire or bomb bursts, for miniature or full-scale effects, can be duplicated harmlessly with low-base explosive material. Concealed in metal containers, they offer a degree of control of speed and direction. Ignited by means of electric contacts or fuses, the controlled charges go off on cue without endangering those in the vicinity of the blast. Subsurface explosions caused by mines and depth charges, for example, are produced in similar fashion. Vibration effects, produced by rocking the camera during filming, can be synchronized with the burst to add realism to the "shot." Several frames of black-and-white leader or preexposed positive and negative film, intercut immediately after the explosion, can add to the visual effectiveness of the burst. As an alternative, these few frames can be "burnt in" optically so that the scene is only partially visible.

On-stage recording microphones, in scenes that include explosions or blasts from artillery fire, must be shielded by wind screens. The

TABLE E-1 Exciter Lamp Specifications

ASA Code	Type number	Amperes	Volts	Base type	Bulb type
BSW	20A/T5SCP	0.2	7	S. C. Pref.	T-5
BTD	20A/T5/2DCP	0.2	7	D. C. Pref.	T-5
BAK	75A/T5SCP/1	0.75	4	S. C. Pref.	T-5
BFT	75A/S8SC	0.75	4	S. C. Bay.	S-8
BFX	75A/S8SCP	0.75	4	S. C. Pref.	S-8
BGB	75A/S8DCP	0.75	4	D. C. Pref.	S-8
BRD	75A/T5SC	0.75	4	S. C. Bay.	T-5
BRS	75A/T5SCP	0.75	4	S. C. Pref.	T-5
BRK	75A/T5/1	0.75	4	S. C. Pref.	T-5
BRX	1A/T5/5	1.0	6	D. C. Pref.	T-5
BSK	1A/T5/6	1.0	6	S. C. Pref.	T-5
BSS	1A/T5/7	1.0	6	S. C. Pref.	T-5
BXB	4A/T8/86	4.0	8.5	S. C. Pref.	T-8
BXE	7.5A/T8/SCB	7.5	10	S. C. Bay	T-8

Courtesy of Sylvania Electric Products, Inc.
Specifications and packing effective Mar. 1, 1966.
Life when lamp is held at rated amperes.

wire-mesh strainer-type coverings use fine silk to reduce the level of the sound and protect the vibrating diaphragm in the sensitive microphone.

Professional as well as amateur cameramen should check with local authorities before embarking on any filming venture that requires the use of explosive materials. In many areas, possession or use of even the most harmless type of explosive is forbidden by law.

expose (1) The process of producing a latent image on film by allowing light from the subject matter to expose the film's emulsion layer for a predetermined interval. The formula for exposure may be indicated in the following manner.

$$\text{Exposure} = \text{intensity of light} \times \text{length of exposure}$$

(2) The listing of the animated drawings or cels in their proper sequence on the exposure sheets. The exposure sheets are subsequently used as a guide by the animation cameraman during the filming process.

exposure Good print quality is not achieved with hit-or-miss arbitrary evaluations and assessments. Neither is exposure a matter of judgment or opinion. In photography, the number of influential

ated fe, ours	Light center length	Maximum overall length	Rated lumens	Carton quan.	Burning position	Bulb finish
50	1⅛	2⅜	13	10	Any	Clear
50	1⅛	2⅜	15	10	Any	Clear
00	1⅛	2	30	10	Any	Clear
50	1¼	2	32	10	Any	Clear
50	1⅛	2	32	10	Any	Clear
50	1⅛	2	32	10	Any	Clear
50	1¼	2	32	10	Any	Clear
50	1⅛	2⅜	32	10	Any	Clear
50	1⅛	2⅜	32	10	Any	Clear
00	1⅛	2⅜	70	10	Any	Clear
00	1⅛	2⅜	70	10	Any	Clear
00	1⅛	2⅜	70	10	Any	Clear
00	1⅝	3⅛	680	10	Any	Clear
00	1¾	3⅛	—	10	Base down	Clear

factors that directly affect the results makes each imprudent decision not only risky but costly as well. Experience is not an adequate substitute for methods and devices that remove the guesswork and accompanying risk from the photographic process.

Light meters, for example, available to the amateur and professional alike, assure maximum control and predictable results. Proper lens settings can be made simply and with a minimum of mathematical computation by merely pointing the sensitive instruments at light sources or subject matter.

One type of exposure meter, generally used for obtaining incident-light readings, measures the intensity of the light falling on the subject matter rather than the brightness of the subject matter itself. A second type of exposure meter provides the cameraman with reflected-light readings or the amount of light reflected from the subject matter when the meter is pointed in that specific direction. With the first type, generally referred to as an incident-light meter, the measuring device is held in front of the subject with the photocell facing the camera.

Although these sensitive instruments measure the intensity of light with amazing accuracy by means of photoelectric cells, readings are made on the basis of candlepower. Originally, this standard of illumination used the amount of light produced by a candle that burned exactly two grains of wax per minute. A footcandle is equal to the amount of light reflected from a 12- by 12-inch card placed so that every part of its surface is exactly one foot from the source of light.

The term *candle* has now been supplanted by the term *candela.* By definition, the candela is the luminous intensity of $\frac{1}{60}$ square centimeter of projected area of a black-body radiator operating at the temperature of the solidification point of platinum. The lumen, a unit of measurement, is based on this formula.

Obviously, objects that are closer to a light source are clearer and more distinguishable than objects that are farther away. Although the light source remains in the same position, the intensity of illumination on the object decreases proportionately as the object is moved away from the light. The formula that defines this relationship between light source and subject matter is known as the *law of inverse square.* Mathematically, the intensity of illumination varies inversely as the square of the distance from the light source.

Stated simply, an object exactly 1 foot from a light source receives four times the amount of illumination received by the same object at a distance of 2 feet from the light source. Subject matter that is 3 feet from a light source receives $\frac{1}{9}$th the amount of illumination, and at 4 feet only $\frac{1}{16}$th. In order to compensate for this geometric reduction in illumination, the object 2 feet away from the light source needs four times as much light as the object only 1 foot from the light, in order to provide the same exposure-meter reading. This basic guide for using and measuring illumination applies to both still and motion picture photography.

Although the terms overexposure and underexposure are usually associated with errors in lens settings, some special effects can be achieved by deliberate variations from normal settings. Flashes and simulating distant lightning can be easily produced by stopping the camera at irregular intervals during the filming process. Because of the extra time taken by the shutter during its buildup to normal running speed, the first few frames of film, after the camera has been started, receive more exposure than the frames that follow and are, therefore, lighter. When projected at normal viewing speeds, these lighter frames create the desired illusory effect.

Obviously, "hot" scenes can be exaggerated through overexposure. Jungle or desert scenes, for example, can be brightened considerably and made to appear uncomfortably hot and humid by simply opening the lens to a point one or two stops over normal. Similarly, underexposure can be used to produce night effects with daylight filming conditions. In other instances, underexposure is quite valuable for creating dramatic effects that help emphasize tense or menacing moods. See Figure E-7, DAY-FOR-NIGHT PHOTOGRAPHY, and EXPOSURE METERS.

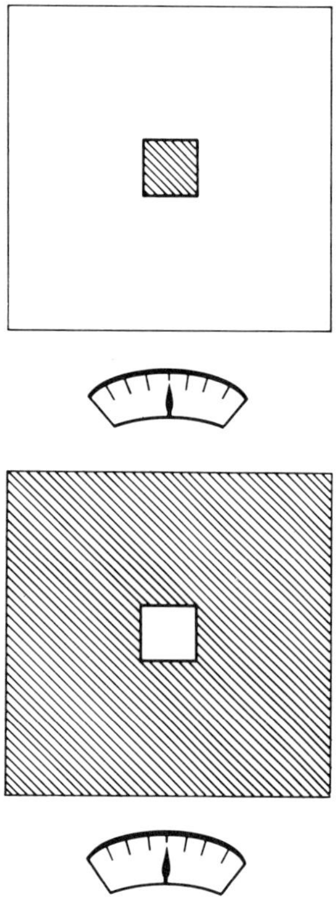

Fig. E-7 Exposure. Under similar lighting conditions, the meter will indicate the same reading regardless of the background color behind the subject matter.

TABLE E-2 Exposure Decrease

Shutter setting, degrees	Decrease in exposure, *f*-stops
180–170	Full exposure
165	
160	
155	
150	
145	¼
140	
135	
130	
125	
120	½
115	
110	
105	
100	
95	
90	
85	1
80	
75	
70	1¼
65	
60	1½
55	
50	
45	
40	2
35	2¼
30	2½
25	
20	3
15	3½
10	4
5	
0	Closed

exposure decrease When exposure must be decreased (either because there is too much light in the scene or because the film in the camera is too fast), and the lens aperture cannot be closed down sufficiently, the variable shutter can be used to obtain additional decreases in exposure. If, for example, the smallest lens aperture available is *f*/22 and the exposure-meter reading indicates a setting of *f*/32, the 170-degree variable shutter can be set at 85 degrees to obtain a full *f*-stop decrease in exposure.

Most variable shutters have a maximum opening between 170 and 180 degrees. Table E-2 is applicable within this range and shows the degree of shutter opening in relation to *f*-stops.

exposure factor Any influencing element that necessitates changes from normal exposure.

From a practical viewpoint, good exposure is the result of proper shutter speeds and correct lens-aperture settings. The first of these makes possible good print quality and stops movement, thus eliminating blur. The second increases the depth of field. The use of filters, the choice of lighting made necessary by extremely wide brightness ranges in color photography, and the several types of processing procedures used by different laboratories that influence the method of exposure are some of the factors that must be taken into consideration before filming begins.

exposure formulas

$$\text{Exposure time} = \frac{\text{shutter opening, degrees}}{\text{frames per second} \times 360 \text{ degrees}}$$

$$\text{Shutter opening, degrees} = \text{exposure time} \times \text{frames per second} \times 360 \text{ degrees}$$

As an example, if the shutter opening is 180 degrees at 16 frames per second, the exposure time is given by

$$\text{Exposure time} = \frac{180 \text{ degrees}}{(16)\ (360)} = \frac{1}{32} \text{ second}$$

exposure-index ratings Exposure-index ratings indicate the sensitivity of a film's emulsion. Arrived at by sensitometric measurements of a film's emulsion, the recommended ratings made by film manufacturers are based on standards set by the United States of America Standards Institute (the old American Standards Association, known by the initials as ASA). Because of the many methods used for processing the various types of film, it is often advisable to use the ratings only as a basis for exposure tests.

Since other systems for determining film-speed ratings use their own standards as a criteria, it is almost impossible to establish a mathematical relationship among the various systems. The figures in the table closely approximate the relationship. See Table E-3.

exposure meter—the principles Exposure meters are light-sensitive instruments capable of measuring the intensity of illumination falling on or being reflected from the subject matter being photographed.

All exposure meters are designed to make comparisons of density and scene brightness. Of the many types of metering devices available to the cinematographer (sensitized paper, calculators of various design, visual extinction and photoelectric types), the most favored generally is the photoelectric cell meter. This type of meter provides mathematically accurate measurements of the light striking the meter's sensitive cell. (The Weston V meter is used as an example, in the paragraphs that follow, for describing the functions of exposure meters generally.)

Measured in footcandles on an indicating dial, the brightness of the scene is translated into practical, usable terms of exposure by a second or calculating dial. This second dial is adjustable and can be set to provide accurate readings for different film types of any speed.

TABLE E-3 Exposure-index Ratings (Also Emulsion-speed Values)

ASA*	Weston (older meters)	General Electric	Scheiner	Hurter & Driffield	DIN†
1	0.7	1	11	17.5	1/10
1.2	1.0	1.5	12	25	2/10
1.6	1.2	2	13	30	3/10
2	1.5	2.5	14	38	4/10
2.5	2	3	15	50	5/10
3	2.5	4	16	63	6/10
4	3	4.5	17	75	7/10
5	4	6	18	100	8/10
6	5	8	19	125	9/10
8	6	10	20	150	10/10
10	8	12	21	200	11/10
12	10	16	22	250	12/10
16	12	20	23	300	13/10
20	16	24	24	400	14/10
25	20	32	25	500	15/10
32	24	40	26	600	16/10
40	32	48	27	800	17/10
50	40	64	28	1,000	18/10
64	50	80	29	1,250	19/10
80	64	100	30	1,600	20/10
100	80	125	31	2,000	21/10
125	100	150	32	2,500	22/10
160	125	200	33	3,120	23/10
200	160	250	34	4,000	24/10
250	200	300	35	5,000	25/10
320	250	400	36	6,250	26/10
400	320	500	37	8,000	27/10
500	400	600	38	10,000	28/10
650	500	800	39	12,500	29/10
800	650	900	40	16,250	30/10
1,000	800	1,000	41	20,000	31/10

*ASA ratings are now universally used and are the same as the BSI (British Standards Institute), Kodak, and Ansco exposure values.

†DIN is the abbreviation for Deutsche Industrie Norm (German Standards Organization).

Although meters vary greatly in design, the procedure for taking readings is similar, generally. Most meters have two sliding light-value scales that provide correct readings under extremely bright or very dim light conditions. Hinged baffles automatically change scales. When the baffle is closed, the high-light scale moves into position. If the light reading is below a certain point on that scale, the baffle should be opened for more accurate readings on the extended low-light scale.

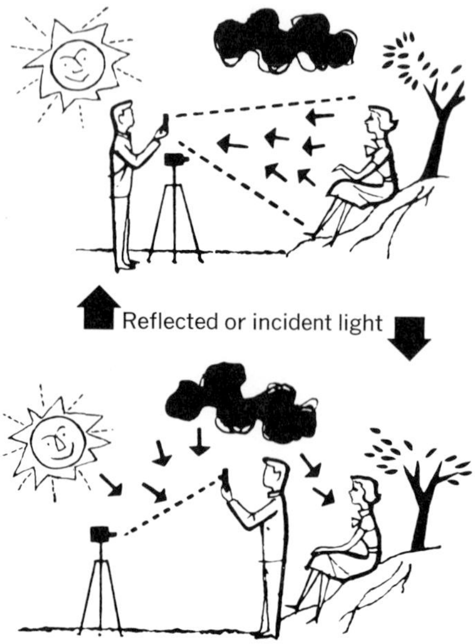

Fig. E-8

reflected or incident light. Reflected light is light reflected from the subject to the camera. Incident light is the light that falls on the subject. Measurement of either reflected or incident light can be used to determine correct exposure. With the addition of an Invercone attachment (in the Weston meter), or hemispheres or plates in other meters, reflected-light meters can be used to measure incident light. For certain types of photography, such as portrait or copy work, incident light will be found quite convenient, but for the majority of photographic assignments, reflected light is more suitable. See Figure E-8.

aiming the meter. Obviously, when the meter is being aimed at the subject matter, the meter's photocell must not be obstructed either by the holder's fingers or by a dangling neck cord. In outdoor scenes, when the reading is taken from the camera position, the meter should be tilted at an angle slightly downward so that the line of sight passes over the front edge of the meter. This positioning excludes sky areas which tend to inflate the reading and cause underexposure.

camera position method for determining exposure. The arrow on the exposure control dial should be set to the light reading obtained. Any combination of *f*-stop and shutter speed opposite each other may be selected.

close-up method for determining exposure. This method should be used for scenes where there is but one subject of interest and the background is of relatively little importance. In general, the meter reading should be taken from a position close to the subject matter. The normal arrow on the exposure control dial is set to this reading. Any combination of *f*-stop and shutter speed opposite each other may be selected. When the meter reading is taken from a person's face, the C (for contrast) position on the dial is set to the reading instead of the normal arrow. The camera settings are then selected in the usual manner.

brightness-range method for determining exposure. This is the most accurate method for determining the correct exposure of scenes consisting of a wide range of bright and dark light values. Two close-up readings should be taken: one for the darkest object and one for the brightest. In color photography, black and white are not considered colors and should not be measured. The normal arrow is set midway between the two values measured. This gives an average exposure.

When necessary, the person taking the reading may use the palm of his hand as a substitute for the subject's face or skin tones. Under those circumstances, the arrow on the exposure control dial is set to the light reading. The meter's C position is used when reading flesh tones.

For correct exposure, the cameraman must first determine the exposure index of the film from the manufacturer's data sheet, and the constant shutter speed of the camera at the normal framing speed. The exposure index is set in the exposure-index window. The meter is then aimed at the scene, and the reading is noted. The meter reading is the point at which the meter arrow is set on the light scale of the exposure control dial. The correct *f*-stop is found just above the camera shutter speed.

Assume, for example, that the camera operates at $\frac{1}{30}$ second at 16 frames per second, an exposure index of 10 and a light reading of 13. With the meter arrow set at 13, the correct *f*-stop (*f*/8) is found directly above the $\frac{1}{30}$ second on the shutter-speed dial. Therefore, the scene should be filmed at 16 frames per second at *f*/8.

increasing or decreasing exposure. It may be necessary either to increase or to decrease the indicated exposure for special effects or under unusual light conditions. Since the *f*-stop scale markings are one full *f*-stop apart, the exposure can be increased one full *f*-stop by turning the outer control dial to the next lower number. To decrease the exposure, the dial should be turned to the next higher number.

exposure for special effects and unusual conditions

Snow, Beach and Water Scenes. Take readings of the brightest and darkest objects, and use the *f*-stop midway between, or take a reading from the palm of the hand and use the C position on the control dial. Best snow texture results when the snow is back or cross-lighted.

Copy Work. When copying pages of a book or photographs in black and white or color, take a reading from a white card placed over the subject. Divide the exposure index by 5, and set this value in the exposure-index window. Point the arrow at the light reading obtained, and select the camera settings.

Television. Adjust the television screen for contrasty black and whites. Set the shutter at $\frac{1}{25}$ second, and focus on the lines across the screen. Dim the room lights. Take a close-up average reading, holding the meter about 6 inches from the screen. Set the meter arrow at this reading, and select the camera settings.

Aerial Pictures. To prevent the sky from inflating the reading, aim the meter down toward the ground. In general, below 1,000 feet use the meter reading indicated; from 1,000 to 2,000 feet stop down $\frac{1}{3}$ *f*-stop; from 2,000 to 4,000 feet close down $\frac{1}{2}$ *f*-stop; above 4,000 feet close the camera aperture one full stop from the reading. A skylight filter should be used with daylight color film. No exposure correction is required.

Overcast Scene. This type of scene frequently lacks contrast owing to diffused light and the absence of shadows. In such situations, black-and-white pictures can be improved by using the A position (A is for absence of contrast) on the meter, rather than the normal arrow when scene brightness is measured from the camera position.

Key and Fill Light. For normal results in studio photography, the lighting contrast range should be limited to 8:1 for color, and 64:1 for black and white. This means that the key light for color should produce no more than 8 times the light falling on the subject than the fill lights, and for black and white the key light should produce no more than 64 times the fill light.

To determine the ratio of key light to fill light, insert the Invercone in the exposure meter, stand at the subject's position, and measure the light produced by the key light by aiming the Invercone at this light, with all other lights turned off. Then, turn off the key light, and measure the light produced by the fill light. If the ratio is greater than the suggested limits, move the lamps near to or farther away from the subject, as required.

Contrasty Scenes. Occasionally, scenes metered from the camera position will be too contrasty. Exposure for black-and-white pictures can be improved by doubling the exposure indicated by placing the C position opposite the brightness reading.

Absence of Contrast. For scenes with little or no contrast, metered for black-and-white film from the camera position, set the A on the meter opposite the reading obtained. This will halve the exposure and result in a better print.

Available Light. Occasionally, scene brightness will be too low to obtain a reading from the camera position. In those situations, take a close-up reading from the brightest object in the scene. If no object in the scene is bright enough to produce a reading, substitute a piece of white paper or a handkerchief adjacent to the brightest object, and take a close-up reading from those objects. Set the O position to the light reading obtained, and select the camera settings.

Scene Depth. Proper lens settings and corresponding shutter speeds can be combined for the purpose of achieving scene depth. As an alternative, a suitable shutter speed needed for stopping an action can be selected and then used to determine the lens setting. Settings of *f*/11 at $1/25$ second, for example, will provide exposures similar to those produced with a lens setting of *f*/8 at $1/50$ second. Although greater scene depth will be achieved with the *f*/11 setting at $1/25$ second, the *f*/8 setting at $1/50$ second will stop faster action.

Most of the professional cinematographers measure the light falling on the subject (incident light) by aiming the meter at the source of light—especially when the level of lighting is low. In order to determine the footcandles of light falling on a relatively small area of a set or stage, a reading should be taken from a piece of white matte paper placed over that portion of the set. Table E-4 can be used to convert the readings.

R Readings
B Brightness equivalent, candles per square foot
FC Approximate footcandles

TABLE E-4 Weston V Exposure-meter Readings

R	B	FC	R	B	FC	R	B	FC
16	1,600	6,400	11	50	200	6	1.6	6.4
—	1,300	5,200	—	40	160	—	1.3	5.2
—	1,000	4,000	—	32	128	—	1	4
15	800	3,200	10	25	100	5	.8	3.2
—	650	2,600	—	20	80	—	.65	2.6
—	500	2,000	—	16	64	—	.5	2
14	400	1,600	9	13	52	4	.4	1.6
—	320	1,280	—	10	40	—	.32	1.28
—	250	1,000	—	8	32	—	.25	1
13	200	800	8	6.5	26	3	.2	.8
—	160	640	—	5	20	—	.16	.64
—	130	520	—	4	16	—	.13	.52
12	100	400	7	3.2	12.8	2	.1	.4
—	80	320	—	2.5	10	—	.05	.2
—	65	260	—	2	8	—		

Some other Weston exposure meters have the brightness values expressed directly in candles per square foot. For those meters, the brightness values should be multiplied by 4 in order to obtain the approximate footcandle values. Note that in the above table the numbers in the brightness column when multiplied by 4 equal the amounts in the footcandles column.

Elwood Foto-Meter. This meter is designed to read very small areas of light on the subject from any distance or position. It does not measure wide angles of scattered light and is unaffected by background lighting. It is powered by miniature mercury batteries which have an approximate two-year life.

EPOI Spotron CdS spot exposure meter. The spot meters produced by the Spotron Division, Ehrenreich Photo-Optical Industries, measure the brightness of small areas, even at a distance, for determining accurate exposure and subject contrast. The Pentaview has a 10-degree acceptance angle, equal to the angle of a 250mm telelens on a 35mm camera and provides accurate sensitivity over a wide range of light values. Prism finders show the subject matter upright and unreversed.

EPOI's professional model has an ultraselective 2-degree acceptance angle. At 25 feet, the meter measures an area only $10\frac{1}{2}$ inches in diameter. Independent high-low sensitivity circuits cover the extreme range of light values.

Fairchild high-speed photography meter. Designed for high-speed photography, this meter is of the reflectance type with an acceptance angle of $1\frac{1}{2}$ degrees. Meter focus is set for nominal subject distance of 6 feet. Exposure ratings for black-and-white and color film are

TABLE E-5 General Information on Illumination Meters and Footlambert Meter

Model	Type	Size, inches	Scale length, inches	Accuracy (See Note 1)	
				<1,000 ft-c	>1,000 f
614	Illumination	3⅝ W 7⅜ L 2¼ D	2.37	±7.5	10
703	Illumination	2⅝ W 4⅝ L 1 5/16 D	1.43	±7.5	10
756	Illumination	5¼ W 6⅛ L 3½ D	4.06	±5	8
759	Luminance	5¼ W 6⅛ L 3½ D	4.06	See Note 2	

Courtesy of Weston Instruments Inc.

W-width; L-length; D-diameter.

Notes:

1 Accuracy expressed in % full scale.

2 Luminance 600 ft-L or less ±5% of full scale except ±3% of reading correction curve. Luminance greater than 600 ft-L ±8% of full scale except reading with correction curve.

3 These light levels not recommended for prolonged exposure.

read directly on the calculator in terms of lens *f*-stop openings as related to picture-taking rates (frames per second). The calculator has an integrated light-value scale calibrated for subject incident illumination levels of 0 to 500,000 footcandles. A high and low scale separately calibrated for 0 to 25,000 and 0 to 500,000 footcandles is provided. ASA film-index values of 2.5 to 20,000 and picture-taking rates from 100 to 20,000 pictures per second are incorporated in the calculator.

Fastax high-speed photography meter. This meter is made for Wollensak by Weston Electric Instrument Corporation. Because of the extreme speed of high-speed cameras, conventional exposure meters cannot be used. This special high-speed meter is made exclusively for use with high-speed cameras. The meter has ranges of 0 to 3,000, 0 to 30,000, and 0 to 300,000 footcandles. These ranges cover all light conditions from normal sunlight to extremely high artificial-light levels, such as are obtained with multiple banks of high-intensity spotlights.

General Electric Golden Crown. The PR-3 meter reads reflected light. An incident-light attachment is optional. Direct reading is in *f*-stops. The pointer may be locked on the reading. The Dynacell

aximum no. of anges	Minimum ft-c ranges available				Maximum ft-c (See Note 3)
	Glass or quartz	Viscor	Cosine	Viscor and cosine	
3	10	25	30	100	15,000
1	5	10	20	50	1,500
9	2	5	7.5	20	15,000
2		60 ft-L			1,200 ft-L

light booster may be attached for very dim light. Incident-light scale equivalents may be transposed to footcandles.

Gossen Luna-Pro CdS exposure meter. This meter extends the capabilities of the Lunasix by mastering exposure problems on location and in the studio, darkroom, and laboratory. Four optional lock-on attachments convert the Luna-Pro from a reflected-and-incident meter to a variable-angle spot meter, enlarging meter, photomicrography meter, and ground-glass reading meter. The meter measures reflected and incident light (with a built-in hemispheric diffuser) and has a 30-degree light-acceptance angle, a single seesaw selector button for high and low range, and an automatic memory needle lock.

Gossen Lunasix Electronic. The Lunasix measures reflected and incident light (with a built-in hemispheric diffuser). The meter has a 30-degree light-acceptance angle, two-button brightness-range system, automatic needle lock, built-in battery tester, and external zero adjustment. It is calibrated for both motion picture and still cameras. Its performance is attained through an electronic circuit which incorporates a specially designed photoconductive cadmium sulfide cell powered by a long-life mercury battery.

Gossen Super Pilot CdS exposure meter. This meter rates next to the Luna-Pro in range and sensitivity. The meter has two full ranges—normal to low light, ASA 6 to 12,500, *f*/1 to *f*/45, 1/1,000 second to 2 hours; EV −6 to EV +24, Cine from 8 to 128 frames per second.

Gossen Sixticolor temperature meter. This meter can be used outdoors or under artificial light. It reads from 2600 to 20,000°K directly in Kelvin degrees or in microreciprocal degrees (mired values). The meter may be preset for any type of color film. The color-temperature reading is locked in and automatically indicates the filter necessary for color correction.

Gossen Pilot precision exposure meter. The meter has a $1\frac{1}{2}$-ounce jeweled movement. The meter reads incident or reflected light and has a built-in diffuser. ASA 6 to 6400, 1 minute to 1/1,000 second, *f*/1.4 to *f*/22, EV 1 to 18.

Gossen Tri-Lux footcandle meter. The Tri-Lux linear scale assures accurate readings over a range of 0 to 12,000 footcandles—lowest scale reading is 0.2 footcandle. It is indispensable for the measurement of incident light to determine light distribution, to achieve even illumination, to measure lighting contrast, or to calculate exact exposure. The ultrasensitive photoelectric cell is housed in a convenient probe with a handgrip, and connects to the microgalvanometer by an extension cord for flexibility and measuring ease.

Gossen Tri-Lux model C. This meter reads incident light (the reading is taken with a flat disk on the remote handle). The flat disk is wired into the meter case. The direct-reading footcandle scale has two rows of numbers.

Gossen Sixtomat X-2. This meter reads reflected light. Incident light is read with a built-in diffusion screen. Reflected or incident light can also be read directly.

Jaro-Salford. This British meter reads reflected or incident light. (Component parts are interchanged.) Direct incident-light readings are in footcandles: 0 to 200, 0 to 2,000, and 0 to 20,000. A computer is used to read reflected light. The meter may be used in several combinations by varying the assembly of the component parts: meter unit, cell unit, B/1 unit, range-change hood, and extension unit.

Kelvilux color-temperature meter. The meter evaluates Kelvin readings for proper filter selection from 2000 to 10,000°K, indoors or outdoors. The meter is in two units. The main body has a built-in zero galvanometer connected with the measuring head by an extension cable. The measuring head has two photoelectric cells which are exposed through an opal glass and colored filters. The three-position switch permits adapting the instrument to prevailing light intensities.

Norwood Super Director, Walz. The meter reads incident light with a three-dimensional Heliosphere and reflected light with a Heliogrid. Direct reading is in *f*-stops for either incident or reflected light. The Heliovalve may be set for any combination of ASA exposure index and shutter speed. Two Heliovalves are available for covering all photographic situations. Valve AA-K handles average light conditions. Valve L-P is used with ultrasensitive film and very low light levels.

Rebikoff color-temperature meter. The meter can be used outdoors or under artificial light and reads degrees Kelvin from 2000 to 10,000. The photoelectric cells are covered by red and blue filters coupled to microammeter with needle. Direct mechanical reading from the light source. Degrees Kelvin is indicated by an arrow. The filter is selected from the filter chart.

Sekonic studio exposure meter. The meter reads both reflected and incident light. It is equipped with a three-dimensional light collector and three direct-reading slides and attachments. Readings from 4 to 32,000 footcandles are possible with a white disk. ASA 0.1 to 12,000, *f*/1 to *f*/45, 60 seconds to 1/1,000 second, and direct-reading exposure dial.

Sekonic Microlite exposure meter L-88. The meter utilizes a match-head-sized cadmium sulfide photoresistor cell and a dime-sized mercury battery that provides service for well over a year. The meter gives accurate readings under very low light levels. A dual-range system provides accuracy under the entire range of lighting conditions from dimly lit rooms to sun-drenched beaches. ASA indexes from 6 to 12,000, *f*/1 to *f*/32, 16 to 1/2,000 seconds, direct-reading movie speeds from 8 to 64 frames per second, and unit-value film speeds from 1 to 12 degrees.

Spectra Professional Gold Seal. The meter reads incident light with a three-dimensional Photosphere. Incident light is read in footcandles with a flat Photodisk and reflected light with a Photogrid. Direct reading is in either *f*-stops or footcandles (dual scales). Each meter is individually calibrated. 13 ASA exposure-index slides (calibrated in $\frac{1}{50}$ second or any other desired shutter speed). Computer-scale operation is also provided.

Spectra Combi-500 professional exposure meter. The meter combines a selenium cell and an ultrasensitive photoconductive cell. Its sensitivity is 500 times that of the average selenium cell meter, 100 times more than the highly sensitive Spectra professional meter. Exposures are indicated for extremely low light levels. Under normal lighting conditions, the standard selenium cell operates without the aid of the battery. The Spectra professional meter can be converted to the Combi 500. See Figures E-9*a* and *b*.

Fig. E-9*a* The Spectra Photospot.

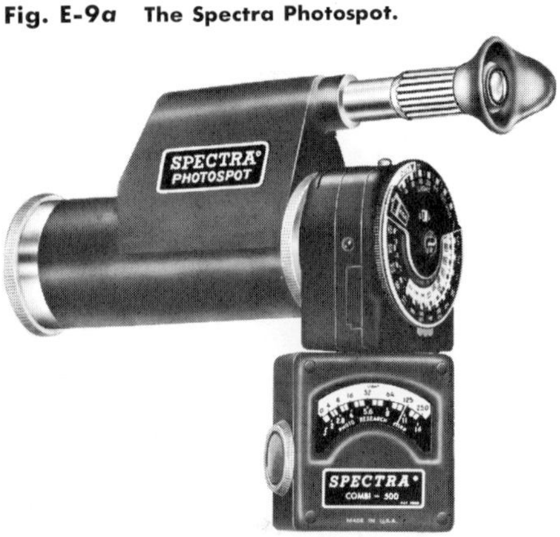

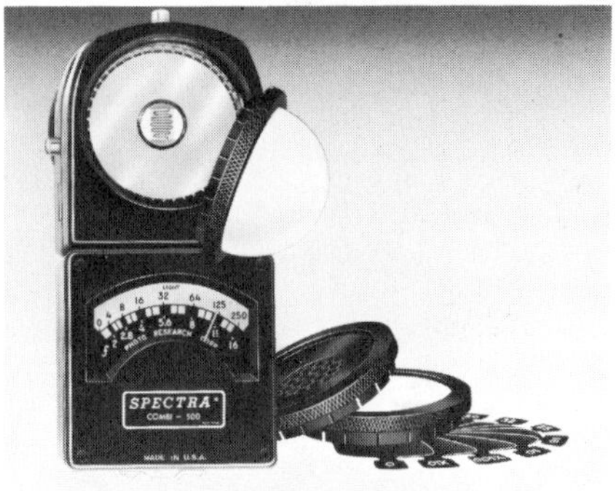

Fig. E-9*b* The Spectra Combi-500.

Spectra brightness spot meter. This meter measures the brightness of a small spot from a remote location. This self-contained instrument reads directly in footlamberts having a spectral response closely approximating that of the CIE luminosity curve. The meter also provides accurate data for many specialized applications such as measuring the uniformity of illumination on projection screens from any distance or angle (independent of color response of the observer).

The objective lens of the meter may be focused from 4 feet to infinity. The unit can be mounted on 1/4-20 tripod socket. Full-scale readings of 50, 500, 5,000, and 50,000 footlamberts. Scale multipliers: ×1, ×10, ×100, and ×1,000.

Spectra brightness spot meter C-3. The C-3 is designed for portable use and speed in measurement at a lower cost. It reads directly in footlamberts having a spectral response closely approximating that of the CIE luminosity curve. Human error is entirely eliminated.

Spectra three-color meter. For indoor and outdoor use, the meter reads directly on two scales: blue-red and red-green in Spectra index units. The meter has a table for conversion from 2000 to 30,000°K, or can be read in direct Kelvin degrees (on optional scale). A calculator indicates the filters required for correction for color temperature and additional green correction. It may be used with all types of color film. The meter has a pistol-grip housing and an inclined dial for easier reading. Models covering shorter ranges with direct-reading Kelvin scales are available.

Spectra micro- and millicandela. These footcandle meters measure the incident light falling on a cosine-corrected receptor disk. The seven full-scale measuring ranges are controlled by a switch. The basic difference between the microcandela and the millicandela is in the degree of sensitivity, the microcandela being 1,000 times more sensitive.

Spectra candela X-100. This pocket-sized footcandle meter incorporates a selenium photovoltaic cell and a CdS photoconductive cell. The photovoltaic cell measures normal and high light levels. The CdS photoconductive cell measures low light levels. A switch built into the side of the meter makes changing from high- to low-level ranges an instantaneous process. The meter provides six full-scale measuring ranges—from 1 to 50,000 footcandles.

Spectra screen-brightness meter. This metering device measures the brightness of a projection screen in footlamberts. The spot meter provides a 5-degree viewing angle and has three ranges—0 to 25, 0 to 250, and 0 to 2,500 footlamberts. It is powered by batteries, and meter readings are taken from a position beside the projector.

Spectra TV Optoliner. For use in production facilities, quality-control operations, or standard labs, the Optoliner contains a constant, adjustable light source and a special meter to indicate the exact illuminance and color temperature falling on the face of the camera tube.

This high-resolution, precision TV camera tester replaces the elaborate test equipment formerly used for this operation. The Optoliner attaches directly into the camera lens mount and provides microscopic alignment (within 0.002 inch) of the slide-mounted test patterns to the center of the camera lens.

Weston Master IV. The model 745 meter reads reflected light. Incident light is read with the supplied Invercone attachment. A swinging baffle automatically changes high- and low-light scales. An on/off button locks the pointer and the meter's four positions, U,

O, A, and C, cover brightness range and contrast situations with color and black-and-white film.

Weston footcandle meter. The model 614 reads incident light only with a flat disk. The meter has low, medium, and high ranges. Direct reading is in footcandles only. The Photronic cell is mounted on a hinge so that it can be set at an angle for easy reading. The meter is supplied calibrated in any desired range.

Weston sight meter. The model 703 is a pocket-size direct-reading incident-light meter. Use of the sight meter multiplier disk extends the range to 10 times its normal full-scale value.

Weston Master V. The versatile model 748 exposure meter can be used to measure reflected light as well as incident light. (The latter reading is made with the Invercone attachment.)

The combination of sensitive selenium photoelectric cells and meter movements gives measurements in any usable light conditions. Low- and high-light scales allow accurate measurements from 0.05 to 1,600 candles per square foot. A cell baffle automatically changes the scales. Shutter speed and *f*-stop combinations for all cameras indicate limits of black-and-white and color films, permit negative density control, and aid in compensations for the light-reducing effect of filters.

Although the meter can be used conventionally, with free pointer movement, an on/off button switch permits locking the pointer when taking incident-light readings, in extremely low light locations or whenever required. A one-piece, snap-on light converter comes as standard equipment for incident-light measurements. Exposure indexes cover ASA to 16,000, DIN to 42, and exposure values (Polaroid) to 18.

Zoomar SEI. A versatile exposure meter that measures reflected light by the comparison method. Direct reading can be made in footlamberts without the necessity of conversion into footcandles. The meter has a brightness range from 1 to 1,000,000. Standard brightness for comparison is controlled by the photoelectric cell and microammeter.

exposure sheets, animation Exposure sheets contain a record of every frame of film in an animated scene. These sheets show the order in which every cel in the scene is to be photographed, the position of the background, the camera effects, and the dialogue in the sound track. The information, listed originally by the animator, is used as a guide by the animation cameraman during the filming process. See Figure E-10.

extension tubes A cylindrical tube or fitting that increases the distance between the camera and the lens. It is especially useful with telephoto lenses for producing larger images.

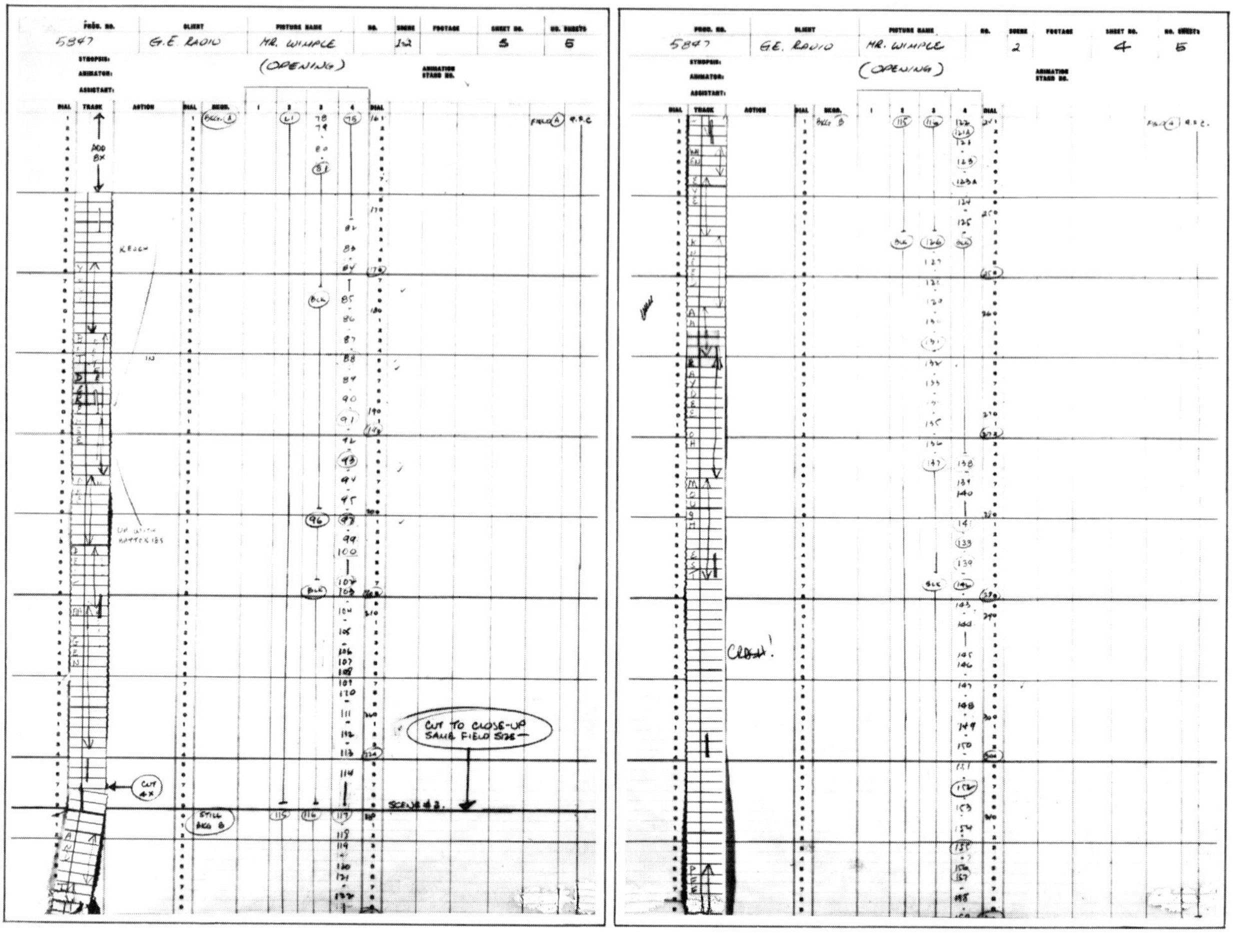

Fig. E-10 **Exposure sheets.**

Extension tubes permit extreme close-ups with regular fixed-focal-length lenses. Extension tubes are more versatile than close-up lenses since they provide greater magnification without decreasing the optical quality of the lens. Extension tubes can be used with any focal-length lens, but their use is usually limited to standard and telelenses.

The same magnification can frequently be obtained with different focal-length lenses by changing the length of the extension. For that purpose, the choice of lens should be based on the desired working distance. The longer the focal length of the lens, the longer the distance from the camera to the subject.

exterior shot A scene photographed either on location or on a studio stage that is designed to simulate an outdoor setting.

extinction meter The principle behind this type of metering device is that detail in the scene being viewed is lost or becomes extinct at certain lens settings. Exposure readings are determined by viewing the scene through a calibrated density wedge which is positioned between the eyepiece and camera lens or by gradually closing the diaphragm until the scene's detail, seen on a small screen in the metering device, is no longer visible. The lens opening, at the point of extinction, is used as the basis for estimating* exposure (in conjunction with a conversion table).

extra A performer best described as a part of the background.

extremes The key drawings in an animated action. These drawings contain spacing guides and other instructions that serve as a guide for the assistant animator or in-betweener who make the intermediate drawings that are needed to complete the animated action.

eye light Subtle illumination designed to produce reflections from the performer's eyes or teeth without affecting the general lighting on the set.

eyepiece An attachment linked to the camera's optical system that permits the cameraman to appraise the composition of a scene before filming begins.

*The words *estimating exposure* are used because it is entirely possible for two cameramen to view the same scene and obtain completely different readings. For this reason, the cameraman using this type of meter should make his own calibration scale based on his visual acuity.

F

f The symbol for focal-length factor. It is an index of a lens aperture in relation to its focal length.

f-stops Settings that indicate the amount of light transmitted through a lens. Increasing the aperture by one *f*-stop doubles the amount of light transmitted through a lens. Decreasing the aperture by one *f*-stop halves the amount of transmitted light.

f-stop calibrations are computed by dividing the focal length by the aperture diameter of the lens. Established by the International Congress of Photography around 1900, the *f*-stop system represented the first attempt at standardization. The computations used for developing this system, however, did not take into account the amount of light lost during exposure by reflection and absorption. Neither does the system take into consideration the fact that it is possible for one lens to transmit a greater percentage of light, at the same *f*-stop, than another lens (with the same general characteristics) is capable of transmitting. Many professional cameramen score their own *f*-stop calibrations on lenses based on tests which indicate the actual amount of light that is transmitted by that particular lens.

A lens rated at *f*/1 indicates that the lens will have a diameter of 1 inch for every inch of focal length. A 6-inch lens with a rating of *f*/1 will, therefore, have a diameter of 6 inches. Conversely, a 6-inch

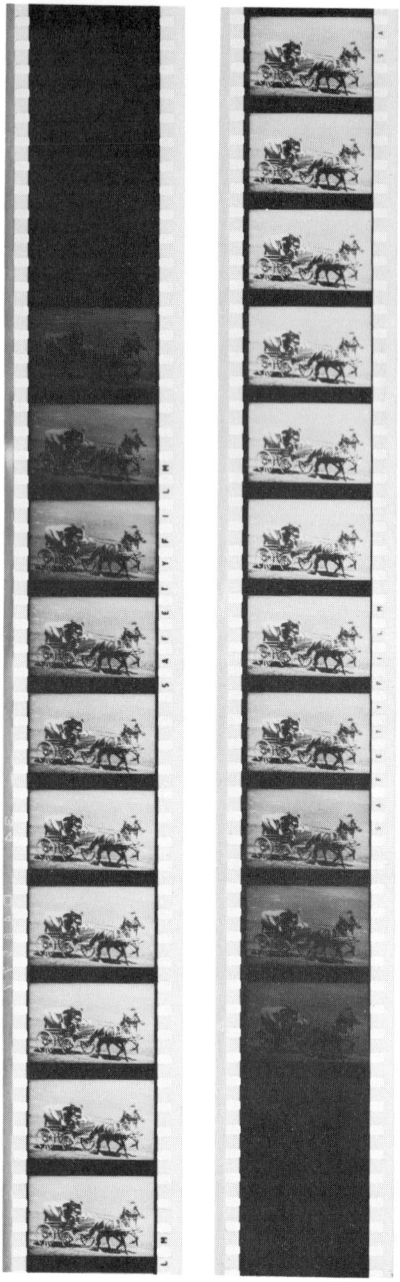

Fig. F-1 Fades. A 12-frame fade-in (left) and a 12-frame fade-out (right). Note that the frame preceding the actual fade-in is completely black, as is the frame that immediately follows the last frame of the fade-out.

lens with a rating of *f*/2 will have a 3-inch diameter (2 inches of focal length for each inch of diameter). Although the diameter of the lens has no influence on the size of the image, lenses with a smaller diameter will produce sharper but proportionately less contrasty pictures. Lenses with larger diameters will produce brighter pictures with less definition.

factor, DIN A German standard for rating the sensitivity or the speed of a film's emulsion. The term "DIN" is derived from the initial letters of the phrase **D**as **I**st **N**ormal. See EXPOSURE-INDEX RATINGS.

factor, H & D A basis for measuring the inertia of a film's emulsion. The findings are subsequently translated in terms of gamma. H & D are the initials of Hurter and Driffield, the scientists who developed the system. See EXPOSURE-INDEX RATINGS.

factor, Scheiner A universal system for designating the degree of sensitivity or the speed of a film's emulsion. *See* EXPOSURE-INDEX RATINGS.

factor, Weston A numerical system for rating the speed of an emulsion. Based on tests using Weston exposure meters, these speed ratings can be related to other numerical systems used for this purpose. See EXPOSURE-INDEX RATINGS.

fade An optical or camera transitional effect in which each successive frame of film receives proportionately less exposure until the final frame of the scene is completely black. For a fade-in, each frame receives proportionate increases in exposure until the scene's density and tonal values duplicate those in the original subject matter. See Figure F-1 and TRANSITIONAL EFFECTS.

Fahrenheit thermometer An instrument used for measuring temperature. The freezing point of water on the Fahrenheit scale is 32 degrees. The boiling point of water is 212 degrees. Compare CENTIGRADE THERMOMETER and KELVIN DEGREES.

Farmer's reducer Potassium ferricyanide. A reducer used in laboratory processing. It is named after its originator, E. H. Farmer.

Farnsworth dichotomous test A method of determining the degree of color blindness in individuals. It detects both functional color blindness and moderate deficiencies in color perception.

Farnsworth-Munsell 100-hue test The 100-hue method of testing for determining the degree of color blindness is designed for those individuals who are expected to make accurate, practical, everyday decisions and judgments about color. The test is especially useful for art directors, scenic designers, and graphic artists.

favoring shot A scene in which one of the actors in a group is photographed facing the camera.

featurette Also called **short**. Relatively short motion pictures with a running time of 5 to 20 minutes, approximately. Such films cover a wide variety of subjects and usually precede the showing of the featured motion picture. The popular travelogue series is included in the general category.

feedback The distortion caused by sound waves from an interconnected loudspeaker system reaching a microphone and being reamplified and rebroadcast. If the condition is sufficiently serious, the circuit begins to oscillate, or "howl." Reducing the volume of amplification usually improves the situation.

feed reel The metal or plastic reel on which the film or tape is originally wound.

ferrotype Also referred to as **tintypes**, an outdated photographic process that produced images on black enameled-iron plates. The pictures were reversed—right and left—as in a mirror image.

fidelity The accuracy with which sound or picture is reproduced. Fidelity, with respect to sound, requires reproduction of the entire range of audible frequencies present in the original signal, with the same relative relationship among them, a minimum of distortion, and a reasonable re-creation of the original dynamic range.

In motion pictures, fidelity refers to the accurate reproduction of the full range of tonal values in a black-and-white image, from black through intermediate shades of gray to white. In color photography, fidelity requires the accurate reproduction of each of the colors and hues found in the original subject matter.

fiducial marking An optical system used between the camera and the taking lens that records, simultaneously, the subject and reference marks (usually cross hairs) on each frame of film. The image from the objective lens is focused on the planosurface of a collective lens on which the reticle cross-hair pattern is engaged. A 1:1 lens system then relays the image and reticle pattern to the film plane. (Special reticle patterns are available from Fastax.) These systems are not recommended for use with Fastax lenses shorter than 1-inch focal length.

In addition to their value in survey work, these cross-hair reference points have become familiar symbols in photographs of the moon that have been taken from satellites and transmitted back to earth.

In continuous projectors that are used for projecting motion picture film for television scanning, the continuously moving film is projected onto the target of a vidicon tube at the rate of 30 frames per second. Light from xenon flash lamps, directed onto the target area, is used to illuminate the picture being projected. The short duration of the flash, triggered by a transparent fiducial mark on the film, stops the motion of the projected picture to permit scanning.

field, animation In animation, a numbered guide on a transparent sheet of Celluloid is used by the cameraman for composition purposes as well as for positioning animation or artwork in relation to the camera. The abbreviation F (field) is usually preceded by a number that indicates the size of the field. 9F, for example, indicates that the width of the field is 9 inches.

field, lens The horizontal limits (measured in degrees) a lens is capable of "seeing" and, consequently, the area that will be exposed on the film in the camera. Depth of field, by comparison, is the range of distances from the camera within which objects are in acceptable focus. See FIELD OF VIEW.

field, television One-half of the transmitted television picture. Two interlaced fields are equal to one frame of picture.

During the scanning process, the odd-numbered lines of the 525 lines that make up the complete picture are scanned first by the electron beam. The electron gun swings back to the top of the screen during the blanking interval and then scans each of the even-numbered lines in turn to complete the composite picture. See ODD-LINE INTERLACING SYSTEM and Figure 0-1.

field angle In lighting, the field angle of a light source refers to the angle between those points of a candlepower curve where the candlepower is 10 percent of the maximum candlepower output of the source. See COVERAGE, LIGHTING.

field coverage See COVERAGE, FIELD.

field frequency The rate at which a complete field is scanned electronically, actually one-half of a composite picture. The field-frequency rate in the United States is 60 fields per second. The standard rate in Europe is 50 fields per second. See FIELD, TELEVISION.

field lens The lens in a camera eyepiece that is nearest to the objective. The lens closest to the eye is referred to as the *eye lens.*

field lumens See LUMENS, FIELD.

field pulse Magnetized impulses, along the edge of a length of videotape, that are used as a splicing guide by the editor. Compare TALKING CLOCK.

field of view The area "seen" by a lens, expressed in degrees. Wide-angle lenses, for example, cover a wider expanse of scene than the field of view covered by other lenses.

Photographic lenses are manufactured for every possible type of filming assignment and are designed to cover fields ranging from an angle of approximately 5 degrees to the 180-degree coverage offered by a fisheye lens. Lenses designed to give best results for specific fields can, of course, also be used for other fields. The results, however, are not always satisfactory. A lens designed for a 50-degree angular field, for example, will give good results over a 30-degree field. A 30-degree-angle lens will not, by comparison, provide satisfactory results when used over a 50-degree field. See ANGLE OF VIEW.

Lens	*Angular coverage, degrees*
Microscope and telescope lenses	5–15
Telephoto lenses	15–30
Motion picture lenses	20–50
Wide-angle lenses	60–100
Peripheral vision (human eye)	120
Ultra-wide-angle lenses	150
Fisheye lenses	180

filament A metal conductor, usually tungsten, through which an electric current is passed, thereby causing it to become incandescent. These filaments are housed in glass envelopes. See Figure F-2.

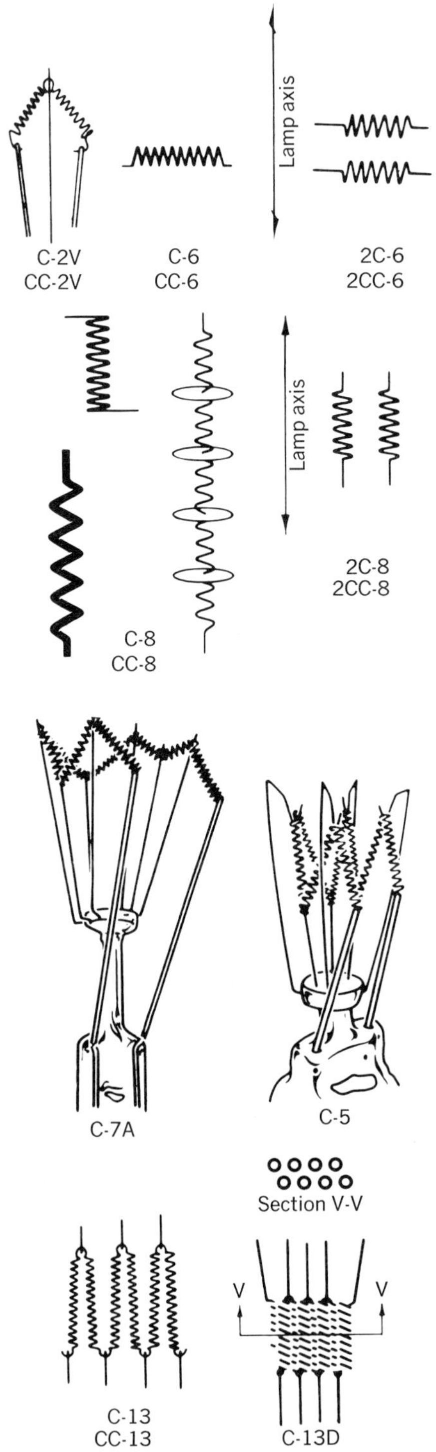

Fig. F-2 Common incandescent filament forms—not to scale. (*Sylvania Electric Products, Inc.*) Filament forms are identified by a two-part designation. The first part indicates that the filament is coiled (C), or that the coil is formed into a second larger coil before mounting, i.e., a coiled-coil (CC). If there is more than one basic filament section, the number of sections is sometimes used as a prefix. The second part of the designation is an arbitrary letter or number describing the configuration of the filament and support in relation to the lamp axis. Most of the filaments shown in the diagram can be either of C or CC construction. The mounting details and proportions can vary considerably for each designation.

fill light The supplementary lighting for the filming area. Fill light, in motion picture photography, is used to supply illumination of medium intensity and eliminate or minimize shadows.

Fill light has been referred to as a "wrinkle remover" in close-up shots. Fill-light sources are often mounted in the vicinity of the camera or attached to the camera itself so that the light units move with the camera during zooms and tracking shots. See KEY LIGHT and LIGHTING CONTRAST.

film A cellulose base, in roll or sheet form, coated with a photographically sensitive emulsion. Emulsion, in turn, consists of a thin layer of dispersions of light-sensitive materials in a colloidal medium, usually gelatin.

To the motion picture producer, the rolls of unprocessed film are a means to an end. As a necessary budgetary item, the cost of the raw stock or unexposed film is negligible compared with the large sums of money that are expended annually for the purpose of exposing their coated, sensitized surfaces.

To the director, the cellulose rolls are a permanent record of accomplishment. His interest is in talent, drama, emotional impact, and, of course, recognition in the form of an Oscar, Emmy, or Clio.

For the cameraman, and the army of dedicated technicians whose efforts are reflected on giant motion picture screens throughout the world, film has magical properties—a fact also discovered by a French magician named George Melies who amazed viewers with his startling illusory effects before the turn of the century.

Before 1950, most of the cans of motion picture film delivered to production studios contained a built-in hazard in the form of a nitrate base. In spite of the highly flammable support, and the need for precautionary measures in handling and storage, nitrate stock was the overwhelming choice of the professional film producer. Safety film, available since the industry's pioneer days, failed to match the physical properties, printing characteristics, or popularity of the nitrate variety.

Recognizing the ever-growing need for a safety film that would satisfy the requirements of the rapidly growing industry, the film manufacturers eventually developed cellulose triacetate and acetate propionate bases. These supports, with their comparatively thin but tough safety bases, gained the industry's almost immediate approval and acceptance.

Since considerable amounts of nitrate stock are still in circulation, however, precautionary measures must be taken with all film. Nitrate negative, both picture and sound, along with duplicate negatives, masters, and old prints, will remain a hazard until chemical and natural changes cause deterioration and render them harmless.

While Father Time can be expected to help remove that major hazard, Mother Nature is inclined to be less cooperative. Showing no favoritism, she makes few exceptions in her general treatment of all things, film included. Following the grand pattern, film expands in warm weather, contracts in cold, swells when wet, and shrinks when dry.

Motion picture film is supplied in a variety of widths and as many different sprocket-hole arrangements for registration purposes. Varying in size from the 70mm width, with 10 perforations on each side of the film frame, to the 8mm width, with its single sprocket hole for each frame, every film type is designed to produce professional results. See EMULSION, COLOR FILM, and MULTILAYER FILM.

In normal developing, black-and-white film produces a negative image, or one in which the tonal values are exactly opposite to those in the original subject matter at the time of exposure. In the negative reproduction, the lighter areas have proportionately greater silver deposits than the darker areas. Blacks, therefore, appear as clear areas in the negative; and whites, or lighter areas, as the denser portions of the film frame.

In reversal processing, the in-camera raw stock is used to produce a direct positive. Although the silver halide deposits are similar to the granular formations in negative film, with the buildup greatest where the density is lowest, it is the unused silver halide in the emulsion that produces the tonal values of the positive image. Reversal film is first developed so that the silver in the emulsion is exposed. This exposed silver is bleached away by chemical action, leaving the unexposed developed silver. The unexposed silver is reexposed, redeveloped, and fixed to produce the positive image. See REVERSAL PROCESSING.

The reproduction of color values is made possible by a chemical reaction described as *coupler development.* During the developing process, the exposed silver halide, in the first stage of the reaction, forms metallic silver. As the developing agent oxidizes, it combines with the chemical coupler to produce dyes in direct proportion to the amount of metallic silver formed during the initial reaction. The insoluble dyes duplicate the color values in the original subject matter. The negative images are approximately complementary to the color values in the original. See COUPLER DEVELOPMENT.

In color-reversal processing, the positive images are produced by developing each of the three layers of emulsion separately if the coupler components must be extracted from the developing solutions; or by developing the original film in a single solution if the couplers were added to the layers of emulsion by the manufacturer. Either way, the basic process is the same. See KODACHROME and KODACOLOR.

Developed in a black-and-white bath to produce a negative image, the silver halide in each of the emulsion layers is reexposed. The resultant positive dye images, produced by the chemical action of the couplers, duplicate the color values in the original subject. Including the developing, coupling, bleaching, fixing, and washing stages, the exposed film passes through as many as 10 separate baths before it is dryed and ready for projection. Unlike color negative which can be easily corrected during the printing process, adjustments in the color balance of reversal film are, at best, limited.

With few exceptions, most film bases have gray dye or jet-black coatings to prevent undesirable halation. Exercising no effect on the quality of the print, these coatings prevent light from passing through the emulsion, striking the acetate film base and reflecting back to the sensitive emulsion to cause halation or light flares around bright objects. See Figure F-3.

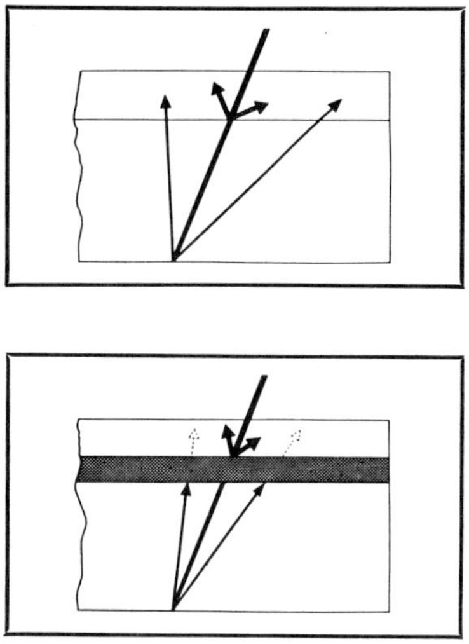

Fig. F-3 A strong beam of light coming from the subject area goes through the lens and the emulsion. Some light is reflected from the surface between the emulsion and the film base and scatters to spread the image. Much of this light, however, goes right on through the base and is reflected from the back of the film base, bouncing back to the emulsion again, spreading the halo still further.

Today, manufacturers include a special dye layer, between emulsion and base, that absorbs a large part of the light that gets through the emulsion. This dye layer also absorbs the small amount of light that could still make its way back toward the emulsion. Requiring no special treatment, the dye layer disappears completely during the processing procedure.

Today, almost all the camera, editing, optical, processing, and projection equipment is designed to accommodate the film widths generally favored by professional producers—35 and 16mm. Except for size and relative cost, differences between them are negligible. Both widths, under similar photographic conditions, will produce equally professional results. And, frequently, motion pictures exposed on the wider 35mm stock are released on the narrower 16mm film. Since many of the motion pictures produced for entertainment purposes contain multiple exposures and special effects that require precise registration during the filming and optical processes, the 35mm film size is favored for theatrical, television, and commercial purposes. The narrower 16mm film size, with one sprocket hole for each frame instead of the four perforations on each side of the 35mm frame, is less expensive and the more popular choice for the non-theatrical producer with limited funds or low budgets. Paradoxically, even the producer with unlimited funds must, on occasion, take the lighter, portable equipment and the narrower 16mm rolls of film to locations where the use of heavier 35mm equipment would either be inadvisable or even hazardous.

During the filming process, a roll of film unwinds from its core in the first of two compartments in a lightproof container known as a *magazine,* passes through the camera, and is rewound on the take-up reel in the second compartment. In the camera, the film passes over sprocketed wheels that engage the perforations and bring each frame into exposing position. Perfectly registered during the interval of exposure, each frame of film moves intermittently as it passes through the camera. See EMULSION, STANDARD POSITION.

Sprocket holes, like the motion picture film in which they are made, can be either negative or positive, depending on the shape of the perforations. Negative perforations are designed for in-camera use and provide the needed precision during the exposure interval.

Positive-type perforations, by comparison, are shaped to withstand the stress and strain of repeated projection.

Both types of perforations are available in either short or long pitch sizes. *Pitch,* in motion picture terminology, refers to the distance between perforations. Measured from the leading edge of one sprocket hole to the leading edge of the next perforation, the short pitch is used during the photographic process while the long pitch, as explained below, is generally used for laboratory printing purposes. The distance from the leading edge of one sprocket hole to the leading edge of the next is 0.1866 inch for the short pitch and 0.1870 inch for the long pitch.

During the lab printing process, the in-camera negative and the raw stock on which the images are copied pass over the same sprocketed wheel, simultaneously, with the print film over the negative film. Measured from the center of the sprocketed wheel, the difference in distance between that point and each of the two lengths of film can, obviously, be no greater than the thickness of one length of film. The amount of compensation for that extra thickness is 0.0004 inch (the difference between 0.1866 and 0.1870 inch). This minute amount assures precise registration and eliminates the inherent danger of damaged sprocket holes.

The perforations used in high-speed cameras approximate the size and shape of 35mm positive stock. The difference between the two lies in the extremely critical tolerances of the high-speed perforations.

Like the wider lengths of 35mm film, some rolls of 16mm film are perforated for in-camera use while others are perforated for duplicating or printing purposes. The rolls of raw stock used in "silent" cameras have two rows of perforations, one row along each edge of the film. Single-system cameras, by comparison, record sound and picture simultaneously and use film with only one row of perforations. The sound track, in the standard emulsion position, appears on the right side of the frame and replaces the second row of perforations. See A & B WINDINGS and A & B ROLLS—EMULSION POSITIONS.

In order to take the guesswork out of film identification, the industry has developed a relatively simple system of symbols and abbreviations. The letters BH (Bell & Howell), for example, are used to indicate negative-type perforations. Similarly, KS (Kodak Standard) indicates that the film has positive-type perforations. DH (Dubray-Howell) identifies the perforations used for color-print film. CS (CinemaScope) is associated with the smaller perforations that provide the additional space needed for the multiple sound tracks in the release prints used for CinemaScope projection. 1R and 2R indicate the number of rows of perforations in 16mm film. Footage numbers, for editing purposes, are latent-image-printed along the edges of most 35 and 16mm rolls of raw stock. Horizontal markings, after every fourth perforation, indicate frame lines on negative as well as positive lengths of film.

Numbers that follow the letter designations, in ASA descriptions, indicate the pitch. BH 1866, for example, means that the film has negative-type perforations and that the pitch, or distance from the leading edge of one perforation to the leading edge of the next sprocket hole, is 0.1866 inch. KS 1870, by comparison, describes a film having positive-type perforations with a long pitch, or a distance of 0.1870 inch between sprocket holes. In all cases, labels on film cans and cartons clearly indicate the type of film in the container as well as the emulsion and roll number.

To the manufacturers who must anticipate trends while filling current needs, the production of film is a challenging assignment. In an industry where change is the only constant, manufacturers must maintain standards that are measured in millimeters while conducting research and keeping pace with new equipment and production techniques.

Conscientiously aware of the industry's needs and problems, film manufacturers have combined scientific skills with engineering principles to produce film supports with lower shrinkage characteristics, lower granularity, higher emulsion speeds, increased sensitivity, greater resolving power, better definition, and processing properties that parallel improved laboratory techniques. In addition to in-camera black-and-white, color, negative, and reversal-type films, catalogs place at the producer's disposal a film for every specific need and every imaginable application. The list includes film types that are sensitive to colors beyond the range of the human eye, film especially suitable for television recording, film for recording sound, high-contrast film for titling and matte work, high-speed rapid-reversal film for photographing under difficult lighting conditions, and leader stock for editing purposes.

The ability of an emulsion to produce sharp images is of relatively negligible importance if the cameraman must expose test lengths from each roll of raw stock delivered to him. The amateur cameraman who buys one roll of film at a local store and the purchasing agent for a major studio who orders thousands of feet of film directly from the manufacturer share the assurance and knowledge that each new roll of a specific film type will have the same characteristics and degree of sensitivity as the last roll exposed.

In order to maintain the high standards outlined in the specification sheets prepared by the ASA, manufacturers use a number of devices for controlling the quality and consistency of their collective product. Sensitometers, capable of measuring the response of an emulsion to a light source, and densitometers, for measuring the densities of a length of developed film in terms of logarithmic units, are but two of the many instruments that remove the guesswork from film manufacturing processes and assure the required degree of consistency.

Even as the editors prepare these pages for the printer, new literature reaches us describing a film type that virtually eliminates the need for adjusting aperture settings and changing shutter speeds. The inventor, Dr. Harold Edgerton, is also known as the father of electronic or stroboscopic flash.

The new film type is known as XR with the letters standing for *extended range.* The film's tricolor emulsion is capable of photographing "impossible" subject matter with identical shutter speeds and aperture settings. The full-range black-and-white photographs, achieved by using different filters during processing, contain such picture information as the inside of a blast furnace, rocket exhaust, or explosion. Most of these effects in the past, with ordinary film, would look like lightstruck overexposures.

Like most new film types, XR film is at the present time available for still cameras only. In time, however, it is safe to assume that the new film will be used for exposing images in the motion picture camera as well.

Agfa-Gevaert duplicating films

Duplicating Positive, Type 3.62. A yellow dyed film of extremely fine grain, intended for the printing of master positives of high quality. The use of this film, in conjunction with duplicating negative pan film, type 4.64, meets the most exacting requirements of duping.

Color sensitivity:	Blue-sensitive only
Recommended gamma:	1.7
Resolving power:	250 lines per mm
Edge marking:	The letter D just before the footage number

Duplicating Negative, Type 4.51. Type 4.51 is intended for preparing dupe negatives. This low-contrast material has a fine grain and high resolving power, ensuring an excellent rendering of detail.

Color sensitivity:	Blue-sensitive only
Recommended gamma:	0.65
Resolving power:	180 lines per mm
Edge marking:	The letter E just before the footage number.

Duplicating Negative Pan, Type 4.64. Type 4.64 produces dupe negatives practically indistinguishable in gradation and grain size from the original negative. It is printed from master positives made on duplicating positive film. The extreme fineness of grain of this film and its high resolving power are associated with a relatively high sensitivity, owing to the panchromatic sensitization of the emulsion, which appreciably increases the sensitivity to the incandescent light of the printers.

Color sensitivity:	Panchromatic
Recommended gamma:	0.65
Resolving power:	165 lines per mm
Edge marking:	The letter G just before the footage number

Duplicating Reversal, Type 8.91. A reversal film with a very fine grain, intended for making prints from original reversal films or positive duplicates. Exposure must be determined by testing. The relative exposure coefficient of this film is 25 with respect to positive fine grain, type 5.61, taken as 100.

Sensitization:	Blue-sensitive emulsion
Recommended gamma:	1.1
Resolution:	165 lines per mm
Edge marking:	Latent-image name Agfa-Gevaert S between perforations

Agfa-Gevaert GevaColor films

GevaColor Positive, Type 9.54. A multilayer color film for printing release copies from color negatives, with or without a mask, either directly or via duplicating stock. Type 9.54 consists basically of three separate emulsions, each containing an appropriate color coupler. Color development produces positive pictures in natural colors. The film has a black antihalo layer on the back of the film which is removed by a preliminary treatment. In order to improve definition, the film contains dyes which disappear during processing. Because of the very different characteristics of printing machines, the practical speed of the film must be ascertained experimentally for the type of printing machine used.

Sensitometric wedges indicate the following gamma values: blue-sensitive layer—3.20, green—2.90, red—2.90. See Table F-1.

GevaColor Reversal Print, Type 9.02. A multilayer color-reversal film for printing release copies from reversal color originals. It may also be used for 16mm reduction from 35mm color prints. This film consists basically of three separate emulsions which produce images in the same natural colors as the original. It cannot be used for picture taking in the camera. The film has a black antihalo layer on the back of the film which is removed by a preliminary treatment. Dyes, disappearing during processing, have been added to the emulsions to improve definition.

GevaColor reversal print film is balanced for the light normally used in conventional printers. Slight deviations in the color balance can be adjusted by using the appropriate color-correction filters.

Because of the very different characteristics of printers, the sensitivity of the film must be ascertained experimentally for each type of printer used.

Recommended gammas: blue-sensitive layer—1.50, green—1.40, red—1.30.

Agfa-Gevaert reversal camera films

GevaPan 30 *Reversal, Type* 8.63. A panchromatic reversal film intended for filming documentaries, educational films, newsreels, television productions, and for all work under normal lighting conditions, both by daylight and artificial light. The gamma has been chosen so that normal contrast is obtained on prints, together with a relatively high maximum density. GevaPan 30 reversal, type 8.63, can be processed in developers with either a high or a low alkaline content. The base has a light-blue tint.

Meter settings	*Daylight*	*Tungsten*
ASA	64	40
DIN	19	17

Color sensitivity:	Panchromatic emulsion
Recommended gamma:	1.2
Resolution:	145 lines per mm
Edge marking:	Latent-image footage numbers between perforations preceded by the letter N, as well as the name Agfa-Gevaert S

GevaPan 36 *Reversal, Type* 8.80. A high-speed panchromatic reversal film intended for filming under unfavorable conditions in either daylight or tungsten light. Its low contrast permits working by illumination of very high contrast so that the contrast of a release print still lies within acceptable limits. Type 8.80 is also very suitable for high-speed cinematography. GevaPan 36 reversal film, type 8.80, may be processed in developers with high or low alkaline content. The base is tinted light blue.

Meter settings	*Daylight*	*Tungsten*
ASA	320	200
DIN	26	24

Color sensitivity:	Panchromatic emulsion
Recommended gamma:	1.2
Resolution:	115 lines per mm
Edge marking:	Latent-image footage number preceded by the letter H between perforations, as well as the name Agfa-Gevaert S

TABLE F-1 GevaColor Positive Film, Type 9.54

Arrangement of the Layers

Before processing	After color development	After bleaching and fixing
	Protective layer	
Green-sensitive silver-chloride emulsion with magenta coupler	Partial image consisting of silver and magenta dye	Magenta dye image
Red-sensitive silver-chloride emulsion with cyan coupler	Partial image consisting of silver and cyan dye	Cyan dye image
Blue-sensitive silver-bromide emulsion with yellow coupler	Partial image consisting of silver and yellow dye	Yellow dye image
	Clear cellulose triacetate safety base	
Antihalation backing layer		

Agfa-Gevaert negative camera films

GevaPan 30, *Type* 1.65. A black-and-white camera negative for all work under normal exterior lighting conditions and for interiors where an adequate lighting level is available. Because of the fine grain and the very high resolving power of its emulsion, the film gives sharp images. GevaPan 30 film possesses all the traditional qualities of the ideal studio film.

Meter settings	*Daylight*	*Tungsten*
ASA	80	64
DIN	20	19

Color sensitivity:	Panchromatic emulsion
Recommended gamma:	0.65
Resolving power:	130 lines per mm
Edge marking:	The letter N just before the footage number

GevaPan 36, *Type* 1.92. A panchromatic negative of extremely high sensitivity intended especially for newsreel and for all cinematography under poor lighting conditions. This film can also be used under normal lighting conditions, thus permitting a smaller stop to be used when a greater depth of field is desired. An extremely high speed has been obtained with due regard for relatively fine grain.

Meter settings	*Daylight*	*Tungsten*
ASA	250	200
DIN	25	24

Color sensitivity:	Panchromatic emulsion
Recommended gamma:	0.65
Resolving power:	115 lines per mm
Edge marking:	The letter H just before the footage number

Agfa-Gevaert positive films

Positive Fine Grain, Type 5.61. A positive film with extremely fine grain, intended principally for printing high-quality release copies. The fine grain of the emulsion retains the utmost quality of the picture and sound. This film can also be used for titling and other work. Its speed has been taken as a basis of comparison for the other Agfa-Gevaert films with the exception of the negative and reversal camera films.

Color sensitivity:	Blue-sensitive only
Recommended gamma:	2.45
Resolving power:	165 lines per mm

GevaPrint Film for TV, Type 5.64. Type 5.64 has been specially formulated for making copies for television transmission. The brightness-reproduction characteristics meet the demands made by telecine techniques. It has a sensitivity equal to that of positive fine-grain film, type 5.61, and can be developed in the same way.

Type 5.64 ensures optimum stability of the picture as well as excellent quality in reception of picture and sound. The graininess of the film is finer than that of ordinary positive films. Consequently, the level of background noise is kept to an absolute minimum. Furthermore, the fineness of the grain exerts a favorable influence on the quality of the picture and ensures optimum reproduction of sound from an optical sound track.

Color sensitivity:	Blue-sensitive only
Recommended gamma:	1.7–1.8
Resolving power:	165 lines per mm

Positive Super Contrast, Type 5.53. A positive film of high contrast, yielding vigorous blacks and good transparent whites, intended for titling and similar work requiring a very contrasty emulsion.

Color sensitivity:	Blue-sensitive only
Recommended gamma:	4
Resolving power:	220 lines per mm

Agfa-Gevaert telerecording film

Telerecording Reversal, Type 8.50. A reversal film for photographing the television screen. The emulsion is blue-sensitive and has a very fine grain. The base is tinted light blue. Exposure-meter settings vary with the wavelength of the screen illumination. Test exposures are essential.

Sensitization:	Blue-sensitive emulsion. The camera can therefore be loaded by the light of a red safelight.
Recommended gamma:	1.2
Maximum density:	2.6
Resolution:	150 lines per mm
Edge marking:	Latent-image footage numbers preceded by the letter M between perforations, as well as the name Agfa-Gevaert S

Agfa-Gevaert leader film, type 5.00. A transparent and clear cellulose triacetate film base on which no emulsion is coated. It is very resistant to impact, and its thickness is 0.2mm. The leader film is used as a protective leader and trailer for films as well as for leader in processing machines. Both sides of type 5.00 take the usual ball-point pen, so that it is easy to write notes and references on it.

DuPont negative films. See Table F-2.

DuPont reversal films. See Table F-3.

DuPont laboratory stock. See Table F-4.

DuPont triacetate-base films. See Table F-5.

DuPont Cronar films. Table F-6.

DuPont cathode-ray tube exposure data. See Table F-7.

Eastman camera film. See Table F-8.

Eastman duplicating films. See Table F-9.

Eastman release-print films. See Table F-10.

Eastman leader films. See Table F-11.

Eastman specialized types. See Table F-12 and Figure F-4.

Eastman standard roll lengths. See Table F-13.

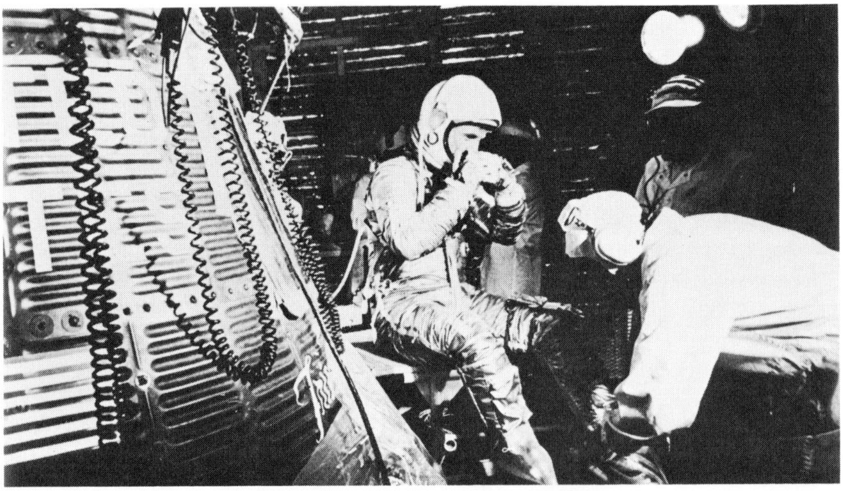

Fig. F-4 ***(NASA photo.)*** For Colonel John Glenn's orbital flight, the Eastman Kodak Company produced not only a special camera that could fit within the limited confines of the tiny capsule but a new, thin-base film as well. This high-speed color film known as Ektachrome ER, with an ASA exposure index of 160, is slightly thicker than the cellophane wrapper on a cigarette pack. Designed for use under low light levels, 590 feet of this film type was loaded on a normal spool with a 400-foot capacity. To right overhead are special ColorTran lights with heat filters, providing illumination for film and TV coverage.

Eastman winding and perforating specifications. To meet the requirements of various types of equipment and operating procedures in trade use and to ensure optimum performance characteristics, specific types of windings and perforations are provided for different products. For detailed information, refer to the individual specification sheets which are available on request.

Eastman filters for color films. See Table F-14.

Eastman exposure and filter compensation for color films. See Table F-15.

GAF negative films

VersaPan Negative Film: Type 2531, 16mm; *Type* 5531, 35mm. High-quality, medium-speed panchromatic film. Fine grain, high resolution, and long tone scale make it ideal for general motion picture work in daylight and under artificial illumination. In addition, the excellent quality of release prints made from the negatives makes VersaPan negative film suitable for such specialized applications as medical photography, time-lapse and motion studies, as well as for all uses where sharpest rendering and finest detail are desired.

	Film speed		
	Daylight	*Tungsten*	
	80	64	
Type 2531	16mm	Regular perforations	2R 2994
		Sound B perforations	1R 2994
		High-speed perforations	2R 3000
Type 5531	35mm	BH 1866 perforations	

Super HyPan Negative Film: Type 2671, 16mm; *Type* 5671, 35mm; *Type* 7671, 70mm. High-quality panchromatic film. It is ideal for general motion picture work under artificial illumination and in low-level daylight situations. Super HyPan negative film yields high-quality release prints.

	Film speed		
	Daylight	*Tungsten*	
	200	160	
Type 2671	16mm	Regular perforations	2R 2994
		Sound B perforations	1R 2994
		High-speed perforations	2R 3000
Type 5671	35mm	BH 1866 perforations	
		KS 1870 perforations	

GAF reversal films

Medium-speed Reversal Film, Type 2955, 16mm. High-quality panchromatic film. With reversal processing it yields a positive image for direct projection that has excellent pictorial quality, good sharpness, and fine grain.

TABLE F-2 DuPont Negative Films

Specifications	Cronar fine-grain Superior 2, type 136	Cronar high-contras fine-grain negative type 140
Description	A very fine grain, medium-speed, wide-latitude, extended-red-sensitivity instrumentation negative	For photographing missiles against a sk background. Extenc red sensitivity
Speeds		
Negative, daylight	Low contrast: 125 High contrast: 200	
Negative, tungsten	Low contrast: 100 High contrast: 160 Missile tracking: 800	
Reversal, daylight	—	
Reversal, tungsten	—	
Relative speed CRT recording (P-11 phosphor)	400 460	300 375
Safelight	Total darkness. Wratten series 3, or equivalent, when necessary	Total darkness. Wratten series 3, o equivalent, when necessary
Recommended developers		
Negative	Low contrast: 6-D High contrast: 16-D or D-19	Low contrast: 6-D High contrast: 16-I or D-19
Reversal	—	—
Approximate development time at 68°F (machine)		
Negative	Low contrast: 4–6 minutes High contrast: 3–5 minutes	Low contrast: 4–6 minutes High contrast: 5–7 minutes
Reversal	—	—
Control gamma negative	Low contrast: 0.7 High contrast: 1.2	Low contrast: 1.0 High contrast: 2.0
Identifying character in latent-image stencil	V	VY

Superior is DuPont's registered trademark for its motion picture films.
Cronar is DuPont's registered trademark for its polyester film base.

Superior 2, type 936	Superior 3, type 937	Superior 4, type 928
ry fine grain, edium-speed, wide- titude negative film r both exterior and terior work	Fine-grain high-speed, wide-latitude negative film for both interior and exterior use under limited illumination	High-speed, wide-latitude negative film for both interior and exterior use under limited illumination
125	250	320
100	200	250
—	—	—
—	—	—
460	1,500	750
750	1,700	1,050
al darkness. ratten series 3 en necessary	Total darkness. Wratten series 3 when necessary	Total darkness. Wratten series 3 when necessary
6-D	6-D	6-D
—	—	—
5 minutes	$6\frac{1}{2}$ minutes	7 minutes
—	—	—
0.7	0.7	0.7
V	XY	X

TABLE F-3 DuPont Reversal Films

Specifications	Rapid reversal, type 930	High-speed rapid reversal, type 931	Ultraspeed rapid reversal, type 932
Description	Medium-speed film for rapid reversal processing. Widely used for sports and news coverage in daylight	High-speed, reversal film. Widely used for night sports and news coverage under incandescent lights	Very high speed, reversal film, especially for difficult, low-level illumination of sports and TV news events
Speeds			
Negative, daylight	32	80	125
Negative, tungsten	25	64	100
Reversal, daylight	64	160	320
Reversal, tungsten	50	125	250
Relative speed CRT recording (P-11 phosphor)	—	—	—
Safelight	Total darkness. Wratten series 3 when necessary	Total darkness. Wratten series 3 when necessary	Total darkness. Wratten series 3 when necessary
Recommended developers			
Negative	6-D	6-D	6-D
Reversal	1st: 107-D 2d: 129-D	1st: 107-D 2d: 129-D	1st: 107-D 2d: 129-D
Approximate development time at 68°F (machine)			
Negative	$4\frac{1}{2}$ minutes	4 minutes	$3\frac{1}{2}$ minutes
Reversal	1st: 2 minutes 2d: 50 seconds	1st: 2 minutes 2d: 50 seconds	1st: 2 minutes 2d: 50 seconds
Control gamma negative	0.7	0.7	0.7
Identifying character in latent-image stencil	Y	Z	UY

TABLE F-4 DuPont Laboratory Stock

Specifications	Fine-grain duplicating negative, type 908	Reversal duplicating, type 910	Cronar fine-grain positive, type 225
Description	Exceptionally fine grain high-resolution film designed specifically for duplicating work. Fully panchromatic. Its wide available gamma range allows it to be used with master positives of low as well as normal contrast.	Fine-grain panchromatic film designed to yield high-quality reversed duplicates when printed from either color or black-and-white reversed camera films or positive prints	For general release work and dubbing prints requiring optimum picture and sound quality. Also positive transparencies for projection viewing or direct reading. Dimensional stability makes it an excellent companion to Cronar negatives. Yields blue-black image.
Tungsten speed	4	20	2
Safelight	Wratten series 3 or equivalent	Wratten series 3 or equivalent	DuPont S-55X or equivalent
Recommended developers	6-D	1st: 107-D 2d: 129-D	123-D
Approximate development time at 68°F (machine)	5½ minutes	1st: 2 minutes 2d: 50 seconds	3 minutes
Control gamma	0.7	—	2.40
Identifying character in latent image stencil	U	UZ	J

TABLE F-5 DuPont Triacetate-base Films

Film width	Film name	Film type	Type of perforations	American standard	Film length, feet	Spool or core type
	Standard perforations				Roll lengths, core and spool types	
16mm	Superior 2 Superior 3 Superior 4	936A 937A 928A	1R 2994, winding A or B 2R 2994 2R 3000 2R 1500 (8mm perf.)	PH 22.109 PH 22.110 PH 22.5 PH 22.17	100 D/L* 200 D/L* 400 D/L* 400 L/P 1200 L/P	44 49 60 32 32
	Rapid reversal High-speed rapid reversal Ultraspeed rapid reversal	930A 931A 932A	1R 2994, winding A or B 2R 2994 2R 3000 2R 1500 (8mm perf.)	PH 22.109 PH 22.110 PH 22.5 PH 22.17	100 D/L* 200 D/L* 400 D/L* 400 L/P 600 L/P 1200 L/P	44 49 60 32 32 32
	Panchromatic reversal duplicating	910A (SL353)	1R 3000 winding A 2R 3000 2R 1500 (8mm perf.)	PH 22.12 PH 22.5 PH 22.17	400 L/P 1200 L/P 2000 L/P	32 32 32
35mm	Superior 2 Superior 3 Superior 4	936B 937B 928B	BH 1866	PH 22.93	100 D/L 200 L/P 400 L/P 1000 L/P	16 4 or 24 25 4 or 24
	High-speed rapid reversal Ultraspeed rapid reversal	931B 932B	KS 1870	PH 22.36	100 D/L 500 L/P 1000 L/P	16 24 4 or 24
	Duplicating negative	908B	BH 1866	PH 22.93	400 L/P 1000 L/P	24 24

Magnetic striping is available on the following films:

930A 1R 2994, winding B.
931A 1R 2994, winding B.
932A 1R 2994, winding B.
936A 1R 2994, winding B.
937A 1R 2994, winding B.

*Not available 1R 2994, winding A.

TABLE F-6 DuPont Cronar Films

Perforations				Roll lengths, core and spool types	
Film width	DuPont film type	Type of perforations	American Standard	Film length, feet	Spool or core type
16mm	136, 140	2R 3000 2R 2994 1R 2994, winding A or B	PH 22.5 PH 22.110 PH 22.109	125 D/L* 250 D/L† 450 D/L 500 L/P 1250 L/P	44 49 60 32 32
	225	2R 3000 1R 3000, winding A or B Unperforated	PH 22.5 PH 22.12	1250 L/P	32
35mm	136, 140	BH 1870 BH 1866 KS 1870 Unperforated	PH 22.34 PH 22.93 PH 22.36	125 D/L* 250 L/P 500 L/P 1250 L/P	16 4, 24 24 24
	225	KS 1870 Unperforated	PH 22.36	1250 L/P	24
70mm	136, 140, SL-360	Type I (0.2340) Type II (0.1870) Unperforated	PH 1.20 PH 1.20 PH 1.20	125 D/L* 125 L/P 500 L/P 1250 L/P	58 11 11, 59 59
	225	Type I (0.2340) Type II (0.1870) Unperforated	PH 1.20 PH 1.20 PH 1.20	125 L/P 1250 L/P	11 11, 59
5.481 inches	136, 140	Unperforated		125 L/P	68

*137-foot film, self-leader and trailer.
†267-foot film, self-leader and trailer.

TABLE F-7 Cathode-ray Tube Exposure Data

The following relative CRT speeds are calculated at a net density of 1.0.
Exposure: wedge-modulated cathode-ray tube raster.
Development: 6 minutes in D-19 at 68°F.

Film type	Relative CRT speed	
	Phosphor P-11 (blue)	Phosphor P-31 (green)
131	575	1,060
136	400	460
140	300	375
928	750	1,050
936	460	750
937	1,500	1,700

Courtesy of DuPont.

TABLE F-8 Eastman Camera Films

Name of film	Type no.		General description
	(35mm)	(16mm)	
XT Panchromatic negative (*replaces background-X types 5230 and 7230*)	5220	7220	A very fine grain, low-speed n tive film for general exterior tography, background work optical work where high defini and low granularity are requis
Plus-X Panchromatic negative	4231	7231	High-speed, fine-grain neg film for general production both outdoors and in the st
Double-X Panchromatic negative	5222	7222	A high-speed negative film resenting the latest advanc speed-granularity ratio. Sui for exterior and interior ph raphy under difficult lig conditions.
4X Panchromatic negative (*replaces Tri-X negative types 5233 and 7233*)	5224	7224	An extremely high speed neg film of medium graininess. I an exceptional ability to repro shadow detail. It is well suite newsreel work and all ph raphy where the lighting c tions are adverse.
RP Panchromatic negative		7229	A high-speed, prehardened especially designed for proc in the Eastman Viscomat p sor.
Color negative	5251		A multilayer color film su for both exterior and in photography. Balanced for 32 tungsten lamps but may be in daylight with Kodak W filter no. 85.
Ektachrome commercial		7255	A low-contrast, reversal-type, layer color film for general color production work bot doors and in the studio. R mended for originals from it is desired to make one o color release prints.
Ektachrome ER (daylight type)	5257	7257	A high-speed, daylight-ba color-reversal film design use under very low illumi conditions or high-speed tography applications wher cient exposure cannot be ob with slower-speed color-r films.

xposure index		Code letter identification* (precedes footage number)	End markings (on 16mm 100-ft and 200-ft spools)	Footage numbered†	Frame-line marked*	Magnetic pre-striping (16mm only)
light	Tungsten					
5	20	F	XTN	Yes	Yes	Yes
0	64	H	PXN	Yes	Yes	Yes
0	200	C	DXN	Yes	Yes	Yes
	400	G	4XN	Yes	Yes	Yes
	200			Yes	No	Yes
en	50	A(L.I.)		Yes	No	
en	25		EKT-C	Yes	No	Yes
			EKT	Yes	No	Yes

TABLE F-8 Eastman Camera Films (cont.)

Name of film	Type no.		General description
	(35mm)	(16mm)	
Ektachrome ER (type B)	5258	7258	This film, balanced for tung illumination, serves as a com ion to Ektachrome ER film, light type, and is useful for ing industrial pictures u existing plant illumination nighttime sports photogra and for newsreel work.
Ektachrome MS (daylight type)	5256	7256	A medium-speed, daylight anced color film processe reversal to give a positive inte primarily for direct proje Color prints may be made the original and, while not in every respect, are acceptab many purposes. The film pro excellent color rendition and images having low grainines good sharpness. It finds application in data and neering analysis, high- camera use, and sports p raphy under low daylight illu tion conditions.
Plus-X reversal		7276	Fine-grain, medium-speed re film for general motion p production use where re films are indicated.
Tri-X reversal		7278	High-speed reversal film of m granularity provides ex halation control. It is use photography under difficul ing conditions both outdoo in the studio. Especially s for television newsreel wo sports photography.

*On 35mm films only.

†All 35mm films are ink footage-numbered except types 5251, 5256 and 5258, which are latent-image footage-numbered.

‡Exposure to tungsten light not recommended except under emerg conditions. For 3200°K tungsten lamps and with Kodak Wratten filte plus 82A, index is 16.

posure index		Code letter identification* (precedes footage number)	End markings (on 16mm 100-ft and 200-ft spools)	Footage numbered†	Frame-line marked*	Magnetic pre-striping (16mm only)
ight	Tungsten					
0 n tten 85B r)	125		EKT-B	Yes	No	Yes
4	‡See note at bottom of page.		EMS	Yes	No	Yes
	40		PXR	Yes	No	Yes
	160		TXR	Yes	No	Yes

TABLE F-9 Eastman Duplicating Films

Name of film	Type no.		General description	Code letter identification (precedes footage number)*	Footage numbered†	Frameline marked
	35mm	16mm				
Fine-grain panchromatic duplicating negative	5234	7254	A low-speed, extremely fine grain duplicating negative film. May be used to make duplicate negatives from master positive on types 5365 and 7365 or from reversal originals.	D	Yes	Yes
Fine-grain duplicating positive	5366	7366	A blue-sensitive film with extremely low granularity and very high resolution. Intended for making positives from camera negatives. Used as a companion to types 5234 and 7234.	D	Yes(35) No (16)	Yes
Panchromatic separation	5235		A very fine grain film intended for making black-and-white separation positives from color negatives, such as those made on color negative film type 5251.	D	Yes	Yes
Color intermediate	5253	7253	A multilayer color film suitable for use in preparing both color master positives and color duplicate negatives from originals on color negative film, type 5251. Eliminates the need for making black-and-white separation positives except where protection of the original against fading is desired. May also be used for making color duplicate negatives from black-and-white separation positives.		Yes	No
Color internegative	5270	7270	A multilayer color film designed for making color internegatives from color-reversal originals. The film is designed for printing onto Eastman color print film types 5385 and 7385.		Yes	No

*On 35mm films only.

†All 35mm films are ink footage-numbered except types 5253 and 5270, which are latent-image footage-numbered.

TABLE F-10 Eastman Release-print Films

Name of film	Type no.		General description	Footage numbered*	Frameline marked*
	35mm	16mm			
Fine-grain release positive	5302	7302	A film for general black-and-white production release printing. Also useful for making negative and positive titles, dubbing prints for sound, and kinescope recordings from negative tube images.	Note 1, 2	Yes*
Fine-grain release positive		7303	A very fine grained release print film. Used in conjunction with XT panchromatic negative film, it will yield the finest 16mm black-and-white prints.	Note 2	
High-speed positive	5305		A positive film of higher speed than that of type 5302. Useful for making release prints, negative and positive titles, variable-density and variable-area sound recording, and other uses.	Yes	Note 1
High-contrast positive	5362	7362	A high-contrast material useful for making negative and positive titles, silhouette mattes for process work, and travelling mattes for printer light control.	Notes 1, 2	Yes
Color print	5385	7385	A multilayer color film designed for general color release printing from color negatives and color duplicate negatives. May be used in conjunction with types 5251, 5270 and 7270, and 5253 and 7253.	No	Yes*
Reversal color print		7387	A multilayer, reversal-type color film designed for making prints from 16mm Ektachrome commercial, Ektachrome ER and MS, and Kodachrome II, daylight and type A films. Requires special processing similar but not identical to that used for Kodachrome films.	No	No
Ektachrome reversal print film	5386	7386	A reversal-type multilayer color film used for making prints from reversal color originals, particularly those made on Ektachrome ER films. However, satisfactory prints may also be obtained from originals made on Ektachrome commercial film. Processing of Ektachrome reversal print film is identical with that used for Ektachrome ER films.	No (35) Yes(16)	No
Reversal duplicating		7361	A black-and-white reversal film intended for making prints from black-and-white or color positive originals. Suitable for making work prints for editing purposes. Can be processed in same solutions as used for other black-and-white reversal films.	No	No

*35mm only.

Notes:

1 All 35mm with positive perforations is not footage-numbered; all 35mm with negative perforations is footage-numbered.

2 16mm is neither footage-numbered nor frame-line-marked.

TABLE F-11 Eastman Leader Stocks

Name	Type no.	Description	Thickness	Length	Perforations
No. 4 clear leader	5980	Clear, transparent, uncoated support	0.0050 in.	35mm × 1,015 and 2,015 ft 16mm × 1,220 ft	Negative or positive or unperforated Perforated one edge, two edges, or unperforated
Title leader	5983	Clear, transparent, uncoated support	0.0050 in.	35mm × 1,000 ft	Negative
No. 6 black-and-white opaque leader	5981	Opaque, black on one side and white on other side. Satisfactory for use in various types of processing machines	0.0083 in.	35mm × 800 ft 16mm × 800 ft	Oversize positive or unperforated Oversize perforated two edges, perforated one edge, or unperforated
No. 6 green leader	5982	Green, transparent, uncoated support. Recommended for use in processing machines where long leader life is desired	0.0095 in.	35mm × 800 ft 16mm × 800 ft	Oversize positive or unperforated Oversize positive perforated on two edges, perforated on one edge, or unperforated

TABLE F-12 Eastman Specialized-type Films

Film		Color sensitivity	Applications and special properties	Base thickness and type	Backing	Photo-recording sensitivity and ASA speed
2490 RAR (Estar base)		Blue	For oscillography (mercury, xenon); modulated glow-tube (argon) recording; seismic recording (variable area or density); and CRT recording with blue-emitting phosphors (P5, P11, P16).	4-mil polyester	Dyed gel	64
2492 RAR (Estar base)		Ortho	For oscillography (tungsten); fluoroscopy; hydrogen-bubble chamber photography; high-speed photography; and CRT recording with green-emitting phosphors (P1, P2, P15, P24, P31).	4-mil polyester	Clear gel	160
2479 RAR (Estar—AH base)		Ext. red Pan	For outdoor and artificial-light (all sources) photorecording; oscillography (all sources); modulated glow-tube (neon) recording; and CRT recording with all types of phosphors.	4-mil polyester	Clear, fast-drying	650 (ASA 400)
2496 RAR	Reversal processing	Ext.	For artificial-light (all sources) and outdoor photorecording; cloud-chamber photography; modulated glow-tube (neon) recording; and CRT	4-mil polyester	Clear, fast-	160

(Estar—AH base) (5498 RAR)	Negative processing	Pan	low. Reversal-processed films have very good projection quality. 2498 RAR film has an antihalation coating under the emulsion to enhance sharpness and minimize halation. 5498 RAR film has medium resolving power, extremely fine granularity, and very high acutance.	polyester (5498: $5\frac{1}{4}$-mil triacetate)	Clear, fast-drying	250) / 250 (ASA 250)
Linagraph recording (Estar base and Estar thin base)		Blue	For oscillography (mercury, xenon); seismic and modulated glow-tube (argon) recording; well logging; and CRT recording (P5, P11, P16). Thin base for maximum footage per load.	$2\frac{1}{2}$- and 4-mil polyester	Dyed gel	32
Linagraph drift survey		Ortho	For drift-survey cameras and photoclinometers. Will withstand 300°F temperature for up to 2 hours during the exposure run.	$5\frac{1}{4}$-mil triacetate	None	6
Linagraph ortho		Ortho	For oscillography (tungsten); fluoroscopy; hydrogen-bubble-chamber photography; and CRT recording with green-emitting phosphors (P1, P2, P15, P24, P31). Also available on gray base.	$5\frac{1}{4}$-mil triacetate	None	400
Linagraph Shellburst (Estar—AH base)		Ext. red Pan	For cinetheodolite photography of aerial objects against a blue- or white-sky background and for cloud-chamber and hydrogen-bubble-chamber photography.	4-mil polyester	Clear, fast-drying	200
2475 recording (Estar—AH base)		Ext. red Pan	For all types of photorecording and police science photography when weak illumination, extremely short exposure times, or high writing speeds require an extremely high speed film.	4-mil polyester	Clear, fast-drying	1000 (ASA 800)
2485 high-speed recording		Ext. red Pan	Extremely fast film for all types of photorecording and police science photography. Film speeds and contrasts can be varied over a wide range by selecting the proper developer-time-temperature condition for a particular situation.	4-mil polyester	Clear gel	2500
High-speed infrared		1R	For all types of recording between 700 and 900 mμ, such as thermography, infrared luminescence, and high-speed photography. Requires a Wratten filter no. 25, 29, 70, 87, 87C, 88A, or 89B.	$5\frac{1}{4}$-mil triacetate	None	320 (ASA 160; No. 25 filter)

TABLE F-13 Eastman Standard Roll Lengths (in feet)

Films	35mm		16mm	
	Black-and-white	Color	Black-and-white	Color
Camera*	100, 200, 400, 1,000, and in some cases, 2,000	100, 200, 400, 1,000 (special order, 2,000)	100, 200, 400, 1,200	100, 200, 400, 1,200
Release print	1,000, 2,000 (special order, 3,000)	1,000, 2,000 (special order, 3,000)	400, 1,200, 2,000	1,200, 2,000
Sound recording and television recording	1,000 (special order, 3,000)		400, 1,200 (special order, 2,400)	
Duplicating	1,000, 2,000 (special order, 4,000)	1,000, 2,000	400, 1,200, 2,000	1,200, 2,000

*Camera films are normally supplied on spools for loading in subdued light when purchased in 35mm × 100-ft and 16mm × 200-ft lengths. All other lengths of camera films and other film types are supplied on cores.

TABLE F-14 Filters for Kodak Color Films

Light source	Daylight: Kodachrome, Ektachrome, Kodacolor-X, Ektacolor Professional, Type S	Type A: Kodachrome II Professional, Type A	Type B: Ektachrome	Type L: Ektacolor Professional, Type L
Daylight	None*	85	85B	85
Electronic flash	None†	Not recommended	Not recommended‡	Not recommended
Blue flashbulbs	None	Not recommended	Not recommended‡	Not recommended
Clear flashbulbs	80C or 80D¶	81C	81C	Not recommended
Photoflood lamps	80B§	None	81A	81A
3200°K lamps	80A§	82A	None	None

*With reversal color films, you can use a skylight filter (no. 1A) to reduce excessive bluishness of pictures made in open shade or on overcast days, or pictures of distant scenes, such as mountain and aerial views.

†With Kodacolor-X film, use a no. 81B filter if your prints are consistently too blue.

‡With Kodak high-speed Ektachrome film, type B, use a no. 85B filter.

¶Use a no. 80D filter with zirconium-filled clear flashbulbs such as AG-1, M3, and M5.

§This light source is not recommended with Ektachrome film, daylight type (process E-3).

TABLE F-15 Exposure* and Filter Compensation for the Reciprocity Characteristics of Kodak Color Films

Film	Exposure time, seconds					
	1/1,000	1/100	1/10	1	10	100
Kodacolor-X film	None No filter	None No filter	None No filter	None No filter	+1 stop No filter	+2 stops No filter
Kodak Ektacolor professional film, type S (short exposure)	None No filter	None No filter	None No filter	Not recommended		
Kodak Ektacolor professional film, type L (long exposure)	Not recommended		See film instructions for speeds. No filter from 1/10 to 60 seconds.			
Kodak Ektachrome film, daylight type (process E-3)	None No filter	None No filter	+½ stop CC10B	+1 stop CC30B	+2 stops CC40B+ CC10M	+3⅓ stops CC50B+ CC10M
Kodak Ektachrome film, type B (process E-3)	None No filter	None No filter	None No filter	None No filter	+⅓ stop No filter	+1 stop No filter
Kodak Ektachrome-X film, daylight type	None No filter	None No filter	+⅓ stop CC10C	+1 stop CC20C	+2 stops CC30C+ CC05G	+3 stops CC30C+ CC10G
Kodak high-speed Ektachrome film, daylight type	+⅓ stop No filter	None No filter	+⅓ stop CC10B	+⅔ stop CC10B	+1⅓ stops CC10B	+2⅔ stops CC10B
Kodak high-speed Ektachrome film, type B	None No filter	None No filter	None No filter	+⅓ stop No filter	+1⅓ stops CC10R	+2⅓ stops CC20R
Kodachrome II professional film, type A	None No filter	None No filter	+⅓ stop CC05R	+⅔ stop CC10R	+1⅓ stops CC20R	+2 stops CC25R
Kodachrome II film for daylight	None No filter	None No filter	None No filter	+⅓ stop No filter	+1 stop CC10G	+1⅔ stops CC20G
Kodachrome-X film, daylight type	None No filter	None No filter	None No filter	+⅓ stop No filter	+⅔ stop No filter	+1⅔ stops No filter

*The exposure increase, given in lens stops, includes the adjustment required by any filter or filters suggested. The data for each film in the table apply only to the type of illumination for which that film is balanced, and the figures are based on average emulsions.

The medium speed of this film permits movie making under bright daylight illumination, without the neutral-density filters or variable-shutter adjustments that are often needed with higher-speed films. In addition, the fine grain, high resolution, excellent gradation, and wide exposure latitude of medium-speed reversal film, type 2955, make it ideal for television news and sports photography, for general industrial and scientific work, as well as for all applications where a high-quality, medium-speed reversal film is required.

		Film speed	
	Daylight 64	*Tungsten* 50	
Type 2955	16mm	Regular perforations	2R 3000
		Sound B perforations	1R 3000
		Prestriped for magnetic sound	1R 3000

High-speed Reversal Film, Type 2961, 16mm. High-quality panchromatic film. With reversal processing, it yields a positive image for direct projection that has excellent pictorial quality, good sharpness, and medium grain. In addition to its usefulness in daylight situations, the high speed renders it suitable for filming with average television studio lighting, with portable newsreel lights, and with the lighting found in most indoor sports arenas. This film also yields excellent results outdoors at night, under lighting found at football and baseball fields.

		Film speed	
	Daylight 200	*Tungsten* 160	
Type 2961	16mm	Regular perforations	2R 3000
		Sound B perforations	1R 3000
		High-speed perforations	2R 3000
		Twin 8 perforations	2R 3000
		Prestriped for magnetic sound	1R 3000

GAF Anscochrome films

Anscochrome D/64. Reversal color film balanced for exposure by daylight. It provides high-quality positive originals for direct projection. The medium speed and outstanding color characteristics of this film render it an excellent choice for movies under normal daylight conditions. Anscochrome D/64 film is recommended for motion pictures in the areas of sports, documentaries, industry, education, and news and television reporting, as well as in other fields where medium speed, fine grain, and faithful color rendering are required.

	Film speed		
	Daylight 6000°K 64	*Tungsten* 3400°K 32[a] 3200°K 16[b]	
D/64	16mm	Regular perforations	2R 3000
		Sound B perforations	1R 3000
		High-speed perforations	2R 3000
		Twin 8 perforations	2R 1500
D/64	35mm	BH 1870 perforations	

Anscochrome T/100: *Type* 2240, 16mm; *Type* 5240, 35mm; *Type* 7240, 70mm. Reversal color film provides high-quality positive originals for direct projection. Anscochrome T/100 film is balanced for tungsten light (3200°K). Because of its high speed and outstanding color characteristics, this film is the best choice for a wide variety of professional applications. It is recommended for motion analysis, data recording, stress studies, available-light news and television reporting; and for industrial, scientific, military, medical, engineering, and other work where a combination of high speed and color fidelity is required.

	Film speed		
	Daylight 6000°K 64[c]	*Tungsten* 3400°K 80[d] 3200°K 100	
Type 2240	16mm	Regular perforations	2R 3000
		Sound B perforations	1R 3000
		High-speed perforations	2R 3000
		Twin 8 perforations	2R 1500
		Prestriped for magnetic sound	1R 3000
Type 5240	35mm	BH 1870 and KS 1870 perforations	
Type 7240	70mm	Type I and type II perforations	

Anscochrome D/200: *Type* 2230, 16mm; *Type* 5230, 35mm; *Type* 7230, 70mm. Reversal color film balanced for exposure by daylight. It provides high-quality positive originals for direct projection. The extremely high speed and outstanding color characteristics of this film make it the best choice for movies under low-level daylight illumination. It is also excellent for applications where short exposures are encountered. Anscochrome D/200 film is recommended for high-speed motion analysis, data recording, stress studies, available-light news and television reporting; and for industrial, scientific, military, medical, engineering, and other work where a combination of high speed and color fidelity is required.

[a] With 80B filter.
[b] With 80A filter.
[c] With 85B filter.
[d] With 81A filter.

	Film speed			
	Daylight	*Tungsten*		
	6000°K	3400°K 3200°K		
	200	100[a] 50[b]		
Type 2230	16mm (acetate base)	Regular perforations	2R 3000
		Sound B perforations	1R 3000
		High-speed perforations	2R 3000
		Twin 8 perforations	2R 1500
Type 2230	16mm (polyester base)	High-speed perforations	2R 3000
Type 5230	35mm (acetate base)	BH 1870 perforations	
		KS 1870 perforations	
Gafstar film	35mm (polyester base)	BH 1870 perforations	
		KS 1870 perforations	
Type 7230	70mm (acetate base)	Type I and type II perforations	
Gafstar film	70mm (polyester base)	Type I and type II perforations	

Anscochrome D/500. Extremely high speed, reversal color film balanced for exposure by daylight. It is excellent for applications where low light levels or short exposures are encountered. Anscochrome D/500 film is recommended for high-speed motion analysis, data recording, stress studies, and similar work. It is also useful for available-light news and television reporting; and for industrial, scientific, military, medical, engineering, and other work where a combination of extremely high speed and color fidelity is required.

	Film speed		
	Daylight	*Tungsten*	
	6000°K	3400°K 3200°K	
	500	250[a] 125[b]	
D/500	16mm	High-speed perforations	2R 3000
Gafstar film	35mm	KS 1870 perforations	
	70mm	Type II perforations	

GAF aerial film

AnscoPan T. Medium-speed panchromatic film meets the exacting requirements of modern aerial photography. It has extended red sensitivity to improve haze penetration, fine grain, and high resolution to provide maximum ground detail, and a high gamma to counteract the contrast-lowering effect of atmospheric haze. Contrast can be varied over a wide range by developing-control to get optimum results from different brightness ranges. The thin, highly hardened emulsion permits rapid development at elevated temperatures.

[a] With 80B filter.
[b] With 80A filter.

Coated on a low-shrink acetate base, AnscoPan T aerial film is recommended for topographic work and general aerial photography. The film is supplied on special order in all standard lengths, in the following widths: 70mm perforated, 5 inch perforated, 5 inch unperforated, and 9½ inch unperforated.

GAF instrumentation and recording films

Hyscan Gafstar Recording Film and Hyscan Pan Gafstar Recording Film. High-resolution, high-contrast films designed for high-temperature and rapid-access processing in photoinstrumentation systems. Hyscan Gafstar film has a sensitivity range which yields maximum response to blue-emitting phosphors (P5 and P16). Hyscan Gafstar film is sensitive to P11, P12, and P19 phosphors, as well as to light sources outside the blue range.

The thin, highly hardened emulsions of Hyscan Gafstar and Hyscan Pan Gafstar film are coated on 0.004-inch polyester base. These films are recommended for use in cathode-ray-tube recording, glow-modulator-tube recording, oscilloscope recording, and airborne instrumentation photography. The film is available in 16, 35, and 70mm, and 5 inch perforated and unperforated rolls, and 9½ inch unperforated rolls.

	Film speed		
	Gafstar	*Pan Gafstar*	
	25	64	
Hyscan Gafstar film	16mm	High-speed perforations	2R 3000
	35mm	BH 1870 perforations	
	70mm	Type II perforations	
Hyscan Pan Gafstar film	16mm	Unperforated	
	35mm	Unperforated	
	70mm	Unperforated	

Ilford films

Ilford Fine-grain Safety Positive. A very fine grain, blue-sensitive emulsion coated on 0.0005-inch safety base which may be either clear or dyed gray. This film is primarily intended for release printing but has many additional applications including high-contrast line and diagram animation in title cameras and the preparation of traveling matte titles.

The film may be handled and processed by the illumination provided by an Ilford S safelight no. 902 (light brown) in an Ilford darkroom lamp with the specified bulb.

The recommended control gamma is 2.2 to 2.4 in standard motion picture positive developer.

Ilford Mark V. A very fast fine-grain panchromatic film coated on 0.0005-inch gray antihalation safety base. This film has finer grain than HP3 (a very fast panchromatic negative film) and a wider exposure latitude. These characteristics, together with its high speed, make the film suitable for the production of pictures of very high quality under a wide variety of lighting conditions. Newsreel photography in poor light, filming at night, and television work generally are situations in which this film is particularly useful. In the studio, the high speed of the film makes great depth of field possible without the need for exceptionally powerful lighting units.

Meter settings	*Daylight*	*Tungsten*
ASA	250–500	200–400
DIN	25–28	24–27

The recommended control gamma is 0.6 to 0.7 in standard motion picture negative developers. This film is exceptional in that considerable speed gains are possible by careful processing techniques. Speeds as high as 1300 ASA have been realized without loss of image quality, using Ilford Microphen fine-grain developer.

Ilford HRT. A 16mm blue-sensitive film primarily intended to be used in fast pulldown television film-recording cameras. The film has a 0.0005-inch gray antihalation safety base. The gray dye helps to increase image sharpness by reducing flare. It also prevents excessive transmission in the highlights.

The emulsion has very fine grain and, because it is blue-sensitive, gives maximum speed when used for filming blue tube phosphors. The high-frequency response necessary for a telerecording film is achieved with a frequency-signal curve having a decided rising characteristic in the upper region. The signal response up to 3 megacycles per second is exceptionally good. The emulsion has a specially lubricated supercoat to overcome the mechanical problems and danger of scratching, commonly associated with the high-speed operating necessary in television film recording. The film base has exceptionally high tensile strength to withstand the severe stresses occurring during the fast transport of the film. The film can be handled and processed by the illumination provided by an Ilford F safelight no. 904 (dark brown) in an Ilford darkroom lamp with the specified bulb.

The recommended control gamma is 1.0 to 1.2 in standard laboratory motion picture negative developer. When used to derive a positive image from tubes displaying phase-reversed pictures, a gamma of about 1.6 is commonly employed.

Ilford Pan F Negative. A medium-speed, extremely fine grain panchromatic film coated on 0.0005-inch gray antihalation safety base. It is recommended for cinematography of all types of subject by daylight, except when lighting conditions are poor. It is also suitable for use in title work when a panchromatic film is necessary.

Pan F is capable of being developed to relatively high gammas in standard negative developers and is therefore suitable for applications requiring high contrast such as traveling mattes.

Meter settings	*Daylight*	*Tungsten*
ASA	25–50	20–40
DIN	15–18	14–17

The recommended control gamma is 0.6 to 0.7 in standard motion picture negative developers.

Ilford FP3 Negative. FP3 is a fast, fine-grain, panchromatic negative film coated on 0.0005-inch gray antihalation safety base. It is intended for cinematography by daylight for all subjects and may also be used for work in artificial light in good lighting conditions. It is widely used for television coverage.

Meter settings	*Daylight*	*Tungsten*
ASA	80–160	64–125
DIN	20–23	19–22

The recommended control gamma is 0.6 to 0.7 in standard motion picture negative developers.

Ilford Newsreel Positive Film. A positive film coated on 0.0005-inch acetate safety base. It is available in 35mm widths only.

Ilford Clear-base Film. A 0.0005-inch clear acetate safety-base film. Available in 16mm widths only.

Ilford Leader Film. Leaders are available in 35 and 16mm widths and are clear base, coated, or opaque. Clear and coated leaders have a 0.0005-inch base, and opaque leaders have a 0.0008-inch base.

Metro/Kalvar film. Metro/Kalvar film is the product of a new and basically different system of photography which requires no darkroom or chemical processing. In use, the film is exposed to ultraviolet light to form a latent photographic image and heated to develop the image permanently. The finished print is unique in that the opaque area of the heat-developed image is composed of light-scattering centers, rather than light-absorbing grains or dyes as in conventional films.

Metro/Kalvar raw stock does not require darkroom or safelight handling and may be used under ordinary levels of visible light. This unique operational freedom should not be overextended. The raw stock has a useful shelf life of at least five years and may be stored and transported in normal ambient temperatures and humidity. The raw stock should, however, be kept in its ultraviolet protective wrapper.

The film is not edge- or frame-numbered. Types 233 and 253 are available with preapplied magnetic sound-recording stripe at additional cost. Spectral sensitivity for all four Metro/Kalvar films is in the near ultraviolet range—3400 to 4400 Å. All Metro/Kalvar's polyester-based films require tape splicing.

Operating under normal room light at speeds up to 100 frames per minute, the Metro/Kalvar printer-processors produce ready-to-project prints from laboratory standard negatives in a single, completely dry operation. Their model 216 features separate sound and picture-printing heads to provide high-speed release printing of 16mm sound motion pictures. The 135/16 model which is convertible to either 35 or 16mm, is designed to produce filmstrips as well as motion picture work prints.

Metro/Kalvar films are composed of a tough Saran plastic emulsion coated on a base of high-strength, dimensionally stable polyester; and feature scratch resistance and long life. The properly exposed, developed, and fixed image is one of the most stable of all photographic images, with high resistance to environmental change and mechanical stresses. Oils and greases, including fingerprints, are easily removed.

See THERMOPLASTIC RECORDING and the KALVAR PHOTOGRAPHIC PROCESS.

Metro/Kalvar Type 233. 16 or 35mm television release positive for television release printing of black-and-white motion pictures.

Base:	3-mil polyester
Roll lengths:	16mm: 800, 1,200, and 2,000 feet
Cores and winding:	ASA 16mm, 2-inch cores in B winding
Perforations:	16mm ASA: PH 22.109-1958 (1R 2994)

Metro/Kalvar Type 253. 16 or 35mm cine-release positive is manufactured for cine-release printing of black-and-white motion pictures.

Base:	3-mil polyester
Roll lengths:	16mm: 800, 1,200 and 2,000 feet
Cores and winding:	ASA 16mm, 2-inch cores in B winding
Perforations:	16mm ASA: PH 22.109–1958 (1R 2994)

Metro/Kalvar Type 18. 35mm high-contrast filmstrip print film.

Base:	3-, 4-, and 5-mil polyester
Roll lengths:	35mm: 300 to 1,000 feet in increments of 100 feet
Cores and winding:	ASA 35mm, 2-inch cores—keyway slotted through entire width of core
Perforations:	35mm ASA: PH 22.34-1956 (BH 1870)

Metro/Kalvar Type 28. 35mm continuous-tone filmstrip print film.

Base:	3-, 4-, and 5-mil polyester
Roll lengths:	35mm: 300 to 1,000 feet in increments of 100 feet
Cores and winding:	ASA 35mm, 2-inch cores—keyway slotted through entire width of core
Perforations:	35mm ASA: PH 22.34–1956 (BH 1870)

film base The plastic material, or support, that carries the light-sensitive photographic emulsion. See FILM.

film chain The equipment that is used to convert film information to electronic information and, subsequently, to tape. Included are 35mm slide projectors and 35 and 16mm motion picture projectors.

film conversion See Table F-16.

Super 8 film.* The frame produced on Super 8 film is 50 percent larger in size than the frame in standard 8mm film. The print, in turn, is considerably sharper and brighter than the picture in the standard 8mm width. Running times for duplicate lengths of film, Super 8 and standard 8mm, are quite different, as shown by the following figures:

Film size	Inches per second		Feet per minute	
	18 fps† (silent speed)	24 fps† (sound speed)	18 fps† (silent speed)	24 fps† (sound speed)
8mm	2.7	3.6	13.5	18
Super 8	3.0	4.0	15.0	20

See Tables F-16 and F-17.

65mm film. At sound speed of 24 frames per second, 1 foot = 12.8 frames.

Feet	*Frames*
1/8	1 3/5
5/16	4
1/2	6 2/5
5/8	8
3/4	10
1	12 4/5

65/70mm film. See Table F-18.

filmograph Also called **slide motion film**. Basically, a filmstrip in which the illusion of movement is produced by camera movement. Still photographs and other nonanimated graphic material are positioned on the compound table of the animation stand, and zoom and panning movements are added during the filming process.

film perforations See PERFORATIONS and FILM.

*72 frames per foot.
†Frames per second.

TABLE F-16 Film Conversion

Time	Frames	Footage			Words
		8mm	16mm	35mm	
Seconds					
1	24	0.3	0.6	1.5	2
2	48	0.6	1.2	3.0	4
3	72	0.9	1.8	4.5	7
4	96	1.2	2.4	6.0	9
5	120	1.5	3.0	7.5	11
6	144	1.8	3.6	9.0	13
7	168	2.1	4.2	10.5	16
8	192	2.4	4.8	12.0	18
9	216	2.7	5.4	13.5	20
10	240	3.0	6.0	15.0	22
11	264	3.3	6.6	16.5	24
12	288	3.6	7.2	18.0	26
13	312	3.9	7.8	19.5	28
14	336	4.2	8.4	21.0	30
15	360	4.5	9.0	22.5	33
16	384	4.8	9.6	24.0	35
17	408	5.1	10.2	25.5	37
18	432	5.4	10.8	27.0	39
19	456	5.7	11.4	28.5	41
20	480	6.0	12.0	30.0	44
21	504	6.3	12.6	31.5	46
22	528	6.6	13.2	33.0	48
23	552	6.9	13.8	34.5	50
24	576	7.2	14.4	36.0	52
25	600	7.5	15.0	37.5	54
26	624	7.8	15.6	39.0	56
27	648	8.1	16.2	40.5	59
28	672	8.4	16.8	42.0	61
29	696	8.7	17.4	43.5	63
30	720	9.0	18.0	45.0	65
31	744	9.3	18.6	46.5	67
32	768	9.6	19.2	48.0	70
33	792	9.9	19.8	49.5	72
34	816	10.2	20.4	51.0	74
35	840	10.5	21.0	52.5	76
36	864	10.8	21.6	54.0	78
37	888	11.1	22.2	55.5	80
38	912	11.4	22.8	57.0	82
39	936	11.7	23.4	58.5	85
40	960	12.0	24.0	60.0	87
41	984	12.3	24.6	61.5	89
42	1,008	12.6	25.2	63.0	92
43	1,032	12.9	25.8	64.5	94
44	1,056	13.2	26.4	66.0	96
45	1,080	13.5	27.0	67.5	98

TABLE F-16 Film Conversion (cont.)

Time	Frames	Footage			Words
		8mm	16mm	35mm	
46	1,104	13.8	27.6	69.0	100
47	1,128	14.1	28.2	70.5	102
48	1,152	14.4	28.8	72.0	104
49	1,176	14.7	29.4	73.5	106
50	1,200	15.0	30.0	75.0	108
51	1,224	15.3	30.6	76.5	110
52	1,248	15.6	31.2	78.0	112
53	1,272	15.9	31.8	79.5	115
54	1,296	16.2	32.4	81.0	117
55	1,320	16.5	33.0	82.5	119
56	1,344	16.8	33.6	84.0	122
57	1,368	17.1	34.2	85.5	124
58	1,392	17.4	34.8	87.0	126
59	1,416	17.7	35.4	88.5	128
60	1,440	18.0	36.0	90.0	130
Minutes					
2	2,880	36.0	72.0	180.0	260
3	4,320	54.0	108.0	270.0	390
4	5,760	72.0	144.0	360.0	520
5	7,200	90.0	180.0	450.0	650
10	14,400	180.0	360.0	900.0	1,300
20	28,800	360.0	720.0	1,800.0	2,600
30	43,200	540.0	1,080.0	2,700.0	3,900
60	86,400	1,080.0	2,160.0	5,400.0	7,800

TABLE F-17 Super 8 Conversion

Film footage	18 frames per second (silent speed) minutes:seconds	24 frames per second (sound speed) minutes:seconds
1	0:04	0:03
2	0:08	0:06
3	0:12	0:09
4	0:16	0:12
5	0:20	0:15
10	0:40	0:30
15	1:00	0:45
20	1:20	1:00
25	1:40	1:15
30	2:00	1:30
50	3:20	2:30
100	6:40	5:00
150	10:00	7:30
300	20:00	15:00
500	33:20	25:00
1,000	66:40	50:00

TABLE F-18 65/70mm Film Conversion

At sound speed of 24 frames per second

Time	Feet and	Frames
Seconds		
1	1	11
2	3	$9\frac{2}{5}$
3	5	$7\frac{4}{5}$
4	7	$6\frac{1}{5}$
5	9	$4\frac{2}{5}$
10	18	$8\frac{4}{5}$
15	28	$\frac{3}{5}$
20	37	5
25	46	$9\frac{3}{5}$
30	56	$1\frac{1}{5}$
35	65	$5\frac{4}{5}$
40	74	$10\frac{1}{5}$
45	84	$1\frac{4}{5}$
50	93	$6\frac{2}{5}$
55	102	$10\frac{4}{5}$
Minutes		
1	112	$2\frac{2}{5}$
2	224	5
3	336	$7\frac{3}{5}$
4	448	$10\frac{1}{5}$
5	561	
10	1,122	
20	2,244	
30	3,366	
40	4,488	
50	5,610	
1 hour	6,732	

film speed The degree of sensitivity of an emulsion. A film's emulsion is rated in terms of its sensitivity to light. That degree of sensitivity is determined by varying the frame-to-frame exposure over a length of film that is selected for testing. Exposed under precise standards of illumination, the film is then developed in a controlled solution. The degree of blackness in the negative (the rate of development) is plotted in relation to exposure to produce the characteristic curve of the film. This information is used to state the speed of the film numerically.

Basically, if emulsion *A* is rated twice as fast as emulsion *B*, emulsion *A* needs half the exposure time given to emulsion *B* in order to provide the same degree of density in the resultant image. See CHARACTERISTIC CURVE.

film storage Raw stock (unexposed film) should be kept in its original packing and not be subjected to high temperatures or serious temperature fluctuations in order to assure a reasonably long shelf life. A storage temperature of 50°F with a relative humidity of approximately 55 percent will ensure that unexposed film will maintain the highest quality for more than six months. Raw stock should always be stored with the film roll flat to avoid stress marking.

Extremely long storage periods will bring about changes in the emulsion. The photographic sensitivity will gradually diminish, and the fog level, particularly of extremely fast emulsions, will increase, causing an overall loss of image quality and contrast. These changes can also be caused by storage temperatures which are unacceptably high. All unexposed motion picture films may benefit from storage at low temperatures, but it is important to restore the stock to normal operating temperatures over a period of some hours before the original packing is opened.

The most important requirement is that the temperature during storage should be constant, since it is only in this way that the image quality, which is carefully standardized in manufacture, can be maintained from reel to reel. Significant delays between exposure and processing should be avoided, since storage conditions during this period are particularly critical, and very long delays may cause fading of the latent image.

Exposed and processed motion picture film achieves archival permanence only if the storage conditions are the same as those for unexposed stock. The permanence of the image, however, depends not only on the storage conditions but also on the adequacy of the processing techniques—particularly during the final washing stage.

filmstrip Also called **slide film**. A sequence of related still photographs, illustrations, or other graphic material on motion picture film. Unlike standard motion picture projection processes, each frame of film is advanced individually and remains focused on

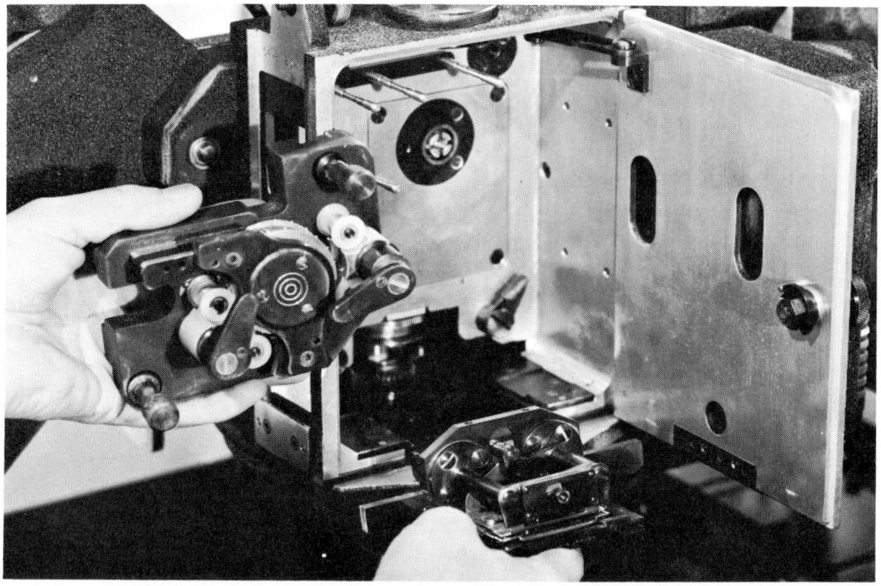

Fig. F-5 The film transport—the heart of the camera. *(Animation Equipment Corp.)*

the screen for the length of the accompanying narration. The sound tracks that complement the picture information are recorded on disks or tape.

Early filmstrips were also known as *bell* pictures. The reason for that designation was based on the method used for maintaining synchronization between the picture information in each frame and the sound track. During projection, a soft bell sound that had been incorporated in the sound track cued the operator who, when hearing the sound, advanced the film manually and brought the next picture into projection position. In recent years, the practice has been to record sound tracks on magnetic tape. These tapes have built-in cueing devices that trigger the projector and bring each frame into projection position automatically.

film transport The transport mechanism in the camera that controls the movement of film—from the feed-reel compartment in the magazine, through the shuttle for exposure, and on to the magazine's take-up compartment. See Figure F-5 and INTERMITTENT MOVEMENT.

filters Transparent gelatin or glass designed to alter the colors of the subject matter during exposure or processing.

Filters are available in a variety of colors. For convenient and efficient handling, they are protected against dust and fingerprints by thin pieces of optical glass. These are placed in a device known as an *adapter ring* and positioned in front of the camera lens.

The camera lens collects reflected light rays from the subject matter and transmits them to light-sensitive film. If the camera lens opening, shutter speed, and focus are all properly adjusted, the processed film should contain an image that accurately duplicates the color values in the original subject matter.

Consistent results, due to variations in the sensitivity of different types of film, are not possible. Filters, however, provide the means for restoring the color balance and duplicating the original tonal values seen by the cameraman at the time of filming. These are *corrective* filters. Other filters, used mainly for achieving special effects, either distort or exaggerate the color values in the original subject matter.

In general, both types permit the passage of light rays of their own color but hold back or absorb the light rays of complementary colors. Yellow filters, for example, permit colors in that range of the spectrum to be exposed on the film in the camera. Complementary colors are absorbed by these selective yellow filters. See Tables F-19 to F-27.

Mr. H. K. Harrison, an authority on the subject, offers this interesting comment: "Most people would know and recognize a filter when they see it. But, what actually is a filter? Saying that a filter is made of solid glass, laminated glass, gelatin, or gelatin between glasses is not enough. A filter is a colored object. It remains a colored object, and only that, until it is placed between two unknowns—a sender and a receiver. At that time, the filter becomes an instrument. In photography, the unknowns are light and film. In television, they may be the light and a TV camera or, in the alternative, light and videotape. When either of the two unknowns is changed, the function of the filter begins."

Selective filters, as the name implies, are sensitive only to the colors within their own spectral region. They reflect or absorb complementary colors. Subtractive filters, by comparison, are sensitive to practically all colors of the spectrum with the exception of the filter color itself.

Theoretically, every selective filter can be negated or offset by a complementary subtractive filter. Selective, or separation, filters are also used for making black-and-white separation negatives from material originally photographed on color film. These negatives are

TABLE F-19 Filters and Their Effect on Colors

Color of filter	*Light absorbed*
Red	Blue and green
Blue	Red and green
Green	Red and blue
Yellow	Blue
Magenta	Green
Cyan	Red
Black	Red, green, and blue
White	None
Gray	Equal amounts of red, green, and blue

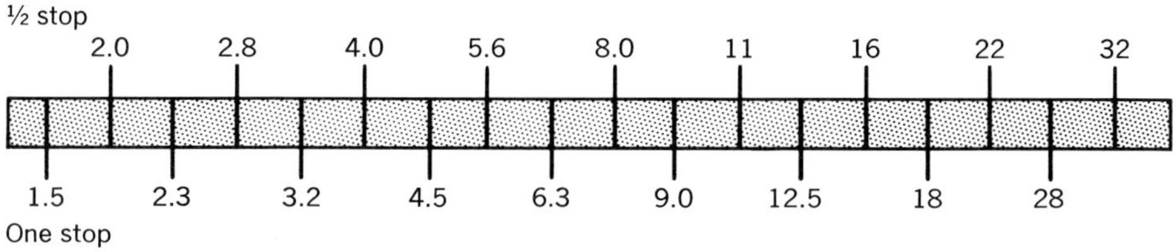

TABLE F-20 Stop Rule, Filter Data, and Exposure Data

Film type	131	136	140	928	930	931	932	936	937
Speed:									
Daylight	80	200	100	320	64	160	320	125	250
Tungsten	64	160	80	250	50	125	250	100	200
Filter	*f*-stop increase								
3 (Aero-1)	½	½	½	½	½	½	½	½	½
Aero-2	1	1	½	1	1	1	½	1	1
6 (K1)	1	½	½	1	1	1	½	1	1
8 (K2)	1	1	1	1	1	1	1	1	1
11 (X1)	2	2½	2½	2	2	2	2	2	2
12	1	1	1	1	1	1	1	1	1
13 (X2)	2½	3	3	2½	2½	2½	2	2	2
15 (G)	1½	1	1	1½	1½	1½	1½	1½	1½
21	1½	1	1½	2	1½	1½	1½	1½	1½
23A	2½	1½	1½	2½	2½	2½	2½	2	2
25 (A)	3½	2	2	3	3	3½	3	2½	3
29 (F)	5	2½	2½	4½	4	5	4	4	4
47 (C5)	2½	3	3	2½	2½	2½	3½	2½	2½
56	2	2	2	2	2	2	1½	2	2
58 (B)	3	3½	3½	3	3	3	2½	3	3
3N5	2½	2½	2½	2½	2½	2½	2½	2½	2½
8N5	3	2½	2½	3	3	3	3	3	3
30 ND	1	1	1	1	1	1	1	1	1
60 ND	2	2	2	2	2	2	2	2	2
90 ND	3	3	3	3	3	3	3	3	3

Courtesy of DuPont.

used for the subsequent reproduction of the original subject matter on color film.

Compensating, correction, and orthochromatic filters provide the means for duplicating the color values in the processed film so that they appear similar to the colors seen by the cameraman at the time of filming. In black-and-white photography, yellow or light-green filters assure the faithful reproduction of the equivalent values in the subject matter. In color photography, true tonal values can be obtained with light-blue, cyan, magenta, or yellow filters.

By comparison, contrast filters, in black-and-white photography, are used to lighten or darken a specific color without affecting the other colors in the subject matter. Filters of the same color are used for making colors appear lighter. Filters of a complementary color have the opposite effect.

TABLE F-21 Descriptions and Uses of Filters

Type	Number	Description and use
Colorless	0	For compensating thickness of other gelatin filters in optical systems.
	1A	Kodak skylight filter. Reduces excess bluishness in outdoor color photographs in open shade under a clear, blue sky.
Yellows	2A	Absorbs radiant energy below 405 mμ.
	2B	Absorbs ultraviolet below 390 mμ.
	2C	Absorbs ultraviolet below 385 mμ.
	2E	Absorbs radiant energy below 415 mμ.
	3	Light yellow (Aero no. 1).
	3N5	No. 3 plus 0.5 neutral density.
	4	Yellow. Approximate correction on panchromatic materials for outdoor scenes, including sky.
	6	K1—Light yellow. Partial correction outdoors.
	8	K2—Yellow. Full correction outdoors on type B panchromatic materials. Widely used for proper sky, cloud, and foliage rendering.
	8N5	No. 8 plus 0.5 neutral density.
	9	K3—Deep yellow. Moderate contrast in outdoor photography (with black-and-white films).
	11	X1—Yellowish-green. Correction for tungsten light on type B panchromatic materials; also for daylight correction with type C panchromatic materials in making outdoor portraits, darkening skies, or lightening foliage.
	12	Minus blue. Haze cutting in aerial photography.
	13	X2—Dark yellowish-green. Correction for type C panchromatic materials in tungsten light.
	15	G—Deep yellow. Overcorrection in landscape photography. Contrast control in copying and in aerial infrared photography.
	16	Blue absorption.
	18A	Transmits ultraviolet and infrared only (glass).
Oranges and reds	21	Blue and blue-green absorption.
	22	Yellow-orange. For increasing contrast in blue preparations in microscopy. Mercury yellow.
	23A	Light red. Two-color projection—contrast effects.
	24	Red for two-color photography (daylight or tungsten). White-flame-arc tricolor projection.
	25	A—Tricolor red for direct color separation. Contrast effects in commercial photography and in outdoor scenes. Two-color general viewing. Aerial infrared photography and haze cutting.
	26	Stereo red.

TABLE F-21 Descriptions and Uses of Filters (*cont.*)

Type	Number	Description and use
Oranges and reds	29	F—Deep red. Red color separation from transparencies. Strong contrast effects. Copying blueprints. Tungsten tricolor projection.
	92	Red. For densitometric measurement of color films and papers.
Magentas and violets	30	Green absorption.
	31	Green absorption.
	32	Minus green.
	33	Strong green absorption.
	34	Violet.
	34A	Blue separation.
	35	Contrast in microscopy.
	36	Dark violet.
Blues and blue-greens	38	Red absorption.
	38A	Red absorption. Increasing contrast in visual microscopy.
	39	Contrast control in printing motion picture duplicates (glass).
	40	Green for two-color photography (tungsten).
	44	Minus red. Two-color general viewing.
	44A	Minus red.
	45	Contrast in microscopy.
	45A	Blue-green. Highest resolving power in visual microscopy.
	46	Blue projection (experimental).
	47	C5—Tricolor blue for direct color separation and from Kodak Ektacolor film for dye transfer. Contrast effects in commercial photography. Tungsten and white-flame-arc tricolor projection.
	47A	Light blue. Used for medical photography.
	47B	Tricolor blue for direct color separation, separation from transparencies and from Kodak Ektacolor film for graphic arts.
	48	Green and red absorption.
	48A	Green and red absorption.
	49	C4—Dark blue.
	49B	Dark blue.
	50	Very dark blue. Mercury violet.
	94	Blue. For densitometric measurement of color films and papers.
	98	Blue. For use with Kodak Ektacolor paper—equivalent to a no. 47B plus a 2B filter.
Greens	52	Light green.
	53	Medium green.
	54	Very dark green.
	55	Stereo green.
	56	Very light green.
	57	Green for two-color photography (daylight).
	57A	Light green.

TABLE F-21 Descriptions and Uses of Filters (*cont.*)

Type	Number	Description and use
Greens	58	B—Tricolor green for direct color separation. Contrast effects in commercial photography and microscopy.
	59	Green for tricolor projection (white-flame-arc).
	59A	Very light green.
	60	Green for two-color photography (tungsten).
	61	N—Green color separation from transparencies and Kodak Ektacolor film. Tricolor projection (tungsten).
	64	Red absorption (light).
	65	Red absorption.
	65A	Red absorption.
	66	Light green. Contrast effects in microscopy and medical photography.
	93	Green. For densitometric measurement of color films and papers.
	99	Green. For use with Kodak Ektacolor paper—equivalent to a no. 61 plus a no. 16 filter
Narrow-band	70	Dark red. For use with Kodak Ektacolor paper and color separation from Kodak Ektacolor film (with tungsten).
	72B	Dark orange-yellow.
	73	Dark yellow-green.
	74	Dark green. Mercury green.
	75	Dark blue-green.
	76	Dark violet (compound filter).
	77	Transmits 546 $m\mu$ mercury line (glass plus gelatin).
	77A	Transmits 546 $m\mu$ mercury line (glass plus gelatin).
Photometrics	78	Bluish. Photometric filter (visual).
	78AA	Bluish. Photometric filter (visual).
	78A	Bluish. Photometric filter (visual).
	78B	Bluish. Photometric filter (visual).
	78C	Bluish. Photometric filter (visual).
	86	Yellowish. Photometric filter (visual).
	86A	Yellowish. Photometric filter (visual).
	86B	Yellowish. Photometric filter (visual).
	86C	Yellowish. Photometric filter (visual).
Light balancing	80A	Blue. Kodak 3200°K filter for Kodak daylight-type color films.
	80B	Blue. Kodak photoflood filter for Kodak daylight-type color films.
	80C	Blue Kodak photoflash filter for Kodak daylight-type color films and clear flashbulbs.
	80D	Blue. Kodak photoflash filter for daylight-type color films and AG-1 (clear) flashbulbs.
	81	Yellowish. For warmer color rendering.
	81A	Yellowish. For Kodak type B color films with photoflood lamps.
	81B	Yellowish. For warmer color rendering.

TABLE F-21 Descriptions and Uses of Filters (cont.)

Type	Number	Description and use
Light balancing	81C	Yellowish. For Kodak type A and type B color films with clear flashbulbs.
	81D	Yellowish.
	81EF	Yellowish.
	82	Bluish. For cooler color rendering.
	82A	Bluish. For Kodachrome II film type A with 3200°K lamps.
	82B	Bluish. For cooler color rendering.
	82C	Bluish. For cooler color rendering.
	85	Orange. Kodak daylight filter for Kodak type A color films.
	85B	Orange. Kodak daylight filter for Kodak type B color films.
	85C	Light orange. Kodak daylight filter for Kodak type F color films.
	85N3	For daylight exposure of Eastman color negative film at relatively large apertures.
	85N6	For daylight exposure of Eastman color negative film at relatively large apertures.
Miscellaneous	79	Photographic sensitometry. Corrects 2360°K to 5500°K.
	87	For infrared photography. Absorbs visual.
	87A	Absorbs visual and near infrared, transmits infrared.
	87B	Absorbs visual and near infrared, transmits infrared.
	87C	Absorbs visual, transmits infrared.
	88A	For infrared photography. Absorbs visual.
	89B	For infrared photography.
	90	Monochromatic viewing filter.
	96	Neutral filters for controlling luminance.
	97	Dichroic absorption.
	102	Conversion filter for Barrier-layer photocell (to luminosity).
	106	Conversion filter for S-4 type photocell (to luminosity).

Courtesy of Eastman Kodak.

In color photography, red filters deepen reds, oranges, and yellows. Blue and green colors, similarly, reproduce considerably darker after being photographed through the same red filters. The indiscriminate use of combinations of color filters is inadvisable since the results would be more or less unpredictable.

Paradoxically, for simulating night effects in daylight photography, combinations of filters are used to help achieve the desired effects. Not quite as desirable, however, are the light flesh tones caused by the use of red filters in day-for-night filming. This problem is resolved quite simply by the application of darker makeup. The

TABLE F-22 Filter Recommendations for Commercial Subjects

With Panchromatic Material unless Otherwise Specified

Subject	Effect desired	Suggested filter
Architecture		
Light-colored buildings against blue sky	Separation between building and sky	No. 8 (K2)
	Greater contrast between building and sky	No. 15 (G)
	Dark sky	No. 25 (A)
	Darker sky with any of above filters	Polarizing screen
Red-brick buildings	Lightening of tone to show texture	No. 15 (G)
Reflections from roofs and windows	To subdue	Polarizing screen
Automobiles		
Dark colors	Lightening of tone to show detail	Filter similar to color of car
Lettering on trucks, buses, etc.	Contrast to increase legibility	G, A, B, or C5
Reflections from lacquered surfaces	To subdue	Polarizing screen
Colored china, glass, plastics, pottery, tile, mosaics	Natural rendering	No. 8 (K2) or No. 11 (X1)
	Contrast to show pattern or indicate color contrast	G, A, B, or C5
Reflections	To subdue	Polarizing screens at lens and lights
Fabrics, clothing, tapestries, upholstered furniture	Natural rendering	No. 8 (K2) or No. 11 (X1)
	Contrast to show pattern or indicate color contrast	G, A, B, or C5
Dark colors	Lightening of tone to show texture	Filter similar to color of material
Furniture		
Red woods, as mahogany and rosewood	Best rendering of wood grain	No. 25 (A)
Yellow woods and others, as maple, walnut, oak, etc.	Best rendering of wood grain	No. 15 (G) with Kodak Tri-X Ortho film (Estar thick base)
Wood furniture with other colors, as in a room setting	Best grain with good rendering of other colors	No. 8 (K2) or No. 15 (G)
Leather, linoleum, wallpaper	Natural rendering	No. 8 (K2) or No. 11 (X1)
	Contrast to show pattern or indicate color contrast	G, A, B, or C5

TABLE F-22 Filter Recommendations for Commercial Subjects (cont.)

Subject	Effect desired	Suggested filter
Colored leather with chromium	Lightening of leather so image of metal detail will print	Filter similar to color of leather
Reflections from linoleum	Subdue reflections that obscure pattern	Polarizing screen
Packaging	To separate light reds from other colors	No. 11 (X1) or No. 8 (K2)
	To make lettering more readable or to separate various colored parts of package design	G, A, B, or C5
	Penetration of colored acetate sheeting to show material or detail within	Filter similar to color of sheeting
	To subdue reflections from Kodapak sheet or other clear sheeting	Polarizing screen

Courtesy of Eastman Kodak.

TABLE F-23 Descriptive Index of Filters and Light-correction Disks

Filters for	Spectro-Wedge numbers
Color photography	
Absorption of ultraviolet	1–7
Raising color temperature	31.1–32.4
Warmer color rendering	79–80
Lowering color temperature	79–80
Cooler color rendering	31.1–32.4
Color compensation	10, 25, 40, 55, 70, 85
Film conversion	30, 33, 77.5, 78, 81
1A (skylight)	84
UV15, 16, 17, and 2B series	1–7
81 to 81EF yellowish series	79
C$\frac{1}{8}$ to C14 coralite series	80
82 to 82C bluish series	32.4
B$\frac{1}{8}$ to B8 blue series	31.1
83, 85B, 85, type A, conversion 11	77–84
Conversion 10, 80B, 80C	29, 33
Black and white photography	
Cloud emphasis in black and white	63–71
Foliage emphasis in black and white	58–63
Contrast emphasis in black and white	83–92
RD1 to RD8 series	83–92
YL1 to YL9 series	69–74
GR1 to GR15 series	58–63
Blue 10 to 12 series	37
OR1 to OR3 series	76–80.5

TABLE F-24 Neutral-density Filters

Density	*% of transmission*
0.10	80
0.20	63
0.30	50
0.35	45
0.40	40
0.50	32
0.60	25
0.70	20
0.80	16
0.90	13
1.00	10
1.20	6.3
1.30	5
1.50	3.2
2.00	1
2.50	0.32
3.00	0.10
3.50	0.032
4.00	0.010

horses in western films, do not share these makeup problems. With them, it is merely a horse of another color.

Some filters, in both black-and-white and color photography, are used for reducing exposure. Other filters provide the means for exaggerating, distorting, or eliminating the light rays that are transmitted to the light-sensitive film in the camera.

Neutral-density filters are used mainly for reducing exposure. They are extremely useful when the amount of light falling on the subject matter is too great to allow even the smallest lens opening to provide the proper exposure. These filters make it possible to avoid the use of a small lens opening with the accompanying, and often undesirable, increase in the depth of field. In day-for-night filming, graduated neutral-density filters are used to darken sky areas without affecting the tonal values in the remainder of the scene.

Pola screens, or Polaroid filters, are used for controlling glare and brightness. They are especially useful for photographing large areas of sky and water in strong sunlight. Rotating the filter, in relation to the lens, causes such scenes to appear progressively lighter or darker, depending on the position of the filter. Like the neutral-density filters, they are not used for the purpose of altering color values but rather for eliminating glare and reflections.

Ultraviolet and infrared filters are sensitive only to the invisible spectrum. Their ability to absorb visible light makes them especially important to dual-film, traveling-matte photographic processes. See SELF-MATTEING PROCESSES.

TABLE F-25 Harrison Duraline Filters

Color or type	Spectro-Wedge number and common description	Use
Colorless	1 (Plain glass) 2B (2B) 2C (2C) 5 (UV15) 6 (UV16) 6A (2A) 7 (UV17) 7E (2E)	For focusing. Absorbs Ultraviolet below 390 mμ. Absorbs Ultraviolet below 385 mμ. Absorbs blue and violet only. Absorbs ultraviolet below 400 mμ. Absorbs ultraviolet below 405 mμ. Absorbs ultraviolet below 410 mμ. Absorbs ultraviolet below 415 mμ.
Magenta primary	10C (CC05M) 10D (CC10M) 10F (CC20M) 10H (CC30M) 10K (CC40M) 10L (CC50M) 10M (CC60M)* 10N (CC70M)* 10P (CC80M)* 10R (CC90M)* 10S (CC100M)*	Magenta color compensation series: Absorbs green for reducing greenish cast in color photography, printing and projection. *Also available in ½ steps of density.
Bluish magenta	13W (RD9) 13W (70)	For Kodak Ektacolor paper and color separation from Kodak Ektacolor film with tungsten. Also available in densities K to W.
Magentish blue	23W	Pan viewing glass, shows filter results for pan film.
Blue primary	25D (CC05B) 25E (CC10B) 25H (CC20B) 25K (CC30B) 25M (CC40B) 25N (CC50B) 25R (CC60B)* 25S (CC70B)* 25ST (CC80B)* 25T (CC90B)* 25U (CC100B)*	Blue color compensation series: Absorbs red and green for reducing yellow cast in color photography, printing, and projection. *Also available in ½ steps of density.
Cyanish blue	29M (Conversion 10)	Ansco daylight film with tungsten light. Also available in densities D to S.
	32D (82) 32E (82A) 32G (82B) 32H (82C)	For cooler color rendering. For Kodak type A film with tungsten 3200°K lamps. For cooler color rendering. For Kodak type F. S. film with tungsten 3200°K lamps. Above 4 filters also available in densities H to S.

Density Scale

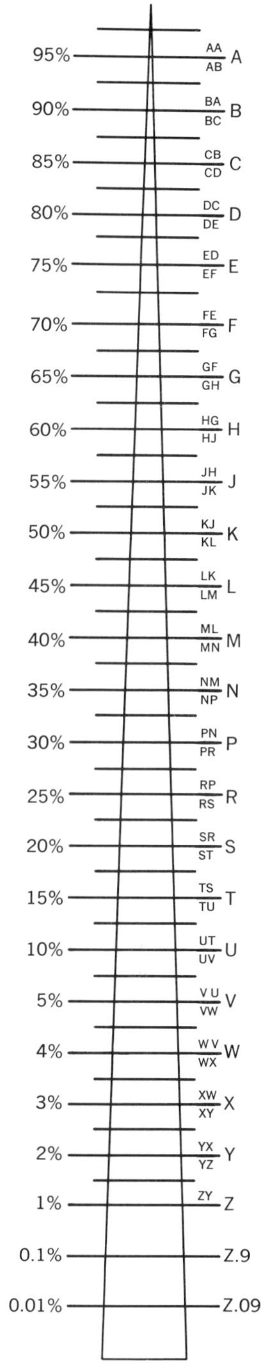

Transmission equivalent of the Spectro-Wedge letters following each color number. Example: 10D Letter D equals 80% total transmission.

TABLE F-25 Harrison Duraline Filters (cont.)

Color or type	Spectro-Wedge number and common description	Use
	33E (B$\frac{1}{8}$) 33F (B$\frac{1}{4}$) 33G (B$\frac{1}{2}$) 33H (B1) 33J (B2) 33K (B3) 33L (B4) 33P (B5) 33PR (B6) 33R (B7) 33S (B8)	Blue light corrector series: For correcting reddish light that is incorrect for the color film being used by raising the color temperature in specific steps of degrees Kelvin. See color temperature chart (Table C-3).
	33J (80C) 33L (80B)	Kodak daylight film with flash. Kodak daylight film with photoflood.
Bluish cyan	37VW (BL10) 37VW (47) 37WW (BL11) 37WW (48) 37W (BL12) 37W (49)	Standard tricolor, ortho viewing for ortho films. Experimental tricolor. Experimental tricolor.
Cyan primary	40C (CCO5C) 40D (CC10C) 40E (CC20C) 40F (CC30C) 40G (CC40C) 40H (CC50C) 40K (CC60C)* 40L (CC70C)* 40LM (CC80C)* 40M (CC90C)* 40N (CC100C)*	Cyan color compensation series: Absorbs red for reducing red cast in color photography, printing, and projection. *Also available in $\frac{1}{2}$ steps of density.
Greenish cyan	52R (GR11) 52R (61)	Color separation. Also available in densities D to R.
Green primary	55C (CC05G) 55D (CC10G) 55E (CC20G) 55F (CC30G) 55G (CC40G) 55H (CC50G) 55K (CC60G)* 55L (CC70G)* 55M (CC80G)* 55MN (CC90G)* 55N (CC100G)*	Green color compensation series: Absorbs blue and red for reducing bluish red cast in color photography, printing, and projection. *Also available in $\frac{1}{2}$ steps of density.
Yellowish green	58M (GR9) 58M (59) 58N (GR10) 58N (60)	Tricolor projection. Also available in densities D to T. Two-color photography. Also available in densities D to T.

TABLE F-25 Harrison Duraline Filters (cont.)

Color or type	Spectro-Wedge number and common description	Use
	60D (GR1)	Extra light correction.
	60F (GR2)	Very light correction.
	60M (GR6)	Two-color photography (daylight).
	60M (57)	
	60N (GR5)	Indoor tungsten (portraiture).
	60N (X2)	
	60R (GR8)	Tricolor separation.
	60R (58A)	
	62L (GR4)	Darkening skies, lightening foliage. Also available in densities D to P.
	62L (X1)	
	62P (GR7)	Standard color separation. Also available in densities D to P.
	62P (58)	
Greenish yellow	63G (GR3)	Light correction. Also available in densities D to VW.
	63G (56)	
	63VW (GR15)	Mitchell monoviewing glass. Also available in densities D to VW.
	68A (GY1)	Light aerial haze correction.
	68BA (GY2)	Light aerial haze correction.
	68B (GY3)	Medium aerial haze correction.
	68C (GY4)	Heavy aerial haze correction.
	P68B (YN3)	Very light yellow + 0.5ND.
	P68B (3N5)	
	P69C (YN8)	Medium yellow + 0.5ND.
	P69C (8N5)	
	69B (YL1)	Partial correction.
	69B (K1)	
	69CB (YL2)	Medium correction.
	69CB (K1½)	
	69C (YL3)	Full correction.
	69C (K2)	
Yellow primary	70BA (CC05Y)	Yellow color compensation series: Absorbs blue for reducing bluish cast in color photography, printing, and projection. *Also available in ½ steps of density.
	70B (CC10Y)	
	70BC (CC20Y)	
	70CB (CC30Y)	
	70C (CC40Y)	
	70CD (CC50Y)	
	70DC (CC60Y)*	
	70D (CC70Y)*	
	70DE (CC80Y)*	
	70ED (CC90Y)*	
	70E (CC100Y)*	
Reddish yellow	71D (YL4)	Moderate contrast.
	71D (K3)	
	71E (YL5)	Haze cutting in aerial photography.
	71E (Minus blue)	
	71G (YL6)	Overcorrection for beach scenes.
	71G (G15)	
	74U (YL9)	Monoviewing.
	74U (90)	

TABLE F-25 Harrison Duraline Filters (cont.)

Color or type	Spectro-Wedge number and common description	Use
	76K (OR1)	Blue and blue-green absorption.
	76K (21)	
	76N (OR2)	For increasing contrast in blue preparations in microscopy.
	76N (22)	
	76U (YL7)	Night effects for black-and-white film.
	76W (YL8)	Moonlight effects for black-and-white film.
	76W (72)	
Orange	77G (Conversion 11)	Ansco tungsten color to daylight exposure.
Yellowish red	78G (83)	Kodachrome commercial with daylight exposure.
	79BA (81)	For warmer color rendering.
	79B (81A)	For Kodak Ektachrome film, type B with photofloods.
	79BC (81B)	For warmer color rendering.
	79C (81C)	For Kodachrome film, type A with photoflash lamps.
	79CD (81D)	For Kodachrome film, type A with photoflash lamps.
	79DC (81EF)	For Kodak Ektachrome film, type B with photoflash lamps.
	79H (85B)	For Kodak Ektachrome film, type B with daylight exposure.
	80A (C$\frac{1}{8}$)	Coral light corrector series: For correcting bluish light that is incorrect for the color film being used by lowering the temperature in specific steps in degrees Kelvin. Refer to color temperature chart (Table C-3).
	80B (C$\frac{1}{4}$)	
	80C (C$\frac{1}{2}$)	
	80D (C1)	
	80E (C2)	
	80EF (C3)	
	80F (C4)	
	80G (C5)	
	80H (C6)	
	80J (C7)	
	80JK (C8)	
	80K (C9)	
	80KL (C10)	
	80L (C11)	
	80LM (C12)	
	80ML (C13)	
	80M (C14)	
	81E (85C)	Kodak type F color film with daylight exposure.
	81F (85)	Kodachrome type A film with daylight exposure.
	83H (RD1)	Very light contrast.
	83K (RD2)	Light contrast.
	83N (RD3)	Medium contrast.
	83P (RD4)	Two-color projection.

TABLE F-25 Harrison Duraline Filters (cont.)

Color or type	Spectro-Wedge number and common description	Use
	84B (Skylight) 84S (RD5) 84S (25A)	Absorbs violet blue and green. Standard tricolor red.
Red primary	85C (5R) 85D (10R) 85F (20R) 85H (30R) 85K (40R) 85L (50R) 85N (60R)* 85P (70R)* 85R (80R)* 85RS (90R)* 85S (100R)*	Red color compensation series: Absorbs blue and green for reducing bluish green cast in color photography, printing, and projection. *Also available in ½ steps of density.
Reddish magenta	92S (RD6) 92S (26) 92T (RD7) 92T (27) 92U (RD8) 92U (29)	Experimental projection red. Two-color viewing. Copying, strong contrast.

Haze filters are capable of penetrating and thus minimizing the effect of atmospheric haze by absorbing ultraviolet light which would normally "fog" the film in the camera. The ever-present light haze, in extreme long shots, can be penetrated with yellow filters in black-and-white photography. For denser concentrations, red filters are more effective. These filters cancel out some of the haze by reducing the blue in the water droplets that create the condition. The aero number 1 and 2 filters are made specifically for that purpose.

Color-temperature corrective filters, another specialized type, are used for balancing the color temperature of the light source with the sensitivity of the film.

When viewed under different lighting conditions (such as artificial light and daylight), the color of familiar objects such as food and clothing does not appear the same to the eye. Even when an object is examined under light sources that appear to be similar, the eye can easily be deceived because it has the ability to compensate for differences in the intensity of the light sources. Since the camera lens cannot make the same allowances, the color of the light source must be adjusted to suit the sensitivity of the film type.

Color-temperature meters make it possible for the cameraman to control the quality of the light so that accurate color results can be obtained in the print. (Color-temperature meters and exposure meters are not used for the same purpose. The latter measures levels of

TABLE F-26 Tiffen Filters for Black-and-white Film

Filter no.	Color or name	Purpose	Ortho Day-light	Ortho Tung-sten	Pan Day-light	Pan Tung-sten
			f-stop increase			
3	Aero 1	Aerial photography, haze penetration.	1	$\frac{2}{3}$	$\frac{2}{3}$	—
3N5	Yellow	Combines no. 3 with 0.5 neutral density.	$2\frac{2}{3}$	2	2	$1\frac{2}{3}$
6	Yellow 1	For all black-and-white films, absorbs excess blue, outdoors, thereby darkening sky slightly, emphasizing the clouds.	1	$\frac{2}{3}$	$\frac{2}{3}$	$\frac{2}{3}$
8	Yellow 2	For all black-and-white films, most accurate tonal correction outdoors with panchromatic films. Produces greater contrast in clouds against blue skies, and foliage. Can be used for special effects with color film.	$1\frac{1}{3}$	1	1	$\frac{2}{3}$
8N5	Yellow	Combines no. 8 with 0.5 neutral density, for control of high-speed films with same application of no. 8 (yellow 2) filter.	3	$2\frac{2}{3}$	$2\frac{1}{3}$	$2\frac{1}{3}$
9	Yellow 3	Deep yellow for stronger cloud contrast.	$1\frac{1}{3}$	1	1	$\frac{2}{3}$
11	Green 1	For all pan films. Ideal outdoor filter where more pleasing flesh tones are desired in portraits against the sky than can be obtained with yellow filter. Also renders beautiful black-and-white photos of landscapes, flowers, blossoms, and natural sky appearance.	—	—	2	$1\frac{2}{3}$
12	Yellow	Minus blue cuts haze in aerial work, excess blue of full moon in astrophotography.	$1\frac{2}{3}$	$1\frac{1}{3}$	1	$\frac{2}{3}$
13	Green 2	For male portraits in tungsten light, lightens foliage, with pan film only.	—	—	$2\frac{1}{3}$	2

TABLE F-26 Tiffen Filters for Black-and-white Film (cont.)

Filter no.	Color or name	Purpose	Ortho Day-light	Ortho Tung-sten	Pan Day-light	Pan Tung-sten
15	Deep yellow	For all black-and-white films. Renders dramatic dark skies, marine scenes, aerial photography. Contrast in copying.	$2\frac{1}{3}$	$1\frac{2}{3}$	$1\frac{2}{3}$	1
16	Orange	Deeper than no. 15. With pan film only.	—	—	$1\frac{2}{3}$	$1\frac{2}{3}$
21	Orange	Absorbs blues and blue-greens. Renders blue tones darker such as marine scenes. With pan film only.	—	—	$2\frac{1}{3}$	2
23A	Light red	Contrast effects, darkens sky and water, architectural photography. Not recommended for flesh tones. With pan film only.	—	—	$2\frac{2}{3}$	$1\frac{2}{3}$
25A	Red 1	Use with pan films to create dramatic sky effects, simulated moonlight scenes in midday (by slight underexposure). Excellent copying filter for blue prints. Use with infrared film for extreme contrast in sky, turns foliage white, cuts through fog, haze and mist. Used in scientific photography.	—	—	3	$2\frac{2}{3}$
			Filter factor			
29	Dark red	For strong contrasts; copying blueprints.	—	—	$4\frac{1}{3}$	2
47	Dark blue	Accentuates haze and fog. Used for dye-transfer and contrast effects.	—	—	$2\frac{1}{3}$	3
47B	Dark blue	Lightens same color for detail.	$2\frac{2}{3}$	3	3	4
56	Light green	Darkens sky, good flesh tones. With pan film only.	—	—	$2\frac{2}{3}$	$2\frac{2}{3}$
58	Dark green	Contrast effects in micro copy, produces very light foliage.	3	$2\frac{1}{3}$	3	3
61	Dark green	Extreme lightening of foliage.	—	—	$3\frac{1}{3}$	$3\frac{1}{3}$

TABLE F-26 Tiffen Filters for Black-and-white Film (*cont.*)

Filter no.	Color or name	Purpose	Ortho Daylight	Ortho Tungsten	Pan Daylight	Pan Tungsten
64	Green	Absorbs red.				
70	Dark red	Commercial darkroom work only.				
72B	Dark orange	Commercial darkroom work only.				
74	Dark green	Mercury lamp, commercial darkroom work.				
23A + 56		Creates night effects in daylight, with pan film only.				
87		For infrared film only, no visual transmission.				
87C		For infrared film only, no visual transmission.				
113		For infrared film only, no visual transmission.				
Neutral density	For all black-and-white films	For uniform reduction of light with high-speed films for still and movie cameras. No change of color value.	Available .1, .2, .3, .4, .5, .6, .7, .8, .9, 1.00 neutral densities. Available in all series sizes.			
Polarizer	For all black-and-white films	Eliminates surface reflections, unwanted glare, or hot spots from any light source.	2	2	2	2

Exposure depends on hour of the day and amount of light reflected from object photographed after the glare has been penetrated. If picture is taken through glass or water, additional exposure will be required.

Handy Filter Conversion Chart

Filter factor	If normal *f*-stop without filter is: 1.4	1.8	2.0	2.5	2.8	3.5	4.5	5.6	6.3	8	11	16	22
	Then here's *f*-stop with filter. No calculation necessary!												
1.5	1.1	1.3	1.6	2.0	2.3	2.9	3.7	4.6	5.1	6.6	9.2	12.5	18
2		1.2	1.4	1.8	2.5	2.8	3.5	4.5	5.6	6.3	8.0	11.3	16
2.5			1.2	1.6	2.3	2.5	3.2	3.6	4.0	5.2	7.1	10.1	14
3				1.4	1.6	2.0	3.0	3.3	3.6	4.6	6.6	9.2	13.1
4				1.2	1.4	1.7	2.5	2.8	3.2	4.0	5.6	8.0	11.3
5					1.3	1.6	2.2	2.5	2.9	3.6	5.0	7.2	10.1
6						1.4	2.0	2.3	2.6	3.3	4.5	6.3	9.2
7						1.3	1.8	2.1	2.4	3.0	4.2	6.1	8.5
8							1.6	2.0	2.2	2.8	4.0	5.6	8.0
10							1.4	1.8	2.0	2.5	3.5	5.0	7.1

TABLE F-27 Tiffen Filters for Color Film

Filter no.	Film type	Lighting	*f*-stop increase	Comments
Sky 1A	Daylight	Daylight	—	Use at all times, outdoors, to reduce blue and add warmth to scene. Also in open shade.
Haze 1	Daylight	Daylight	—	Reduce excess blue caused by haze and ultraviolet rays. Ideal for mountain, aerial, and marine scenes.
Haze 2A	Daylight	Daylight	—	Greater ultraviolet correction than Haze no. 1 filter and adds some warmth to the visible colors.
UV15	Agfachrome and Anscochrome daylight	Daylight	—	Haze filter, also for Anscochrome 3200°K film with photofloods.
UV16	Daylight	Electronic flash Daylight	—	Reduces excessive blue in electronic flash, also may be used for haze correction.
UV17	Daylight	Daylight	—	Greater haze correction, reduces blue in shade.
80A	Daylight	3200°K lamps	2⅓	Converts daylight film for use with 3200°K lamps.
80B	Daylight	3400°K floods	2	Converts daylight films to use indoors with 3400°K flood lamps.
1D-Flash	Daylight	Outdoor flash	1	With all flash (except SM and SF) outdoors or in deep shade where flash is main source of light. Cooler than no. 80C.
80C	Daylight	Early A.M., late P.M. Outdoor flash	1	Outdoor in early A.M. or late afternoon, also where flash is main source of light.
80D	Daylight	AG-1 clear flashbulbs	—	Balances clear AG-1 for daylight color film.
81	Type F Type F negative	M2 flash	⅓	Yellowish, warming filter.
81A	Type B	Electronic flash 3400°K floods	⅓	Balances daylight films to electronic flash. Corrects type B films for use with 3400°K lamps. Prevents excessive blue.
81B	Type B	3500°K lamps	⅓	Permits using spotlights or white fluorescent lamps (3500°K) with 3200°K films.
81C	Ektachrome type B (E-3) H.S. Ektachrome type B Kodachrome II type A	Clear flash (except SM or SF)	⅓	Permits the use of clear flash lamps except SM or SF.
81D	Anscochrome T Ektachrome type B (E-3) H.S. Ektachrome type B Kodachrome II type A	Clear flash (except SM or SF)	⅔	Permits the use of clear flash except SM or SF. Will render warmer results than no. 81C filter.

TABLE F-27 Tiffen Filters for Color Film (*cont.*)

Filter no.	Film type	Lighting	*f*-stop increase	Comments
81EF	Type A (3400°K)	M2 flash	⅔	For any 650°K drop, flash lamps with Ektachrome type B.
82	Daylight	Daylight	⅓	For any 100° increase in Kelvin temperature for cooler renderings.
82A	Type A (3400°K) Daylight, daylight negative Type F, type F negative	3200°K lamps Early A.M., late P.M. 3400°K floods	⅓	With daylight and daylight negative films use in early A.M. or late P.M. to reduce the excessive red of the light. When using type A (3400°K), films under 3200°K lamps. For type F and type F negative films when shooting under 3400°K photofloods.
82B	Type F Type F negative	SM, SF flash	⅔	For cooler results.
82C	Daylight negative Type F negative	3200°K lamps	⅔	For cooler results or when using 3200°K lamps.
83	Commercial		—	For commercial processing only.
85	Type A	Daylight Electronic flash	⅔	Converts indoor film to outdoor use. Corrects electronic flash when using daylight negative films.
85N3	Type A daylight negative	Daylight Electronic flash	1⅔	Combines no. 85 and 0.3 neutral density into one filter. Use same as no. 85. Will reduce light output of electronic flash when shooting close to subject.
85N4	Type A daylight negative	Daylight Electronic flash	1⅔	No. 85 with 0.4 neutral density combined into one filter for movie cameras with electric eye. Produces same exposure on Kodachrome II (ASA 25/40) as on the discontinued Kodachrome (ASA 10/16).
85N6	Type A daylight negative	Daylight Electronic flash	2⅔	No. 85 with 0.6 neutral density combined into one filter. Reduces light source more than 85N4.
85B	H.S. Ektachrome type B Anscochrome T-100 Ektachrome type B (E3) Ektacolor type L	Daylight	⅔	Converts indoor film to outdoor use. Also acts as haze filter.
85C	Type F Kodacolor Ektacolor AFA color CN 14, CN 17	Daylight	⅓	For negative film to be printed at home or in a custom laboratory. Type F film used in daylight.
CC30-R	Daylight films	Daylight Top window Side window Underwater	⅔	For underwater photography, to correct color. Also, to compensate for distortion of color when shooting through transparent plastic windows similar to vistadomes provided by railroads for camera fans.
820B	Polacolor	3200°K	—	For use with MP-3 camera to balance 3200°K light to Polacolor film.

TABLE F-27 Tiffen Filters for Color Film (*cont.*)

Filter no.	Film type	Lighting	f-stop increase	Comments
840B	Polacolor	75W or 150W floods (2900°K)	—	For use with MP-3 camera to balance lights with Polacolor film.
Neutral density	All film types Color or black and white	All light sources		For uniform reduction of light with high-speed films for still and movie cameras. No change of color value.
Polarizer	All film types Color or black and white	All light sources	2	Eliminates surface reflections, unwanted glare or hot spots from any light source. The only filter that will darken a blue sky.

Exposure depends on hour of the day and amount of light reflected from object photographed after the glare has been penetrated. If picture is taken through glass or water, additional exposure will be required.

Note: All exposure increases shown are approximate, and should be varied with differences in light and subject. All Exposure Indexes are ASA.

illumination.) See COLOR-TEMPERATURE METER and KELVIN DEGREES. Also see Table C-4 under COLOR TEMPERATURE.

The Spectro-Wedge color system is an ingenious method that is designed for the specific purpose of removing the guesswork from the selection of filters. With this standard guide, even the color-blind photographer can select the correct filter, or combination of filters, and be assured of achieving the desired results. In this system of positive identification, each color is assigned a definite number while the density of each color is identified by a letter of the alphabet.

Obviously, even colorless filters prevent a certain amount of light from passing through a camera lens. In order to compensate for this loss, even though in some cases it may be almost negligible, the cameraman can either increase the amount of exposure time or use a larger lens opening. This compensation, or the amount of increase over the normal exposure, is known as the *filter factor.*

Each filter is given an identifying number. See FILTER FACTOR. The accompanying exhibits show how the factor numbers are used for the purpose of computing the proper lens stop. See Figures F-6*a* and *b* and Tables F-20 to F-27.

filter factors Because filters absorb a percentage of the light passing through them, a proportionate increase in exposure, referred to as the filter factor, is made necessary because of their use.

This filter factor is a measure of the difference between normal exposure and the exposure compensation required by the particular filter being used. A filter with a factor of 2 indicates an increase in exposure equal to one *f*-stop. An added *f*-stop increase in exposure is necessary every time the factor doubles. A factor of 4 necessitates an exposure increase equal to two stops. A factor of 8, in turn, indicates an increase in exposure of three *f*-stops. See Tables F-28 and F-29.

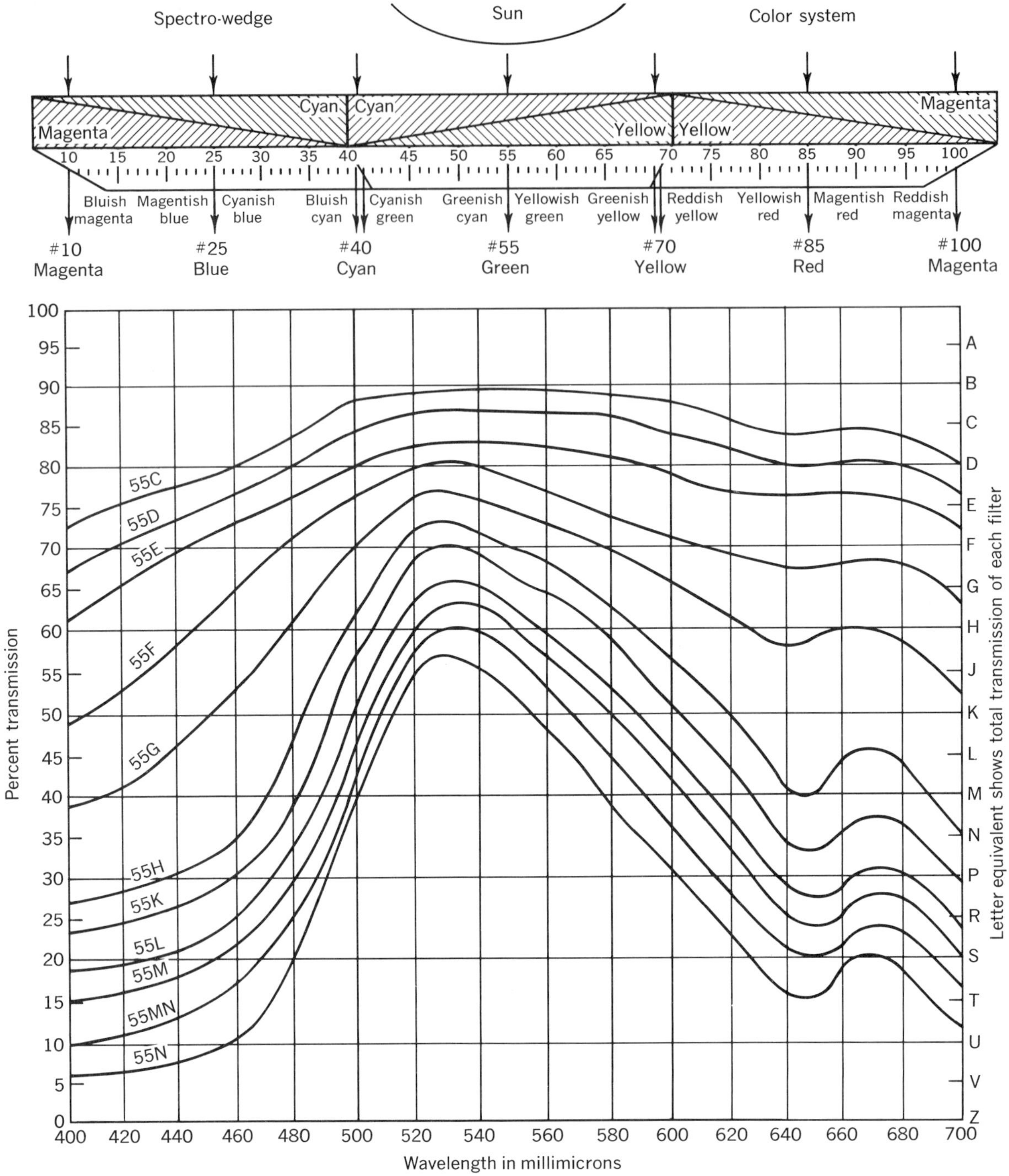

Fig. F-6a Spectro-Graphic curves. The Spectro-Graphic transmission curves graphically illustrate the transmission characteristics of each color filter with corresponding density letters. The transmission curves are grouped together in one graph to show how the density changes the position of the curve without changing its general shape. The number 55 green, for example, shows that each filter step from number 55B to number 55U has a different light characteristic ranging from 400 to 700 mμ. The 400-to-450 and the 640-to-700 range are always lower in transmission than the 450-to-640 range. Since each group of curves is taken from different sections of the color, the density steps for each individual filter can be selected with accuracy.

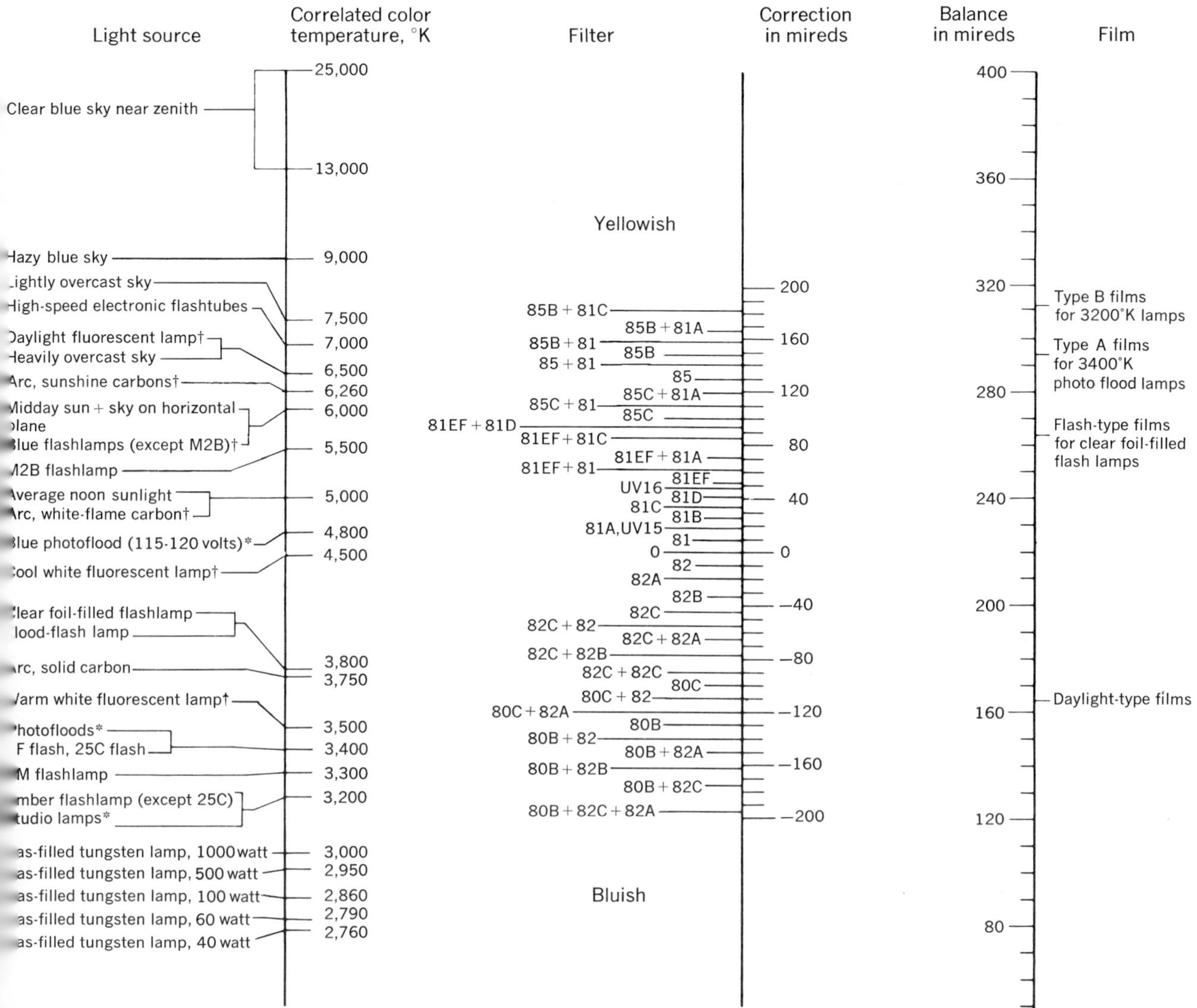

*The correlated color temperature of these lamps increases about 11°K for each volt increase applied potential, in the neighborhood of 115 volts. As lamps are used, the correlated color mperature (at a given voltage) decreases, often from 50°K above to 50°K below the rated value ıring the life of the lamp.

†Color temperature is only an approximate specification of these sources.

Fig. F-6*b* Color filter nomograph. (*Chart prepared by C. S. McCamy, NBS Photographic Research Section.*) One of the problems facing the color photographer is matching his color film to the light he is using to take the picture. Color films are usually balanced for a particular kind of light, classified as daylight, flash, 3200°K, or photoflood. When a film is used in light with a different color balance, filters are used over the camera lens to bring the combination back into agreement. This chart provides a quick and easy way to select the right filter for almost any combination of light source and color film. A straightedge laid across the three scales so that it connects the light source and the color film will cross the center line at a point corresponding to the proper filter. Notations consisting of a series of filter designations (e.g., 80B + 82C + 82A) mean that all these filters should be used at once for proper correction. When color-correcting filters are used, the exposure must be increased by the filter factor assigned by the filter manufacturer. See FILTER FACTOR.

TABLE F-28 Filter Factor Conversion

Filter factor number	*Lens stop opening increase*
1.0	0
1.5	½
2.0	1
2.5	1¼
3.0	1½
3.5	1¾
4.0	2
4.5	2⅛
5.0	2¼
6.0	2½
7.0	2¾
8.0	3
10.0	3¼
12.0	3½
14.0	3¾
16.0	4
20.0	4¼
24.0	4½
28.0	4¾
32.0	5
40.0	5¼
48.0	5½
56.0	5¾
64.0	6

Example A

Normal stop	1 stop open	2 stops open	3 stops open	4 stops open	5 stops open	6 stops open
f/16	*f*/11	*f*/8	*f*/5.6	*f*/4	*f*/2.8	*f*/2

If the normal exposure is *f*/16 and a filter having a factor of 4 is used, the lens stop must be opened to *f*/8, or two stops open from normal.

Example B

Normal stop	1 stop open	2 stops open	3 stops open
f/11	*f*/8	*f*/5.6	*f*/4

If the normal exposure is *f*/11 and a filter having a factor of 8 is used, the lens stop must be opened to *f*/4 or three stops from normal.

Example C

Normal stop	1 stop open	2 stops open	3 stops open	4 stops open	5 stops open
f/16	*f*/11	*f*/8	*f*/5.6	*f*/4	*f*/2.8

If the normal exposure is *f*/16 and a filter having a factor of 32 is used, the lens stop must be opened to *f*/2.8, or five stops open from normal.

TABLE F-29 Filter Factors and Lens Settings for Kodak Films and Plates

*Filter Factors for General Classes of Films and Plates**

Filter number	Color of filter	Non-color-sensitized		Ortho-chromatic		Pan-chromatic	
		Daylight	Tungsten	Daylight	Tungsten	Daylight	Tungsten
3	Light yellow	4	3	2	1.5	1.5	...
4	Yellow	8	5	2	1.5	1.5	1.5
6	K1, light yellow	4	3	2	1.5	1.5	1.5
8	K2, yellow	12	10	2	2	2	1.5
9	K3, deep yellow	20	16	2.5	2	2	1.5
11	X1, yellowish green	...	...	...	...	4	3
12	Yellow	...	...	3	2.5	2	1.5
13	X2, dark yellowish green	...	...	...	...	5	4
15	G, deep yellow	...	...	3	3	3	2
23A	Light red	...	...	...	...	6	3
50	Very dark blue	...	...	...	...	20	40
25	A, red	...	...	...	...	8	4
58	B, green	...	...	6	5	8	8
47	C5, blue	...	...	5	8	6	12
29	F, deep red	...	...	...	...	16	8
61	N, green	...	...	...	...	12	12
47B	Blue	...	...	6	8	8	16
Polarizing screen	Gray	For dark sky effects, use a factor of 2.5 (in addition to the exposure increase required for sidelighting), or increase the exposure four times as compared to the exposure for the same subject with frontlighting and without a polarizing screen.					

Lens Settings for Use with Filters (Average subject in bright sunlight, shutter at 1/100 or 1/125 second)

Kodak Film: Roll and 135	Filter			
	No. 8 (K2)	No. 11 (X1)	No. 15 (G)	No. 25 (A)
Panatomic-X	*f*/5.6	*f*/4	*f*/4-5.6	*f*/2.8
Verichrome Pan	*f*/11	*f*/8	*f*/8-11	*f*/5.6
Plus-X Pan	*f*/11	*f*/8	*f*/8-11	*f*/5.6
Tri-X Pan	*f*/22	*f*/16	*f*/16-22	*f*/11

Courtesy of Eastman Kodak.

*The filter factors for a specific film or plate may vary somewhat from this listing. See individual product instructions.

filter, videotape The filter, in videotape recorders, provides the precisely controlled overall response characteristic required for optimum signal-to-noise ratio and frequency response of the tape-playback system.

filter, visual A visual filter is not a filter in the true photographic sense but rather a colored glass that is used as a guide by the cameraman or director for the purpose of scanning the scene to see how the tonal values and colors will reproduce in a black-and-white or color print.

finder Also called **viewfinder**. A sighting device which is either attached to or built into the camera. In addition to its value for composition purposes, the finder shows the exact area that will be exposed on the film during the photographic process. The finder also makes it possible for the cameraman to focus with accuracy and correct for parallax.

In some turret mounts, the cameraman merely brings the desired lens into photographic position and sights through a matching lens that automatically comes into place in the viewfinder. In other cameras, built-in coupled range finders make precise focusing a relatively simple and routine operation. In recent years, camera manufacturers have incorporated mirrored shutters which allow the cameraman to view the scene while the camera is in operation. This innovation makes it possible for the camera operator to adjust focus or correct framing during the photographic process.

An *Albada* finder (named after its inventor) has a negative lens which is transparent from only one direction. The fields covered by the camera are outlined by rectangular markings.

The *compensating* finder, with its movable axis, allows the cameraman to set the finder so that it corresponds with the field covered by the lens.

The *iconograph* finder is perhaps the oldest, simplest, and best-known finder. Basically, the finder is a wire frame that is attached to the camera. It indicates the approximate boundaries of the scene.

The *Newtonian* finder is also known as a direct finder. Like the Albada finder, it has a short-focal-length negative lens which allows the subject to be viewed directly through the lens.

A *range* finder is used to view the subject matter from two different positions in order to find the exact distance between the subject matter and the camera. This type of finder eliminates the guesswork with regard to distance and makes it possible for the cameraman to focus properly.

A *reflecting* finder is a sophisticated viewfinder that houses a lens, mirror, and viewing surface. The viewing surface may be either a ground glass with engraved reticle markings or a lens. The latter is referred to as a *brilliant* finder. See Figure F-7.

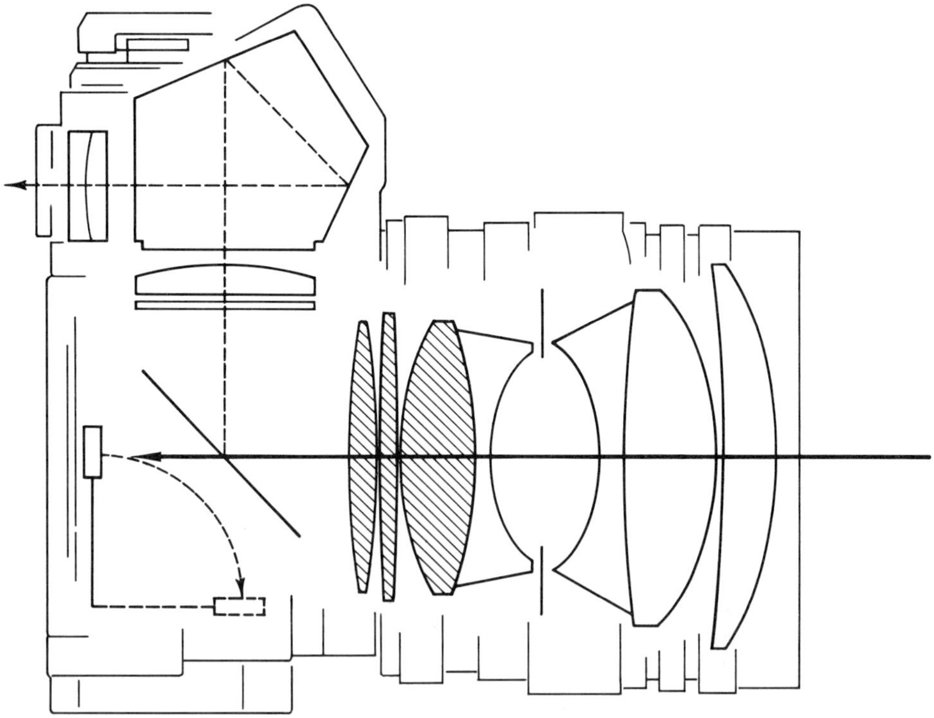

Fig. F-7 Reflecting finder. The Bell & Howell Canon Pellix camera shown here has a microthin two-way stationary pellicle instead of a conventional flip-up mirror. This stationary mirror transmits light directly to the film and reflects light up to the viewfinder for composing and focusing.

fine cut A number of scenes in the preoptical stage, spliced together in the editor's work print to approximate the continuity of the final version.

fine grain Film in which the minute silver halide grains in the emulsion are smaller and thus produce a granular texture that is finer than those usually found in most in-camera raw stock. Fine-grain film is used mainly for optical and duplicating work. The print material used for duping is relatively transparent and less grainy than the print material used for release purposes. Fine-grain prints are sometimes referred to as lavender prints.

finger A translucent diffusing device that is positioned in front of a light source for the purpose of reducing the intensity of the light originating from it.

fire effects The staging of full-scale conflagrations and lesser pyrotechnic displays is often not only impractical but dangerous. Wherever possible, illusion and suggestion should be substituted for reality.

Usually limited to fireplaces, most indoor blazes can be confined and controlled by merely regulating the flow of gas through perforated pipes. Capable of producing a flickering flame or a roaring fire, these pipes are camouflaged and concealed wherever needed on the studio set. "Burning logs" and other combustible materials are actually shaped wires that act as the skeletons for the asbestos covering that is applied to their outer surfaces.

As an alternative for optically superimposed stock shots of fires, wood alcohol or paraffin can be ignited in exterior locations with relative safety to produce the required blaze. To augment the effect, flames from small fires, in asbestos containers that are positioned directly in front of and below the camera lens, blend these foreground flames with the blaze in the background to create a composite "four-alarmer." Billowing smoke, produced by pumping petroleum products (kerosine and high-grade diesel fuel) through heated pipes, helps lend credibility to the effect.

Obviously, the effects of the fire should be reflected in the scenes that follow. "Burnt" and damaged areas should present a discolored or charred appearance, hardly a challenging assignment for the talented background artist.

The film maker's power of suggestion and the viewer's imagination can both be combined effectively to produce adequate results and a reasonably acceptable substitute for actual fire. The illusion is created by tacking several short lengths of linen or fine curtain material to a wood slat that is subsequently positioned in front of a light source. Their slightest movement casts flickering shadows on studio walls to provide amazingly realistic flame effects. They can also be used to suggest watery reflections.

first run As the term implies, the first public showing of a motion picture at one of the larger theaters in an area. After an indefinite "run" (determined by the success of the picture as measured by box office receipts), the film is released for showing at neighborhood theaters.

The first-run picture differs from the preview and premiere in that the audience at a preview is usually made up of studio personnel along with those guests who may have either a financial interest in the production or are potential distributors. The audience at a premiere, by comparison, consists of members of the cast and their escorts, name performers from other studios, critics (sometimes), and VIPs generally.

first trial composite The first composite print from picture and sound-release negatives. It is used for checking picture and sound quality and, generally, the assembly of the picture in terms of optical continuity and scene-to-scene color balance. This composite print is in projection synchronism. See flow charts under PROCESSING.

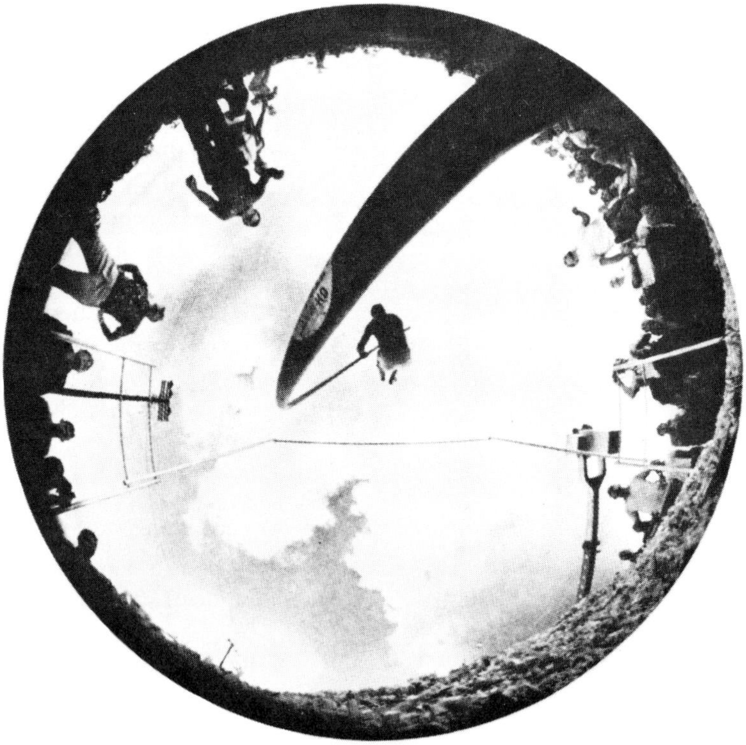

Fig. F-8*a* The fisheye lens—180-degree coverage. (*Nikon, Inc.*)

fisheye lens An ultra-wide-angle lens with 180-degree coverage. With this type of lens, a photograph taken from a doorway would show every item in the room including the door frame itself. Its depth of field extends from inches to infinity, and in spite of its extreme angle of coverage, its pictures are amazingly sharp. The image formed by these lenses is, of course, distorted. See Figures F-8*a* and *b*.

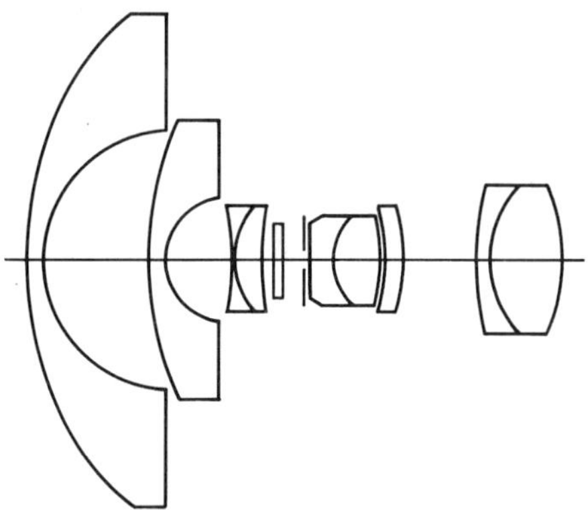

Fig. F-8*b* Fisheye lens components.

fish pole A device similar in design and function to a microphone boom. A microphone is suspended from the end of the "fish pole" so that it can be positioned over the acting area.

fixed-focus lenses and settings A fixed-focus lens is one that is positioned on a camera so that it remains at a fixed distance from the film plane. Although these small-aperture lenses have no provision for focusing, they usually have great depth of field and will keep all objects—between approximately 6 feet from the camera to infinity—in sharp focus.

Lenses in a focusing mount are most desirable for close-up work. Fixed-focus lenses can be used as long as the subjects are not closer than the hyperfocal distances for the various diaphragm openings. The depth of field at each *f*-stop opening is from infinity to half this distance, measured from the film plane. Fixed-focus lenses can also be used together with close-up lenses or extension tubes for extreme close-ups.

In general filming, focusing-mount lenses are occasionally set to a so-called "fixed-focus" setting, which is the hyperfocal distance, rather than focused each time for the exact camera-to-subject distance. The general recommendation is to focus the lens for each shot, regardless of the focal length of the lens.

fixing Also called **fixation**. A laboratory process in which residual sensitive silver salts are removed from developed film in order to render the image permanent. In the process, the developed image is hardened or "fixed."

flag Also called **gobo** and **mask**. A square sheet of metal, or opaque screen, that is used to shield the camera lens from stray lighting that could cause reflections during the filming process. It is usually attached to a metal support that is positioned near the camera. See Figure L-9 under LIGHTING ACCESSORIES.

flare An undersirable light streak or halation in the processed film, usually caused by light reflected from the surface of a faulty lens.

flash The process of reexposing an already exposed but undeveloped negative by a relatively weak printing light for a predetermined amount of time. Flashing strengthens the overall image and reduces the contrast between light and dark areas, thus making it possible to bring the negative into acceptable printing range. See LATENSIFICATION.

flashback An insert scene in a dramatic presentation showing an action or event supposed to have taken place in the past. The flashback is designed to establish a relationship with current events taking place in the film.

flash box A device that produces a brilliant flash of light and an ominous puff of white smoke. The device is used to create special effects for use in motion pictures as well as in television and stage productions. The flash box (and flash powder) is available through S.O.S. Photo-Cine-Optics.

flashbulb A small, incandescent lamp used in still photography in which the filament is usually in the form of crumpled foil or lengths of thin, snarled wire. One intense white flash, lasting a small fraction of a second, and its period of usefulness is over. See Table F-30.

TABLE F-30 Flashbulbs and Photofloods for Color Film

Film	Daylight		Any clear flashbulb		3400°K photoflood		3200°K photoflood		Any blue flashbulb or daylight (blue) photoflood	
	Film speed	Filter	Film speed	Filter	Film speed	Filter	Film speed	Filter	Film speed	Filter
Anscochrome D/50	50	None	25	80C	16	80B	10	80A	50	None
Anscochrome D/100	100	None	50	80C	32	80B	20	80A	100	None
Anscochrome D/200	200	None	100	80C	64	80B	40	80A	200	None
Anscochrome T/100	64	85B	64	81D	80	81A	100	None	64	85B
Kodachrome II A	25	85	30	81C	40	None	32	82A	25	85
Kodachrome II daylight	25	None	12	80C	8	80B	6	80A	25	None
Kodachrome-X	64	None	32	80C	20	80B	16	80A	64	None
Kodacolor-X	80	None	32	80C	25	80B	20	80A	80	None
Ektacolor S	100	None	40	80C	25	80B	20	80A	100	None
Ektacolor L	—	—	—	—	64	81A	100	None	64	85B
Ektachrome-X	64	None	32	80C	20	80B	16	80A	64	None
Ektachrome professional daylight	50	None	25	80C	16	80B	12	80A	50	None
Ektachrome professional type B	25	85B	20	81C	25	81A	32	None	25	85B
Ektachrome high-speed daylight	160	None	80	80C	50	80B	40	80A	160	None
Ektachrome high-speed type B	80	85B	80	81C	100	81A	125	None	80	85B
Polacolor	75	None	75	*	12	80B + CC20B	12	80A	75	None

Courtesy of General Electric.

Note: Flashbulbs have no true color temperature. However, for purposes of comparison, the following approximate color-temperature ratings may be used: clear foil-filled flashbulbs, 3800°K; blue filtered flashbulbs, 5500°K.

*Use clear M3 with no. 268, 269, 270, or J5 reflectors. With other reflectors use proper G.E. blue flashbulb.

flash frame A deliberately overexposed frame that is intended to serve as a cue mark for the film editor or optical layout man.

Also, the first few frames of a scene that may be overexposed while the camera reaches its correct filming speed.

flash pan Also called **zip pan**. A transitional effect produced by panning the camera away from the subject matter being filmed. The quick panning action, during which no detail is distinguishable, is generally used to indicate a change in location and usually leads to the shot that follows.

flat (1) A negative or positive in which the colors or tonal values in the gray scale are low in brilliance and contrast. (2) A painted background supported by a wooden frame which, along with other such sections, is lashed together and placed in position on a stage to form the setting for a scene.

flat response The reproduction of sound with such fidelity that a graph of the output gain over the entire audio spectrum is essentially horizontal, or flat. A variation of ± 1 decibel in the range from 30 to 15,000 cycles per second would qualify as flat.

flight analyzer A flight analyzer* records photographically on a single record the space trajectory of a moving object. This is accomplished by the operator tracking the object manually with the aid of attached binoculars and keeping an aperture aligned with the object, exposing the sensitized plate in a series of pictures abutting each other. The result is 58 images of the target appearing at equally spaced positions along the trajectory. At the same time that the position of the object in space is photographed, the instantaneous time of exposure is recorded on the base of each strip.

The analyzer record is a true undistorted graph, self-plotted by the aircraft at constant scale, showing attitude, position, and relative time.

floating pegs In animation, a flat metal plate with standard pegs for registration purposes. Positioned in movable tracks or mounted wherever needed on the animation compound table, artwork that is registered to floating pegs remains in a fixed position, regardless of background and camera movement.

flood-flash flashbulb A high-intensity, long-duration flash lamp with an average flash duration of 1.75 seconds. Ideal for use in high-speed motion picture studies and other applications requiring a self-contained, brilliant, long-duration flash.

It may be used with a polished reflector, with the lamp in a vertical position, base down. One FF-33 (Sylvania) will expose film to a density equal to that of three R-34 or RSF-2 photofloods. It requires 4.5 to 45 volts for flashing. A single flood-flash lamp will expose 37 feet of 16mm film at 1,000 frames per second, and 75 feet of 16mm film at 2,000 frames per second. At slower camera speeds, several lamps may be flashed in sequence of 1.75 seconds. If more light is required, two or more lamps may be flashed simultaneously. The color temperature of the lamp is 3800°K. Compare FLASHBULB.

flood lamps Incandescent lamps with heavier filaments than the ones found in conventional bulbs used for home lighting. Because they are purposely used at approximately twice the voltage for which such filaments would normally be rated, their useful life-span is considerably less than that of other types of incandescent bulbs.

floodlight A relatively large unfocused light source.

*The Fairchild is used here as an example of flight analyzers generally.

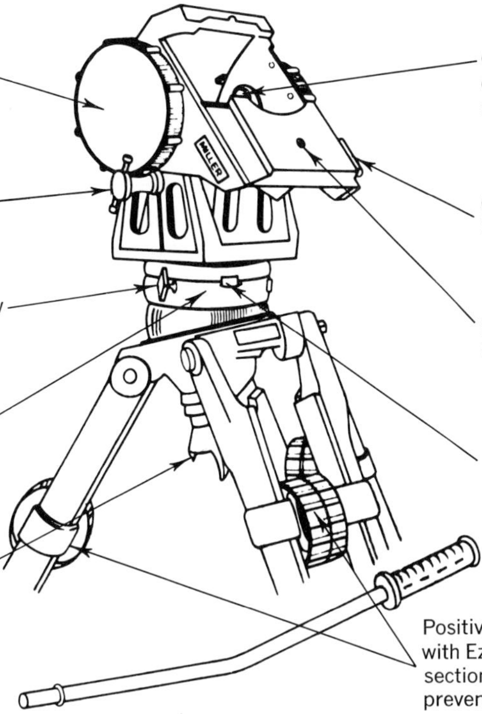

Fig. F-9 The Miller professional fluid head.

fluff Mistakes made during the filming process either in actions or in the lines of dialogue spoken by the performers.

fluid head A camera mount. With a fluid-action tripod head (Miller), the camera's weight rides on the fluid, and all tension adjustments take place within the fluid chambers. This arrangement provides smooth camera movement with no slack, bounce, backlash, or jitter. The professional model is designed for cameras weighing up to 35 pounds. The model D, with ball-leveling adjustment, is designed for cameras weighing up to 14 pounds. See Figure F-9.

fluorescent light Coated, tube-shaped light sources that produce an intense yet soft, cold light. An electric discharge causes the neon, argon, xenon, or krypton gas within the tube to activate the fluorescent lining. Using little current and producing practically no heat, they are capable of emitting a variety of colors, depending on the tube's coating. See Table F-31.

fluorescent screen The chemically coated target or screen at the end of the cathode-ray tube. Fluorescent particles in the screen glow when struck by the electron beam with an intensity that is proportional to the strength of the beam, thus creating the dark and light patterns that duplicate the tonal values in the original subject matter. See TELEVISION.

TABLE F-31 Filters for Fluorescent Light

Type of fluorescent lamp	Type of Kodak color film		
	Daylight type, Kodacolor-X, and type S	Type B and type L	Type A
Daylight	40M + 30Y + 1 stop	No. 85B + 30M + 10Y + 1 stop	No. 85 + 30M + 10Y + 1 stop
White	20C + 30M + 1 stop	40M + 40Y + 1 stop	40M + 30Y + 1 stop
Warm white	40C + 40M + 1⅓ stop	30M + 20Y + 1 stop	30M + 10Y + 1 stop
Warm white deluxe	60C + 30M + 1⅔ stop	10Y + ⅓ stop	No Filter None
Cool white	30M + ⅔ stop	50M + 60Y + 1⅓ stop	50M + 50Y + 1⅓ stop
Cool white deluxe	30C + 20M + 1 stop	10M + 30Y + ⅔ stop	10M + 20Y + ⅔ stop

Note: Increase exposure by amount shown in table.

C Cyan.
M Magenta.
Y Yellow.

flutter An undesirable fluctuation of sound caused by variations or inconsistencies in the speed of the transport mechanism in recording equipment. These are similar to wows which occur at slower intervals. See Figure F-10.

flux The basis of measurement for luminous flux (light flow) is the *lumen.* Individual light sources are also rated by the amount of flux they generate.

The amount of light that is emitted by a source equal to 1 candela provides a luminous flux of 12.57 lumens.

flying spot A television scanner which transmits a beam of light through projected motion picture film during the transmission process. As each frame of film reaches the scanning area in the continuous-motion projectors that are used for this type of transmission system, the beam from the flying spot scans the film and, in passing through it, activates a photoelectric cell in the transmission chain. The film is scanned at the standard television transmission rate of 30 frames per second. See CONTINUOUS PROJECTORS and MULTIPLEXED TELECINE SYSTEM.

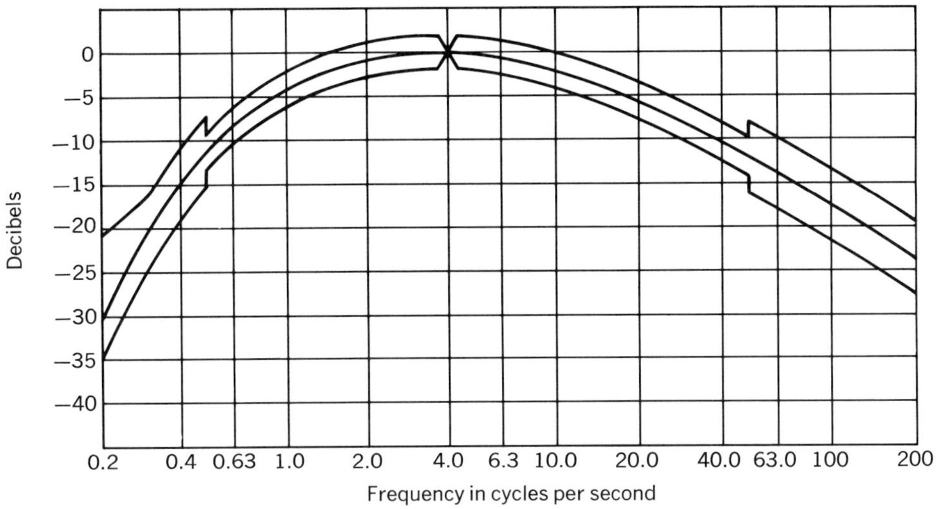

Fig. F-10 Weighting curves for weighted flutter measurements. (*National Association of Broadcasters.*)

Frequency	Decibels	Tolerance
0.2	−30.6	+10, −4 db
0.315	−19.7	0.315 to 0.5 cps
0.4	−15.0	(±4 db)
0.63	−8.4	0.5 to
0.8	−6.0	less than
1.0	−4.2	4 cps
1.6	−1.8	(±2 db)
2.0	−0.9	
4.0	0	±0 db
6.3	−0.9	Greater than
10	−2.1	4 cps to
20	−5.9	50 cps
40	−10.4	(±2 db)
63	−14.2	50 to
100	−17.3	200 cps
200	−23.0	(±4 db)

FM The abbreviation for **frequency modulation**. FM is a method of broadcasting through the variation of the frequency waves of radio rather than the amplitude waves as in AM broadcasts. Since static is caused by electric charges in the atmosphere and these charges do not distort frequency waves, FM broadcasting is free from static and provides sound of higher quality. The system was authorized by the Federal Communications Commission in 1941.

FM reference FM reference, in videotape recorders, provides a white reference frequency keyed into the vertical blanking interval of the signal for checking FM deviation. FM reference also provides the crystal-controlled reference frequency for correct setting of deviation standards. It also provides the noise test facility for head-wheel adjustment.

focal distance The distance between the lens and the subject matter or point of focus.

focal length The distance between the optical center of a lens and the focal point, the film's light-sensitive layer of emulsion, when the lens is focused on infinity.

The size of an image at a predetermined distance from the lens is governed by the focal length of the lens. The size of the image increases in proportion with the focal length of a lens. See Figure F-11.

To convert the focal length into diopters, divide 1 by the focal length. For instance, a lens engraved with a focal length of 1 meter is a 1-diopter lens ($1/1 = 1$). An 0.5-meter close-up lens equals 2 diopters ($1/0.5 = 2$).

The focal length of a lens in meters is obtained by dividing 1 by the diopter power of the lens. For example, $1/4 = 0.25$, meaning a 4-diopter lens has a focal length of 0.25 meter (25 centimeters or about 10 inches). Note that the focal length must always be expressed in meters to make these conversions.

focal plane The plane in which the image formed by the camera lens is in sharp focus. Actually, the light-sensitive layer of emulsion.

focal-plane shutter A shutter designed to operate in a plane close to the film's light-sensitive layer of emulsion. The speed with which the shutter moves, and the width of the slit in its surface, are the factors that determine the length of the interval of exposure. See FOCAL-PLANE SHUTTERING CAMERA.

focal-plane shuttering camera A type of camera used for high-speed photography. It is designed around a dual optical system similar to

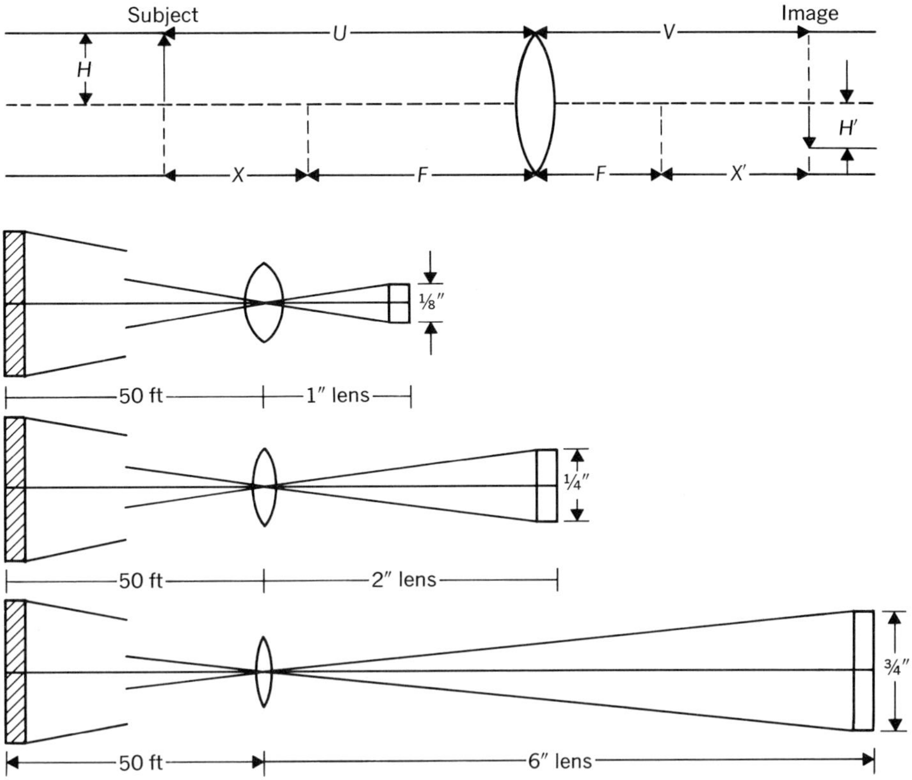

Fig. F-11 Focal length.

F = focal length
X' = distance of image from focal point, or distance lens is moved from infinity setting
X = distance of subject to focal point
U = subject distance from lens
V = image distance
H = height of subject
H' = height of image

those found in framing cameras. Electrooptical shuttering (0.1 to 0.001 microsecond), using Kerr or Faraday light-modulating magnetooptic cells, provides short exposure times and frame rates in excess of 1 million per second. See FRAMING CAMERA and HIGH-SPEED PHOTOGRAPHY.

focal point The point at which rays of light (reflected from the subject matter) intersect after being transmitted through the lens.

foci, conjugate Focal points on either side of the lens. One point of focus is the subject matter. The point of focus on the other side of the lens is the film in the camera.

focimeter An instrument used in optics for finding the focus of a lens.

focus The point at which light rays converge after being reflected or refracted.

focusing mount The housing that accommodates the lens and permits adjustments in focus by allowing the lens to be moved closer to or farther away from the film plane.

focusing scale The markings, in terms of distance, on the camera's focusing mechanism. This timesaving device enables the cameraman to set the reference line at a point opposite a specific distance marking and be assured that the image, at that setting, will be sharply focused.

focusing screen Any of the devices built in or attached to the camera that are used for viewing and focusing—a ground glass, for example. The image that is seen on the focusing screen is identical to the image that will subsequently be exposed on the film in the camera. The reticle markings etched on the surface of these screens are extremely useful for composition purposes.

focusing tube (camera) The focusing tube is used for accurate focusing and framing through the camera lens. Its aperture is sized and positioned to correspond exactly with the camera aperture. With the camera box racked over to the right, the focusing tube is positioned behind the lens. Thus the scene may be framed through the lens, and the lens may be focused. When ready for the take, the camera box is racked over to the left, placing the camera aperture in position without disturbing the focus or position of the lens.

The focusing tube has an eyepiece adjustable to individual eyesight. This eyepiece should be adjusted so that the crosslines in the optical system are sharp to the individual eye before focusing the lens on the object. The image is seen on a ground glass, upright and correct as to right and left. The magnification ratio can be adjusted to 5 or 10 times by sliding a knurled knob at the bottom of the tube. With 10× magnification, the entire field is not visible. Two viewing filters, one green and one amber, are built into the tube to aid in judging exposure and contrast. The green filter is used with orthochromatic film; the amber filter is used with panchromatic film. Each filter is placed in position by turning a knurled knob on the side of the tube. A sliding door permits access to the matte slot and to the ground glass for cleaning. The filters on the BNC camera are placed in the light path by pushing one or the other of two buttons near the focusing tube eyepiece in the outer case. See Figures F-12*a* and *b*.

fog An undesirable graying of the image caused by light striking the film before it has been developed.

fog-effect filters, half-and-half Half-and-half fog-effect screens (Harrison), with half of the filter simulating fog and the other half

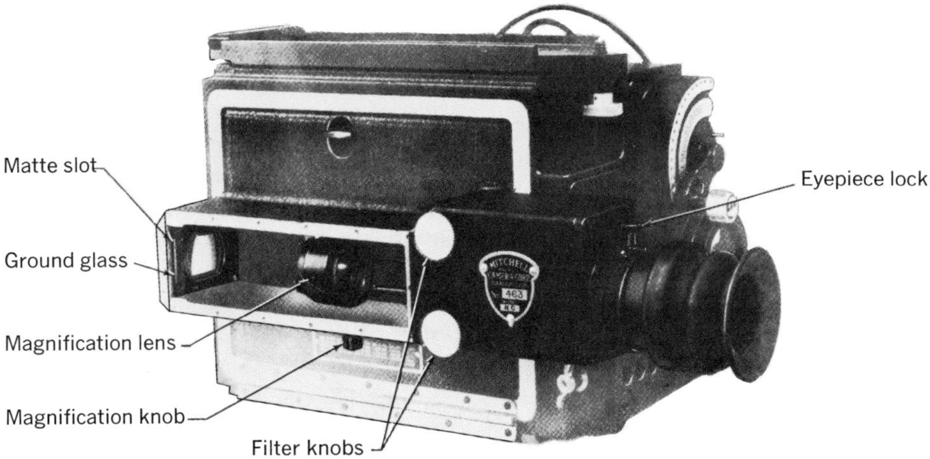

Fig. F-12*a* **Focus tube (Mitchell NC).**

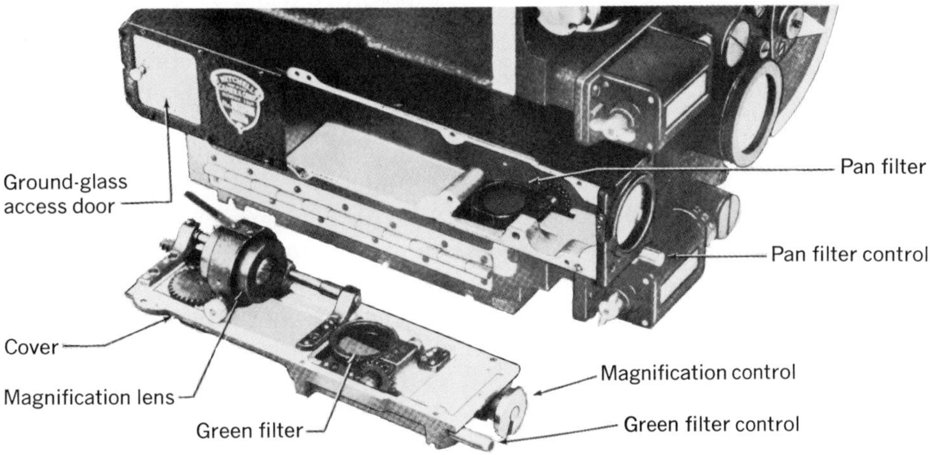

Fig. F-12*b* **Focus tube (Mitchell BNC).**

clear, enables the photographer to create fog effects in one-half of the scene without affecting the subject matter in the remaining part of the frame. These filters can be used for creating atmosphere effects on indoor sets as well as for simulating fog and haze in location filming, while maintaining clear foreground areas. Similarly, in day-to-day location filming, where it is necessary to match the previous day's atmospheric conditions, these filters are especially important.

While the range of five filters covers practically every type of fog condition, additional effects can be created by varying the focal length of the lens, the lens opening, or the distance between the filter and the lens.

Half-and-half fog filters are also useful where it is necessary to keep part of the subject matter completely clear while creating a diffusion or fog effect over the remainder of the film frame. Regardless of the usage, however, exposure compensations are unnecessary since the effect created by their use, a lowering of contrast, is similar to the exposure used for filming natural fog. As with other filters, viewing glasses help the cameraman "see" and evaluate the results that will subsequently be produced on film.

fog-effect filters, regular The principles of diffraction and dispersion of light by small droplets of water that create natural fog are also applied in its simulation by fog-effect filters. As the filters have a different density from that of air and a higher dispersion of light, they create a bluish halo around every bright object in the scene.

In order to duplicate the range of natural fog conditions, there are 10 equally graduated filters with as many degrees of diffusion. In addition to the filters, image size, lens contrast, lighting contrast, exposure level, and development—in either black-and-white or color—can influence the degree of fog. Similarly, the visual halo or degree of fog sometimes appears to be greater than the actual photographic result. Contrast-viewing glasses (either pan glass for black-and-white filming or a color-viewing glass for color) are excellent guides for determining the photographic effect that is most desirable.

With photographic variables applying to fog filters in a manner similar to diffusion disks, a set of four fog filters is considered adequate for most filming assignments. The number 2, 4, 6, and 8 filters, ranging from light to heavy, are the ones recommended for average conditions.

Fog-effect filters tend to lower contrast by scattering light from highlights over the shadow areas. This produces a grayish appearance in relatively dark areas. The reduced contrast gives an impression of increased exposure in the overall picture, so that no exposure compensations are required for fog filters.

Since fog-effect filters flare the highlights into the shadow areas, the effects created in printing from negative to positive do not have a natural appearance because the shadow areas flare into the highlights and create a reverse feeling that is, at best, unnatural. In any reversal process that goes from positive to positive, however, the flaring of the highlights into the shadows and the overall fog effect that results can be almost as natural as the conditions found in the original scene. Here again, a set of four filters is necessary for creating the desired effect. See Figure F-13.

fogged track The deliberate exposure of the film's sound-track area at the laboratory. This is done in order to prevent undesirable noises from interfering with the projection of silent films or, in motion picture presentations where the accompanying sound originates with an off-screen announcer. See VOICE OVER.

foil In videotape recording, small pieces of aluminum foil that are applied to the film* by the editor to indicate cut points in the work print. These pieces of foil are used as a guide when the picture information is transferred to videotape. As the transfer is made, a pickup device reads the location of the aluminum foil on the film and places an electronic cue on the videotape. This permits precise editing of the tape.

following shot A photographic effect in which the camera is positioned in front of, behind, or alongside a moving performer or vehicle. The relationship or distance between the camera and the

*Immediately after videotaping is completed, the selected takes of each scene are transferred to 35 or 16mm direct-positive black-and-white film. The live audio track from the videotape is transferred to 35mm magnetic sound-recording film.

Fig. F-13 Fog artificially produced with the Mole-Richardson type 1983 Mister Fogmaker Moleffect.

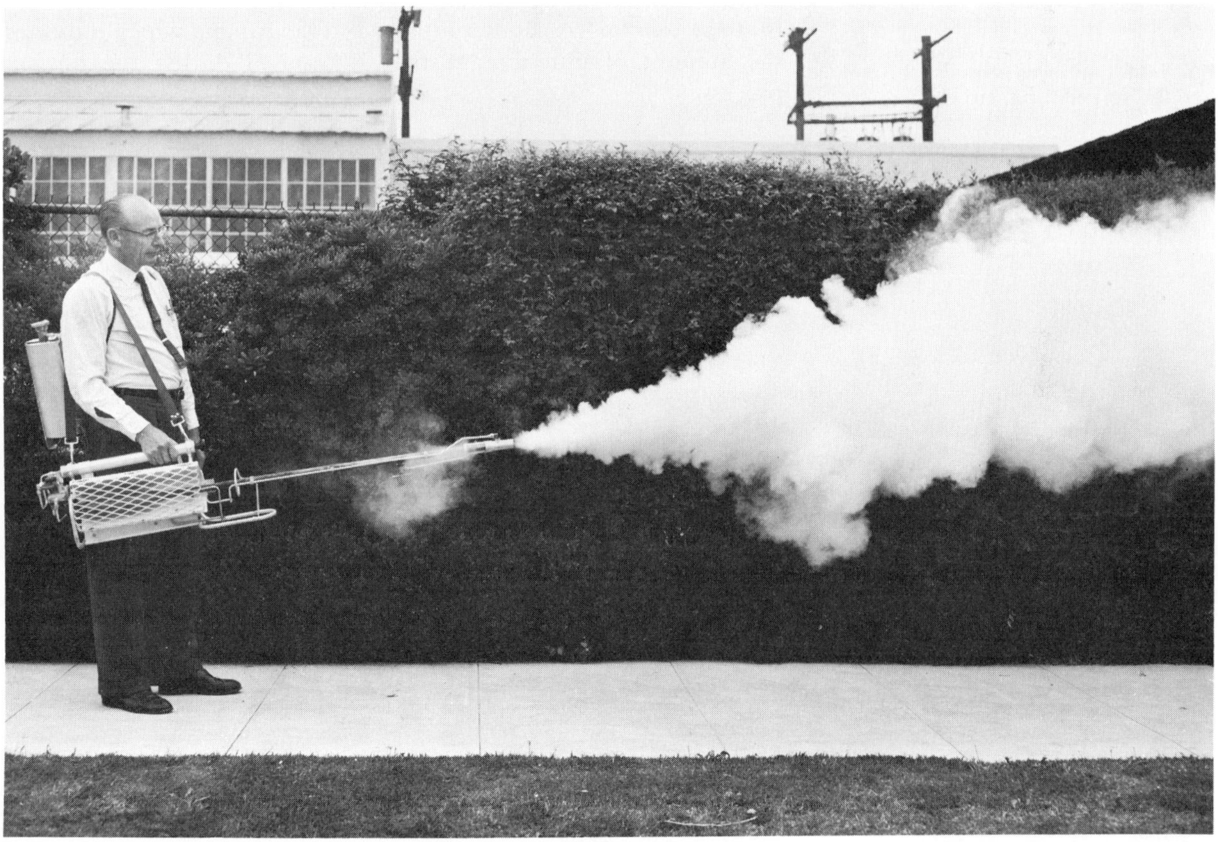

subject matter remains the same during the entire shot in spite of the fact that the background is continually changing. See Figure F-14.

follow spot The follow spot (ColorTran) provides up to 400,000 center-beam candlepower for television, motion picture, and theatrical lighting. The 1,000-watt PAR 64 3200°K quartz-iodine lamp produces a soft, uniformly illuminated field, free of objectionable filament patterns or other variations.

It operates directly from 120 volts a-c or d-c and maintains constant intensity and color temperature throughout its rated life of 150 hours. Through the use of mattes, a variety of light patterns can be produced. Two 8-inch planoconvex lenses are employed. One, fixed, at the rear of the housing, focuses parallel rays from the lamp. The front lens is for image focusing. The iris is used for controlling image size without affecting intensity. It is variable from $3\frac{1}{8}$ to $\frac{3}{16}$ inches, and incorporates a "douser" for completely blanking off the light. Iris and focusing controls are conveniently located for ease of operation in coordinating effects. See Figure F-15.

footage counter A counter attached to the camera which indicates the amount of film footage that is exposed during the filming process.

Fig. F-14 Following shot. Look, ma. No hands. No pedaling, either.

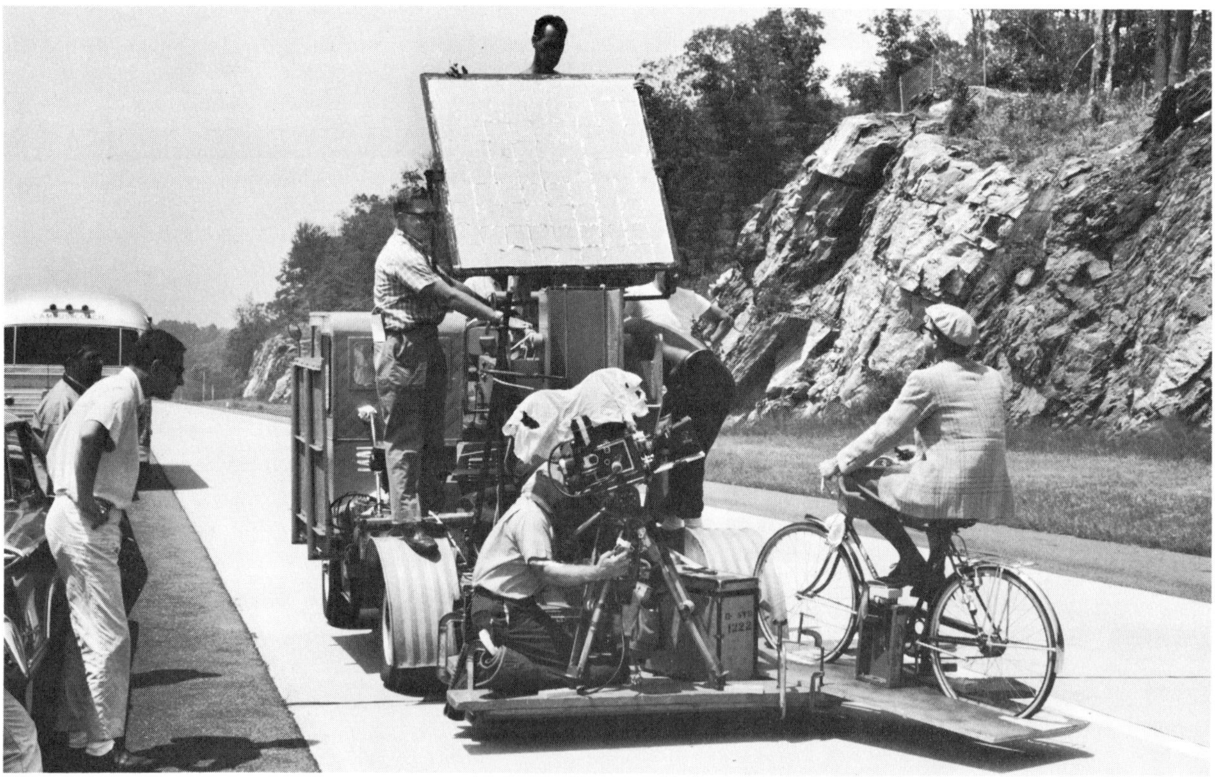

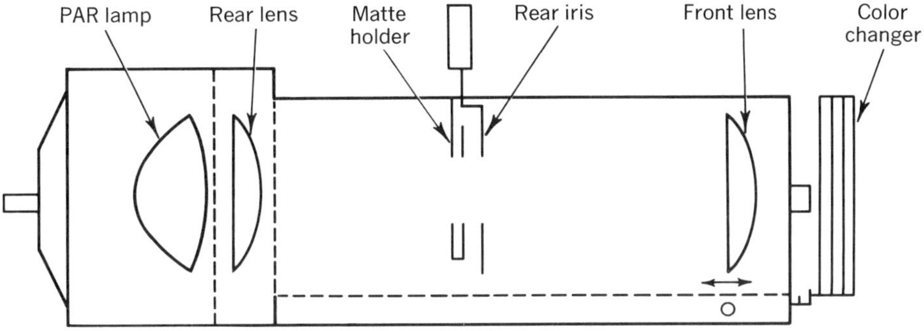

Fig. F-15 ColorTran follow spot. A three-point screw adjustment system permits alignment of the PAR 64 lamp axis for maximum output and efficiency.

footcandle An arbitrary standard measure of light intensity. The amount of light produced on a surface one foot from the standard source equals one footcandle.

The footcandle is a linear measure. For example, if the standard source produces one footcandle at one foot, then the equivalent of 10 standard sources will produce 10 footcandles at one foot. In general, when a table indicates the maximum intensity of a fixture, it specifies the maximum footcandle level at the most commonly used distance for that particular fixture. This maximum intensity is generally measured at the center of the light pattern. The footcandle is expressed, in practice, as an "average" footcandle which is the average intensity within a defined coverage.

On a one-square-foot area, the amount of light falling on that surface would equal one lumen. Incident-light exposure meters, capable of measuring light coming directly from a source, are used for determining footcandle readings.

$$\text{Footcandles} \times \text{area of surface in square feet} = \text{total lumens}$$

footlambert A unit of measurement indicating the brightness of a specific area. Light reflected at the rate of one lumen per square foot, from a more or less perfect diffusing surface, would have a brightness equivalent of one footlambert in all directions.

footlambert meter An instrument for general brightness measurements. Its cell is equipped with a tubular baffle which restricts the oblique light and provides a basic range of 0 to 60 footlamberts. A removable multiplier disk provides the second range of 0 to 300 footlamberts.

footlights As the name implies, a row of lights that border a stage in front of the acting area. Unlike borderlights which are mounted immediately behind the proscenium arch, footlights are positioned either on the floor or in recessed grooves along the edge of the stage.

foreground The area between the camera and the subject matter.

foreign-language sound-release negative A sound negative that contains all the re-recorded sound elements with the exception of the dialogue track. A new dialogue track, in the language for which the release is being prepared, is recorded in synchronization with the corresponding picture information and substituted for the original dialogue track.

foreign picture-release negative A release negative, usually a dupe negative, used for producing the picture material needed for foreign-version release prints.

foreign-version release print A composite print which has the dialogue in a language other than the one used during the original production. The foreign dialogue either is dubbed in, in place of the original sound track, or appears in the form of superimposed captions.

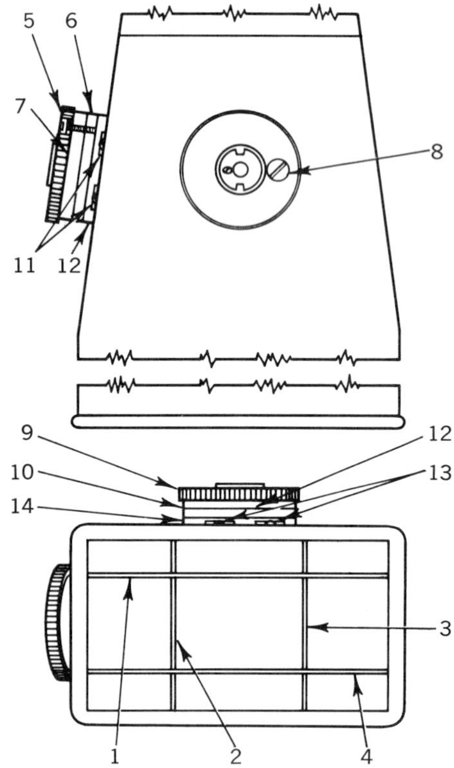

Fig. F-16 Adjusting four-way viewfinder mattes (Mitchell BNC). (1) Upper horizontal matte; (2) left vertical matte; (3) right vertical matte; (4) lower horizontal matte; (5, 8, 11, 13) screws; (6, 10, 12, 14) rings; (7, 9) knobs.

Formalin A gas (formaldehyde) in the form of a water solution. In photography, it is used as a preservative for stabilizing the dyes in color prints. It can also be used as an alkali in certain developers (hydroquinone, for example).

format The screen proportions of the projected picture. The term *flat* indicates that the projected picture has an aspect ratio of 1.33:1. Conversely, the term *wide-screen* format is an all-inclusive designation for any of the wide-screen or "scope" processes—CinemaScope, for example.

Also, a term used to describe the general pattern of a television show that is seen regularly.

four-way mattes (camera) Four-way mattes are built inside the upright section of the camera base and are operated from the outside by means of adjusting knobs. The mattes may be operated in pairs, or each may be operated independently. These mattes are so placed that there is no danger of reflection from backlight rays. The outer door must be opened to gain access to the four-way matte adjusting knobs on the BNC camera. See Figure F-16.

frame In motion pictures, one of a series of pictures in a length of film. Each picture is separated from the others by a narrow black border that is referred to as a *frame line.*

In television, one complete television picture made up of two interlaced fields. Unlike motion picture film, with a projection rate of 24 frames per second, the television transmission rate is 30 frames per second.

frame counter A counter attached to the camera which indicates the number of film frames that have been exposed during the filming process.

framing camera A type of high-speed camera in which the image is first received on a primary focal plane (by means of moving optical elements) and relayed to the film at the secondary plane where the image is recorded. The light reflected from the image to either a rotating mirror or rotating prism is the primary image. This image is transmitted through refocusing lenses to the film plane and is referred to as the secondary image. These cameras are capable of speeds up to several million frames per second. See HIGH-SPEED PHOTOGRAPHY.

framing control In television, settings for controlling and adjusting the width and height of the televised picture.

freeze frame Also called **hold frame**. An optical term used to describe the reexposure of a specific frame of film (from the scene being duplicated) for a predetermined number of frames. This is usually done to slow down an action or to extend the footage of the scene.

frequency The rate at which electric impulses, or sound and light waves, are repeated in a specific time period. In sound, frequency is measured in cycles per second, but in broadcasting and electronics, the same unit has been renamed *hertz*.

frequency, field The rate at which a complete field is scanned electronically, actually one-half of a composite picture. The field-frequency rate in the United States is 60 fields per second. The standard rate in Europe is 50 fields per second.

frequency, frame The rate at which television frames are scanned. The United States standard rate is 30 frames per second. The European rate is 25 frames per second. Compare FREQUENCY, FIELD. See INTERLACED SCANNING.

frequency, infrasonic Frequencies ranging below 15 cycles per second. Compare FREQUENCY, ULTRASONIC.

frequency, supersonic See FREQUENCY, ULTRASONIC.

frequency, ultrasonic Sound energy of frequencies that are above the range of the human ear. Frequencies above 15,000 cycles per second are in this category. Compare FREQUENCY, INFRASONIC.

frequency interleaving The process of fitting (interleaving) both the monochrome and color subcarriers into the same bandwidth. The color subcarrier frequency (3.58 megacycles) fits into regularly spaced intervals, between each of the component frequencies that form the composite monochrome signal.

The interval between each monochrome and color component is 7,875 cycles. The interval between any two monochrome components is 15,750 cycles. Thus, signal components, monochrome and color, are alternated within the 4-megacycle band.

frequency range The spread between the lowest- and highest-pitched sounds that can be reproduced at a usable level. Conventionally, this is taken to be approximately 20 to 16,000 cycles per second. The fundamental tones of a standard piano keyboard range from approximately 30 to 4,200 cycles per second. The range of a base singer is from 90 to about 400, and for a tenor 130 to 530 cycles per second. An alto's range is between 190 and about 800 cycles per second. The notes reached by a soprano range from about 250 to approximately 1,200 cycles per second.

frequency response The range of tones or musical notes that a sound system will record and reproduce faithfully. The human ear can hear frequencies from about 20 up to approximately 16,000 cycles per second. See FREQUENCY-RESPONSE CHARACTERISTICS TABLE.

frequency-response characteristics table The frequency-response characteristics that are generally recommended for $\frac{1}{4}$-inch magnetic tape recorders operating at different recording speeds are listed here.

Frequency response, kilocycles per second	*Recording speed, inches per second*
2.5–0.3	$\frac{15}{16}$
50–7.5	$3\frac{3}{4}$
40–10	$7\frac{1}{2}$
50–15	15
30–15	30

frequency-response function A term used in connection with lenses to describe the interrelationship between contrast and resolution. Lenses having extremely high resolution do not necessarily provide good contrast. Characteristic curves of lens performance are plotted to produce visible readings on the face of an oscilloscope showing the balance between resolution and contrast. The patterns are used to evaluate the performance of a lens as well as its inherent aberrations.

Fresnel lens A glass lens, with raised concentric rings, positioned in front of a lamp housing and used for focusing the light beam originating from it. See Figure F-17.

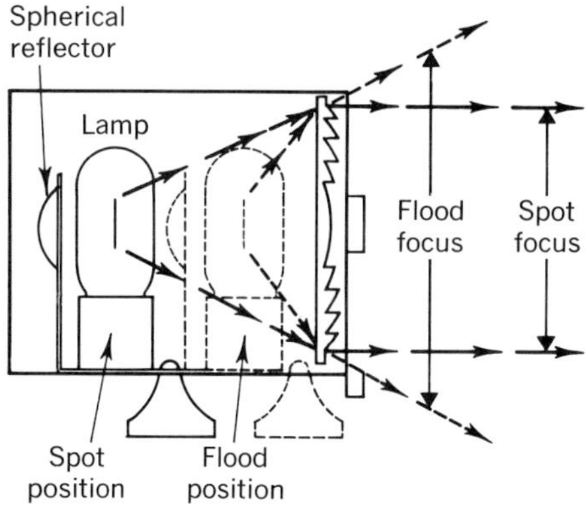

Fig. F-17 The Fresnel system. (*Sylvania Electric Products, Inc.*)

friction head A camera mounting in which disks, rather than gears or flywheels, help achieve smooth movement during pans or tilts.

frilling Because of excessive heat or the use of warm developing solutions, the emulsion will sometimes part company with the film base along the edges of the film. This separation is known as frilling.

fringing A halo or glare, usually around light objects, caused by improper registration or undesirable light reflections.

front porch The time interval between the beginning of the television blanking signal and the beginning of the synchronizing signal immediately following.

front-screen projection Also, **Alekan-Gerard process, Jenkins process, Scotchlite process**. A photographic process in which a background scene or plate is projected on a reflecting-type screen with the performers or subject matter positioned in front of the projected background. The process is used to produce composite effects on the in-camera negative.

Usually, the development of a technique precedes the introduction of new equipment and materials that are designed for the purpose of improving the process or making it more efficient. In the case of front-screen projection, the tail wagged the dog. A new material introduced in 1949, Scotchlite sheeting, led to the development of the technique.

Less than a year after its introduction, the Motion Picture Research Council and the Stanford Research Institute issued a detailed report describing front-screen projection. In 1952, a Virginian named Will Jenkins applied for the first patent. Two Frenchmen, Alekan and Gerard, also working independently, developed a similar process for which they were granted English patents approximately five years later.

Designed for the producer with both budget and space limitations, the front-screen projection process makes possible the same variety of effects that rear-screen processes are capable of producing. Although the procedures for the production of composite effects are generally similar for both processes, there are also basic differences.

In front-screen projection, for example, the print or plate that serves as the background for the scene being filmed is projected from a position either in front of or from one side of the screen. But, unlike the diffused-translucent screen on which the background is projected in rear-screen processes, the screen used for front projection is of the reflecting type. The projectors and cameras, in both systems, are driven by synchronous, interlock motors.

The Scotchlite sheeting, a product of the 3-M Company, is the same type of material that can be purchased in roll form, cut into strips, and taped to the bumpers of automobiles for use as reflectors. For motion picture production, the wider 3-foot rolls are applied to the surface of flats or large sheets of plywood. With millions of tiny, round beads of glass embedded in the tape's silver surface, the most suitable color for the process, almost all the projected light that strikes the surface of the screen is reflected back to the projector. Having an extremely narrow angle of reflection, the sheeting material is preferable to aluminized or flat-white material used for screens.

If desirable, set props matching items in the projected background (doors, for example) may also be covered with the reflective sheeting and treated as a part of the background. These matching props make it possible for a performer to walk behind them, thus creating the illusion that he is disappearing through the door on the projected background.

The beam combiner that is positioned between the camera and screen can be a two-way mirror, thin sheets of optical glass, or a transparent plastic material approximately 0.0005 inch thick. The last-mentioned is referred to as a pellicle. If the two-way mirror is used, its reflective surface should be positioned so that it faces the projector. The transmitting surface of the mirror faces the camera lens. Optically clear glass, unless it is very thin, can turn the projected image into a ghost image by reflecting off both surfaces. Also, because of the low reflective qualities of the glass, stronger light sources must be used in the projector lamp housing. The lighter-weight pellicle, by comparison, does not have the reflective properties of the two-way mirror nor the comparative sturdiness of optical glass, but since it is easier to handle and eliminates the possibility of off-register ghosts, it is used wherever possible. Regardless of the type of material used, the beam combiner should be positioned so that the distance between its surfaces and the projector and camera lenses are equal. Different focal-length lenses will, of course, change the relationship between the background and subject matter.

If, while setting up the shot, the performer's shadow or what appears to be a matte line around him is visible in the viewfinder, the cameraman must make slight adjustments in the position of the camera. If the performer must move from his set position, during the filming, that movement must be rehearsed. What might appear to be a good shot in the original setup might produce undesirable shadows or matte lines with the introduction of movement. It must be realized that, at best, because of the size of the projected backgrounds, movement is obviously limited.

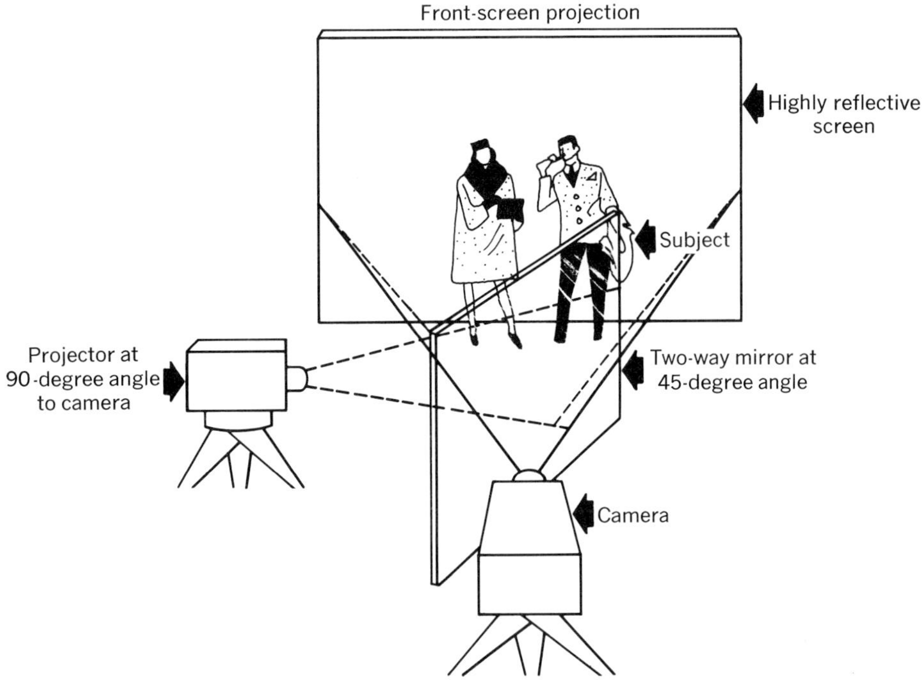

Fig. F-18 **Front-screen projection.**

The front-projection process permits more latitude and flexibility in the positioning of the light sources than rear-screen processes. Key light and fill light can be used creatively as on the conventional set. Finally, by keeping scenes achieved with this process relatively short, the audience does not have an opportunity to properly appraise the composite effect with respect to related scenes that precede and follow the front-screen shot. See Figure F-18. Compare REAR-SCREEN PROJECTION.

frying pan A screenlike unit that is positioned in front of a light source to soften the light originating from it.

full apple A large box that is used for shipping or storing lighting equipment.

full-coat magnetic film Perforated film that has the ferromagnetic coating covering the entire film base.

full shot Also called **long shot**. A large area photographed from a distant point. Indicating the general setting, this type of shot helps establish a relationship between the photographed area and subsequent scenes or actions taking place within it.

Also, a shot in which the full figures of the performers are shown in the frame of film.

full track A magnetic coating that covers the entire sound-track area in 16mm film. This arrangement eliminates the possibility of mixing magnetic sound with optical sound as is the case with the half-track. Compare HALF-TRACK.

funnel Also called **snoot**. As the name implies, funnel-shaped metal tubes that are positioned in front of spotlights in order to control the beam of light originating from the source.

G

gaffer A studio term for an electrician. The head gaffer may, on occasion, be the person responsible for lighting the motion picture or television stage.

gain Basically, the control of volume during the recording and reproduction processes. Gain or volume controls are used to keep the amplitude of the sound signal within the acceptable limits of the equipment for faithful reproduction.

galvanometer An electrical device, actually a light modulator, that makes it possible to convert sound waves into light waves. When magnetic sound is being transferred to optical sound, for example, the galvanometer is used to produce the irregular patterns and gradations in the variable-area and variable-density sound tracks.

gamma The term used for indicating the characteristics of a film with respect to contrast. Represented on H & D curves by the slope of the straight line, gamma expresses changes in density in relation to exposure.

Black-and-white prints prepared for theater projection are developed to a gamma of approximately 2.60. For television usage, print material is developed to a gamma between 1.80 and 2.0. The gamma for fine-grain material is between 1.60 and 1.80.

gamma strip A length of film, with varying tonal values ranging from white through intermediate shades of gray to black, exposed in a sensitometric printer and used by laboratory technicians as a reference for controlling gamma. Gamma strips are also known as **control strips**. See GAMMA.

gap Indicated in mils or microns, the gap represents the distance between opposite magnetic poles in record or erase heads in tape-recording equipment.

gate, camera A part of the camera film channel that permits loading of the film into the shuttle. During the filming process, the gate is that area in the shuttle where each frame of film is held for a predetermined amount of time in order to allow the light passing through the camera lens to expose it. Compare GATE, PROJECTION.

gate, projection That part of a projector where each frame of film is held against the aperture plate long enough to allow the projection light to pass through it so that the picture information is subsequently focused on the screen.

gate mask An auxiliary device in the projector gate that masks or cuts off a portion of the projected picture. Masks, with cutoffs that indicate the amount of picture information that will be visible in the television safety field, are invaluable in reviewing film that will subsequently be televised. See SAFETY AREA.

gauze A loosely woven material that is used to diffuse light being emitted from a source.

gear head A type of camera mounting in which pan and tilt movements are made possible by a built-in gear arrangement. Compare FRICTION HEAD and FLUID HEAD.

gelatin With general characteristics similar to refined glue, gelatin is used as a vehicle for the sensitive silver salts in emulsions and in the manufacture of filters. Gelatin is hard when dry, swells in cold water, and dissolves in hot water.

gelatin emulsion The mixture of gelatin and sensitive silver salts. See EMULSION and MULTILAYER FILM.

gemini tape-film system A method of production during which a motion picture negative (or positive) and videotape recording are obtained simultaneously. Gemini is a mechanical-optical method of combining a television and a 16mm film camera so that images seen by the television camera can be simultaneously recorded directly on 16mm film.

The gemini unit is in essence an adapter comprised of a base plate, common lens, beam splitter, optical relay system, and a 16mm camera. See Figures G-1*a* to *d*.

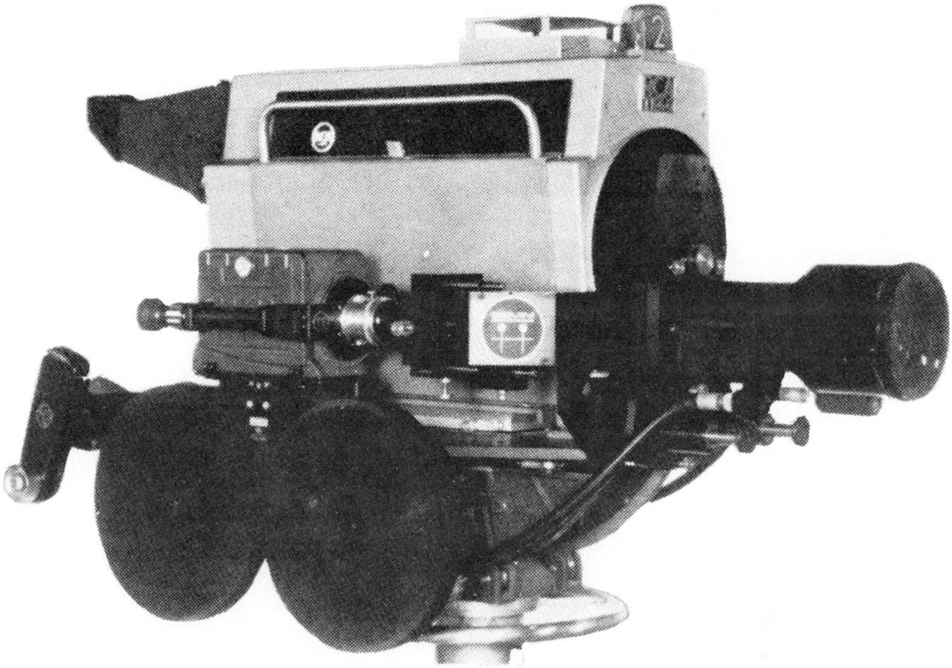

Fig. G-1*a* **Gemini unit: a TV camera (RCA TK 60) and a 16mm film camera combined by a common taking lens. (*M-G-M Telestudios.*)**

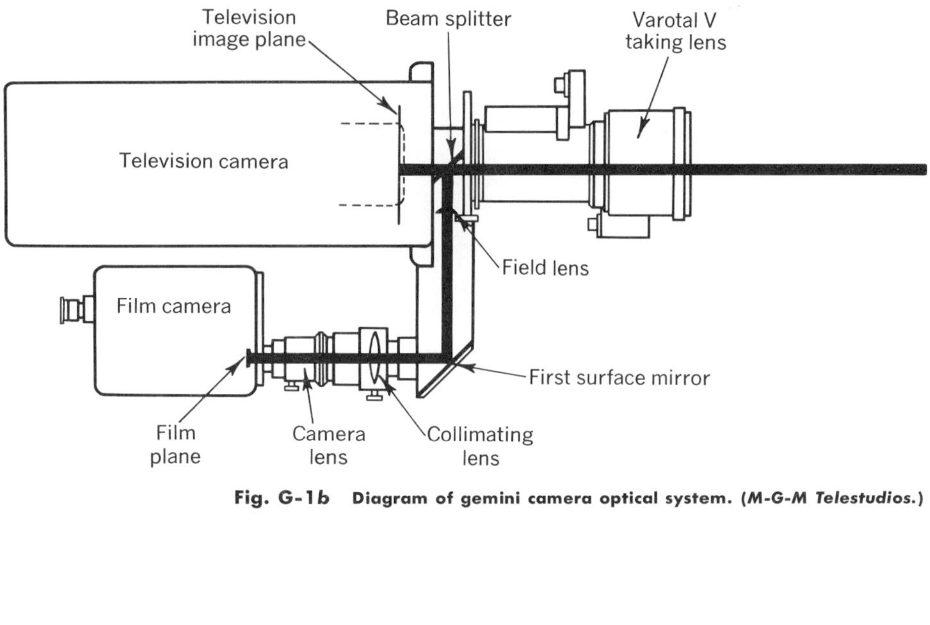

Fig. G-1*b* **Diagram of gemini camera optical system. (*M-G-M Telestudios.*)**

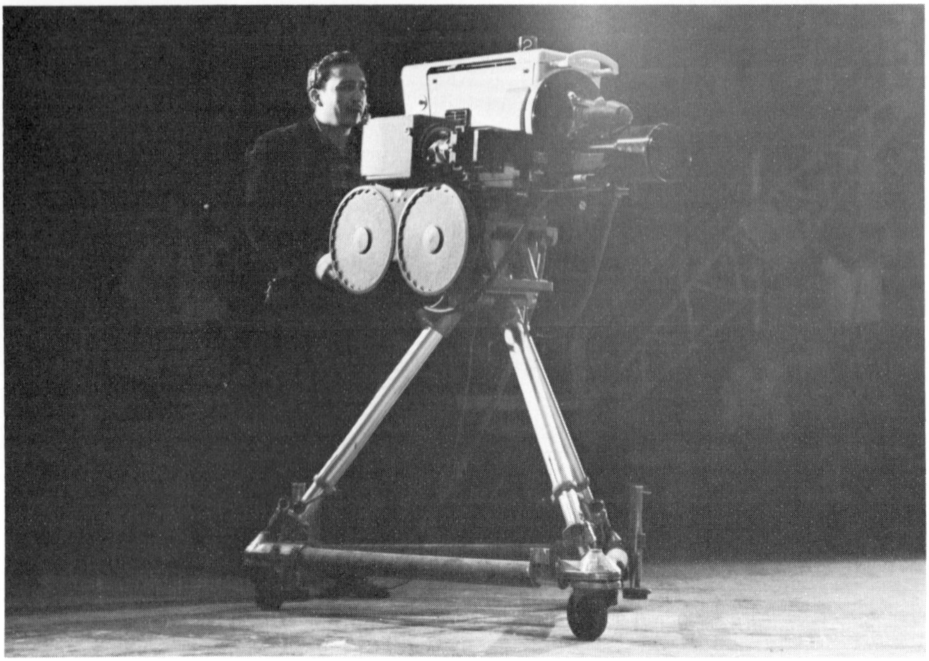

Fig. G-1c The gemini tape-film system—mediums combined. (*M-G-M Telestudios.*)

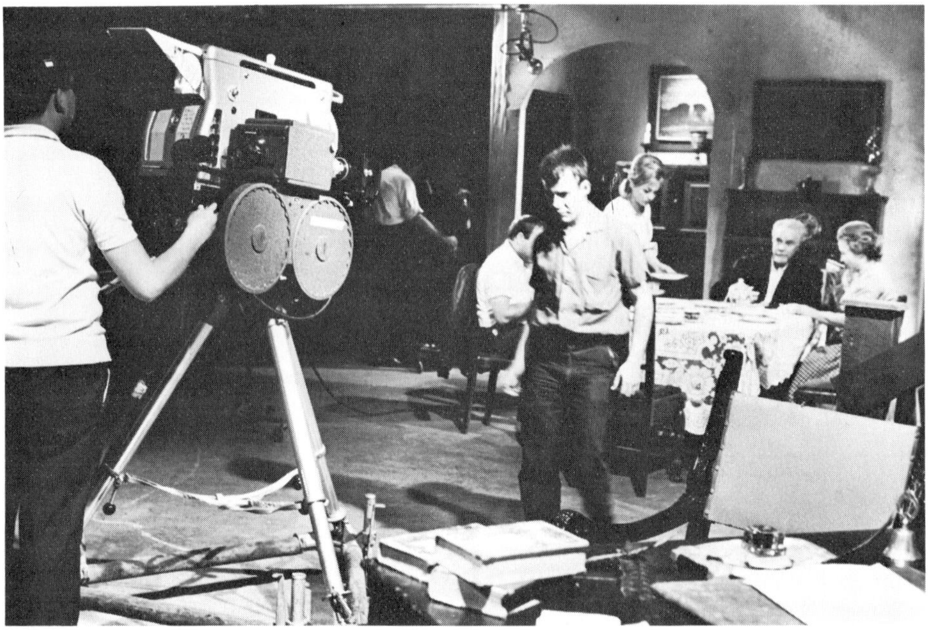

Fig. G-1*d* The motion picture camera is mounted immediately to the right of the TV camera. (*M-G-M Telestudios.*)

generation In film production or videotape recording, a term used to indicate original material as opposed to successive duplicates. First-generation material is referred to as an original or master. A duplicate negative or a dub of a master is known as a second-generation negative, and so on.

ghost A shadowy or weak image in the received picture, offset either to the left or to the right of the primary image—the result of transmission conditions which create secondary signals that are received earlier or later than the main or primary signal. A ghost displaced to the left of the image is designated as *following* or lagging. When the tonal variations of the ghost are the same as the primary image, it is designated as *positive,* and when it is the reverse, it is designated as *negative.*

gimmicks Shortcuts used by animators for the purpose of achieving effects without having to prepare the usual mountainous pile of drawings.

gladiator A carbon-arc follow spotlight (produced by the Strong Electric Corporation) with an output of 26,500 lumens. The unit has a two-element variable-focal-length optical system which employs a silvered glass reflector and provides a single-lever means of changing the size of the spot. The automatic carbon feed maintains a constant uniform arc gap to assure a steady light free from hiss or flicker.

glare Brilliant reflections or scattered light rays that have a graying effect on the print.

glass shots Glass has been defined as a hard, brittle, transparent substance produced by fusion, usually consisting of mutually dissolved silica and silicates. It can be shaped, cut, ground, polished, coated—and painted with opaque colors* for use on motion picture sets. For the motion picture producer, glass shots, or backgrounds painted on glass, provide a comparatively "paneless" method of producing composite effects. Economically advantageous, these opaque renderings on the flat surface of a sheet of glass are used in conjunction with auxiliary backgrounds and eliminate expensive set-construction costs from tight budgets. Since the glass paintings are photographed simultaneously with the action and secondary background, costly matteing and time-consuming optical camera work are also unnecessary.

During the photographic process, the glass, with its opaque rendering of a portion of the background, is mounted in a wooden frame

*Miller Brush Colors and Grumbacher Paint Company carry black-and-white and color opaque paints.

and positioned in front of the camera. The size of the glass and the painting depends on the scene's specific requirements. Precise relationship between the glass painting and the portion of the set that is visible through the clear area of the glass can easily be established before filming begins. Similarly, necessary changes in composition and framing, with the performers in the action area between the glass and the studio background, are immediately apparent in the camera's viewfinder.

Relatively small when compared with the size of the auxiliary sets, the paintings act as a mask and may include one or more renderings that are positioned where needed on the surface of the glass. The perspective, in the paintings, must blend with and match the coloring and detail in the stage sets. The rendered illustrations may depict upper or lower portions of structures, mountains or wooded areas, ocean, sky, or foreground detail. When projected, even the most critical observer finds it extremely difficult to point out where the background scene and glass painting are joined in the composite print.

The painted glass and the scene beyond it may be lit from a central point or from several separate light sources, depending on the desired dramatic effect and the position of the performers during the photographic process. Unlike rear-screen process shots and the lighting problems associated with them, glass paintings do not limit the performers to an action area directly in front of the projected backgrounds. With glass shots, and controlled lighting on each of the various planes, action may be staged anywhere between the opaque paintings and the scene's backing behind them. Obviously, greater scene depth is possible with the glass shot than with rear-screen projection.

Although cameras can be panned or tilted while filming, they should be mounted on a special panhead with the lens centered over the nodal or pivot point of the tripod. In this position, a constant relationship can be maintained between the glass painting and the set beyond it. Errors in parallax, the differences between the photographed area and that shown through the viewfinder, are minimized when the camera lens is centered over the horizontal and vertical axes.

glass shots and composite matteing With minor variations during the photographic process, glass paintings can also be combined with mattes to produce amazingly realistic composite effects. When well planned and precisely executed, these composite-matte shots can be used to simulate "remote" settings on the studio stage with a minimum of set construction. As a time and budget saver, this type of matte shot further proves its versatility by making a few stars and a handful of extras seem like a cast of hundreds. Limited only by the collective imaginations of the director and set designer, glass paintings combined with composite-matte filming processes can be used to produce a variety of illusory effects in relatively small areas with equally small, but mobile, casts.

In the planning stages, each of the component sets needed to achieve the composite effect are combined in a single drawing. Perspective, in relation to the position of the camera, along with proportional sizes of props and performers are established in these detailed layouts.

Quite popular for economic reasons, the specially designed sets are divided into small shooting areas in which related actions are subsequently filmed. Set construction is limited to the essential detail in these specific areas with glass paintings and mattes concealing the supporting framework and undesirable areas of the stage itself. All parts of the set, with the exception of the area being photographed, are, in turn, masked or matted out of the composite frame. In order to ensure maximum registration and a precise match between the separate sections, the camera remains in a fixed position until the action in each of the sets has been photographed. Similarly, light

Fig. G-2 Glass shots and composite matteing.

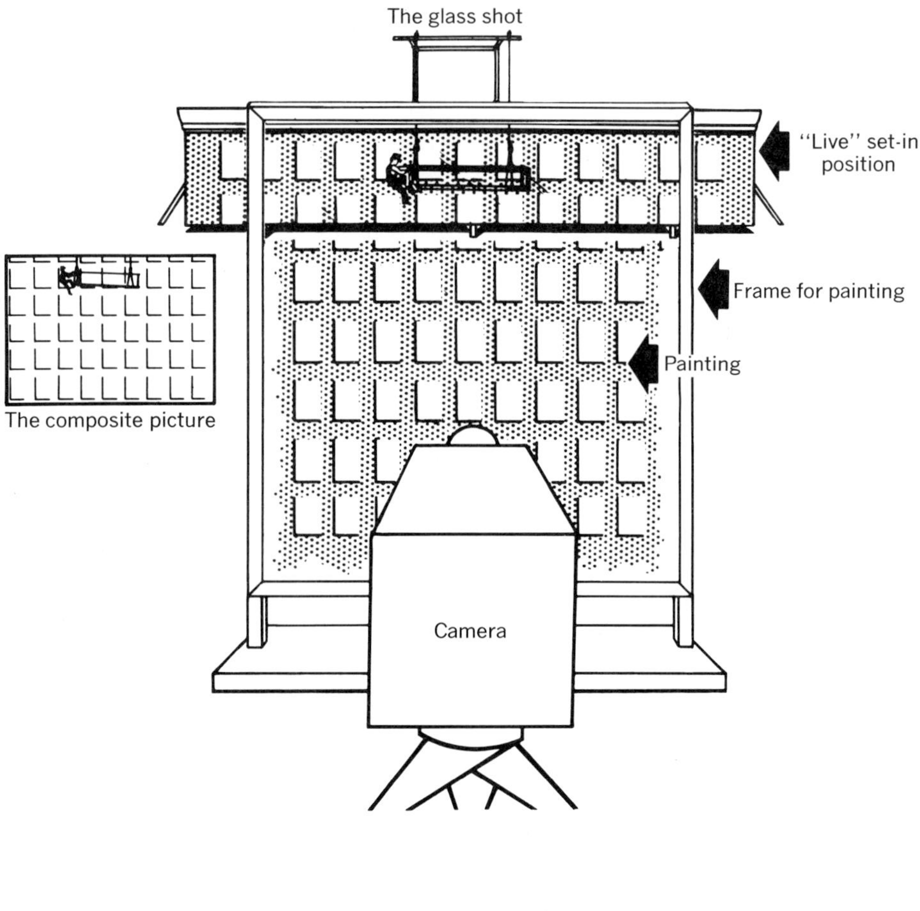

sources and shadow patterns must be consistent throughout the filming.

Because each portion of the scene is photographed separately, large, spectacular settings or remote locations can be staged within a relatively small area. The stars and extras, as each component section is photographed, can be moved from one action area to the next filming location just a few steps away. Since overlapping actions would create undesirable double-exposure effects in adjacent areas, the action on each set may not extend beyond the particular section being filmed.

Similar effects once achieved by bipacking, multiple exposures, and the use of matte boards that served as a printing light source can now be accomplished with optical camera work. Differences in print quality, because of duping and optical combinations, are negligible.

Matched like the pieces of a giant jigsaw puzzle, each of the component parts of the frame help form a composite picture that tests the powers of observation of the most discerning audience. See Figure G-2.

glass shots, paneless In addition to their many and varied uses in motion picture production, large sheets of unbreakable glass, positioned between the camera and the subject matter, provide a shield for the unprotected lens and vulnerable cameraman.

An integral part of the pioneer production, these panes stopped many meringue and cream pies that were aimed directly at the camera. For the film editor, however, the "wipe" effect that followed was a natural transition between scenes.

As a protective device, the transparent sheets of optically clear glass are especially useful to newsreel crews on location. Exposed to falling debris, water, and other equally hazardous conditions, cameras remain undamaged and the "panestaking" filming assignment becomes relatively harmless.

glitch A studio term indicating a random disturbance in the videotape signal.

globe An inclusive term for the glass enclosures that house lamp filaments or gaseous elements. When positioned in contact with a supply of electric current, the globes produce an intense light that is capable of illuminating relatively large areas.

glossy print A photographic print produced on smooth-surfaced paper. Unlike prints made on matte paper, glossy prints have no surface texture and show detail sharply. This type of print is best for reproduction.

glycine A powdery substance used as a developing agent. Although it will not dissolve in water, the white powder will mix readily in alkaline solutions.

gobo An absorbent material that is used for screening out harsh sounds during the recording process. A length of cheesecloth or other loosely woven material serves the purpose quite adequately when either is wrapped around or draped over a microphone to shield it during location filming.

Also, a square sheet of metal or opaque screen that is used to shield the camera lens from studio lights.

Goldberg wedge A density wedge in which a colored or tinted layer of gelatin, varying in thickness, is positioned between two sheets of optically clear glass. Continuously increasing in density from one end to the other, the wedge is used for sensitometric purposes. See SENSITOMETRY.

goose control A unit that permits the synchronization of the camera with the event to be photographed, including regulation of the camera speed and remote control of the camera or event. A four-position selector switch controls the camera and event start-stop cycle so that the camera and the event can be started together, or the camera started before the event by any predetermined time interval, or, conversely, the event started first with the camera start followed by any predetermined time interval. The event is actuated through a selected open or closed relay.

The camera speed is controlled by a variable a-c autotransformer with maximum output of 280 volts. The camera control circuit also incorporates a 70-millisecond delay timer when operating above 130 volts. (Fastax Goose Control units are available through Gordon Enterprises.)

gradation The range of tonal values in a print.

gradient A portion of a characteristic curve and the corresponding degree of image contrast over that specific portion. See CHARACTERISTIC CURVE.

grading The process of balancing exposure, on a scene-to-scene basis, throughout a motion picture.

graduated wedges—color or neutral density Where it is necessary to reduce exposure in some areas of the scene without affecting the other areas, or where a color must be altered without changing the transmittance or illumination level, neutral-density wedges can be used to produce an even level of illumination from one end of the wedge to the other.

Harrison color and neutral-density wedges are available in widths up to 4 inches and lengths up to 15 inches. Since the color changes uniformly from dense at one end of the wedge to light at the other end, the exact density of a given color for any particular requirement is assured.

The wedges are available in all the colors and densities listed in the Spectro-Wedge scale. The length of the wedge determines the degree of color change—the longer the wedge, the more gradual the change in color. See Figure F-7*a* and Table F-25 under Filters.

Where a large uniform area is desirable, in order to increase the working aperture, the wedges should be as long as possible, to provide a minimum of change over the greatest distance. Two wedges of the same color can be overlapped and moved together in equal amounts until a larger uniform area of the same color is obtained over the entire length of the wedges.

Since the wedges are made by laminating two pieces of glass together with colored adhesive, the accuracy of the wedges is determined by the flatness of the glass supports—$\frac{1}{8}$ to $\frac{1}{4}$ inch thick. The overall thickness of the wedge is $\frac{1}{4}$ to $\frac{1}{2}$ inch.

grain Microscopic silver grains in developed film that impart the tonal values to an image. Granular groupings, in varying densities, create the effect. Compare FINE GRAIN.

granularity The characteristic composition of an emulsion. The film, after development, may show an objectionable grain structure. See Figure G-3.

graphics Illustrative material, other than cell animation, that is photographed on the animation stand.

graphidone A developing agent used in the processing of original black-and-white negative film.

graphol A developing agent that is used to produce low-contrast results in the original black-and-white negative film.

graticule Also called **reticule**. The etched fine lines on the surface of the ground glass in camera viewfinders that are used for framing and composing.

grease paint A slang term generally used for makeup. Grease paint is actually applied to the face and serves as a foundation or base for powder, rouge, or other cosmetics.

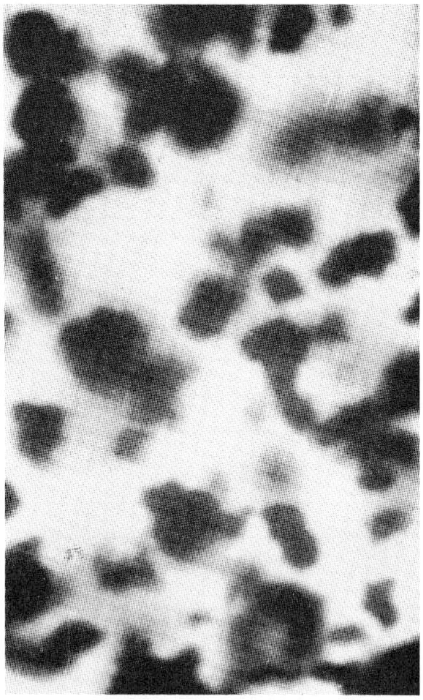

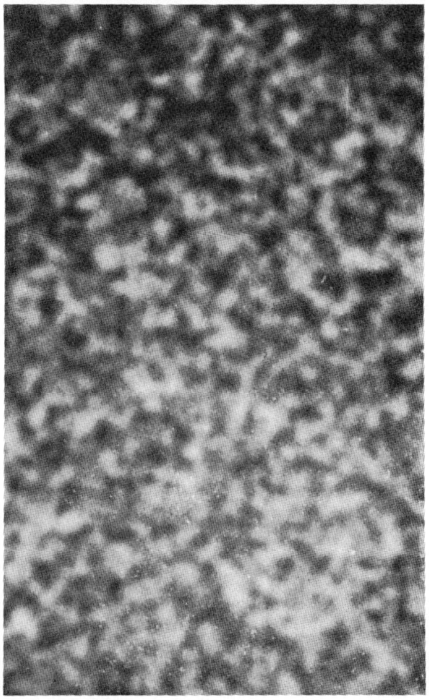

Fig. G-3 Grain structure. LEFT: magnification approximately 15,000×; RIGHT: magnification approximately 5,000×.

green film Freshly processed film. During projection, such film may warp and, subsequently, cause a buildup of emulsion on the aperture plate and result in the scratching of the film's picture area.

grip A stagehand.

ground glass Glass with a surface that has been roughened by grinding or etching. Ground glass, in camera viewfinders, is used for focusing and composing the subject matter seen on their surfaces. The material is also useful for light-diffusion purposes. See Figure G-4 and VIEWFINDER.

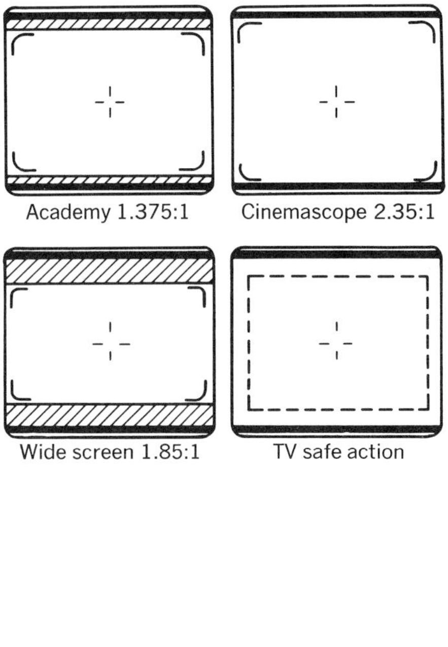

Fig. G-4 Interchangeable ground glass. The Arri 35 2C viewfinder shows the full cinemascope field and provides for a complete program of ground glasses—all quickly and easily interchangeable by the cameraman. The series of interchangeable ground glasses are each precisely etched to show camera aperture, projector aperture, and center cross for: Academy 1.375:1; cinemascope 2.35:1; wide screen 1.85:1; and TV safe action 0.565 × 0.759 inch.

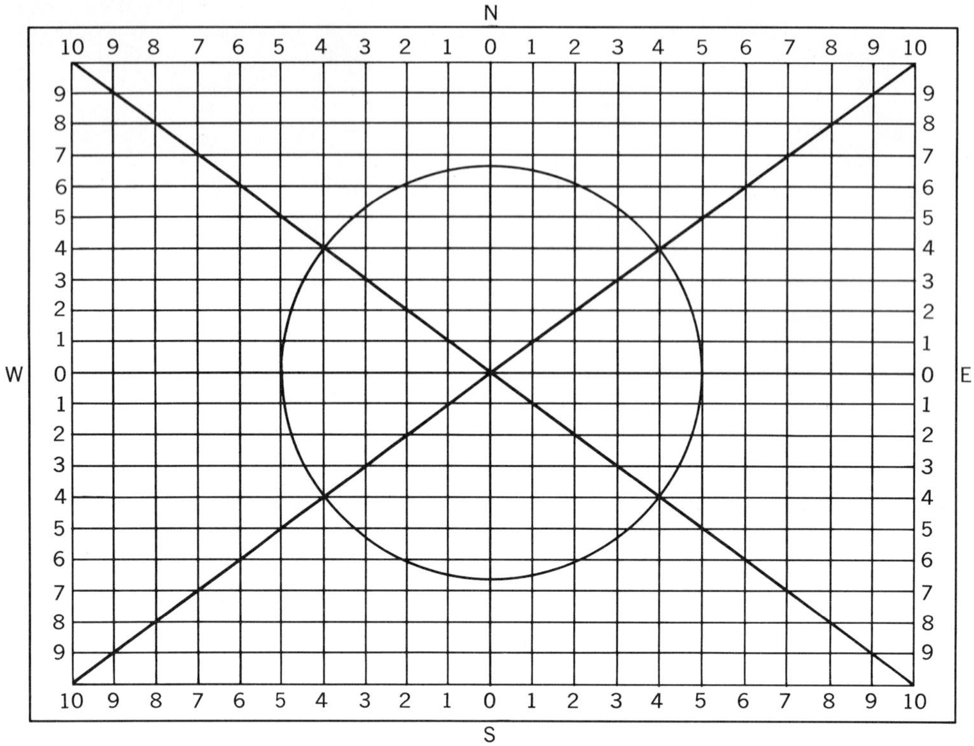

Fig. G-5 The field guide used for photographing animation (not actual size).

ground noise Undesirable sound caused by imperfections in the recording material. It should not be confused with system noise which is usually at a much higher frequency and is due to the electrical characteristics of the equipment. See SYSTEM NOISE.

guide In animation, a layout used by the animation cameraman for moving cutouts and other graphic material, as well as for the filming of scratch-offs and other effects. See Figure G-5.

guide rollers Fixed wheels at various points in cameras, projectors, and laboratory equipment that keep the film precisely aligned as it travels through the transport mechanism.

guide servo The guide servo is an accessory for videotape recorders which accurately positions the vacuum guide to eliminate picture skewing.

guide track A temporary sound track that is used as a guide for synchronizing purposes or to time subsequent recordings.

gun cameras Originally mounted on the leading edges of a combat airplane's wings, the cameras were activated when the airplane's guns were fired. The cameras were focused on anything within the range of the lens and traced the paths taken by the bullets. Today, with only minor modifications, these compact, rugged cameras have many applications in motion picture photography. Formerly produced by the Bell & Howell Company, these cameras are now manufactured by the Consolidated Systems Corporation and distributed by the Traid Corporation.

KS-27B, 16mm. This gun camera has a photocell for measuring light levels and a shutter which is set automatically by a computer. This feature is especially useful where constant attention cannot be given to camera settings and controls. If desirable, the photocell and computer may be installed in locations remote from the camera. The camera accepts C- or GSAP-mount lenses and has rates of 16, 32, and 64 frames per second, and film capacities of 50 or 100 feet.

KB-3A, 16mm. This camera is adaptable for practically any movie-making application, including sky diving. The compact camera has filming speeds of 16, 32, and 64 frames per second, accepts C- and GSAP-mount lenses, and has a variable shutter.

KD-7, 16 mm. This pulse-data-recording camera operates in either a pulse mode up to 10 frames per second or at a motion picture filming speed of 12 frames per second. The camera offers three modes: open-to-open shutter, closed-to-closed shutter, and alternately open to closed. The camera is compact and lightweight, has a film capacity up to 100 feet, and accepts either C- or GSAP-mount lenses.

gyro head A type of camera mounting in which pan and tilt movements are made possible by a built-in fly wheel arrangement. Compare FLUID HEAD and FRICTION HEAD.

gyro stabilizer The gyro stabilizer (Kenyon) is an ingenious device for stabilizing the longitudinal axis of any hand-held camera or optical instrument such as sextants or binoculars. It will dampen vibration from a moving platform or involuntary movement of arms. The stabilizer provides a firm resistance through application of gyroscopic principles. Known in professional motion picture circles as the "invisible tripod," it provides rock-steady hand-held pans and tilts, even when shooting from airplanes, helicopters, boats, and moving cars. An aluminum alloy housing encloses hermetically sealed, electrically driven gyroscopes. The stabilizer has precision ball bearings on all moving parts.

H

H The optical designation for **hyperfocal distance**.

H & D curve Also called **characteristic curve**. A method of plotting and determining, by graphs, the sensitivity and development characteristics of an emulsion. It is named after the two scientists who were responsible for developing the present system of photometric evaluation—Ferdinand Hurter of Switzerland and Vero C. Driffield of England. With the H & D system, a film with a speed rating of 1,000 would be twice as fast as another film type that is rated at 500.

halation Also called **halo** and **flare**. Hot spots, background ghosts, and angling streaks that become visible in the processed print. The undesirable effect is caused when strong beams of light pass through the lens, go through the film's layer of emulsion, and reflect back from the film base to the emulsion. Special antihalation backings between the emulsion and base absorb most of the light passing through the layer of emulsion. The antihalation backing is completely removed during processing.

half-and-half diffusion screen See DIFFUSION SCREENS, HALF-AND-HALF.

half-and-half fog-effect filter See FOG-EFFECT FILTER, HALF-AND-HALF.

half broad A 1,000-watt light source. Compare BROAD.

halftones The intermediate shades of gray in print material.

half-track A magnetic coating that covers half the area usually occupied by the optical track in 16mm sound film. This arrangement makes it possible to use optical or magnetic tracks alternately. See FULL TRACK.

halo A light flare or streak around objects in the scene. The undesirable effect is caused by light striking the base of the film, after passing through the emulsion layer, and reflecting back through it during the filming process. Compare HALATION.

In television, most commonly, a dark area surrounding an unusually bright object caused by overloading of the camera tube. The reflection of studio lights from a piece of jewelry, for example, might cause this effect. With certain camera-tube operating adjustments, a white area may surround dark objects.

halogen Any of the chemical elements of the bromine, chlorine, fluorine, iodine family. They are used in combination with the silver in a film's emulsion. During the developing process, the silver which has been exposed to light is converted into pure metallic silver. The halogen, during the same developing process, is dissolved.

Also, these elements in vapor form, sealed within lamp housings.

halogen-cycle lamps When halogen lamps were introduced, they were manufactured from quartz and contained an atmosphere of iodine. Today, with new developments in lamp technology, material such as Vycor (Sylvania) is used to replace the quartz; gases other than iodine in the halogen family are used. Therefore, it is more accurate and more meaningful to refer to them as halogen lamps.

The halogen gas sealed in the lamp reduces blackening and maintains the color temperature for the life of the lamp. Quartz, or the special high-silica-glass Vycor has a very high melting point and a low coefficient of expansion, permitting bulb-wall temperature up to 600°C.

In operation, tungsten particles evaporate and collide with halogen gas particles. The tungsten combines chemically with halogen gas and is deposited on the lamp's filament. The halogen gas is released ready to repeat the chemical process.

hard A laboratory term for a contrasty negative.

hardener Any of the chemicals (chrome alum or potash alum, for example) that are used in hardening or fixing baths for the purpose of making developed film more resistant to physical damage such as abrasions or scratches.

hard light A light source that produces sharply defined shadows.

harmonics The specific branch of acoustics that deals with musical sounds. Also, partial vibrations produced by a secondary tone accompanying a fundamental or pure tone.

harmonic distortion Distortion characterized by the appearance in the output of harmonics of the fundamental frequency when the input wave is sinusoidal.

haze The scattering of light by dust and other particles in the atmosphere, which creates a cloudy or foggy effect in distant views. Since the scattering is strongest in the blue and ultraviolet regions, and photographic emulsions are generally sensitive to ultraviolet light, haze will often affect pictures even when the view appears clear to the eye. Haze filters, which absorb blue and ultraviolet light, can reduce this effect.

haze filter A type of filter capable of reducing the effects of atmospheric haze by absorbing blue and ultraviolet light.

Hazeltine Also called **Hazel**. Before printing, scenes in a negative are checked to determine the proper printing light by the Hazeltine, an electronic color-film analyzer which scans the color negative and projects a color-positive image for direct visual analysis. See Figure H-1.

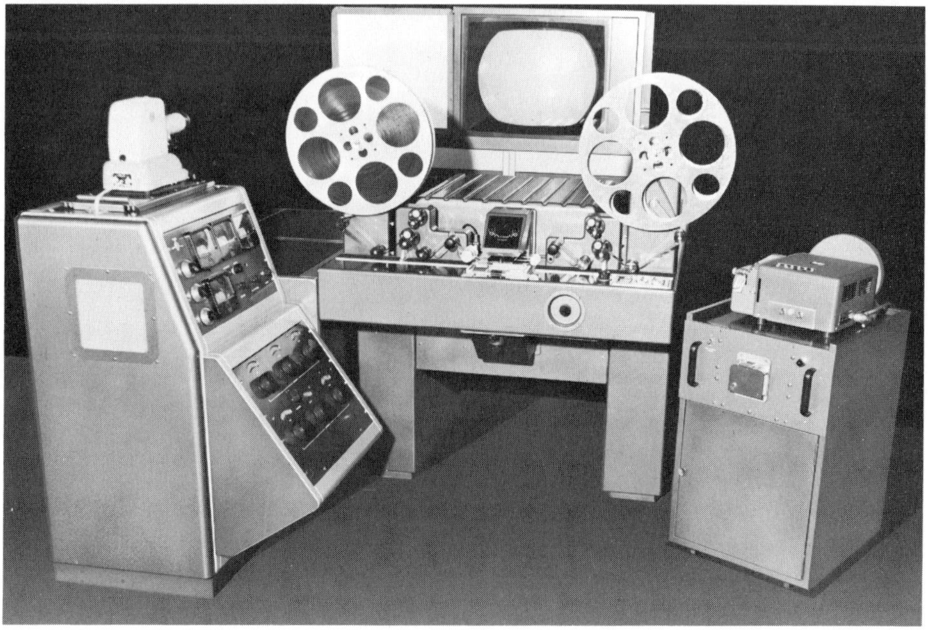

Fig. H-1 Hazeltine electronic color-film analyzer.

head The mechanism in a projector or moviola that contains a source of illumination and the sprocketed transport facilities needed for continuous viewing at proper projection speeds.

In magnetic recorders, the electromagnetic devices which magnetize the tape's iron oxide coating as the tape passes across them during the recording process.

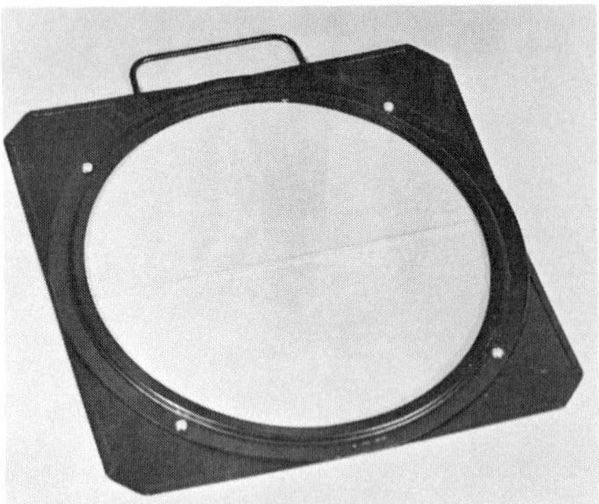

Fig. H-2 ColorTran heat filter.

head alignment The positioning of the record and reproduce heads on a tape recorder so that their gaps are mutually parallel and perpendicular to the path of travel of the tape. See GAP.

head out The position of film on a reel. The leader, when the film is wound head out, is on the outside of the roll in position for threading in a projector or moviola.

heat filter Heat filters, when placed over lights, keep a motion picture set and the actors cool and comfortable. These filters also make it possible to photograph delicate plants and other subject matter which would normally be affected by the heat from conventional illumination. Similarly, a high-speed photographic process under intense light is also a more comfortable procedure with these filters. Since the color temperature of the light is raised only 100°K, the heat filter is ideal for all work involving color photography without affecting the resulting color values.

Since the heat filter transmits a major portion of the actinic* rays produced by all light sources, it can be used with all types of illuminants such as arc, quartz-iodine, tungsten, and pulsed xenon. Actual applications include backlighting of transparencies, contact printing of color and black-and-white material, and enlarged transparency copying. See Figure H-2.

heavy A term synonymous with the "bad guy" or villain.

helical scan A method of recording video on tape in which the signal is positioned diagonally on the tape in adjacent strips.

helicopter mount Also called **Aero-Vision**. An antivibration mount designed for use in helicopters, airplanes, or other moving vehicles. The mount includes a bucket seat, safety belt, and circular turning bar. See AERO-VISION.

*The main actinic light of an arc source is above 3000 angstroms. Very little of this energy is absorbed by the filter.

helmet cameras When a close-up of an instrument panel, or a panoramic shot of the countryside is needed, an unusual bird's-eye view can be obtained with cameras that are built into helmets. These helmets* have been successful in obtaining hard-to-get participants' views of an action. The actions most frequently photographed with the helmet cameras are water skiing, snow skiing, motorcycling, automobile racing, and sky diving. These helmets were originally designed by Bell Toptex, manufacturer of safety helmets. The first camera helmet was used for filming scenes of the television series "Ripcord."

hemisphere An attachment used with certain types of exposure meters to obtain incident-light readings with reflected-light meters. The Invercone attachment, which relates specifically to Weston exposure meters, serves a basically similar purpose.

hertz unit In broadcasting and electronics, frequency is measured in terms of hertz units rather than in cycles per second as it is for sound. Named after Heinrich Rudolph Hertz, the German physicist, the unit of electromagnetic wave frequency is equal to one cycle per second. See FREQUENCY.

high-band recording A videotape recording made with a high-frequency carrier (10 megacycles), yielding a very high signal-to-noise ratio. The higher FM deviation frequencies eliminate moiré and noise. See MOIRÉ.

high-contrast film A contrasty film used for titles, matte work, and effects photography. The processed lengths of high-contrast film are subsequently used in combination with other film footage and reexposed in the optical printer to produce the desired composite effects.

high frequency Also called **highs**. Frequencies of 15,000 cycles per second—minimum. In sound recording, the electric impulses and modulations produced from high-pitched sounds. Although the audible frequency range is between 15 and 20,000 cycles per second, the term high frequency usually implies frequencies above the audible range.

high-frequency interference Interference effects that occur at high electrical frequencies. Interference in a signal transmission path, generally, is extraneous energy which tends to interfere with the reception of the desired signals.

*Gordon-Bell cameras available through Gordon Enterprises: the GB 35 mm, GB 16 mm, and GB 35 DF (double frame).

high key The general arrangement of set lighting that produces gradations between white and intermediate shades of gray in the print material. Brightly lit scenes, with almost no dark areas, are referred to as high-key scenes.

highlight The lightest parts of the print in a positive image. Conversely, the darkest or densest parts of the negative image.

high-speed photography As a time magnifier, high-speed photography, paradoxically, produces slow-motion effects. Movements of any kind, at speeds far greater than the eye can follow, become almost stationary when projected at normal viewing speeds. Aside from the achievement of unusual illusory effects, the uses for this filming technique are as many as they are varied.

Scientists and engineers use the filming technique for observing action and reaction. Movement at fantastic rates of speed can be stopped in its tracks for analysis. The motion of a missile, or the reaction time of chemicals, can be easily determined by projection and measured in milliseconds or microseconds, if necessary.

On the lighter side, high-speed photography makes possible the study of other types of projectiles. The high standard of play and the mass attack on sports record books can be traced directly to the use of the filming technique and its effect on coaching practices. Following the movements of 22 men and 1 ball traveling in as many directions is a herculean task for any sideline spectator. For the football coach, preoccupied with other details, it is an impossible assignment. With the high-speed filming technique, however, it is possible not only to follow the movement of each player and observe the skill with which assignments are fulfilled but also to count the stitches on the ball during its flight to a downfield receiver. The postgame study of the film helps eliminate future mistakes and improve the overall quality of play in subsequent games.

In commercial use, high-speed techniques make it possible for an advertiser to demonstrate his wares in many ways. The effectiveness of these demonstrations lowers the resistance of the most skeptical consumer, a fact that is reflected by commercial surveys.

Statistics, relative to high-speed photography, stagger the imagination. Specially designed reimaging cameras, for example, are capable of taking approximately 30 million pictures per second. If these separate images were projected at the rate of 24 per second, normal sound speed, the viewing time for the one-second sequence would be almost $2\frac{1}{2}$ weeks.

The exposure of motion pictures, over the standard projection rate of 24 frames per second, does not automatically qualify the effort or entitle it to a high-speed classification. By present standards, the term *high speed* applies only to exposure rates that are greater than those that can be obtained with an intermittent camera. Since some intermittent cameras can take as many as 400 pictures per second, the high-speed technique must, obviously, use that figure as the minimum requirement for the classification. Statistically, even this minimum figure is equal to 16.66 times the normal rate of projection.

Two types of cameras are used for filming high-speed subject matter. In the first group, the film moves continuously, and exposure is made possible with the help of rotating prisms, lenses, mirrors, and synchronized flash lamps. In the second group, the camera components move continuously in relation to the stationary film.

With either camera, the operator has at his disposal a variety of film types including panchromatic, orthochromatic, infrared, and color. Reversal film, both black-and-white as well as color, widens the choice and makes available a film type ideally suited for every purpose. Tungsten lamps, high-pressure mercury arcs, synchronized electronic flash and spark systems, and photoflash lamps are all capable of providing the high-intensity light source that is needed for high-speed photographic processes.

The number of pictures required per second, or the *taking rate,* is determined by the type of study and analyses needed. If, for example, the image or subject matter moves from one point to another at a predetermined speed, the number of pictures needed for the study is the important consideration. Thus, if 100 pictures are deemed sufficient for the study, the taking rate would be equal to the number of pictures divided by the time of the action.

$$\text{Number of pictures} \div \text{time of action} = \text{taking rate}$$
$$100 \text{ pictures} \div 0.5 \text{ second} = 200 \text{ frames per second}$$

Projection time is determined by the following formula;

$$\text{Taking rate} \div \text{projection rate} = \text{projection time}$$
$$200 \text{ fps} \div 24 \text{ fps} = 8\tfrac{1}{3} \text{ seconds}$$

Changes in the direction of the moving object, toward or away from the film plane, slow the motion down in proportion to the angle of motion. The slowdown is relatively little when the object's movement is almost parallel to the film plane. The amount of slowdown, caused by any movement that takes the object directly away from the camera or brings it directly to the camera, is negligible. See Figure H-3.

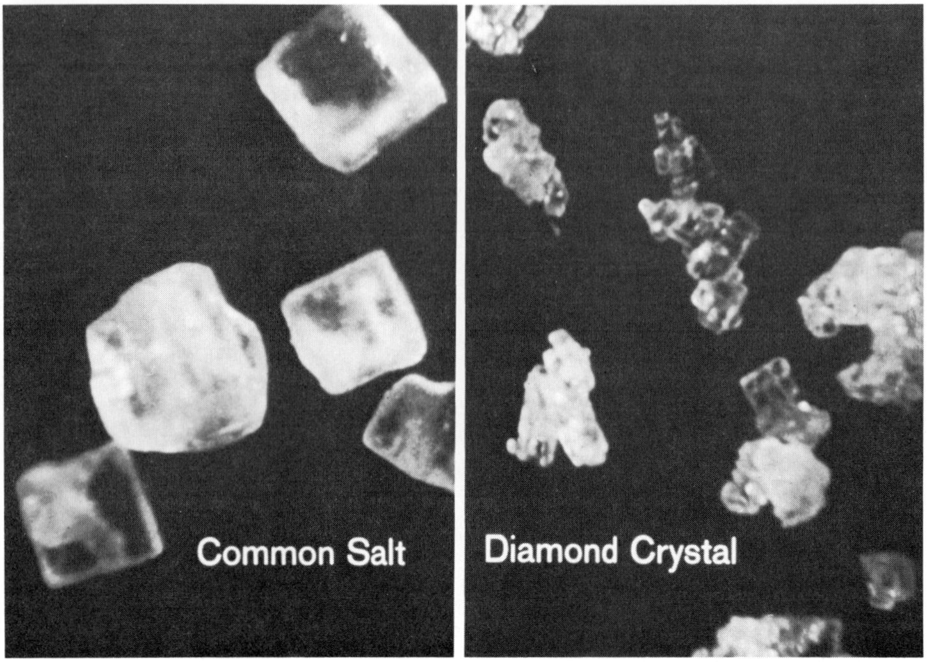

Fig. H-3 High-speed photography. In order to produce the effect shown here, the manufacturers of Diamond Crystal Salt combined high-speed photographic techniques with microcinematography. High-speed photosonic cameras, capable of speeds up to 4,000 frames per second, were used for filming the falling salt on Eastman high-speed Tri-X film with positive perforations.

Because of the high-intensity lights that were used to light the set, cameramen and technicians wore dark glasses. The tricky optical effect turned out to be a "sweet" job in spite of the subject matter.

high-speed cameras High-speed motion picture cameras differ from ordinary motion picture cameras in that the film is pulled past the aperture in a continuous motion by means of a drive and a take-up motor. These higher framing-rate systems expose each frame through a rotating prism compensator which moves the subject image in synchronism with the film. This method permits speeds up to 18,000 frames per second. An integral sleeve with windows enveloping the prism functions as a shutter. Other cameras, in addition to the prism, have either drum or disk shutters with which it is possible to vary shutter speed.

Cameras of 8, 16, and 35mm with 100-, 400-, 500-, and 1,200-foot capacities, and speeds from 150 to 18,000 frames per second, are available in straight motion picture, oscillographic, or combined motion picture–oscillographic models. See Figure H-4.

Beckman & Whitley model 333 Magnifax, 16mm. The Magnifax is a rotating-prism-type camera designed to operate from 1,000 to 3,200 frames per second. A rheostat controls framing speeds of a 115-volt a-c/d-c motor. The camera has a built-in event synchronizing switch to coordinate subject and camera action electrically as well as a built-in timing light. A light-tight door is attached to the camera body by a single knurled knob. The camera has a built-in boresighting and focusing finder.

Bensen-Lehner photographic recorder model HS-16D2, 16mm. This intermittent-type high-speed camera is designed to operate up to 400 frames per second while mounted on aircraft, missiles, and sleds, or in any location where photographic information is required for test data or documentary purposes. Register pins on both sides of the film ensure precise registration at the time of exposure and eliminate the danger of film loop loss.

The camera utilizes coaxially mounted daylight-loading spools of 200-foot film capacity and has a standard 16mm single-frame aperture. The camera uses standard C-mount lenses and has a specially designed universal motor with four-speed manually operated transmission, capable of operating on 25 volts d-c, 115 volts d-c, or 115 volts a-c, 50 to 1,000 cycles per second, single phase.

Fig. H-4 Hycam optical schematic diagram.

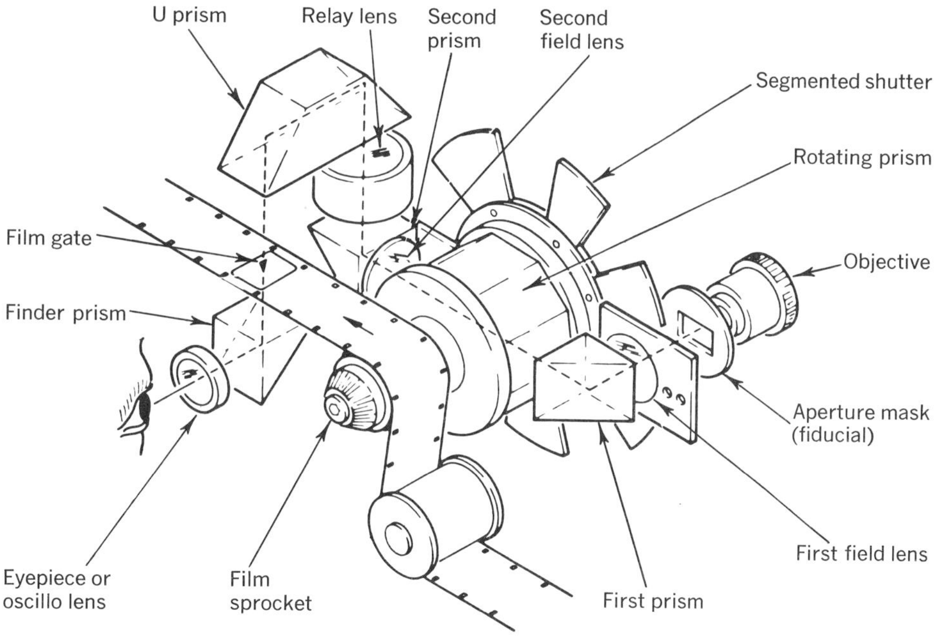

Eastman, 16mm. The Eastman camera can film at speeds from 1,000 to 3,200 frames per second. In addition to a built-in 115-volt a-c/d-c motor with rheostat control, the camera has specially hardened gears, high-speed shafts with sealed precision bearings, and nylon belts. The camera is permanently lubricated to provide trouble-free operation. It is available with built-in camera timing lights and event synchronizer. Focusing is accomplished through a magnifying eyepiece.

Fastair, 16mm. The camera is built to withstand acceleration forces up to 100*g* in four principal mounting attitudes. It exposes from 12 to 600 frames per second and reaches top speed almost instantaneously. The camera maintains uniform speed throughout the entire film run. An NE 2*a* neon lamp exposes pips or marks on the film for reference. Fastair offers a choice of three film magazines; 50-foot daylight-loading spool, 100-foot daylight-loading spool, and 200-foot darkroom load on plastic core. The camera can be started and stopped any time during the run at speeds under 400 frames per second. Thermostatically controlled heaters in the camera body and magazines allow operation from 40 to −65°F. The camera is supplied with a 12½, 17, or 25mm *f*/2.3 Pro-Raptar lens.

Fastax. These cameras are not complex instruments. Anyone with a general knowledge of photography can operate them and produce excellent results with a little training and experience. Fastax cameras differ from ordinary motion picture cameras in that the film is pulled past the aperture in a continuous motion. Instead of a shutter exposing each individual frame during a stationary cycle, the camera exposes frames continuously through a rotating prism traveling at the same speed as the film. By this method, speeds up to18,000 frames per second are possible. Fastax cameras are available in 8, 16, and 35mm, with speeds from 150 to 18,000 frames per second.

Various models are available for straight high-speed motion pictures, oscillographic recording, and combination motion picture–oscillographic work.

The Fastax cameras incorporate an integral viewfinder for critical focusing and viewing through the camera aperture, double timing lights, automatic cutoff switch, and two motors (one for driving the film and one for film take-up). Both motors are powered by 115-volt a-c/d-c 60-cycle current. All models are equipped with the Fastax lens mount. See Figure H-5.

Gordent 200, 16mm. A compact camera designed for photoinstrumentation applications where space is limited. The camera operates at 50, 100, and 200 frames per second and has a 150-degree shutter opening. The film capacity is 50 feet using preloaded magazines. The camera accepts standard C-mount lenses. Speed is achieved by incorporating precision bearings and a high-torque 28-volt d-c governor-controlled motor. An automatic end-of-run switch is supplied to prevent excessive camera operation on remote locations. A special lubrication system allows extended operation. (Cameras are available with timing and correlation marker lights.)

Hulcher 70. The Hulcher's versatility and ability to provide exacting details of fast-moving action makes it ideally suited for recording sports, military, research, and industrial assignments. Hulcher 70mm rapid-sequence still cameras are capable of photographing $2\frac{1}{4}$- by $2\frac{1}{2}$-inch or $2\frac{1}{4}$- by 5-inch format on 100-foot internal spools or 400- or 1,000-foot external magazines. The cameras accommodate a wide variety of interchangeable lenses from wide angle to extreme telephoto. The camera's unique film-transport design completely eliminates all reciprocating parts, thus reducing vibration, jamming, and weight. All the Hulcher cameras have a double-rotating disk-shutter arrangement operating close to the focal plane.

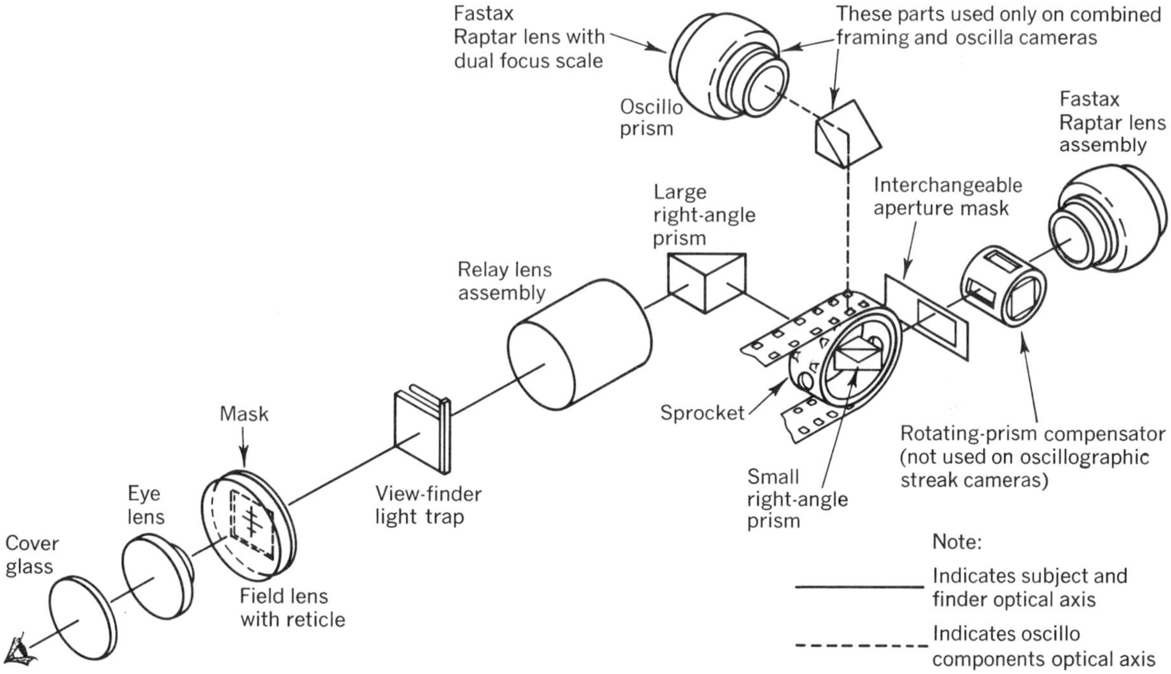

Fig. H-5 Fastax camera. In the Fastax camera, light from the subject passes through the taking lens, through a rotating prism, and then on to the film. Two motors are used: one to drive the film-sprocket-prism mechanism, and the other to take up the exposed film.

In the combined motion picture–oscillographic camera, observation of the moving subject, as well as visual representation of other data peculiar to the operation of the subject, is superimposed over the motion picture and is available for a correlated study.

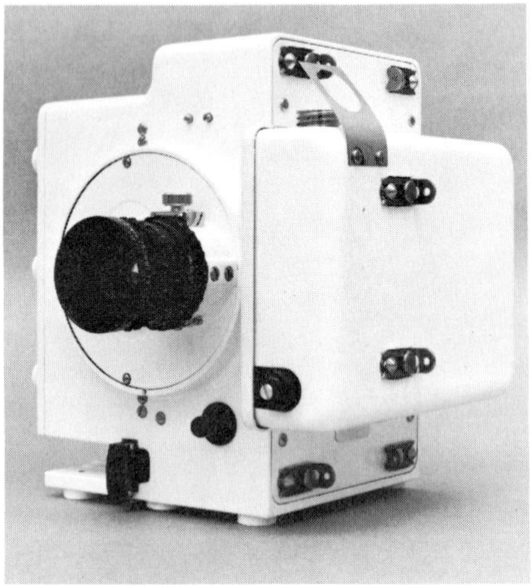

Fig. H-6a Flight research model IV-E 35mm Multidata camera.

35mm model IV-E Multidata flight research camera. The flight research 35mm Multidata camera is available with an auxiliary data chamber which records supporting and supplementary data directly on each film frame.

Since the supplementary data are recorded at the same time the main "target" picture is taken, all the pertinent supplementary data can be compiled for each individual frame. Commonly recorded as supplementary data are time, date, frame number, run number, project number, and the like. Coding lamps or digital counters can be provided to permit the recording of supplementary material and numerical data obtained from other instruments being used in a test.

The auxiliary data chamber permits the combining of target and the photopanel information on a single strip of film. The two coding lights located where the film is traveling at constant speed can be used as event markers or to add additional supplementary data. Space is available for mounting a number of different components. A standard configuration including a 24-hour clock with sweep-second hand, four-digit counter, and data slate is also provided.

The optical path incorporates a front-surface mirror, a short-focal-length lens, and a specially designed prism to bring the data chamber image to the face of the film. The optical path passes behind the camera shutter so that the data-recording exposure is independent of shutter speed. Lamps are incorporated in the data chamber and are triggered at each frame by the standard camera output pulse. The camera is available with either the variable-shutter or the standard fixed-shutter front plate. See Figures H-6*a* and *b*.

Photo-Sonics 1-B, 16mm. The 1-B high-speed camera is capable of taking from 12 to 1,000 pictures per second with one motor. Exposure rates from 1/60 to 1/40,000 second provide even exposure over the entire frame. The use of a variable rotating-disk shutter synchronized with the rotating four-sided prism provides excellent image resolution. The camera accommodates film magazines with 100-, 200-, 400-, and 1,200-foot film capacities—with start-stop capability at any frame rate. The camera also has two timing lights, a run-out switch, and heaters that are effective to −20°F. Illuminated fiducial markers may be installed as an optional accessory.

Photo-Sonics 1-B/a-c, 16mm. The 1-B/a-c camera provides the same high image resolution as the 1-B model, but also offers the convenience of a 110-volt a-c drive motor. Operating speeds are 200, 400, 600, 800, and 1,000 frames per second. The governor-controlled motor assures very rapid rise time and constant speeds.

An optional gearbox provides for slower frame rates of 14, 25, 50, 75, 100, 125, 150, and 250 frames per second, all using the single-drive motor. This camera uses the same film-capacity magazines as the 1-B model.

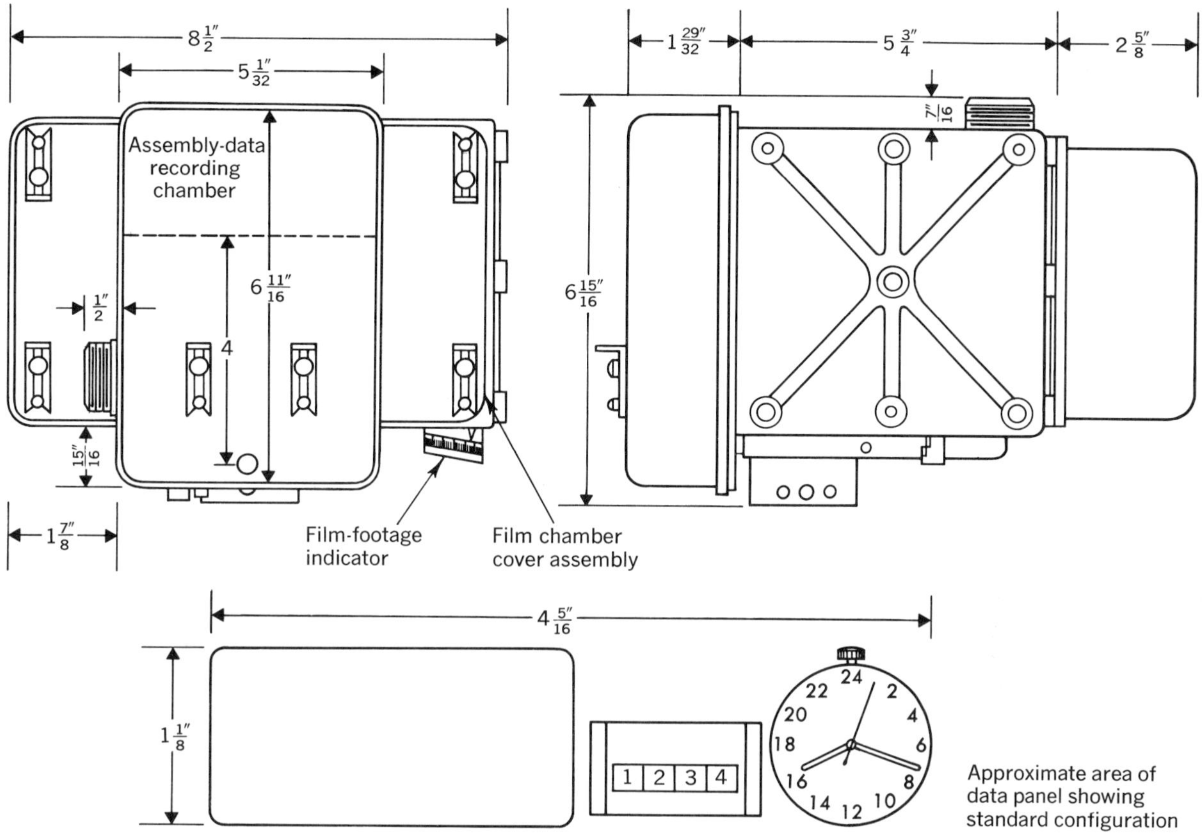

Fig. H-6*b* Dimensional drawing of 35mm model IV-E Multidata camera.

Traid 200, 16mm. Traid cameras are available in several versions with numerous optional features such as 100- or 200-foot capacity, with or without timing systems, single or variable speeds, standard C mount or special lens mounts, with or without heat element and thermostats, standard 72-degree shutter or other opening, boresight, and pulsing unit. The camera has successfully withstood approximately 25*g* in guided-missile filming. Its operating speed is 200 frames per second. The camera is controlled by a shunt-type, governor-controlled motor. The camera can be supplied to fit individual requirements.

hi-hat A relatively small camera mount that can be fastened on any substantial tripod or other suitable support. The mount can be placed on the floor for low-angle or worm's-eye-view shots, or secured to a cartop, station wagon tailgate, or dolly.

Hinelight A lighting unit in which high-intensity, circular fluorescent tubes provide the illumination. The Hinelight unit has an opening

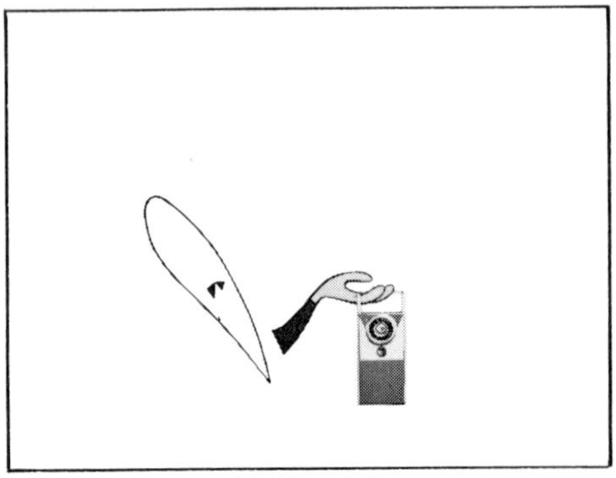

Hold cel 1

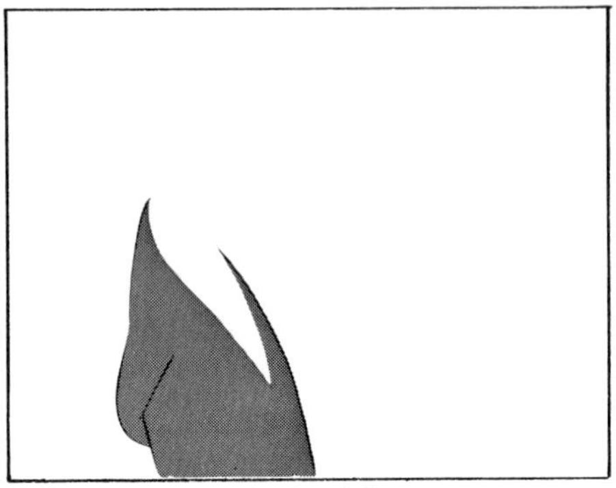

Hold cel 2

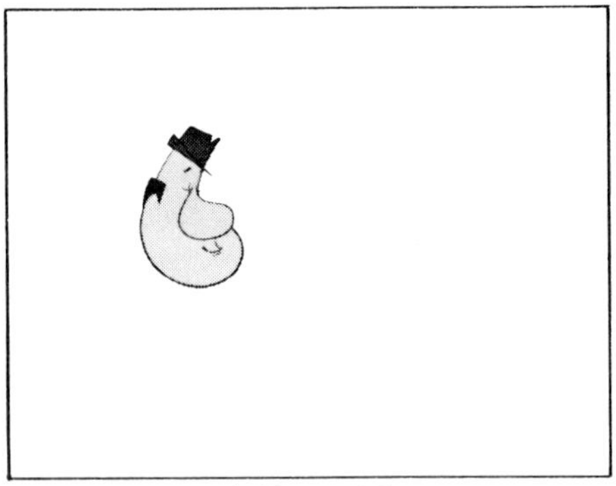

Action cel 3

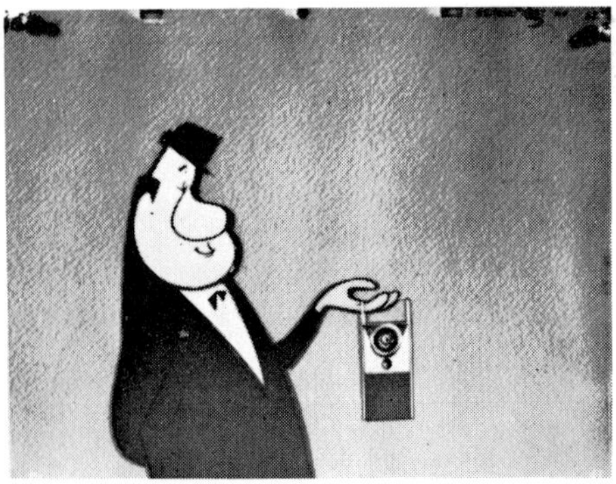

The three cels placed in position over the background, as they will appear on the finished film.

in the center of its reflector that fits around the camera lens, thus providing an even field of flat illumination. This type of lighting unit is used mainly for recording scientific data.

hi-spot An extremely high-output light source for high-speed and instrumentation photography. Capable of producing a relatively long throw, the unit uses PAR 64 quartz-iodine lamps.

hiss High-frequency noise, evident during the playback, caused by imperfections in the recording equipment.

hitchhike A short commercial announcement that follows the program material being televised. Compare COWCATCHER.

hold In animation, a term used to indicate a "hold" or stop position for an action or pan background.

A cel that is exposed for a number of frames while other cels carry the action is referred to as a *hold cel.* If a cartoon character is speaking, for example, the head may be drawn as a hold cel while the mouth movements are inked and opaqued on separate cels that overlay the cel containing the cartoon character's head. This eliminates the necessity of tracing the head for every mouth movement drawn by the animator. See Figures H-7*a* and *b*.

holography A technique—still experimental—of photographically recording the wave patterns of objects illumined by laser-generated coherent light to produce, under subsequent laser illumination, a three-dimensional image of the object.

The recording of the subject matter is made possible by illumination from a laser or other light source capable of providing continuous waves of coherent light—that is, light confined to a single frequency or wavelength. A mirror, adjacent to the subject, reflects part of the incident light falling on it from the laser beam back to the photographic plate, in which the light patterns are recorded.

The wavefront reconstruction of the image is accomplished by placing the resulting hologram in the path of another beam of coherent light. The diffracted waves produce both real and virtual images. These are optically perfect reproductions of the original subject matter.

And although the hologram image duplicates the proportions found in the subject, changes in the perspective of the image can be achieved easily by shifting the viewing position. With this image-recording and wavefront-reconstruction technique, the three-dimensional images can be viewed without Polaroid glasses or other stereo devices.

Fig. H-7*a* **The hold cel. (*From Levitan, "Animation Art in the Commercial Film," Reinhold Publishing Corporation, New York.*)**

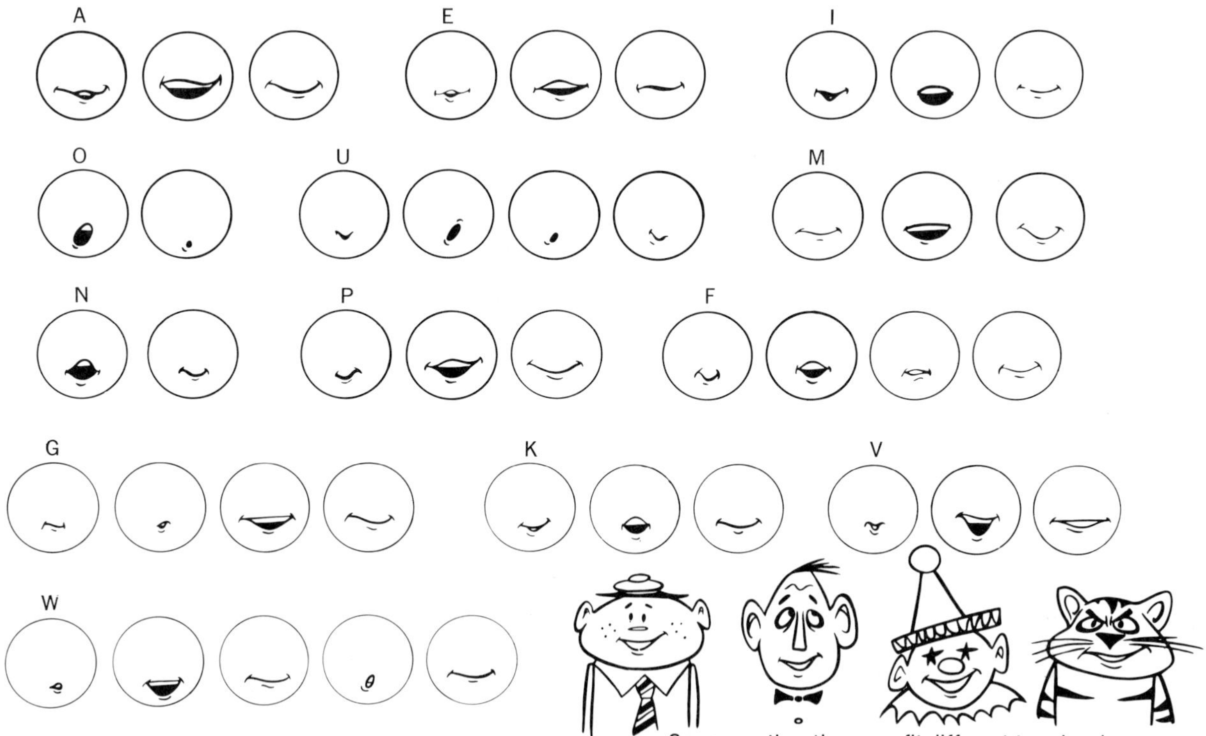

Same mouth actions can fit different-type heads.

Fig. H-7*b* Mouth-action chart. The illustration shows the positions of the mouth during the formation of different sounds. Mouth-action charts are used as guides by animators.

Mouth action	*Also used to form*
A	R
E	C
I	L
O	Y
U	Q
F	S
G	T, Z
K	D, H, X
P	B, J

The technique can best be described by assuming that the photographic plate is a window with the memory capabilities of a tape recorder. All the light waves passing through the window are recorded and can be played back at any time.

In ordinary photographic techniques, light from two-dimensional images passes through the lens system and is exposed on film. Exposure increases in direct proportion to the brightness of the subject matter. More important, however, is the fact that any specific point or component part of the object or scene being filmed is exposed on only one specific point of the photographic plate.

In the holography technique, by comparison, every component part of the scene or subject matter is exposed on *every* part of the photographic plate. Looking through the window (photographic plate), a viewer could change the perspective of the scene by changing his viewing position. Thus, a person could see behind objects by merely moving slightly and looking around them. The effect is similar to that produced when a person moves from one side of a relatively wide picture window to a position on the other side. The overall scene that is visible through the window does not change. The perspective does.

The name for the technique is taken from the Greek word *halos,* meaning whole. As applied to the technique, the inference would be "whole picture."

The first experiments that led to the development of the technique were conducted in 1801 by Thomas Young. The Englishman, in one of a series of experiments, found that patterns of dark and light values could be produced on a screen by splitting a pencil-like beam of ordered or coherent light into two separate beams by means of a mirror. The converging light waves, where they augmented each other, produced light areas. In those places where the light waves were out of phase or fought each other, the brightness was reduced, thus producing dark areas on the screen. Mr. Young had succeeded in stopping light—no easy accomplishment, considering that the previously unobservable waves travel at speeds of approximately 186,000 miles per second. More important, he had discovered a method for generating interference patterns—the basis for the holography technique.

Applying the same interference principles, Gabriel Lippman produced the first color photos in 1891 and, subsequently, received the Nobel prize for his efforts. Coating his mercury-backed plates with a thick emulsion to form a reflective surface, Lippman allowed light from the subject matter to pass through the thick layer of emulsion and reflect back through it after striking the mercury-backed plate. Both sets of light waves met, interacted, and formed interference patterns within the emulsion layer. The different wavelengths in the ordinary white light that fell on the plate, after development, responded to corresponding interference patterns in the emulsion. The colors thus produced duplicated those in the original subject matter.

Approximately one half-century later, Dr. Dennis Gabor of the Imperial College of Science and Technology in London duplicated Thomas Young's experiments. Dr. Gabor's original aim had been the achievement of enormous magnification through the development of an electron microscope. The key to that plan called for the generation of interference patterns that would expose, on a photographic plate, the characteristics of light waves emanating from the subject matter.

In Dr. Gabor's experiments, two sets of light waves originated from a mercury-vapor lamp. Some rays fell on the subject matter. Other rays bypassed the subject completely. Both sets, however, met on the recording emulsion to set up overlapping interference patterns which were frozen into the film's layer of emulsion.

After development, the image could be reconstructed by passing a similar beam of coherent light through the photographic plate. Affected by the interference patterns in the plate, the beam reflected the same wave characteristics that had emanated from the subject at the time of exposure.

In 1962, Dr. Yu N. Denisyuk of the Soviet Union proposed a variation of the technique used by Lippman. Applying the same interference principles advanced by the Nobel Prize winner, Dr. Denisyuk's plans included a method for making holograms visible under ordinary light.

That same year, Dr. Emmett Leith and Juris Upatnieks recognized the possibilities offered by the new laser beam as a light source for the photographic technique. The beam, they concluded, had properties that made it an excellent source of light. It was extremely intense, of a single wavelength, and highly coherent.

Using a laser, the two University of Michigan scientists utilized one part of the beam to illuminate the subject matter and expose the photographic plate. Light from the second part of the split beam fell on the plate at an angle after being diverted by mirrors. The hologram on the emulsion's surface was the result of interference set up between the two beams.

Using a similar laser beam for the playback, the viewer could almost touch the dimensional image that seemed suspended in space. The image had all of the directional and brightness characteristics found in the original subject matter.

In the experimental stage, or already tested and proved applicable with the holography technique are:

1. Three-dimensional radar systems.
2. Scanners capable of mapping planets in distant space.
3. Methods for observing stress patterns in objects moving at extremely high speeds.

4. Computer studies of three-dimensional objects that exist only in imaginative form on computer tapes.

5. Methods for photographing fog and dust clouds for analytical study.

6. Methods for the processing and sorting of great numbers of photographs and other visual material of similar content.

7. Methods that enable aircraft to visualize cloud-covered ground installations via three-dimensional radar systems.

8. Microscopic studies showing the three-dimensional structure of molecules composed of millions of atoms. These studies could lead to the dramatic development of new drugs.

See Figures H-8*a* to *c*.

hookup In animation, a term indicating that the first and last drawings of an animated action, or the first and last positions of a pan background, are the same and interchangeable. This is a great time and budget saver—repeat actions and pan backgrounds may be exposed for as much film footage as is needed.

hot Extreme brightness, either in print material or in lighting setups.

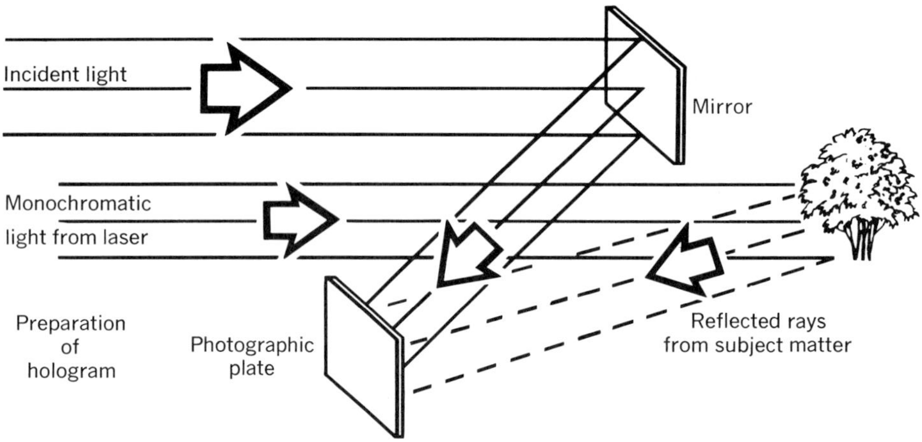

Fig. H-8a Leith-Upatnieks hologram—interference patterns on the surface of the photographic plate. With the Leith-Upatnieks technique for making holograms, the interference patterns are produced on the surface of a photographic plate by the convergence of highly coherent light beams originating in a laser. One beam, the illuminating beam, is reflected from the subject matter. The second, or reference beam, is reflected from a mirror surface to the photographic plate.

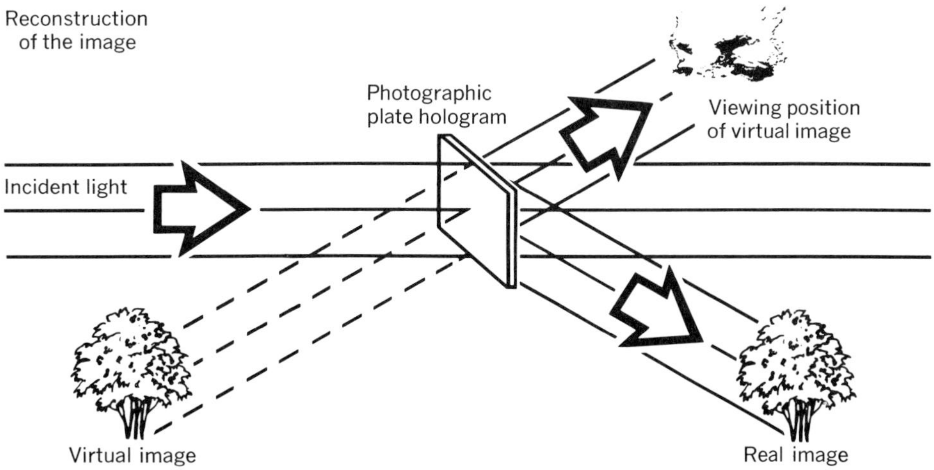

Fig. H-8*b* Leith-Upatnieks hologram. In order to reconstruct the dimensional image, light originating from a laser strikes the surface of the photographic plate, following the same path as the original reference beam. The light waves, altered by the interference patterns on the surface of the photographic plate, duplicate all the characteristics found in the original subject matter. An observer, by changing viewing positions, is able to see the dimensional image as well as objects behind it.

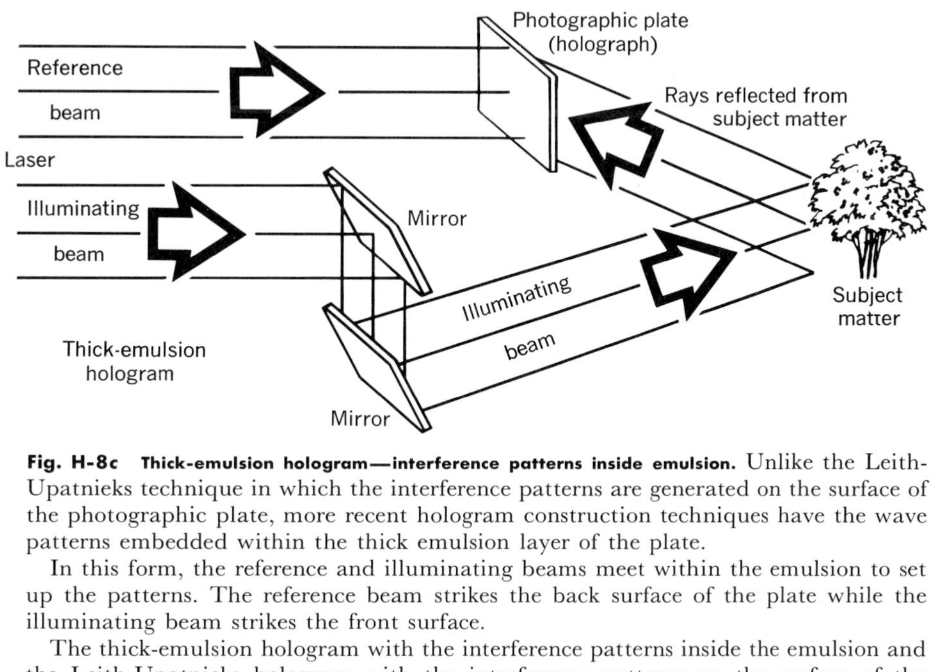

Fig. H-8*c* Thick-emulsion hologram—interference patterns inside emulsion. Unlike the Leith-Upatnieks technique in which the interference patterns are generated on the surface of the photographic plate, more recent hologram construction techniques have the wave patterns embedded within the thick emulsion layer of the plate.

In this form, the reference and illuminating beams meet within the emulsion to set up the patterns. The reference beam strikes the back surface of the plate while the illuminating beam strikes the front surface.

The thick-emulsion hologram with the interference patterns inside the emulsion and the Leith-Upatnieks hologram with the interference patterns on the surface of the photographic plate are both viewed in the same manner.

hot splicer A hot splicer is used to obtain precision in splicing. The stainless steel blocks on these splicers are fitted with tungsten carbide inserts for cutting. The attached scrapers are adjustable for depth—fitting the jig incorporated in the splicer base. Tempered steel springs provide even pressure, positioning the film firmly during the scraping and splicing process. Thermostatically controlled heaters maintain constant 100° temperatures. Many hot splicers have two sets of pilot pins which permit splicing with the perforations either toward or away from the operator, eliminating the necessity of looping the film around the splicer. This is particularly important when using A & B rolls.

hot spot An area that is more brightly lit than any other area on the set.

houselights A general lighting system permanently installed in the audience area of either a theater or a studio which provides sufficient illumination for safe movement in the area.

hue Any of the colors in the spectrum. Hue refers only to color, or wavelength. It is distinguished from *value,* which describes the brightness of the color, and *intensity* or *chroma,* which describes the white components in it.

hydrochloric acid A colorless solution used in bleaching baths. It has a rather strong, pungent odor.

hydrofluid tripod A medium-sized tripod with adjustable pan-and-tilt tension. Pan-and-tilt action utilizes the silicone dampening effect for smoothness of movement. The ball-type adjustment permits fast leveling for repositioning. The tripod is designed for the 16 Mitchell, Eclair, Maurer, Camerette, Arri 16/35, Cine special, Eyemo, Filmo, and Auricon (CineVoice 660-1200). Compare FLUID HEAD, FRICTION HEAD, and GEAR HEAD.

hydroquinone A developing agent used in processing. Also known as **quinol**, it is a relatively slow working, contrasty developer.

hydrotype process See IMBIBITION.

hygroscopic Describing any substance that absorbs moisture from the air. These substances are useful for storing exposed film in arctic and tropical climates, prior to their development. See DESSICATOR.

hyperfocal distance With the lens focused at infinity, its hyperfocal distance is the distance between the lens and the nearest object to it that is in acceptable focus. Focused on hyperfocal, all objects will be in sharp focus from a point approximately halfway between the lens and that nearest object and infinity.

hyperfocal distance formula.

$$\text{Hyperfocal distance} = \frac{(\text{focal length of lens})^2}{f\text{-stop} \times \text{circle of confusion}}$$

Circle of confusion in 16mm cameras = 0.001 or 1/1,000 inch
Circle of confusion in 35mm cameras = 0.002 or 1/500 inch

hypersensitize Any of the several processes used for increasing the emulsion speed of film prior to exposure. Immersion in alkaline solutions containing diluted ammonia is one method. Exposure to the vapors produced by mercury or sulfurous acid solutions is another. Flashing, or brief exposure of the film to light is also an effective method for increasing the speed of an emulsion. In all cases, the film must be used as soon as possible after having been treated. See LATENSIFICATION.

hypo A laboratory term for sodium thiosulfate—a fixing agent. *Hypo* is taken from hyposulfite of soda.

hysteresis synchronous motor A motor in which the speed is governed by the alternations in the voltage. Assuming that the power supply is at the usual 60 cycles per second, the motor will run at a constant 3,600 revolutions per minute, without variations in speed. It is used to control the speeds of magnetic recorders and for other applications where constant speed is crucial.

ice cream cone A studio term for a relatively bright light source, 750 to 5,000 watts, with a white, reflective back surface.

ICI Standards Also, **CIE Standards**. ICI stands for International Commission of Illumination, the body that sets the standards used for describing colors.

Iconoscope Also known as **Ike**, an Iconoscope is a television camera pickup tube (developed by RCA) that is capable of converting visual images into electric signals for subsequent transmission.

identification, film See FILM.

idiot card Also called **cue card**. A method of prompting by positioning cards (containing lines of dialogue or lyrics to a song) in the vicinity of the camera. These cards permit the performer to face the camera normally while taking cues from the information printed on the cards. Compare TELEPROMPTER.

idle rollers Freely revolving wheels at various points in such film-handling devices as cameras, and projection and laboratory equipment, that reduce friction and keep the film precisely aligned as it travels through the transport mechanism.

i-f amplifier The intermediate frequency amplifier in receiving equipment that provides gain control and prevents overloading. The incoming video and sound signals are converted (in stages) and sent, in turn, to the video detector and video amplifier where they are again amplified to the required reception level. See VIDEO DETECTOR and VIDEO AMPLIFIER.

illumination Any light with enough intensity to affect the film's sensitive layer(s) of emulsion. Lighting, by comparison, is controlled illumination.

illumination meter A portable instrument for light measurements requiring a high degree of accuracy over a broad range of light values. Ideal for research and routine work or as an instrument for checking other light-measuring instruments and devices. All types of commercial lighting, regardless of color composition, can be read directly without correction filters.

In the Weston model 756 meter, a Viscor filter corrects for color and a cosine filter for angle of incidence. The unusually long mirror form of scale provides for accurate readings and eliminates parallax errors. Calibration is by direct comparison to accurate reference standard lamps.

The ranges in these instruments are selected by a convenient range-changing switch, providing exceptionally good readability on low light values. The light target, housing two Photronic cells, connects to the color-coded binding posts of the instrument. The Weston model is standardized at a color temperature of 2700°K for stock ranges listed:

0–20/100/500 footcandles (with Viscor and cosine correction filters)
0–20/50/500 footcandles (with Viscor filter)

illumination meter, sunlight This light-measuring instrument incorporates ranges of 1–120/1,200/12,000 footcandles and requires only one Photronic cell equipped with a quartz filter so that the ultraviolet portion of the spectrum is included in the light measured. The extra-high-range coverage is ideally suited for making comparisons of natural sunlight intensities, or for any survey in which high intensities of light must be measured with accuracy. Agricultural bureaus, schools, and industry find this meter indispensable for studying the effect of sunlight on paint, textiles, and numerous other products.

illumination meter, television studio This light-measuring instrument was designed to provide full coverage for the varied requirements of television studio lighting. Special ranges of 0 to 300 and

0 to 1,500 footcandles are suited both for black and white and for color. A special filter is provided to match the instrument spectral response to the no. 1850A Iconoscope used for television film transmission.

The spectral response of the no. 5820 and no. 5826 image orthicons used as television camera pickup tubes are close enough to the standard luminosity curve so that the Viscor filter is recommended. The meter is standardized at a color temperature of 3000°K.

image A photographically produced likeness, exposed on film, and rendered visible by chemical processing. A *latent* image is an invisible image, chemically present, on film that has been exposed but not developed. The image becomes visible in negative form after the developing process. A *real* image is an image produced with normal photographic processes and, subsequently reproduced on film. A *virtual* image is an unreal or illusionary image which cannot be brought to a focus or projected because of its divergent rays.

image enhancer An accessory trademarked by CBS Laboratories which improves the resolution of the video picture.

image orthicon A highly sensitive television camera pickup tube that is especially effective for low-key shots, or when the available illumination is insufficient and would normally make exposure difficult.

image-to-eye ratio The apparent size of an object in relation to the size of the image on film.

image-to-object ratio The actual size of an object in relation to its size on film.

imbibition Also called **hydrotype process**. Imbibition is a process used for printing color films. The images are produced by successive contact with three separation matrices (relief images), dyed in appropriate colors. These colors, the photographic primaries, are transferred to the print stock in direct proportion with the amount of color on the relief images or matrices, to produce the composite color image. See TECHNICOLOR.

immersion bath A laboratory method of processing in which the film passes through baths containing the various developing and fixing solutions. Compare SPRAY BATH. See PROCESSING.

impedance In sound recording, impedance is expressed in ohms. The impedance of subsystems like microphones or speakers, for example, must be "matched" to connecting networks to avoid distortion in recordings and their subsequent reproduction.

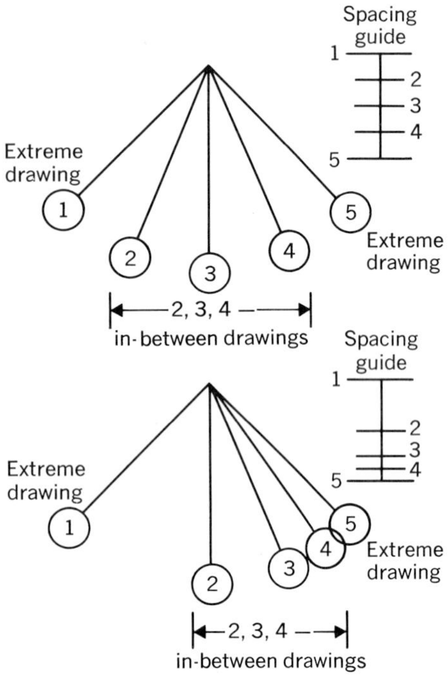

Fig. I-1 In-betweens.

in-betweens In animation, the intermediate drawings known as in-betweens that are needed to complete an animated action. The animator first draws the extremes or key positions of an action. These drawings are turned over to the in-betweener who assists the animator and is responsible for making the missing drawings. See Figure I-1.

incandescence The production and emission of light by heated objects—as when an electric current is passed through a lamp filament.

incident light Light rays falling on an object or surface directly as opposed to reflected or refracted light rays. See Table I-1.

incident-light formula

$$\text{Footcandles} = \frac{25 \times f\text{-stop}^2}{\text{ASA exposure index} \times \text{exposure time}}$$

incident-light meter A light meter that measures the intensity of the light being used to illuminate the subject matter. This type of meter does not measure the light being reflected from it. The readings are guides for the lens settings used during the filming process.

TABLE I-1 Incident-light Factors

Shutter Speed: About 1/50th second or 24 frames per second

ASA rating (exposure index)	Number of footcandles required (for lux values multiply by 10)			
	f/2	*f*/2.8	*f*/4	*f*/5.6
10	500	1,000	2,000	4,000
12	400	800	1,600	3,200
16	320	640	1,250	2,500
20	250	500	1,000	2,000
25	200	400	800	1,600
32	160	320	640	1,250
40	125	250	500	1,000
50	100	200	400	800
64	80	160	320	640
80	64	125	250	500
100	50	100	200	400
125	40	80	160	320
160	32	64	125	250
200	25	50	100	200
250	20	40	80	160
320	16	32	64	125

index of refraction The degree of change in the direction of a light ray passing from one transparent medium to another. The changes are caused by the differing densities of the two mediums and the increases or decreases in the velocity of the light ray.

indirect floodlight A rectangular-shaped overhead lighting source that is suspended from scaffolding and uses six 1,000-watt pear-shaped bulbs. Equally useful for both black-and-white and color photography, the unit produces even, shadowless light over a relatively large area.

induction motor A motor used to drive cameras in single-system sound filming or projection equipment. The a-c motors are useful where fractional gains or losses in running speed are relatively unimportant, as in projection equipment, for example, where the sound track and picture exist on the same length of film, and the synchronization between them cannot be affected by changes in the operating speed of the equipment.

induction period That interval, measurable from the time when an exposed length of film is immersed in a developing solution to the time when the image begins to form.

industrial films Motion pictures of varying lengths produced for and sponsored by large industrial organizations. The subject matter varies considerably. The film usually contains a message that comes under the heading of Public Service.

infinity As used in photography, infinity does not imply immeasurable distance, but merely that distance from a lens at which rays of light seem to be parallel and do not vary.

infrared filter A type of filter capable of transmitting only those wavelengths in the infrared portion of the spectrum. These filters are used to reduce the effects of fog, haze, or mist.

infrared light The position of infrared wavelengths in the spectrum is just beyond visible red: 7000 to 150,000 angstroms. Although the rays are invisible, their radiation is felt by heat. Motion picture film exposed by infrared light must respond to a minimum wavelength of approximately 8000 angstroms. As the film is also sensitive to blue and red light, filters must be positioned in front of either the camera lens or the lighting units in order to avoid contamination from undesirable wavelengths. Kodak Wratten filter 25A is recommended for use with daylight illumination along with filters 29F, 70, 87, 88A, and 89B. Under artificial lighting conditions, a 15G filter is suggested.

Tungsten lamps (red glass bulbs, for example), ordinary floods dipped in infrared gelatin dyes, electronic flashbulbs, and infrared heat lamps with infrared glass filters produce illumination almost free of the visible spectral light and are, therefore, quite useful as sources of illumination for infrared photography.

infrared photography Infrared is used in many unusual ways to fill many unusual needs. It is especially useful where the level of illumination is too low to make other filming methods effective. Because of its sensitivity to colors in that invisible region of the spectrum, infrared can even be used for filming subject matter in apparently complete darkness.

Infrared film literally turns day into night in the production of western movies. Bright moonlight effects, filmed on the range in broad daylight, are easily achieved by using blue-extinction filters with this extremely sensitive film. These day-for-night scenes, printed darker than normal with laboratory processes, are amazingly realistic as nighttime effects. For matteing purposes, as in the infrared self-matteing process, the combination of infrared film and infrared lighting provides excellent results that help make many unusual effects possible.

For scientific observation and analysis, infrared has no equal. Because many materials look altogether different in visible light from what they do when viewed under infrared lighting, objects that previously defied scientific observation can now be studied conveniently. The difficult filming conditions usually associated with aerial photography are also offset by the use of this unusual film type. Bodies of water, for example, appear quite dark in contrast with land masses. The sky, by comparison, is considerably darker than cloud formations, open fields, and grassy areas. This high degree of contrast is, of course, most desirable for many types of studies. Sharper definition and better picture quality, both equally difficult to achieve under extreme haze conditions, are completely possible with infrared film.

Because of the different amounts of infrared rays that are absorbed or reflected by different objects, the results of infrared filming are not always predictable. The dissimilarity in the wavelengths of infrared radiation and visible light rays make focusing adjustments necessary when this film type is being used. Best results are usually achieved by using the smallest lens opening permissible and focusing at a point closer than the actual distance between the camera and the subject matter. The difference in focal points can best be determined by pretesting. All the laboratory processes and solutions used for developing and fixing, may also be used for film exposed with infrared lighting.

Unusual effects can be achieved with infrared photography—the

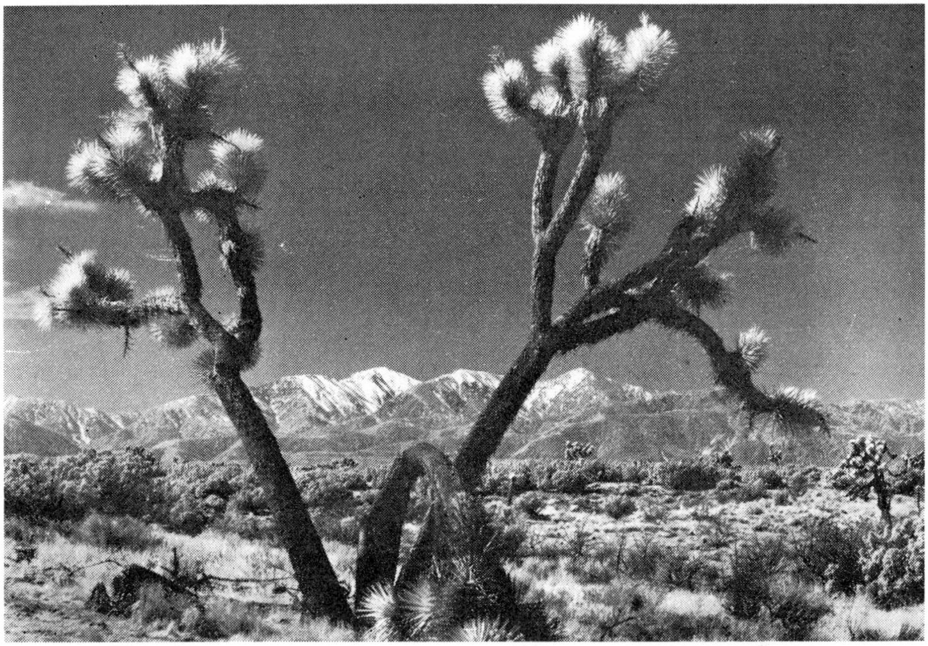

Fig. I-2 Infrared photography. (*Eastman Kodak.*) A little imagination, infrared film, and a Kodak No. 25A Wratten filter all combined to produce this spectacular shot.

extent of these effects being limited only by the cinematographer's imagination and his willingness to experiment. When used for close-up work or portrait photography under studio lighting conditions, for example, flesh tones are usually lighter and have a translucent appearance, depending on individual characteristics. Hair and eyes appear quite dark in contrast with extremely light or white lips. In some cases, beard growth beneath the skin level, and veins close to the surface, will cause unusual effects in the subsequent print material. Certainly, not the kind of portrait one would wish to frame and hang over a fireplace. See Figure I-2.

infrared self-matteing system See SELF MATTEING and PICKLEY PROCESS.

infrasonic frequencies Frequencies ranging below 15 cycles per second. The audible frequency range is between 15 and 20,000 cycles per second.

inker (animation) The artist who traces the lines of the animator's drawings on cels (acetates).

inky dink A comparatively small (250-watt) light source which, in combination with a snoot, is used to minimize lines in a performer's face and bring a sparkle to his eyes.

inlay (videotape) Also called **keyed insertion** and **static electronic matte**. An electronic insertion process that makes it possible to combine picture information from separate sources to form a composite. The success of the inlay electronic-picture-insertion system depends mainly on the accuracy of the specially prepared mattes that are needed for inserting picture information into a corresponding area of a second picture.

in-line heads Also called **stacked heads**. The position of the recording heads in magnetic recorders. In-line heads are arranged so that the head gaps are in a vertical position—one being directly over the other.

in phase A term indicating the synchronous operation of shutter and transport mechanism in motion picture equipment.

Also, the matched recurrence of pulses or peaks from separate but interrelated light waves, sound waves, or electric currents.

input The delivery of current or signal voltage to an electric device such as an amplifier.

ips Inches per second. A unit of measurement indicating the tape speed of a recording. Generally, slower speeds like $^{15}\!/_{16}$, $1\,^{7}\!/_{8}$, and $3\,^{3}\!/_{4}$ ips are used for recording data or speech information. When fidelity is important, faster speeds like $7\,^{1}\!/_{2}$ or 15 ips are used to record the widest possible range of frequencies and, therefore, the most "sound" information.

insert scene A short scene, usually a close-up, intercut between two main scenes to help the visual continuity of a sequence. An insert scene might show only the performer's hands holding a letter or newspaper, or a close-up showing the movement of the eyes as they scan the page. In commercial applications, an insert scene usually shows the client's product. See Figure I-3.

instrument A name applied to theatrical-type lighting fixtures.

integral tripack Film with three layers of emulsion, each of which is sensitive to one of the photographic primaries. This type of film is used for the production of separation negatives. See MULTILAYER FILM and SEPARATION NEGATIVES.

intensification A process used for increasing the density or contrast of an image. The process is especially useful for improving the quality of negatives that have been underdeveloped. See LATENSIFICATION.

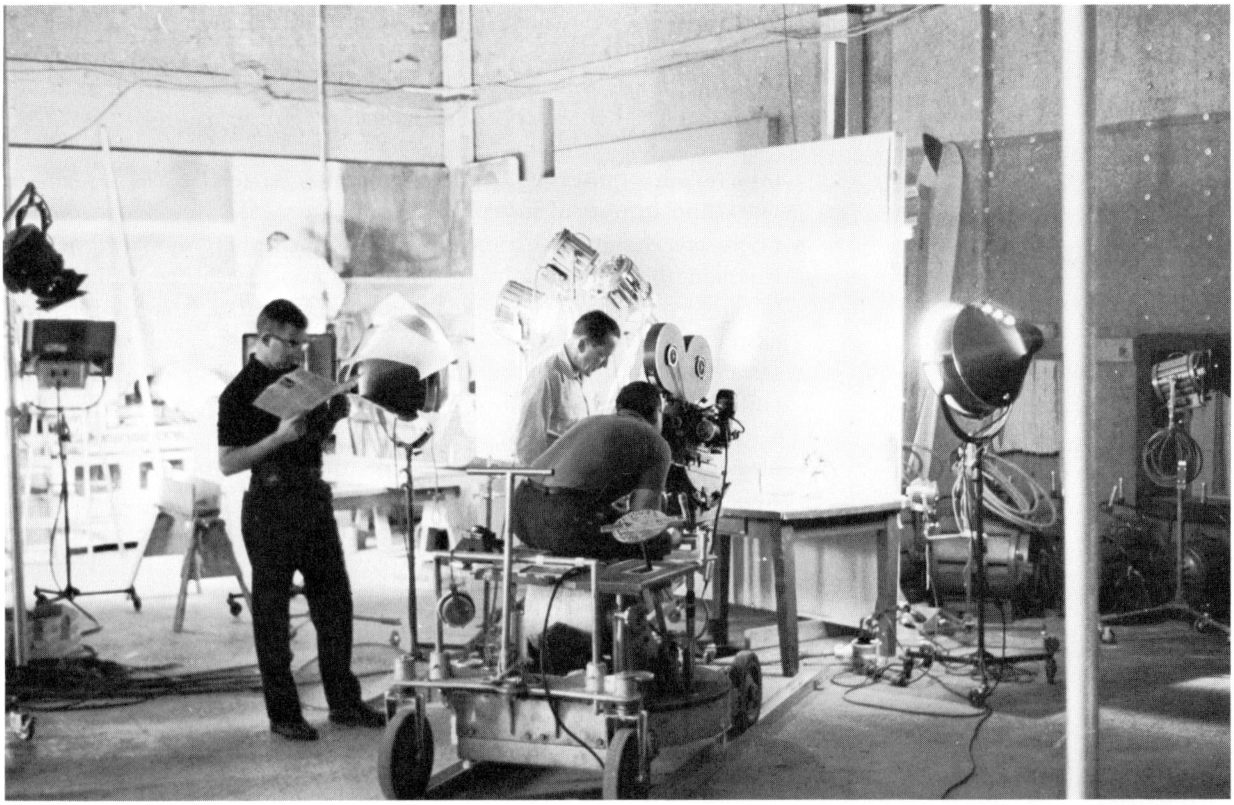

Fig. I-3 **Insert shooting requires the same amount of preparation and skill as the feature motion picture. The product "stars" in this television commercial production. (*MPO Videotronics.*)**

intensity A term used to indicate the power of a light source. It is also a measure of the "strength" of a light source in a particular direction. Intensity is independent of the distance from the source. The unit used for measuring intensity is the *candela.*

intensity distribution A graph or diagram that describes how the intensity of a light source varies with direction.

intensity scale A series of frames of film, each receiving the same amount of exposure but different degrees of light intensity. This is the procedure followed in exposing a step wedge. Other scales are made by varying the amount of exposure, for each frame of the test strip, while keeping the intensity of the light at a constant level. See STEP WEDGE.

intercutting A term used to describe an editing technique in which scenes from two related or separate actions are alternated to create the impression that both actions are taking place simultaneously.

interference The merging of sound waves, some as a result of reflection, with each originating from a different point. The results, usually undesirable, vary with the waveforms that are constantly changing in their relationship with one another.

interference filter A type of filter that is especially useful for laser work and in optical instrumentation. These nearly monochromatic filters are designed to produce interference phenomena at specific wavelengths, thus permitting the rejection of unwanted radiation by selective reflection. High passband transmittance or rejection, and extremely narrow bandwidths, are achieved at any desired center wavelength. These special filters are available through Barnett Associates in Hauppauge, New York.

interior shot Scenes photographed either on location or on a studio stage in which indoor settings are reproduced. The lighting arrangement, in either case, must be in keeping with the type of lighting normally associated with those areas.

interlaced scanning An electronic picture-scanning process in which each adjacent scan line belongs to the alternate field. See ODD-LINE INTERLACING SYSTEM.

interlock The process wherein the picture reel and the sound-track reel are kept in phase during projection for the purpose of checking synchronization. A moviola is used for similar purposes.

interlock motor Also called **selsyn motor**. Specially designed motors that drive related equipment (the projector and camera in rear-screen projection shots, for example), keeping both in synchronization during the filming process.

intermittent effect The difference between the density of an image that is photographed "on the run" and the density of the same image photographed with single-frame, stop-motion processes. Although the image in both instances receives the same amount of exposure, the single-frame image is denser—by an almost insignificant amount.

intermittent movement The stop-and-go movement by which film advances from one point to another in motion picture cameras and projectors. The film is held motionless in the shuttle during the interval of exposure (or projection) and is advanced, while the shutter is in its closed position, to bring the next frame of film into position in the shuttle. This process takes advantage of the "afterimage" effect by which the eye retains an image for the fraction of a second before the next image is projected, thus giving an illusion of continuous viewing.

intermittent prism Also called **rotating prism**. Lens arrangements that are used in some high-speed cameras to take the place of the shutter. These lenses, in projection equipment that is used with certain types of telecine systems, make it possible to project a flickerless picture. See HIGH-SPEED PHOTOGRAPHY and MULTIPLEXED TELECINE OPERATION.

internegative A negative produced directly from an original color-reversal film. Second- and third-generation negatives produced from film other than reversal films are referred to as *duplicate color negatives.*

interpositive Interpositives used in color optical and duplicating work are used in the same manner that fine-grain material is used for black-and-white optical and duplicating work. Basically, an interpositive is a denser-than-normal print that is made from the original camera negative or from a duplicate negative. (Dupe negatives can also be made from interpositives.) Unlike the physical appearance of conventional color prints, the interpositive has an orange-colored base or mask.

Inter-Sync Inter-Sync is an ampex device that permits combining the output of two or more videotape recorders, line for line and field for field, with complete synchronization. Replacing the standard drum servo unit in a videotape recorder, Inter-Sync also synchronizes the playback from one or more recorders with the output of any television signal source—live-camera chains, film chains, slide projectors, or network feeds.

intervalometer A timing device used for time-lapse photography. Interval timers provide frame intervals of 1, 2, 4, and 6 seconds up to 25 minutes. Exposure timers provide time exposure from 1 to 10 seconds. The device has a built-in controlled 8-volt d-c power supply. It requires a single-frame motor modified for double-pulse operation; one pulse to open the camera shutter and one pulse to close it. Master controls can be set to control the operation of the camera, lights, and auxiliary shutter at predetermined intervals.

Invercone An attachment that makes it possible to convert reflected-light exposure meters into incident-light meters. Its basic function is to scatter the light evenly over the photocell area.

inverse-square law The intensity of the light falling on an object varies inversely as the square of the distance of the subject from the source of light. This rule provides a measure of the intensity of the light source in relation to the amount of illumination it produces over a given area.

In practical terms, it means that illumination of an object falls

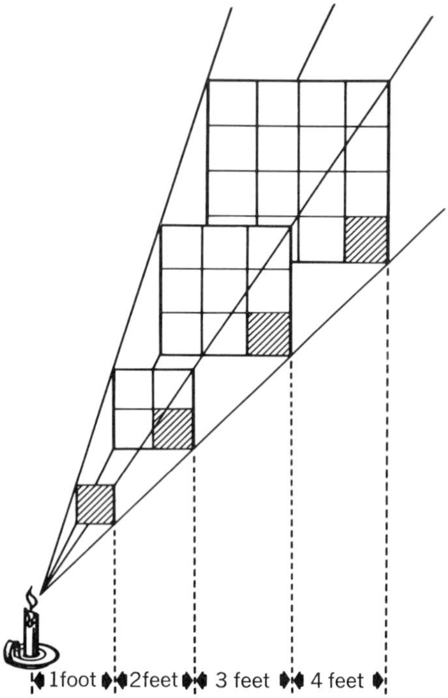

Fig. I-4 **Law of inverse squares.**

off very sharply as the object is removed farther from the light source. At twice the distance, the object receives one-fourth the illumination; at three times the distance, one-ninth; and so on. See Figure I-4.

inversion The optical phenomenon that causes images to be inverted when formed by a photographic lens. The rights and lefts are also interchanged in these upside-down images.

inverted telephoto lens A short-focal-length lens consisting of a series of converging and diverging lenses—the arrangement giving the lens a long back focus. This type of lens is useful where, because of the inner construction or optical arrangement of cameras, additional room is needed between the camera lens and film plane.

invisible radiation Invisible radiation above and below the range of visible light. Above the violet rays at the top of the visible spectrum are the ultraviolet rays, x-rays, and gamma rays. Below the red rays at the other end of the visible spectrum are the infrared and heat rays. Ultraviolet, infrared, x-rays, and gamma rays are useful in photography as they are all capable of forming an image on film.

invisible splices Splices made by the film editor on film that is being prepared for A & B roll printing. To produce this type of splice, the film is positioned in a splicer so that one of the cuts is made on the frame line. The second cut, on the film that is to be joined to the first length, is made in the picture area. When the two lengths of film are joined and lined up correctly with the leader strip, the splices should be invisible in the subsequent print. These splices must be made carefully in order to avoid scrape markings that would be reproduced in the print. See Figure I-5.

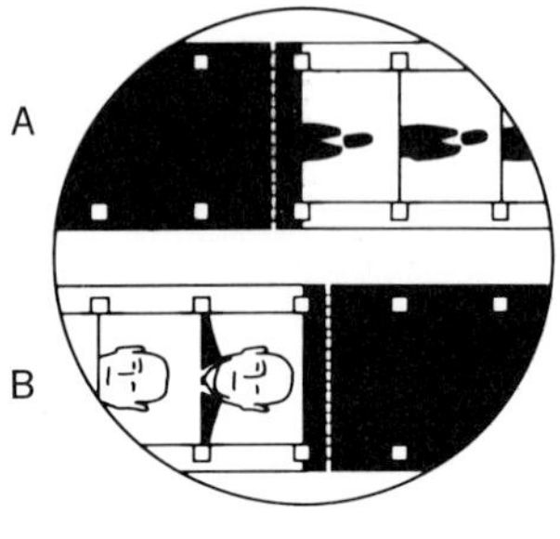

Fig. I-5 **Invisible splices. Scrape picture only —never scrape emulsion from black leader. (*Color Service Co.*)**

ion traps Methods of controlling the flow of ions in a television picture tube. Produced when electrons come in contact with extraneous gases within the tube, ions strike the fluorescent screen and, in time, cause a brown spot to appear on its surface.

Two methods or traps are used to eliminate the cause of the ion spot. In the first, a thin aluminum coating is applied to the phosphor screen to prevent ion penetration. In the second method, a magnetic field (outside the electron gun) directs the ions to the anode and, at the same time, bends the electron beam away from it.

iridescence The ability of a surface to display specific colors that vary in both hue and pattern. The changes in color are dependent on the direction of the light falling on the reflective surface.

iris, lens Also called **diaphragm, stop, aperture**. The adjustable part of the camera that controls the amount of light passing through the lens to expose the film. Overlapping blades form a circular opening that can be increased or decreased in size to permit more or less exposure.

iris, lighting In lighting, an adjustable device that is used for controlling the size of the circle of illumination. It is positioned in front of a lamp housing.

iris in (iris out) A circular wipe effect in which the picture information is seen within a circle. The area around the circle is usually black. The image, as the circle increases in size, subsequently fills the entire film frame.

In an iris out, the image and the circle both diminish in size until the frame is completely black.

irradiation The scattering of light rays in a film's layer of emulsion. As the light rays pass through and are reflected from the silver halide crystals in the emulsion, they are scattered, causing a blurring or loss of definition in the resulting image. This undesirable effect is more likely to occur in thick rather than in thin emulsions. The blurring effects caused by halation are more definite than those caused by irradiation. See HALATION.

Also, the process of directing light to a surface, or the intensity of radiation falling on a surface at any specific time.

J

jack An electric receptacle that accommodates the plug connector leading to either the input or output circuits in recording equipment.

jelly A studio term for gelatins that are positioned in front of a light source in order to correct the color value of the source.

Jena glass Produced in Jena, Germany, the glass named after that village was the first to be ground to predetermined specifications. The lower degree of dispersion in these glass lenses, as related to the refractive index, was a radical departure as compared with the design of other lenses manufactured during that period. The first lenses were produced by Otto Schott and Ernst Abbe. Carl Zeiss joined them during the 1880s, and they monopolized the field until after World War I when England began producing optical lenses on a commercial basis.

jenny A studio term for a generator which is used for supplying electric power in location filming.

jitter A tendency toward lack of synchronization of the picture. It may refer to individual scanning lines in the picture or to the entire field of view. As the name implies, the picture jitters (usually in short vertical or horizontal movements) until the controls are adjusted and synchronization is restored.

jump cut The process of eliminating a number of frames from a scene without affecting the continuity of the action.

junction box Also called **spider box**. A conveniently located enclosure that houses electric cable terminals and receptacles and sockets for the various lighting sources on a motion picture set or television stage.

junior Solarspot An all-purpose spot-flood combination lighting unit with a $9\frac{7}{8}$-inch Fresnel lens that accommodates bulbs rated between 1,000 and 2,000 watts.

juvenile A young male or female performer.

K

kaleidoscope A combination of three Greek words: **kalos** meaning beautiful, **eidos** for form, and **skopeo**, the equivalent of viewing or to view. By any other name, the kaleidoscope is a beautiful, eye-arresting optical effect that is produced by the reflection of colors and symmetrical forms from several mirror surfaces.

The effect is produced by positioning the subject matter between two mirrors at angles to each other. With this type of arrangement, the subject matter or pattern is reflected in a circular manner as many times as is desirable. The number of images is determined by the angle of the mirrors. An angle of 90 degrees, for example, will produce four images in a circular pattern. Similarly, mirrors placed at 45-degree angles to each other will produce eight images. The camera is positioned so that the lens is aimed at a point between the two angled mirrors.

The method used for determining the angle of the mirrors in relation to the number of images desired is expressed by the formula

$$360 \text{ degrees} \div \text{number of images desired} = \text{angle of mirrors}$$

See Figure K-1.

Kalvar photographic process See THERMOPLASTIC RECORDING.

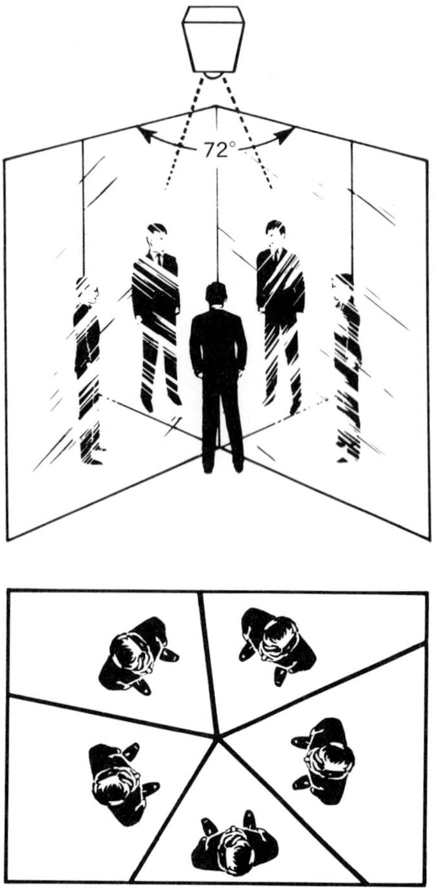

Fig. K-1 Kaleidoscope. The number of images produced on film varies with the angle of the mirrors. In this illustration, the subject was positioned between two mirrors at a 72-degree angle to each other thus producing five images in the composite.

keg light A spotlight in a cylindrical housing. The lighting unit accommodates 500- and 750-watt bulbs.

Kelvin degrees A unit of measurement indicating the color temperature of a light source. Expressed in degrees Kelvin (abbreviated by the letter **K**), 273°K is the equivalent of 0° on the centigrade scale. Each one-volt increase in the color temperature of a lamp is equal to approximately 10° Kelvin. The unit of measurement is named after the originator of the system, W. T. Kelvin. See COLOR TEMPERATURE.

kelvinometer A light-sensitive instrument used to measure the color temperatures of light sources. Readings derived from it may not always be completely accurate owing to uneven spectral emission of the light source.

Kerr cell shutter Designed for high speeds, the Kerr cell shutter has two crossed polarizing screens that have liquid mounted between them. A high-voltage current changes the polarization of the liquid and allows light to be transmitted through it. This shutter makes possible exposure times of one-millionth of a second.

key The basic lighting setup in relation to the desired results. See HIGH KEY and LOW KEY.

key insert signal An electronically controlled matteing process in which a portion of the picture information previously recorded on one length of videotape can subsequently be inserted into a corresponding portion of a second recording to produce a composite of the combined material.

key light The principal source of illumination for a motion picture set. The intensity of the key light should take the performers' comfort into consideration. A reasonable working formula for determining key-light intensity might be

Key-light intensity = light power: size of set: "comfort of talent"

keystone A distortion of the image caused by an improper relationship between the camera and the subject matter during the filming process, or between the projector and the screen. If the subject matter is a square, for example, and the camera is not aligned correctly in relation to the plane on which that square appears, the vertical (or horizontal) lines would not be parallel. The lines might be further apart at the top and closer together than they should normally be at the bottom of the square. The amount of keystone depends on the angle or degree of tilt in the camera or projector during filming or projection.

As a general rule, the effect of keystoning can be eliminated or at least be minimized by aligning the plane surfaces of the subject matter so that they are parallel to the axis of the camera lens.

kicker Also called **eye light**. Illumination that is used to produce a specular reflection from the eyes or teeth of the performer without adding a significant increase of light on the subject.

kilocycle Abbreviated **kc**. A unit of frequency equal to 1,000 cycles per second.

kinescope Also called **telerecording** and **teletranscription**. The process used for transferring a televised picture to motion picture film.

kinescope recording In motion picture production, images are recorded on photographic film, combined with a sound track, and processed to produce a composite print that contains both elements. The recorded roll of videotape, by comparison, includes both picture information and sound track, and can be viewed or played back immediately—without processing. Unlike the visible images on processed film, recorded and unrecorded reels of tape look alike—brown plastic ribbons. In either medium, however, picture and accompanying sound can be preserved, thus providing a permanent record of the photographed information.

Although the all-seeing eye of the electronic television camera can provide visible information for instantaneous transmission, the medium, unfortunately, has no "memory." For repeating the information in subsequent broadcasts as well as for "remembering," generally, the televised images are recorded on film. The photographic technique that preserves these images for posterity is known as kinescope recording.

As in videotape transmission, the interlaced picture information from two fields forms one composite frame. Each half picture or field thus contains a $262\frac{1}{2}$-line image, or half of the 525-line total that is visible on television screens for $\frac{1}{30}$ second. Since the kinescope recording is subsequently projected on motion picture equipment, the 30 television images transmitted each second must be photographed to conform with the projection rate of film, or 24 frames per second. The conversion from television scanning time to motion picture projection standards is accomplished by exposing each of the scanned television images for exactly $\frac{1}{30}$ second.

The electronic or mechanical shutters in the TV recording cameras, in order to compensate for the differences in exposure time, remain in an open position for $\frac{1}{30}$ second and in their closed position for $\frac{1}{120}$ second. During the photographic process, and while the camera shutter is in its open position, two fields in the TV scanning cycle are exposed on each frame of motion picture film. The camera shutter then remains in its closed position for $\frac{1}{120}$ second, or the interval during which one-half of the next field would have been projected. The total time for the complete cycle or two shutter positions, open and closed, equals $\frac{1}{24}$ second—the standard sound motion picture projection rate.

Kinescopes are recorded directly from the face of the picture tube in the monitoring set. Because the curvature of the glass faceplate would tend to distort the outer areas of the picture, the coated, short-focal-length camera lens is focused on a smaller area in the center of the picture tube. The governing factor, with respect to the sharpness of the photographed image is the size of the electron scanning spot—the larger the size of the transmitted image, the clearer the picture.

To minimize the cost of the raw stock and processing, most kinescopes are made on 16mm film. Since the processed film will, quite likely, be spliced together with commercial material and other film footage, the kine prints must conform with the standard emulsion position—emulsion side of the film facing the screen during projection. The standard emulsion position for 35mm film is exactly opposite to the 16mm standard. 35mm prints are projected with the emulsion side of the film away from the screen.

Although negative stock, high-contrast positive, and reversal film may be used for photographing the televised images from the monitor tube, the kinescope print is, at best, inferior when compared with the quality of standard motion picture prints.

Depending on the type of in-camera film and its subsequent usage, the transmitted tube picture may be either positive or negative, with the latter images produced by reversing the polarity of the video signal. The nonstandard emulsion position, in the print made from negative film, can be corrected by changing the horizontal scanning direction of the transmitted picture. The resultant mirror image, reversing right and left, allows the final print to be read correctly during projection. Negative tube images, photographed on high-contrast positive stock, will produce positive film images. Prints made from the dupe negative will have the emulsion in its proper projection position. As an alternative method, positive tube images may be photographed with reversal stock to produce an original positive. Here again, the emulsion side of the film, in the prints made from the dupe negative, will be in correct viewing position.

Correct exposure, for film sensitivity and laboratory processing, is determined by light-meter readings. The maximum kinescope transmission brightness, before beam-resolving power decreases, and the minimum signal needed for transmitting dark portions (the "black" level) provide the kinescope range and light intensity needed for the exposure-meter readings.

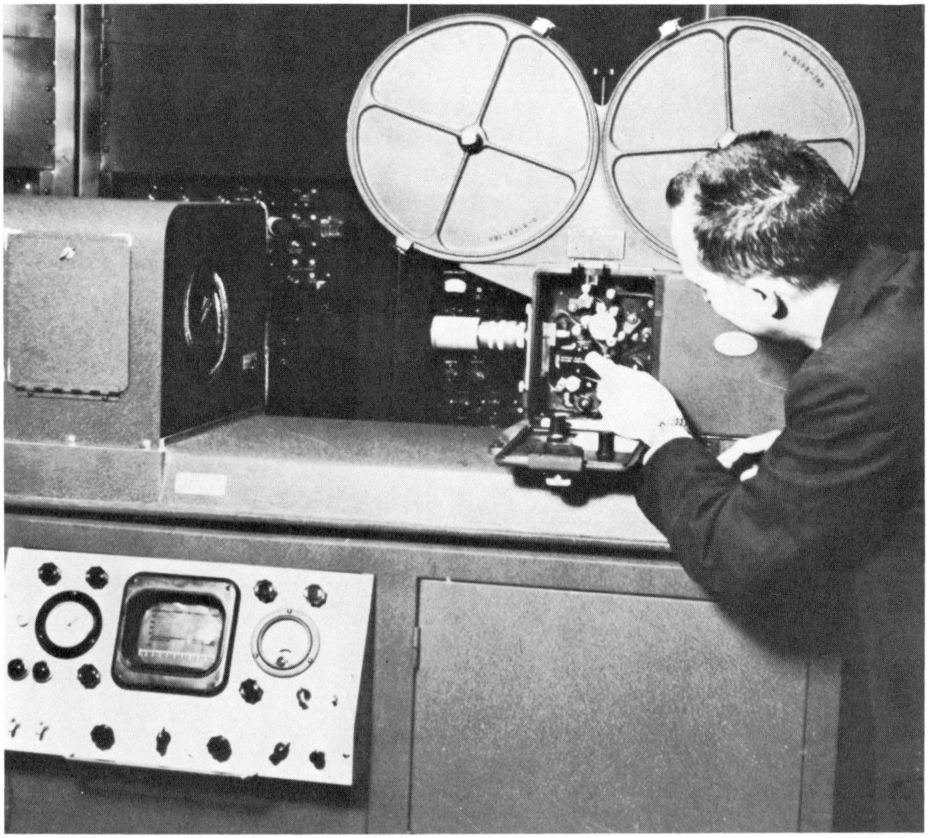

Fig. K-2 The kinescope.

Sound, as in motion picture production, is reproduced either with single- or double-system recording. In the single-system recording, both the sound-track and picture information appear on the same length of film. With the latter system, the sound track is recorded separately, on equipment that is synchronized with the camera, and later edited and combined with the film that contains the accompanying picture information.

In many cases, kinescopes are made from previously recorded lengths of videotape in order to provide an editor with a "visible" record of the picture information. The prints obtained from the kinescope recording are cut to form a work print that can be screened for client approval before the composite videotape is produced. By any other name, video recording, television recording, teletranscription, or video photography, the kinescope provides the only available method for preserving the images produced by the television cameras. See Figure K-2 and GEMINI-TAPE-FILM SYSTEM.

kleiglights Open-arc lights named after their designer.

Kodachrome A reversal color film produced by the Eastman Kodak Company, from which color prints are obtained by the subtractive process. The layers of emulsion in this integral tripack film do not contain color formers. This type of film, after being exposed, is developed in black-and-white developers. The resultant negative silver image is then reexposed to fog the remaining silver halide in the emulsion and render it developable. The dyes needed for producing the color values (coupler development) are added in the next step of the processing procedure. See INTEGRAL TRIPACK and COUPLER DEVELOPMENT. See Table F-15.

Kodacolor An additive color film when first introduced by the Eastman Kodak Company, Kodacolor is now a negative-positive film from which prints are produced by the subtractive process. Kodacolor film is a multilayer film and is used to make color negatives. The coupler components of the dyes are added to the film's emulsion during manufacture.

After exposure, the film is developed in a color developer to produce both a dye image and a silver image in each of the layers of emulsion. The silver image is then removed by bleaching and fixing. The remaining dye images are negative and thus complementary to the colors in the original subject matter. See COUPLER DEVELOPMENT.

L

laboratory The place where rolls of sensitized emulsions on cellulose acetate bases are magically transformed into lengths of film that contain picture information that can be projected for viewing by large groups of people. In nontechnical language, the place equipped to develop and print film. See PROCESSING.

laced process, MGM Also called **tri-lace process**. A method of projection used in the production of wide-screen process shots. With this rear-projection system, a composite background is projected on a translucent screen by two or more synchronized projectors. Each of the different portions of the background is overlapped or partially superimposed over the portion next to it. The advantage of this system over others used for rear projection lies in the increased size and added brightness of the projected backgrounds.

lambert A unit of measurement for determining the degree of luminance or the intensity of light that is reflected by an illuminated surface. A reflecting surface with a 100 percent reflective capability, receiving 1 lumen per square foot, reflects 1 footlambert. 1 lumen per square centimeter equals 10,000 lumens per square meter. See FOOTLAMBERT.

lamp An all-inclusive term for any device (globe, bulb, tube, for example) capable of providing light for illumination. See Figure L-1.

lamp charts Most of the fixtures used in motion picture and television production accept a wide variety of lamps. Lamp charts, therefore, include each lamp's maximum intensity in a particular fixture, along with the more usual nominal information relating to lamps

Fig. L-1 Common incandescent lamp bases—not to scale. (*Sylvar Electric Products, Inc.*) The illustration shows some of the mo common lamp bases. Double-ended lamps are specified terms of maximum overall length (MOL). The filament loc tion of single-ended lamps is specified by the light cent length (LCL) measured from the center of the filament to t bottom of the base for screw bases, to the top of base pins f bayonet bases, to the top of fins for prefocus bases, to t shoulder of posts for mogul bipost bases, and to the bottom bulb base for medium bipost bases. Certain other bases ha special references as indicated in the illustration. The ma mum overall length (MOL) of a single-ended lamp is measu from the top of the bulb to the bottom of the base.

Cand	Candelabra
DC Bay	Double-contact bayonet candelabra
DC Pf	Double-contact prefocus candelabra
EMEP	Extended mogul end prong
F	Ferrule contact
L Tf	Locking trufocus (also: locking four pin)
Mc	Minican
Med	Medium
Med Bp	Medium bipost
Med Pf	Medium prefocus
Med Skt	Medium skirted
Med 2P	Medium two pin
MEP	Mogul end prong (also: extended mogul end pron
Mog	Mogul
Mog Bp	Mogul bipost
Mog Pf	Mogul prefocus
MS	Miniature screw (with reference shoulder)
MSP	Medium side prong
Rect RSC	Rectangular recessed single contact
RSC	Recessed single contact (also: single contact reces
S	Metal sleeve
SC Bay	Single-contact bayonet candelabra
ST	Screw terminal
Tf	Trufocus (also: four pin)
TP	Two pin
TrP	Trupin (also: miniature two pin)
V 2P	Vented two pin
2B	Two button
3P	Three prong

Note: R indicates special reference point for LCL.

(voltage, wattage, color temperature, and other pertinent information). When lamps are interchanged in a fixture, the maximum intensity, lumens, and candlepower change, while coverage and light distribution usually remain the same.

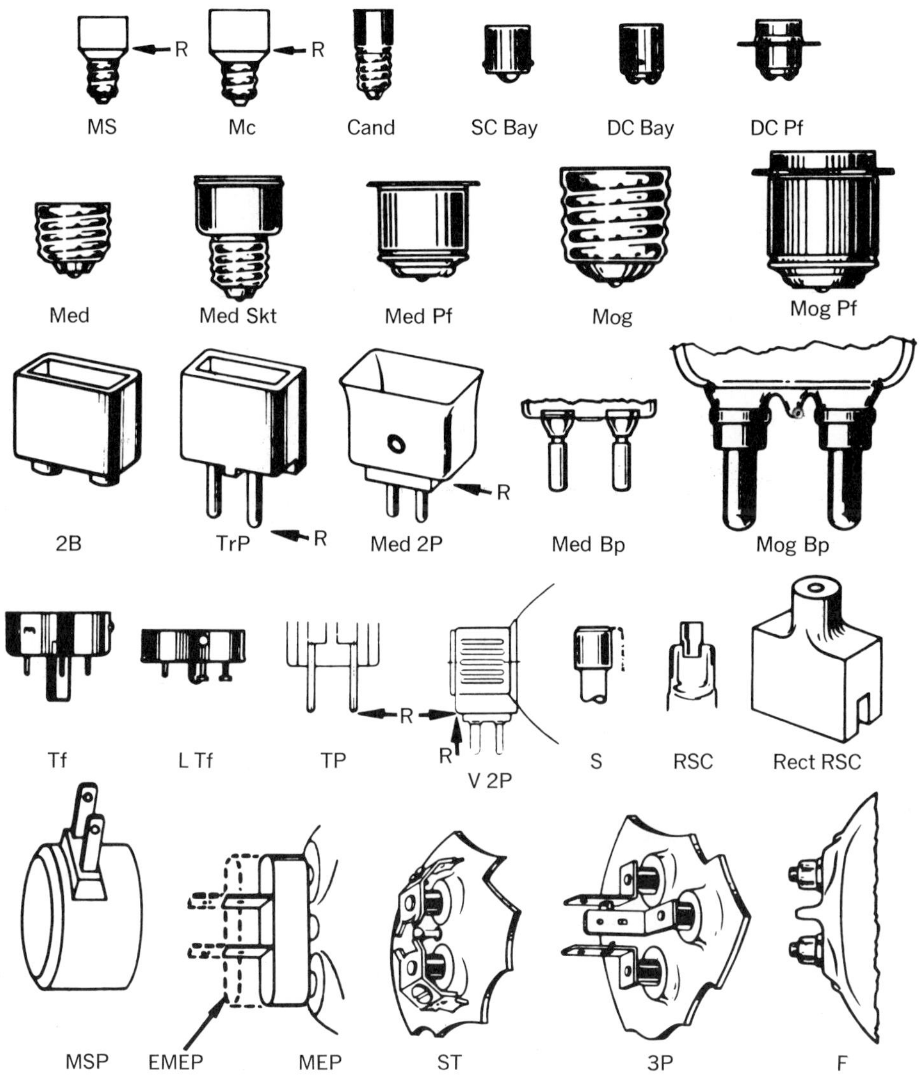

lamphouse In projectors, the housing for the light source.

lamp lumens The total amount of light available from any particular lamp.

landscape A tract of open country or a natural scene or setting that provides a suitable background for filming.

lantern slide projector An antiquated device for projecting transparencies. The lantern slide projector made it possible for many people to view a picture simultaneously as compared with the even older method in which a picture was passed from one person to another and viewed individually. The projection device consisted of a light source in a glass housing and a holder in which the transparency (mounted between two thin pieces of glass) was positioned during projection.

lap dissolve Also called **cross-dissolve**. Lap is the abbreviation for the overlapping of picture information (over the same film footage) during the transitional effect.

laser The letters in the word laser stand for **L**ight **A**mplification by **S**timulated **E**mission of **R**adiation. Its immediate acceptance as a light source for medical, scientific, industrial, and communication purposes is based on its extraordinary ability to focus narrow beams of coherent, or monochromatic, light over amazingly long distances.

Many times hotter than the surface of the sun, a single laser beam can burn holes in diamonds, perform ultradelicate surgical feats, weld minute electric circuits, and transmit great numbers of telephone, radio, and television messages simultaneously.

How this concentrated source of light will be used by the motion picture industry is, at this writing, a matter for speculation. Certainly, its ability to travel over relatively great distances and to cut its way through dense fog should make it extremely useful as a light source in both celestial and underwater photography. See Figure L-2 and HOLOGRAPHY.

latensification A method for increasing the speed of negative emulsions when filming under difficult conditions. Before developing, the exposed negative is reexposed to a rather weak printing light for a predetermined amount of time. The reexposure produces a slight amount of fog in either weakly exposed or completely unexposed areas.

Parts of the granular texture of the film's emulsion not affected during the filming process are affected by this fogging and strengthen the overall image. This process also reduces the contrast between light

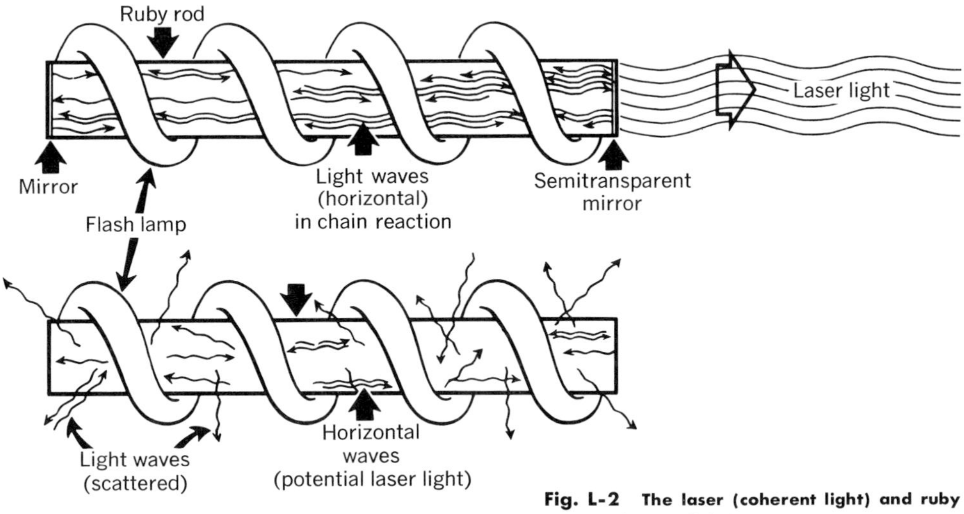

Fig. L-2 **The laser (coherent light) and ruby rod (incoherent light).**

and dark areas, thus making it possible to bring the negative into acceptable printing range.

latent image An image exposed, but not yet developed, on film.

lateral color A form of aberration that causes fringing around the images in color photography and a blurring or fuzziness around the subject matter in black-and-white photography. These undesirable effects, which can occur even in lenses corrected for chromatic aberration, are due to the varying degrees of refraction of the different colors. Stopping down (or using smaller apertures) does not reduce this effect.

lateral inversion The term is used to describe a mirror image in which the subject matter is reversed with respect to right and left. A developed negative shows this type of image although the orientation is corrected in the subsequent print. The images seen in some viewfinders (in reflex cameras, for example) are also reversed—right and left.

latitude The range of lens apertures and shutter speeds in which an emulsion will receive proper exposure and, consequently, a negative from which good-quality prints can be produced. In some cases with color films or emulsions that are relatively slow, the cameraman has little latitude in his choice of lens settings. Faster films, by comparison, give the cameraman a much greater latitude in the selection of lens stops.

lavalier microphone A relatively small microphone that can be worn around a person's neck. This type of microphone is ideally suited for announcers who must hold or demonstrate a product, as well as for performers who find it necessary to have their hands free while acting.

lavender print A fine-grain print used for duplicating purposes. It is especially useful for reducing contrast.

layout (1) In animation, a guide made by the animator in which the scene's action is plotted. (2) In motion picture and television production, the plotting of action, camera angles, and lighting arrangements prior to filming. (3) In opticals, timings and frame counts prepared for the optical cameraman from information on the work print supplied by the film editor.

lead Also called **star**. The starring part in a film or the performer playing the leading role.

leaders Lengths of film at the beginning and end of a reel. These blank lengths are used for threading purposes, thus eliminating the need for handling the picture portions of the film, or wasting portions of them when projection equipment is started. In the composite print, the leaders contain numbers, in reverse sequence, that are printed in the center of the frame at one-foot intervals for the guidance of the projectionist. See Figures L-3*a* to *d*.

lead sheet Also called **bar sheet**. A cue sheet that provides a visual synopsis of an animated sequence. Prepared by the animation director, the bar sheet shows the number of frames allotted to an action, the position of each musical beat, and the specific frames in which lip sync or mouth movement occurs.

lenses A lens is a shaped, transparent substance, usually glass, which acts to focus light rays.

Lenses are divided into two main groups, positive and negative, and manufactured in a number of shapes or types. Positive lenses converge light rays and are thicker through the center portion than at the edges. Negative lenses produce divergence of light rays and are thicker at the edges than through the center portion of the lens. The most frequently used lenses for photographic purposes are the plano-convex (converging), double-convex or biconvex (converging), converging concavo-convex, diverging concavo-convex, plano-concave (diverging), and double-concave or biconcave (diverging).

When light waves pass from one transparent medium into another at an oblique angle, their direction and speed are both subject to

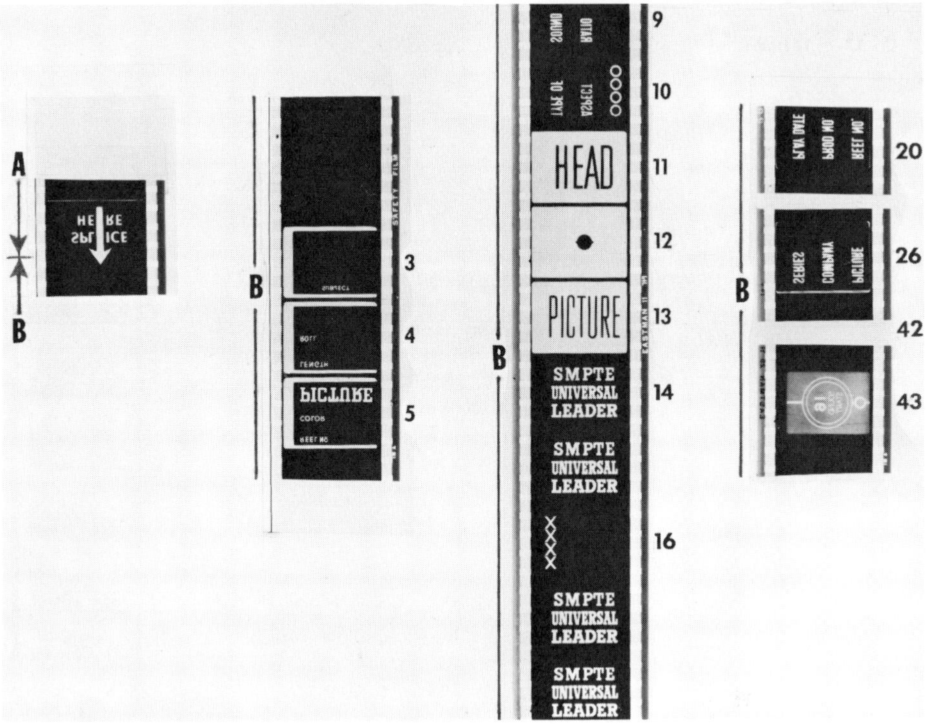

Fig. L-3*a* Academy leader—head. (*A*) Protective section: 6 to 8 feet. (*B*) Identification section: 2 feet and 10 frames.

Fig. L-3*b* Academy leader—head. (*C*) Synchronizing section: 13 feet and 10 frames.

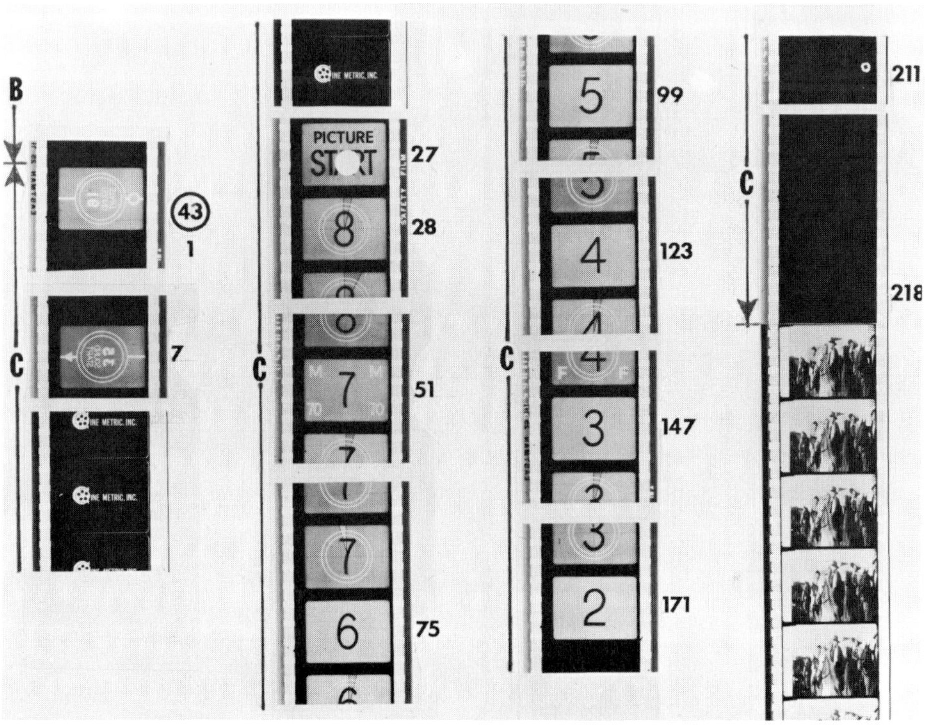

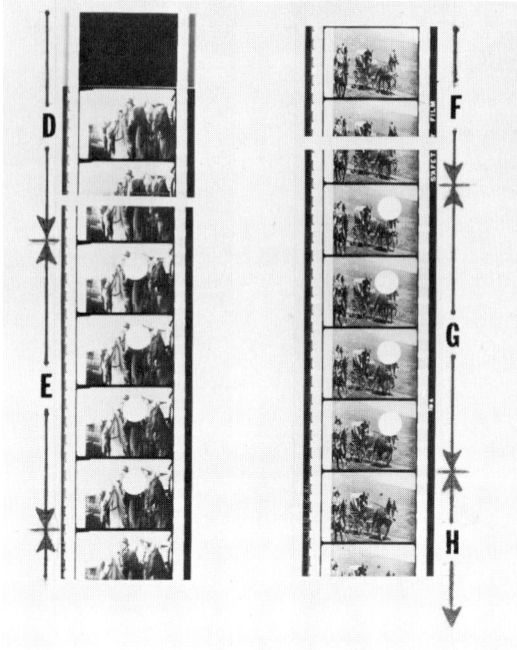

Fig. L-3c Academy leader—tail. (*D*) Picture, head. (*E*) Motor cue: 4 frames. (*F*) 168 frames (7 seconds) between motor cue and changeover cue. (*G*) Changeover cue: 4 frames. (*H*) Runout section: 24 frames to last frame of picture.

Fig. L-3d Academy leader—tail. (*I*) Runout section: 5 feet and 7 frames. (*J*) Identification section: 3 feet and 5 frames. (*K*) Protective section: 6 to 8 feet.

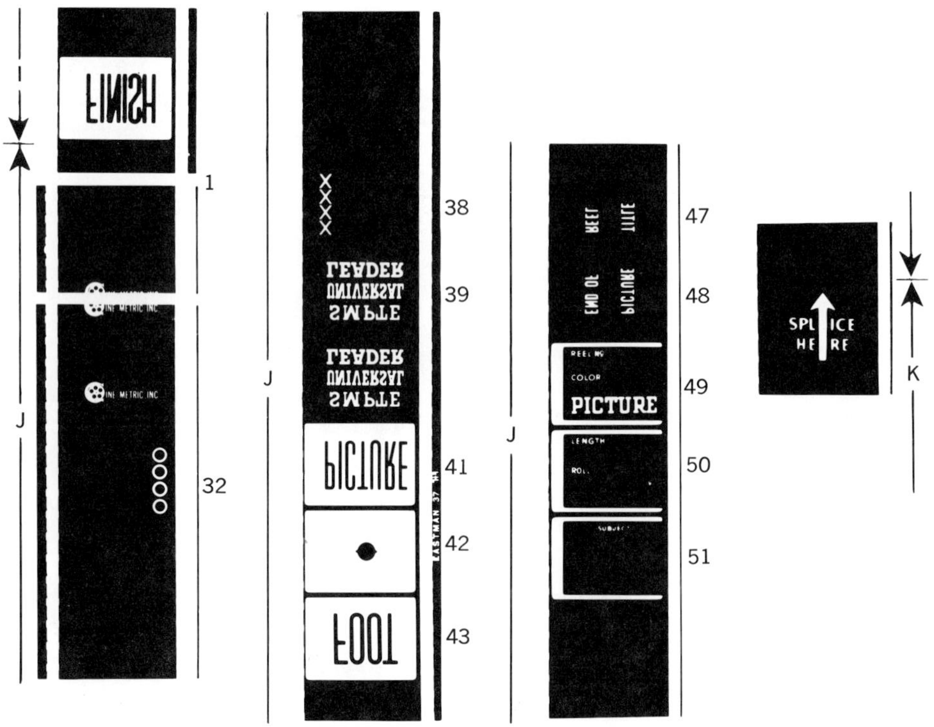

change. This change in direction, or bending of light, is known as *refraction.* Light waves passing through a glass window, for example, will continue in a straight line if the light wave strikes perpendicular, or at right angles to the pane. If the light wave strikes the transparent surface at an angle, however, the light wave will be refracted and change direction. Although the light wave will subsequently emerge at the same angle, its original path is deflected in proportion with the thickness of the glass pane. The difference between the angle of entry and the emerging angle is known as *displacement.*

Early experimenters learned that light rays, passing through a curved lens, could be concentrated and focused. In the absence of matches, the converging light rays could be used for starting fires.

Further investigation led to the observation that lenses, when placed in front of an object, made the object appear larger. But, if the viewer looked at the same object through a lens, from a distant point, the image seen through the lens would not only become considerably smaller but would also be upside down, or inverted. Images could also be formed by placing a lens between a light source and a sheet of paper. These images could be made clearer and sharper by changing the distance or relationship between the lens and the paper. Although the images remained in sharp focus at only one specific point, they remained upside down.

This basic information was subsequently studied, appraised and translated into appropriate scientific terminology. To optical glossaries were added definitions for terms such as reflection, refraction, displacement, divergence, and convergence. Accompanying diagrams showed that parallel rays from an object entered the lens, were refracted, and then met at the focal point beyond the lens. Properly captioned, the image remained upside down.

The ability of a lens to form images that are both clear and sharp, by concentrating or dispersing the refracted rays of light from an object, is its primary function in photography. This faculty is referred to as *defining power* or *definition.* During the photographic process, light passes through a lens and the images are exposed on the film in the camera.

The eye, in many ways, performs similarly with the pupil acting as the aperture and the retina taking the place of the film. In dimly lit surroundings, the pupil enlarges. In strong sunlight, the pupil makes its own compensation and becomes considerably smaller. Muscles within the eye control the focal length and make automatic adjustments for differences in lighting conditions and distances. These reactions can be compared with the lens adjustments that must be made in order to expose the film properly.

The eye sees in dimension and perspective. Unless stereo equipment is used, the lens is limited to two dimensions. Depth can only be assumed by the relationship, size, and focus of the subject matter relative to other objects and the background.

The first lenses were made of glass. In recent years, lenses made of plastic and other transparent materials have been used successfully without any appreciable sacrifice in performance or quality. Molds guide the skilled craftsmen in the shaping and grinding processes. Polishing the mathematically perfect, curved surfaces of a lens is a time-consuming as well as painstaking task. Some lenses, specially prepared for use in high-power telescopes, take many months to complete. Any appreciable nick or scratch immediately voids its usefulness as an optical instrument.

In spite of the delicate handling and grinding processes, aberrations or disturbances that prevent the proper focusing of light rays are present in almost all lenses. Theoretically (although no longer a positive actuality), these minute aberrations in even the most perfectly ground lenses virtually eliminate the possibility of a geometrically perfect focus. Spherical aberrations, for example, prevent all light rays from converging at the exact focal point of the lens. These and similar imperfections, however, do not seriously affect the general image-formation properties of lenses.

Lenses can be used singly (simple lenses), or in combination with other lenses (compound lenses).

Not all the light that enters the outer surface of a lens reaches the film in the camera. Some light is lost by reflection. Additional light losses are caused by the absorption qualities of the lenses themselves. In compound lenses, the light loss may amount to 50 percent of the total amount of light transmitted. Coatings on each of the separate lens surfaces reduce the amount of light lost by reflection to a mere 1 percent instead of the approximate 4 percent loss from uncoated surfaces.

Convex, or converging lenses concentrate light waves to a central point. The image formed by this type of lens is inverted. Concave lenses spread or disperse light waves (divergence). Double-concave lenses make an image appear smaller than normal—but right side up.

Differences in the relative positions and curvature of lenses cause variations in focal length—the distance between the optical center of the lens and the focal plane on which the image is exposed. The size of an image, at a predetermined distance from the lens, is governed by the focal length of the lens. Increases in focal length cause images to become proportionately larger, just as a projected image increases with the distance from the screen to the projector. Thus, objects placed at similar distances from lenses with the same focal length appear equal in size. (A 6-inch-focal-length lens would, therefore, form an image six times the size of an image formed by a 1-inch-focal-length lens, assuming that the subject matter remains in the same position for both.)

Because the subject matter, in motion picture and television production, is usually positioned at relatively greater distances from the camera than in other types of photography, lenses with longer focal lengths are better suited than lenses with shorter focal lengths. The average field covered by these lenses is approximately 30 degrees. By comparison, lenses used for still photography usually cover a 50-degree angle while those used in microscopes and telescopes have an angular coverage of about 10 degrees.

The size of the image formed by a lens is proportional to the size of the field (diagonal of the negative) that is recorded in the individual frames of film. Since longer-focal-length lenses produce proportionately larger images, the selection of a particular lens is governed by the size of the field that the cinematographer wishes to include in the film frame.

Lenses having focal lengths shorter than the picture diagonal are referred to as wide-angle lenses. Telephoto or long-focus lenses have focal lengths that are considerably longer than the picture diagonal. Zoom lenses merit a special category because of their unique ability to vary focal length while keeping the image in focus. In motion picture photography, their many uses easily outnumber the more than 20 separate lens elements in their complicated optical systems.

The depth of field is the distance in front of and behind the subject matter in which detail remains in acceptable focus. This distance decreases proportionately as the subject matter is moved closer to the lens and increases at greater distances from it. Smaller lens stops and short-focal-length lenses increase the depth of field, whereas larger lens openings and long-focal-length lenses have the opposite effect. Objects outside this limited area appear as blurred circles and are aptly referred to as *circles of confusion.*

Depth of focus, often confused with depth of field, is measured within the camera, between the lens and the film plane. In theory, the image transmitted through the lens is sharpest at the focal plane on which it is exposed. If the film is not precisely positioned in the camera during exposure, the resultant buckling or breathing caused by improper registration produces images that lack good definition. The distortion increases with the use of short-focus lenses.

The hyperfocal distance of a lens is the distance, when the lens is focused at "infinity," beyond which all objects are in relatively sharp focus. When focused on a point at the hyperfocal distance, objects remain sharp and clear from infinity to approximately half the distance between the hyperfocal distance and the lens. If the hyperfocal distance is 100 feet, for example, a lens focused at 100 feet will have all objects from 50 feet to infinity in sharp focus. In fixed-focus cameras, all objects beyond a relatively short distance from the camera are in sharp focus. The hyperfocal distance decreases proportionately with corresponding reductions of the lens aperture and increases at larger lens openings. *f*-numbers on lens diaphragms

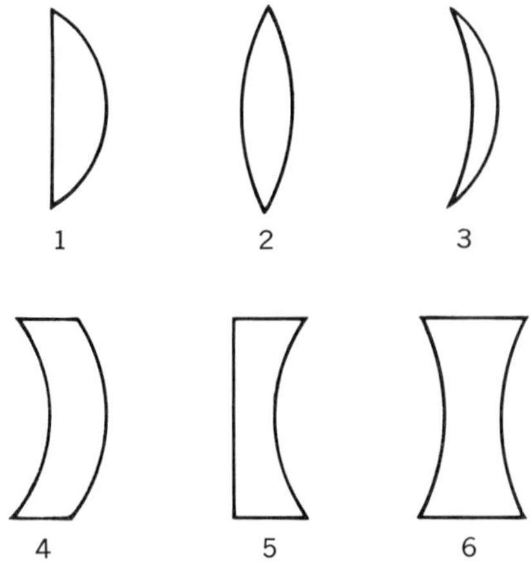

Fig. L-4*a* Lens types. (1) Plano-convex—converging. (2) Double-convex (bi-convex)—converging. (3) Concavo-convex—converging. (4) Concavo-convex—diverging. (5) Plano-concave—diverging. (6) Double-concave (bi-concave)—diverging.

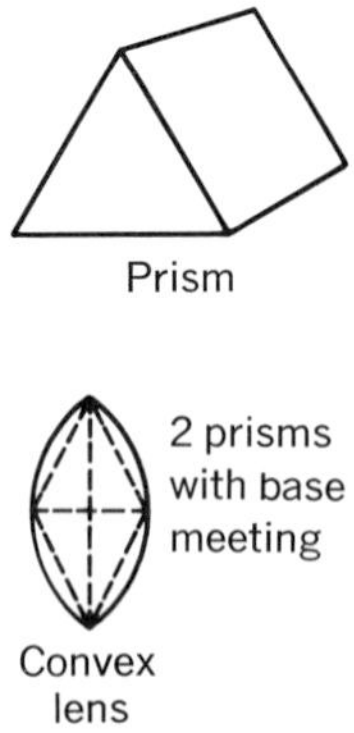

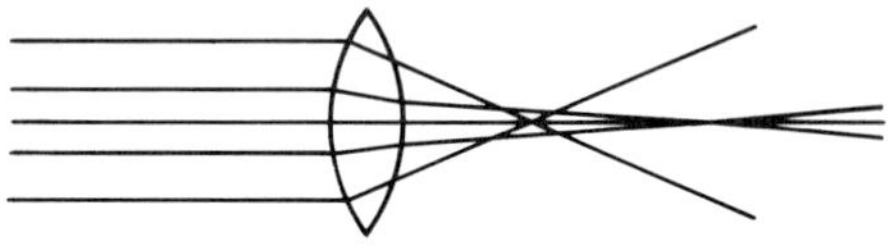

Spherical aberration
Rays through central portion of lens and rays through edges do not converge at same point.

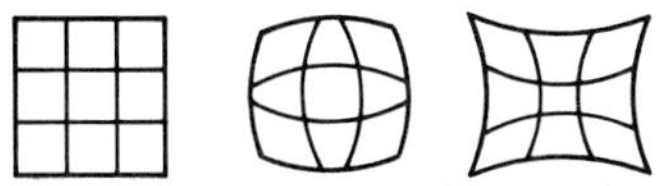

Subject Images distorted
Image is accurate through center only. Distortion occurs from center outward.

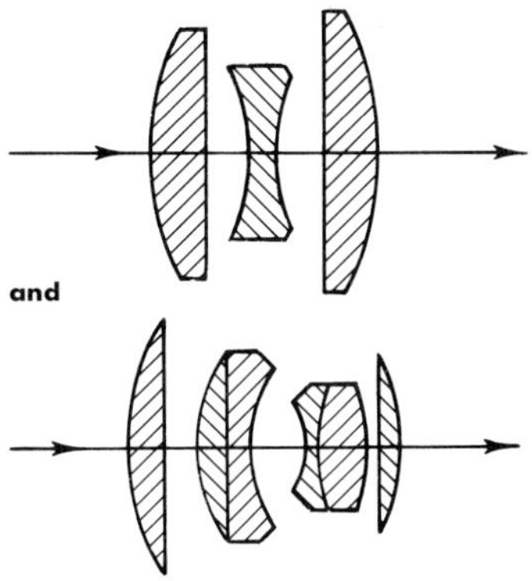

Fig. L-4*b* Examples of compound lenses and their components.

indicate the amount of light that is transmitted through the lens. Each *f*-stop, or aperture setting, increases or decreases the amount of transmitted light by one-half of the amount at the adjacent *f*-stop.

An aperture of *f*/4 allows only one-half as much light to pass through the lens as the amount transmitted at the next lens setting of *f*/2.8. A lens setting of *f*/8, by comparison, doubles the amount of exposure at the *f*/11 setting, the next diaphragm marking. See Figures L-4*a* to *h*.

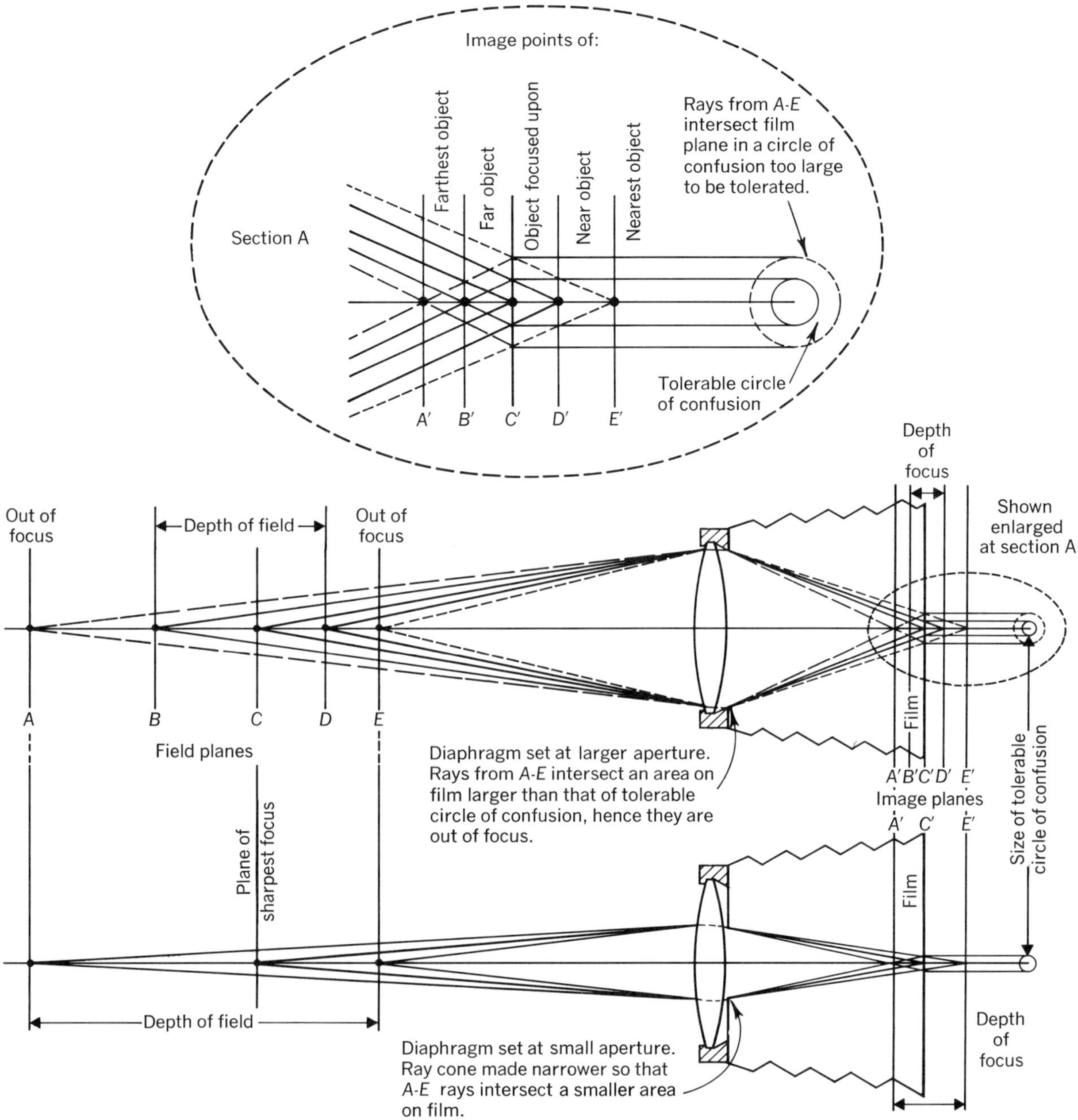

Fig. L-4c The effect of lens apertures on depth of field.

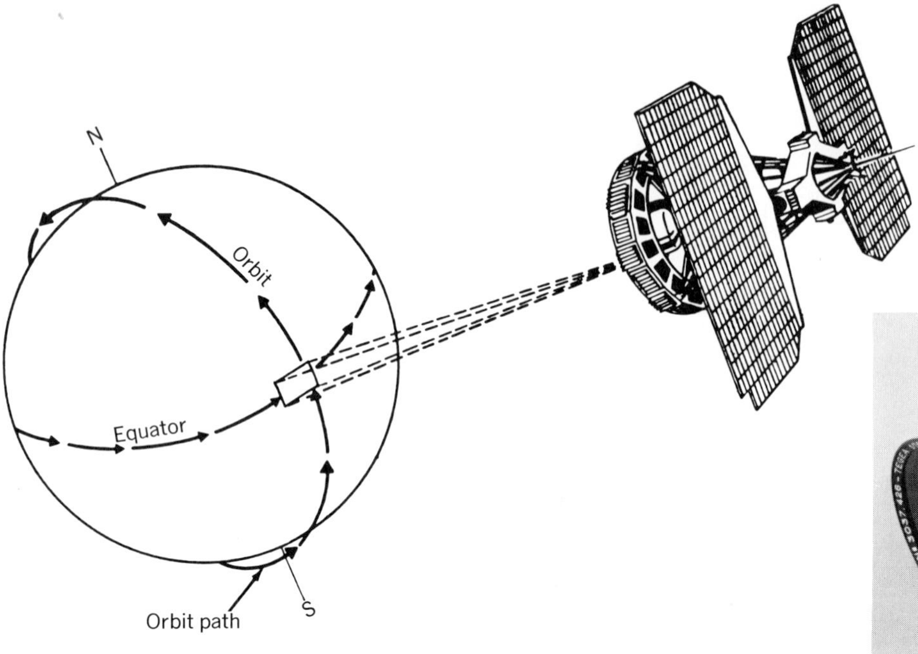

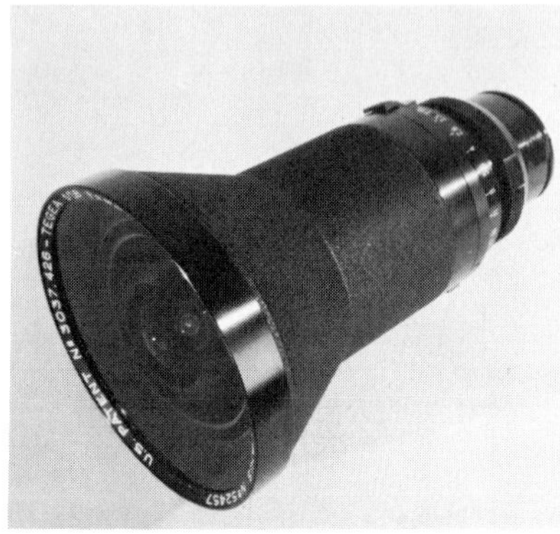

Fig. L-4*d* Schematic diagram of automatic picture transmission coverage by Nimbus weather satellite incorporating a Kinoptik 5.7mm *f*/1.8 Tegea super-wide-angle lens (shown at right). This lens, supplied by Karl Heitz, Inc., was used in the Nimbus C weather satellite launched by NASA. Mounted on an automatic picture transmission (APT) vidicon camera, the lens takes 200 pictures of the earth every day from an orbital height of approximately 680 miles. Especially protected against radiation, each lens covers a field of 108 degrees, equaling an earth area of 2 million square miles. A 40-millisecond exposure with an electromagnetic shutter produces an 800-scan line picture on the photosensitive surface of a special one-inch-diameter vidicon tube for automatic picture transmission to receiving stations on earth. Such line resolution makes landscapes two miles across clearly visible, giving detailed weather information of the smallest part of the earth.

Kinoptik lenses have played an important role in earlier space achievements. Most previous weather satellites, such as the Tirus and Nimbus A, were equipped with Kinoptik 5.7mm f/1.8 Tegea super-wide-angle lens.

Portions of the historic rendezvous between Gemini VI and VII were also filmed with Kinoptik 18mm *f*/1.8 Apochromat, as were sequences of the first hookup attempt between Gemini VIII and an Agena rocket.

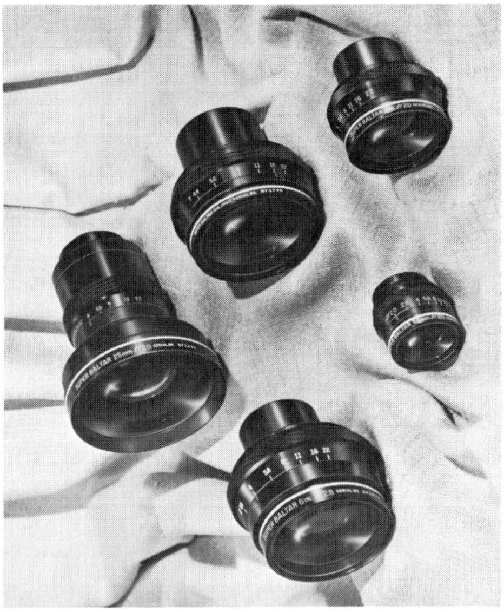

Fig. L-4e Bausch & Lomb Super Baltar professional motion picture lenses.

Fig. L-4f The Camart optical FX unit and rotator lens.

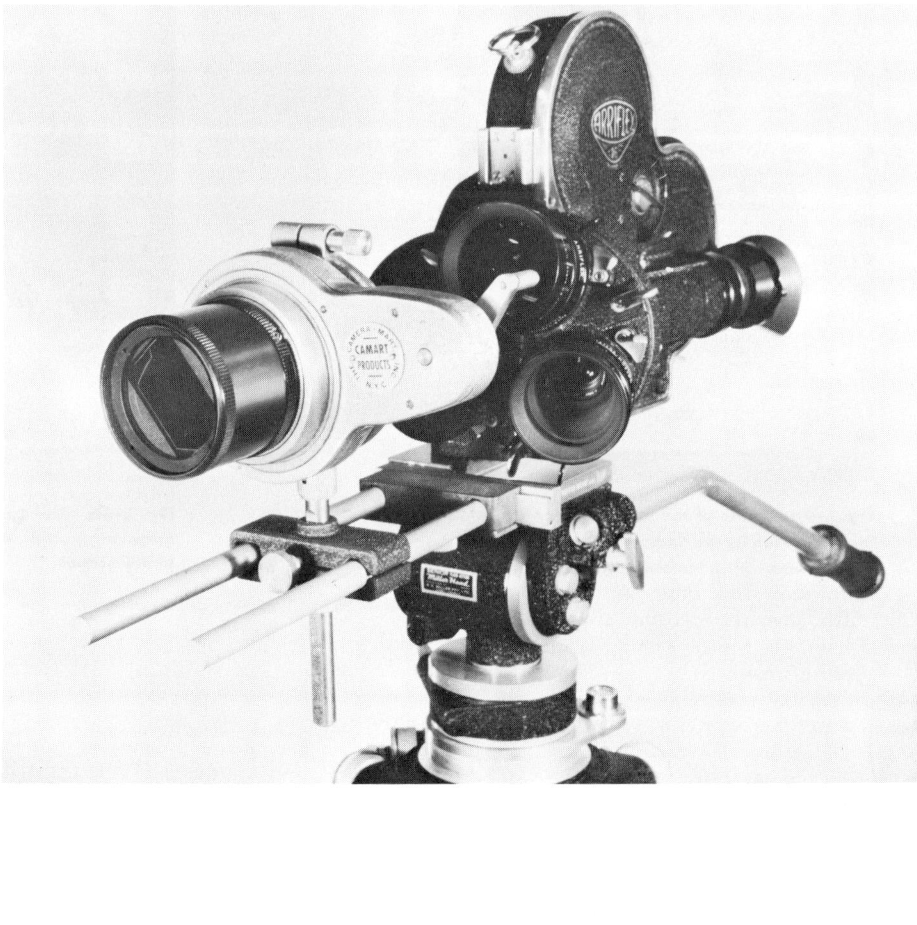

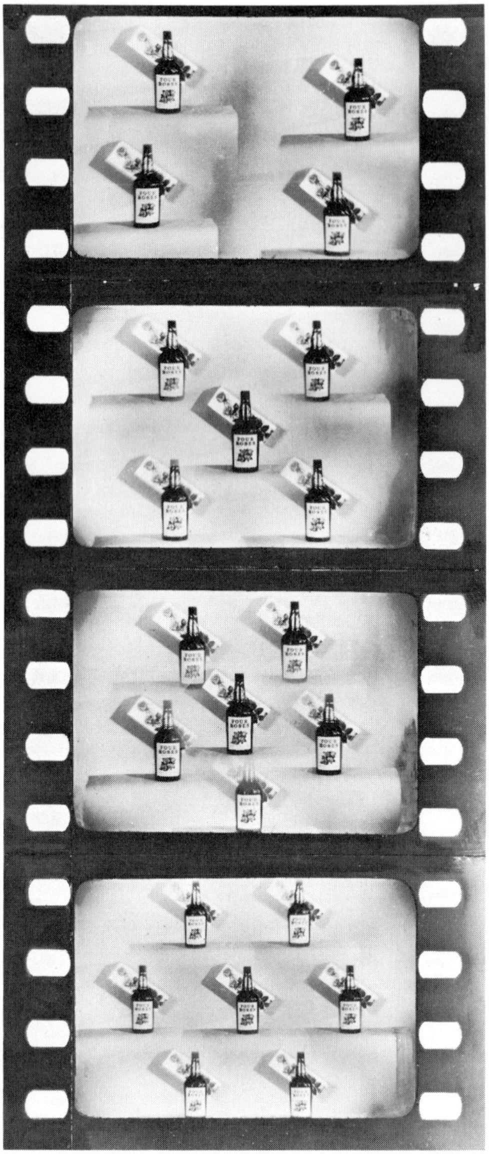

Fig. L-4*g* Some of the interesting optical effects produced by the Camart optical FX unit and rotator lens. Top to bottom: four images in rotation; four images rotate around a "fifth"; five images rotate around a sixth; six images rotate around a stationary seventh image.

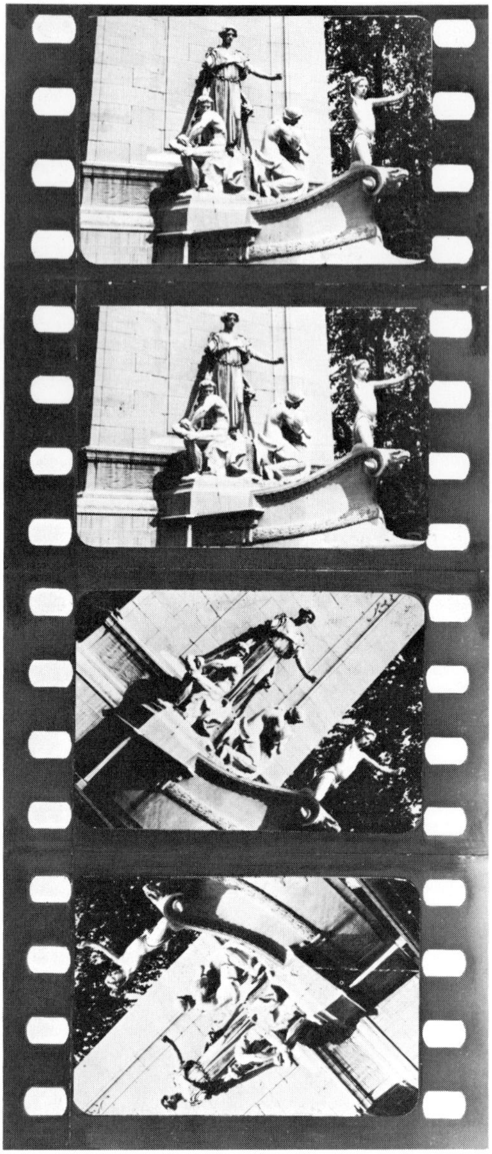

Fig. L-4*h* The Camart rotator lens is used for achieving multiple-image shots and for rotating entire scenes.

A comprehensive listing of motion picture and photo instrumentation lenses has been compiled and designed by Gordon Enterprises of Hollywood (Figure L-4*i*). The charts show which lenses are available from stock, along with those that can be custom-mounted for a specific camera.

	16MM Arriflex A	16MM Bolex Reflex B	16MM Standard "C" C	16/35MM Camerette D	16MM Eastman Bayonet "S" E	16MM Eastman H.S. & Magnifax F	16MM Eastman Reflex "R" G	16MM Fairchild H.S. H	16MM Fastair or Photo-Sonic I	16/35MM Fastax J	16MM G.S.A.P. (N-9, KB-3) K	16MM Mitchell L	35MM Arriflex M	35MM Bell & Howell #2709 N	35MM Eyemo "C" O	35MM Mitchell Std., H.S., & N.C. P	35MM Mitchell BNC Q	35MM Mitchell Reflex R	70MM Flight Research S	70MM Hulcher 100 & 101 T	70MM Hulcher 102 U	70MM Hulcher 103 V	70MM Mitchell W	70MM Photo-Sonic X
NGENIEUX																								
1001 9.5mm f/2.2			S4								C5	S5												
1002 10mm f/1.8			S4								C5	S5												
1003 14.5mm f/3.5											C5		C	C	C	C	C							
1004 15mm f/1.3			S								C5	S5												
1005 18.5mm f/2.2	S		C								C5		C	C	C	C	C							
1006 24mm f/2.2	C										C5		C	C	C	C	C							
1007 25mm f/0.95			S								C5	S5												
1008 25mm f/1.4			S								C5	S5												
1009 28mm f/1.8	C										C5		C	C	C	C	C							
1010 32mm f/1.8	C	C	C	C							C5	C	C	C	C	C	C							
1011 40mm f/1.8	C	C	C	C							C5	C	C	C	C	C	C	C						
1012 50mm f/1.8	C	C	C	C							C5	C	C	C	C	C	C	C						
1013 75mm f/1.8	C	C	C	C							C5	C	C	C	C	C	C	C						
1014 75mm f/2.5	C	S	S								C5	S5												
1015 100mm f/2	C	C	C	C	C	C	C	C	C	C	C5	C	C	C	C	C	C	C						C
Bausch & Lomb																								
ALTARS																								
1501 15mm f/2.3 (T2.5)																								
1502 17.5mm f/2.3 (T2.5)											C5	C												
1503 20mm f/2.3 (T2.5)											C5	C												
1504 25mm f/2.3 (T2.5)		C	C								C5	C		S	S	S	S							
1505 30mm f/2.3 (T2.5)	C	C	C								C5	C	C	S	S	S	S							
1506 35mm f/2.3 (T2.5)	C	C	C	C							C5	C	S	S	S	S	S							
1507 40mm f/2.3 (T2.5)	C	C	C	C							C5	C	C	S	S	S	S							
1508 50mm f/2.3 (T2.5)	C	C	C	C							C5	C	S	S	S	S	S							
1509 75mm f/2.3 (T2.5)	C	C	C	C	C	C	C	C			C5	C	S	S	S	S	S	C						
1510 100mm f/2.3 (T2.5)	C	C	C	C	C	C	C	C	C	C	C5	C	C	S	S	S	S	C						
1511 152mm f/2.7 (T2.8)	C	C	C	C	C	C	C	C	C	C	C5	C	C	S	S	S	S	C	C		C		C	C
JPER BALTARS																								
1512 20mm T2.3											C5			C	C	C	C	S						
1513 25mm T2.3	C		C								C5	C	C	C	C	C	C	S						
1514 35mm T2.3	C		C		C						C5	C	C	C	C	C	C	S						
1515 50mm T2.3	C	C	C	C	C	C	C	C	C	C	C5	C	C	C	C	C	C	S						
1516 75mm T2.3	C	C	C	C	C	C	C	C	C	C	C5	C	C	C	C	C	C	S						
1517 100mm T2.3	C	C	C	C	C	C	C	C	C	C	C5	C	C	C	C	C	C	S						
1518 150mm T2.8	C	C	C	C	C	C	C	C	C	C	C5	C	C	C	C	C	C	S						
1519 225mm T4	C	C	C	C	C	C	C	C	C	C	C5	C	C	C	C	C	C	S						
BALTARS																								
1520 15mm f/2.5			S4								C4	C4												

TV BALTARS (continued)	16MM Arriflex A	16MM Bolex Reflex B	16MM Standard "C" C	16/35MM Camerette D	16MM Eastman Bayonet "S" E	16MM Eastman H.S. & Magnifax F	16MM Eastman Reflex "R" G	16MM Fairchild H.S. H	16MM Fastair or Photo-Sonic I	16/35MM Fastax J	16MM G.S.A.P. (N-9, KB-3) K	16MM Mitchell L	35MM Arriflex M	35MM Bell & Howell #2709 N	35MM Eyemo "C" O	35MM Mitchell Std., H.S., & N.C. P	35MM Mitchell BNC Q	35MM Mitchell Reflex R	70MM Flight Research S	70MM Hulcher 100 & 101 T	70MM Hulcher 102 U	70MM Hulcher 103 V	70MM Mitchell W	70MM Photo-Sonic X
L1521 20mm f/3.5			S4								C4	C4												
L1522 25mm f/3.5			S4								C4	C4												
L1523 25mm f/2.5			S4								C4	C4												
L1524 35mm f/3.5			S4								C4	C4												
L1525 35mm f/2.5		S	S								C5	S5												
L1526 50mm f/3.5		S	S								C5	S5												
L1527 50mm f/3.5		S	S4								C4	S5												
L1528 50mm f/2.5		S	S								C5	S5												
L1529 75mm f/3.5		S	S4								C4	C4												
L1530 75mm f/3.5		S	S								C5	S5												
L1531 100mm f/3.5		S	S4	C	C	C		C	C	C	C4	C4												
L1532 100mm f/3.5		S	S	C	C	C		C	C	C	C5	S5												
Bell & Howell																								
MILTARS																								
L2001 25mm f/1.9														S	S	C								
L2002 35mm f/1.9														S	S	C	C							
L2003 50mm f/1.9	C													S	S	C	C							
L2004 75mm f/1.9	C	C	C	C	C	C	C	C	C	C	C	C	C	C	S	C	C	C						
L2005 100mm f/1.9	C	C	C	C	C	C	C	C	C	C	C	C	C	C	S	C	C	C						
L2006 6 inch f/3.5	C	C	C	C	C	C	C	C	C	C	C	C	C	C	S	C	C	C				C	C	C
L2007 10 inch f/4	C	C	C	C	C	C	C	C	C	C	C	C	C	C	S	C	C	C				C	C	C
CANON																								
L3001 50mm f/1.2		C	C									C5			C									
L3002 50mm f/1.8		C	C		C	C	C	C	C	C	C5	C5			C									
L3003 85mm f/1.8		C	C		C	C	C	C	C	C	C5	C5		C	C	C	C							
L3004 100mm f/2	C	C	C	C	C	C	C	C	C	C	C5	C5	C	C	C	C	C	C						
L3005 135mm f/3.5	C	C	C	C	C	C	C	C	C	C	C5	C5	C	C	C	C	C	C	C					
L3006 135mm f/2.5	C	C	C	C	C	C	C	C	C	C	C5	C5	C	C	C	C	C	C	C					
L3007 200mm f/3.5	C	C	C	C	C	C	C	C	C	C	C5	C5	C	C	C	C	C	C	C					
L3008 300mm f/4	C	C	C	C	C	C	C	C	C	C	C5	C5	C	C	C	C	C	C	C	C				
L3009 400mm f/4.5	C	C	C	C	C	C	C	C	C	C	C5	C5	C	C	C	C	C	C	C	C				
L3010 600mm f/5.6	C	C	C	C	C	C	C	C	C	C	C5	C5	C	C	C	C	C	C	C	C				
CENTURY PRECISION																								
L3501 100mm f/2	S	S	S		S	C	C	C	C	C	C5	S5												
L3502 150mm f/2.2	S	S	S		S	C	C	C	C	C	C5	S5												
L3503 150mm f/2.8	S	S	S	C	S	C	C	C	C	C	C5	S5												
L3504 230mm f/2.8	S	S	S	C	S	C	C	C	C	C	C5	S5	S	C	C	C	C	C						
L3505 230mm f/3.8	S	S	S	C	S	C	C	C	C	C	C5	S5	S	C	C	C	C	C						
L3506 300mm f/3.5	S	S	S	C	S	C	C	C	C	C	C5	S5	S	C	C	C	C	C						
L3507 300mm f/4.5	S	S	S	C	S	C	C	C	C	C	C5	S5	S	C	C	C	C	C						

Fig. L-4*i* Comprehensive listing of motion picture and photo instrumentation lenses. (*Copyright © 1962 by Gordon Enterprises.*)

C1, S1 Will not cover full 16mm frame.
C2, S2 Will not cover full 35mm frame.
C3, S6 16mm format only.
C4, S4 Fixed focus only.
C5, S5 Requires C-mount adapter.

	16MM Arriflex	16MM Bolex Reflex	16MM Standard "C"	16/35MM Camerette	16MM Eastman Bayonet "S"	16MM Eastman H.S. & Magnifax	16MM Eastman Reflex "R"	16MM Fairchild H.S.	16MM Fastair or Photo-Sonic	16/35MM Fastax	16MM G.S.A.P. (N-9, KB-3)	16MM Mitchell	35MM Arriflex	35MM Bell & Howell #2709	35MM Eyemo "C"	35MM Mitchell Std., H.S., & N.C.	35MM Mitchell BNC	35MM Mitchell Reflex	70MM Flight Research	70MM Hulcher 100 & 101	70MM Hulcher 102	70MM Hulcher 103	70MM Mitchell	70MM Photo-Sonic
CENTURY (cont.)	A	B	C	D	E	F	G	H	I	J	K	L	M	N	O	P	Q	R	S	T	U	V	W	X
L3508 385mm f/4.5	S	S	S	C	S	C	C	C	C	C	C	S	S	C	C	C	C	C						
L3509 400mm f/4.5	S	S	S	C	S	C	C	C	C	C	C	S	S	C	C	C	C	C						
L3510 500mm f/5.6	S	S	S	C	S	C	C	C	C	C	C	S	S	C	C	C	C	C						
L3511 500mm f/4.5	S	S	S	C	S	C	C	C	C	C	C	S	S	C	C	C	C	C						
L3512 600mm f/4.8	S	S	S	C	S	C	C	C	C	C	C	S	S	C	C	C	C	C						
L3513 1000mm f/7	S	S	S	C	S	C	C	C	C	C	C	S	S	C	C	C	C	C						
Eastman																								
CINE EKTARS																								
L4001 15mm f/2.5			S		S		S				C5	S												
L4002 25mm f/1.9		S	S		S		S				C5	S												
L4003 25mm f/1.4		S	S		S		S				C5	S												
L4004 50mm f/1.9		S	S		S	S	S				C5	S												
L4005 63mm f/2		S5	S5		S	S	S				C5	S												
L4006 102mm f/2.7		S5	S5		S	S	S	C	C	C3	C5	S												
L4007 152mm f/4		S5	S5		S	S	S	C	C	C3	C5	S												
CINE EKTONS																								
L4008 10mm f/1.8			C				S				C5	C5												
L4009 15mm f/1.3			C				S				C5	C5												
L4010 25mm f/0.95			C				S				C5	C5												
L4011 25mm f/1.4			C				S				C5	C5												
L4012 50mm f/1.5		C	C				S				C5	C5												
L4013 75mm f/2.5	C	C	C			C	S				C5	C5												
L4014 100mm f/2.5	C	C	C			C	S	C	C	C3	C5	C5												
L4015 150mm f/2.7	C	C	C			C	S	C	C	C3	C5	C5												
ELGEET																								
L4501 3.7mm f/1.5			S1					S			S1	C1												
L4502 5mm f/1.5			S1					S			S1	C1												
L4503 8mm f/1.5			S1					S			S1	C1												
L4504 12mm f/1.2			S								C5	C												
L4505 13mm f/2.5			S								C5	C												
L4506 13mm f/1.5			S		S	S		S	S		C5	C												
L4507 25mm f/1.9			S			S		S	S		C5	C												
L4508 25mm f/1.5			S								C5	C												
L4509 50mm f/1.5		S	S		S						C5	C												
L4510 76mm f/1.9		S	S		S	S	C	S	S		C5	C												
L4511 76mm f/2.9		S	S		S		C	C			C5	C												
L4512 102mm f/2.7		S	S		S	S	C	S	S		C5	C												
L4513 152mm f/3.8		S	S		S	C	C	S	C		C5	C												
L4514 259mm f/3.5	C			C	C	C	C	C	C	C	C5	C												

Fig. L-4*i* (*continued*).

	16MM Arriflex	16MM Bolex Reflex	16MM Standard "C"	16/35MM Camerette	16MM Eastman Bayonet "S"	16MM Eastman H.S. & Magnifax	16MM Eastman Reflex "R"	16MM Fairchild H.S.	16MM Fastair or Photo-Sonic	16/35MM Fastax	16MM G.S.A.P. (N-9, KB-3)	16MM Mitchell	35MM Arriflex	35MM Bell & Howell #2709	35MM Eyemo "C"	35MM Mitchell Std., H.S., & N.C.	35MM Mitchell BNC	35MM Mitchell Reflex	70MM Flight Research	70MM Hulcher 100 & 101	70MM Hulcher 102	70MM Hulcher 103	70MM Mitchell	70MM Photo-Sonic
	A	B	C	D	E	F	G	H	I	J	K	L	M	N	O	P	Q	R	S	T	U	V	W	X
Elitar																								
SOLIGORS																								
L2501 10mm f/1.8			S								C5	C5												
L2502 13mm f/1.5			S								C5	C5												
L2503 15mm f/1.9			S								C5	C5												
L2504 17mm f/2.5			S								C5	C5												
L2505 25mm f/0.95			S								C5	C5												
L2506 25mm f/1.1			S								C5	C5												
L2507 25mm f/1.5			S								C5	C5												
L2508 50mm f/1.9		S	S								C5	C5												
L2509 75mm f/1.5		S	S								C5	C5												
L2510 75mm f/1.9		S	S								C5	C5												
L2511 75mm f/2.5		S	S								C5	C5												
L2512 150mm f/3.5	C	S	S		C						C5	C5												
TV SOLIGORS																								
L2521 12.5mm f/1.4			S								C5	C5												
L2522 15.8mm f/1.4			S								C5	C5												
L2523 25mm f/1.4			S								C5	C5												
L2525 50mm f/1.4		S	S								C5	C5												
L2521 12.5mm f/1.4		S	S								C5	C5												
L2526 75mm f/4.5		S	S		C			C	C	C3	C5	C5												
Gordon Enterprises																								
VIDEOTELS																								
L5001 3 inch f/1.5		S	S								C5													
L5002 6 inch f/2.5		S	S		C	C	C	C	C	C3	C5	C												
CYCLOGONS																								
L5003 3.7mm f/1.5			C1								C1	C1												
L5004 17mm f/2.7			C								S	C												
CYCLOTARS																								
L5005 3 inch f/2.8		C	C								C5	C									C	C	C	
L5006 4 inch f/2.5	C	C	C	C3	C	C	C	C	C	C	C5	C												
L5007 6 inch f/2.8	S	S	S	C3	S	C	C	C	C	C	C5	C												
L5008 6 inch f/2.5	C	C	C	C3	S	C	C	C	C	C	C5	C									C	C	C	
L5009 7 inch f/2.5	C	C	C	C	S	C	C	C	C	C	C5	C	S	C	C	C	C	C	C	C	C	C	C	
L5010 9 inch f/3.8	S	S	S	C	S	C	C	C	C	C	C5	C	S	C	C	C	C	C						
L5011 10 inch f/4.5	S	S	S	C	S	C	C	C	C	C	C5	C	S	C	C	C	C	C						
L5012 12 inch f/4.5	S	S	S	C	S	C	C	C	C	C	C5	C	S	C	C	C	C	C						
L5013 12 inch f/2.5													C					C	C	C	C	C	C	
L5014 15 inch f/4.5	S	S	S	C	S	C	C	C	C	C	C5	C	S	C	S	C	C	C						
L5015 20 inch f/5.6	S	S	S	C	S	C	C	C	C	C	C5	C	S	C	S	C	C	C	C	C	S	S	C	

	16MM Arriflex	16MM Bolex Reflex	16MM Standard "C"	16/35MM Camerette	16MM Eastman Bayonet "S"	16MM Eastman H.S. & Magnifax	16MM Eastman Reflex "R"	16MM Fairchild H.S.	16MM Fastair or Photo-Sonic	16/35MM Fastax	16MM G.S.A.P. (N-9, KB-3)	16MM Mitchell	35MM Arriflex	35MM Bell & Howell #2709	35MM Eyemo "C"	35MM Mitchell Std., H.S., & N.C.	35MM Mitchell BNC	35MM Mitchell Reflex	70MM Flight Research	70MM Hulcher 100 & 101	70MM Hulcher 102	70MM Hulcher 103	70MM Mitchell	70MM Photo-Sonic
LOTARS (continued)	**A**	**B**	**C**	**D**	**E**	**F**	**G**	**H**	**I**	**J**	**K**	**L**	**M**	**N**	**O**	**P**	**Q**	**R**	**S**	**T**	**U**	**V**	**W**	**X**
16 24 inch f/6	C	C	C	C	C	C	C	C	C	C	C5	C	C	C	C	C	C	C	C	C	S	S	C	C
17 36 inch f/8	C	C	C	C	C	C	C	C	C	C	C	C	C	C	C	C	C	C	C	C	C	C	C	C
18 40 inch f/8	C	C	C	C	C	C	C	C	C	C	C5	C	C	C	C	C	C	C	C	C	S	S	C	C
aillard																								
TARS																								
01 10mm f/1.6		S	S								C5	C5												
02 16mm f/1.8		S	S								C5	C5												
03 25mm f/1.4		S	S								C5	C5												
04 50mm f/1.4		S	S								C5	C5												
05 75mm f/1.9		S	S								C5	C5												
AR																								
06 26mm f/1.9		S	S								C5	C5												
RS																								
07 16mm f/2.8		S	S								C5	C5												
08 75mm f/2.8		S	S								C5	C5												
09 100mm f/3.3		S	S								C5	C5												
10 150mm f/4		S	S								C5	C5												
ilfitt																								
KRO-KILARS																								
01 40mm f/2.8	S	C	C								C5	C5	S		C									
02 90mm f/2.8	S	C	C	C	C	C	C	C	C	C	C5	C5	S	C	C	C	C	C						C
E-KILARS																								
03 150mm f/3.5	S	S	S	C	S	C	C	C	C	C	C5	C5	S	C	C	C	C	C	C	C	C	C	C	C
04 300mm f/5.6	S	S	S	C	S	C	C	C	C	C	C5	C5	S	C	C	C	C	C	C	C	C	C	C	C
05 300mm f/4	S	S	S	C	S	C	C	C	C	C	C5	C5	S	C	C	C	C	C	C	C	C	C	C	C
N-KILARS																								
06 400mm f/5.6	S	S	S	C	S	C	C	C	C	C	C5	C5	S	C	C	C	C	C	C					C
07 400mm f/4	S	S	S	C	S	C	C	C	C	C	C5	C5	S	C	C	C	C	C	C	C	C	C	C	C
08 600mm f/5.6	S	S	S	C	S	C	C	C	C	C	C5	C5	S	C	C	C	C	C	C					C
inoptik																								
CHROMATS																								
01 1.9mm f/1.9	S1		S1								C1	C1												
02 5.7mm f/1.8	S4		S4				C4				C4	C4												
03 9.8mm f/1.9	C										C5	C5												
04 12.5mm f/2.5	S		S				C				C5	C5												
05 18mm f/2	S		S	C			C				C5	C5	S	C	C	C								
06 20mm f/1.9	S		S	C			C				C5	C5	S	C	C	C								
07 25mm f/2	S		S	C			C				C5	C5	S	C	C	C	C							
08 28mm f/2	S		C	C			C				C5	C5	S	C	C	C	C							
09 32mm f/1.9	S		C	C			C				C5	C5	C	C	C	C	C							

Fig. L-4*i* (*continued*).

	16MM Arriflex	16MM Bolex Reflex	16MM Standard "C"	16/35MM Camerette	16MM Eastman Bayonet "S"	16MM Eastman H.S. & Magnifax	16MM Eastman Reflex "R"	16MM Fairchild H.S.	16MM Fastair or Photo-Sonic	16/35MM Fastax	16MM G.S.A.P. (N-9, KB-3)	16MM Mitchell	35MM Arriflex	35MM Bell & Howell #2709	35MM Eyemo "C"	35MM Mitchell Std., H.S., & N.C.	35MM Mitchell BNC	35MM Mitchell Reflex	70MM Flight Research	70MM Hulcher 100 & 101	70MM Hulcher 102	70MM Hulcher 103	70MM Mitchell	70MM Photo-Sonic
APOCHROMATS (cont.)	**A**	**B**	**C**	**D**	**E**	**F**	**G**	**H**	**I**	**J**	**K**	**L**	**M**	**N**	**O**	**P**	**Q**	**R**	**S**	**T**	**U**	**V**	**W**	**X**
L6510 50mm f/2	S	S	S	C	C	C	C				C5	C5	S	C	C	C	C	C						
L6511 75mm f/2	S	S	S	C	C	C	C	C		C	C5	C5	S	C	C	C	C	C						
L6512 100mm f/2	S	S	S	C	C	C	C	C	C	C	C5	C5	S	C	C	C	C	C						
L6513 150mm f/2.5	S	S	S	C	C	C	C	C	C	C	C5	C5	S	C	C	C	C	C						
L6514 210mm f/2.8	S	S	S	C	C	C	C	C	C	C	C5	C5	S	C	C	C	C	C						
L6515 300mm f/3.5	S	S	S	C	C	C	C	C	C	C	C5	C5	S	C	C	C	C	C						
L6516 500mm f/5.6	S	S	S	C	C	C	C	C	C	C	C5	C5	S	C	C	C	C	C						
FULGIOR																								
L6517 50mm f/1.3	S	S	S								C5	C5	S		C	C								
EXTRA-CORRECTED APOCHROMATS																								
L6518 32mm f/2.8	S	S	S	C	C		C				C5	C5	S		C	C	C	C						
L6519 40mm f/2.8	S	S	S	C	C		C				C5	C5	S		C	C	C	C						
L6520 50mm f/2.8	S	S	S	C	C		C				C5	C5	S		C	C	C	C						
LEITZ																								
L7001 50mm f/1.4		C	C								C5	C5												
L7002 50mm f/2		C	C								C5	C5												
L7003 85mm f/1.5		C	C				C				C5	C5												
L7004 90mm f/2	C	C	C	C	C	C	C	C	C	C	C5	C5	C	C	C	C	C	C						
L7005 90mm f/2.8	C	C	C	C	C	C	C	C	C	C	C5	C5	C	C	C	C	C	C						
L7006 90mm f/4	C	C	C	C	C	C	C	C	C	C	C5	C5	C	C	C	C	C	C						
L7007 125mm f/2.5	C	C	C	C	C	C	C	C	C	C	C5	C5	C	C	C	C	C	C						
L7008 135mm f/4	C	C	C	C	C	C	C	C	C	C	C5	C5	C	C	C	C	C	C						
L7009 200mm f/4	C	C	C	C	C	C	C	C	C	C	C5	C5	C	C	C	C	C	C						
L7010 280mm f/4.8	C	C	C	C	C	C	C	C	C	C	C5	C5	C	C	C	C	C	C						
L7011 400mm f/5	C	C	C	C	C	C	C	C	C	C	C5	C5	C	C	C	C	C	C						
Nikon																								
NIKKORS																								
L7501 50mm f/2		C	C		C	C	C	C	C	C	C5	C5			C									
L7502 58mm f/1.4		C	C		C	C	C	C	C	C	C5	C5			C									
L7503 105mm f/2.5	C	C	C		C	C	C	C	C	C	C5	C5	C	C	C	C								
L7504 135mm f/3.5	C	C	C	C	C	C	C	C	C	C	C5	C5	C	C	C	C	C	C						
L7505 135mm f/2	C	C	C	C	C	C	C	C	C	C	C5	C5	C	C	C	C	C	C						
L7506 180mm f/2.5	C	C	C	C	C	C	C	C	C	C	C5	C5	C	C	C	C	C	C						
L7507 200mm f/4	C	C	C	C	C	C	C	C	C	C	C5	C5	C	C	C	C	C	C						
L7508 250mm f/4	C	C	C	C	C	C	C	C	C	C	C5	C5	C	C	C	C	C	C	C	C	C	C	C	C
L7509 350mm f/4.5	C	C	C	C	C	C	C	C	C	C	C5	C5	C	C	C	C	C	C	C	C	C	C	C	C
L7510 500mm f/5	C	C	C	C	C	C	C	C	C	C	C5	C5	C	C	C	C	C	C	C	C	C	C	C	C
L7511 500mm f/5.5	C	C	C	C	C	C	C	C	C	C	C5	C5	C	C	C	C	C	C	C	C	C	C	C	C
L7512 1000mm f/6.3	C	C	C	C	C	C	C	C	C	C	C5	C5	C	C	C	C	C	C	C	C	C	C	C	C

	A 16MM Arriflex	B 16MM Bolex Reflex	C 16MM Standard "C"	D 16/35MM Camerette	E 16MM Eastman Bayonet "S"	F 16MM Eastman H.S. & Magnifax	G 16MM Eastman Reflex "R"	H 16MM Fairchild H.S.	I 16MM Fastair or Photo-Sonic	J 16/35MM Fastax	K 16MM G.S.A.P. (N-9, KB-3)	L 16MM Mitchell	M 35MM Arriflex	N 35MM Bell & Howell #2709	O 35MM Eyemo "C"	P 35MM Mitchell Std., H.S., & N.C.	Q 35MM Mitchell BNC	R 35MM Mitchell Reflex	S 70MM Flight Research	T 70MM Hulcher 100 & 101	U 70MM Hulcher 102	V 70MM Hulcher 103	W 70MM Mitchell	X 70MM Photo-Sonic
PACIFIC OPTICAL																								
L7601 3.45mm f/1.5			C1								C5													
L7602 3.45mm f/1.5			C1								C5													
L7603 3.45mm f/1.9			C1								C5													
L7604 4mm f/1.5			C1								C5													
L7605 6.51mm f/6.3			C1								C5													
Rodenstock																								
HELIGON RAINBOW																								
L8001 16mm f/2 (T2.2)	S		C																					
L8002 25mm f/1.5 (T1.6)	S		C																					
L8003 50mm f/2 (T2.2)	S		C																					
Schneider																								
CINEGON																								
L8501 10mm f/1.8	S										C4	C4												
CINE-XENONS																								
L8502 16mm f/1.9	S	S	S								C5	C5												
L8503 25mm f/1.4	S	S	S								C5	C5												
L8504 28mm f/2	S		C	C							C5	C5	S	C	C	C								
L8505 35mm f/2	S		C	C							C5	C5	S	C	C	C								
L8506 50mm f/2	S	S	S	C	C	C	C				C5	C5	S	C	C	C	C							
L8507 75mm f/2	S	C	C	C	C	C	C	C	C	C	C5	C5	S	C	C	C	C	C						
L8508 100mm f/2	S	C	C	C	C	C	C	C	C	C	C5	C5	S	C	C	C	C	C						
CINE-TELE-XENARS																								
L8509 75mm f/2.8		S	S								C5	C5												
L8510 100mm f/2.8		S4	S4								C4	C4												
L8511 150mm f/4		S4	S4								C4	C4												
XENONS																								
L8512 20mm f/2	S		C	C							C5	C5	S	C	C	C								
L8513 40mm f/2	S		C	C							C5	C5	S	C	C	C	C							
L8514 50mm f/2		C	C	C	C	C	C				C5	C5	S	C	C	C	C							
L8515 125mm f/2		C	C	C	C	C	C	C	C	C	C5	C5	S	C	C	C	C	C						
TELE-XENARS																								
L8516 90mm f/3.5		C	C		C	C	C	C	C	C	C5	C5	C	C	C	C								
L8517 135mm f/3.5	C	C	C	C	C	C	C	C	C	C	C5	C5	C	C	C	C								
L8518 200mm f/5.5	C	C	C	C	C	C	C	C	C	C	C5	C5	C	C	C	C								
L8519 360mm f/5.5	C	C	C	C	C	C	C	C	C	C	C5	C5	C	C	C	C	C	C	C		C	C	C	C
XENOTARS																								
L8520 80mm f/2.8	C	C	C	C	C	C	C	C	C	C	C5	C5	C	C	C	C	C	C	C		S	C	C	C
L8521 105mm f/2.8	C	C	C	C	C	C	C	C	C	C	C5	C5	C	C	C	C	C	C	C		C	C	C	C
L8522 135mm f/2.	C	C	C	C	C	C	C	C	C	C	C5	C5	C	C	C	C	C	C	C		C	C	C	C

Fig. L-4*i* (*continued*).

XENOTARS (continued)	A 16MM Arriflex	B 16MM Bolex Reflex	C 16MM Standard "C"	D 16/35MM Camerette	E 16MM Eastman Bayonet "S"	F 16MM Eastman H.S. & Magnifax	G 16MM Eastman Reflex "R"	H 16MM Fairchild H.S.	I 16MM Fastair or Photo-Sonic	J 16/35MM Fastax	K 16MM G.S.A.P. (N-9, KB-3)	L 16MM Mitchell	M 35MM Arriflex	N 35MM Bell & Howell #2709	O 35MM Eyemo "C"	P 35MM Mitchell Std., H.S., & N.C.	Q 35MM Mitchell BNC	R 35MM Mitchell Reflex	S 70MM Flight Research	T 70MM Hulcher 100 & 101	U 70MM Hulcher 102	V 70MM Hulcher 103	W 70MM Mitchell
L8523 150mm f/2.8	C	C	C	C	C	C	C	C	C	C	C5	C5	C	C	C	C	C	C	C	C	C	C	C
TELE-ARTONS																							
L8524 180mm f/5.5	C	C	C	C	C	C	C	C	C	C	C5	C5	C	C	C	C	C	C	C		C	C	C
L8525 240mm f/5.5	C	C	C	C	C	C	C	C	C	C	C5	C5	C	C	C	C	C	C	C	C	C	C	C
L8526 270mm f/5.5	C	C	C	C	C	C	C	C	C	C	C5	C5	C	C	C	C	C	C	C	C	C	C	C
Taylor-Hobson																							
SPEED PANCHROS																							
L9001 18mm f/1.7 (T2)	S										C5	C5	S	S	S	S	S						
L9002 25mm f/1.8 (T2.2)	S										C5	C5	S	S	S	S	S						
L9003 32mm f/2 (T2.3)	S		C	C							C5	C5	S	S	S	S	S	C					
L9004 40mm f/2 (T2.3)	S		C	C							C5	C5	S	S	S	S	S	C					
L9005 50mm f/2 (T2.3)	S	C	C	C							C5	C5	S	S	S	S	S	C					
L9006 75mm f/2 (T2.3)	S	C	C	C			C	C	C	C	C5	C5	S	S	S	S	S	C					
L9007 100mm f/2.5 (T2.8	S	C	C	C	C	C	C	C	C	C	C5	C5	S	S	S	S	S	C					
KINETALS																							
L9008 9mm f/1.8 (T2)	S		C								C5	C5											
L9009 12.5mm f/1.8 (T2)	S		C								C5	C5											
L9010 17.5mm f/1.8 (T2)	S		C								C5	C5											
L9011 25mm f/1.8 (T2)	S		C								C5	C5											
L9012 37.5mm f/1.8 (T2)	S		C								C5	C5											
L9013 50mm f/1.8 (T2)	S	C	C								C5	C5											
L9014 75mm f/2.6 (T2.8)	S	C	C								C5	C5											
L9015 100mm f/2.6(T2.8	S	C	C		C	C	C	C	C	C	C5	C5											
L9016 150mm f/3.8 (T4)	S	C	C		C	C	C	C	C	C	C5	C5											
Wollensak																							
L9500 5.3mm f/1.5			S1								C1	C1											
CINE RAPTARS																							
L9501 13mm f/1.5			S								C5	C5											
L9502 13mm f/1.2			S								C5	C5											
L9503 17mm f/2.7			S								C5	C5											
L9504 25mm f/1.4			S								C5	C5											
L9505 25mm f/1.5			S								C5	C5											
L9506 25mm f/1.9			S								C5	C5											
L9507 40mm f/1.5			S								C5	C5											
L9508 50mm f/1.5		S	S								C5	C5											
L9509 50mm f/3.5		S	S								C5	C5											
L9510 63mm f/2.5		S	S								C5	C5											
L9511 75mm f/1.9		S	S								C5	C5											
L9512 75mm f/2.5		S	S								C5	C5											
L9513 75mm f/2.8		S	S								C5	C5											

The charts may be utilized in two ways. (1) Select the camera at the top of the chart. Immediately below, in the vertical column, find which lenses are available. (2) Select the required lens in the left-hand column, and read across to find whether it is available in the particular mount for the camera selected for filming. An S designates that the lens is available in a stock mount supplied by the manufacturer. A C designates that the lens may be custom-mounted to function with that particular mount and camera.

CINE RAPTARS (cont.)	16MM Arriflex A	16MM Bolex Reflex B	16MM Standard "C" C	16/35MM Camerette D	16MM Eastman Bayonet "S" E	16MM Eastman H.S. & Magnifax F	16MM Eastman Reflex "R" G	16MM Fairchild H.S. H	16MM Fastair or Photo-Sonic I	16/35MM Fastax J	16MM G.S.A.P. (N-9, KB-3) K	16MM Mitchell L	35MM Arriflex M	35MM Bell & Howell #2709 N	35MM Eyemo "C" O	35MM Mitchell Std., H.S., & N.C. P	35MM Mitchell BNC Q	35MM Mitchell Reflex R	70MM Flight Research S	70MM Hulcher 100 & 101 T	70MM Hulcher 102 U	70MM Hulcher 103 V	70MM Mitchell W	70MM Photo-Sonic X
L9514 75mm f/4		S	S								C5	C5												
L9515 101mm f/2.5		S	S						S	S	C5	C5												
L9516 101mm f/4.5		S	S								C5	C5												
L9517 152mm f/4.5		S	S						S	S	C5	C5												
L9518 152mm f/3.8		S	S								C5	C5												
PRO RAPTARS																								
L9521 25mm f/2.3										S2				S	S	S	S							
L9522 35mm f/2.3		C	C							S2	C5	C5	C	S	S	S	S							
L9523 50mm f/2.3		C	C	C						S	C5	C5	C	S	S	S	S							
L9524 75mm f/2.3	C	C	C	C	C		C			S	C5	C5	C	S	S	S	S							
L9525 100mm f/2.3	C	C	C	C	C	C	C	C	C	C	C5	C5	C	S	S	S	S							
L9526 152mm f/2.7	C	C	C	C	C	C	C	C	C	C	C5	C5	C	S	S	S	S							
RAPTARS																								
L9530 90mm f/4.5	C	C	C	C	C	C	C	C	C	C	C	C	C	C	C	C	C	C	C	C	S	C		
L9531 101mm f/4.5	C	C	C	C	C	C	C	C	C	C	C	C	C	C	C	C	C	C	C	C	S	C		
L9532 127mm f/4.5	C	C	C	C	C	C	C	C	C	C	C	C	C	C	C	C	C	C	C	C	S	C		
TELEPHOTO RAPTARS																								
L9540 254mm f/5.6	C	C	C	C	C	C	C	C	C	C	C	C	C	C	C	C	C	C	C	C	S	C		
L9541 380mm f/5.6	C	C	C	C	C	C	C	C	C	C	C	C	C	C	C	C	C	C	C	C	S	C		
Carl Zeiss																								
PLANARS																								
L9601 32mm f/2	S		C								C5	C5	S											
L9602 50mm f/1.5	S	C	C								C5	C5	S											
SONNARS																								
D9603 85mm f/2	S	C	C	C							C5	C5	S	C		C								
L9604 135mm f/4	C	C	C	C	C	C	C	C	C	C	C5	C5	C	C	C	C	C							
L9605 180mm f/2.8	C	C	C	C	C	C	C	C	C	C	C5	C5	C	C	C	C	C	C						
L9606 250mm f/5.6	C	C	C	C	C	C	C	C	C	C	C5	C5	C	C	C	C	C	C						
L9607 300mm f/4	C	C	C	C	C	C	C	C	C	C	C5	C5	C	C	C	C	C	C						
Wollensak																								
MIRROTELS																								
L9527 20 inch f/6.3	C	C	C	C	C	C	C	C	C	C	C5	C5	C	S	S	S	S	C						
L9528 40 inch f/8	C	C	C	C	C	C	C	C	C	C	C5	C5	C	C	C	C	C	C	C	C	C	C	C	
L9529 80 inch f/14	C	C	C	C	C	C	C	C	C	C	C5	C5	C	C	C	C	C	C	C	C	C	C	C	
Zoomar																								
REFLECTARS																								
L9701 20 inch f/5.6	C	C	C	C	C	C		C	C	C	C5	C5	C	C	C	C	C	C		C	C	C	C	C
L9702 25 inch f/6.3	C	C	C	C	C	C	C	C	C	C	C5	C5	C	C	C	C	C			C	S	C	C	C
L9703 40 inch f/8	C	C	C	C	C	C	C	C	C	C	C5	C5	C	C	C	C	C	C	C					C
L9704 80 inch f/15	C	C	C	C	C	C	C	C	C	C	C5	C5	C	C	C	C	C	C	C	C	C	C	C	C

Fig. L-4*i* (*continued*).

REFLECTARS (continued)	16MM Arriflex A	16MM Bolex Reflex B	16MM Standard "C" C	16/35MM Camerette D	16MM Eastman Bayonet "S" E	16MM Eastman H.S. & Magnifax F	16MM Eastman Reflex "R" G	16MM Fairchild H.S. H	16MM Fastair or Photo-Sonic I	16/35MM Fastax J	16MM G.S.A.P. (N-9, KB-3) K	16MM Mitchell L	35MM Arriflex M	35MM Bell & Howell #2709 N	35MM Eyemo "C" O	35MM Mitchell Std., H.S., & N.C. P	35MM Mitchell BNC Q	35MM Mitchell Reflex R	70MM Flight Research S	70MM Hulcher 100 & 101 T	70MM Hulcher 102 U	70MM Hulcher 103 V	70MM Mitchell W	70MM Photo-Sonic X
L9705 100 inch f/20	C	C	C	C	C	C	C	C	C	C	C5	C5	C	C	C	C	C	C	C	C	C	C	C	C
L9706 150 inch f/25	C	C	C	C	C	C	C	C	C	C	C5	C5	C	C	C	C	C	C	C	C	C	C	C	C
ZOOM LENSES																								
ANGENIEUX																								
L9799 12 to 120mm f/2.2	S	S	S								C5	C5												
L9800 17 to 68mm f/2.2		S	S								C5	C5												
L9801 17.5 to 70mm f/2.2	S	S	S								C5	C5												
L9802 20 to 80mm f/2.5		S	S								C5	C5												
L8903 35 to 140mm f/3.5	S	C	C	C	C	C	C				C5	C5	S	C	C	C	C	C						
L9804 35 to 140mm f/2.2	C	C	C	C	C	C	C				C5	C5												
Berthiot																								
PAN CINORS																								
L9805 17.5 to 70mm f/2.4	S	S	S	C3	S						C5	C5												
L9806 17.5 to 85mm f/2	S	S	S	C3	S						C5	C5												
L9807 25 to 100mm f/3.5	S	S	S	C3	S		C				C5	C5												
L9808 38 to 150mm f/3.8	S	C	C	C3	C		C				C5	C5	S	C	S	S	S	C						
L9809 60 to 240mm f/5.9	S	C	C	C3	C		C				C5	C5	S	C	S	S	S	C						
CANON																								
L9810 25 to 100mm f/1.8	S	S	S		C						C5	C5												
Elgeet																								
ZOOM NAVITAR																								
L9811 20-80mm f/1.8	C	S	S								C5	C5												
NIKKOR ZOOM																								
L9812 25 to 85mm f/2.8					C	C					C5	C5												
L9813 85 to 250mm f/4	C	C	C		C	C					C5	C5	C	C	C	C								
L9814 200 to 600mm f/9.5	C	C	C		C	C					C5	C5	C	C	C	C								
Schneider																								
VARIOGON																								
L9815 20 to 80mm f/2	S	S	S								C5	C5												
L9816 30 to 120mm f/2.8	S	S	S								C5	C5												
Wollensak																								
VARI-FOCUS RAPTAR																								
L9817 20 to 60mm f/1.8		S	S								C5	C5												
Zoomar																								
TV VIDICON ZOOM																								
L9822 17 to 70mm f/2		S	S								C5	C5												
L9823 25 to 150mm f/3.5		S	S								C5	C5												
L9824 30 to 125mm f/2.8											C5	C5												
L9825 40 to 240mm f/5.6		S	S								C5	C5												

lens adapter A device attached to the front of a camera that makes it possible for the cinematographer to interchange lenses quickly and easily.

lens barrel The housing in which the lens elements are mounted. See LENS MOUNT.

lens cap A covering that protects the lens when it is not in use.

lens coatings Magnesium fluoride coatings in lenses. A part of the light passing through a lens is reflected on each of the element surfaces. The light bounces back and forth, within the lens, and finally emerges to expose the film in the camera.

Normally, some loss of light, definition, and contrast, as well as "flare" around light objects, can be anticipated. Lens coatings, however, neutralize the reflected light. The different colors of the coatings vary with the index of refraction and the transmission characteristics of the glass used in the manufacture of the lens. Any light lost during transmission, because of the colored coatings, can be regarded as negligible.

lens hood An attachment that shields the camera lens and prevents undesirable light from striking its surface. See MATTE BOX.

lens mount The metal housing in which each of the component glasses making up the lens are positioned.

In *between-lens* shutter mounts, the lens and shutter blades can be regarded as a composite unit since the shutter blades and leaves of the iris diaphragm are positioned between the front and rear lens elements. The shutter in these lens mounts is especially practical for longer exposures since it can be accurately synchronized with flash lamps.

Fig. L-5 **Super Baltar lens mount.**

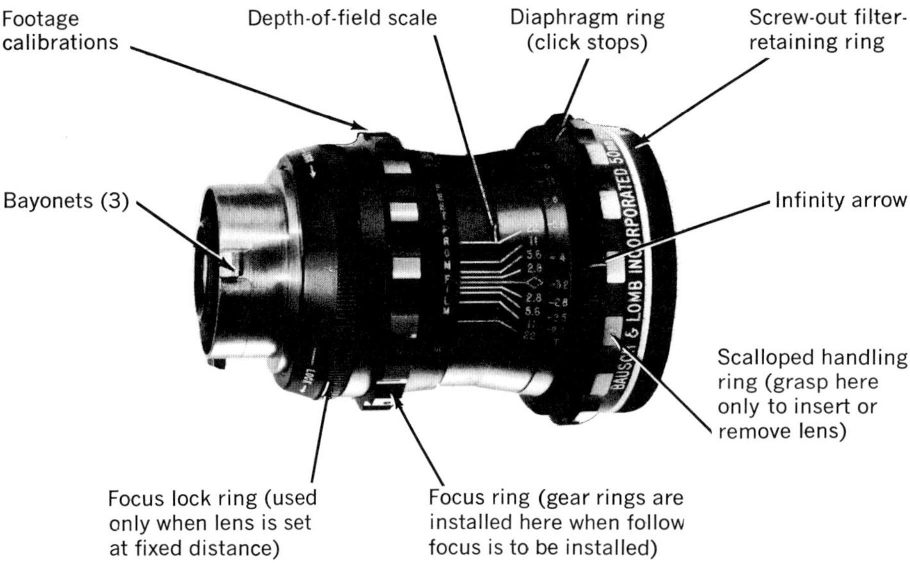

A *focusing* mount, as the name implies, contains the focusing mechanism. In these mounts, the lens may be turned or slid into the mount in order to make adjustments in focus.

A *micrometer* lens mount has a fine thread adjustment that permits critical focusing.

Spiral lens mounts have a coarse screw thread adjustment for rapid focusing.

In *standard* mounts, the lens is screwed into the threaded flange near the back of the lens barrel. The major portion of the lens, in this type of mount, projects from the front panel.

Sunk mounts, by comparison, have the flange thread close to the front of the lens barrel so that only a small portion of the lens is projected from the front panel. See Figure L-5, BAYONET LOCK, and C MOUNTS.

lens sizes See Tables L-1 and L-2.

lens speed The larger the aperture of a lens, the faster the speed of the lens. Lenses that work at *f*/2 or faster are known as *high-aperture lenses*. The maximum so far developed is a Zeiss product with a speed of *f*/0.85. High-aperture lenses are capable of filming images under difficult lighting conditions and at relatively high shutter speeds.

TABLE L-1 Lens Sizes*

Millimeters	Inches	Millimeters	Inches
12½	½	90	3⅗
15	⅗	95	3⅘
20	⅘	100	4
25	1	125	5
28	1⅛	150	6
30	1⅕	175	7
32	1¼	200	8
35	1⅖	225	9
38	1½	250	10
40	1⅗	275	11
45	1⅘	300	12
50	2	325	13
55	2⅕	350	14
60	2⅖	375	15
65	2⅗	400	16
70	2⅘	425	17
75	3	450	18
80	3⅕	475	19
85	3⅖	500	20

*The lens sizes stated in inches are approximate. A 100mm lens, for example, is actually 3.937 inches, and not 4 inches as stated.

TABLE L-2 Horizontal and Vertical Angular Coverage of Different-sized Lenses and the Aspect Ratios Obtainable with 35mm Film

35mm frame size, inches	Aspect ratio	Lens sizes: 20mm $\frac{4}{5}$ in.	25mm 1 in.	35mm $1\frac{2}{5}$ in.	50mm 2 in.	75mm 3 in.	100mm 4 in.	150mm 6 in.
0.980 × 0.735	Full-							
Horizontal angle	screen	63.8	52.9	39.2	27.9	18.8	14.1	9.4
Vertical angle	aperture	50.0	41.0	29.8	21.1	14.2	10.6	7.1
0.868 × 0.631	Academy							
Horizontal angle	aperture	57.7	47.5	34.9	24.8	16.7	12.5	8.4
Vertical angle	1.33:1	43.6	35.5	25.7	18.2	12.1	9.1	6.1
0.825 × 0.600	Wide							
Horizontal angle	screen	55.3	45.5	33.3	23.7	15.9	12.0	8.0
Vertical angle	1.33:1	41.7	33.9	24.5	17.3	11.6	8.7	5.8
0.825 × 0.497	Wide							
Horizontal angle	screen	55.3	45.5	33.3	23.7	15.9	12.0	8.0
Vertical angle	1.66:1	35.0	28.3	20.4	14.4	9.6	7.2	4.8
0.825 × 0.472	Wide							
Horizontal angle	screen	55.3	45.5	33.3	23.7	15.9	12.0	8.0
Vertical angle	1.75:1	33.3	26.9	19.4	13.6	9.1	6.8	4.5
0.825 × 0.446	Wide							
Horizontal angle	screen	55.3	45.5	33.3	23.7	15.9	12.0	8.0
Vertical angle	1.85:1	31.7	25.5	18.4	12.9	8.6	6.5	4.3

lens spotlight A lighting fixture embodying a lamp and a simple lens (plano-convex or biconvex), with or without a reflector, which has variable field and beam angles obtained by changing the spacing between the lamp and the lens.

lens squeeze ratio The degree of horizontal squeeze or compression of a film image produced by a lens used for anamorphic photography. Many of the wide-screen "scope" pictures have a 2:1 squeeze ratio. This would indicate that an object in the original negative would be only half as wide, in the horizontal meridian, as it would be if photographed with a normal lens. See ANAMORPHIC LENS.

lens-stop computer See Table L-3.

lens symbols and formulas See Figure L-6.

O = size of object
D = distance between object and camera lens
F = focal length of lens
A = size of aperture

$$D = \frac{O \times F}{A} \qquad F = \frac{D \times A}{O}$$

$$O = \frac{D \times A}{F} \qquad A = \frac{F \times O}{D}$$

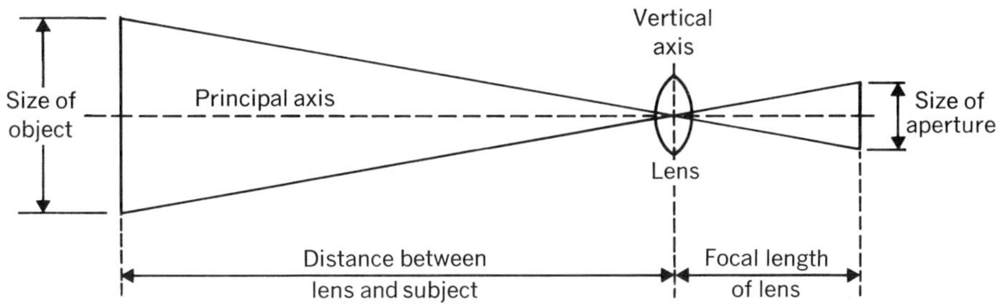

Fig. L-6 **Lens symbols and formulas.**

TABLE L-3 Lens-stop Computer

f-stop setting*	1/4 *f*-stop closed	1/2 *f*-stop closed	3/4 *f*-stop closed	1 *f*-stop closed	*f*-stop setting*	1/4 *f*-stop closed	1/2 *f*-stop closed	3/4 *f*-stop closed	1 *f*-stop closed
1.2	1.3	1.4	1.5	1.6	8.5	9.1	10.0	11.0	12.0
1.3	1.4	1.5	1.6	1.8	9.1	10.0	11.0	12.0	12.5
1.4	1.5	1.6	1.8	2.0	10.0	11.0	12.0	12.5	14.0
1.5	1.6	1.8	2.0	2.1	11.0	12.0	12.5	14.0	16.0
1.6	1.8	2.0	2.1	2.3	12.0	12.5	14.0	16.0	17.0
1.8	2.0	2.1	2.3	2.5	12.5	14.0	16.0	17.0	18.0
2.0	2.1	2.3	2.5	2.8	14.0	16.0	17.0	18.0	20.0
2.1	2.3	2.5	2.8	3.0	16.0	17.0	18.0	20.0	22.0
2.3	2.5	2.8	3.0	3.2	17.0	18.0	20.0	22.0	24.0
2.5	2.8	3.0	3.2	3.6	18.0	20.0	22.0	24.0	25.0
2.8	3.0	3.2	3.6	4.0	20.0	22.0	24.0	25.0	28.0
3.0	3.2	3.6	4.0	4.2	22.0	24.0	25.0	28.0	32.0
3.2	3.6	4.0	4.2	4.5	24.0	25.0	28.0	32.0	34.0
3.6	4.0	4.2	4.5	5.0	25.0	28.0	32.0	34.0	36.0
4.0	4.2	4.5	5.0	5.6	28.0	32.0	34.0	36.0	40.0
4.2	4.5	5.0	5.6	6.0	32.0	34.0	36.0	40.0	45.0
4.5	5.0	5.6	6.0	6.3	34.0	36.0	40.0	45.0	50.0
5.0	5.6	6.0	6.3	7.2	36.0	40.0	45.0	50.0	
5.6	6.0	6.3	7.2	8.0	40.0	45.0	50.0		
6.0	6.3	7.2	8.0	8.5	45.0	50.0			
6.3	7.2	8.0	8.5	9.1	50.0				
7.2	8.0	8.5	9.1	10.0					
8.0	8.5	9.1	10.0	11.0					

To find comparable settings for fractional lens openings, read from right to left on the same line.

*Each *f*-number indicates the focal length divided by the diameter of the diaphragm. Each of the underlined numbers provides twice as much exposure as the next underlined number (stop). *f*/5.6, for example, would give twice as much exposure as *f*/8; *f*/4 would give twice as much exposure as *f*/5.6; etc.

lens turret The rotating lens turret is mounted on the front plate of the camera base and usually accommodates four lenses. By rotating the turret, each lens can be easily and quickly shifted to position before the camera aperture. A locating pin locks the turret when the lens is correctly placed. The locating-pin assembly can be adjusted so that the lens will be centered exactly over the camera aperture (when the zero marks are aligned), or positioned so that the lens will be centered above or below the center of the camera aperture. The latter positioning produces the effect of a rising and falling front to produce an adjustment equivalent to a 15-degree tilt of the tripod without the usual distortion due to photographing at an angle. Each lens is secured in the turret apertures by several screws.

lenticular Pertaining to lenses, or lens-shaped. More specifically, anything having the form of a double-convex lens. See LENTICULAR FILM and LENTICULAR SCREEN.

lenticular film A panchromatic emulsion on a textured film base. The surface of the film base has vertical lenticular ribs or ridges, forming a series of cylindrical lenses. Exposed through the base, the right and left sides of the image are on different sides of each ridge. When projected, each part of the image is seen on a corresponding surface of the lenticular-type screen that is used for stereoscopic projection.

Used basically for stereo effects, a filter with bands of the three primary colors is positioned in front of the lens during both the photographic and projection processes so that each of the cylindrical lenses in the film's base focuses each of the colors individually. See Figure L-7 and CYLINDRICAL LENS.

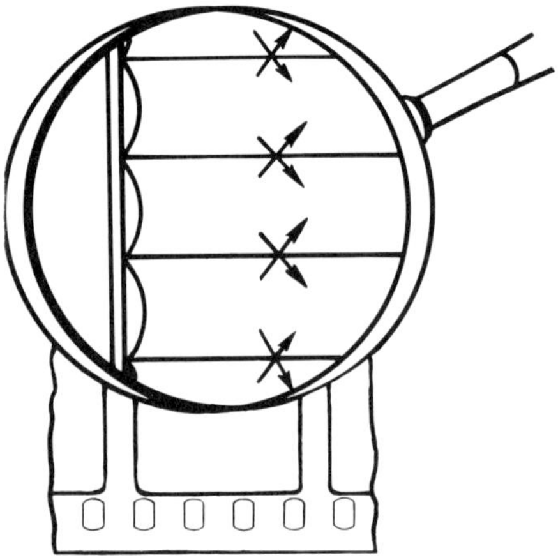

Fig. L-7 Lenticular film.

lenticular screen A matte white screen with heavy embossing for maximum diffusion. Used mainly for stereoscopic projection, a lenticular screen provides a relatively small viewing angle. The composite picture is seen as a series of interlaced images, the right and left images being focused side by side on the screen. This follows the pattern in the positive film base on which the image is recorded. The surface of the film base has vertical lenticular ribs or ridges, forming a series of cylindrical lenses, which are capable of projecting the right and left images to their corresponding screen positions. The viewer does not have to wear polarized spectacles with this type of screen in order to get the full stereo effect. See LENTICULAR FILM and CYLINDRICAL LENS.

level, camera The spirit level indicates to the cameraman, while setting up the equipment, that the camera is properly positioned with respect to the horizontal plane. It is usually located on the rear of the camera base.

level, recording The amplitude of the electronic signal; it is used for determining balance during the recording process.

library footage Also called **stock shot**. Scenes or lengths of film, photographed for use in previous motion pictures, that are used as inserts by the film editor in subsequent productions.

light A form of electromagnetic, radiant energy. Light travels in a straight line at a speed of approximately 186,000 miles per second in air. The speed varies, depending on the density of the medium through which it is traveling.

Light can be reflected and refracted. Its rays can be dispersed in all directions or polarized and confined to a single plane. Within the range of wavelengths between 400 and 700 millimicrons, light rays are visible. On either side of this range are the ultraviolet and infrared portions of the spectrum which, although they can affect photographic emulsions, are invisible to the human eye.

light, artificial Any light reaching us from the sun, moon, or stars, whether directly or by reflection, is referred to as natural light. Light received from the moon is reflected light (from the sun). Electricity and other forms of man-made lighting, are classified as artificial.

Film emulsions are more sensitive to natural light (daylight, for example) than they are to most light sources used in photography generally. Because of this, the speed ratings for a film are usually stated separately for daylight and artificial light.

light, diffused Lighting that is softened by the placement of different types of material in front of the light source. Diffusion tends to minimize or eliminate shadows.

light, hard The opposite of diffused light. Hard light is intense light which produces dark, sharply defined shadows.

light, polarized Unlike ordinary light which emanates from its source in all directions, polarized light is confined to a single plane.

light, soft Subdued or diffused light that is used for minimizing shadows.

light-balancing filters Filters positioned in front of a camera lens and used to cool or warm specific colors. Blue filters, for example, are used for cooling, and yellow filters for warming.

light distribution The degree of uniformity or consistency of illumination falling on a specific area.

Many of the data sheets contain graphs showing light-distribution curves. A distribution curve is obtained by placing the fixture at what is considered a commonly used distance, and taking footcandle readings, at uniform intervals, along a flat wall. The intensity changes shown on the resultant curve indicate the rate of falloff for the fixture (whether the light from the fixture is a very flat uniform distribution, for example, a gradually tapering distribution, or a very sharp peaked distribution). Any of these distributions are quite useful for particular lighting setups. For example, a very peaked distribution would be easy to use when only a particular subject is to be illuminated, without illuminating the surrounding background. A very slow tapering distribution lends itself for use for base floodlighting.

cable power-distribution data. How to use Exhibits A, B, C, and D:

1. Know the total amount of power at your supply point.
2. Be sure your supply has ample fusing.
3. Compute the amperage load (Exhibit A) for the maximum units of lighting equipment you expect to use, and be sure it does not exceed your total supply. (The quantity and variety of lights owned should be sufficient to provide flexibility in all photographic situations, and since all will not be in use at one time, only the maximum number of lights to be used at one time will form the basis for computing amperage load.)
4. Relate your available power to the current-carrying capacities in Exhibit B, and decide on adequate cables.
5. Determine whether you have a two- or three-wire system, and use either Exhibit C or D as a guide in laying out your distributing system.
6. Select components and accessories.

EXHIBIT A Amperage Consumption*

Incandescent lamps at 120 *volts*

Globe wattage	*Globe amperage*
200	1.7
500	4.2
750	6.3
1,000	8.3
1,500	12.5
2,000	16.7
5,000	41.6
10,000	83.3

Arc lamps at 115–120 *volts*

Type number	*Amperage*
40	40
90	120
170	150
450	225

*Courtesy of Mole-Richardson.

EXHIBIT B Amperage Carrying Capacities of Cable

Cable		Ampere ratings			
		Per conductor		Total load	
AWG size of conductors	Number of conductors	Continuous	Intermittent*	Continuous	Intermittent*
8	2	45	60	45	60
6	2	65	85	65	85
4	2	85	115	85	115
4	3	85	115	170†	230†
2	2	115	155	115	155
2	3	115	155	230†	310†
4/0	1	230	310	230‡	310‡

*Ampere ratings usually found to be satisfactory for intermittent service normally encountered in motion picture studios.

†Maximum load amperes for balanced condition on a three-wire system.

‡Additional capacity may be secured by paralleling as many feeders as are necessary to carry the load.

EXHIBIT C Typical Three-wire 240/120-volt Distribution System

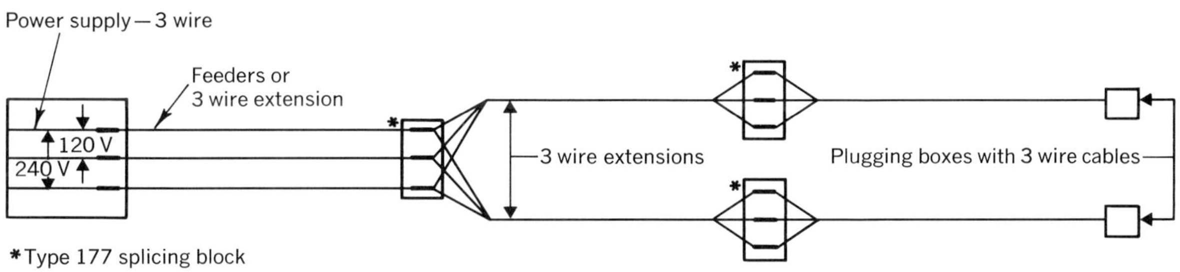

EXHIBIT D Typical Two-wire 120-volt Distribution System

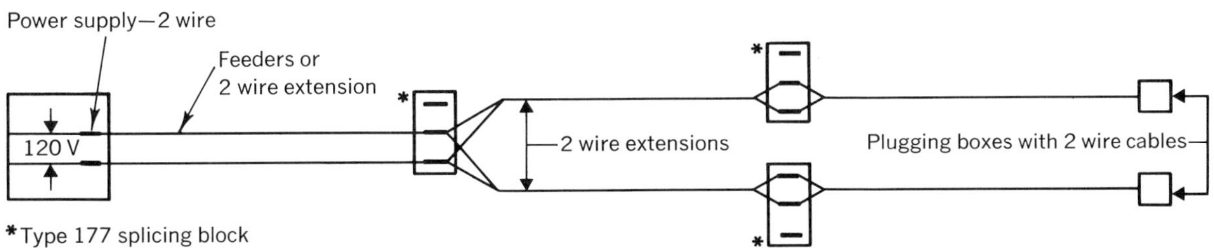

lighting If the natural light sources that normally illuminate an area were replaced by artificial light that exactly duplicated the original highlight and shadow patterns, the lighting director could look with justifiable pride on the achievement. This, briefly, is the basis for successful motion picture lighting for both interior and exterior locations.

Like little children who should be seen rather than heard, motion picture lighting setups should not take the spotlight from the on-stage performers. If the studio lighting competes with the subject matter, or attracts attention to itself, it has failed to accomplish its assigned task. Obviously, then, the broad rule would dictate the use of artificial lighting arrangements that simulate the light pattern that is most natural for a specific setting. The same ruling applies to mood and special-effects lighting that exaggerates normal light patterns. Oddly enough, even under strong sunlight in outdoor photography, natural light must be augmented by artificial lighting sources. Reflectors, in location photography, direct the natural light to those places where it is needed.

Basically, studio lighting falls into three distinct categories. Each grouping has a specific purpose and must be well planned in relation to the other sources of light. The three groupings, when blended together, should provide a composite effect that is both pleasing and believable.

The key light establishes the overall lighting pattern and is used for obtaining exposure settings. The intensity of the key light is governed by the size of the set and should remain constant during the entire photographic process. In keeping with the size of their assignment, the lighting sources are aptly termed *brutes*.

Unless shadows are needed for establishing a mood or creating an effect, they are distracting and, at times, objectionable. The *fill,* or secondary light source, reduces the shadows cast by performers to acceptable proportions. Scoops, cone lights, or DuArcs, and diffusers where necessary, are best for this purpose. This supplemental lighting source should not be strong enough to influence or affect the exposure reading taken from the key light.

The third grouping, the *kicker* and *backlight,* add depth and dimension to the overall scene. Easily controlled spotlights are best for this purpose and provide the desirable separation between the subject matter and the background. Greater scene depth can also be achieved by allowing the background to become soft or go out of focus. Positioning the subject matter closer to the camera will also add depth to a scene.

While professional lighting directors use established patterns as a guide for the placement of lights, the illumination for each set varies and requires individual planning. Key lights are usually positioned away from the camera and to the side of the subject matter. The keys, kickers, and background lights, wherever possible, should be suspended in order to facilitate the movement of cameras and sound equipment. Fill lights, however, should be positioned at camera level, and the backlights should be above and behind the subject matter or performer.

Unusual lighting arrangements, for achieving special effects, should be given particular consideration. Established patterns are usually discarded when a desired mood or atmosphere is needed to create an immediate impression. Shadows, usually objectionable in normal lighting setups, may be quite tolerable in such scenes.

Generally, the key light should furnish approximately twice as much illumination as the fill light. Stark whites should be toned down or avoided. Similarly, black props or costumes have a tendency to lose detail, and their use should be discouraged wherever possible. Obviously, the color of the background should not be similar to the color of the clothing worn by the performers. Uniform density in large areas should be broken up with soft-focus "cookie" patterns, or textured effects. Finally, lighting should be planned so that scenes, when edited in their correct viewing order, are balanced and do not alternate between extremes of light and dark, high key and low key. The latter effect, unless it is used for a specific reason, can be quite disturbing to the viewer.

Although the general information in these paragraphs may be used as a guide for interior and location photography, lighting should be regarded as a creative effort. As such, its relative importance in the overall production pattern entitles it to all the privileges and variations enjoyed by other processes. See Figure L-8 and Table L-4; also COVERAGE, FILL LIGHT, KEY LIGHT, and LIGHTING CONTRAST.

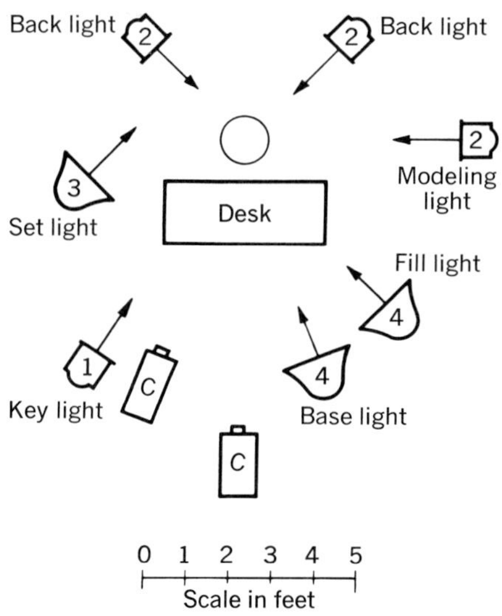

Fig. L-8 Typical TV news set lighting layout. (*Sylvania Electric Products Inc.*) (*C*) Camera. (1) 2,000-watt Fresnel. (2) 1,000-watt Fresnel. (3) 1,500-watt Scoop. (4) 1,000-watt Scoop.

TABLE L-4 Professional Photo Lamps Ordering and Specification Guide

ASA Code	Suggested list price	Carton quantity	Case quantity	(10) Voltage	Bulb	Watts	Base	Special description	Filament	(1) Maximum life, hours	Average light center length, inches	Maximum overall length, inches	Approximate initial lumens, candlepower	Approximate mean color temperature, °K	Notes
Photoflood spot and reflector															
BBA	$0.31	6	60	115–120	A-21	250	Medium	Inside frost	C-9	3	. . .	$4\frac{15}{16}$	8,650	3400	
EBV	0.55	6	24	115–120	PS-25	500	Medium	Inside frost	C-9	6	. . .	$6\frac{15}{16}$	17,000	3400	
DXR	1.60	1	24	115–120	PS-35	1,000	Mogul	Inside frost	C-7A	10	. . .	$9\frac{3}{8}$	33,500	3400	
BCA	0.45	6	60	115–120	A-21	250	Medium	Inside frost	C-9	3	. . .	$4\frac{15}{16}$	5,600	4800	
EBW	0.85	6	24	115–120	PS-25	500	Medium	Inside frost	C-9	6	. . .	$6\frac{15}{16}$	11,000	4800	
BEP	0.93	1	24	115–120	R-30	300	Medium	Inside frost	C-9	4	. . .	$5\frac{3}{16}$	12,000	3400	2
EBR	1.15	1	24	115–120	R-30	357	Medium	Inside frost	C-9	4	. . .	$5\frac{3}{16}$	13,000	3400	2
BFA	1.25	1	12	115–120	R-40	375	Medium	Inside frost	C-9	4	. . .	$6\frac{1}{2}$	13,000	3400	2
DXC	1.75	1	12	115–120	R-40	500	Medium	Inside frost	CC-2V	6	. . .	$6\frac{1}{2}$	4,500	3400	2
DXB	2.40	1	12	115–120	R-40	500	Medium	Inside frost	CC-2V	6	. . .	$6\frac{1}{2}$	50,000	3400	2
DXY	4.95	1	8	115–120	PAR-36★	650	Screw terminal	Medium beam	CC-6	8	. . .	$2\frac{3}{4}$	25,000	3400	
DWA	4.95	1	8	115–120	PAR-36★	650	Ferrule contact	Medium beam	CC-6	8	. . .	$2\frac{7}{16}$	25,000	3400	
FDL	4.95	1	8	115–120	PAR-36★	650	Two prong	Medium beam	CC-6	8	. . .	$2\frac{3}{4}$	25,000	3400	

Studio lighting															
BFE	$10.75	1	12	120	T-20★	750	Medium prefocus	. . .	C-13	200	$2\frac{3}{16}$	$5\frac{3}{4}$	16,950	3000	3
BFK	8.50	1	12	120	T-20★	750	Medium prefocus	Blue top	C-13	200	$2\frac{3}{16}$	$5\frac{3}{4}$	16,950	3000	3
DSN	9.50	1	12	120	T-20★	750	Medium bipost	. . .	C-13	200	$2\frac{1}{2}$	$6\frac{1}{2}$	16,950	3000	3
EAG	4.20	1	6	120	PS-52	1,000	Mogul	White	C-7A	1,000	$9\frac{1}{2}$	$13\frac{1}{16}$	21,500	2850	4
DKL	6.25	1	6	120	PS-52	1,500	Mogul	White	C-7A	1,000	$9\frac{1}{2}$	$13\frac{1}{16}$	32,100	2900	4
EDN	40.00	1	1	120	T-64★	5,000	Mogul bipost	. . .	C-13	75	$6\frac{1}{2}$	$13\frac{3}{8}$	165,000	3350	5
ECP	108.00	1	1	120	G-96★	10,000	Mogul bipost	. . .	C-13	75	10	$17\frac{3}{8}$	330,000	3350	5
3200° *Kelvin*															
ECA	$ 0.45	6	120	120	A-23	250	Medium	Inside frost	C-9	20	$4\frac{3}{8}$	$6\frac{1}{16}$	6,500	3200	4
ECT	0.80	6	60	120	PS-25	500	Medium	Inside frost	C-9	60	$5\frac{1}{4}$	$6\frac{15}{16}$	13,500	3200	4
EDB	5.50	1	12	120	T-20★	500	Medium prefocus	. . .	C-13	30	$2\frac{3}{16}$	$5\frac{3}{4}$	13,400	3200	
EDG	7.20	1	12	120	T-24★	750	Medium bipost	. . .	C-13	30	$2\frac{1}{2}$	$6\frac{1}{2}$	20,625	3200	
FAT	13.00	1	8	120	PAR-56★	1,000	Two prong	Widebeam 70° spread	CC-2V	100	. . .	5	13,500	3200	7, 8, 9
ECV	3.00	1	24	120	PS-40	1,000	Mogul	Inside frost	C-7A	60	7	$9\frac{3}{4}$	26,500	3200	4
ECX	4.60	1	6	120	PS-52	1,000	Mogul	. . .	C-7A	75	$9\frac{1}{2}$	$13\frac{1}{16}$	26,500	3200	4
ECY	5.75	1	6	120	PS-52	1,500	Mogul	. . .	C-7A	100	$9\frac{1}{2}$	$13\frac{1}{16}$	41,000	3200	4
DVF	15.00	1	6	120	G-48★	2,000	Mogul bipost	. . .	C-13	100	5	$9\frac{3}{8}$	57,500	3200	
ECN	39.00	1	1	120	G-64★	5,000	Mogul bipost	. . .	C-13	150	$6\frac{1}{2}$	$11\frac{7}{8}$	141,000	3200	5
DPK	48.00	1	1	120	T-64★	5,000	Mogul bipost	. . .	C-13	150	$6\frac{1}{2}$	$13\frac{3}{8}$	141,000	3200	
Ultra-high-speed photoflood															
DVP	3.15	1	24	115–120	R-30	300	Medium	Reflector	CC-2V	4	. . .	$5\frac{3}{16}$	. . .	3400	4, 10, 12
DXA	3.30	1	24	28	R-30	300	Medium	Reflector	CC-6	4	. . .	$5\frac{3}{16}$	. . .	3400	4, 12

TABLE L-4 Professional Photo Lamps Ordering and Specification Guide (cont.)

ASA Code	Suggested list price	Carton quantity	Case quantity	(10) Voltage	Bulb	Watts	Base	Special description	Filament	(1) Maximum life, hours	Average light center length, inches	Maximum overall length, inches	Approximate initial lumens (CP = candlepower)	Approximate mean color temperature, °K	Notes
								Darkroom							
BPG	$2.70	6	60	115–125	S-14	10	Medium	*Ruby	C-9	1,500	. . .	$3\frac{1}{2}$	. . .	. . .	4
BAS	2.90	6	60	115–125	A-19	25	Medium	*Ruby	C-9	1,000	. . .	$3\frac{15}{16}$	. . .	. . .	4
BBX	3.00	6	60	115–125	A-19	40	Medium	*Ruby	C-9	1,000	. . .	$4\frac{1}{4}$	. . .	. . .	4
BCJ	3.00	6	60	115–125	A-19	60	Medium	*Ruby	CC-6	1,000	. . .	$4\frac{7}{16}$	. . .	. . .	4
								Halogen-cycle photoflood and projection							
DWY	7.95	1	12	120	T-4	650	Recessed S.C.	Photoflood†	CC-8	25	$1\frac{1}{2}$	$3\frac{1}{16}$	20,000	3400	4
DWZ	9.55	1	12	30	T-4	375	Recessed S.C.	Bowl score†	CC-8	1,000	$1\frac{1}{2}$	$3\frac{1}{8}$	7,500	2900	4
FAL	7.95	1	12	120	T-4	420	Recessed S.C.	Projection†	CC-8	75	. .	$2\frac{5}{8}$	11,000	3200	4
DVY	5.95	1	12	120	GT-6	650	Two pin	Photoflood†	CC-6	25	$1\frac{11}{16}$	$2\frac{3}{8}$	20,000	3400	3
								Film viewer							
BKP	0.80	6	120	115–125	S-11	30	Candelabra	Clear	CC-2V	50	$1\frac{5}{8}$	$2\frac{1}{4}$	400	. . .	3
BKV	0.85	6	120	115–125	S-11	30	S.C. bayonet	Clear	CC-2V	50	$1\frac{3}{8}$	$2\frac{3}{8}$	400	. . .	3, 11
BLC	0.95	6	120	115–125	S-11	30	D.C. bayonet	Clear	CC-2V	50	$1\frac{3}{8}$	$2\frac{3}{8}$	400	. . .	3, 11
EEC	1.05	6	120	115–125	S-11	15	Candelabra	Inside frost	CC-2V	50	$1\frac{5}{8}$	$2\frac{1}{4}$	205	. . .	3

Enlarger

Lamp number & ordering abbreviation															
PH/111	1.20	6	120	115–125	S-11	75	S.C. bayonet	White	CC-2V	25	$1\frac{3}{8}$	$2\frac{3}{8}$	1,125	2900	3, 6, 10, 11
PH/111A	1.20	6	120	125	S-11	75	S.C. bayonet	White	CC-2V	25	$1\frac{3}{8}$	$2\frac{3}{8}$	1,125	2900	3, 6, 11

Sound reproducer

ASA Code						Amperes									
BFX	0.80	10	100	4	S-8	0.75	S.C. prefocus	Clear	C-6	50	$1\frac{1}{8}$	2	32	. . .	4
BGB	0.90	10	100	4	S-8	0.75	D.C. prefocus	Clear	C-6	50	$1\frac{1}{8}$	2	32	. . .	4
BRK	1.30	10	100	4	T-5	0.75	S.C. prefocus	Clear	C-8	50	$1\frac{1}{8}$	$2\frac{3}{8}$	32	. . .	4
BRD	1.05	10	100	4	T-5	0.75	S.C. bayonet	Clear	C-6	50	$1\frac{1}{4}$	2	30	. . .	4
BRS	1.15	10	100	4	T-5	0.75	S.C. prefocus	Clear	C-6	50	$1\frac{1}{8}$	2	30	. . .	4
BVK-BVS	1.40	6	24	5	T-8	6.5	S.C. prefocus	Clear	C-8	50	$1\frac{5}{8}$	$2\frac{5}{8}$	365	. . .	3
BSB	1.30	10	100	6	T-5	1.0	S.C. prefocus	Clear	C-8	15	$1\frac{1}{8}$	$2\frac{3}{8}$	90	. . .	4
BRX	1.25	10	100	6	T-5	1.0	D.C. prefocus	Clear	C-6	100	$1\frac{1}{8}$	2	70	. . .	4
BSK	1.15	10	100	6	T-5	1.0	S.C. prefocus	Clear	C-6	100	$1\frac{1}{8}$	2	70	. . .	4
BSS	1.30	10	100	6	T-5	1.0	S.C. prefocus	Clear	C-8	100	$1\frac{1}{8}$	$2\frac{3}{8}$	70	. . .	4
BSW	1.35	10	100	7	T-5	.20	S.C. prefocus	Clear	C-8	50	$1\frac{1}{8}$	$2\frac{3}{8}$	15	. . .	4
BTD	1.20	10	100	7	T-5	.20	D.C. prefocus	Clear	C-6	50	$1\frac{1}{8}$	2	15	. . .	4
BXC	1.35	6	24	8.5	T-8	4.0	S.C. bayonet	Clear	C-6	100	$1\frac{3}{4}$	$3\frac{1}{8}$	. . .	. . .	
BXM	1.55	6	24	9	T-8	4.0	S.C. prefocus	Clear	C-6	500	$1\frac{15}{32}$	$3\frac{1}{8}$	. . .	. . .	
BXR	1.35	6	24	10	T-8	5.0	S.C. bayonet	Clear	C-6	100	$1\frac{19}{32}$	$3\frac{1}{8}$	1,000	. . .	
BXN	1.45	6	24	10	T-8	5.0	S.C. prefocus	Clear	C-6	100	$1\frac{15}{32}$	$3\frac{1}{8}$	1,000	. . .	
BXE	1.55	6	24	10	T-8	7.5	S.C. bayonet	Clear	C-8	100	$1\frac{3}{4}$	$3\frac{1}{8}$	. . .	. . .	
BXG	1.40	6	24	10	T-8	7.5	S.C. bayonet	Clear	C-6	100	$1\frac{19}{32}$	$3\frac{1}{8}$	1,600	. . .	
BXA	1.55	6	24	10	T-8	7.5	S.C. prefocus	Clear	C-6	100	$1\frac{15}{32}$	$3\frac{1}{8}$	1,600	. . .	

Courtesy of Westinghouse Electric Corporation.

Notes: All lamps burned base down unless otherwise specified by footnote.

1. Life under specified test conditions.
2. Maximum beam candlepower at 10 feet.

★ Heat-resisting glass bulb.

3. Burn base down to horizontal.
4. Burn any position.
5. Tungsten powder cleaner.
6. Use only equipment that provides adequate ventilation.
7. Internal dichroic reflector. Collector grid. For use with heat-resisting connectors.
8. Burn base 45° to horizontal.
9. Support bulb at rim. Use only in a fixture with well-ventilated back and high-temperature wire insulation.
10. Design volts 120.
11. Base pins parallel to plane of filament.
12. For intermittent burning. Use porcelain sockets.

*Natural color lamps are not safe for use with all sensitized materials. For critical applications refer to photographic manufacturer's recommendation for proper filter to be used with standard clear or inside frosted lamp in safelight fixtures.

† Iodine cycle.

S.C. Single contact.

lighting accessories See Figures L-9*a* and *b*.

lighting contrast The relationship between the light reflected from the subject and the total amount of illumination used for lighting both the subject and the set. The ratio between those two elements is used to determine the proportionate amounts of key and fill light that are needed to ensure subject delineation and good lighting contrast.

Assuming that the key light is three times as intense as the fill light, the total number of light units falling on the subject would be four, three from the key light and the one from the fill light. Since lighting ratio is the combined total of the key light and fill light to the fill light only, the lighting ratio for the above example would be four to one. The higher the ratio, of course, the greater the contrast.

Key + fill light (measured in footcandles)	Fill light	Ratio
10	5	2:1
50	16	3:1 approximately
100	25	4:1
200	35	6:1 approximately

lighting equipment for color photography See Table L-5 and Figures L-10*a* and *b*.

lighting focus The process of varying a beam of light being emitted from a light source—from its wide flood position, for example, to a relatively narrow, focused beam or spot.

lighting grid A steel or aluminum structure (which may sometimes consist of long lengths of pipes joined together to form a scaffolding), that is positioned over a stage or motion picture set. Light units are secured or suspended from the structure as needed. See Figure L-11.

lighting requirements for TV and film studios and theaters See Table L-6.

light intensity See Tables L-7 to L-9.

light meter Also called **exposure meter**. A device for measuring the amount of illumination falling on the subject matter or being reflected from it. See EXPOSURE and EXPOSURE METER—THE PRINCIPLES.

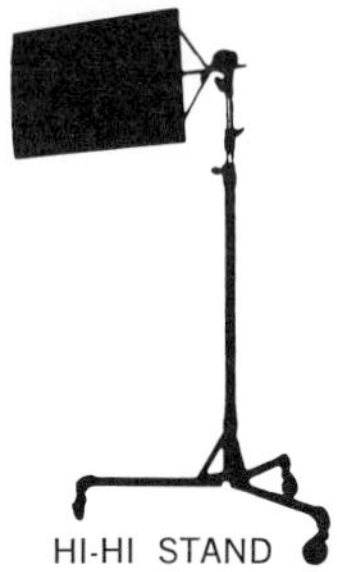

HI-HI STAND
One of the sturdiest stands made. Extra heavy- duty construction. Large full-swivel casters. Extends to 20 feet high.

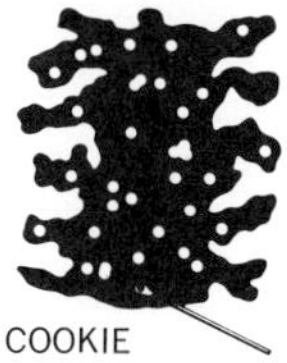

COOKIE
Available in a wide variety of sizes and patterns. Provides shadow background pattern when placed in front of light.

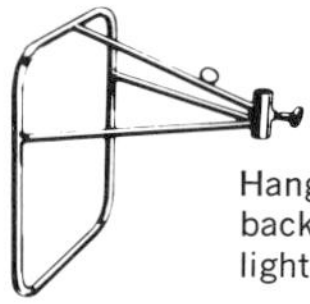

WALL SLED
Hangs over top of set or background cyc to support light on set wall.

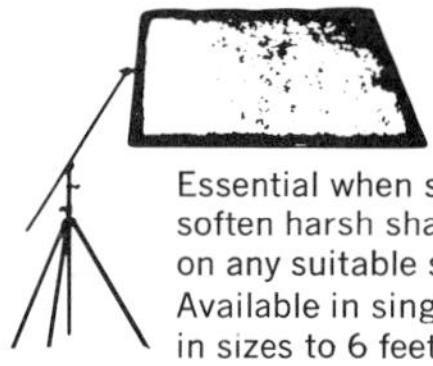

BUTTERFLY
Essential when shooting outdoors to soften harsh shadows. Can be mounted on any suitable stand or pedestal. Available in single or double thickness, in sizes to 6 feet square.

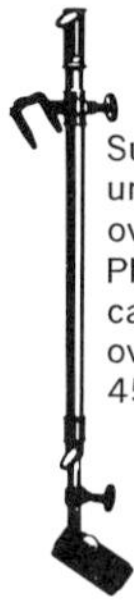

BAZOOKA
Supports lighting units over edge of overhead catwalk. Plate is nailed to catwalk, and fits over hand rail at 45° angle.

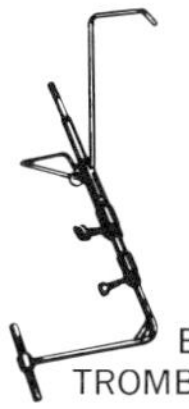

BABY TROMBONE
Hangs over top of set to support lights on set wall. Can be placed over any supporting beam or railing.

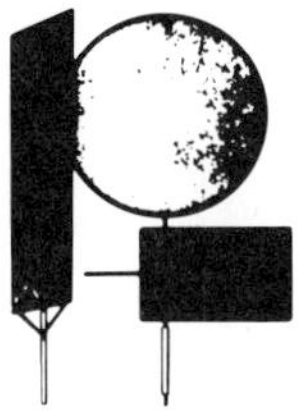

SCRIMS, FLAGS AND CUTTERS
Available in a variety of sizes and shapes for control of light and to eliminate "hot spots." Scrims are available single or double thickness.

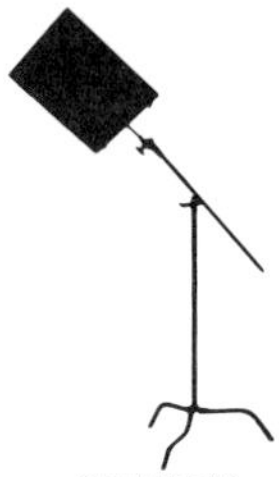

CENTURY STANDS
Lightweight, heavy duty stands for supporting lights, flags, scrims, etc. on the set or on location. 40″ closed; extends to 100″. Folds for convenient transportation or storage.

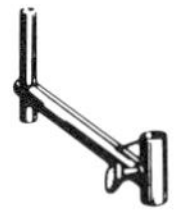

LIGHT STAND OFFSET
Permits mounting additional lights on stand. Also allows more flexibility in positioning of lights.

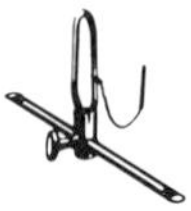

TRAPEZE
Allows suspension of lights from overhead ropes or chains. Can be guyed to prevent swinging.

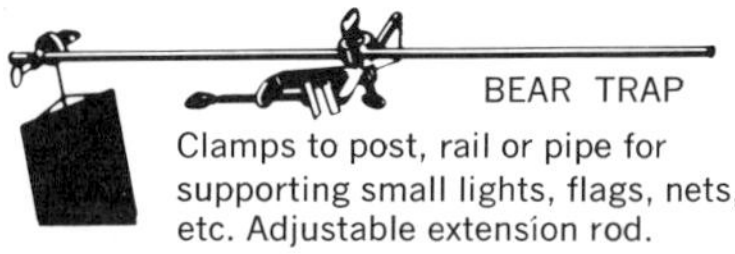

BEAR TRAP
Clamps to post, rail or pipe for supporting small lights, flags, nets, etc. Adjustable extension rod.

Fig. L-9*a* Lighting accessories. (*Motion Picture Equipment Distributing Corporation.*)

Fig. L-9*b* Lighting accessories.

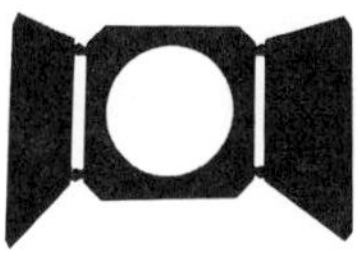

2-leaf barn door

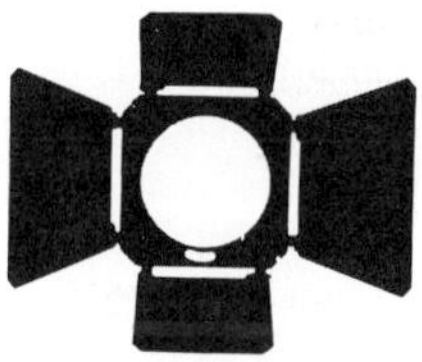

4-leaf barn door

Used to cut off light from the camera and also to provide special lighting effects.

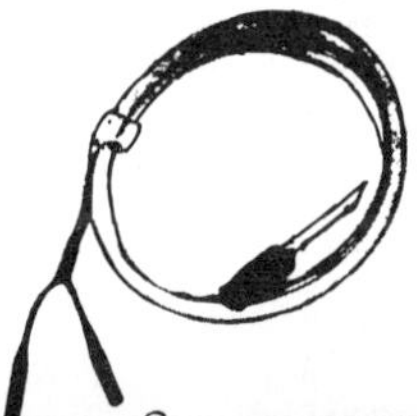

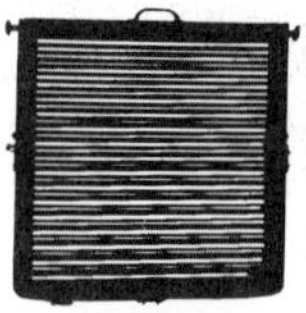

Shutter
Controls light intensity especially for color photography.

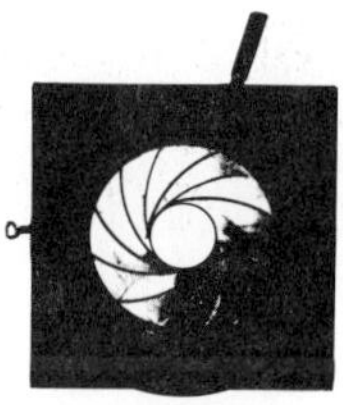

Iris
Quick and easy way to control size of circle of illumination.

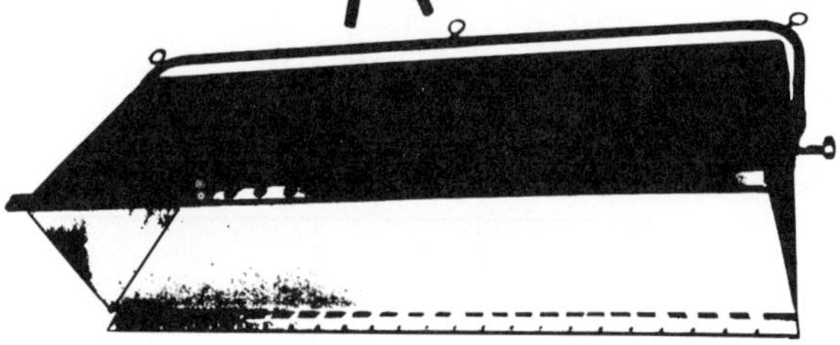

Five-light overhead strip

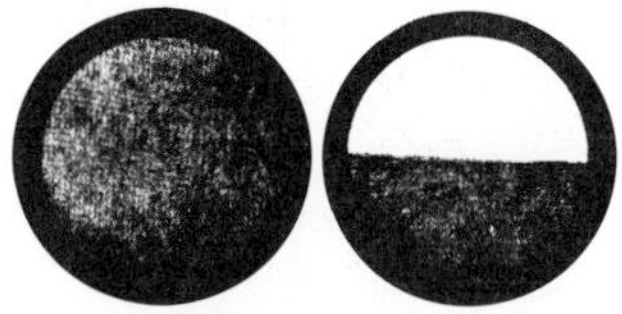

Lamp scrims
Diffusing medium available in single, double, half-single and half-double thickness.

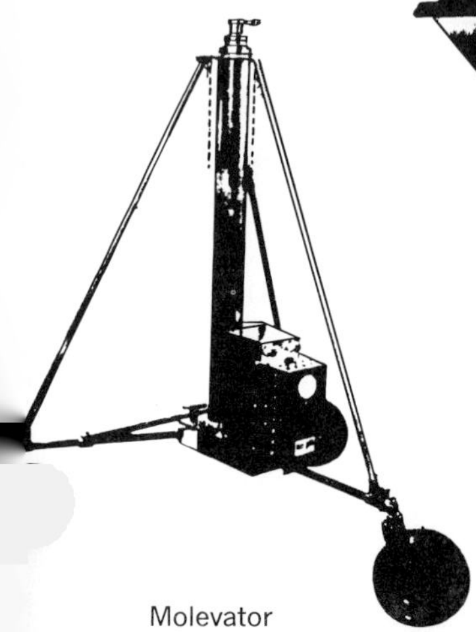

Molevator

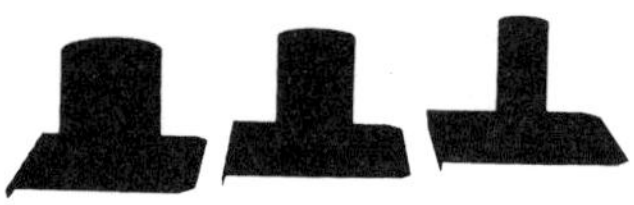

Snoots
Sets of three sizes for directing light to a particular spot.

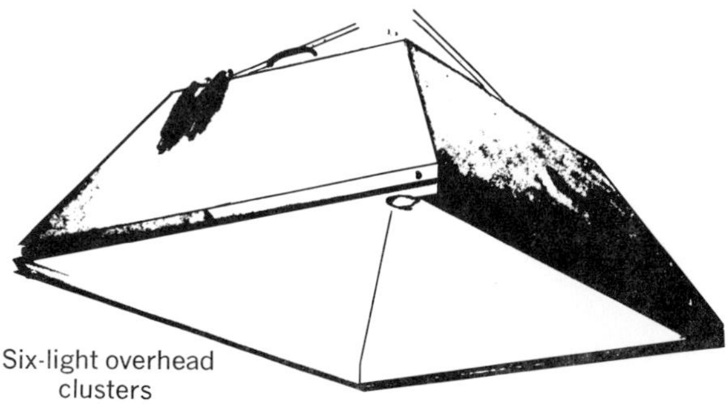

Six-light overhead clusters

TABLE L-5 Lighting Equipment for Color Photography

Arc lamps with white-flame carbons	Gelatin filters: For film balanced for daylight	Gelatin filters: For film balanced for 3350 or 3400°K
MR type 450, 225-ampere Molarc	Y-1	MT-2 + Y-1*
MR type 170, 150-ampere Molarc	Y-1	MT-2 + Y-1
MR type 90, 120-ampere Molarc	Y-1	MT-2 + Y-1
MR type 40, 40-ampere Duarc		MT-2

Incandescent lamps	Name	Globes	MacBeth filter
	For film balanced for daylight		
Type 415†	Senior	2–3‡	415-B
Type 412 or 2001†	Junior	6–7‡	412-B
Type 407 or 751†	Baby	20–24‡	407-B
	For film balanced for 3350 *or* 3400°*K*		
Type 416	Tener	1	
Type 415	Senior	2–3	
Type 412 or 2001	Junior	6–7	
Type 407 or 751	Baby	20–24	
Type 16	Cinelite	13	
Type 22	Single broad	13	
Type 23	Double broad	13	
Type 1000	Bell	13	
Type 245	Skylite	2–3	
	For film balanced for 3200°*K*§		
Type 415	Senior	4–5	
Type 412 or 2001	Junior	8	
Type 407 or 751	Baby	21–27	
Type 1000	Bell	12–17–18	
Type 22	Single broad	12–17–18	
Type 23	Double broad	12–17–18	
Type 16	Cinelite	18–28	
Type 245	Skylite	4–5	

Courtesy of Mole-Richardson.

*A yellow-flame carbon balanced for this Kelvin rating is available. This carbon requires a minimum of filtering.

†For other equipment insert no. 26 gelatin, in single or double thickness, in a diffuser frame.

‡A filter must be used with these CP (color photography) globes for matching to daylight.

§A special line of globes having this rating has been designed, but many other globes are usable since they are in the range specified. Only the special line is listed here.

TABLE L-6 Power Requirements for Typical Television and Film Studios and Theaters

Service Power in Kilowatts for Required Footcandle Level

NPA,* sq ft	Footcandles								
	50	100	110†	200	250	300‡	350	400	500§
500	8	17	20	34	42	50	58	68	84
1,000	17	33	40	66	83	100	117	132	166
1,500	25	50	60	100	125	150	175	200	250
2,000	34	67	80	134	168	200	234	268	336
2,500	42	83	100	166	208	250	292	332	416
3,000	50	100	120	200	250	300	350	400	500
3,500	59	117	140	234	293	350	409	468	586
4,000	67	133	160	266	333	400	467	532	666
4,500	75	150	180	300	375	450	525	600	750
5,000	84	167	200	334	418	500	584	668	836
6,000	100	200	240	400	500	600	700	800	1,000
7,000	117	233	280	466	583	700	817	932	1,166

Courtesy of Sylvania Electric Products, Inc.

*NPA (net production area): the unobstructed area within the confines of the cyclorama.

†Typical average footcandle requirements are 110 ft-c for monochrome and 300 ft-c for color.

‡Typical average footcandle requirement for proscenium-type stages is 300 ft-c.

§Typical average footcandle requirement for major film studios is 500 ft-c.

Air-conditioning requirements—lighting only. Multiply above listed power-service kilowatt requirements by 0.14. E.g.:

$$300 \text{ kw} \times 0.14 = 42 \text{ tons}$$

For service ampere rating divide kilowatts by

0.12 for 120 volts a-c
0.24 for 120/240 volts a-c
0.36 for 120/208 volts a-c, three-phase

E.g., for a 300-ft-c average-level 3,000-sq ft studio:

$$\frac{300 \text{ kw}}{0.120} = 2{,}500 \text{ amperes at 120 volts a-c}$$

or

$$\frac{300 \text{ kw}}{0.360} = 835 \text{ amperes at 120/208 volts a-c, three-phase}$$

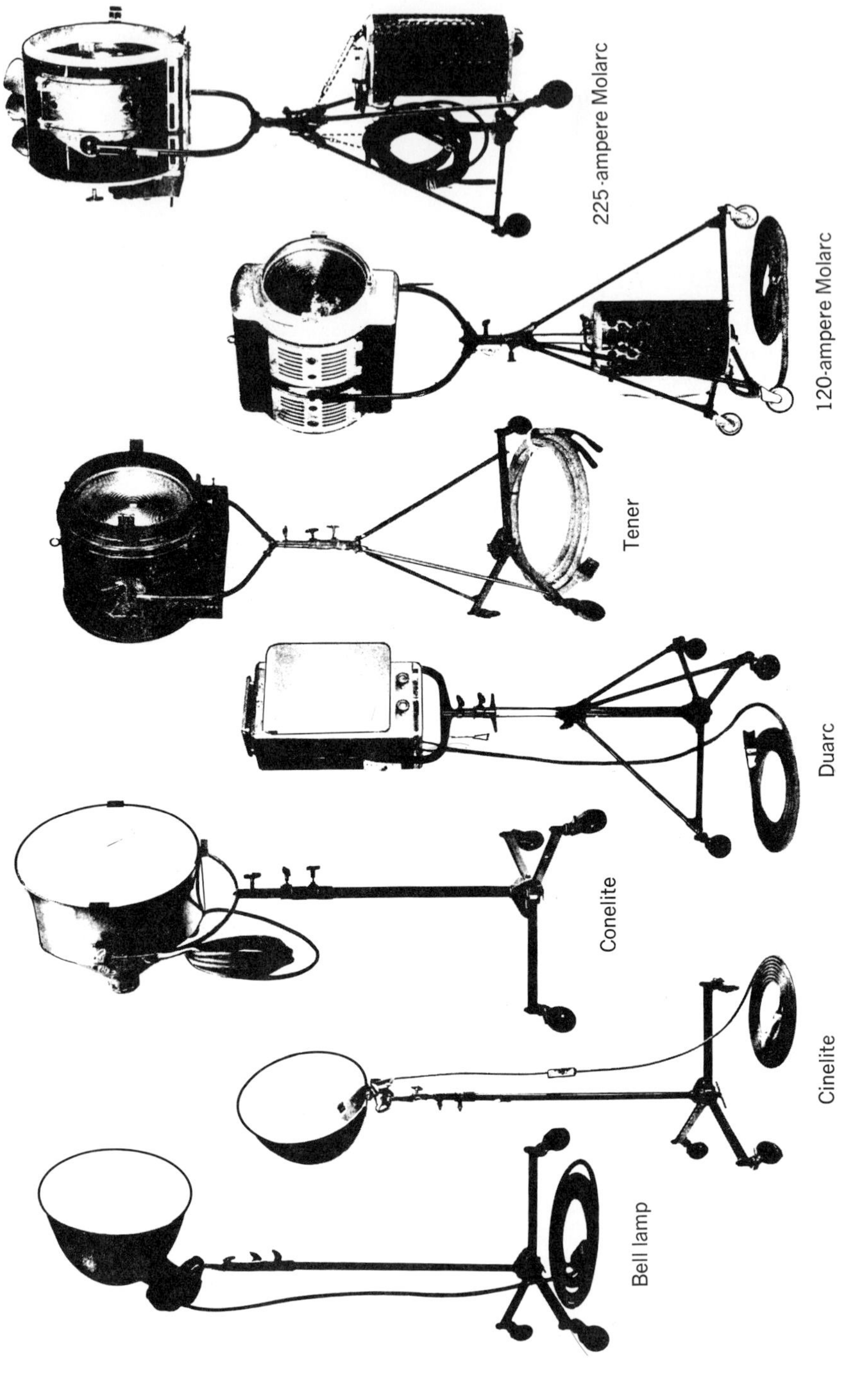

Fig. L-10a Lamps for color photography.

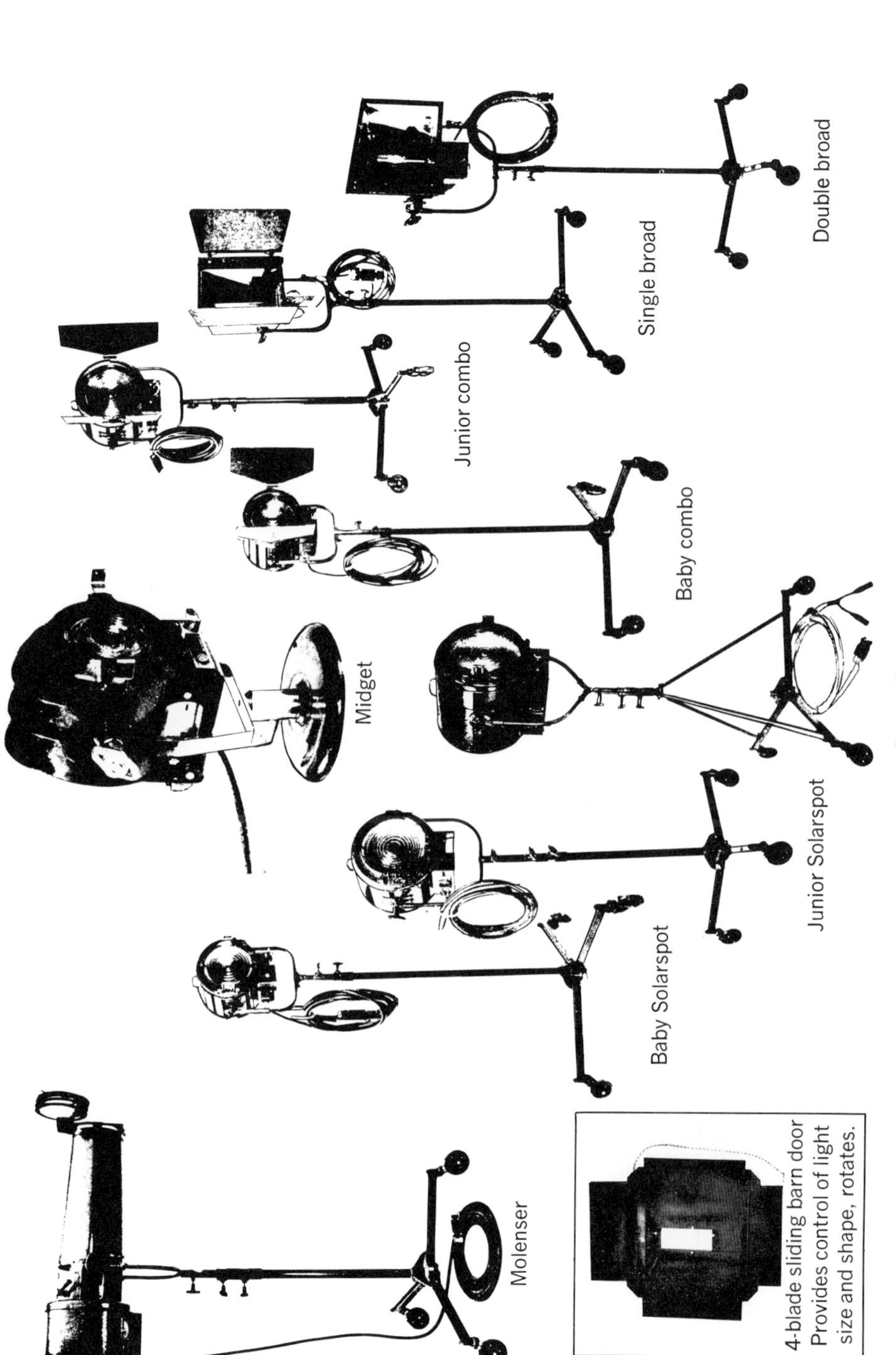

Fig. L-10b **Lamps for color photography.**

TABLE L-7 Spot Lamps

*Light Intensity in Footcandles (FC) and Spot Diameter (DIA) at Various Distances**

Lamp type	Watts	Globe number	Amperes	Distance in feet: lamp to subject																	
				5		10		15		20		30		40		50		75		100	
				FC	DIA	FC	DIA	FC	DIA	FC	DIA	FC	DIA	FC	DIA	FC	DIA	FC	DIA	FC	DIA
450†			225											450	25	300	32	130	48	70	64
														7,000	3	5,000	4	2,100	6	1,200	8
170†			150									370	22.5	200	30	130	38	55	57	30	72
												5,500	2	3,000	3	2,000	4	900	6	500	8
90†			120							320	15	170	22	90	30	55	38	25	57		
										3,200	2	1,700	3	900	4	550	5	250	7		
416	10,000	1	83.3							800	11	400	16	220	22	140	28	60	42		
										4,200	3	2,000	5	1,100	7	700	9	320	13		
415‡	5,000§	2-3	41.6					330	15	190	20	80	30	45	40	30	50				
								3,500	2.5	2,000	3.5	850	5	480	7	320	9				
412‡	2,000§	6–7	16.7			480	7	210	11	120	15	50	22	30	30						
						4,200	1.5	1,900	2	1,000	3	460	4	260	6						
	1,000	14	8.3			160	7	70	11	40	15	18	22								
						2,300	1	1,000	1.5	580	2	260	3								

407‡	750§	20	6.3	580	4	150	8.5	65	13	35	17										
				5,800	0.75	1,500	1.5	650	2	350	3										
	500§	24	4.2	380	4	95	8.5	42	13	23	17										
				3,800	0.5	950	1	420	1.5	230	2										
1,404	200	30	1.7	60	4	15	8														
				700	0.5	180	1														

Courtesy of Mole-Richardson.

*The FC and DIA values represent only those that are typical at the limits of focus. The upper values are measured at maximum flood, and the lower values at minimum spot. Intermediate corresponding values of FC and DIA cannot be accurately interpolated from the table since their relation is not directly proportional.

†With gelatin filter, loss: Y-1, 10%; MT-2 plus Y-1, 40%.

‡With Whiterlite filter, loss: 64%.

§Measured with 3350°K globe. Reduce 15% for 3100–3250°K globes.

TABLE L-8 Flood Lamps

Light Intensity in Footcandles (FC) and Spot Diameter (DIA) at Various Distances

Lamp type	Watts	Globe number	Globe size	Degrees Kelvin	Distance in feet: lamp to subject													
					4		6		8		10		12		14		16	
					FC	DIA	FC	DIA	FC	DIA	FC	DIA	FC	DIA	FC	DIA	FC	DIA
40*	ARC			4700			670	10	380	13	240	16	170	19	120	22	95	26
245	5,000	2–3	T or G-64	3350							700	14	500	17	370	20	280	23
	5,000	4–5	T or G-64	3200							600	14	420	17	310	20	240	23
16	1,000	13	PS-35	3400			600	7	340	9	220	12	150	14	110	17	85	19
	1,000	18	PS-40	3200			500	7	280	9	180	12	130	14	90	17	70	19
	500	28	PS-25	3200	560	5	250	7	140	9	90	12	65	14	45	17	35	19
	500	25–26	PS-40	3000	420	5	180	7	100	9	65	12	45	14	35	17		19
1,000	1,500	12	PS-52	3200	675	8	300	12	170	16	110	20	75	24	55	28	40	32
	1,500	10–11	PS-52	3050	560	8	250	12	140	16	90	20	65	24	45	28	35	32
	1,000	13	PS-35	3400	560	8	250	12	140	16	90	20	65	24	45	28	35	32
	1,000	18	PS-40	3200	470	8	210	12	120	16	75	20	50	24	40	28		32
	1,000	17	PS-52	3200	470	8	210	12	120	16	75	20	50	24	40	28		32
	1,000	15–16	PS-52	3050	380	8	160	12	90	16	60	20	40	24	30	28		32
	750	22–23	PS-52	3000	260	8	120	12	65	16	40	20	30	24		28		32
	500	25–26	PS-40	3000	150	8	65	12	35	16		20		24		28		32

23 (2 globes)	1,500	12	PS-52	3200	675	12	300	18	170	24	110	30	75	36	55	42	40	48
	1,500	10–11	PS-52	3050	560	12	250	18	140	24	90	30	65	36	45	42	35	48
	1,000	13	PS-35	3400	560	12	250	18	140	24	90	30	65	36	45	42	35	48
	1,000	18	PS-40	3200	470	12	220	18	120	24	75	30	50	36	40	42	30	48
	1,000	17	PS-52	3200	470	12	220	18	120	24	75	30	50	36	40	42	30	48
	1,000	15–16	PS-52	3050	380	12	160	18	90	24	60	30	40	36	30	42		48
	750	22–23	PS-52	3000	260	12	120	18	65	24	40	30	30	36		42		48
	500	25–26	PS-40	3000	150	12	65	18	35	24		30		36		42		48
22	1,000	13	PS-35	3400	300	8	140	12	75	16	50	20	35	24				
	1,000	18	PS-40	3200	260	8	120	12	65	16	40	20	30	24				
	1,000	17	PS-52	3200	260	8	120	12	65	16	40	20	30	24				
	1,000	15–16	PS-52	3050	210	8	95	12	55	16	35	20		24				
	750	22–23	PS-52	3000	150	8	65	12	35	16		20		24				
	500	25–26	PS-40	3000	75	8	35	12		16		20		24				

Courtesy of Mole-Richardson.
*With MT-2 gelatin filter, loss: 35%.

TABLE L-9 Globes

Item no.	Watts	Ordering code	Bulb	Base	Filament	Initial lumens	Hours of life	Degrees Kelvin	Amperes at 120 volts
1	10,000	10KG96	G-96	Mog. Bip.	C-13	325,000	75	3350	83.3
2	5,000	5M/G64/3	G-64	Mog. Bip.	C-13	165,000	75	3350	41.6
3	5,000	5M/T64/1	T-64	Mog. Bip.	C-13	165,000	75	3350	41.6
4	5,000	5M/G64/7	G-64	Mog. Bip.	C-13	140,000	150	3200	41.6
5	5,000	5M/T64/3	T-64	Mog. Bip.	C-13	140,000	150	3200	41.6
6	2,000	2M/G48/14	G-48	Mog. Bip.	C-13	65,500	25	3350	16.7
7	2,000	2M/T48/4	T-48	Mog. Bip.	C-13	65,500	25	3350	16.7
8	2,000	2M/G48/18	G-48	Mog. Bip.	C-13	56,000	100	3200	16.7
9	2,000	2M/T48/1	T-48	Mog. Bip.	C-13	57,000	100	3265	16.7
9A	2,000	2M/G48/17	G-48	Mog. Bip.	C-13	53,000	200	3050	16.7
10	1,500	1500(Clear)	PS-52	Mog. Sc.	C-7A	33,000	1,000	3050	12.5
11	1,500	1500/IF	PS-52	Mog. Sc.	C-7A	33,000	1,000	3050	12.5
12	1,500	1500PS52/78	PS-52	Mog. Sc.	C-7A	40,000	100	3200	12.5
13	1,000*	PH/4 Photoflood	PS-35	Mog. Sc.		33,500	10	3400	8.3
14	1,000	IM/G48/8	G-48	Mog. Bip.	C-13	25,000	100	3150	8.3
15	1,000	1000(Clear)	PS-52	Mog. Sc.	C-7A	21,500	1,000	3050	8.3
16	1,000	1000/IF	PS-52	Mog. Sc.	C-7A	21,500	1,000	3050	8.3
17	1,000	IM/PS52/77	PS-52	Mog. Sc.	C-7A	26,500	75	3200	8.3
18	1,000*	PH/IM/PS40/1	PS-40IF	Mog. Sc.	C-7A	26,500	60	3200	8.3
18A	1,000	1000/SBIF	PS-52	Mog. Sc.	C-7A	20,000	1,000	3050	8.3
19	750	750/T24/5	T-24	Med. Bip.	C-13	15,000	200	3080	6.3
20	750	750/T24/13	T-24	Med. Bip.	C-13	24,500	12	3350	6.3
21	750	750/T24/16	T-24	Med. Bip.	C-13	20,500	50	3200	6.3
22	750	750(Clear)	PS-52	Mog. Sc.	C-7A	15,600	1,000	3000	6.3
23	750	750/IF	PS-52	Mog. Sc.	C-7A	15,600	1,000	3000	6.3
24	500	500/T20/60	T-20	Med. Bip.	C-13	16,000	8	3350	4.2
25	500*	500(Clear)	PS-40	Mog. Sc.	C-9	9,900	1,000	3000	4.2
26	500*	500/IF	PS-40	Mog. Sc.	C-9	9,900	1,000	3000	4.2
27	500	500/T20/63	T-20	Med. Bip.	C-13	13,000	35	3200	4.2
28	500†	PH500/PS25/5	PS-25IF	Med. Sc.	C-9	13,000	60	3200	4.2
29	200‡	PH200/T10/5DC	T-10	D.C. Bay.	2CC-8	4,700	25	3100	1.7
30	200	CCM	T-8	D.C. Bay.	2CC-8	4,600	25	3075	1.7
31	150	CGP	T-8	D.C. Bay.	2CC-8	3,500	25	3085	1.3
32	100	100T8/DC	T-8	D.C. Bay.	CC-13	1,950	50	2955	0.8
33	75	CBX	T-8	D.C. Bay.	CC-13	1,300	50	2920	0.6
34	50	PH50T8/22DC	T-8	D.C. Bay.	CC-13	775	50	2870	0.4

Courtesy of Mole-Richardson.
*This globe may be used in types 22, 23, and 1,000 if a mogul-to-mogul extension is inserted in the socket.
†This globe may be used in type 16 by inserting a mogul-to-medium reducer in the socket.
‡Used in type 404 midget.
Mog. Mogul.
Bip. Bipost.
Med. Medium.
Sc. Screw.
D.C. Bay. Double-contact bayonet.

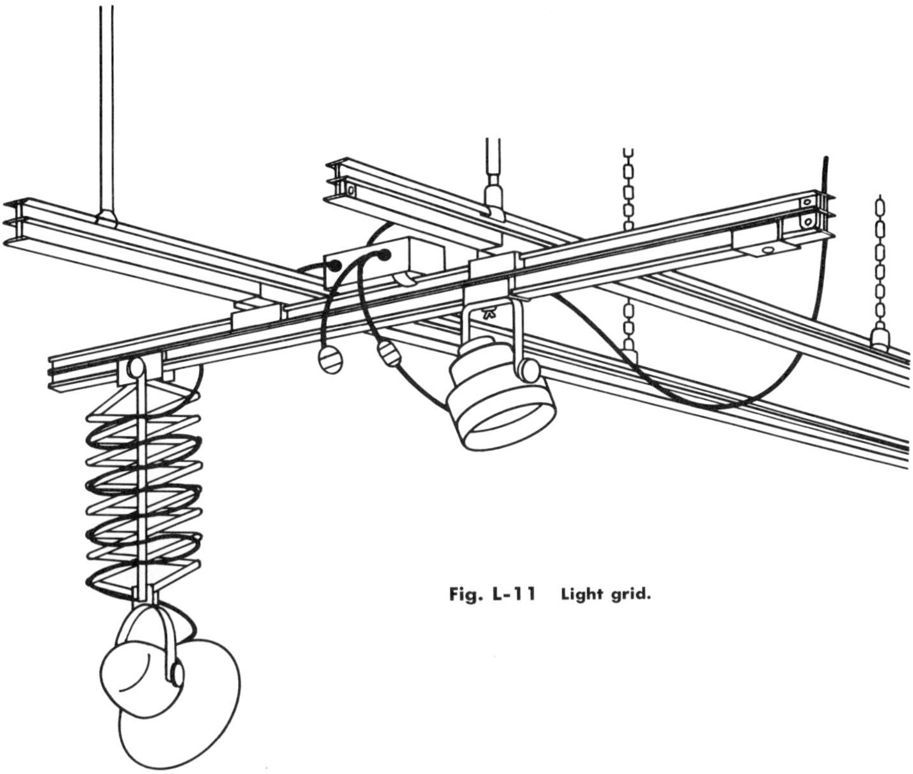

Fig. L-11 Light grid.

lightning According to statistics, less than 20 electrical storms occur in any one area during a 12-month period. More than half of this number take place during daylight hours. If a motion picture camera was set up so that its lens pointed in the general direction of a storm, the chances of getting some reasonably good, usable footage of a stroke of lightning would still be almost negligible. With the shutter in its closed position for 50 percent of the total exposure time, the odds against getting a good shot of lightning become almost prohibitive.

The ambitious and patient cinematographer who loads his camera with film and the hopes of arresting a stray lightning bolt is, indeed, an optimistic optimist. Each of the separate "elements," the length of the flash, the time of its occurrence, its pattern and relative position in the film frame at the time of exposure, would all have to be completely compatible before the effort could be called reasonably successful.

As an alternative for a good case of the sniffles, the more likely result, the cinematographer has several choices open to him. The first and perhaps the most practical option lies in the use of library or stock footage.* Commercially available, these lengths of film provide a variety of day or night bolts and flashes that can fill the most specific need. Intercut with other scenes, superimposed, or combined optically with previously exposed footage, these stock shots offer a practical solution to the cornered cameraman and panicky producer.

Cloudy sky scenes that serve as backgrounds for the selected stock footage should be photographed without filters and underexposed by approximately three to four stops. Clear sky areas, in scenes where other subject matter is also visible in the film frame, can be darkened by using a red filter. These scenes do not require any exposure compensation. The combination of landscape and cloudy sky area, by comparison, needs no correction filters but must be underexposed by approximately three to four stops.

Studio machines and other electrical devices that cause visible flashes of current to span the open areas between contact points can also be used to produce realistic lightning effects. The amount of preparation and the subsequent cost of opticals, for superimposing the effect over a scene, makes this method somewhat impractical. In the hands of the special-effects technicians, these electrical devices serve a more practical purpose in some mad scientist's laboratory during the filming of a horror picture.

Graphic lightning effects can be produced by photographing a rendered bolt on an animation or titling stand. To achieve the effect, a jagged white line, representing the shape of the bolt, is drawn on a gray card. In its filming position on the animation stand, a similar gray card is placed over the card on which the jagged line has been drawn. For each frame of the stop-motion filming, the covering gray card is repositioned to reveal a portion of the drawn line. The line should be visible, in its entirety, within three or four frames. At normal viewing speeds, the projection time for the graphic effect is $\frac{1}{8}$ second, considerably longer than an actual flash which travels at approximately 20,000 miles per second, or about one-tenth the speed of light.

If more than one flash is needed for the scene, the jagged pattern should be changed. The first "graphic" flash, if repeated, may be recognized for what it is, an artificial bolt, thus spoiling the illusion when the footage is combined optically with the background scene.

Although most material for optical superimposure is photographed on a black card, the gray card is preferable for these effects because it lightens the entire scene during the length of the flash—naturally. An overexposure of one or two stops will also soften the edges of

*In many cases, these shots are obtained through the efforts of free-lance cameramen who sell the film to organizations who rent stock footage or to studios who catalog the film and add the shots to their libraries for future use.

the drawn line and cause some desirable halation—an added touch that lends realism to the effect.

In an emergency, the effect can be achieved by intercutting several opaque black and one or two frames of clear film or preexposed frames of positive and negative film. Spaced at irregular intervals, these black and clear frames, when preceded by the sound of thunder in the accompanying sound track, can be substituted for the required flash or used immediately after an explosion to add to the effectiveness of the shot. Cut into the scene during the editing process, the few frames needed for the effect can be used in their original form or burnt in optically at approximately 50 percent exposure so that the background scene is partially visible. Repeated at regular intervals, the flash pattern may become monotonous and spoil the illusion. Lightning isn't supposed to strike more than once in the same way—or place.

light traps Devices or arrangements designed to prevent actinic light from striking the film during photography or processing. In photographic equipment, light traps in film magazines prevent stray light rays from exposing the film. In laboratories, the light trap might be in the form of an anteroom leading to the area where the film is being processed.

light valve In processing equipment, a slit that opens and closes on commands dictated by the perforations in a control tape or in response to electromagnetic impulses. The slit or opening regulates the amount of light passing through it as in the additive color system where the valves regulate the amount of light reaching the film.

In sound-recording units, the valves control the amount of light reaching the sound-track area of the film. See MODULATION.

lily Cards containing the primary and intermediate colors or the tonal values in the gray scale. Photographed over a representative scene (on the animation stand) before the film is unloaded from the camera, the lily is used as a guide by the laboratory technician or as a standard for checking the accuracy of the colors and tonal values in the processed film. See Figure L-12 in the color insert.

lim A unit of measurement that is used to indicate frequency-range changes (in sound-reproducing equipment such as a loudspeaker) that are noticeable to a listener.

Since there are no standards for frequency limits, the range is divided into four groups to provide a basis for comparison as well as a method of classification. The lim scale adopted by the Jensen Manufacturing Company covers a range of frequencies from 3,000 cycles per second (+1 lim) to 20,000 cycles per second (+8 lims).

The low range (+1 lim) includes the minimum acceptable range of musical tones. The medium and high ranges (+2 to approximately +6 lims) includes most of the sounds within the audible range. The audibility limits or ultra-high range, up to 20,000 cycles per second, are rated at +7 to +8 lims.

limbo A shot in which the background has no distinguishable detail. This type of background is used mainly for product shots in television commercials.

limited animation A technique designed to produce the maximum in movement with a minimum amount of drawings or artwork. With this technique, a cartoon character's head and body may be opaqued on one cel level while other cels containing the animating mouths or eyes carry the action and are overlaid to form the complete figure. See HOLD CEL.

line In television, one of the 525 lines that are used to form the composite picture on home receivers. The colors and tonal values are reproduced on these lines during transmission by the action of the scanning beam. See INTERLACED SCANNING.

linear distortion A defect or fault in lenses that causes straight lines to curve, in one direction or another, away from normal.

line feed A coaxial cable connecting a point where shooting is taking place to a remote videotaping facility. The cable is supplied by the telephone company on a rental basis.

line frequency In television, the number of horizontal scans per second of a television frame, nominally 15,750.

line lighting Back- and sidelighting that produces highlights around the subject matter being photographed, thus effectively outlining it.

linelock The linelock in videotape recorders locks the machine to local horizontal and vertical sync signals to permit the use of special effects. As in motion picture production, where precise registration is needed for the achievement of composite effects (matte shots, for example), the linelock serves a similar purpose in that it provides that registration electronically.

lip synchronization In animation, the drawings containing the mouthing of dialogue by cartoon characters and their subsequent synchronization with the dialogue in the sound track recorded by live performers. As in other forms of dubbing, the film containing the animation is projected onto screens and the dialogue is recorded in synchronization with the lip movements of the cartoon characters. See POSTSYNCHRONIZATION and PRESYNCHRONIZATION.

In motion pictures, the editorial synchronization between the movements of the performer's lips (in those scenes containing dialogue) and the corresponding sound track. See DUBBING.

liquid gate Also called **wet gate**. An attachment used by laboratories and in optical printers through which processed film passes before entering the projection shuttle. Before exposure, it coats the film used for duplicating with a special solution to minimize scratches or abrasions.

live recording The simultaneous recording of sound and picture during the actual filming process.

loading The placement of film in motion picture magazines prior to filming. The positioning of film in the camera, projector, or moviola's transport mechanism is referred to as *threading*. See Figure L-13.

location An area, not within studio grounds, selected for filming. Quite obviously, if the mountain cannot be brought to Mohammed, then Mohammed is left with no alternative. The academic question revolves around the choice of routes leading to the high country. In spite of the motion picture industry's well-deserved reputation for achieving the unusual, bringing a mountain to a studio sound stage could be judged a neat trick, if not a minor miracle. The problems connected with transporting small armies of technicians and performers to distant locations, along with cameras and lighting equipment needed for filming, are not all economic ones. At best, it is a hercu-

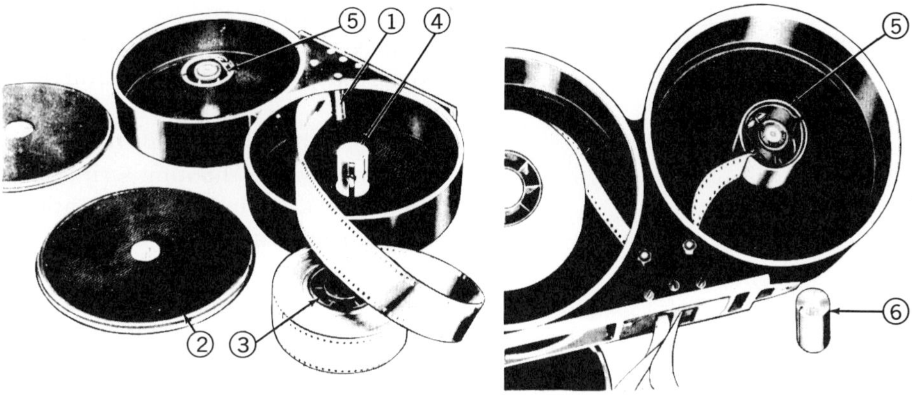

Fig. L-13 Loading magazine. (*Mitchell Camera Corp.*) (1) Idle roller. (2) Magazine cover. (3) Film core. (4) Spindle. (5) Retractable spool. (6) Solid spool.

lean task—and one that any producer would rather avoid, if at all possible.

Going "on location," unfortunately, is not the fanciful flight to faraway places with romantic settings that it is assumed to be by the readers of motion picture magazines. Some visualize an expedition in which long lines of native bearers carry large packages of photographic equipment on their collective heads while hacking a path through thick underbrush. Others think of desert settings, with nearby beaches, where the performers and crew pause frequently for a refreshing dip in the foamy surf.

Although disillusioning, the fact remains that most "location" crews seldom, if ever, leave the country. Quite often, the photographic process takes place within city limits and sometimes, just around the corner from the studio itself. The word *remote,* as applied to television or videotape productions, is also quite deceiving. In these mediums, the term applies loosely to any location away from the actual studio. And here, too, the electronic cameras may be focused on subject matter in the lobby of the building that houses the television studio.

Although location sites are usually associated with open country and exterior settings, more often than not a good portion of the filming will take place under a sheltering roof or other protective covering. Location filming, in spite of the implication, includes interior as well as exterior staging.

In most western thrillers, for example, the location site or the "home on the range" might be an open area where blue skies and multicolored mountains provide ideal backgrounds. In these appropriate settings, both the "good guy" (white horse) and the "bad guy" (black) have ample room for riding, roping, and taking turns trying to do each other in. Sooner or later, however, the cowboys will break off the chase, seek shelter, and eventually dismount in front of an abandoned shack where they can rest their tired horses and bed down for the night.

Indoors, the clapsticks click and filming continues. And although the cameras, actors, and technicians are no closer to the home studio than they were in the previous shot, the interior of the shack has become the new location site. Interior or exterior, location is location.

Obviously, the selected site must visually conform with the written descriptions in the shooting script. Since budgetary limitations are almost always a primary consideration, the availability and accessibility of nearby alternate sites, as opposed to more authentic but distant locations, should always be taken into account. Similarly, the climatic conditions at the contemplated site must also be a matter for review. Torrential downpours in the middle of a rainy season may be quite welcome if called for in the script but disastrous, economically, if the scene must be filmed under bright skies. The

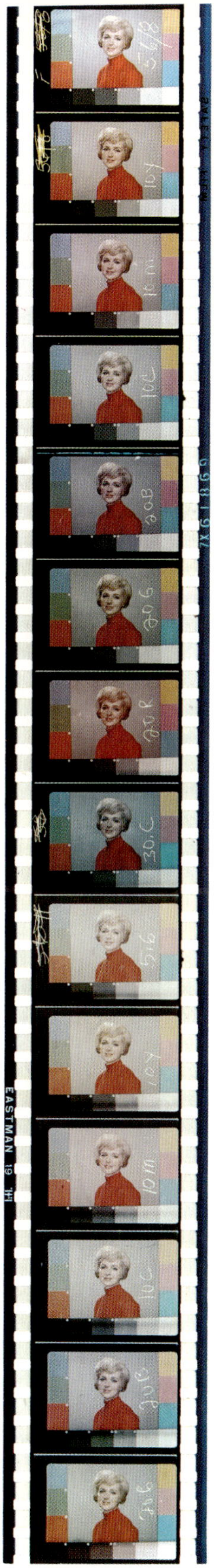

Fig. C-11 China girl.

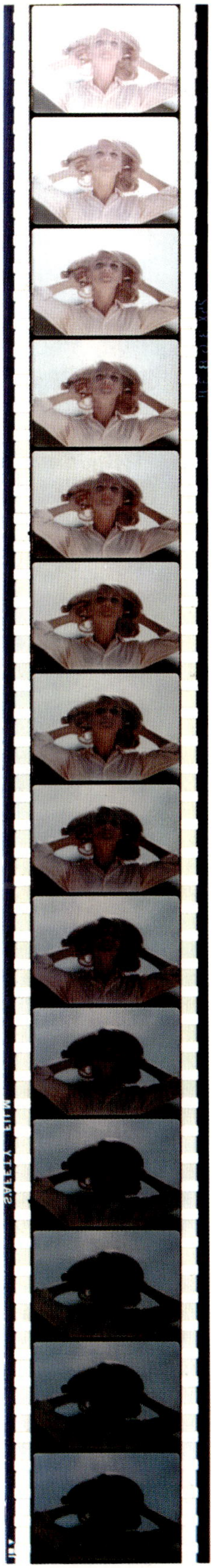

Fig. C-18 Color scene tester Cinex.

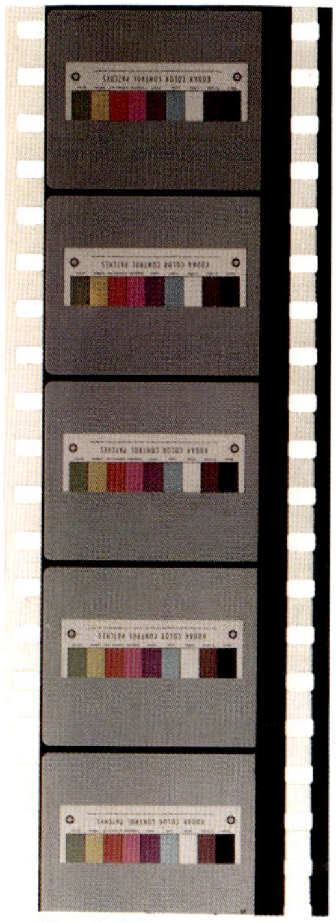

Fig. L-12 The lily.

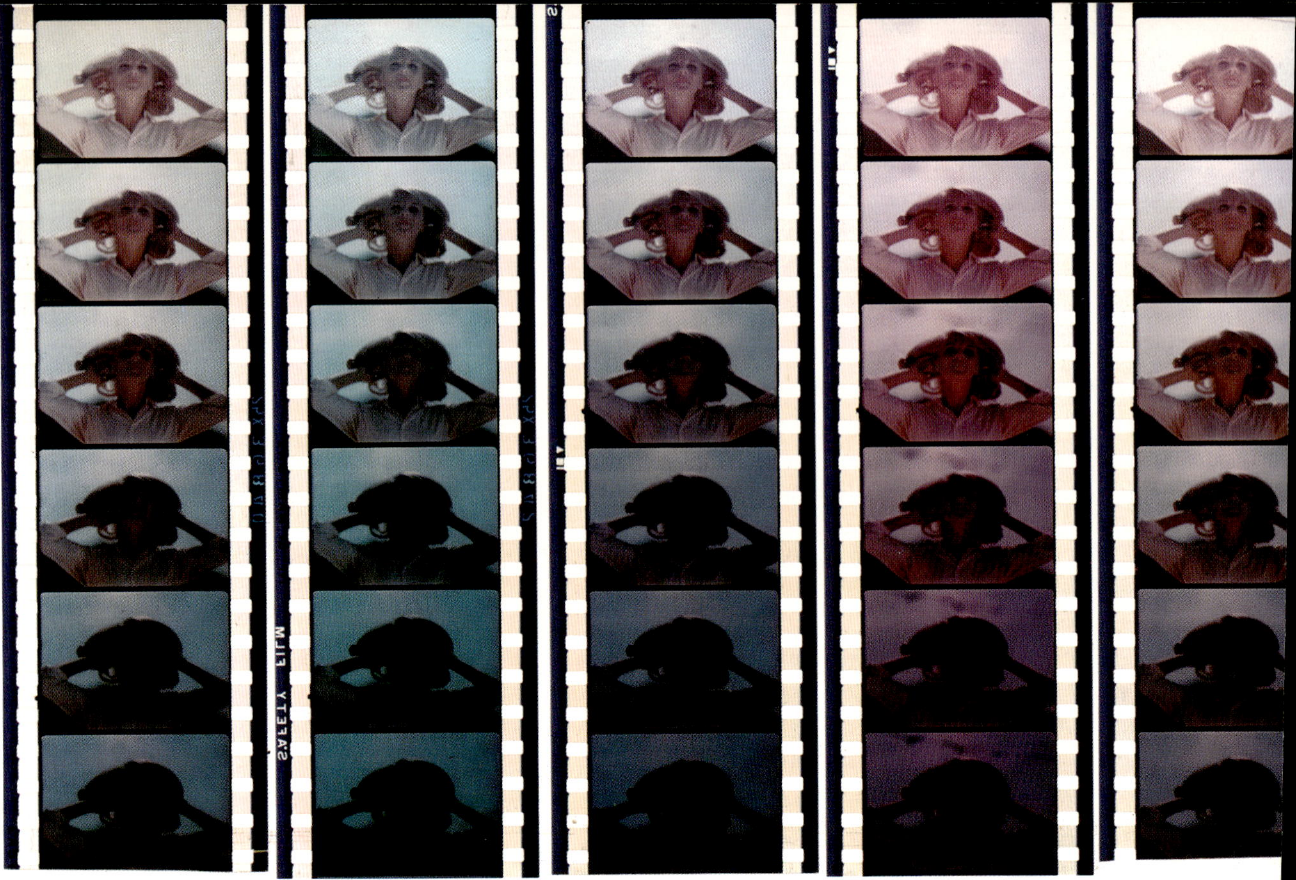

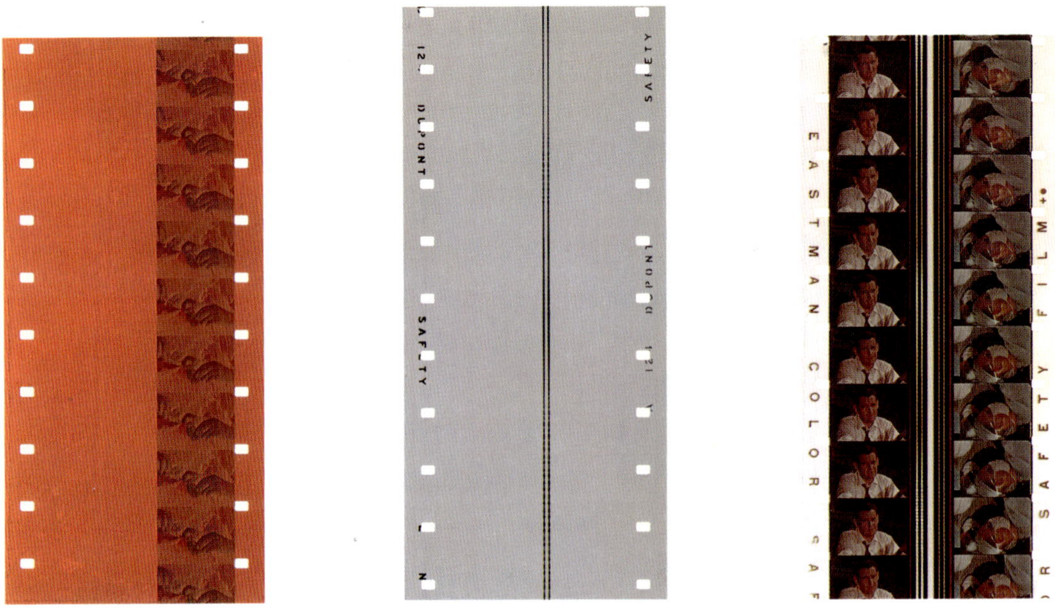

A B C

Fig. P-16a General Film Laboratories 35/32mm printing process. (*A*) Negative picture. (*B*) Negative track. (*C*) 35/32 composite.

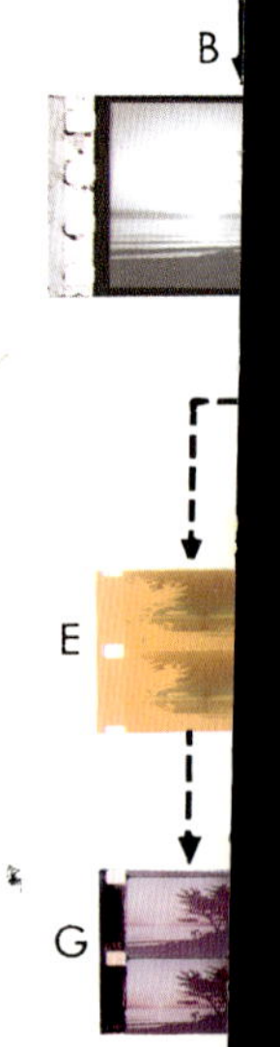

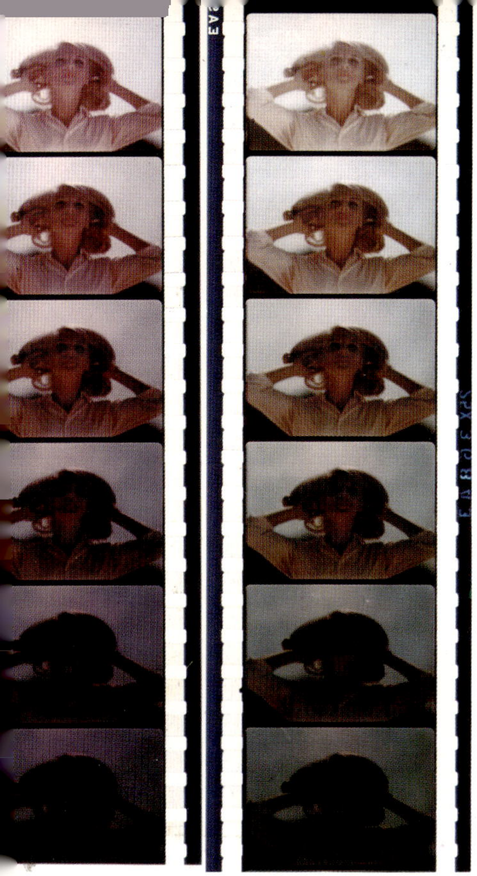

Fig. C-13 **Color wheel Cinex.**

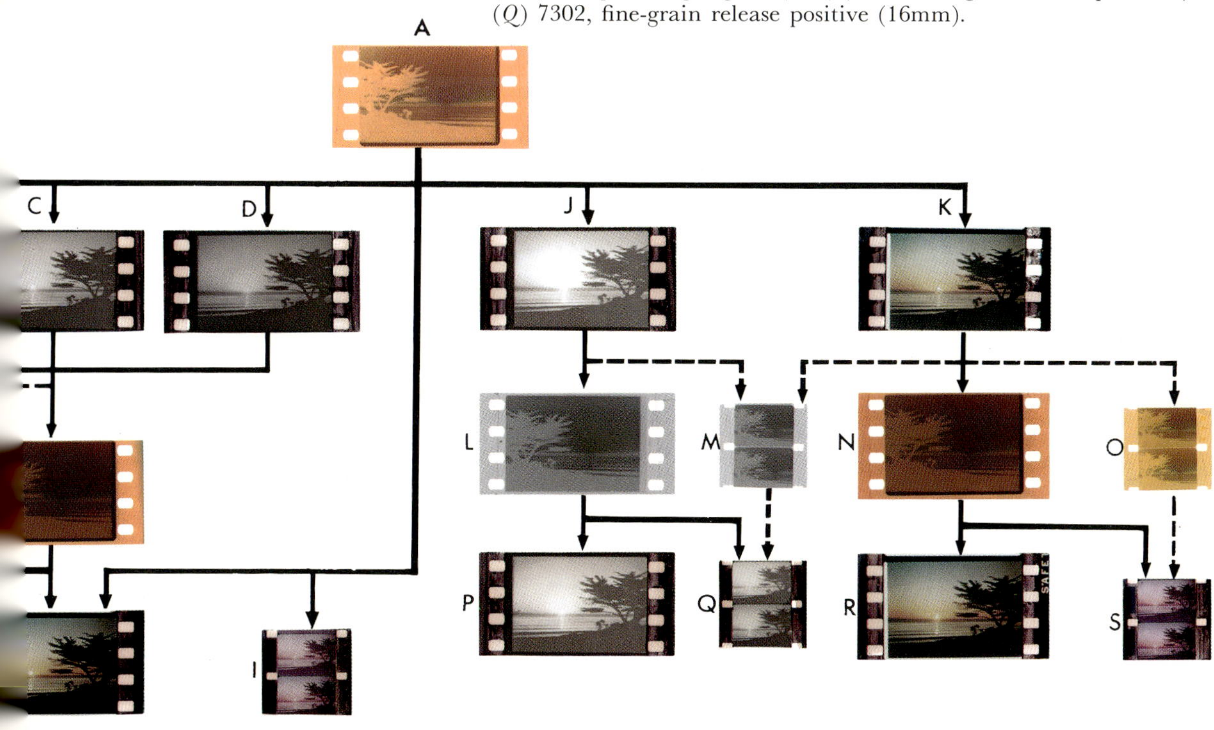

Fig. P-15e **Schemes of printing 35 and 16mm prints from 35mm negatives.** (*A*) 5251, Eastman color negative (original camera negative). (*B*, *C*, *D*, *J*) 5235, panchromatic separation (35mm duplicating film). (*E*, *O*) 7253, color intermediate (16mm duplicating film). (*F*, *K*, *N*) 5253, color intermediate (35mm duplicating film). (*G*, *I*, *S*) 7385, color release print (16mm). (*H*, *R*) 5385, color release print (35mm). (*L*) 5234, fine-grain panchromatic (35mm duplicating negative). (*M*) 7234, fine-grain panchromatic (16mm duplicating negative). (*P*) 5302, fine-grain release positive (35mm). (*Q*) 7302, fine-grain release positive (16mm).

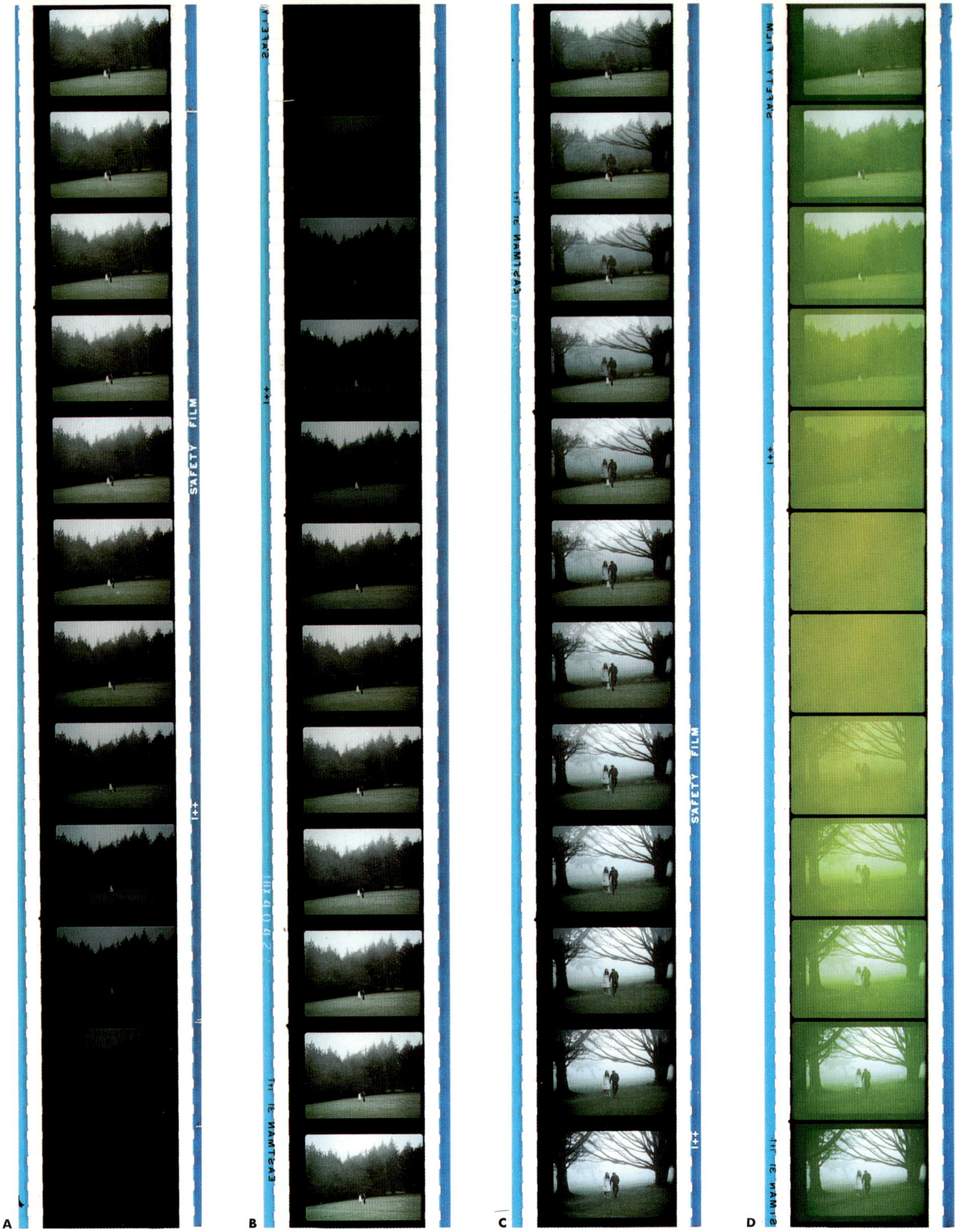

Fig. T-17 **Transitional effects.** (*A*) 12-frame fade-out. (*B*) 12-frame fade-in. (*C*) 12-frame cross-dissolve. (*D*) 12-frame cross-fade.

producer should also remember that seasonal weather conditions do not necessarily approximate those in the film maker's own immediate area. With the sun at its hottest and air conditioners going full blast at the studio, other far-off places are quite likely to be covered by a thick blanket of snow.

Even in nearby locations, unforeseen annoyances can slow the production process and irritate both technicians and performers. Crowds, attracted by the glamour of the event, can easily become unmanageable and seriously hamper production. Traffic in the immediate vicinity, unless detoured, can be a greater nuisance than the crowds of onlookers. And extraneous noises, such as the shrill toot of a policeman's whistle or the honking horns from passing cars, can easily ruin the good "take" when sound pictures are being filmed. Outside of congested city or suburban areas, shooting sound films is a far less harrowing experience.

Interior locations, by comparison, present their own particular problems. Although filming takes place indoors, the producer who depends on existing outlets for supplying the needed current for cameras and lights is gambling unnecessarily. Rarely, if ever, is the available power supply adequate. Technicians, forced to resort to temporary makeshift arrangements, are seriously handicapped. The economic advantages, if any, are usually reflected in the quality of the production. Generators, a must for outdoor locations, are capable of supplying enough power for operating cameras, sound recorders, and lights, and other electric equipment. Equally valuable as a central power source for indoor locations, the small, portable units take up little room and can be stored in any of the trucks that carry other photographic equipment to the filming site. Since these units can be rented from organizations that supply other photographic accessories, even the low-budget producer can ill afford to go on location without including this important item.

In selecting interior locations, the producer should remember that, in spite of economic arrangements with the owner of the premises, he is still a visitor and, as such, must not wear out his welcome. In most cases, the lessor expects to be inconvenienced as a result of the attendant activity connected with the filming process. This inconvenience can be minimized by preplanning shots and lighting positions. Similarly, the movement of cameras and microphone booms should be plotted in advance from blueprints or diagrams of the premises. Spontaneous maneuvering or ad-libbing close-up and zoom shots, at the time of filming, only creates unanticipated problems during the editing processes. If at all possible, the director and cameraman should visit the site selected for the filming and discuss their plans with the owners of the premises. Such trips usually benefit all concerned.

Lighting, for both exterior and interior locations, must be planned carefully. The same rules that guide the lighting of the studio set also apply to location photography. In open areas, the chore is lightened—naturally. There, the cameraman and director can take advantage of the best, and cheapest, light source available—the sun. The fill and key lighting provided by this natural source of illumination is usually sufficient for most outdoor filming assignments. Auxiliary lighting sources and reflectors that direct the sun's rays to those areas where they are needed are usually adequate for illuminating problem areas and shaded places. Where direct sunlight causes strong contrasts or objectionable light-and-shadow patterns, in wooded areas or city streets, for example, reflectors help soften the undesirable shadows.

Proper exposure is, obviously, as important on location as it is on the studio stage. In open areas, where the subject matter is some distance from the camera, incident-light exposure meters can be read at the camera position with excellent results. In extreme long shots, haze, not always visible to the cameraman, may produce effects similar to those achieved when film is accidentally overexposed. The use of haze filters, when doubtful conditions exist, can do little harm and usually helps to avoid the necessity of a retake. As a general rule, if precise lens settings are difficult to obtain, the cameraman should underexpose rather than overexpose, since it is easier for laboratory technicians to make corrections in film that has been underexposed.

For determining the proper exposure and lens settings for interior locations, meter readings should be made from a gray card rather than from an arbitrary position on the set. Although the basic construction of the setting may make strict observance of the lighting rules impossible, the director and cameraman should try to duplicate the natural lighting of the area. Light entering through doors and windows does not, generally, interfere with the artificial lighting sources on the set. In most cases, the effect lends authenticity to the overall scene and is quite desirable. Wherever possible, lighting setups should be made with the actors in the same positions that they are to occupy during the actual filming.

As a precautionary measure, all equipment should be checked at the studio under conditions that will duplicate those at the selected location site. Lens shades for shielding purposes, Pola screens for eliminating objectionable highlights, and filters for achieving night effects in daylight and reducing haze are all standard items that must accompany the cameraman on location. And, although the location site may not be too far from the studio grounds, arrangements must be made for parking the trucks that carry equipment and personnel to congested areas.

Going on location? Have a nice trip but don't forget to take along some insect repellant and a first aid kit. See Figures L-14*a* and *b*.

Fig. L-14*a* Location is where the camera is. For this MPO production, the motion picture camera is installed in the forward section of a New York Central locomotive. The sequence was filmed by MPO for "The Big Train," produced for New York Central.

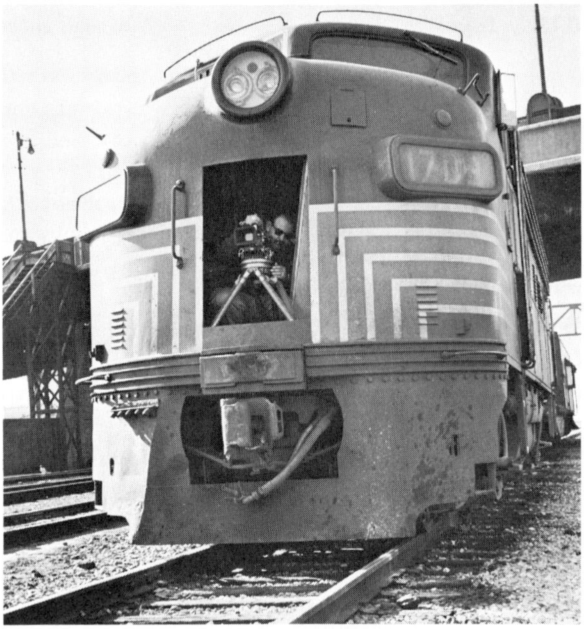

Fig. L-14*b* "West Side Story" and a location background. (*United Artists Corporation.*) A second background will be matted in and added optically in the area behind the camera crane.

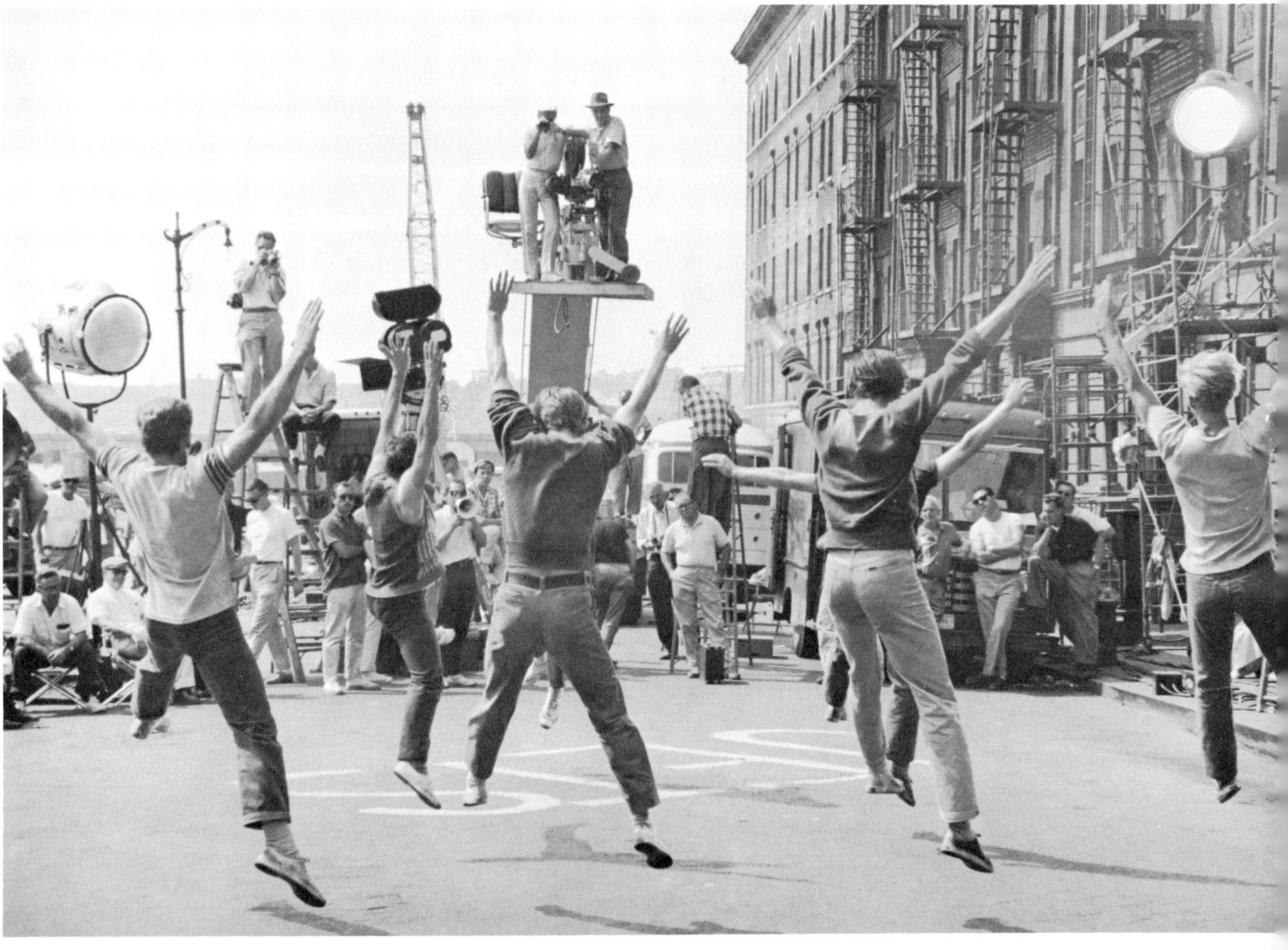

logo A trademark or symbol for an organization or advertised product. The term is used quite regularly in commercial film production. See Figure L-15.

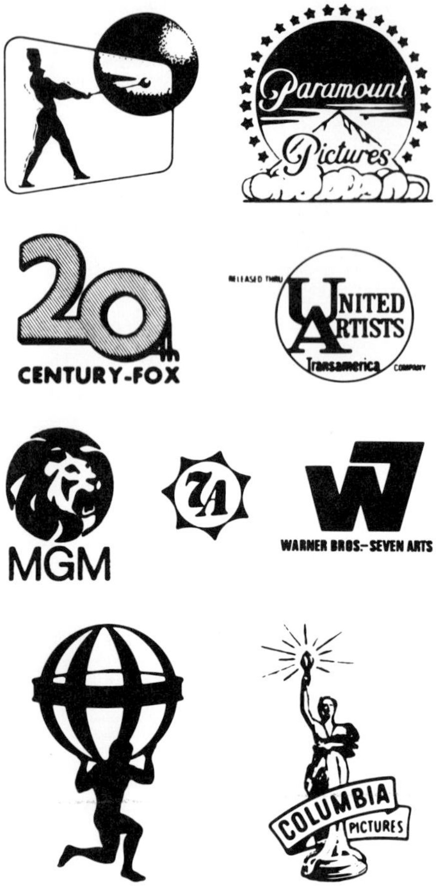

Fig. L-15 **Motion picture logos.**

log sheet Also called **sound report**. A synopsis of the recorded sound on tracks produced during the filming process. Similar in content to the camera reports that are used to keep track of the day's filming activities, log sheets contain information relative to the speed of the transport mechanism during the recording process, extraneous noise, sync source, frequency, and the footage counts indicating which takes are to be transferred to magnetic film.

long-focus lens A type of lens that is extremely useful for location filming and newsreel photography. Unlike wide-angle lenses, long-focus lenses reduce the feeling of distance between camera and subject matter. Long-focus lenses are used to magnify the image on the negative. The focal length of these lenses is longer than that of a normal lens and thus longer than the diagonal of the negative. These lenses are usually considerably heavier and bulkier than telephoto lenses although each will yield images of similar size under more or less similar conditions.

long shot Also, **L.S.** or **establishing shot**. A large area photographed from a distant point. This type of shot helps establish a relationship between a general setting and scenes or actions taking place within it.

loop, film The extra length of film, between sprocket wheels and camera gate, that absorbs the shock of intermittent movement as the film travels through the camera.

loop, sound A length of sound track that is spliced at both ends to form a continuous loop so that the sound it contains is repeated during the mixing or re-recording process.

looping, sound Also called **line replacement** and **dubbing**. A process for replacing the sound track recorded during the filming process with a new sound track that is recorded under studio conditions. This dubbing process is used when the ambient sound or extraneous noise at the filming location made it impossible to record a usable sound track.

The process is called looping because the picture and sound used for cueing purposes at the sound studios (actually the same sound that was recorded on location) are cut into loops so that they can be projected over and over again during the rehearsal period that precedes the re-recording of the sound track.

lot A large area, on studio grounds, that is used for a number of purposes including set construction and filming.

loudspeaker The device by which recorded sound waves are transformed from electric currents back into audible sounds. See LIM.

low-band recording A recording made with a low-frequency carrier (5 megacycles) yielding a mediocre signal-to-noise ratio, with moiré present. Compare HIGH-BAND RECORDING.

low frequencies Also called **lows**. Frequencies below 15 cycles per second—the minimum in the audible frequency range. Electric impulses and modulations produced by low-pitched sounds (thunder, or the base notes struck on a piano keyboard, for example) are in the low-frequency range.

low key The general arrangement of set lighting that produces gradations between the middle shades of gray and black. Overall dark scenes are considered to be low key.

low-speed photography The standard rate of projection for sound motion picture film is 24 frames per second. Actions, in order to appear realistic when projected, must be photographed at the same speed, 24 frames per second.

Camera speeds faster than the standard rate can slow an action to a crawl or, in some cases, make time seem to "stop in its tracks." At extremely high speeds, thousands of pictures per second, for example, the camera lens can record movement that is far too rapid for the eye to follow. The processed film, when projected at normal speeds, not only provides a means for convenient study and analysis but also makes it possible for an observer to time the recorded images in microseconds and milliseconds, if necessary.

Compared with the many practical applications of high-speed photographic processes for scientific, industrial, and commercial purposes, the uses for low-speed techniques are relatively few. Slower camera speeds, generally used for entertainment purposes, help achieve dramatic effects and provide comedy relief for moviegoing audiences.

In a more serious vein, objects that move too slowly to allow accurate study or actions that must be speeded up for demonstration purposes are ideal subjects for low-speed photographic processes. The passage of time by showing the movement of clock hands or the gradual formation of a full-grown flower from a tiny bud or similar actions that take place over relatively long periods of time are, by comparison, best recorded with stop-motion, time-lapse photographic processes.

Actions photographed at camera speeds slower than the standard 24 frames per second appear to be proportionately faster when projected at the normal viewing rate. The speed of an action, regularly requiring 30 feet of film, can be doubled by photographing the action at half speed (12 frames per second), on 15 feet of film. Instead of a projection time of 20 seconds, the accelerated action requires only 10 seconds of projection time. Obviously, the latter action will appear to be twice as fast as a similar action photographed at the normal rate of 24 frames per second.

Variable-speed motors, with speeds controlled by a governor and read on a tachometer, make it possible for a photographer to decrease the camera speed and expose as few as 4 frames of film per second. At that rate, an action would be six times faster than normal when projected at the standard rate.

Since exposure time increases in direct proportion with decreases in camera speeds, the amount of light that exposes the in-camera film must also be reduced. These exposure compensations can be made with mathematical accuracy by closing down the lens to the *f*-stop that best suits the camera speed. Since each lens stop, or aperture setting, increases or decreases the amount of transmitted light by one-half of the total amount, a lens setting of *f*/11, for example, will halve the amount of exposure at *f*/8, the next higher diaphragm setting. Thus, if the normal lens setting for a scene photographed at the standard rate of 24 frames per second is *f*/8, the lens would have to be closed down to the *f*/11 setting for camera speeds of 12 frames per second. Similarly, for camera speeds of 6 frames per second, the lens setting would be *f*/16. (When lighting conditions are difficult and extra exposure is needed, cameramen sometimes take advantage of slower camera speeds and wide-open lens settings to achieve results that would be impossible with other methods.)

Slow camera speeds make it possible for the cameraman to achieve a wide variety of unusual dramatic and comic effects. The natural movement of clouds, for example, can be speeded up to simulate extreme turbulence or herald approaching storms. To add to the general effect, light rains assume the proportions of a cloudburst while swaying shrubs and tall grass whip about violently when the processed film that had been photographed at reduced camera speeds is projected at normal viewing rates.

Similarly, effects that can only be achieved by using miniatures or scale models can be made to look amazingly realistic by photographing the scene at reduced camera speeds. An automobile hurtling over a cliff will fall at a known rate of speed and complete its descent in a calculable amount of time. Scale models on miniature sets, in duplicate actions, will also follow the laws of gravity but will always "fall" far short of being convincing. Their acceleration due to gravity is constant, but their lighter weight makes them more susceptible to being slowed by air resistance. When the action is photographed at reduced camera speeds, however, the timing seems reasonably authentic.

Accidents involving actual cars and trains can be staged safely by slowing the vehicles as well as the camera speed during the photographic process. At quarter speed, or 6 frames per second, vehicles moving at speeds of 15 miles per hour will seem to be traveling at 60 miles per hour when the film is subsequently projected. The drivers, properly protected, will emerge from the wreckage shaken but unhurt after the suspense-filled collision. Similar crashes, filmed while the vehicle was moving at the rate of 60 miles per hour, would be disastrous. Certainly, the dramatic impact of the scene is only slightly less than the staged impact of the cars involved in the accident.

Other actions, within this type of scene, must be slowed down proportionately in order to make the overall effect believable. Performers, for example, must be rehearsed and made to move at one-quarter normal speed if the car in the same scene is also moving at the same reduced speed. When projected, the film will show the actor walking at a normal pace while the car races by and through the scene.

Sequences that include "chases," shots of traffic that show cars and pedestrians at busy intersections, and fast-moving sports events make ideal subject matter for low-speed photographic techniques. Although moviegoing audiences have become more sophisticated since the industry's pioneer days, general reaction to these "gag" shots has remained unchanged. Low-speed photography plus normal projection speeds, in these shots, equals instant laughter.

lumens A measure of an amount of light, one lumen is equal to one footcandle of light covering a one-square-foot area. *Total lumens* is defined as average footcandles, within an area, times that area. Lumens are independent of distance. That is to say, just as many lumens are produced by a lighting fixture at one distance as are produced at any other distance.

The lumen figure is of value to the photographer if he desires to produce a base level of so many footcandles in a certain defined area such as a motion picture set. The photographer may simply determine the number of lumens required for his base level and divide it by the number of lumens available in a fixture of his choice. This will give him the total number of fixtures required. In practice, the fixture mounting distance to the set is normally predetermined; therefore, the intensity of the chosen fixture is known at that distance. For most effective light utilization, he must be certain that his choice of fixture is appropriate—that is, it falls within the following two limits. First, the total number of fixtures times the footcandles of each fixture must provide at least as many footcandles as the set requires. Second, the footcandles of any one fixture cannot be greater than the set requirement.

The following example is given as the lumen approach for determining the number of fixtures required to light a set. Suppose the photographer has a 20- by 20-foot set which he wishes to light to a base level of 500 footcandles. He then chooses the appropriate fixture (within the limits discussed in the previous paragraphs). To obtain the total number of lumens required to light this set, he merely multiplies the 400 square feet (20 by 20 feet) by the 500 footcandles and obtains a total of 200,000 lumens. If his fixture produces 10,000 lumens, he divides the 10,000 lumens into the 200,000 lumens, obtaining 20 as a result. Therefore, he will require 20 fixtures to illuminate a 20- by 20-foot set to a base level of 500 footcandles.

lumens, beam The amount of light available within the area covered to 50 percent of maximum intensity.

lumens, field The amount of light available within the area covered to 10 percent of maximum intensity.

lumens, fixture The amount of light available from the fixture once the lamp has been placed in that fixture. In general, fixture lumens are substantially less than lamp lumens as a direct result of reflector (and lens) light absorption and scattering. There are other light losses in the fixture as a result of inefficiencies due to design and ventilation requirements. The two basic measures of fixture lumens are beam lumens and field lumens.

lumens, lamp The total amount of light available from any particular lamp.

lumens per watt The number of lumens produced by a light source for each watt of electric power supplied to the light source. If a 1,000-watt lamp produces 20,000 lumens, for example, then

$$\frac{20{,}000 \text{ lumens}}{1{,}000 \text{ watts}} = 20 \text{ lumen/watt}$$

lumiere cam An integral part of the film-transport mechanism in early, silent motion picture cameras.

luminaire An all-inclusive term for lighting instruments. The term is intended to include the light housing, the lamp or globe, and the supporting standard.

luminance A term of comparative reference with respect to the brightness or intensity of colors.

In lighting, a measure of the light (flux) per unit area leaving a surface in a particular direction. The unit of measurement is the footlambert. This quantity was formerly known as *brightness.*

luminance analyzer A high-definition spot photometer. Its acceptance angle is $\frac{1}{2}$ degree (a circle 1 inch in diameter at 8 feet). Designed to measure light ratios of more than 1,000 to 1, the readings can be made from small, inaccessible areas such as the inside of a surgical incision as well as from objects at great distances (cloud highlights, for example). Power for the meter is derived from batteries. See Figures L-16*a* and *b*.

luminescent Describing a source capable of producing light without generating heat, such as certain phosphors.

luminous A term indicating the production of light by reflection (from a luminous source, for example), as distinguished from electric sources.

The light emitted by the numbers on the face of a wristwatch is referred to as luminous since the glow produced by them is visible

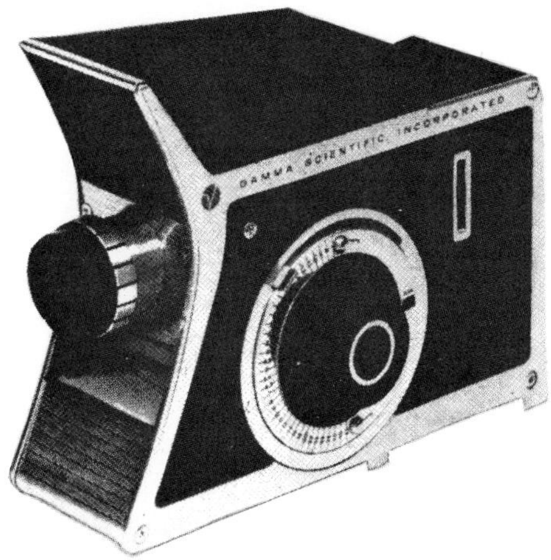

Fig. L-16*a* **Scene luminance analyzer model A-500. (*Gamma Scientific, Inc.*)**

Fig. L-16*b* **Schematic diagram of scene luminance analyzer model A-500. (*Gamma Scientific, Inc.*)**

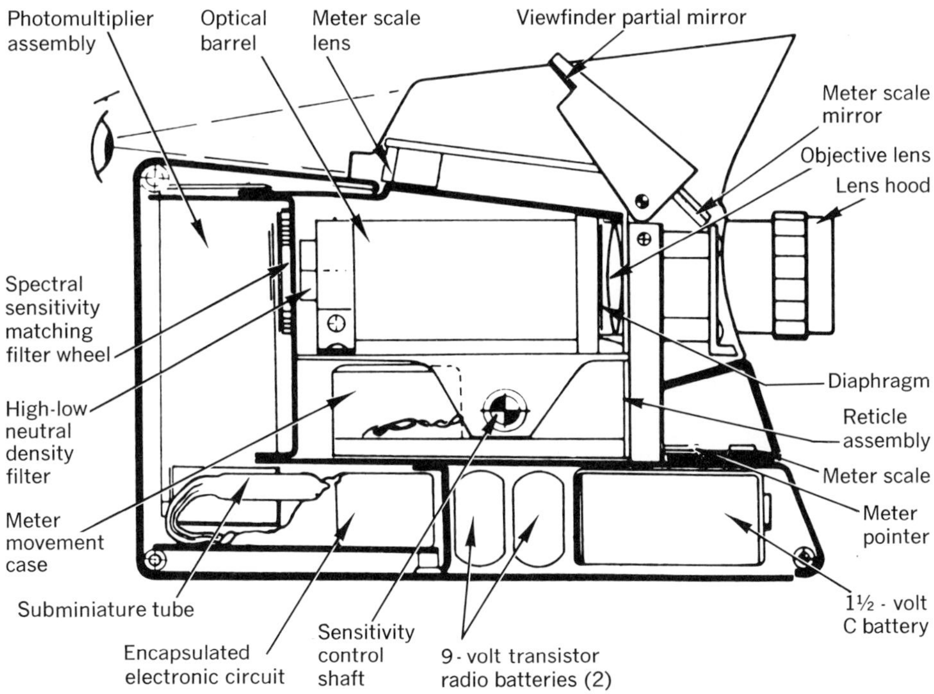

by reflection. The term is usually used in connection with *intensity* or *flux*.

luminous flux The time rate of light flow from a source, measurable in lumens.

luminous intensity The measurable amount of light reflected by an illuminated surface. The degree of luminance of the reflective surface is measured in terms of candles per square inch.

lux The metric equivalent for footcandle. Used internationally, the photometric term indicates the intensity of illumination of a surface. One lux is the equivalent of one lumen of light per square meter of surface.

10.764 square feet = 1 square meter
1 footcandle = 10.8 lux

M

macrolens A lens with a special application in close-up filming. It is equipped with a close-up focusing extension—a built-in extension tube—which permits filming at closer distances than is possible with other lenses of the same focal length.

M and E tracks Sound tracks used for dubbing purposes. Only the music (M) and sound effects (E) are included in these tracks. Dialogue is recorded on a separate track.

M and E tracks are especially prepared for films that are to be distributed internationally. In those cases, the foreign dialogue is recorded separately and combined with the existing music and effects track to form a composite track.

magazine, camera A film container with lightproof compartments. Mounted on the camera, the raw stock is unwound from the roll in one compartment (the feed), passes through the camera where it is exposed, and travels to the next compartment (the take-up) where it is rewound on the core.

Most magazines have a 400- to 1,000-foot capacity. Others, produced for specific cameras, hold 100 to 1,200 feet of film. Usually corduroy- and/or velvet-lined, two lids on the side of the magazine unscrew to permit loading or removal of the film. Both compartments of the magazine are lightproof, and a spring-loaded ball at the center of each lid ensures a snug fit and prevents rattling.

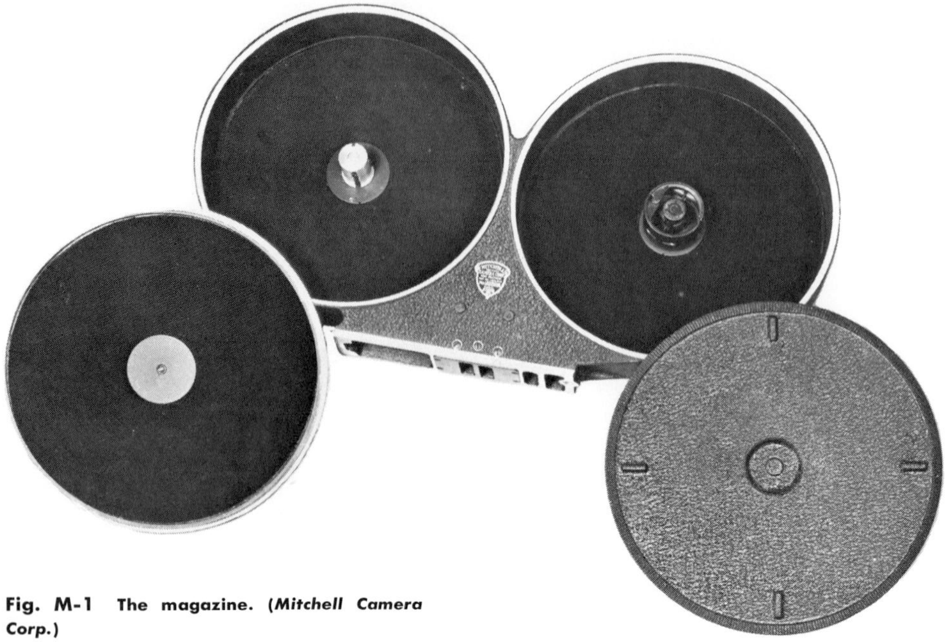

Fig. M-1 **The magazine.** ***(Mitchell Camera Corp.)***

The magazine throat or light trap consists of three velvet-covered rollers held in contact at even pressure by spring tension on the outer rollers. This keeps the magazine lighttight at all times, and the rotation of the rollers as the film is drawn from the magazine prevents film scratches. The loading compartment has a keyway to fit standard film cores, and the take-up spools are contractible.

A leather belt driven by the sprocket-shaft pulley turns the pulley on the magazine take-up spindle. In some cameras, take-up and tension are controlled by small motors.

The magazine is held in place on the camera by a step lip at the front and by a holddown screw at the rear. See Figure M-1 and BIPACK MAGAZINE.

magenta A reddish purple color which is the complementary of green. Dyes of this color, in subtractive systems, produce the images in the green-sensitive layer of emulsion in multilayer film.

Magna-Stripe A process for applying magnetic striping to a film base. The process was developed and introduced by Reeves Soundcraft.

magnetic biasing Magnetic biasing is the simultaneous conditioning of the magnetic recording medium during recording by superposing an additional magnetic field upon the signal magnetic field. In general, magnetic biasing is used to obtain a substantially linear relationship between the amplitude of the signal and the remanent flux intensity in the recording medium. A-c magnetic biasing is biasing accomplished by the use of an alternating current, usually well above the signal frequency range, in the recording head.

magnetic film A sprocketed film base with a magnetic coating for full- or multiple-track audio recordings.

magnetic heads The electromagnetic transducers that are used to convert electric signals to magnetic patterns, or the reverse.

Usually three heads are used in audio systems. The erase head is energized with a very high frequency, well beyond the audio range, designed to destroy any audio signal that may be present on the tape before it passes over the recording head.

The record head is energized with audio frequencies representing sound which is to be recorded. This head receives its signals from a chain of equipment beginning with either a microphone, radio signal, or record player pickup—each of which provides energy that is passed to an amplifying circuit and, in turn, to the recording head. At the recording head, the signal appears as a magnetic force which is alternating at the audio frequency to be recorded. This alternating magnetism is induced into the moving tape.

The playback head is similar to the record head, but it receives magnetic force from the tape as it passes over a small gap in the head. The head converts magnetic signals into small changes in voltage, which are then amplified and fed to a loudspeaker which converts the signals into audible sound.

Properly stored at normal room temperature and kept away from sources that may have a demagnetizing effect, the tape may be played back or reproduced any number of times without loss of signal strength.

magnetic master The edited magnetic sound record that is used for transfer to either magnetic release prints or photographic sound negatives that will subsequently be used for making prints with photographic (optical) sound tracks.

magnetic original The original sound recording on magnetic film or tape.

magnetic recording Although usually associated with the recording of sound and its subsequent playback or reproduction, magnetic recording can also activate mechanisms, instruct computers, and store impulses that can be played back as picture information.

Any one of these functions would have made the technique an extremely valuable one. Its versatility, however, complements other equally amazing developments that are so much a part of this wonderful age of electronic achievement.

Simple factory mechanisms, as well as the more complicated devices that are placed aboard satellites hurtling through space, obey signals on command from a distance of many thousands of miles. Magnetically recorded, the series of signals fed into instrumentation tapes provide a fixed relationship with the receiving equipment that triggers the responding mechanism.

Similar signals have paved the way for the development of an entire industry—computer systems—which, in turn, has influenced the production techniques of almost every other major industry. To these computer tapes the most complicated mathematical problems are child's play. Reduced to simple digital components, the signals recorded on the magnetic tape rearrange themselves in seconds to solve problems that would mean months and years of painstaking computation for a mere mortal.

Computer *logic,* or the process of feeding hypothetical problems to the machine and allowing the computer to make logical assumptions or practical conclusions, gives the machine almost human qualities. Unprejudiced and uncommitted, the computer is able to predict the outcome of an election, for example, based on little more than a minute fraction of the total votes cast and the information that had been previously fed to the computer's memory bank.

Magnetic recording principles are also used to record and store signals that are converted into picture information and transmitted or projected. This system of recording these signals on videotape relies on television cameras which reconstruct the image by scanning procedures similar to the methods used for television transmission.

Basically, variations in light and dark areas of the image produce corresponding variations on the target area in the television camera pickup tube. Converted into electric signals, the various tonal values are then reproduced on the lengths of videotape as invisible magnetic patterns. These patterns, or signals, are fed to a cathode-ray tube where they are scanned, and subsequently, amplified, and transmitted to receiving equipment.

The oxide-coated magnetic tapes that make all these processes and techniques possible are produced in ¼-, ½-, 1-, and 2-inch widths. The mylar or polyester supports for the magnetic coating are extremely flexible and not affected by humidity changes. Unlike motion picture film, magnetic tape has no perforations or sprocket holes that act as guides while the tape travels through a recorder's transport mechanism.

Since magnetic recording tapes do not have to be developed and printed in order to produce visible images, and since the recorded sound or picture information is immediately available for playback, the advantage that tape has over film is obvious. Unlike film, where torn sprocket holes and objectionable surface scratches can be regarded as inevitable, magnetic recordings can be duplicated and played back dozens of times without any appreciable loss of quality.

Sound accompaniment to motion pictures may be achieved in several ways. It may be recorded photographically as variable-area or variable-density tracks, magnetically on $\frac{1}{4}$-inch tape or on 16 or 35mm film coated with ferric oxide over the entire width or over selected bands of the film, or on disks similar to those used with record players. Sound may also be recorded on the same length of film that carries the picture information or on separate lengths of film passing through equipment that is electronically interlocked to operate in synchronism with the motion picture camera.

In recent years, 16mm cameras have been designed essentially to satisfy the requirements of television film services. These all provide the combined features of photographing the scene and recording the sound on one film. This is accomplished by means of a magnetic sound-recording stripe applied to the film stock.

The magnetic stripe consists of a narrow ribbon of ferric oxide capable of magnetic flux in proportion to the frequency and volume of all audible sounds from approximately 40 to 15,000 cycles per second. The magnetic stripe may be applied to the film in three ways: (1) A ready-made independent ribbon is mechanically laminated to the film base. (2) A groove is cut into the film base, and this is filled with liquid oxide. (3) The liquid oxide is coated directly onto the film base.

A magnetic coating can be described as a series of very small bar magnets, aligned head to tail, along the film or tape. During recording, the coated side of the material is passed in contact with the recording head. As the polarity of the recording head is constantly alternating at audio frequencies, the bar magnets become permanently magnetized in a pattern that coincides with the alternating polarity of the recording head.

In the low frequency, the polarity of each bar magnet is indicated by an N for North and an S for South. See Figure M-2*b*. As an alternating current passes through the recording head, one side of the recording gap (or one pole in the head) exhibits maximum north polarity at the peak of a pure waveform but, half a cycle later, the same side of the gap exhibits maximum south polarity. Because of this, the magnetic polarity of the particles throughout the film or tape will also alternate.

This example can only be related to the recording of a pure note. The situation when speech or music is recorded (involving many combinations of frequencies and harmonics) is extremely complex. The highest frequency that can be recorded is limited by the magnetic particle size. Lower frequencies are recorded by several particles acting as one and, in effect, becoming longer bar magnets.

A sync generator associated with the camera drive, creates a synchronizing signal during the filming process that is transmitted by a pair of wires to a tape recorder where it is recorded on the same $\frac{1}{4}$-inch tape as the audio signal. This provides a synchronizing index between picture and sound. The magnetic pulses on the tape are, in effect, magnetic sprocket holes that maintain a constant relationship with the perforated sprocket holes on the camera film.

During reproduction or transfer to sprocket-type magnetic film, the synchronizing signal controls the relative position of the tape with respect to the sprocket-driven film, thus ensuring synchronization. See Figures M-2*a* and *b*.

magnetic recording specifications (NAB) The width of the magnetic track for full-track recordings is 0.238 inch (+0.010 to −0.004 inch). The width of the track for two-track monophonic or stereophonic recordings is 0.082 inch (±0.002 inch), with a center-to-center spacing of 0.156 inch (±0.004 inch). For two-track stereophonic recordings, track 1 carries the recording for the left-hand channel (as heard by the audience), and track 2 carries the recording for the right-hand channel.

The width of the recorded tracks for four-track recordings is 0.043 inch (+0.000 to −0.004 inch). The center-to-center distances between tracks 1 and 3, and between tracks 2 and 4 is 0.134 inch (+0.002 to −0.000 inch). The four tracks are equally disposed across the tape with a tape width of 0.244 inch. The outer edges of tracks 1 and 4 are coincident with the edges of the tape. In four-track monophonic recordings, the track-recording sequence is 1-4-3-2. In four-track stereophonic recordings, tracks 1 and 3 are used simultaneously for one direction of tape travel, and tracks 2 and 4 for the other direction.

Fig. M-2*a* Magnetic recording—the principles.

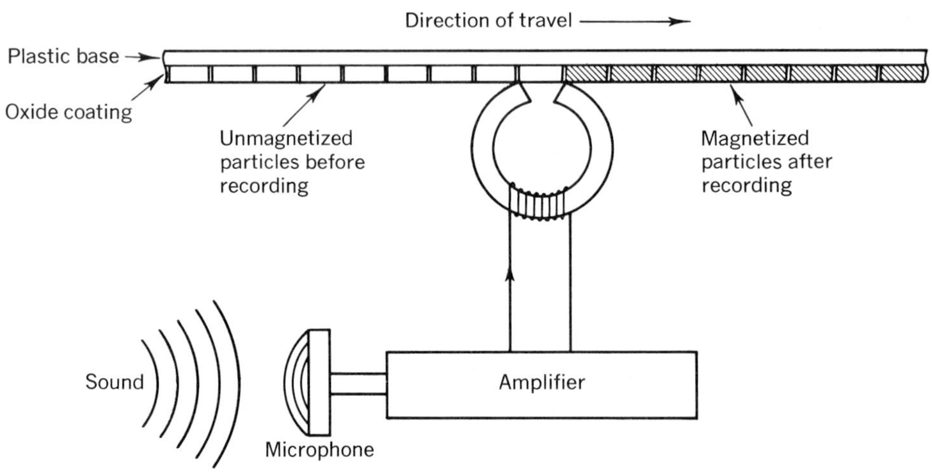

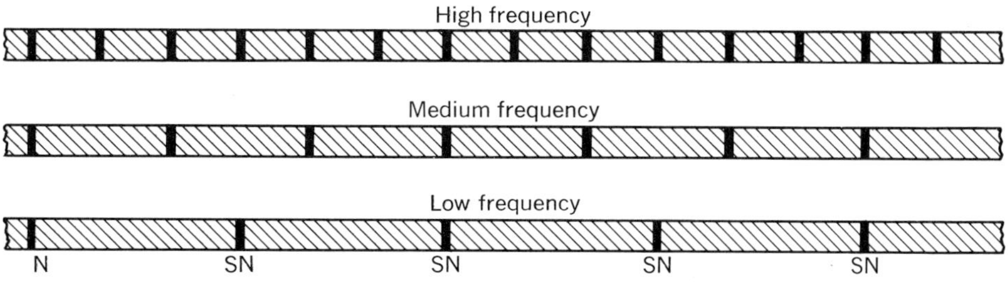

Fig. M-2*b* Magnetic recording—high and low frequencies.

The preferred recording speed (NAB Standards) is 7½ inches per second (ips). The recorded program material should produce the same reference deflection on a standard volume indicator as that produced by a 400-cycle-per-second (cps) sine-wave signal recorded at the NAB standard reference level. The reproduce-system response at 7½ ips from a 7½ NAB 65 test tape should be within the tolerance limits—between 30 cps and 15 kilocycles (kc). The reproduce-system response at 15 ips from a 15 NAB test tape should be within the tolerance limits—between 30 cps and 15 kc. At 3¾ ips from a 3¾ NAB 65 test tape, the reproduce-system response should be between 50 cps and 10 kc. See Figures M-3*a* to *c*.

magnetic recorders

Auricon Filmagnetic. The sound-on-film recording system consists of a twin-head transistor matched Gold Seal camera recording unit with magnetic record and instant monitor heads. The model RPH-11, installed as original equipment in Auricon cameras when ordered (with or without optical sound-on-film in the camera), includes an all-transistorized amplifier model MA-11 with 14 temperature compensated transistors, model E-7 high-fidelity 250-ohm transistor matched microphone, 50 feet of microphone cable, desk

Fig. M-3*a* NAB reproduction standards for recorded magnetic tape, reproducing amplifier output for constant flux in the core of an ideal reproducing head. The letters ips indicate the recording and playback speeds in inches per second.

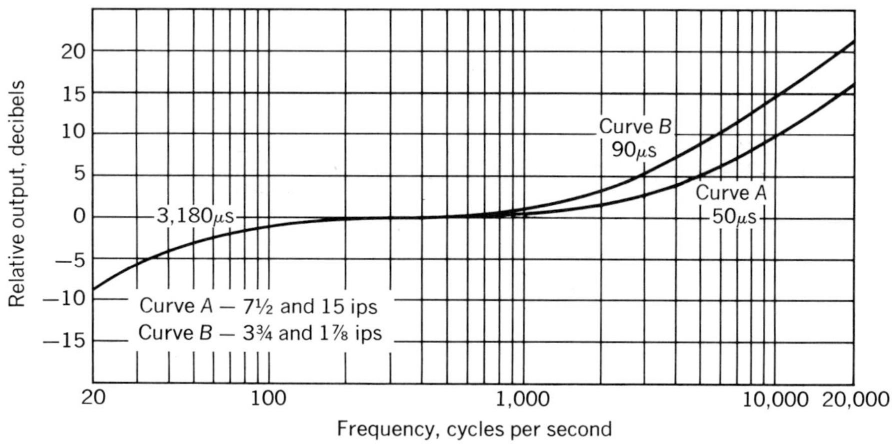

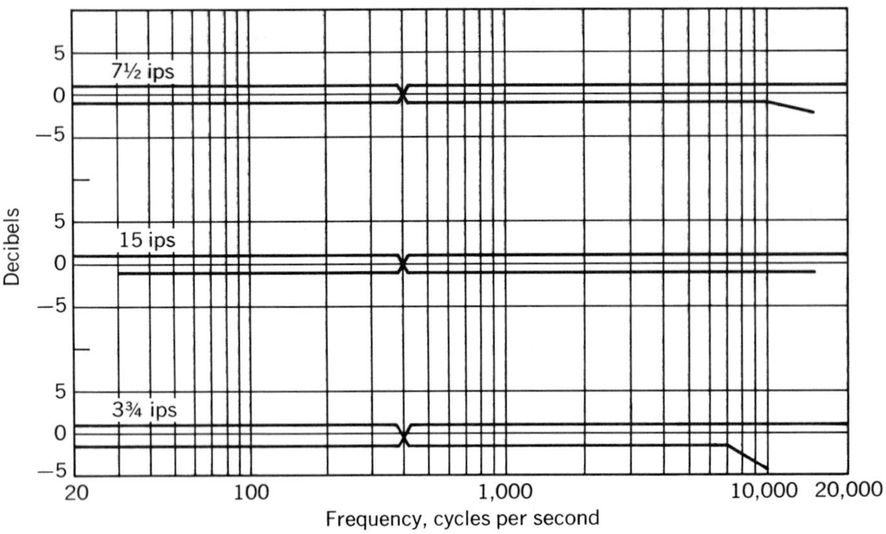

Fig. M-3*b* **NAB standard recorded response limits.**

stand, headphones, and connecting cables. The amplifier operates from a built-in rechargeable battery for complete portability, or from a-c power line current. The unit includes an a-c power pack and battery charger. Model BC-10 operates the MA-11 amplifier from 115-volt 50- or 60-cycle alternating current or the recharged battery. The Filmagnetic outfit is supplied in a fitted dispatch-type carrying case. The transistorized model MA-11 can be used in any Auricon camera that is already equipped to take the former model MA-10 Filmagnetic system.

Fig. M-3*c* **NAB standard reproducing systems response limits.**

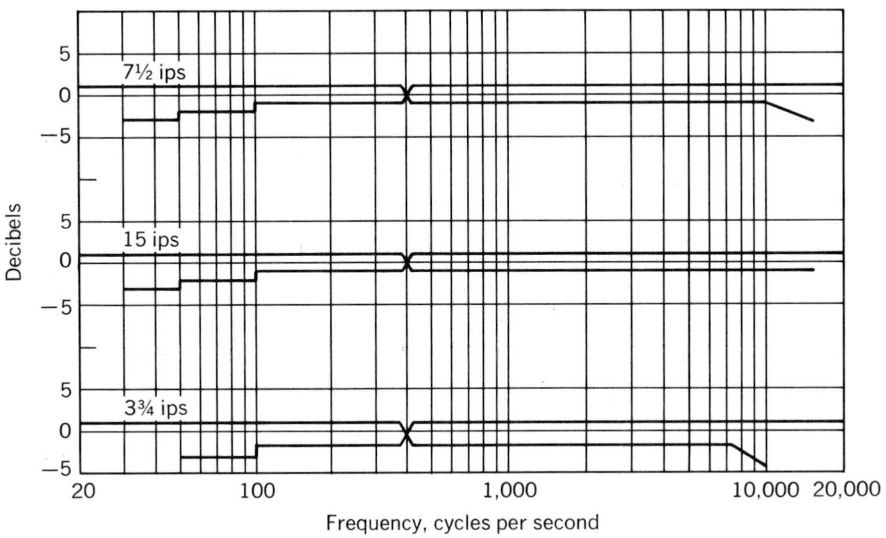

Auricon 1200 model RM-30, 16mm. The 1200 is available for either variable-area or variable-density optical sound-on-film recording. It can be supplied for existing Auricon amplifiers. The unit uses 1,200-foot film magazines (compatible with the Super 1200) for up to 33 minutes of continuous recording. Self-blimped with sponge-rubber interior lining, the unit is completely quiet in operation and can be run within 6 feet of the recording microphone. The recorder will not operate unless the take-up belt is in its correct position and the film is moving normally. The triple-filtered film flow ensures flutter-free and wow-free recording. The 1200 has a push-button control panel with indicator lights that show that the recorder is running film satisfactorily. The recorder is powered with a reversible synchronous motor permitting A- or B-wind emulsion-position sound tracks to be made.

Erlson lip-synchronous recorder-reproducer. Distributed by the Ryder Magnetic Sales Corporation, the recorder features dual-film synchronous recording, instant recorder start and stop with the camera, with frame-to-frame synchronization throughout the recording. The unit uses 1/4-inch sprocket-driven magnetic film, and the recorder's drive speed matches the camera speeds—16, 18, and 24 frames per second. An interlock generator on the camera controls the recorder servo. An auxiliary capstan motor eliminates film stretch. No clapsticks or interscene markings are required. Self-contained Dryfit storage batteries supply the power for the recorder.

Magnasync X-400-E-1. The X-400 consists of two units which may be operated side by side for console operation or stacked on each other for field use. Positive projector interlock can be made with a simple flexible cable gearbox attachment. Frequency response is 50 to 8,000 cycles. The 400-foot film capacity may be increased to 1,200 feet by adding extension arms. The standard unit is furnished with a 110-volt a-c 60-cycle single-phase salient-pole synchronous motor. Other drive motors, such as 50-cycle 3-phase 220-volt or selsyn interlock are available.

A 3½-inch VU meter monitors the input signal at all times. The monitor output will feed any impedance above 500 ohms. The operator can monitor directly through the amplifier or switch to the Film Monitor for simultaneous playback from film approximately 1/5 second after the recording is made. The unit has a dialogue-equalization switch, high-gain low-impedance microphone input, and a high-impedance input. The unit consumes only 60 watts, including amplifier and motor drive.

Magnasync X-400-E-15, 16mm. This recorder will accommodate standard 1,200–foot film reels for 33 minutes of recording. Its frequency response is 50 to 8,000 cycles. A simplified single switch controls all motor functions. The unit features direct-coupled high-resistance rotor torque motors, simultaneous film monitoring through a separate play head, a record indicator lamp which protects recorded takes, and an accurate, chain-coupled, easy-to-read footage counter. The unit is equipped with a 110-volt a-c 60-cycle single-phase salient-pole synchronous motor. A 220-volt 50-cycle three-phase selsyn interlock or other motors are available. A $3\frac{1}{2}$-inch VU meter monitors the input signal.

Magnasync series 2100. The series 2100 recorders and reproducers are available with all standard film speeds and voltages for complete international capability—for 16, 17.5, or 35mm synchronous magnetic film. The modular design permits combination with other existing units such as Magnasync's tube-type equipment, and virtually all other tube equipment, so that one may gradually acquire a complete series 2100 solid-state system at a minimum amount of immediate cost.

Magnecord 1000 series. The 1000 series Magnecorders feature smooth tape transportation and efficient braking; broadcast standard connectors, impedance, and levels; a cueing speaker with separate volume control, and its amplifier to drive an external speaker; provision for earphones, mixing input with separate gain control; and other equally important details.

The one-hand cue system gives the recording engineer complete control. The cue button lifts the tape to the heads, releases the brakes, and puts low even torque on the reels. The recorder can be remote-started and stopped in a preset mode. The operation can be handled with a single-pole single-throw switch. The recorder also features a solid-state system with transistors selected for low heat, high efficiency, and reliability. Power consumed by the electronics is less than 20 watts. The power supply is regulated so it can cope with both line voltage and load variation.

Nagra $\frac{1}{4}$-inch sound recorder. The Nagra is a self-contained, battery-powered, fully transistorized, $\frac{1}{4}$-inch-tape synchronous sound recorder. The Nagra is normally equipped for mixing with a single microphone plus either a line or bridging input. By the addition of a BS preamplifier, the line input can be converted to a second microphone input for two-microphone mixing.

Using 15-inch speeds, the recorder provides a flat frequency response to within 1 decibel from 30 to 18,000 cycles. A high-powered amplifier provides 2 watts of power for undistorted reproduction. The self-contained auxiliary speaker-amplifier provides remote quality monitoring during either recording or reproduction. The auxiliary power unit can be used as a synchronizing transformer when the camera is being driven by a 60- or 50-cycle power supply. The power unit can also be used to provide external power to the Nagra, operating independently of the internal batteries.

The recorder uses a closed-loop servo drive to keep flutter and wow at a minimum. Three speeds are available with no mechanical linkage, as the speed is controlled electronically by tuned resonant circuits. The recorder is equipped with stroboscopic disks for speed comparison with 60- or 50-cycle power sources. The auxiliary speed variator can be used to change the speed of the recorder for purposes of music-pitch change during recording or to correct for improper camera speed during synchronous transfer from $\frac{1}{4}$-inch tape to sprocket-driven magnetic film.

The time-sync generator provides complete portability of the recorder with no cable connections. The camera can be driven separately from a time control or power line. With high-frequency bias, this system provides the trouble-free synchronization. The auxiliary synchronizer is used for synchronous transfer from tape to sprocket-driven film and for synchronous playback of prescored music for lip synchronization during camera action. It can also be used as a source of a synchronizing signal when the camera is being driven by a 60- or 50-cycle power supply.

For the synchronous transfer of $\frac{1}{4}$-inch tape from the Nagra to sprocket-driven magnetic film, a bloop can be introduced on the transfer sound at the point of camera start. Included in the Nagra is a bloop oscillator which can be made to function automatically with the flashing of a hand-held bloop lamp in front of the camera or the flashing of a bloop lamp installed in the camera. See Figure M-4*a*.

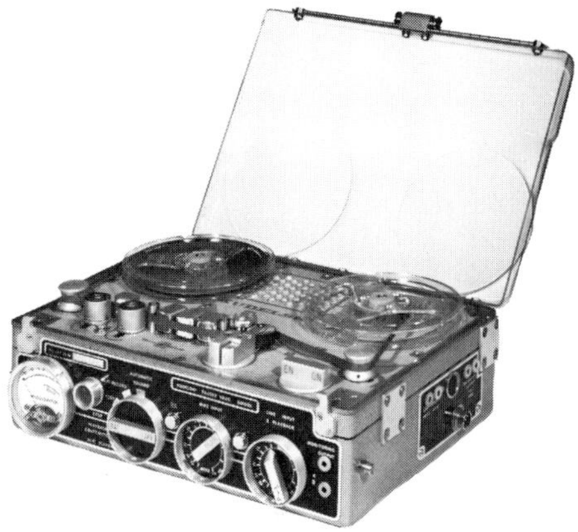

Fig. M-4*a* **The Nagra recorder.**

Nomad Mark 1 recorder-reproducer. A seven-pound, completely transistorized magnetic sound recorder-reproducer. It employs split 16mm sprocketed magnetic film and provides two separate sound tracks (one each for voice and music) which are automatically mixed when played back. This method eliminates the need for sound dubbing and duping of original film where a composite print must be made in order to combine picture and multiple sound tracks at lowest possible cost. The Nomad is mounted beneath the camera. Positive interlock is achieved in either filming or projection by means of a flexible cable. Almost any camera can be converted to a sync-sound camera by employing the Nomad. The unit may be driven with a d-c governor-controlled or a-c synchronous motor which drives both camera and recorder.

The Nomad is supplied complete with remote-control hand mixer, built-in recorder-reproducer amplifier with nickel-cadmium battery, 100-foot roll of magnetic film, take-up reel, and detailed operation handbook.

Perfectone $\frac{1}{4}$-inch magnetic film recorder. A professional, high-quality $\frac{1}{4}$-inch recorder designed to meet the requirements of the industry. Self-contained, including batteries, and lightweight (16 pounds), the equipment is easily portable for location shooting. When used with a sync generator the Perfectone records perfect lip sync.

The EP-6-AR is the standard professional recorder. The EP-6-A2 is the same as the model EP-6-AR with a built-in two-channel mixer.

Rangertone model DC-5, ¼-inch magnetic recorder. A synchronous ¼-inch recorder for studio use. The 4,800-foot reels record one hour at 15 inches per second, or two hours at 7½ ips. It is vacuum-tube-equipped throughout.

Signal-to-noise ratio is 60 decibels against a 1 percent harmonic distortion level. The tape transport comprises a tight loop over a magnetically coupled oversized flywheel. Record and reproduce heads are mounted in close proximity to the flywheel to reduce record-reproduce flutter to a minimum of less than 0.05 percent. The synchronous drive motor is coupled to the tape capstan by a conical drive mechanism. A knob and pointer provide precise tape speed adjustment with a range of ± 0.5 percent to compensate for tapes recorded off speed. A second record head is provided to record the sync control signal in the center of the ¼-inch tape.

The two heads have a gap angle inclined a few degrees to compensate for tape weave. The synchronizer (type S-1008) amplifies the reproduced sync control signal and compares it with the 60–cycle power driving the projector or sound recorder. Correction is supplied to the miniature motor which is coupled through a gear train to the conical tape-drive mechanism. The tape speed is automatically corrected until it is in exact sync with the projector or recorder. A feedback circuit retards the speed correction so that it is inaudible. A frame indicator shows the accumulated error in sync and can be used to correct sound originally recorded out of sync. A small oscilloscope is provided to monitor the recorder's performance.

Sela mixer 2880. For unusually large motion picture and television sets, location recording, multiple camera setups, music recording, and in all instances when several microphones are needed, the Sela mixer is an extension of the Nagra's recording capabilities.

Operating directly from the Nagra, the mixer provides control over four microphones, whether dynamic or condenser, with overload capabilities considerably in excess of those of the Nagra mike input itself and the excellent noise characteristics. The normal Nagra mike input can be used simultaneously, thus making five mikes available. When necessary, several 2880s can be used together. See Figure M-4*b*.

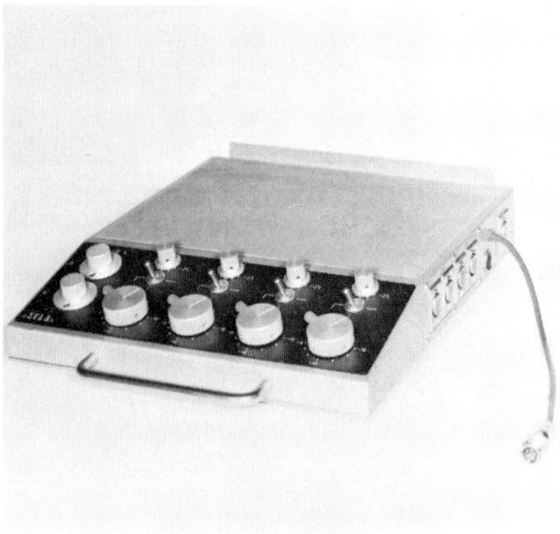

Fig. M-4*b* **Sela mixer 2880.**

Stancil-Hoffman R-70 series. The transport is mated with either integral electronics for a compact, self-contained assembly with preset gain controls or external electronics offering microphone and high-level mixing. With the integral electronics, from one to four tracks of record, playback, automatic gain control, monitor and voice or signal operation are included on a single chassis using the A-70 series of plug-in-modules. All motors and chassis are directly mounted on a rugged ¼-inch dural panel. The basic R-70 transport includes the drive motor, torque motors, push-button controls, counter, monitor switches, and head assemblies.

To handle the various modes and terminations of operation and provide d-c power for the solenoid and dynamic braking, a control-unit chassis is located on the rear side of the transport. When using the integral electronics, control unit CU-76 contains the regulated power supply, bias oscillator, plug-in mode relay and all the automatic control circuitry and connectors. The integral electronics offer the widest applications of magnetic recording. The units may be clock-operated, signal-operated, or remotely operated for delayed broadcast, monaural or stereo, and logging. Complete circuitry is furnished for automatic transfer from one unit to a second unit, pause or reversal.

Switching from high-speed rewind directly into play without stress on the thinnest tape can be accomplished safely. The reels come to a stop and then proceed to the next mode. Differential braking is accomplished by a leaf-spring contact assembly which feels the tape direction to balance the power to either the feed or take-up motors.

Two four-pole torque motors drive the reels for smooth take-up and holdback tension, high-speed winding, and dynamic braking. Designed to handle up to $9\frac{1}{4}$-inch-diameter reels, 6,900 feet of the triple play tape, the R-70 series will record continuously for two days at $\frac{15}{32}$ inches per second or three hours at $7\frac{1}{2}$ inches per second. Using the recorder's interlaced tracks, the tape may be turned over to double this time.

The film transport contains the record and playback heads, the sprocket-drive system, and a motor panel at the rear for mounting either a synchronous motor or an interlock motor. Normally, Stancil-Hoffman motors are supplied with the equipment although sufficient space is provided to accommodate any of the types of motors that are used in motion picture recording equipment.

The motors are coupled to the sprocket system by means of a chain loop. When a change in drive speed is desired, or when changing to a supply voltage of different frequency, the size of the drive sprocket pulley is changed. The film sprocket is coupled to the motor drive by means of a magnetic clutch to permit cueing. The footage counter is coupled directly to the film sprocket so that accurate footage measurements are maintained at all times, including those periods when the sprocket is in free-wheeling condition.

The S-7 model incorporates a mechanical filter system. Two large balanced flywheels provide recordings with negligible flutter, and two self-centering stabilizing arms, one coupled to a pneumatic dashpot, stabilize the tape motion during the start period. The stabilizing arms minimize the effects of splices or damaged film, and also provide an automatic stop in the event of film breakage, or at the end of the film supply.

Stellavox $\frac{1}{4}$-inch magnetic recorder. The recorder derives its power from self-contained flashlight batteries and features both automatic and manual volume controls. It can be used with a sync-signal generator and weighs less than 6 pounds.

Tandberg model 11 Pilotone. The series 11 is a two-track, fully transistorized, battery-driven monaural tape recorder. The recorder has three speeds: $7\frac{1}{2}$, $3\frac{3}{4}$, and $1\frac{7}{8}$ inches per second. One control lever selects fast forward and rewind, and normal forward drive for recording and playback. Selection of recording, playback, and monitor loudspeaker is by means of push buttons, and there is also a push button for automatic regulation of the recording level. Specially developed electronic circuitry ensures precise indication of the recording level on a VU meter.

In addition to the separate erase, record, and playback heads, a fourth head is provided for pilot tone to permit synchronization. Mixing is possible, as there are separate volume controls for microphone and line inputs. A separate control adjusts playback volume. A balanced input with an impedance of 200 ohms is provided for a dynamic microphone, and a balanced output of 60 ohms is used when playing back into an external tape recorder for copying purposes. A monitoring output for 200 ohms headphones is also available. An electronic speed regulator ensures precise speed plus or minus 1 percent. Power is provided by 10 cells of $1\frac{1}{2}$ volts each.

3M $\frac{1}{4}$-inch professional recorder. Featuring Dynatrack electronics and the Isoloop tape transport, the 3M recorder makes possible the recording of the synchronizing or pilot tone from the motion picture camera and sound-track audio on the same quarter-inch tape.

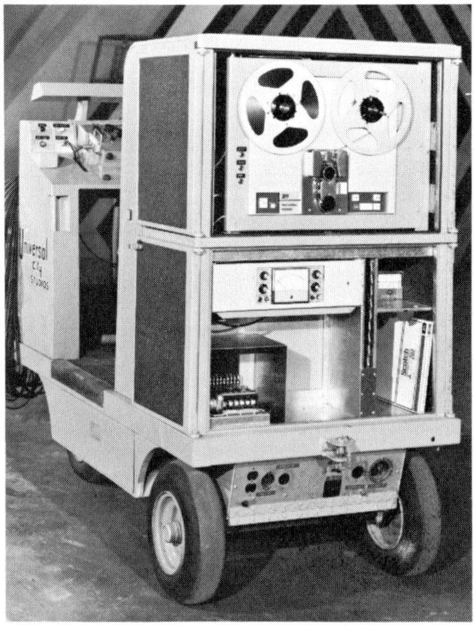

Fig. M-5*a* The 3M film sync recorder. Shown here mounted on a Universal City Studios dolly, it records the synchronizing tone and the sound-track audio on the same quarter-inch-wide magnetic tape.

The recorder consists of a standard quarter-inch two-track transport combined with a Dynatrack electronics module. The Dynatrack system records two low-noise tracks simultaneously with a single signal at different sound levels. In the film-sync module, the NAB audio functions are unaltered, but the second track is converted to an independent record-playback system for the pilot-tone track. Recording a synchronizing or pilot tone from the motion picture camera on the same tape as the sound-track audio to ensure playback synchronization of the film with the voice track, the recorder's capstan motor is controlled by a pilot track to compensate for any changes in camera or recorder speed.

To make the system as simple as possible, some equalization and signal-level compromises are made on the pilot track. The normal NAB low- and high-frequency record-current preemphases are replaced by a flat record current. The reference level of the pilot playback preamp is limited to -20 decibels from a very low impedance output stage to reduce overall power requirement and to eliminate the need for an extra line amplifier. The Isoloop tape transport reduces the unsupported tape path from the usual 12 to $3\frac{1}{2}$ inches, greatly cutting flutter. See Figures M-5*a* and *b*.

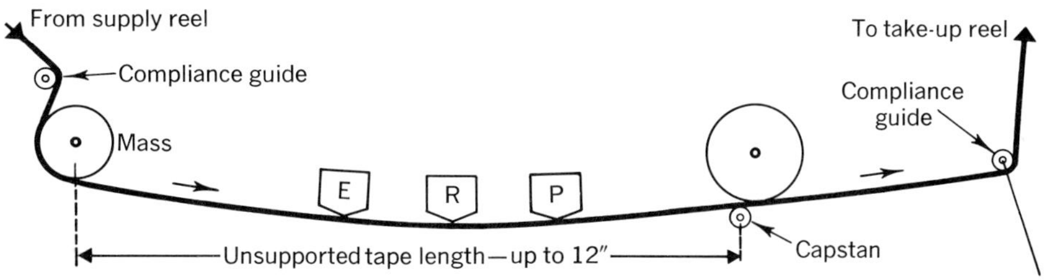

Fig. M-5*b* Conventional tape threading and the Isoloop. (3M Company.)

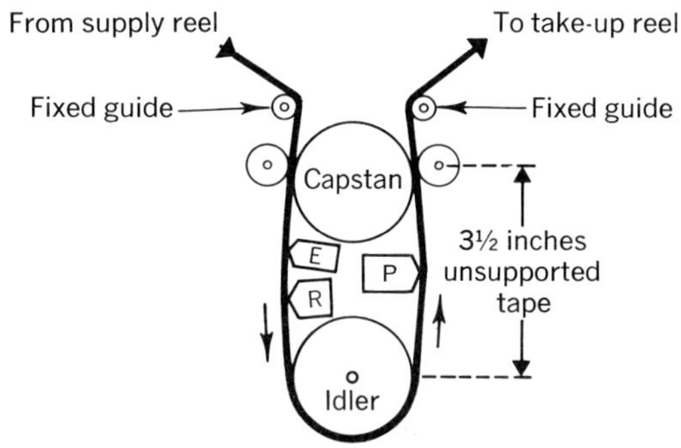

The Isoloop tape transport reduces the unsupported tape path from the usual 12 inches to $3\frac{1}{2}$ inches, greatly cutting flutter.

It has long been recognized that the tape transport is the heart of any professional tape-recording equipment. The transport must move a length of tape over a fixed point at constant speed with extreme precision for the life of the equipment with minimum maintenance. It is also necessary to move the tape gently and rapidly in fast-forward or rewind under complete control to prevent tape spills, stretching, or breakage. More than any other single factor, the performance quality depends on the construction of the tape transport. Unless the transport has the capability of moving the tape smoothly at a constant speed over the heads, no amount of correction in the circuitry can produce top performance.

Uher magnetic tape recorder, $\frac{1}{4}$-inch model S, 4000 series. A fully transistorized professional recorder combining the quality of the fine a-c-powered recorders with the compactness and portability of battery-operated equipment. The four-speed recorder has the following frequency responses: at $\frac{15}{16}$ inches per second (ips), 70 to 5,000 cycles per second (cps); $1\frac{7}{8}$ ips, 50 to 11,000 cps; $3\frac{3}{4}$ ips, 50 to 18,000 cps; and $7\frac{1}{2}$ ips, 50 to 22,000 cps, all at ± 3 decibels.

The recorder accommodates a 5-inch reel under a hinged cover, providing 32 minutes of continuous play on one track of tape at $7\frac{1}{2}$ inches per second. It can be operated with a pulse-sync generator for lip-synchronization recording in the production of sound motion pictures. The recorder is $3\frac{1}{8}$ inches high, $10\frac{1}{2}$ long, and $8\frac{1}{2}$ deep.

Uher 1000 report pilot. A battery-operated portable tape recorder operating at $7\frac{1}{2}$ inches per second. The recorder, featuring sound synchronization for motion picture cameras, is fully transistorized and has a frequency response of 20 to 20,000 hertz. Features include interruptable automatic photoelectric level control and overhead filter, test buttons for pilot-tone level, and battery condition and continuous stroboscopic speed control.

Westrex series 1200. The Westrex series 1200 magnetic recording systems are available for 35, 17.5, or 16mm use. They include a two-position mixer, a magnetic recorder, a set of interconnecting cables, and two headsets.

The mixer (LRA 1592) is a small portable unit which contains audio facilities for operating from two low-impedance microphones.

Each of the two channels has separate dialogue equalization in three steps, provision for monitor, and visual indication of recording and reproducing levels. A two-tone oscillator is included for test purposes.

The mechanism for both the PO 35–17 and PO 16 recorders is mounted within a carrying case. All the controls are located on the front panel. Each of the recorders may be operated in the reverse direction at normal film speed for rewinding. A counter with a manual reset counts the footage in either direction.

The transistorized bias oscillator and monitor amplifier within the recorder are of the plug-in type and can be removed for servicing. The three-phase synchronous motor is controlled by means of a relay-actuated circuit.

Westrex LRA-1552. The LRA-1552 recorder is equipped with torque motors for take-up and fast rewind, and with removable magnetic-head assembly. Modular plug-in, solid-state units with full use of silicon transistors are provided in the electronic circuits. Three types of motors are available on request: synchronous, interlock, and composite, 50 to 60 cycles.

Similar in design to the LRA-1552 recorder, the LRA-1551 reproducer provides high-quality sound reproduction from one or more magnetic tracks.

Westrex series 900. The series 900 recording systems are available for 35 and 16mm film, photographic and magnetic, standard-area negative and direct-positive-area sound-track recording. They include the Westrex RA-1596 transmission cabinet, a choice of mixers to meet the studio's individual requirements, and the Westrex RA-1581 recorder.

The RA-1581-A recorder is designed to record variable-area direct positive or negative standard tracks or double-width push-pull sound tracks on 35mm film in synchronism with the motion picture film.

The RA-1581-C recorder is designed to record variable-area direct positive or negative standard sound tracks on 16mm film in synchronism with motion picture film.

magnetic resolving or transferring magnetic $\frac{1}{4}$-inch tape to sprocket-driven film (sound) There are two methods of resolving or transferring from synchronous magnetic $\frac{1}{4}$-inch tape to sprocket-driven film. In one system, the synchronizing signal is picked up from the tape, amplified, and used to drive a synchronous motor on the sprocket recorder. In this case, the speed of the tape reproducer controls the speed of the sprocket recorder. This system may be used with complete satisfaction on any of several types of recorders as the synchronizing signal recording is done with high-frequency bias.

The second system of resolving is available only with the Nagra or fixed-plant tape decks equipped with servo-loop control. The speed of the Nagra motor drive is under servo-loop control during both recording and reproduction.

In resolving, the speed of the Nagra is controlled by the speed of the sprocket recorder which is driven by a synchronous motor. This system is helpful in major studio operation where most recorders are driven by three-phase 220-volt 60-cycle synchronous motors. It provides the only present system of playing back ¼-inch tape in dual film with television film projectors or ¼-inch tape in dual film with videotape reproduction.

magnetic sensitometry Film "sees." Magnetic stripe "hears."

Magnetic characteristics are plotted in a manner similar to those used for plotting H & D curves for film. Input and output are linear and fidelity is best along the straight-line portion of the graph. A voltage bias is introduced to ensure operation along this linear portion. Changing the bias affects the sensitivity (speed) as well as the shoulder. The dynamic range, similar to the density range in film, is the distance from shoulder to zero-signal noise level.

Overloading results when the signal is driven into the shoulder, the equivalent of blocked highlights in film. Modulation noise can be compared with fog in film. See Figure M-6.

magnetic sound film A film base with perforations along one or both edges with a brown ferromagnetic coating covering the entire film or running in stripes along the film's entire length. The ferromagnetic coating is capable of accepting and reproducing recorded sound images or patterns.

Fig. M-6 Magnetic sensitometry.

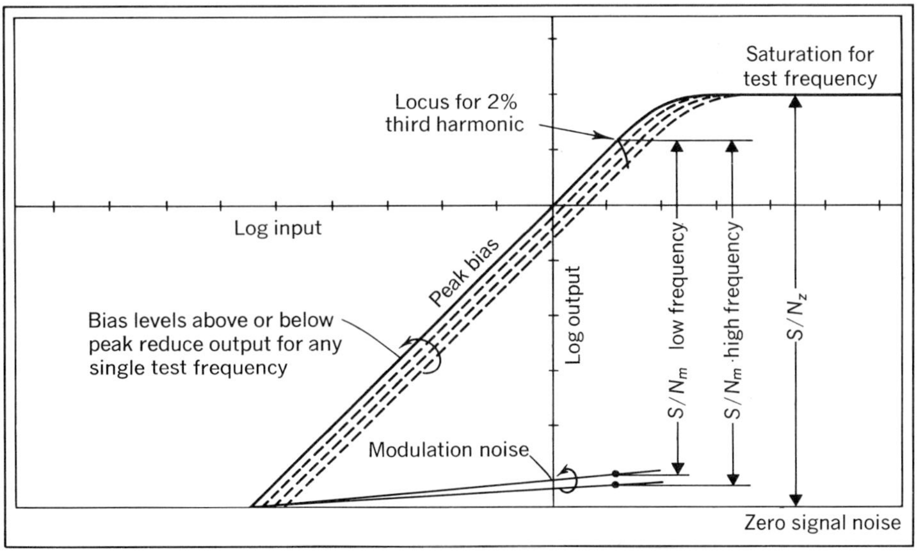

Agfa-Gevaert sound-recording films.

S.T. 4, *Type* 2.54. Type 2.54 is intended for recording by the variable-area method, and is characterized by an extremely fine grain and a steep gradation. The sharpness of the sound track and the absence of ground noise are remarkable. The most suitable densities to assure finest reproduction with a given equipment must be found experimentally and preferably by making cross-modulation tests. In common practice, the combination of a sound negative density of about 2.3 and a black-and-white print density of about 1.5 gives optimum results. For printing on Gevacolor positive film, type 9.53, the following combination is recommended: sound negative density 2.3 and print density 1.7.

Color sensitivity:	Blue-sensitive only
Recommended gamma:	2.7 to 3
Track density:	About 2.3
Resolving power:	180 lines per millimeter
Edge marking:	The letter S just before the footage number

S.T. 6, *Type* 2.56. Type 2.56 is intended for sound recording by the variable-area system, and is characterized by its ultrafine grain and its high contrast. Its sensitivity is lower than that of S.T. 4, and the film is thus suitable for use in more recent recording equipment having a high light intensity. Because of the total absence of light scatter at the densities used in practice, and the high resolving power of the emulsion, the sound track obtained is exceedingly well defined. Reproduction is therefore extremely faithful, especially in the high frequencies.

The most suitable densities to assure finest reproduction with a given equipment must be found experimentally and preferably by making cross-modulation tests. For example, the combination of a sound negative density of about 2.5 and a black-and-white print density of about 1.5 gives optimum results. For printing on Gevacolor positive film, type 9.53, the following combination is recommended: sound negative density 2.8 and print density 1.5.

Color sensitivity:	Blue-sensitive only
Recommended gamma:	3 to 3.5
Track density:	2.5 to 2.8
Resolving power:	200 lines per millimeter
Edge marking:	The letter T just before the footage number

GevaSonor Film, Type 2.01, A perforated triacetate safety film with an extremely sensitive oxide layer whose polished surface gives greatly reduced head wear and remarkable freedom from dropout. In addition, the superior erasure and echo-effect characteristic of these magnetic materials make them the ideal sound-recording medium from both the quality and economic standpoints. In 35mm, it may be supplied fully coated (type 2.01) or striped (type 2.21). Other sizes are only supplied fully coated.

GevaSonor Film, Type 2.02. A perforated triacetate safety film with a lacquer containing very sensitive iron oxide for the magnetic recording of electric signals. Type 2.02 has been designed for carrying out special tasks when a high-sensitive magnetic film is needed. Therefore the thickness of the iron oxide layer has been increased resulting in greater sensitivity (approximately 4 decibels) and a lower total harmonic distortion. GevaSonor film, type 2.02, may be supplied fully coated in 35, 17.5, and 16mm widths.

DuPont sound-recording films. See Table M-1.

Eastman sound-recording films. See Table M-2.

Ilford sound-recording films. All Ilford magnetic sound films are coated on 0.0005-inch triacetate base and have standard perforations (BSI). There are three main groups of 35mm film. The first is fully coated from edge to edge, the second coated only between the two sets of perforations, and the third coated with a narrow ribbon or stripe 0.300 inch wide and an even narrower balancing stripe near the opposite edge of the film. These three basic groups are subdivided into 34 different types. All these 35mm films are used in the same way for original sound recording in the studio and for sound-dubbing and mixing operations. Choice of film is largely a matter of preference related to the following conditions:

1. The edges of fully coated material are tightly wound, and dirt cannot accumulate in the roll.
2. Fully coated material brings the magnetic oxide (which is abrasive) into contact with the sprocket teeth which may cause undue wear.
3. It is sometimes an advantage to have a clear uncoated edge on which instructions or cue marks can be made with a grease pencil, and which also helps to relate sound to picture in editing.

Visible footage numbering can be applied by the film laboratory or the studio editing department to 35mm film coated beween the perforations and to stripe-coated material. Ilford's 16mm magnetic sound-recording film is available as fully coated with single perforations, fully coated with double perforations, and double perforated with one clear edge. The last-mentioned type provides an area in which editors may record cue marks or in which footage numbering may be applied. Also available is a 17.5mm fully coated film that is perforated on one edge. See Figure M-7.

These magnetic films are available with the emulsion wound either in or out, edge dyed in green or blue, and in standard lengths of 1,000, 1,200, 1,500, 2,000 and 2,400 feet.

Ilford magnetic sound-recording tapes. Ilford tapes carry oxide coatings similar to those used in their magnetic films, but they employ much thinner support or base materials. The tapes are available in lengths up to 2,400 feet on spools or up to 7,200 feet on NAB cores.

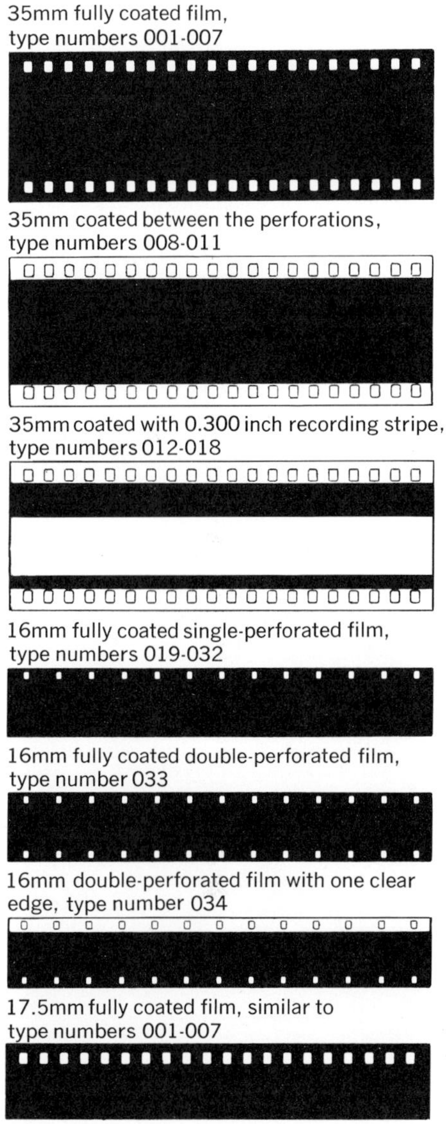

Fig. M-7 The position and size of various magnetic film coatings. ***(Ilford Ltd.)***

TABLE M-1 DuPont Sound-recording and Television-recording Films

Characteristics	Sound recording			Television recording	
		NH			
	Fine grain VA: 833B, 35mm	802B, 35mm	802A, 16mm	834B, 35mm	834A, 16mm
Description	A fine-grain film for variable-area sound-recording using white light. Responds well to high-gamma development. NH-base VA direct recording.	A high-speed variable-area or variable-density recording film for use in 16mm and 35mm recorders. The nonhalation base enhances image sharpness.		A fine-grain low-contrast emulsion designed for photographing either negative or positive images on television monitor tubes. High speed—$2\frac{1}{2}$ times faster than 824.	
DuPont relative speed indices: incandescent	25	45	45	25	25
Recommended developer	125-D	123-D	104-D	123-D	101-D
Approximate development time at 68°F (machine)	5 minutes	7 minutes	$3\frac{1}{2}$–5 minutes	$2\frac{3}{4}$ minutes	7 minutes
Recommended gamma control	3.00	3.00 + VA	0.35–0.50 VD	1.60	1.20
Type of perforations	Positive*	Positive,* P2S (P 8mm)	P1S wind A or B*	Positive, P2S*	P1S wind A or B*
Standard roll lengths, feet; spool and core types	1,000 no. 25	1,000 no. 25	1,200 no. 32	1,000, 3,000 (1 splice) no. 24	1,200 no. 32
Identifying character	CF	C	C	CJ	CJ
Footage-numbered?	Yes	Yes	No	No	No
Safelight	DuPont S-55X or equivalent				

*Short pitch (approximately 0.2 percent less than ASA standard distance between leading edges of any two successive perforations, dimension B).

Note: Some of the above films are available in 32mm, 35mm perforated 32mm, and 70mm.

TABLE M-2 Eastman Sound-recording and Television-recording Films

Name of film	Type no. 35mm	Type no. 16mm	General description	Code letter identification* (precedes footage number)	Frame-line marked*	Footage numbered†
Fine-grain sound recording	5373	7373	A film designed for variable-density sound recording, using a tungsten light source. Can be processed in regular motion picture negative developer.	S	Yes	Yes
Fine-grain sound recording	5375	7375	A film designed for variable-area sound recording, using a tungsten light source. May also be used for direct-playback purposes.	S	Yes	Yes
Television recording	5374	7374	This material is intended for making kinescope recordings of television programs using monitor tubes having either blue (P-11) or ultraviolet (P-16) phosphors. May also be used for making low-contrast prints from either kinescope or regular production negatives when such prints are needed for television transmission.		Yes	Yes
Magnetic sound recording	A704	A704 (also in 17.5mm)	A magnetic sound-recording film designed to give high output, yet with an excellent print-through ratio. The magnetic dispersion used gives unusual abrasion resistance and permits a smooth coating resulting in low distortion. Precise coating techniques and close manufacturing controls provide the ultimate in product uniformity.	Product identified on base side every 15 inches		

*On 35mm films only.
†All 35mm films are ink-footage-numbered; all 16mm films are latent-image footage-numbered.

TABLE M-3 Magnetic Sound-recording Speed

Tape length, feet	*Continuous recording time, minutes, at various recording speeds*		
	$3\frac{3}{4}$ ips*	$7\frac{1}{2}$ ips*	15 ips*
600	32	16	8
900	48	24	12
1,200	64	32	16
1,800	96	48	24
2,400	128	64	32
3,600	192	96	48
7,200	384	192	96

*Inches per second.

magnetic sound-recording level The recording level on a magnetic tape is the level measured by a standard reproducing system with respect to the NAB standard reference level, expressed in decibels.

magnetic sound-recording speed See Table M-3.

magnetic sound-recording tape Professional $\frac{1}{4}$-inch recording tapes have oxide coatings that are similar to those found on magnetic films, but they employ much thinner support or base materials. They are mainly used in the motion picture, television, radio, and music-recording industries where the highest possible sound quality is essential. Tapes are available in lengths up to 2,400 feet on spools, or up to 7,200 feet on NAB cores. (NARTB is an abbreviation for the National Association of Radio and Television Broadcasters. The cores, however, are referred to as NAB cores.)

To obtain the highest possible quality in any recording, high tape speeds are always used. Professional tape recording usually implies a tape speed of 15 inches per second. At that speed, a 7,200-foot roll of tape provides an uninterrupted recording time of 96 minutes.

Continuous-recording times do not necessarily indicate the total recording capacity of a tape. Most tape-recording equipment provides for dual-track recordings—one track is first recorded; then the direction of the tape is reversed and the second track is recorded alongside the first. Recording capacities for dual-track recording are twice as great as the continuous recording times indicated in the table. For example, although a 2,400-foot roll of tape has a continuous-recording time of 64 minutes at $7\frac{1}{2}$ inches per second, it has an effective capacity of 128 minutes of playing time at the same speed.

In addition to their use in conventional sound-recording techniques, $\frac{1}{4}$-inch tapes are also used extensively in conjunction with motion picture cameras and in the production of feature films and television commercials. This is possible only because the motion of the tape can be precisely synchronized to the picture film speed.

One very popular technique for achieving this is sync-pulse generation. With this technique, synchronizing pulses at frequencies much higher than the audible range are generated in the motion picture camera and fed to the tape-recording equipment. On reproduction, these pulses are fed to a control system which automatically governs the tape speed to maintain synchronism with the motion picture film.

magnetic striping Perforated film that has one or more stripes or bands, in varying widths, coated with ferromagnetic compound. These stripes run lengthwise along the film roll.

magnetic transfer A magnetic sound track on film produced by re-recording from an original magnetic sound record. See MAGNETIC RESOLVING.

magnification Abbreviated M, the term magnification comes up frequently in close-up photography. It refers to the proportionate sizes of the image formed by the lens and the actual size of the subject matter. The relationship is expressed by the formula

$$\text{Magnification} = \frac{\text{distance between lens image and film plane}}{\text{distance between subject and camera lens}}$$

If the subject matter on the film appears to be actual size, the image is referred to as lifesize or a 1:1 magnification. If the subject matter is recorded on the film twice its actual size, the magnification is 2.0; if the actual subject is 10 percent larger than its image on the film, the magnification is 0.1.

The image is reduced when the image distance is smaller than the object distance, and proportionately increased when the object distance is reduced and the image distance is increased. Larger images are achieved by using long-focus lenses which increase the image distance.

During projection, an image is normally magnified—its size being proportionate to the distance between the projector and the screen. See PROJECTION FORMULAS.

magnifying power The ability of a lens to enlarge the image of an object. The magnification of an area is the square of the linear magnification. The magnification of a lens is found by dividing the distance of 10 inches by the equivalent focal length of the lens. For example, a 2-inch lens has a linear magnifying power of $10 \div 2 = 5\times$.

magoptical print A print that has optical (variable density or variable area) sound tracks as well as magnetic sound tracks.

magoptical track Lengths of film containing both magnetic and optical sound tracks. Arranged as needed (alternating between each), this type of track is used for editing purposes. See HALF-TRACK.

major conjugate The distance between the subject matter and the front node of a lens.

makeup In 1940, Admiral Chester W. Nimitz explained that "a ship is always referred to as *she* because it costs so much to keep one in paint and powder." Hollywood accountants could be more specific; their ledgers might show that the motion picture industry's annual expenditure for cosmetics was large enough to paint, powder, and equip an entire fleet.

The earliest Biblical references to makeup, generally, and Jezebel, specifically, are found in Ezekiel's charge against Jezebel: "Thou didst wash thyself, paintedst thine eyes and deckedst thyself with ornaments." Historians, however, with sufficient reason for believing that the cosmetician's art originated in the East, assume that makeup, in one form or another, has been in use since man and woman first began noticing each other. Crude, by comparison with any standards, the materials originally used for affecting transformations covered a multitude of sins and, considering the scope of these early efforts, practically everything else. Contrary to generally accepted beliefs, faces and bodies were not smeared with chalk or bedecked with flowers and feathers for the sole purpose of attracting a mate. In many cases, makeup in one form or another was designed to influence and appease the demons and spirits of the day, as well as to help camouflage the hunter and protect the warrior.

The first theater audiences were expected to associate the shape, design, and color of the masks and robes worn by itinerant thespians with the gods of that period. Although each area had its own cast of characters and attendant color schemes, the practice was followed on Chinese, Egyptian, Persian, Italian, and Greek stages. Ancient gods, for example, wore long beards and colorful costumes. Souls of the departed, in direct contrast, made their stage entrances and exits in white, flowing robes. With white paint masking the actors' features, the audiences could easily identify the ghostly roles they portrayed.

The trend toward realism that began in the fourteenth century included some radical departures and touches of ingenuity. Performers discarded their masks and powdered their faces. When necessary, the makeup applied for the opening act was changed, in subsequent scenes, in order to show the character aging. Beards, too, were lengthened between scenes and head coverings added to show progressive stages of baldness.

During the Elizabethan era, the candles and oil lamps that cast their eerie glow on English stages were replaced by illuminating gas. In time, the shadows in the dark corners of the stage disappeared under the incandescent glow from banks of electric light bulbs. Enlightened audiences were no longer satisfied with crude representations. The application of makeup became an art.

With images magnified many times beyond their normal size on motion picture screens, however, the most minute flaws in a performer's makeup were greatly exaggerated and immediately visible. Quick to recognize the need for perfection, the major motion picture studios not only hired experts to hide imperfections but also gave them screen credit as an added inducement. Their knowledge of cosmetics quickly transformed the art into a science.

The new materials and techniques introduced by these artisans were designed to suit any type of complexion, from sallow to ruddy and from dark skin to fair. The many shades of grease paint, mascaras, lip paints, rouges, and face powders, matching the colors on an artist's palette, were applied with the same amount of skill. Minstrel black and clown white, for more specialized requirements, complemented their wide range of colors.

With the emphasis on perfection, wigs were fitted from molds of a performer's head and, where required, included eyebrows and foreheads, complete with wrinkles. Similarly, tresses, beards, and whiskers were applied from long ropes of braided material made of vegetable fiber. Cut and combed out, this crepe hair was pasted to the facial areas with spirit gum and blended with other hirsute areas. If the newly applied hair needed lightening, the makeup man used powder or rice starch for the purpose.

Where necessary, noses and other features are reshaped by applying latex rubber or premolded parts to the performer's face. Covered by makeup, the lines of demarcation between the added and original portions are invisible.

Similarly, fish skin, applied to eyelids, can make unbelievable changes in the shape of those features. Other changes, affecting bodily proportions, are achieved with the aid of sponge rubber, styrofoam, and other plastic materials. Precisely fitted and custom-tailored, artificial protusions and protuberances made from these materials helped transform Walter Hampden's well-shaped nose into the mighty Cyrano proboscis and Frederic March's classic profile into the distorted features of a Dr. Jekyll. On a larger scale, the false back worn by Charles Laughton during his characterization of the misshapen Hunchback of Notre Dame is another excellent example of the makeup man's art and the scope of his assignments.

Best results for the pioneer black-and-white film production were achieved with panchromatic makeup. For the desired monotone effect, tan foundation creams were used as a base and the highlights emphasized by a lighter shade of powder. On color film, however, panchromatic makeup produced a masky appearance—and new problems—magnified and intensified by the sensitive, multilayered emulsion of color film. Makeup artists found the solution in subdued shades of grease paints, rouges, and powders.

Today, with gray foundation cream as a base, black mascara and eyebrow pencils have been discarded in favor of soft browns. Subtle shades of cheek rouge (used only by women) are applied lightly. In keeping with the little-or-none formula, lip paints are applied and almost immediately removed in order to emphasize the outlines and create a natural appearance. By contrast, makeup for television must include strong highlight and shadow areas to prevent features from being washed out during transmission.

Before motion picture cameras grind out their required footage, extras and stars are made up for the roles they have been selected to portray. Traditionally, the bit players are made up first and the stars later. For the more complicated assignments, a performer may spend up to four hours in the makeup chair. The sound of the clapstick does not signal an end to the makeup man's daily activities. Even after filming has begun, the makeup man must remain on the set ready to powder the perspiring performers and repair the damage caused by the hot studio lights.

With the exception of corrective touches that help deemphasize objectionable facial proportions, the rules for applying makeup are few, simple, and not unlike those used by artists who paint portraits on canvas. Unlike the artist who strives for an accurate representation, however, the makeup man is usually concerned with improving the general appearance of the performers.

Similar to the methods used by artists who apply the base or background color with bold strokes of the brush, the makeup man uses a moist sponge to spread the pancake or panstick foundation over the performer's entire face and neck. This is blotted dry, and then powder is patted on with a puff and buffed lightly to smooth the pancake foundation.

To present a more natural appearance, women's eyebrows are shaped with tiny hairlike strokes rather than with bold lines. The eyes are "framed" by lining the upper and lower lids with the same eyebrow pencil. When blended, the soft shadows make the eyes appear larger and more expressive. Men's eyebrows, unless very light or sparse, do not usually need emphasizing or touching up.

Unlike eyelash makeup which is applied generously to the upper lashes and lightly to the lower lashes, eye shadow should be used sparingly. Prominent or protruding bone structures below the eyebrows can be subdued by applying the proper shade of eye shadow or liner.

Light dry rouge buffed on the women's cheeks and dark dry rouge for the men's cheeks help highlight the cheekbones, emphasize the general bone structure, and add sparkle to the eyes. Lip rouge is applied last.

Exposed to the elements or to the heat of studio lights, makeup needs frequent retouching. In most cases, repairs between "takes" can be made by simply applying powder, two or three shades lighter than the original color, to the facial areas that need retouching.

Highlights and shadows, used as corrective measures, can help eliminate a double chin, narrow a nose, or completely cover the heaviest of beards. As a general rule, highlighted areas should be approximately three to four shades lighter than the color used for the foundation. Shadows, by comparison, should be three to four shades darker than the foundation color. The pancake foundation used for shading or highlighting can be applied to large areas with a sponge or to smaller portions with a flat blending brush. If the

TABLE M-4 Makeup for Women

	Black-and-white film and TV		*Color film and TV*	
Makeup	Fair	Medium/dark	Fair	Medium/dark
Pancake or panstick	Film—7N TV-4N, 5N	8N TV-6N	CTV-4W CTV-4W	CTV-4W or 5W CTV-4W or 5W
Powder	Translucent	Translucent	Translucent	Translucent
Eyebrow pencil	Brown	Brown or black	Light brown to brown-black, black	Light brown to brown-black, black
Eyeshadow (liner)	6 (blue-gray) or 22 (brown)	6 (blue-gray) or 22 (brown)	CTV gray or CTV brown	CTV gray or CTV brown
Eyeshadow sticks			Iridescent brown or gray	Iridescent brown or gray
Eyelash makeup	Brownish black	Black	Brown	Black
Mascara	Brownish black	Brownish black	Brown to brown-black	Brown to brown-black
Dry rouge	TV light tech.	TV light tech.	Flame	Flame
Creme rouge (cheeks)	Flame	Flame	CTV creme rouge	CTV creme rouge
Moist rouge (lips)	TV no. 2	TV no. 2	CTV light red	CTV light red
Lipstick	Brilliant red Film-sparkling scarlet	Brilliant red Sparkling scarlet	7-22; 7-22B; 7-28; 7-45	7-22; 7-22B; 7-28; 7-45

TABLE M-5 Makeup for Men

	Black-and-white film and TV		*Color film and TV*	
Makeup	Fair	Medium/dark	Fair	Medium/dark
Pancake or panstick	Film—11N TV 7N	665F Panstick or light Egyptian pancake	CTV-7W	CTV-8W or CTV-9W
Powder	Film tech. special TV tech.	Tech. special TV tech.	Tech. special	Tech. special
Eyebrow pencil	Brown	Brown or black	Brown or brown-black	Brown or brown-black
Eyeshadow (liner)	22 (brown)	22 (brown)	CTV brown	CTV brown
Eyelash (optional)	Brown	Brownish black	Brown	Brownish black
Mascara (optional)	Brownish black	Brownish black	Brownish black	Brownish black
Dry rouge	Dark tech.	Dark tech.	Dark tech.	Dark tech.
Moist rouge (lips)	Film T-3 TV T-3	T-3 T-3	T-2 or T-3	T-2 or T-3

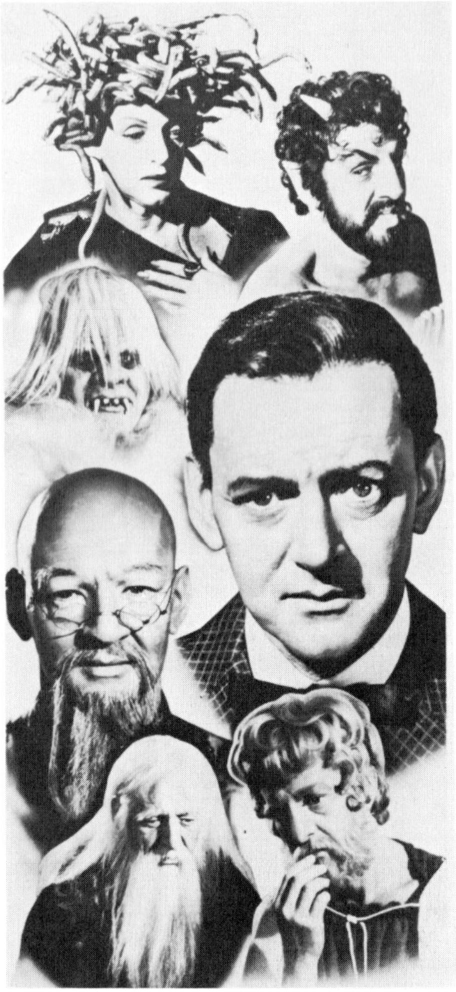

Fig. M-8 Make-up at its finest. Tony Randall in seven characterizations from "The Seven Faces of Dr. Lao."

highlights and shadows are just barely visible in a mirror held at arm's length, the film or television effect will usually be quite adequate.

Makeup in its broadest sense is applied not only to performers but to inanimate objects as well. Because packaged products often play starring roles in television commercials and must be presented at their best, boxes, cans, bottles, tubes, and their identifying labels must be retouched. These color corrections, by members of the studio art department, are complemented by the magical feats of technicians who are expected to make on-camera foods look appetizing and provide bubbles for fizzing liquids. In all cases, objectionable flares and reflections from shiny surfaces must be dulled by spraying or dusting.

Epictetus (circa A. D. 60) noted that "appearances to the mind are of four kinds. Things either are what they appear to be; or they neither are, nor appear to be; or they are, and do not appear to be; or they are not, and yet appear to be. Rightly to aim in all these cases is the wise man's task." Also, the makeup man's goal. See Figure M-8 and Tables M-4 and M-5.

maltese cross A type of mechanism that produces intermittent movement in motion picture cameras. The shaft that drives the camera's transport mechanism is connected to the maltese cross and is geared to bring each frame of film into its precise position for exposure in the camera shuttle.

marriage The synchronization of sound and picture in composite print form.

mask Also called **flag** and **gobo**. A flat, square-shaped sheet of metal, or opaque screen, used to shield the camera lens from stray lighting that could cause reflections during the filming process.

mass An overall tone or value made up of varying tones of the same color or shade of gray. The side of a mountain, for example, may seem to be predominant in one overall color. Closer examination, however, would reveal that the mass is made up of smaller areas of almost similar colors.

master positive The fine-grain print film from which duplicate negatives are produced. See flow charts under PROCESSING.

match In animation, a notation on an animator's drawing indicating that the cartoon character must be matched to some part of the background used for the scene.

matching negative A part of the editing process in which the picture negative is cut and matched to the edited work print.

matrices Singular: *matrix.* Relief images on three separate lengths of film. The thickness of the relief images is proportional to the density of the tonal values in the original image. Dye colors, representing each of the photographic primaries, are absorbed in varying amounts by each of the three lengths of film and used to reproduce the colors in release prints. See TECHNICOLOR and DYE-TRANSFER PROCESS.

matte box An attachment that is positioned around the camera lens to shield it from undesirable light that could possibly interfere with proper exposure. The matte box is also used as a filter or matte holder during the filming process.

To the imaginative film maker, the bellows attached to the front of a motion picture camera is more than a mere sunshade. Classified as an accessory, the indispensable attachment is a versatile device with a wide range of applications.

Although it is also referred to as an effects box, few professional cinematographers would consider the accessory an adequate substitute for the optical printer. The matte box can, however, be used to create many of the simpler optical effects with direct in-camera photographic processes. Matte shots, split screens, superimposures, and wipes are only a few of the effects that overshadow its value as a shade designed to keep the sun out of the camera's eye.

Guide rods, attached to the front of the camera, support the matte box and allow it to be moved forward or backward in relation to the lens. Vertical and horizontal adjustments, for aligning the unit with the optical axis of the camera lens, assure the precise registration needed for achieving effects that require several camera runs. The bellows, also adjustable, serves a dual purpose. Besides accommodating lenses of different focal length, they can be extended to their full length or contracted to change the focus of the mattes or other graphic material positioned in the front or rear frames of the unit. Grooves, at both ends of the bellows, provide a support for mounting slides, gauzes, gelatin filters, diffusion disks, Pola screens, foreground silhouettes, masks, mattes, and other devices for achieving special effects.

Masking effects, in which a scene appears to have been photographed through a window, keyhole, or binoculars, can be achieved easily with direct in-camera photography. The masks, available in sets at camera supply companies or especially cut to fit the individual requirement, are positioned in either the front or rear frames of the matte box, depending on the focal length of the lens being used. Mounted in the front frame, the bellows shade the rear of the mask, ensuring the necessary opacity in the processed print. Simple zoom effects showing the camera moving through windows, doors, or keyholes can be achieved by positioning the cutout masks in the rear frame of the matte box. Moved closer to the taking lens, the outlines of the mask gradually disappear as they go out of camera range, thus revealing the scene beyond in its entirety. With the camera focused on the subject matter only, the silhouetted mask may be kept in soft focus or sharpened to crystal clarity by adjusting the distance between the mask and the lens. Similar effects can be achieved by mounting silhouette cutouts of fences, gates, windows, and other graphic material on glass plates that fit into the front frame of the matte box.

Wipe effects, at the beginning or end of a scene, can be produced by gradually covering the frame in front of the bellows. In order to achieve these transitional effects, pieces of cardboard are cut to the desired size and shape, painted dull black, and then moved manually across the front frame of the unit until it is completely covered. The grooves, at both ends of the bellows, can be used as a guide for registration purposes or for supporting the masking device during the filming. In either case, the direction of the wipe should be consistent with the direction of the scene's action. If a car is moving through the scene from right to left, for example, the wipe effect should duplicate that horizontal movement. The same principle should govern vertical or diagonal actions.

By cutting semicircles, diamond shapes, or other geometric patterns into the leading edges of the two cards used for masking purposes, the cameraman can move them toward each other, from opposite directions, to produce a variety of interesting masking effects. Held in a stationary position, the cutout areas in the cards can be used to highlight any portion of the picture information within the film frame. After running the required footage with the cards in a stationary position, they can be moved away from each other to reveal the subject matter in the previously covered areas, or drawn together to complete the wipe effect. Overlapping wipe effects, in which portions of the outgoing and incoming scenes appear in the same film frame during the transition, require precise, frame-by-frame matching of mask positions, on separate camera runs, and are best left for the optical printer.

Split-screen shots, for various purposes, can be achieved by masking alternate areas of the film frame during successive camera runs. Shots in which a person is shown talking to himself from opposite sides of the frame, for example, can be produced by covering one-half the matte box frame during the first camera run and exposing the previously masked film area during the second camera run. Between the separate exposures, the film must be rewound to its starting position with the shutter closed. The mask in the front matte box frame must be repositioned, for the second camera run, to cover the previously exposed area. If possible, the leading edge of the mask should be positioned along some natural dividing line within the scene. Tree trunks, light poles, building lines, and door frames all provide natural lines of demarcation that help hide the mask's edge in the composite effect. Split horizontally, vertically, diagonally, or in any other desirable manner, this matte-box method of masking can be used to produce a wide variety of novel effects.

For the independent film maker without studio facilities or optical equipment, the matte box has many other applications. An opal glass, positioned in the front frame of the matte box, can convert the camera into a titler, miniature animation stand, or pseudo-optical printer.

Slides or frames of film, mounted in a cardboard frame and positioned in front of the lens, can be copied on the film in the camera by focusing a light source on the outer surface of the opal glass. During the copying process, the slide or film frame can be faded out or cross-dissolved to another slide, frame of film, or to the scene that follows.

Titles and cartoons, mounted on clear glass slides positioned in front of the lens, can also be exposed with the aid of the opal glass, or superimposed over a scene without it. Diffusion and distortion effects can be produced by coating portions of the clear glass slide mounted in front of the camera lens with Vaseline, water glass, or other diffusing or diffracting material. The degree of distortion can be controlled by moving the partially coated glass slide to different positions in relation to the camera lens. The specific uses itemized here merely suggest some of the many effects that can be achieved with the aid of the versatile sunshade known as the matte box. See Figures M-9*a* to M-9*f*.

Fig. M-9a The Mitchell matte box.

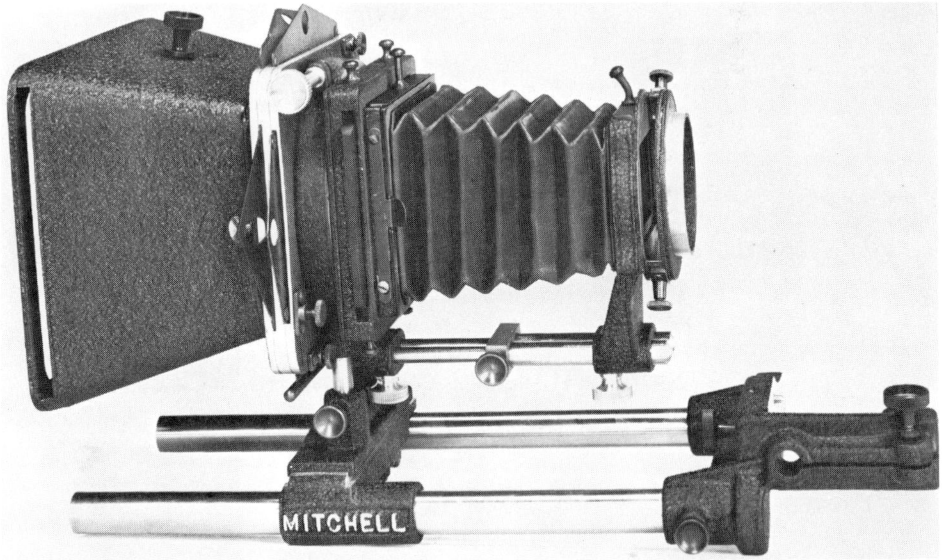

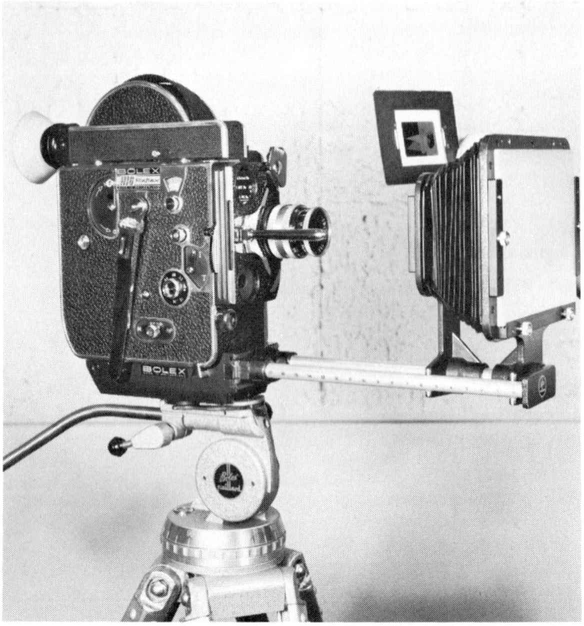

Fig. M-9*b* The matte box and backlighting. The Bolex matte box mounted on sturdy rails before Bolex H Rex 16mm camera, showing method of use for copying 35mm slides. Slides in the front frame and an opal glass in the rear frame for copying purposes.

Fig. M-9*d* The Bolex and the matte box, with mask inserted for a binocular effect.

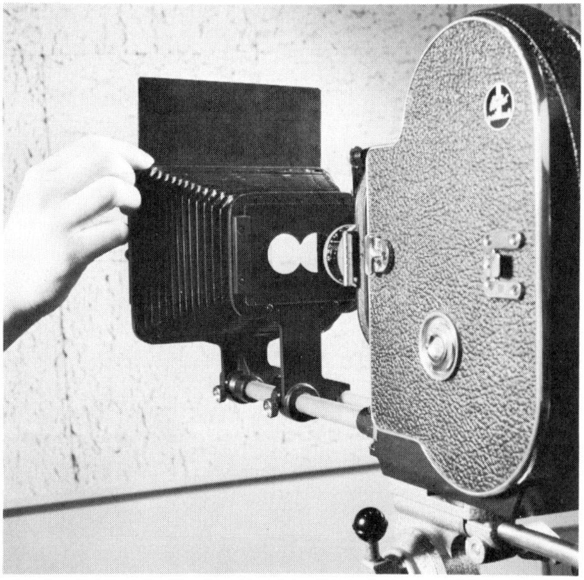

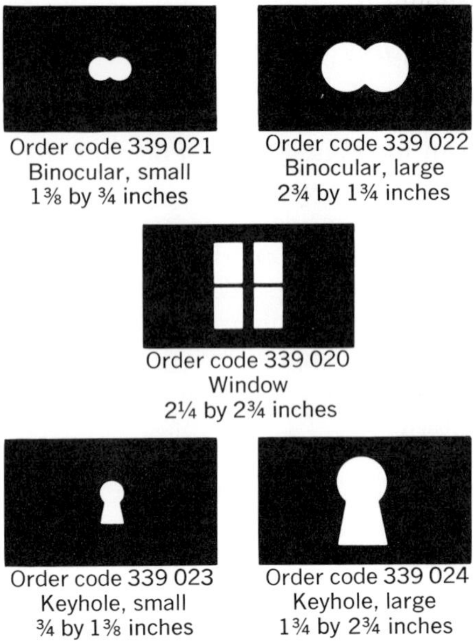

Fig. M-9*c* Special-effects mattes for Arriflex 16 and Arriflex 16M matte box. The mattes illustrated may be used in the matte box of the Arriflex 16 and Arriflex 16M for special effects, thereby increasing the versatility of these cameras. Fabricated from 0.040-inch aluminum alloy, these mattes are well finished and anodized a nonreflective flat black. The mattes slide into the front frame of the Arriflex matte box. Approximate major dimensions and type of aperture are indicated in the description.

Fig. M-9e The matte box and window mask matte.

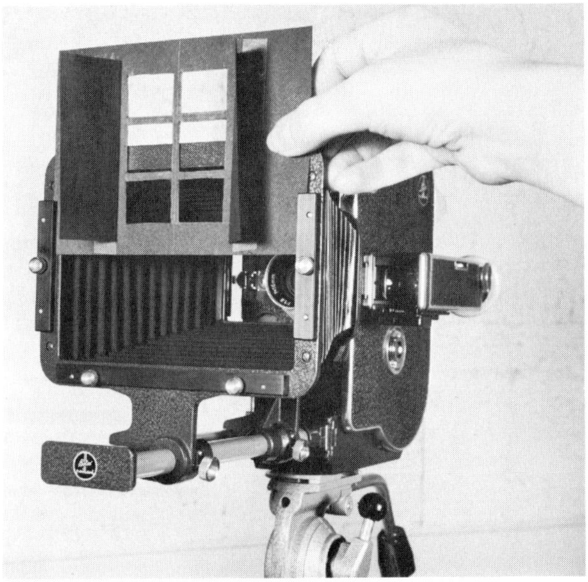

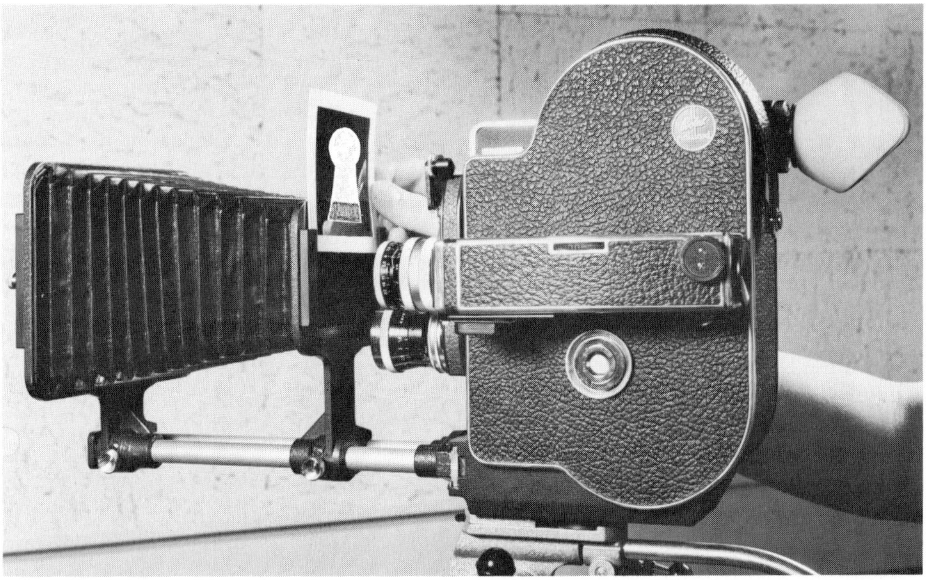

Fig. M-9f For effect of camera moving through keyhole, mask is mounted in rear frame of matte box and moved toward lens while filming.

matte finish A comparatively dull finish on a photographic print. Also, a dull finish similar to the ground-glass surface in a viewfinder.

matteing amplifier See SPECIAL-EFFECTS GENERATOR.

mattes Matteing is a method of masking an area of the film frame during the actual filming or during the optical printing process. On subsequent runs, picture information is inset in the previously matted areas.

Anyone who has ever taken a photograph with any kind of camera realizes that if the film in the camera is not advanced after the taking of each picture, a double exposure or double image will result when the next photograph is taken.

In itself, the double exposure is an effect that is used quite frequently in motion picture production. It is often necessary, however, to combine pictures or lengths of film to achieve a composite effect. In such optical combinations, of course, double exposure is completely undesirable. The elimination of the unwanted effect is made possible through the use of mattes. The matte, or matteing process, in motion picture terminology, refers to either artwork or film which is used for the purpose of concealing or masking an area of the motion picture frame. The concealment is temporary, since picture information will subsequently be inset within the matted area. The mattes, or opaque silhouettes must, therefore, be identical in size, shape, and position with the corresponding picture information. The frequently used term *matteing out* adequately describes the matte's principal function.

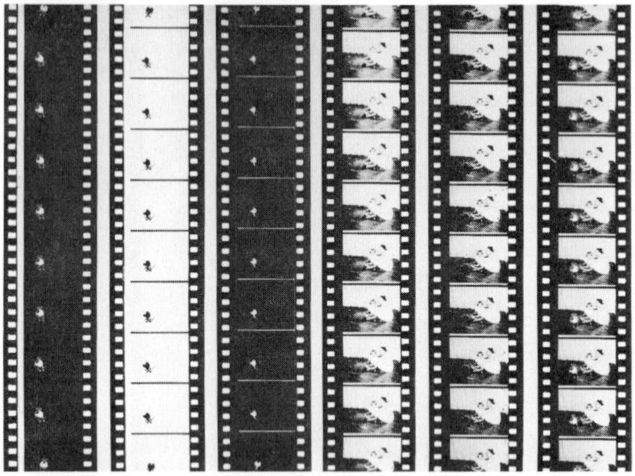

Fig. M-10a Matteing and animation. The animation in column 1 was photographed over a black background. Normal animation stand lighting was used for this exposure. Columns 2 and 3 show the mattes produced by photographing the same animation sequence on high-contrast film, using underlighting. The high-contrast positive is shown in column 2 and the high-contrast negative in column 3. The live-action film is shown in column 4. In column 5, the live-action film has been combined with the high-contrast positive matte. The composite effect is shown in column 6.

In its simplest application, the matte is used for inserting an image, on one length of film, into a previously photographed scene or background plate on another length of film. Assuming that the image remains in one particular position in relation to the film frame, the superimposition of the image over a background plate requires a single matte that occupies the identical position in the frame. Both the negative and positive mattes are used for the optical combination of the insert material with the background.

When the image or subject matter does not remain in a fixed position in the film frame, it is necessary to use separate mattes that correspond to the various movements or positions of the image. These mattes, each in a different position in relation to the frames of film in which they are used, are known as *traveling mattes.*

The earliest methods for making mattes were painstaking, costly, and, at best, unreliable. Homemade devices were used for projecting the frames of film that required matteing. The outlines of the projected subject matter were traced, as accurately as possible, by animators. In the next phase of production, these outlines were retraced on clear sheets of acetate, opaqued to create a silhouette effect, and then photographed on the animation stand. Inaccuracies in tracing showed up only after the mattes had been combined with the other film elements needed for achieving the composite effect.

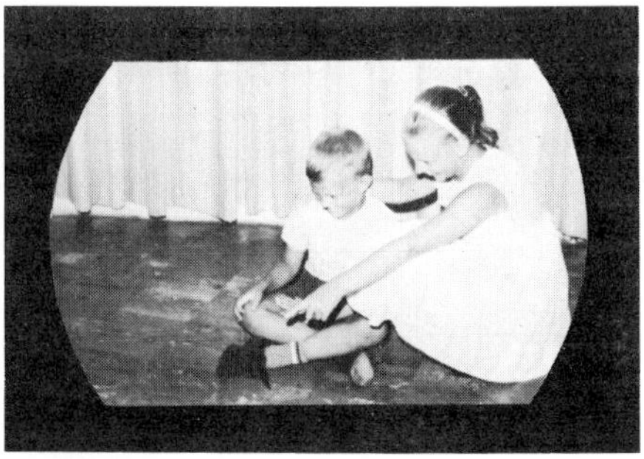

Fig. M-10b The live-action scene and the background for the animation of Fig. M-10a.

Improvements in equipment made it possible to project film directly through the animation camera onto the working surface of the compound table. The punched paper, used for tracing the outlines, fitted the pegs on the animation stand and made the filming of the silhouetted cells a more accurate process. The cameraman, however, could not compensate or be held responsible for inaccuracies in tracing at the time the film was projected.

More precise matteing methods were desperately needed to satisfy the requirements of critical clients who insisted on perfection, and less critical audiences who deserved it. Years of experimentation resulted in *self*-matteing processes that made possible an unlimited variety of effects. See Figures M-10*a* to M-10*e* and SELF-MATTEING and TRAVELING MATTE.

matte white screen A screen with a magnesium oxide surface on which film is projected.

mechanical animation A technique in which three-dimensional models of a mechanism, with moving parts, are used to illustrate the technical subject matter discussed in the film. These setups are photographed on the animation stand with stop-motion processes, and the moving parts repositioned for each frame of the scene.

medium An agency of transmission through which anything can be passed or conveyed. Light waves, for example, can pass through

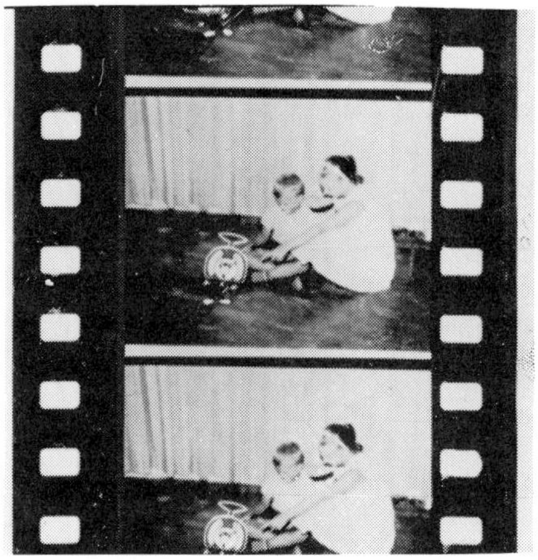

Fig. M-10c The composite effect of Fig. M-10a.

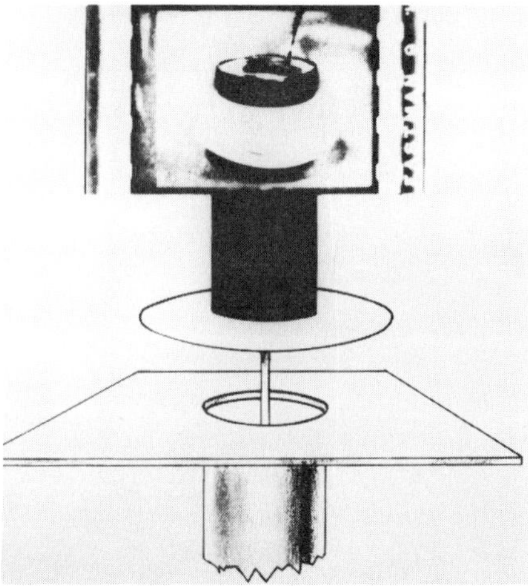

Fig. M-10*d* Maxwell's cup and a half of coffee. The stir created by Instant Maxwell's cup-and-a-half commercial was well worth the effort involved in producing the unusual effect. The fact that a matteing technique was used to achieve the composite effect was immediately obvious to most professional commercial producers. The actual method used, however, had the industry guessing.

The elements needed for the commercial were filmed on two sets, not including the matteing section which was photographed on the animation stand. On the first set, coffee was poured into a bottomless cup. On the second set, a glass cylinder filled with coffee was placed atop a round table.

Magnets positioned on the underside of the table created a swirling effect in the cylinder, thus duplicating the action of coffee being poured into a cup. A hydraulic lift raised the table and cylinder to its correct position in relation to the cup filmed on the first set.

Each of the 200 frames of film needed to show the extra half-cup of coffee had to be matted separately and superimposed over a frame showing the coffee being poured. The mattes were made by projecting the film containing the coffee-pouring action on the animation stand and tracing each position of the cup. The 8-and-⅓-second sequence became an industry conversation piece.

Fig. M-10e Mattes produced by projecting through the animation camera. The frames in column 1 show the Amoco sign at the beginning of the split-screen shot.

Each of the other columns shows the partially completed effect after each of the mattes had been combined with the background on fine-grain film. The composite split-screen effect is shown in column 4.

a transparent or translucent solid substance such as glass. Sound waves, similarly, can travel through such apparent barriers as walls, doors, or windows.

mediums combined Recent developments in motion picture, television, and videotape production techniques have given solid support to the theory that suggests "if one is good, then two must be better."

The Gemini tape-film system, the Mobile video-film system, and the Mitchell system 35 all demonstrate the compatibility of the tape and film mediums. With the Gemini twin system, for example, the simultaneous production of a 16mm motion picture print and videotape recording is easily accomplished. The prints obtained through this unique filming method are appropriately referred to as *Gemprints.*

Developed by a Pittsburgh corporation headed by Warren R. Smith and Robert Ferber, the system named for the zodiacal constellation was devised for the purpose of satisfying the needs of network and major markets as well as those stations in smaller markets that do not have the facilities for transmitting taped commercials. Before the introduction of this filming method, smaller markets, without tape facilities, had to content themselves with kinescopes of comparatively poor quality. The 16mm motion picture print that is used in place of the kinescope assures the local viewer of a picture that is equal in quality to the taped image received by the network viewer.

In operation, a high-resolution television camera is used in conjunction with a 16mm motion picture camera. Although the television camera's viewfinder is used for composition purposes, both cameras use a single lens for recording the tape and film images.

The Mobile video-film system combines the best features of videotape recording with motion picture filming and editing processes. During actual production, two 16mm motion picture cameras are placed on lightweight dollies. A closed-circuit television camera, next to each of the motion picture cameras, feeds the subject matter to the monitors.

The director merely selects the desired picture from the monitoring system and transmits the information to the cameraman. Camera 1, in the video-film system, remains in a stationary position. Camera 2 is rolled from one position to another during the filming process for the purpose of achieving close-up and cover shots.

With this monitoring system, the director not only previews the subject matter in terms of continuity but he is also able to pre-edit the footage during the actual filming. Matching the sound track to the processed film is usually accomplished in little more time than it takes to project the composite print. The magnetic sound track that is recorded during the filming process for protection purposes is seldom needed.

The developers of the system, Snazelle Productions, Inc., have been able to "air" a show within two hours after the film was unloaded from the two cameras. The feeling of a live telecast that is so prevalent in the completed production can be traced directly to the spontaneity of the filming process.

Mediums will undoubtedly continue to join forces, where such cooperation is deemed advisable, in order to assure better, faster, and more economical production. The Gemini tape-film system and the Mobile video-film technique are both excellent examples of cooperative processes. See MITCHELL SYSTEM 35.

medium shot Abbreviated by the letters **M.S.** A scene in which the cameras are positioned to photograph the full figures of the performers and a small portion of the motion picture set beyond them.

megacycle A unit of frequency. One megacycle (abbreviated Mc) equals 1 million cycles per second. A kilocycle (kc) represents 1,000 cycles per second.

meniscus lens Also called **singlet**. A glass lens with one convex and one concave surface. With these lenses, reductions in the size of the aperture minimize the effects of astigmatism. Suitable for the production of contact prints or enlargements, they can also be used as supplementary lenses and are usually found in inexpensive still cameras.

mercuric chloride A white powdery substance used in intensifier solutions. The powder is poisonous and does not dissolve readily in water.

mercuric iodide A red powdery substance used in intensifier solutions. The powder is poisonous and does not dissolve in water.

mercury-vapor lamps Arc lamps whose glowing light is produced in glass vacuum tubes filled with vaporized mercury. These lamps produce a cold bluish light with an abundance of violet rays. The lamps are highly regarded as sources of illumination in studio work—enlarging, or portrait photography, for example. See Table M-6.

meridian A line at right angles to the optical axis of an optical system.

metallography The photographic process used for analyzing the composition of metals. Filming is accomplished by positioning a small piece of the metal to be analyzed under a microscope so that light striking its surface is reflected into the microscope that is used in conjunction with the camera.

TABLE M-6 Filters for Use with Mercury-vapor Lamps

Wratten filter no.	Principal lines transmitted, mμ
22	577 and 578
50	436
74	546
77*	546 (70%)
77A*	546 (55%)
18A	310–390

*Filters 77 and 77A are available only as cemented glass filters.

An alkaline or acid solution is first applied to the surface of the polished piece of metal. The resulting etched surface "exposes" the metal's crystal composition, which takes the form of dark and light areas in the developed negative.

meter Abbreviated by the letter **m**, the meter is a unit of measurement in the metric system. One meter is equal to 39.37 inches.

Metol Trademark for monomethyl-*para*-amino phenol sulfate. A white crystalline powder used as a development agent. The powder dissolves readily in water and must be used carefully since it may cause skin poisoning.

metric system A decimal system of weights and measures in very general use for scientific purposes. The standard unit is the meter (39.37 inches).

1 meter =	10 decimeters	1 dm	= 0.1m
=	100 centimeters	1 cm	= 0.01m
=	1,000 millimeters	1 mm	= 0.001m
=	1,000,000 microns	1 μ	= 10^{-6}m
=	1,000,000,000 mμ	1 mμ	= 10^{-9}m
=	10,000,000,000 angstroms	1 Å	= 10^{-10}m

MGM camera 65 See WIDE-SCREEN PROCESSES.

micro The prefix used to indicate one-millionth in metric units. One microampere, for example, equals 1/1,000,000 ampere. Compare MILLI.

microcinematography A photographic process in which the motion picture camera, set up in combination with a microscope, films subject matter too small for the ordinary camera lens to capture through orthodox photographic processes.

This type of photography permits studying cinematographically

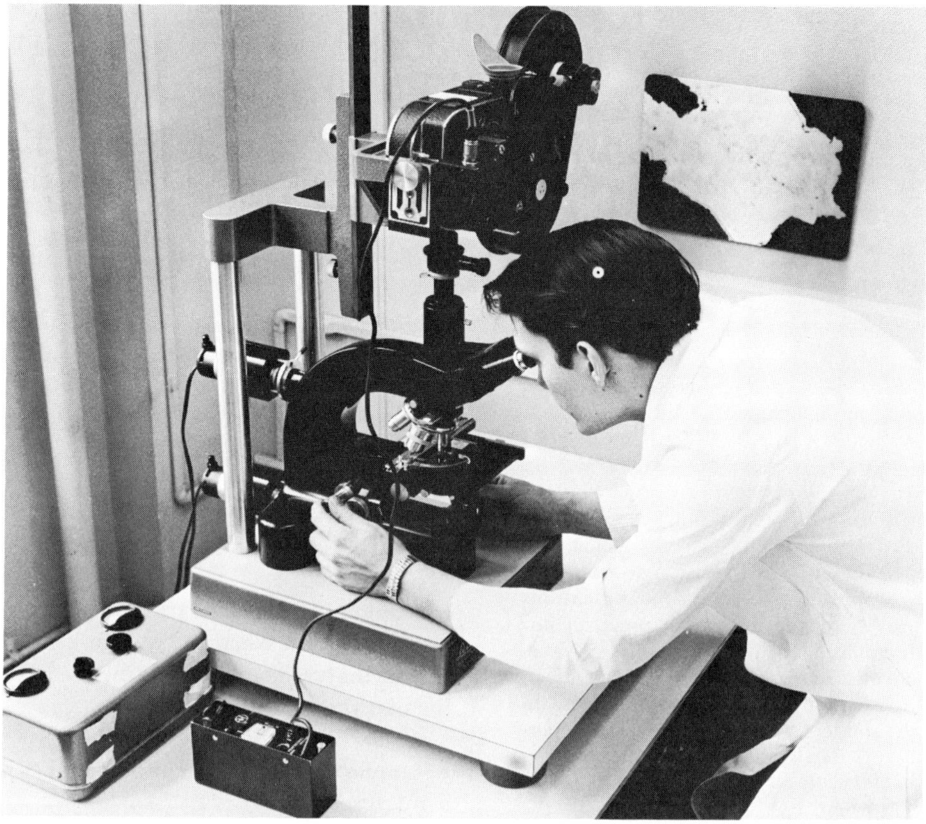

Fig. M-11 Microcinematography. (*Paillard, Inc.*)

the life of microorganisms and other scientific data that can be observed only with the aid of a microscope.

Time-lapse microcinematography is also possible by adding time-lapse units and stop-motion motors. See Figure M-11.

microfilm Microfilming, first used during the early 1930s, is a practical method for photographing records and other information on motion picture film. Each frame, in the processed roll of 16 or 35mm film, is projected individually in filmstrip fashion through special machines.

A common type of microfilm will contain 2,500 images of ordinary $8\frac{1}{2}$- by 11-inch pages at 24 : 1 reduction. See MICROFORMS.

Since the contents of entire newspapers and books can be photographed and stored in small film cans, the space previously needed for storage cabinets can, obviously, be used more advantageously. The growth of microfilming, as a photographic technique, is reflected in the 250 million dollars spent for the preservation of records in just one year.

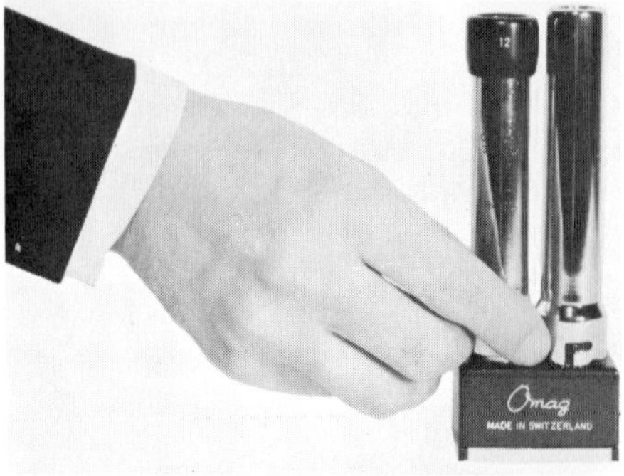

Fig. M-12 Miniature microscope. The OMAG pocket microscope is a Swiss precision instrument which combines compact ($5\frac{3}{4}$ by 2 by 1 inch), lightweight (10 ounce), and shockproof construction with top-quality optics for scrupulous inspection and evaluation.

Fitting easily into a coat pocket, the OMAG, with its precision-ground eyepieces and lenses offers magnifications from 10 to 50 times with precise color fidelity and authentic contrast. A fine-focusing ring provides for critical sharpness at all magnifications. The built-in light renders a brilliant image and operates on two standard 1.4-volt batteries or a 110-volt a-c transformer for indoor or outdoor use. Rigidly designed, the OMAG withstands the most rugged use for lifelong durability.

microforms A general term for microimage types of photographic reproductions of records and printed matter of all kinds. It includes microfilm (see page 463), which generally provides bulk storage of microforms in roll or cartridge film form, micro-opaque cards (such as the trademarked Microcard and Readex Microprint cards), and Microfiche. Microfiche commonly stores up to 96 microimages on a card-size piece of film 105 by 148mm, or approximately 4 by 6 inches. It is provided with a label across the top that is readable to the unaided eye, thus facilitating retrieval and use.

micron μ. A term used to indicate a measurement equal to one-millionth of a meter. There are approximately 25,000 microns to the inch. Also, a prefix used to describe anything that is quite small in size—microfilm, or microflash, for example.

microphone A device used for recording sound. More popularly referred to as a **mike**, it converts sound waves into electric impulses for transmission and recording.

microphonics In video transmission, the mechanical vibration of the elements of an electron tube resulting in a spurious modulation of the normal signal. This usually results in erratically spaced horizontal bars in the picture.

microphot An instrument that makes possible photography through a microscope.

microphotography Unlike photomicrographic and photomacrographic processes in which the image is magnified, microphotography is used for obtaining sharp, precise photographs in which the size of the image is reduced considerably. Microfilming, for example, is used to produce microphotographs.

microscope An optical enlarging instrument which makes possible the observation of minute objects. See Figure M-12 and Table M-7.

microsecond One-millionth of a second.

microwave Wavelengths of less than one meter, or associated with frequencies greater than 300 megacycles, are high-frequency radio waves or microwaves.

microwave link Also called **booster signal** and **relay**. An intermediate point at which the original telecast is received, amplified, and retransmitted to more distant areas.

TABLE M-7 Filters for Microscopy*

Number	Visual color	Spectral transmission
25	Red	From 590 mμ into the infrared
58	Green	From 480 mμ to 620 mμ
47	Blue	From 370 mμ to 510 mμ
35	Purple	From 320 mμ to 470 mμ and from 650 mμ into infrared
22	Yellow-orange	From 550 mμ into the infrared
29	Deep red	From 610 mμ into the infrared
15	Deep yellow	From 510 mμ into the infrared
45	Blue	From 430 mμ to 540 mμ
11	Yellowish green	For correct tone reproduction with tungsten light

Courtesy of Eastman Kodak.

*These filters can be used singly or in pairs. By using them in pairs, the spectrum can be divided into approximately monochromatic portions.

middle-range frequencies Frequencies ranging between 400 and 3,500 cycles per second. The audible frequency range is between 15 and 20,000 cycles per second.

midget A compact Mole-Richardson spotlight that accommodates lamps ranging from 50 to 200 watts. It is used for fill-light purposes.

mil A unit of measurement equal to 1/1,000 inch. The thickness of recording tape is measured in mils.

milk glass Also called **opal glass**. A translucent glass, milky in color, used for diffusing light.

milli The prefix used in the metric system to indicate one-thousandth of a unit. One millimeter, for example, is equal to 1/1,000 meter or approximately $\frac{1}{25}$ inch. Compare MICRO.

miniature Also called **diorama**. See MINIATURES.

miniatures A small-size model of an object, structure, or scene, as opposed to a full-size mock-up. Both are intended, in motion picture practice, to stand in for the original.

To anyone in the property department of a motion picture studio, the word miniature connotes "challenge." And, the challenge is not always on a small scale.

The duplication of large structures and their subsequent demolishment; the simulation of battle scenes on land, sea, and in the air; natural phenomena such as earthquakes, floods, and hurricanes; and many other effects that cannot be duplicated with other production methods are made possible through the use of miniatures. Similarly, actions that have no counterpart in the normal scheme of things become believable through the manipulation of miniatures.

For the accurate reproduction of these scale models, skilled craftsmen use almost every type of construction material available to the builders of the "real thing." Metal, wood, plastics, rubber, glass, leather, papier-mâché, and plaster are only some of the materials that are used for the purpose of achieving the realistic qualities that make a scene or action acceptable to critical audiences.

Obviously, miniatures must be made to look like anything but miniatures in order to be acceptable from a pictorial point of view. The assumption that moviegoers pay little attention to detail is completely erroneous. Audiences, through the years, have become increasingly critical. Their alertness is reflected in the flood of letters that are received by studio technicians after the release of almost any feature-length production. Incorrect details are magnified, by these self-appointed critics, to a point that is completely out of proportion with the minor misrepresentations. These viewers are capable of distinguishing differences between live scenes and miniature reproductions before they are even comfortably settled in their theater seats.

The construction of these minature sets and props, therefore, requires a greater amount of patience and consideration than the preparations that are made for filming live sequences. Perfectly scaled detail, however, is not the only prerequisite for the construction of miniature models. Neatness, of course, is an important and desirable habit, but it can, very often, make the difference between a good model and a believable one. The normal effects of age, wear, and exposure are important considerations in the construction of miniatures. Dust, in those places where it might normally be expected to accumulate, cracked windowpanes in old buildings, and peeling paint on water-soaked vessels are the reasonable results that can be expected under similar conditions. Certainly, these and similar items, add the necessary touches of realism that make the difference between good and bad representations.

The sizes of these dimensional miniatures are in direct proportion to their importance in the shooting script. Naturally, larger miniatures contain a greater amount of detail than smaller models that are used in long shots and are visible for only a few seconds. For both, props or sets can be distorted deliberately and built to a perspective that corresponds with projected backgrounds.

The problems created by the use of miniatures, however, are not always proportionate to their size. Since actual rain, snow, flames, smoke, and similar effects are too large, in relation to the size of the miniatures, it is often necessary to combine these filmed effects optically in order to achieve the desired composite effect. Traveling-matte processes and prephotographed backgrounds or effects are frequently used to establish the proper relationship between normal-sized effects and the scaled-down sets or props.

Any overlooked detail, during the filming process, can make the most realistic reproduction look incongruous. The proportionate size, shape, and speed of miniatures, in relation to the sets, can make the all-important difference between the realistic effect and the unacceptable result. Clouds, waves, smoke, and other effects must be made to conform with the size of the sets or props. Similarly, vehicles on land, ships at sea, and planes in the air must move at believable speeds in relation to their surroundings. The size of the wake from a moving boat, the exhaust from a car, and the smoke cloud billowing from a chimney are other details that help make the composite effect acceptable to audiences.

In feature films, where miniatures are used for script continuity, rather than for their value as eye-arresting effects, the amount of planning and preparation is usually in direct proportion with the projection time of the effect.

For the reenactment of the naval battle in "Air Force," for example, it took approximately 75 men a total of six weeks to plan and construct the 12 ships and 64 miniature airplanes that took part in the action. Fifty men and 10 weeks were required for the actual filming of the sequence at a location where the size of the waves was compatible with the size of the models. To achieve the required effects, explosive charges were built into the models to simulate bomb hits. The effect of near misses was accomplished by setting off other charges below the waterline.

In "Gorgo," a full-length feature produced by King Brothers in London, filming took place in a studio tank approximately 300 feet long. The amazingly realistic set that bordered the huge, water-filled tank was a miniature reproduction of the River Thames. The commercial docks alongside the river, the Tower Bridge, Big Ben, and the Houses of Parliament were re-created to the smallest detail. A 200-foot monster, in relation to the scale of the set, played the part of the "heavy" in the feature film. All the monster's movements, from the slashing tail to the blinking eyes, were controlled mechanically. Altogether, a remarkable effort with truly amazing results.

In another production, a giant squid was constructed by the property department of a major studio. Its steel skeleton was encased in a foam-rubber skin and painted in appropriate colors. Although the original plans had provided for the mechanical operation of the giant figure, manual controls were used during the photographic process. This effort, too, was realistic enough to fool everyone with the possible exception of a real squid.

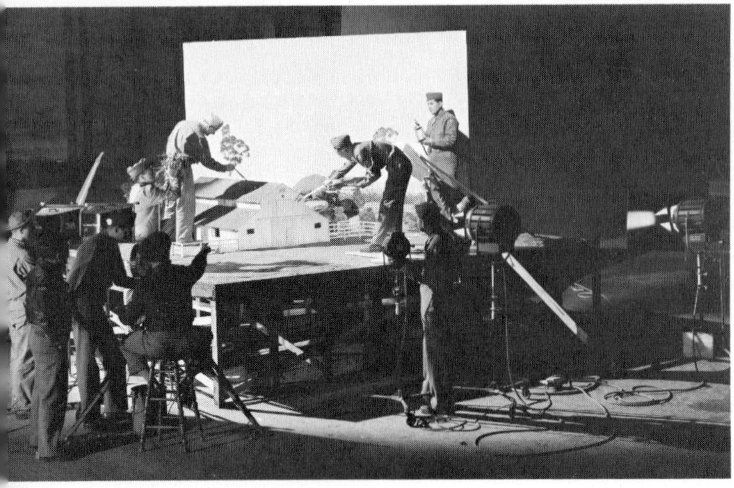

Fig. M-13*a* **The miniature set.**

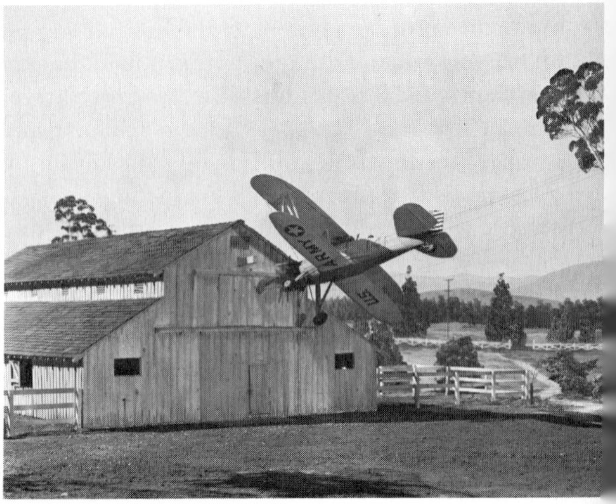

Fig. M-13*b* **A model plane, barn, and trees. Stand by for the . . .**

Obviously, the success of a minature set or prop depends on its credibility and authenticity. But if a miniature looks like a miniature, the motion picture technicians have failed in their attempt to achieve the desired results. See Figures M-13*a* to *i*.

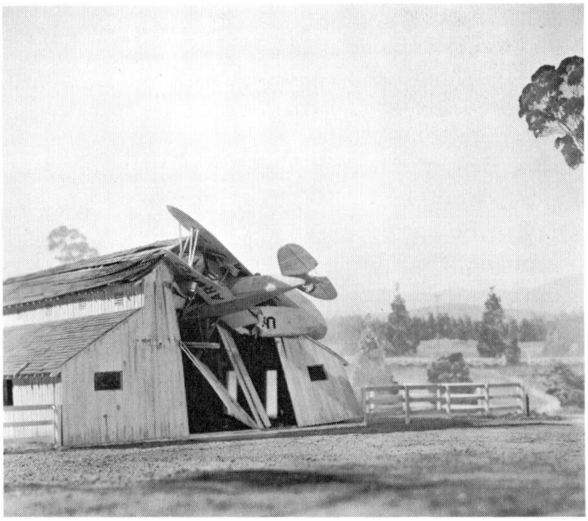

Fig. M-13*c* **Crash!**

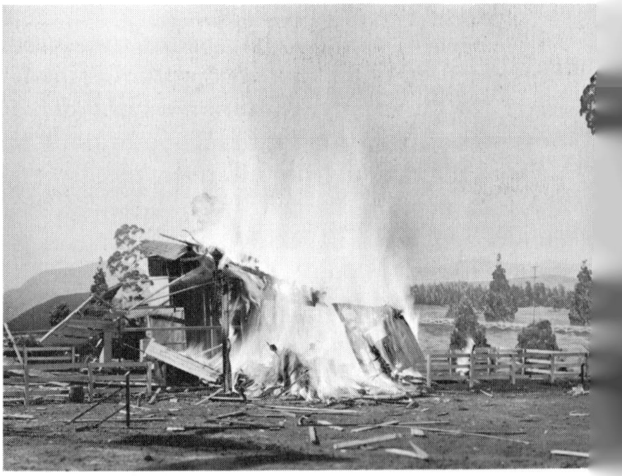

Fig. M-13*d* **How realistic can you get?**

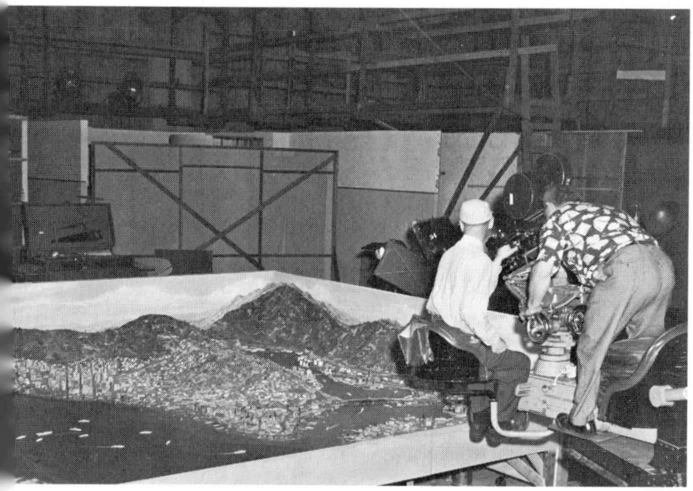

Fig. M-13e A city—in miniature.

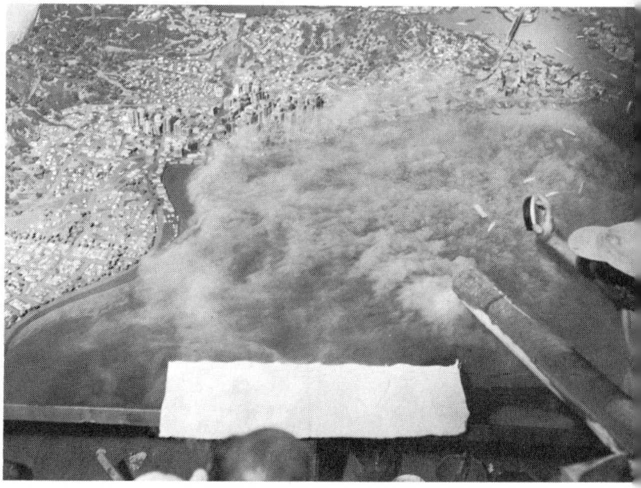

Fig. M-13f Add clouds—and shoot.

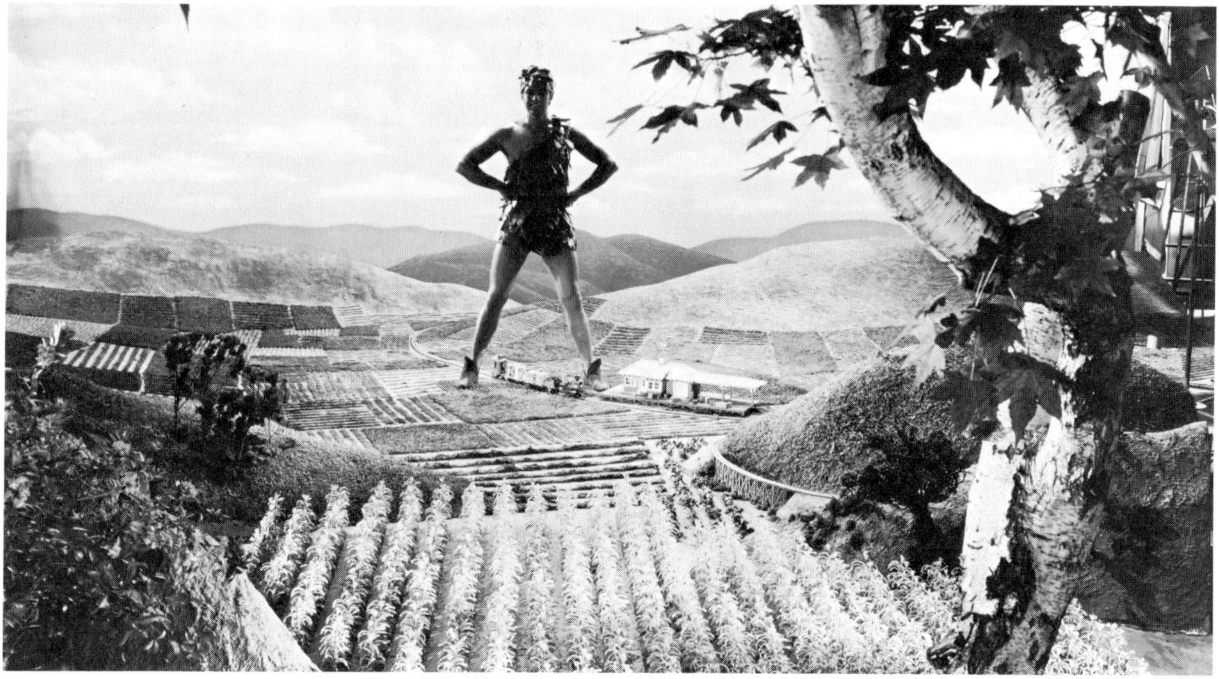

Fig. M-13g Green Giant. Don't fret if you were fooled by this shot. The "giant," as you can see in Fig. M-13*h*, stands under six feet in height.

Quite obviously, it's all a matter of perspective. Other scenes in the commercial use miniature, three-dimensional sets. The animation is photographed separately and combined optically.

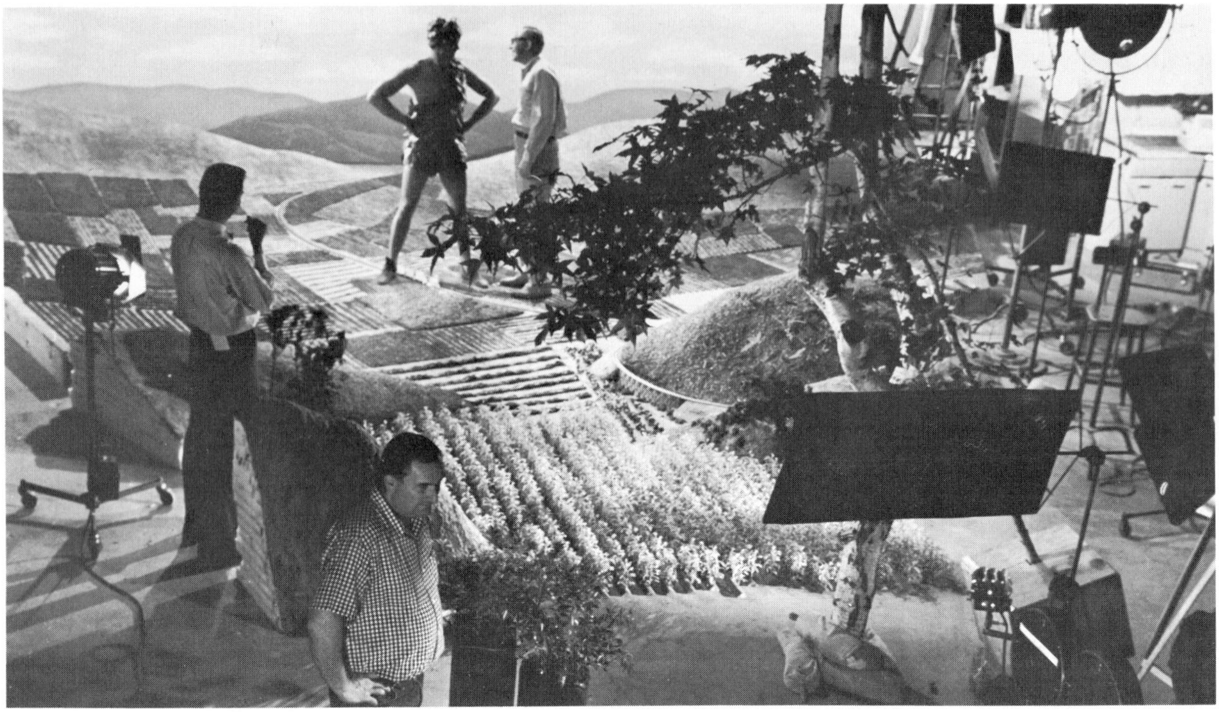

Fig. M-13*h* The Giant shrunk.

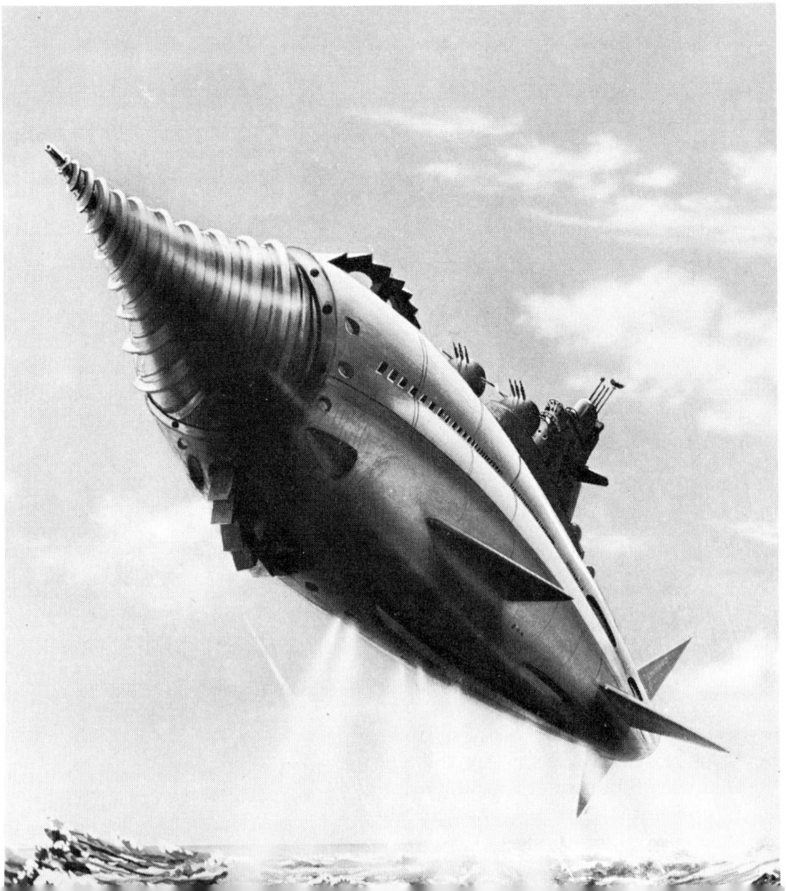

Fig. M-13*i* The Flying Supersub constructed by Japan's Toho Company technicians for "Atoragon," a science-fiction thriller.

formula for determining camera speeds in relation to size of the miniature. Assume that the miniature is one-sixteenth life size—

$$\text{Filming speed} = \text{normal speed} \times \sqrt{\text{reduction}}$$
$$\text{Filming speed} = 24 \text{ frames per second} \times \sqrt{16}$$
$$= 24 \times 4 \text{ or } 96 \text{ frames per second}$$

The camera speed increases when the model is smaller than life size and decreases when it is larger than life size.

Mini-Brute A high-intensity fill light designed to supplement daylight on outdoor locations, as well as for color work in motion picture and television studios. Mini-Brutes accept 650-watt tungsten-halogen quartz (5000°K) PAR-36 lamps with dichroic-coated lenses. Different lamp-switch arrangements provide for various levels of intensity without changing color temperature. See Figure M-14.

Mini-Lite A relatively small, lightweight ColorTran unit, with integral four-leaf barn doors, used for fill lighting. Incorporating a reflector designed to increase the light output by approximately 30 percent, the quartz-iodine lights used with these units provide smooth, broad light patterns with flat intensity curves. Well ventilated, the

Fig. M-14 ColorTran Mini-Brutes on cluster bar.

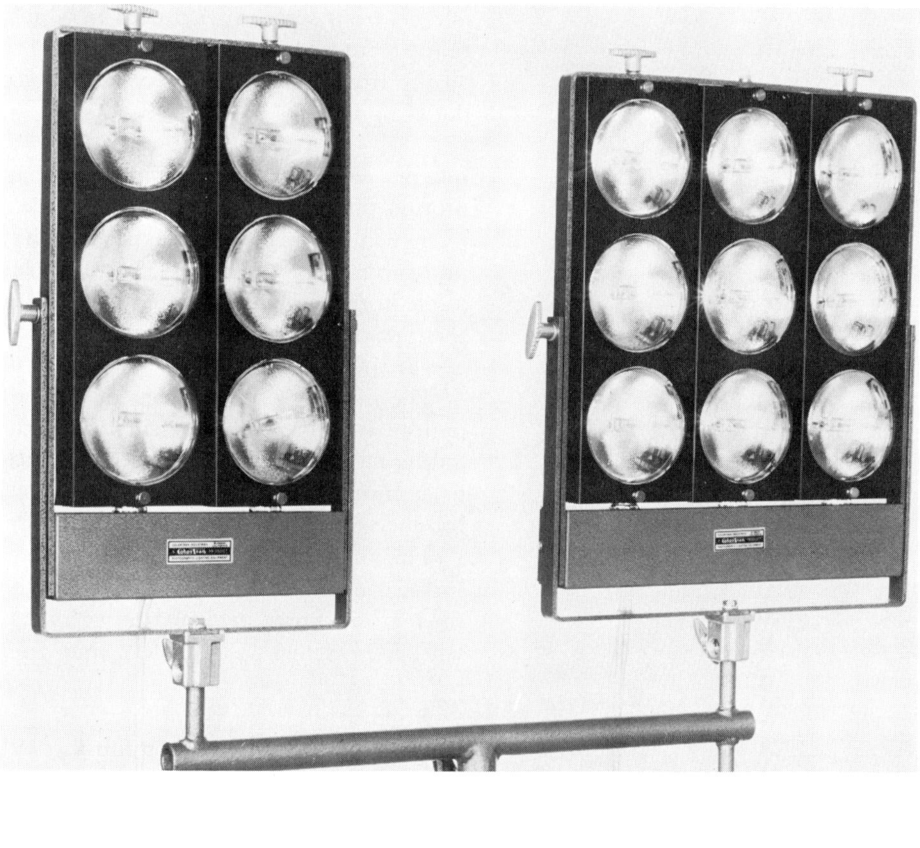

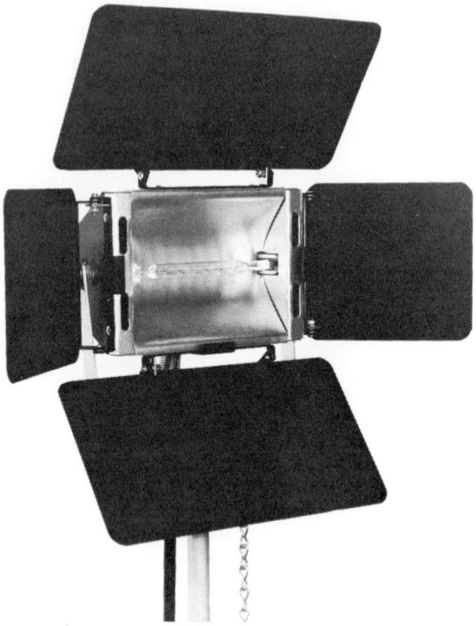

Fig. M-15 ColorTran Mini-Lite with four-way barn doors.

6-inch Mini-Lite takes a 650-watt lamp; the 10-inch Mini-Lite accepts 1,000-watt lamps. The unit operates directly from 120 or 230 volts, a-c or d-c. See Figure M-15.

Mini-Mount A relatively small helicopter mount (Tyler) designed to have the same vibration isolation and smooth-aiming and tracking features found in larger mounts. The unit weighs 88 pounds.

The mount has internal electronics for automatic soft start and stop zoom, fingertip controls for the camera, motorized focus and variable-speed zoom, and a self-contained rechargeable power pack for camera, zoom lens, and focus. The mount can also be used on dollies, boats, cars, or tripods.

minor conjugate In optics, the distance between the subject matter and the rear node of the lens. Compare MAJOR CONJUGATE.

mired values A unit of measurement for indicating the color temperature of a light source. Mired is an abbreviation for **mi**cro **re**ciprocal **d**egrees. Mired values are obtained by dividing 1 million by the color temperature of the light source in degrees Kelvin. See DECAMIRED FILTERS.

mirror, front surface A mirror in which the silver has been applied to the front surface of a glass instead of to the back surface. Silvering on the front surface tends to eliminate undesirable reflections from the glass surface.

mirrors and ghosts The line that separates illusion from reality in motion pictures is quite often no thicker than the layer of emulsion on which the images are exposed. The difference between shadow and substance, in the same art form, usually depends on the camera angle at the time of exposure, the position of the subject matter, and the type of material that is used to reflect an image into the all-seeing eye of the camera.

The materials, large sheets of optically clear glass or transparent panes with silvered backs, each offer ideal reflective surfaces for the manufacture of illusion. The latter substances, more often referred to as mirrors, are used frequently in motion picture production to produce unbelievable effects that border on the supernatural.

The incident- and reflected-light rays that enter and leave the surfaces of these materials follow paths similar to those outlined in the textbooks used in elementary physics courses. Images are reversed, with respect to right and left, and appear to be as far behind the reflective surface as the actual subject is before it.

Because of these peculiar properties, the actual and apparent positions of a camera during filming will vary considerably, depending on the relationship between the camera, the reflective surface, and the subject matter visible in it. If, for example, a mirror is placed over the head of a performer, the camera, in spite of its ground-level position, will actually seem to be looking down on the subject. And, even though a camera is positioned squarely in front of an actor, a mirror angled above and behind the subject will create the impression that the camera was in the position occupied by the mirror during the filming process. Two mirrors, with the first reflecting an image into the second, can be used to produce a wide variety of unusual views.

Depending on the silvery backing and the amount of light falling on the subject, a mirror can be both transparent and reflective at the same time. To the imaginative cinematographer, these properties open a fertile field for experimentation that is ripe with opportunities for achieving unusual illusory effects.

The studio stage, for example, can be made to appear many times larger than it actually is by focusing the camera lens on reflected foreground material whose coloring and perspective blends with the actual background. As an alternative, the camera can be positioned "behind" the background with the concealed lens focused on a large mirror facing the rendered side. The "apparent" background, reflected in the surface of the mirror, will seem to be twice as far from the lens as it would be if the camera faced the background from the position occupied by the mirror.

Since accompanying detail will be reversed, with respect to right and left, all lettering on the background must also be reversed so that it can be read correctly in the processed film.

With the subject matter properly positioned between parallel reflecting surfaces, the "mirror" images can be repeated endlessly with eye-arresting results. In an adaptation of this illusory effect, the same person who acts as the dealer in a card game can also be shown playing each of the dealt hands—simultaneously. Obviously, if one player cheated, all of the participants would, in turn, be guilty of the same type of larceny.

Like the card players, kaleidoscope effects, in which the same image is reflected many times, can be achieved by aiming the camera lens at subject matter that has been positioned between two mirrors. The number of images that will be duplicated in the frame of film varies with the angle of the reflecting mirrors. If the screen is to be filled with four images, the mirrors must be placed at 90-degree angles to each other. Positioned at a 45-degree angle, the same image will be repeated eight times. The proper mirror angle, for any desired kaleidoscope effect, can be determined by dividing the desired number of images into 360 degrees.

More unusual results, in which performers on separate sets are photographed together in a single camera run, owe their success to imaginative cinematographers and mirrors. The Pepper's ghost shot, ideally suited for the production of this type of composite effect, is an old photographic standby and simplicity itself. During the filming process, a sheet of glass, plain or semisilvered, is placed at a 45-degree angle between the camera and the background. The performer, in front of a black, nonreflective background on another set, is reflected on the glass surface and, in turn, into the camera lens. The actual background and the reflected image of the performer, in relation to that background, are both photographed on the same length of film in a single camera run.

The density of the reflected image varies with the amount of light focused on it. As the amount of light falling on the performer is increased, the image becomes more solid than ghostly. Controlled by rheostats or dimmers, gradual changes in light intensity can cause a ghost to materialize magically and mingle with the live performers on the studio set, or disappear before their unbelieving eyes. The possibilities are unlimited.

Sudden appearance, and the ability to disappear just as suddenly into walls and other solid objects, is an accomplishment normally associated with ghosts and ethereal beings. For the studio technician and cameraman, however, the activities of these haunting apparitions are included under the general classification—special effects. The methods used for achieving these startling illusions are as many as they are varied. Some effects, Pepper's ghost, for example, are achieved with transparent or semisilvered sheets of glass. Others are produced either by multiple exposure, in-camera photographic processes or, optically, by double exposure.

If an effect is to be produced with direct in-camera photography, the subject matter in each of the camera runs must be underexposed by one complete stop in order to obtain a composite picture with 100 percent exposure. The photographic process, if the ghost has the scene to himself, is comparatively simple. During the first of the two camera runs, the background is photographed, without the ghost, and underexposed by one full stop. After the film has been rewound, with the shutter in its closed position, the ghost and background are photographed together and underexposed as in the first run. The composite effect, when projected, will show the transparent ghost haunting the solid background, the latter having received a total of 100 percent exposure during the two camera runs.

Ghosts who graciously share the spotlight with live performers and insist on appearing in public can also have their ambitions fulfilled with in-camera photography. Since the live performers must receive full exposure, the cameraman has two alternatives open to him. The first photographic option is, of course, the Pepper's ghost shot. The second alternative gives the cameraman greater latitude for adding illusory effects during the filming process. With the latter method, the performers are photographed with the scene's background at the normal exposure—100 percent. The ghost, during the second run, is photographed in front of a black, nonreflective background and underexposed by a full stop. Illuminated by spotlights, the ghost can move freely from one place to another and mingle with more solid performers exposed during the first camera run.

Unlike his human counterparts, the ghost can be made to float through the air with the greatest of ease, move from one place to another without coming in contact with the studio floor, or rise high above the crowd by merely panning or tilting the camera, to reposition the ghost image in relation to the film frame. Lying in a horizontal position in front of the black background, the camera can be tilted downward so that the apparition seems to be floating in the upper half of the film frame. Standing erect, the ghost can be made to rise high in the air, and even go out of frame, by tilting the camera gradually during the second camera run of the multiple-exposure shot.

For optical double exposures, the ghost image is photographed normally at 100 percent exposure in front of the black background. During the optical combination, the previously exposed scene that is to be used as the background for the ghost's activities, is photographed at the normal exposure. On the second optical run, the length of film on which the ghost image appears, is burnt in to create the desired composite effect. With this latter method, the double-exposed ghost can be made to appear as transparent as is desirable by varying the amount of exposure. The density of the ghost image decreases proportionately as the shutter is closed down.

If necessary, ghosts can be made to appear or disappear by simply fading them in or out of the picture. As an alternative, the ghost image may assume the role of a conscience and appear in the same film frame with his errant counterpart. Similarly, for dream sequences, the performer may be shown sleeping soundly while the ghost, or subconscious image, rises and leaves. Here, again, the possibilities are limited only by the film maker's imagination. See Figures M-16*a* and *b*.

mirror shutter A reflective camera shutter that makes it possible for the cameraman to view the subject matter as it is being exposed. The images are reflected into a prism and, in turn, viewed through a lighttight finder. See Figure M-17.

Mitchell system 35 The Mitchell system 35 uses television techniques as tools for more proficient cinematography. Versatile, compact, and lightweight, the camera can be converted to an inverted magazine configuration for portable and close-quarter filming. For military and photoinstrumentation projects, the camera can be fitted

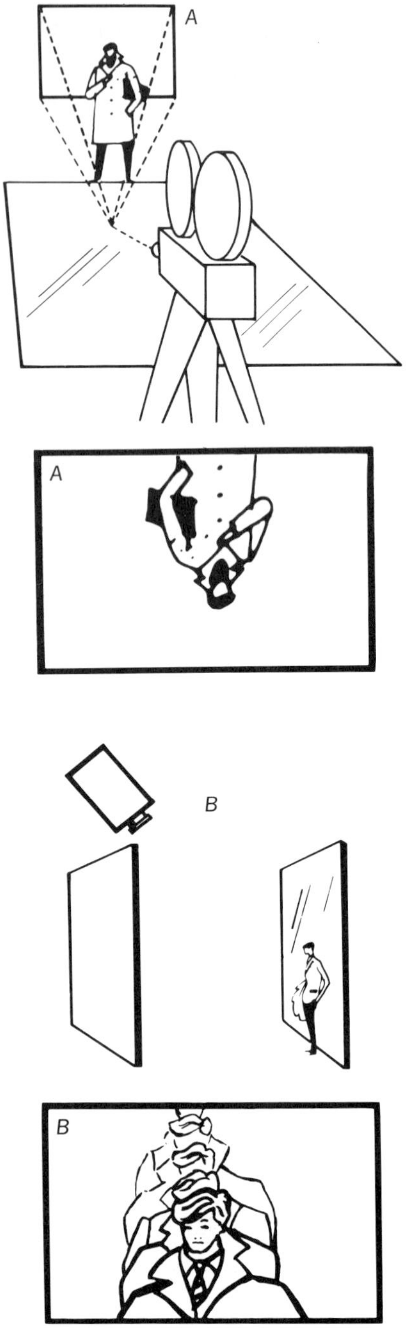

Fig. M-16a Mirrors and images. (*A*) An image can be inverted by aiming the camera, positioned at a low angle, at the reflection in the mirror. (*B*) The subject, positioned between two parallel mirrors, will be repeated over and over again in a long row—with each image gradually decreasing in size. The images, instead of being in a straight line, will curve away from center if the two mirrors are not exactly parallel.

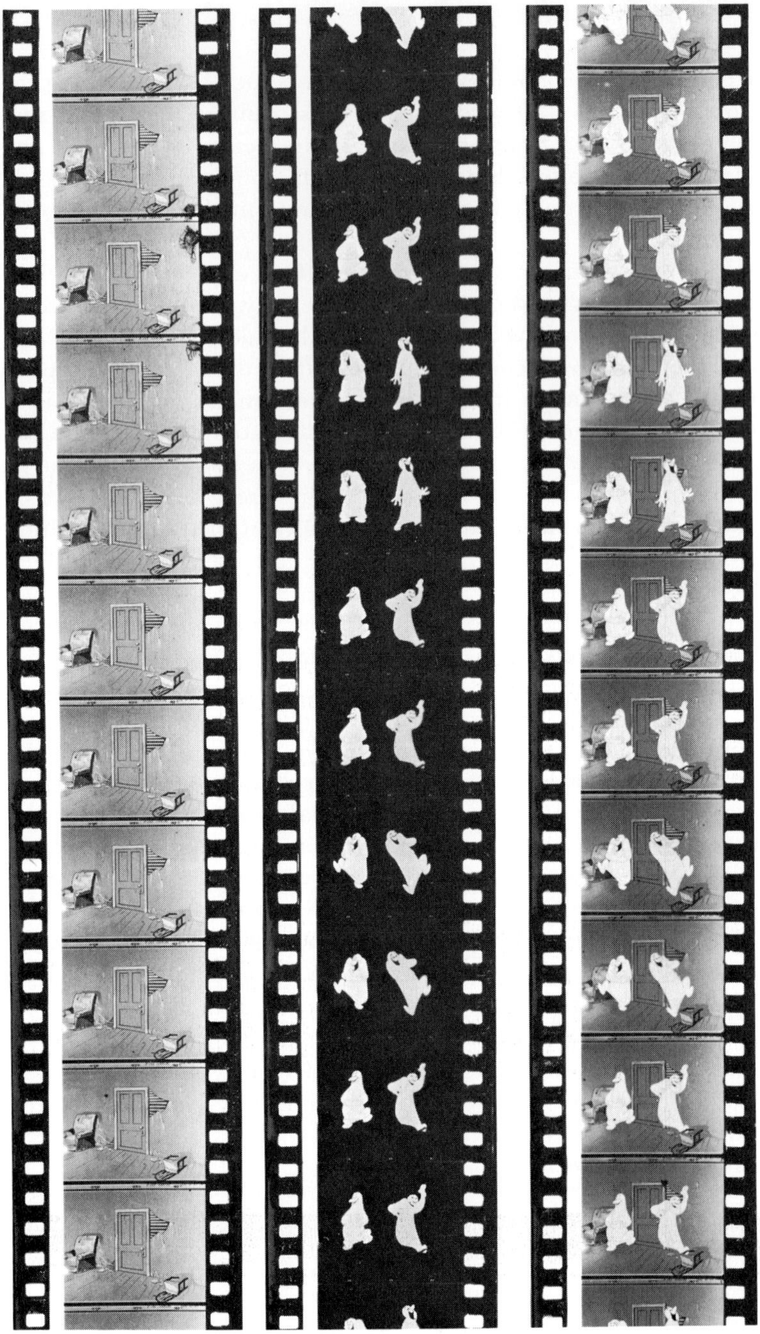

Fig. M-16b Ghosts. In animation, the effect is achieved by photographing the background with the shutter in the open position. The cels containing the animation are then photographed over a black card. On this second run, the shutter remains halfway open. The degree of transparency depends on the shutter opening. The greater the amount of transparency desired, the lower the shutter setting. The composite effect is shown in the third column.

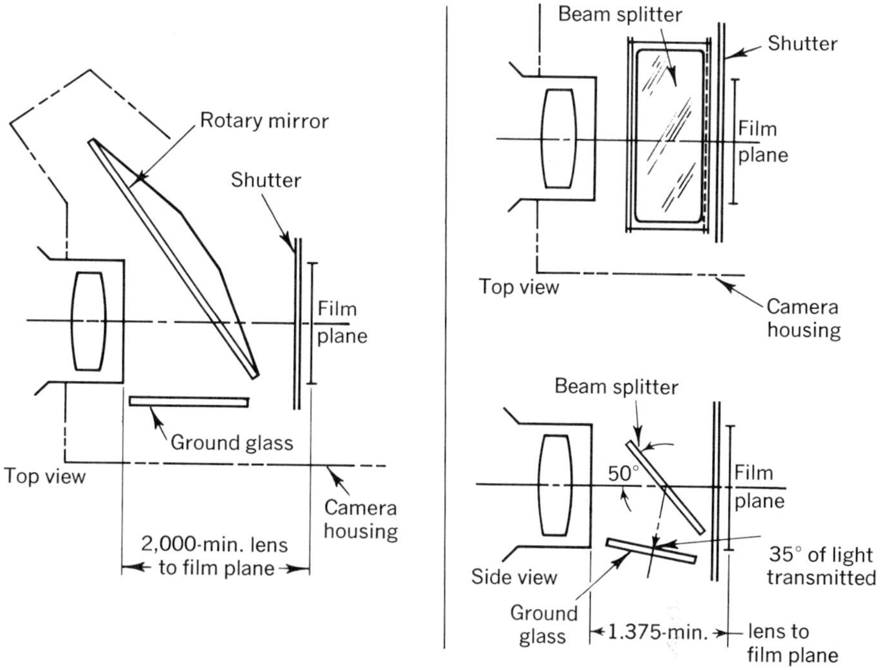

Fig. M-17 Todd-AO AP-65 mirror shutter. LEFT: rotary-mirror viewing system. RIGHT: Beam-splitter viewing system.

with a slant-mounted magazine and equipped for remote control and vidicon viewing. For sound-stage assignments, the camera can be installed in a sound blimp with a reflex-viewing side door. Wired for international use, the "universal" blimp accommodates suitable camera motors for all local power supplies. System 35 is also capable of blending the flexibility of television with multiple-camera shooting, as well as remote-vidicon monitoring of actual camera images and videotape playback for immediate reviewing.

The camera's rotating mirror, through the lens reflex-viewing setup, can be connected to a vidicon camera to provide an integrated cinematographic system with parallax-free, closed-circuit TV viewing of the exact camera/film image, and with videotape recording for immediate previewing of exposed footage. The unit can be controlled and monitored from as far as 2,000 feet away.

For remote operation, the camera is fitted with a vidicon camera, mounted so that it views the on-line image relayed through the camera focus tube. The vidicon unit uses a standard 16mm format, 50mm lens, and a special negative lens cemented to the tube face to correct for field curvature. The vidicon camera produces 700-line horizontal resolution. For extremely critical applications, other vidicon cameras with horizontal resolution as high as 1,250 lines can be substituted.

The vidicon camera scans the reflex image viewed through the camera lens. The image is simultaneously displayed on a blimp-mounted electronic viewfinder as well as on an unlimited number of monitors. In a multiple-camera setup, the director and the director of photography can also monitor and coordinate a series of such cameras from a television-type console. Monitors can also be equipped with switches to compensate for the squeezing effect of anamorphic lenses.

Control of iris adjustment, focal length, and focus—as well as camera orientation—can be handled by a single operator at a remote console. Cameras can also be set up and controlled manually for such range applications as multicamera missile tracking.

mix Also called **re-recording**. The process of combining the various sound elements (dialogue, musical background, and sound effects) into a single sound track that will subsequently be used to produce the composite print.

mm The abbreviation for millimeter. The unit in which film is designated according to width, such as 8, 16, and 35mm.

mobile unit A self-contained television studio on wheels. Usually mounted in a large truck, the television cameras and sound-recording equipment are carried to distant points for remote telecasts.

mobile video-film system A method of production during which a motion picture negative and a videotape recording are produced simultaneously either on a studio set or on location. See GEMINI TAPE-FILM SYSTEM and MEDIUMS COMBINED.

mock-up A specially constructed prop that is built to scale, as opposed to miniatures. See MINIATURES. A mock-up of a building front, for example, will cost less than construction of a complete building, but will still allow actors to be photographed in front of it. In studios involved in the production of television commercials, the building of giant-size models of products are considered routine assignments for the studio property department. See Figure M-18.

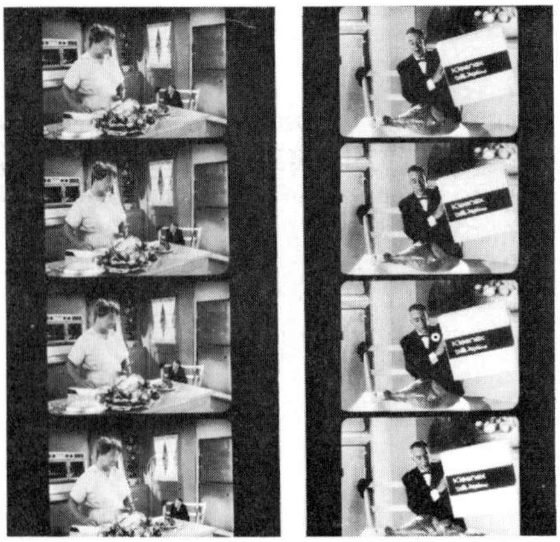

Fig. M-18 Mock-ups, miniatures and Manners the Butler. Looking at the composite effect, the housewife stands head, shoulders, and hips above the pint-sized Manners. Actually, Manners is considerably taller than the picture indicates. The effect is achieved with oversized props, well-planned background perspective, and clever editing.

mode In sound or videotape recorders, the term mode applies generally to settings or methods for positioning controls in order to activate specific circuits in the equipment. A recorder, for example, is said to be in playback mode when it is switched to play back a tape, rather than record new signals.

model A person engaged for the purpose of posing for a specific shot. In motion picture production, for example, a "hand" model might be used to hold a letter apparently held by the star of the motion picture in a preceding shot. To avoid confusion, a small-scale representation of an object is usually referred to as a *miniature* rather than a model. See MINIATURES.

modeling light Illumination on the subject from a direction opposite to that of the key light. It is also referred to as **counter key**. This type of lighting is used to create a three-dimensional quality in the subject matter.

model sheet In animation, a collection of drawings arranged to show the cartoon character in various poses. These sheets serve as a guide for the assistant animators and in-betweeners, thereby assuring a consistency with respect to size, shape, and expression. See IN-BETWEENER.

modulation The patterns of recorded sounds. Invisible on magnetic tracks, they appear as irregular patterns on optical tracks such as the variable area and variable density.

Also, the process of varying the amplitude, frequency, or phase of an electric current in order to impress on it changes representing the signal to be recorded or transmitted.

modulation noise Undesirable noise in the form of a fuzziness in the quality of the recorded sound. The condition is caused by the uneven application and the resultant varying thicknesses of the tape's magnetic coating.

moiré An optical interference pattern caused typically by the imposition of one set of closely spaced lines over another. In television, undesirable moiré can be caused by the near-coincidence of the scanning lines with the pattern of the weave in a man's suit, for example. The result is a wavy or satiny effect that can be highly disturbing.

The undesirable patterns appear as a curving of the lines in the horizontal wedges of the picture and are most pronounced near the center, where the lines forming the wedges converge. A moiré pattern is a natural optical effect when converging lines in the picture are nearly parallel to the scanning lines. The patterns are due to the presence of lines in foreground or background objects that conflict with the evenly spaced scanning lines in receiving equipment. This effect, to a degree, is sometimes due to the characteristics of color picture tubes and of image orthicon pickup tubes. (In the latter, the effect is termed *meshbeat.*) By any other name, in the case of color television, the undesirable dazzling, shimmering patterns can be quite annoying.

Molarc Also called **brute**. A large arc spotlight generally used for key lighting.

Molefay, five-light A light unit that accommodates five FAY tungsten-halogen globes (650 watts PAR-36) capable of producing 3,250 watts of daylight illumination. Spot, medium-flood, or wide-flood illumination is possible with color temperatures ranging from 3200 to 5000°K.

Molenser A versatile light source with a three-lens system. The intensity of the light remains at the same level at a given distance regardless of the size of the spot. Special beam shapes may be projected on the set background with mattes made for that purpose.

Molequartz Mighty-Mole The Mighty-Mole uses 3200 and 3350°K quartz-tungsten-halogen globes in 2,000-, 1,500-, or 1,000-watt ratings. Spot-to-flood control of the variable beam is accomplished by turning a control knob at the rear of the housing to move the reflector. At 10 feet, using the 2,000-watt 3200°K globe, the light intensity may be varied from 2,400 to 280 footcandles.

Molequartz Teenie-Mole A focusable quartz light, the 4051 is rated at 650 watts. At 10 feet, the light beam from an FAD globe can be varied from spot to flood to provide a 9:1 illumination ratio with intensities ranging from 780 to 85 footcandles.

Molevator A power-driven pedestal for large lamps, with a maximum load-carrying capacity of 250 pounds. A mechanical jackscrew provides positive support. A slip clutch protects the mechanism at its travel limits.

Constructed of steel and bronze, the pedestal operates on 120 volts, 5 amperes, a-c/d-c. The unit can be folded for storage or transportation.

monaural Single-channel recordings made with full-track or half-track recording heads. These heads record only one track on the full width of ¼-inch magnetic tape or on a narrower area in the middle of the tape. See BINAURAL and STEREOPHONIC RECORDING SYSTEM.

monitor In sound recording, the process of checking the quality of the sound either while the recording is being made or during the playback. Also, compensating adjustments in volume that are made during the actual recording process.

In video, the term is used for a studio receiver, either in the control room or on the set where the program is actually being televised or taped, that is used for viewing or monitoring purposes.

monitor control A television camera control that is used with a studio switcher to connect the camera chain output to the console monitor.

mono A television camera control that disables the color channels and provides the monochromatic (black and white) signal. See MONOCHROMATIC.

monobath A solution in which the developing and fixing chemicals are combined to provide one-step processing.

monochromatic Gradations or tones of a single color or hue.

monochrome automatic timing corrector In videotape recorder units, the monochrome ATC (automatic timing corrector) performs the function of electronically correcting time-base errors. The correction signal is derived by comparing the time-base errors off the tape to that of a fixed reference. The resultant picture, after correction, contains virtually no geometric errors. For color operation, it is necessary to further reduce the residual timing errors. See COLOR AUTOMATIC TIMING CORRECTOR.

monochrome transmission The transmission of a signal wave (black and white) which represents the brightness values in the picture but not the color (chrominance) values in the picture.

monopack A film with more than one layer of emulsion, each sensitive to one of the photographic primaries. See INTEGRAL TRIPACK and MULTILAYER FILM.

monoscope A cathode-ray tube designed specifically for transmitting still pictures or patterns in order to check receiving equipment. See Figure M-19.

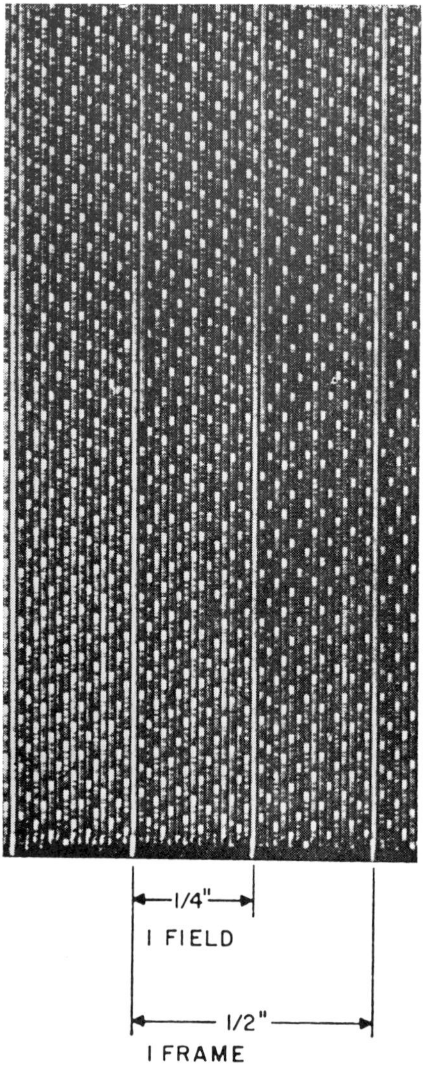

Fig. M-19 A monoscope pattern recorded on videotape. The vertical blanking intervals appear as long lines, while the horizontal sync pulses show as the pattern of dots between them.

montage shot An optical effect in which several images are superimposed or double-exposed in the same film frame to create a specific visual impression.

mood music Also called **background music**. Recorded sound (music and musical effects) used as a background for the dialogue in the sound track. Especially written or scored for specific purposes or, in other cases "canned" (prerecorded for other purposes), the musical background helps establish the mood for the accompanying picture.

mosaic In film, a number of relatively small pieces of picture information which, when assembled and positioned in their proper order, form a single composite of the whole picture.

Mosaics are particularly useful when an area is too large to be covered by a single photograph (as in survey work, for example), or when the amount of detail that is required can best be obtained from a series of individual photographs of smaller sections within the larger area. Usually taken from an airplane, each of the overlapping photographs is identified by either a number or a letter of the alphabet (sometimes both), sorted, and then assembled to form the composite picture.

In television, the sensitive screen in the TV camera tube on which the picture elements are formed after being scanned by the beam from the electron gun.

Motion Analysis Camera (MAC) Motion analysis cameras are high-speed cameras of the rotating-prism type, designed for research and engineering. The Fairchild MAC is used to record from 10 to 8,000 frames per second for study and analysis of phenomena that occur too fast for the human eye to discern. The MAC is a troubleshooter and faultfinder since the action can be photographed at a high frame rate and projected for evaluation and analysis at a much slower rate, thereby magnifying space and time. This facet of photoinstrumentation results in saving valuable hours and cost of troubleshooting, research, repeat testing, and uncertainty. Vibration, shock, flow, and structural change effects may be detected in numerous areas of science and industry.

Constructed for laboratory and field use, the MAC is built to withstand severe acceleration, shock, vibration, and temperature. The main drive consists of a four-sided prism geared to a sprocket. The prism permits photography completely free of fringing while eliminating the need for color-correction filters. To ensure optimum resolution, regardless of camera speed, the film, as it is drawn around the sprocket, is exposed by the rotating prism traveling in synchronization. Power is supplied (except at higher frame rates) on the take-up spool spindle. The film serves as a belt to drive the sprocket that is geared to the rotating prism. For the higher-speed ranges, power is applied to both take-up spindle and sprocket shaft. The motors are series-wound, 28 volts d-c, except for higher-speed motors which are field-wound. The motors are of the high-starting torque type, thereby rapidly reaching maximum velocity.

Timing lights are provided to place "pips" on the film edge so that an accurate time base can be determined when data are evaluated and analyzed. An end-of-run film switch is provided to automatically shut off the camera power when the film supply is exhausted or when threading is incorrect. The detachable open-sight viewfinder can be adjusted for parallax. All lenses used on the MAC are of the positive locking-bayonet-mount type.

Fairchild industrial model HS-101, 16mm. This camera is designed for laboratory or field use and can be hand-held. The camera uses 100-foot-capacity daylight-loading spools. A single motor of the purchaser's selection is supplied with the camera. Speeds range from 10 to 8,000 frames per second, depending on the motor. The camera is provided with three mounting holes for $\frac{3}{8}$-inch no. 16 screws, and a tripod hole for $\frac{1}{4}$-inch no. 20 screws. The camera is 8 inches long by $8\frac{3}{8}$ wide by $7\frac{1}{2}$ high, and weighs 10 pounds.

Fairchild model HS-401, 16mm. The camera has a daylight-load 400-foot film capacity and an exclusive start-stop feature using a magnetic particle brake, enabling the operator to start or stop the camera at any speed. A brake test button and a potentiometer are incorporated with an indicating meter to set the brake adjustment correctly. A single motor of the purchaser's selection is supplied with the camera. Speeds range from 10 to 8,000 frames per second depending on the motor. The camera is provided with three mounting holes for $\frac{3}{8}$-inch no. 16 screws. The camera is $16\frac{13}{16}$ inches long by $12\frac{15}{16}$ high by $9\frac{1}{16}$ wide, and weighs 22 pounds.

motors, camera Various types of motors may be used on cameras. For synchronized sound work, synchronous- or interlock-type motors may be used. A motor adapter door (on the Mitchell equipment described here) is supplied with both the BNC (**B**limped **N**oiseless **C**amera) and the NC (**N**oiseless **C**ameras). This door permits use of a variable-speed or "wild" motor for process work, nonsynchronous work, or special studio laboratory shots where sound insulation is not a factor to be considered. The interlocking or the synchronous types of motors are mounted in insulated soundproof housings which fit directly on the right side of the camera.

These motors are mounted in a horizontal position parallel to the camera, and the camera is driven by means of right-angle gears and flexible coupling. Motor interchangeability is facilitated by the fact that no wiring changes are necessary when different types of motors are mounted, since the electric receptacle connects the motor to the power lines.

NC and BNC motors are not interchangeable, owing to slight variations in electric connections. An external electric terminal is located on the top, right-hand side of the BNC camera outer case. An electric plug is built into the motor assembly and the motor adapter door so that electric connection is made with a socket on the right side of the camera when the motor is mounted. On the NC camera, the electric cord is fastened directly to the motor.

The motor drives on both NC and BNC cameras are virtually identical. The drive is direct to the movement, which in turn, drives the shutter shaft through right-angle helical gears.

The shutter shaft drives by a worm gear, the sprocket, and the magazine take-up pulley, which are mounted on one shaft. The counters and dissolving mechanism on the rear of the camera box are also driven from the shutter shaft.

moving electronic matte Also called **combined montage amplifier.** See BRIGHTNESS SEPARATION OVERLAY.

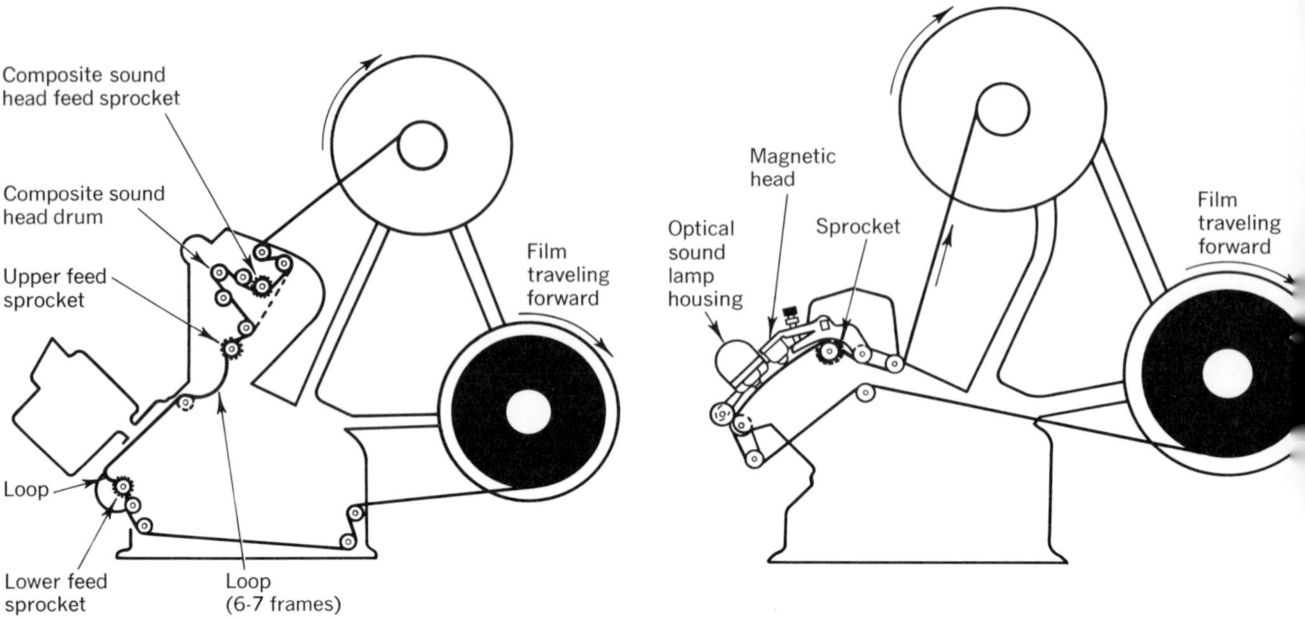

Fig. M-20 Moviola threading diagrams. (*Magnasync/Moviola Corporation.*) LEFT: film threading, 35mm machine. RIGHT: separate sound-head film-threading diagram.

moviola A projection device used by a film editor for synchronizing the picture reel with the sound track.

In operation, the picture reel is threaded so that each frame of film is subsequently visible on the moviola's built-in screen. The sound track, on a separate reel, is threaded so that it passes over the sound head in synchronization with the corresponding picture. The editor can advance either picture or sound track in order to adjust synchronization. The film can be run at sound speed or viewed at slower speeds, whichever of the two is desirable. See Figure M-20.

mμ Millimicron. One millionth of a millimeter.

muddy In sound recording, blurred or fuzzy results in the quality of reproduced sound.

multibeam The light output of the multibeam 650 is continuously variable from 680 footcandles at spot, to 80 footcandles at flood. With the use of a 650-watt quartz-iodine lamp, it is equally effective in the studio or on location. At full spot, the 650 produces 680 footcandles of light, at 10 feet with a 650-watt 3200°K (B6-32) lamp. At full flood, the spread of light exceeds that of a scoop. The reflector design eliminates filament pattern and hot spots throughout the entire focusing range.

The multibeam 1000 is a lightweight, compact, variable-beam quartz-iodine light for a wide range of applications. The beam angle

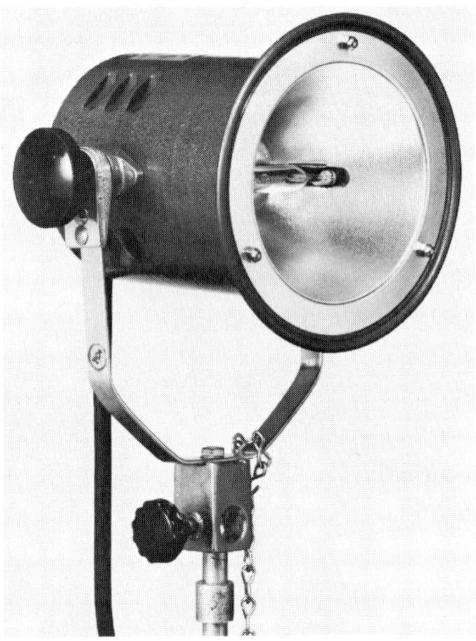

Fig. M-21 ColorTran multibeam 1000.

(to 50 percent of maximum intensity) is continuously variable from 21 degrees at spot position, to 41 degrees at flood. Intensity (at 10 feet) can be varied from 840 to 140 footcandles, providing a usable intensity ratio of 7.3:1. The 1000 uses B10-32E extended quartz-iodine lamps, operating directly from 120 volts a-c/d-c. See Figure M-21.

multibroad This lighting unit is continuously variable from wide- to medium-focus positions, producing 61 to 230 footcandles at 10 feet. Focusing is accomplished with a large knob under the housing. See Figure M-22.

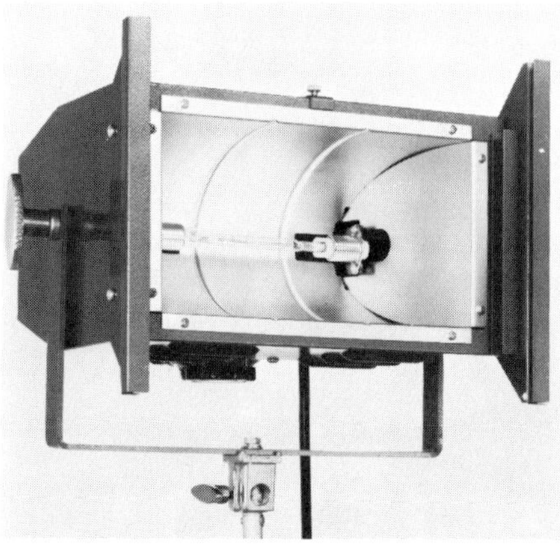

Fig. M-22 **ColorTran multibroad.**

multiduty motor A d-c motor capable of driving a number of cameras and related equipment in synchronization. Because of the frequent need for adjustments in running speed and the annoyances that go hand in hand with the use of battery-driven equipment, this type of motor is gradually being replaced by more sophisticated equipment.

multifrequency film Also called **constant-frequency film** and **sound-transmission film**. A motion picture sound track with frequencies ranging from 40 to 10,000 cycles per second. Used for measuring the sound system's frequency response, each of the changes in frequency on the sound track is identified by an announcement stating the specific frequency.

multilayer film Film with more than one layer of emulsion—with each layer sensitive to its own primary color. Each of the layers contains light-sensitive silver halide along with color formers or dyes of the subtractive type that make it possible to restore colors similar to those found in the original subject matter.

During the developing process, the exposed silver is developed, and the colorant or dye exposed by one of the primary colors is released. The other layers of emulsion react similarly, each releasing its own color dyes. The colors in the developed single-strand negative are complementary to those in the original subject matter. See Figure M-23.

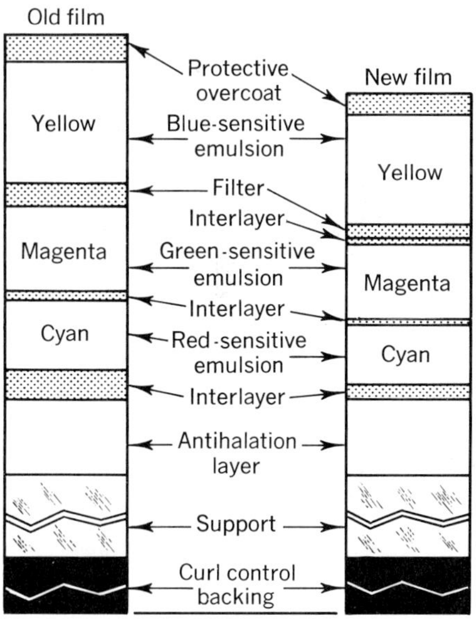

Fig. M-23 **A cross section of multilayer color film.** Thinner: Overall thickness has been reduced by cutting emulsion layers and interlayers. Base thickness remains the same.

Multilock A system developed by Ampex of Canada for synchronizing separately recorded sound with the video picture output of a videotape recorder. Designed primarily as a teleproduction tool for use by networks and production studios, the system may be used whenever separate sound recording is required.

In operation, multiple-channel sound is recorded on a separate audio recorder, and the sound channels are synchronized with video recordings during playback. The system features a fast search mode for the synchronized recorder so that it can be brought rapidly in step with the master recorder.

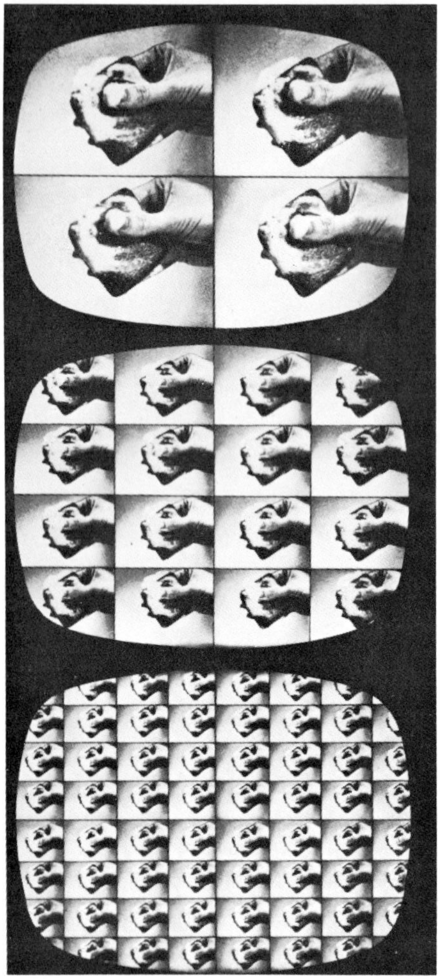

Fig. M-24 The multiple-exposure effect. Produced for the Brillo Manufacturing Company, the effect in this commercial graphically illustrates the effectiveness of their product. Actually, a variation of the split-screen effect, each of the sections is matted in separately in a preoptical. In the final phase of production, the preoptical composites are repositioned to produce the final effect.

multiple-image shot An optical effect in which the same image is repeated a number of times in the same film frame. See Figure M-24.

multiplex Also called **FM stereo**. A method of transmitting two channels of audio signals on the same carrier. The signal received is separated into channels to yield stereophonic or three-dimensional effects when fed to separate loudspeakers.

multiplexed telecine operation Multiplexing is an electronic system that is used for combining signals in the form of picture information, from several sources, for subsequent television transmission. As applied to film transmission, multiplexing consists of an optical arrangement that takes the outputs from a number of projection devices and directs them, either individually or as a composite, to a single camera channel.

In addition to the system's obvious functions, multiplexing offers an economic advantage by freeing other cameras that might ordinarily have to be used for these composite outputs. In operation, the vidicon camera and film projectors must be equipped with lenses in order to increase the working distance between them. Although the film images could be projected directly onto the small target area within the vidicon tube, the projection lens throw would have to be very short. This type of arrangement would require a more critical setup or working relationship between the projector and the camera. As a compromise, the film image is first projected to the plane of a field lens, and this field image is, in turn, directed to the camera lens.

In other multiplexing systems, the film image is directed to the camera lens by way of fully silvered or semisilvered mirrors. Either of those optical arrangements make it possible to combine picture information from several sources to form a single composite. As an alternative, it is also possible to select one particular film image, as needed, from any of the projection devices used in the system. Any light loss caused by the silvered mirrors, due to reflection and transmission, can be regained by boosting the light-source intensity to produce an image of sufficient brightness on the vidicon target.

Vidicon telecines enjoy a production advantage over flying-spot telecines which usually require special optical systems or projection equipment. Vidicon telecines, by comparison, aside from providing a more economical operation, use standard slide and film projection equipment. Differences in the quality of the transmitted picture, in either telecine system, are negligible. See Figure M-25.

Munsell system A color standardization system invented by Albert H. Munsell that arranges colors in a three-dimensional color solid.

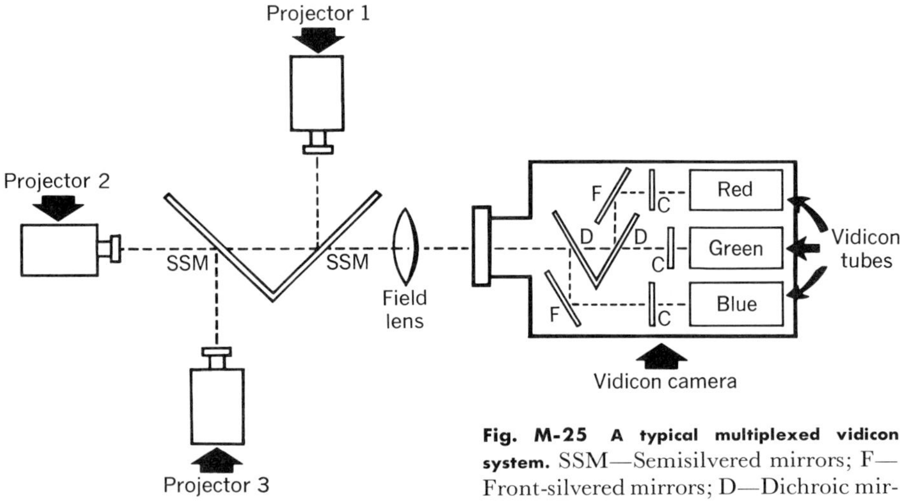

Fig. M-25 A typical multiplexed vidicon system. SSM—Semisilvered mirrors; F—Front-silvered mirrors; D—Dichroic mirrors; C—Filters.

Each color is classified and assigned a position in the system according to its hue, value, and chroma. Spaced at intervals along the equator of the nearly spherical solid are the five principal hues (red, yellow, green, blue, and purple), alternating with the five intermediate hues (yellow-red, green-yellow, blue-green, purple-blue, and red-purple). Each of these 10 major hues is assigned the number 5 in a 10-number series, so that the complete circle of hues numbers 100 colors.

The vertical axis of the color solid is represented by value or reflectance, in a 10-step scale extending from perfect reflectance, or white (number 10) at the top, to no reflectance, or black (number 0) at the bottom.

The horizontal distance from this center axis represents the third element in this classification system. The amount of chroma, or saturation, is indicated along each of the radial spokes for each hue at the various value levels by up to 16 incremental steps. Since colors vary in the amount of chroma or saturation that is possible, the solid is not perfectly symmetrical, and a color like blue-green, for example, has a maximum chroma of only 6. The zero value at the vertical axis represents neutral gray.

mural A relatively large photograph or painting that is usable as a background during the filming process.

mylar base A dimensionally stable material used as a base for the ferromagnetic coating on magnetic tape. Unaffected by changes in temperature and humidity, clear rolls of mylar are also used for splicing lengths of motion picture film.

mylar tape splicer The guillotine splicer is capable of making butt, overlap, or diagonal splices without losing frames. Dissimilar film bases can also be spliced together without difficulty.

In operation, the film is placed on the registration pins in the recessed splicing block, and the applied tape is perforated and cut to conform with the film. The device was invented by Dr. Leo Catozzo and is protected by a United States patent.

N

NAB The abbreviation for the National Association of Broadcasters, now known as the NARTB (National Association of Radio and Television Broadcasters). In many ways similar to the ASA (American Standards Association—now known as the ANSI, American National Standards Institute), the body sets the standards and specifications for broadcasting and transmission with respect to equipment, acceptable frequency response, distortion, and other related items.

NAB curve The standard playback equalization curve set by the National Association of Broadcasters.

narrow-band filters Optical filters capable of isolating relatively narrow spectral bands. They transmit the light from a single color located in a relatively narrow region of the spectrum.

narrow-gauge film Films narrower than the standard 35mm width: 16mm, Super 8, or 8mm, for example.

Naturama See WIDE-SCREEN PROCESSES.

negative Developed film that contains images that are completely opposite with respect to colors and/or tonal values from those in the original subject matter.

negative image An image in which the colors and tonal values are directly reversed in comparison with the values in the original subject matter. Black areas are white, and white areas are black. Intermediate shades of gray are also reversed proportionately. The hues, in color negatives, are complementary compared with the brightness scale in the original subject matter.

negative lens A lens that causes a divergence of light rays. Thinner at the center than along the edges, these lenses are used to correct errors in positive lenses. See LENSES.

neon lamp A glass lamp, usually tubular-shaped, filled with neon gas, and using very little current. The cold glow that is emitted by this type of lamp is of little value for illumination purposes in photography. It is, however, quite applicable in electronic timing systems where it is used as a control for activating other light units employed for illumination purposes.

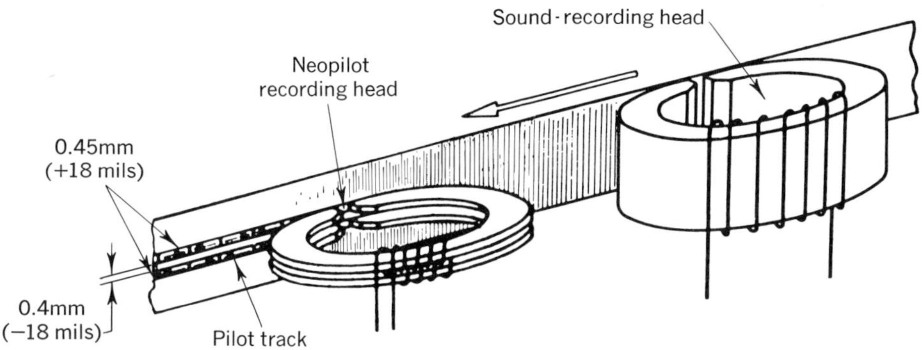

Fig. N-1 Neopilot system. (*Sketch courtesy of K. E. Gondesen.*)

Neo-Pilotton Sync System The Neo-Pilotton system has two standard-type magnetic heads. The slit of each head is about 10 mils in length, and they are separated from each other by about 10 mils. They are in line and parallel to the audio head and placed so as to record down the center of the tape. These heads are operated in push-pull and are balanced so as to give a minimum of crosstalk. See Figure N-1. Compare ECHELON SYNC SYSTEM, PILOTE SYNC SIGNAL, and RANGER SYNC SIGNAL.

network A group of television stations in different locations that share facilities and programs.

neutral-density filter A filter used basically for reducing exposure. See Table N-1 and FILTERS.

neutralization The combining of two lenses of opposite powers so as to produce a resultant without power.

TABLE N-1 Neutral-density Filters

Density	Percent trans- mittance	Exposure compensation (in *f*-stops)	Density	Percent trans- mittance	Exposure compensation (in *f*-stops)
0.10	80	$+\frac{1}{3}$	0.80	16	$+2\frac{2}{3}$
0.20	63	$+\frac{2}{3}$	0.90	13	$+3$
0.30	50	$+1$	1.00	10	$+3\frac{1}{3}$
0.40	40	$+1\frac{1}{3}$	2.00	1.0	$+6\frac{2}{3}$
0.50	32	$+1\frac{2}{3}$	3.00	0.10	$+10$
0.60	25	$+2$	4.00	0.010	$+13\frac{1}{3}$
0.70	20	$+2\frac{1}{3}$			

Courtesy of Eastman Kodak.

neutral scale A gray scale used for reference purposes to check color balance in color-separation negatives.

neutral test card Designed to provide a means of reference in color photography, the gray side of the Kodak neutral test card has an 18 percent reflectance value—the white side, a 90 percent reflectance value. In addition to their use as a method of exposure determination, they are also helpful for adjusting lighting ratios and checking lighting distribution. Light readings are made by holding the card in front of the subject, angled so that the card faces midway between the key light and the camera.

Newton rings Irregularly shaped, colored rings that appear in transparent materials; film, or acetate Celluloid, for example, when contact occurs between two smooth or polished surfaces. Interfering light reflections from both surfaces are responsible for the colored rings.

nitrate film Before 1950 most of the cans of motion picture film delivered to production studios contained a built-in hazard in the form of a nitrate base. In spite of the highly flammable support and the need for precautionary measures in handling and storage, nitrate stock was the overwhelming choice of the professional film producer. Safety film, available since the industry's pioneer days, failed to match the physical properties, printing characteristics, and popularity of the nitrate variety.

Recognizing the need for a safety film that would satisfy the requirements of the rapidly growing industry, the film manufacturers developed cellulose triacetate and acetate propionate bases. These supports, with their comparatively thin but tough safety bases, gained the industry's almost immediate approval and acceptance.

Since considerable amounts of nitrate stock are still in circulation, however, precautionary measures must be taken with all film. Nitrate negative, both picture and sound, along with duplicate negatives, masters and old prints, will remain a hazard until chemical and natural changes cause deterioration and render them harmless.

nodal point Imaginary vertical lines cutting through a lens which indicate the points where light is refracted when entering or leaving the lens. The line where the rays of light are refracted after entering the lens is known as the nodal point of *admission.* The point at which the rays refract when leaving the lens is the nodal point of *emission.*

noise The word noise in video usage, a carry-over from audio practice, describes random electric energy or interference. It may produce a "salt-and-pepper" pattern over the picture. *Heavy* noise is sometimes referred to as *snow.*

noise, audio frequency Any electric disturbance including both hum and hiss introduced from sources extraneous to the signal.

noise, level The amount of ambient or unavoidable sound picked up by the microphone during the recording process.

Fig. N-2 This nomograph shows the amount of recording time that is possible at different recording speeds and the length of the reel (footage) that would be required for the recording. (*DuKane Corporation, St. Charles, Ill.*)

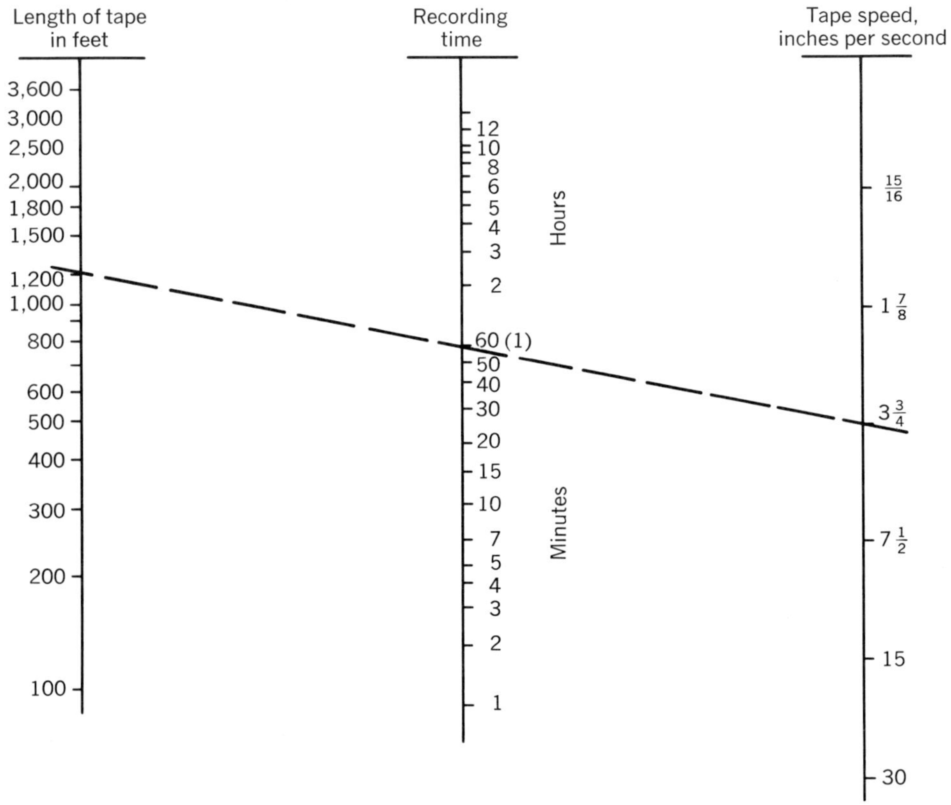

noise, unweighted The noise measured within the audio-frequency passband using a measuring instrument that is uniform in response with respect to frequency over some specified passband. Compare NOISE, WEIGHTED.

noise, weighted The noise measured within the audio-frequency passband using a measuring instrument that has a frequency-selective characteristic. The sensitivity is usually greatest in the frequency range where the ear is most sensitive. Compare NOISE, UNWEIGHTED.

nomograph A graph or chart that permits the user to read off the value of a dependent variable on a graduated scale when a straightedge is laid along the values for independent values as shown on other graduated scales. Nomographs permit rapid determination of values such as the color temperature of a light source without numerical calculation. See Figure N-2.

Norwood analyst A sophisticated viewing glass, actually a meter, that electronically analyzes a scene with respect to overall brightness. The meter indicates the degree of correction needed to bring the lightest and darkest areas within a scene into a proper balance with the average values in the scene. The meter makes it possible for the cinematographer to use special lighting effects while maintaining overall scene balance and proper brightness ratios.

notch A method of identification relied on by still photographers who must identify film types in total darkness. The shape, spacing, and number of notches in the upper right corner of cut films (when the emulsion faces the handler) are clues that tell the cameraman the type of emulsion that he is handling.

Also, a cue mark used by film editors, optical technicians, and laboratory timers, generally, to indicate required changes in printing lights. See Figures N-3*a* and *b*.

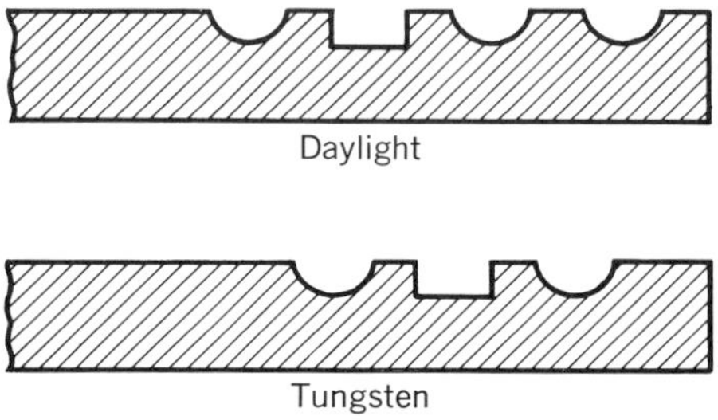

Fig. N-3*b* Anscochrome sheet film is notched as shown, in 3¼ by 4¼ inches and larger. Smaller sizes have a single shallow notch. The emulsion side of the film faces you when the notches are in the upper right-hand corner.

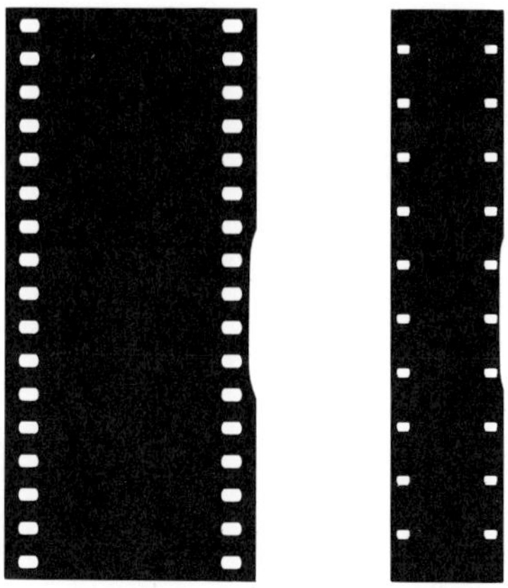

Fig. N-3*a* Notches. In 35mm film, the notch appears on the negative between the sixth and seventh frames of the scene in which the change or correction occurs. In 16mm film, the notch is found between the 15th and 16th frames.

O

objective lenses The primary, light-collecting lens in an optical system. Although the term is more correct when used in connection with telescope and microscope lenses, photographic objectives has become an inclusive term for camera lenses generally.

Included under objective lenses are the singlets (single lenses used in relatively low priced cameras and also referred to as landscape lenses), duplets (usually symmetrical lenses with two components), triplets (three-lens components where each is separated from the others), Petzval or portrait lenses (for use on relatively small fields at high apertures), and Telephoto lenses (with widely spaced negative and positive components).

Since a projector lens is a light-collecting lens, it could, technically, be regarded as an objective lens and thus merit inclusion in this general grouping. The condensing lens in a projector, by comparison, could not be included in the same category since its main function is to concentrate light rays over a specific area.

odd-line interlacing system A scanning system, standard in American television operations, in which the odd-numbered lines are scanned by the electron beam and, subsequently, the even-numbered lines needed to complete the entire picture. See Figure O-1.

The first television pictures were produced on 180-line screens; today 525 scanning lines are used. The European standard is 625 lines.

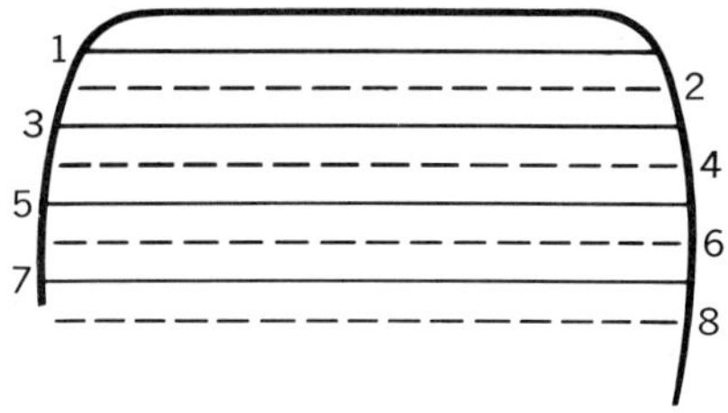

Fig. O-1 **Odd-line interlacing system.**

oersted In sound recording, a unit of magnetizing force abbreviated by cgs (centimeter-gram-second system of measurement). Named after the Danish physicist, Hans C. Oersted, the unit is equal to a force of 1 dyne exerted on a unit magnetic pole.

off register A visual effect, such as a vibration or camera rock, that usually accompanies violent screen actions such as explosions or crashes. Any amount of camera movement, during filming, will produce this effect.

In animation, the effect is achieved by changing the position of the animation camera, in relation to the artwork positioned on the compound table, for each frame of the action.

omnidirectional microphone A microphone capable of picking up sound from every direction, regardless of its placement.

one-shot color camera A type of camera in which three separation negatives are produced with one exposure.

ones, twos, threes In animation terminology, the number of exposures, as indicated by the animator on the exposure sheet, that each of the drawings or cels in an animated action is to receive during the filming process. See Figure O-2.

Fig. O-2 **Exposure sheet: ones, twos, threes.**

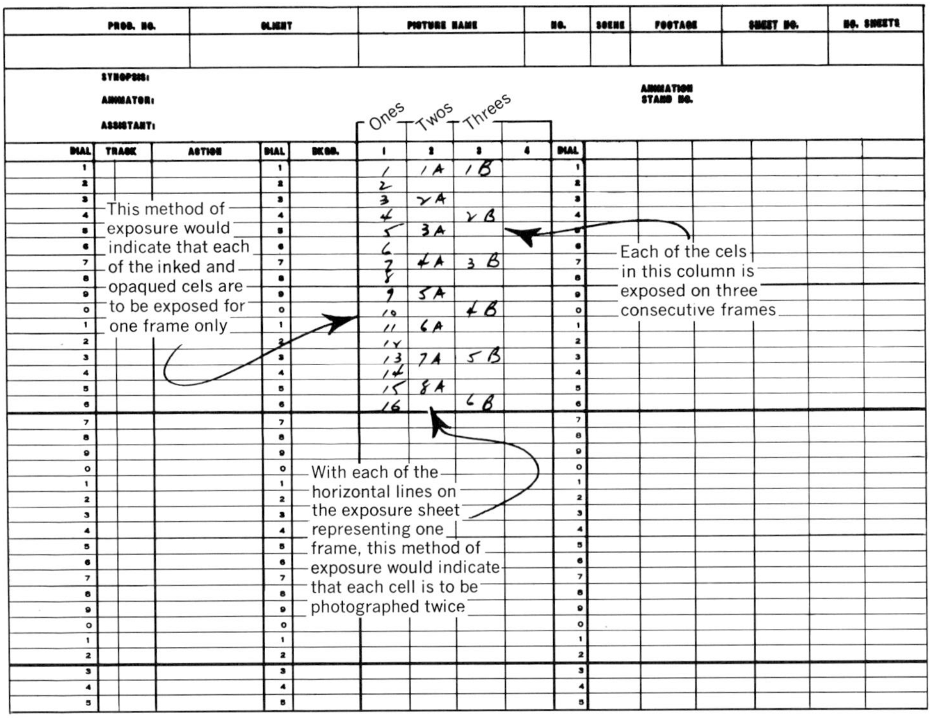

opacity The density of a frame of film in terms of the amount of light that can be transmitted through it.

The formula indicating the relationship between opacity, density, and transmission is

$$\text{Density} = \log^{10} \frac{1}{T}$$

where T is the transmission.

opal glass See MILK GLASS.

opaque Anything incapable of transmitting light is said to be opaque.

opaquer In animation, the artist who applies opaque colors to the various parts of a cel (after the lines of the drawing have been traced on it). The opaque colors are applied (opaquing) to the opposite side of the transparent cel in order to avoid smearing the ink lines that had been traced earlier. Color charts prepared by the animator specify the exact color that is to be used for each area. See Figures O-3*a* to *c*.

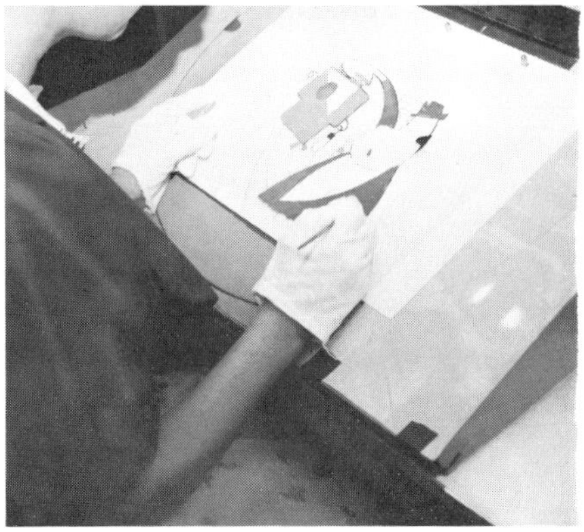

Fig. O-3*a* **The opaquer.**

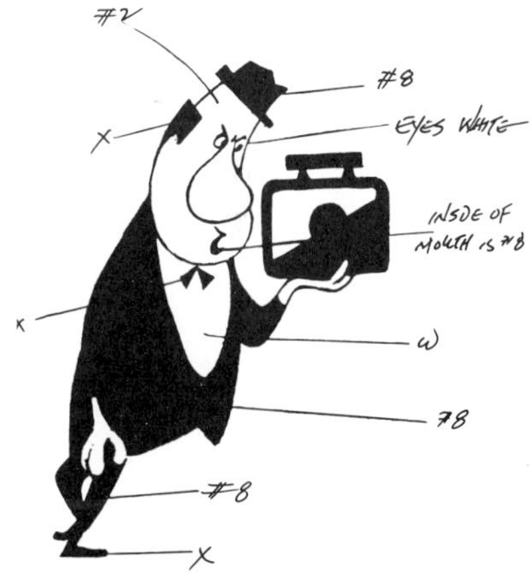

Fig. O-3*b* **An inked and opaqued cel.**

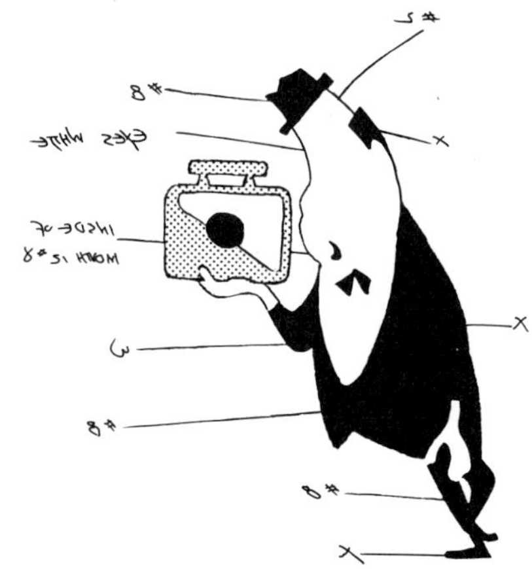

Fig. O-3*c* **The reverse side of an opaqued cel.**

optical axis An imaginary, perpendicular line through the optical center of a lens.

optical camera Also called **process camera**. Facing the optical printer's projection shuttle and the several lengths of processed film passing through it, the picture information in these lengths of film is reexposed on the raw stock in the camera to produce composite effects.

Animation may be combined with live-action scenes, titles may be superimposed over other picture information, and transitional and special effects mattes needed to produce unusual illusory shots may be added where needed. See Figure O-4. See OPTICAL PRINTER.

optical center An imaginary point in a lens through which the secondary and optical axes pass.

optical compensating camera High-speed cameras in which the film runs continuously and optical compensations are used to "freeze" an image during the interval of exposure. In some cameras, framing lenses are synchronized to move with the film, and in others, rotating mirrors or rotating prismatic glass blocks move the images at the same velocity as the moving film. See FRAMING CAMERA and HIGH-SPEED PHOTOGRAPHY.

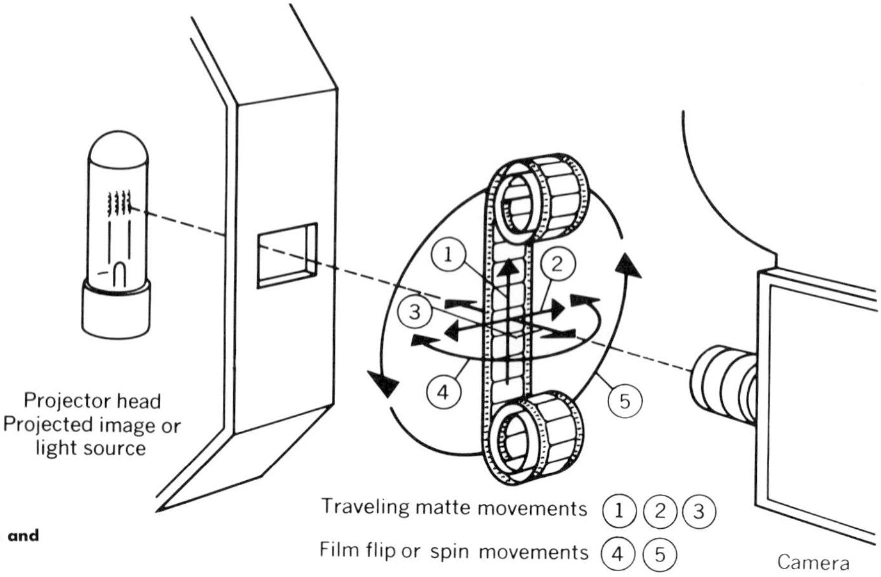

Fig. O-4 Direction of film travel for flip and spin effects. (*Animation Equipment Corp.*)

optical effect An inclusive term for special effects generally and for such transitional effects as **fades**, **cross-dissolves**, and **wipes**. See SPECIAL EFFECTS and TRANSITIONAL EFFECTS.

optical glass Glass ground to predetermined specifications for use in the production of lenses.

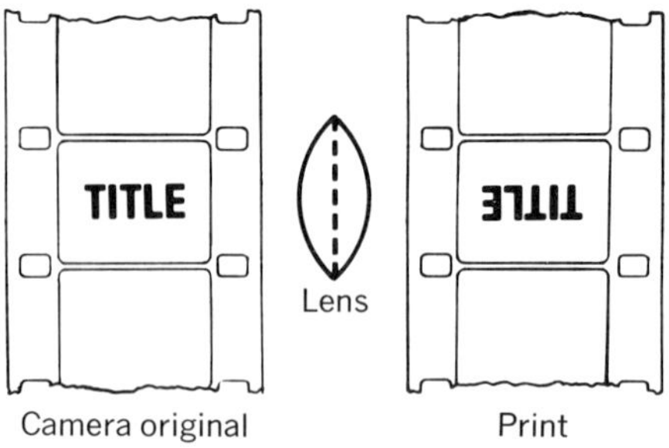

Fig. O-5 Optical printing.

optical print Also called **answer print**. The print produced by optically combining scenes and adding special effects, transitional effects, or titles. If approved, the optical print becomes a guide for the production of the composite print. See Figure O-5. Compare COMPOSITE PRINT.

optical printer An optical, electrical, and mechanical device that combines the basic functions of a camera and a projector. Capable of duplicating the images on previously processed film, the printer is also used to add transitional and other effects in order to produce a film that is complete with respect to visual continuity.

In operation, the film that is to be duplicated, along with mattes, titles, and other film elements, is threaded on the projection side of the printer. The projection shuttle, capable of accommodating several thicknesses of film simultaneously, is synchronized with the shuttle and transport mechanism in the optical camera. Each frame of processed film, as it travels through the projection shuttle, is illuminated by the light source located in the projector housing. In turn, each frame of processed film is copied, or reexposed, on the raw film stock in the optical camera.

Undesirable double-exposure effects that would normally result from the combination of several lengths of film are avoided through the use of mattes. With these opaque film masks on lengths of high-contrast film, it is possible to photograph or copy a portion of the processed film, as it travels through the projection head of the printer, while leaving the other areas of the frame unexposed. On subsequent runs, picture information from other lengths of processed film is inserted in the unexposed or previously matted areas. Animation can thus be combined with film that contains live action—and, to this optical mélange, the producer can add titles and superimposures without risking the possibility of a double exposure.

Many effects that would otherwise be difficult if not impossible to achieve can be accomplished quickly and precisely with the optical printer. Fades and cross-dissolves, transitional effects that can be produced during the actual photographic process, are usually added optically in order to allow the film editor a greater degree of flexibility in the completion process. Wipes, by comparison, are accomplished by combining the closing and opening footage, from successive scenes, with the positive and negative mattes that contain the wipe patterns. Flips and spins, involving entire scenes or individual frames of film, serve a dual purpose. Effective for transitional purposes, flips can also produce a mirror image, reversing right and left.

Capable of enlarging or reducing portions of the picture information within an entire scene or an individual frame of film, the printer can freeze the action, both live and animated, slow it down, speed it up and, where required, even reverse it. Reverse action, guaranteed laugh producers in most cases, enables a diver to defy the laws of gravity by magically retracing the original route between springboard and water.

Many effects that can only be produced by reexposing the same length of film on successive runs, can be achieved without the attendant risk of error and, in many instances, with better print quality by means of the optical printer. Superimposures, double exposures, multiple image, matte and split-screen shots are only a few of the effects that require a number of camera runs.

Designed to fulfill the most imaginative producer's requirements, the newest optical printers incorporate many features that help the cameraman achieve the most difficult of effects with a minimum of time and effort. Variable-speed drives, for example, allow the cameraman to select a convenient filming speed that is best suited for a specific purpose. Automatic skip-frame mechanisms, another recent development designed to slow down or speed up previously exposed actions, can add or drop frames during the optical process at the push of a button. Any number of frames, from 1 to 8, can be added or dropped from the processed film on the projection side of the printer and exposed, in the selected sequence, on the raw stock traveling through the camera.

Newly designed light systems, in air-cooled lamphouses, have replaced the standard condenser lenses in the printer's projection head. Capable of delivering 30 to 50 percent more illumination to the projected frames of film, the achromatic lens system is corrected for color and for spherical and coma aberrations. Influenced by the size of the projected images in wide-screen processes, provision is also made for the installation of anamorphic and other specialized lenses for squeezing images and returning them subsequently to normal proportions.

Zooms, from a 1:3 blowup to a 5:1 reduction, no longer present insurmountable, time-consuming problems for the optical cameraman. Optically centered or off-center zooms with accompanying pan-and-tilt movements from one part of the film frame to another can be made with amazing accuracy and a minimum of calibrations. Automatic follow-focus attachments ensure maximum sharpness throughout the zoom movement. Fades and cross-dissolves, of any length from 8 to 128 frames, can also be made automatically by engaging the dissolve mechanism. With as many as 50 settings available to the cameraman, the mechanism disengages from automatic after the preset number of frames in the effect have been exposed.

Reductions or enlargements, from the standard 35mm width to the 16mm film size, can be accomplished with the aid of interchangeable film-transport mechanisms. Installed in minutes, the different shuttle and sprocket assemblies, for each film size, need only slight compensating adjustments to assure the precise alignment of optical centers.

Matte carriers that advance each of the frames of effect film in stop-motion fashion, superimposure devices for aligning titles and other material in relation to the fine-grain print in the projector head, flip-lens attachments for achieving vertical or horizontal spin effects, tilt mechanisms that permit a scene to be straightened or repositioned, viewfinders that allow the cameraman to check an action for composition, and automatic buckle switches that stop the camera and activate a signal light to indicate jamming or other malfunctions in the transport mechanism—all these are only a few of the many devices that help improve operational efficiency and quality of production.

Similar in many ways to the exposure sheets used by the animation cameraman, the optical layout sheets serve as a guide for duplicating and combining the various film elements that are used during the optical photographic process. Frame and footage counts, indicating the exact order and length of scenes, effects, and transitions, are supplied by the film editor in the form of a work print. The optical layout sheets correspond with the footages indicated on the work print. As many as five counters, three indicating frames and footages

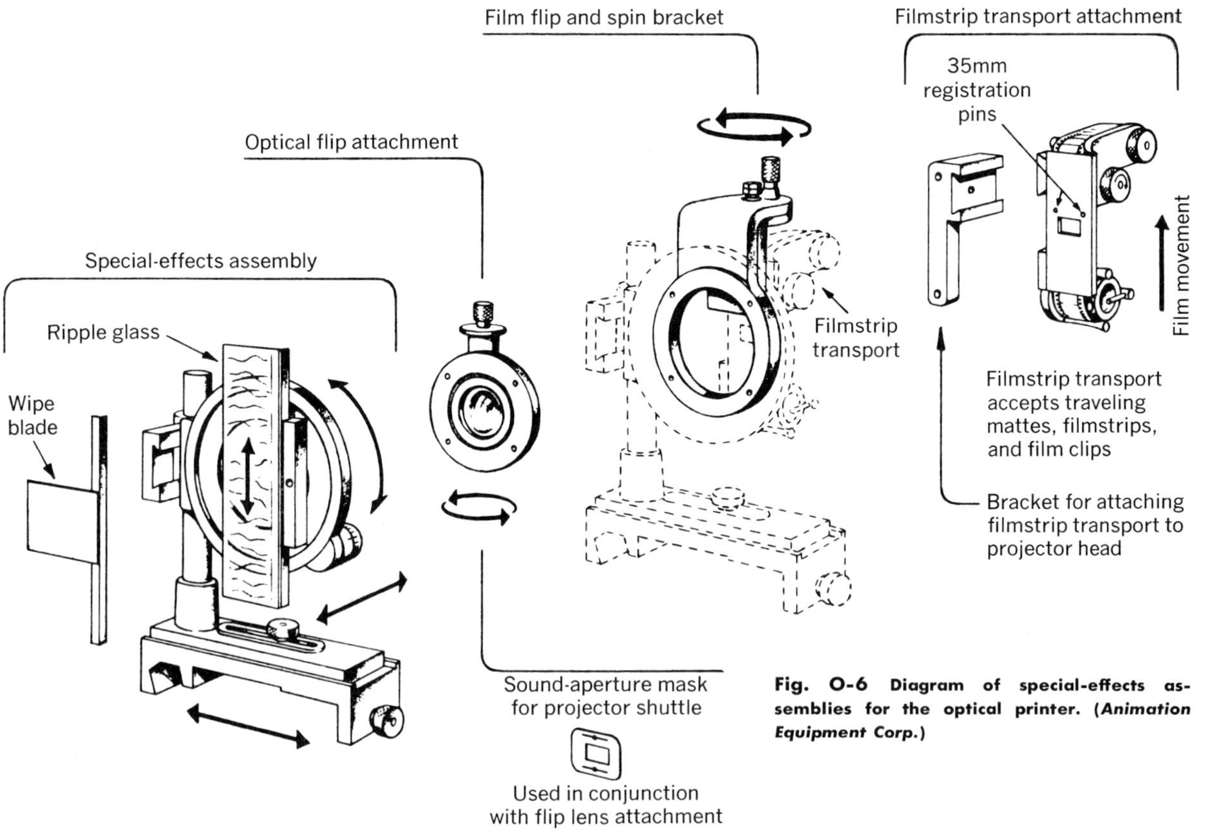

Fig. O-6 Diagram of special-effects assemblies for the optical printer. (*Animation Equipment Corp.*)

and two for positioning purposes, provide the cameraman with a visible guide during every step of the optical filming process.

Often referred to as "one of the most important technical achievements" of the motion picture industry, this optical, electrical, and mechanical device assures the quality and precision generally associated with, and expected in, the final product—the release print. And, since practically every film presentation, from the 1-minute television commercial to the full-length feature picture, includes one or more effects that cannot be achieved with direct in-camera photography, it can be assumed that a large percentage of the film seen by the motion picture or television viewer has been photographed with the optical printer.

Cameras take pictures. Optical printers complete them. See Figure O-6.

Oxberry 1200 series. The Oxberry optical printers have been designed to provide the maximum in effects with a minimum of time and effort. The printers incorporate a number of noteworthy features including automatic follow-focus that is controlled through one lead screw and covers a range from 1:3 blowup through 1:1 and down to 4:1 reduction.

The automatic dissolve provides up to 50 settings ranging from 8 to 120 frames. Interchangeable film-transport mechanisms accommodate 35 and 16mm film in one housing which remains in place on the printer to maintain precision alignment. The printer's tilting mechanism aligns a scene precisely without distortion. The movements permit adjustments of less than 0.001 inch on the lens mount and projector head. The lens and light systems deliver more illumination to the film surface than is possible with straight condensers. Solid steel shafts are used to maintain alignment at all points and permit an extension of lens and camera travel.

The printer has a variable-speed drive which is designed to operate at a number of stop-motion speeds. An electric tachometer indicates the speed at which the printer is operating. A compact clutch system is employed to speed manual work and facilitate skip-frame tasks which add or drop frames to slow down or speed up the action. An auxiliary mechanism can be provided for automatic skip-frame operation.

Working space between camera and projector is approximately 28 inches, allowing the installation of anamorphic and other special lenses. The Oxberry 1200 has five counters, three for film footage and two for positioning purposes. Five dial indicators facilitate interrelated positioning by the operator. A panelboard on the side of the printer includes a safety switch, bypass switch, and other operational controls. Removable plates provide accessibility to electrical and mechanical parts. See Figures O-7*a* and *b*.

Oxberry 1500 series. The Oxberry 35–16mm process camera, the heart of the 1500 model, employs one basic camera housing for both film sizes. Because only the shuttle and sprocket assemblies need be changed, the true optical alignment of the camera is maintained, regardless of film size.

One of the three shutter scales is marked in increments of 10 degrees, from 0 to 170, to indicate the shutter opening. The other two scales are marked for sine dissolves and logarithmic fades. The automatic dissolve operates electrically. The dissolve is driven by an infinitely variable speed unit to permit selection of 8, 9, 10, or any number of frames up to 128.

A feature of the 1500 printer is the spring-loaded pressure plate which maintains proper film contact. It will accommodate two films (plus a splice) without cam adjustment. The loading of film is facilitated by a latched lift button which holds the carrier fork in the up position. The shuttles have fixed pilot-pin registration.

If a double-frame shuttle is ordered for either camera or projector head, it will require a 2:1 gear transmission. If the shuttle is ordered for the camera, it will require a 90-degree swing shutter. If it is ordered

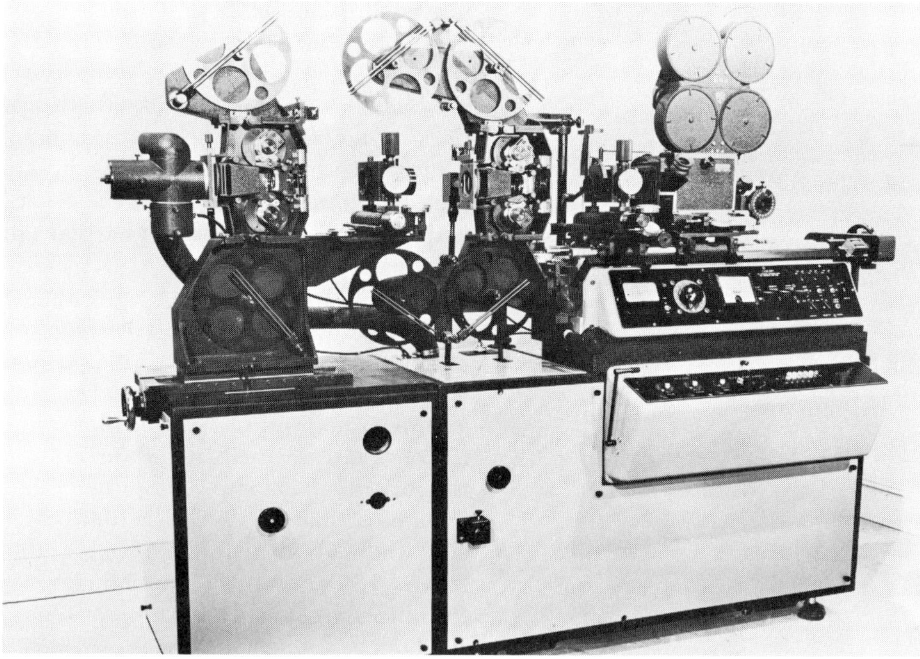

Fig. O-7*a* **Oxberry 1200 series optical printer with aerial image unit.**

Fig. O-7*b* **Oxberry 1200 series optical printer, showing control panel, projector head, and camera.**

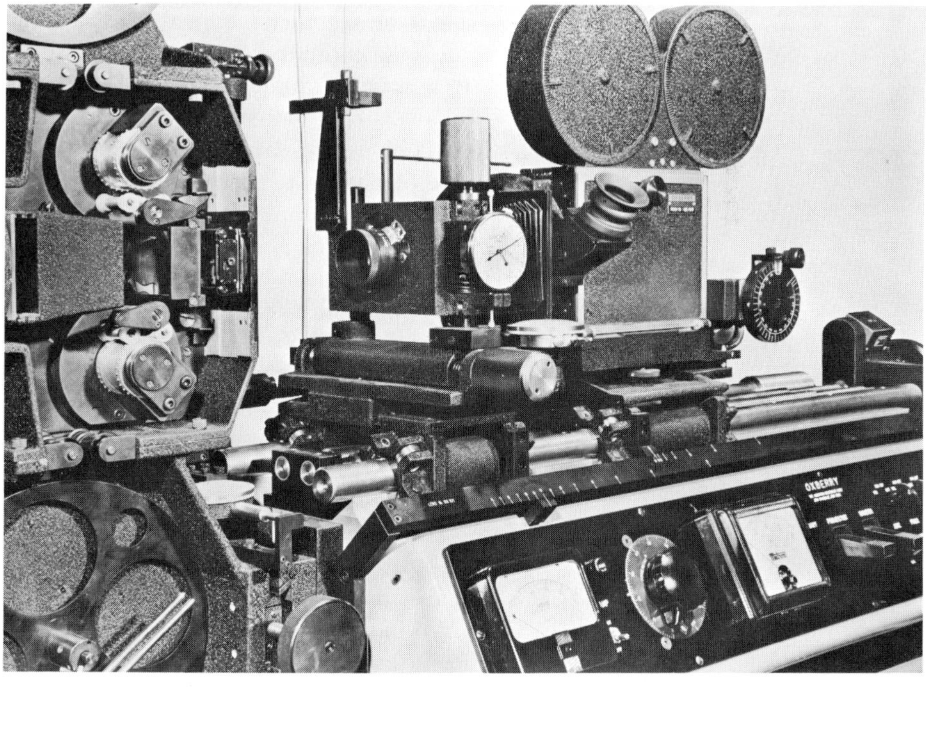

for the projector head, provision must be made for placing the head in a horizontal position, rotated 90 degrees from standard.

The 1500 printer has a reflex-type viewfinder. An adjustable eye-piece compensates for individual focusing. The viewfinder permits the viewing of film for checking action and composing. The viewing device has a removable, dovetailed plate with two fixed registration pins and ground reticle glass. The centers on the reticle are oriented with the registration pins. The plate pins correspond to the fixed shuttle pins.

The 1500 printer also features a focusing scope which permits straight-through sighting. This type of boresight viewing (no parallax) is essential for critical matte lineup work.

The automatic take-up attaches directly to the camera and has a torque motor on each of the film spindles. In operation, only those motors that are used in the direction of film travel are energized for rotational take-up. The other motors are partly energized for dynamic braking to keep the film taut. The take-up accommodates 400- and 1,000-foot rolls, as well as 400-foot bipack magazines, for 35 and 16mm film. The tension control may be adjusted for torque to compensate for film take-up (such as less torque for 16 and more for 35mm). A buckle switch is provided for fast shutoff of the camera drive if a buckle occurs with either 35 or 16mm film. A buckle signal light and reset button are located on the control panel.

The 1500 printer has one reset-type frame counter for totalizing scene-to-scene footage with provision for single-frame adjustment. This counter can be set to add or subtract, regardless of film direction. A second counter is available for totalizing film in the magazine only. If a double-frame shuttle is ordered, a special counter is required. That counter will show one for each double frame.

The camera can be tilted mechanically up to 10 degrees either side of normal. The degree of tilt is registered by a dial indicator.

The lens mount is ball-bearing mounted. A bellows connects the lens and the camera, permitting the greatest latitude in positioning. The lens mount accommodates 100mm *f*/4.5 Ektar enlarging lenses, 103mm *f*/2.8 Ektar printing lenses, 90mm *f*/3.8 Ektar lenses, and anamorphic lenses for wide-screen photography.

Instead of standard condensers in the lamphouse, the 1500 printer has a lens condensing system which permits straight-through printing with a very lightly ground diffusion glass. This system can deliver 22,000 footcandles with a 500-watt lamp without any diffusing system. See Figures O-7*c* and *d*.

Oxberry 1000 series. The 1000 printer is a special-effects optical step-projection printer with a projector head for 35 and 16mm film, and a combination 35–16mm camera. The machine is designed for continuous step-projection printing and a-frame-at-a-time printing from one picture to another.

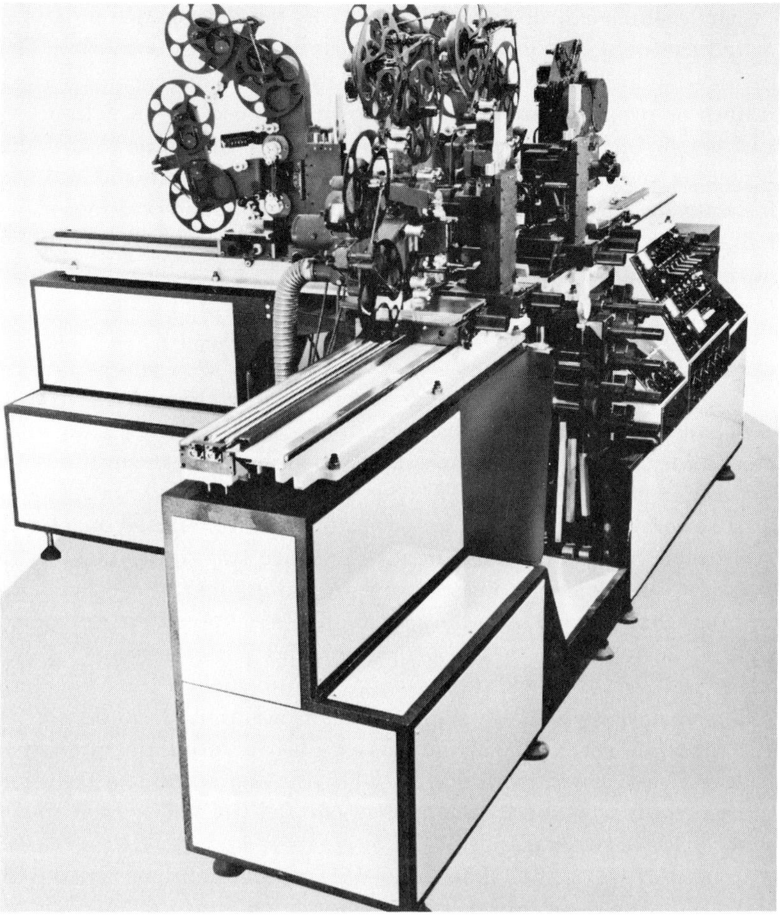

Fig. O-7*c* Oxberry 1500 series optical printer.

Fig. O-7*d* Diagram of Oxberry 1500 series modular optical printer. *A*—Camera and viewer; *B*—Main projector; *C*—Aerial image projector; *D*—Beam-splitter projector; *E*—Beam-splitter aerial image projector; x = horizontal axis; y = vertical axis; z = optical axis.

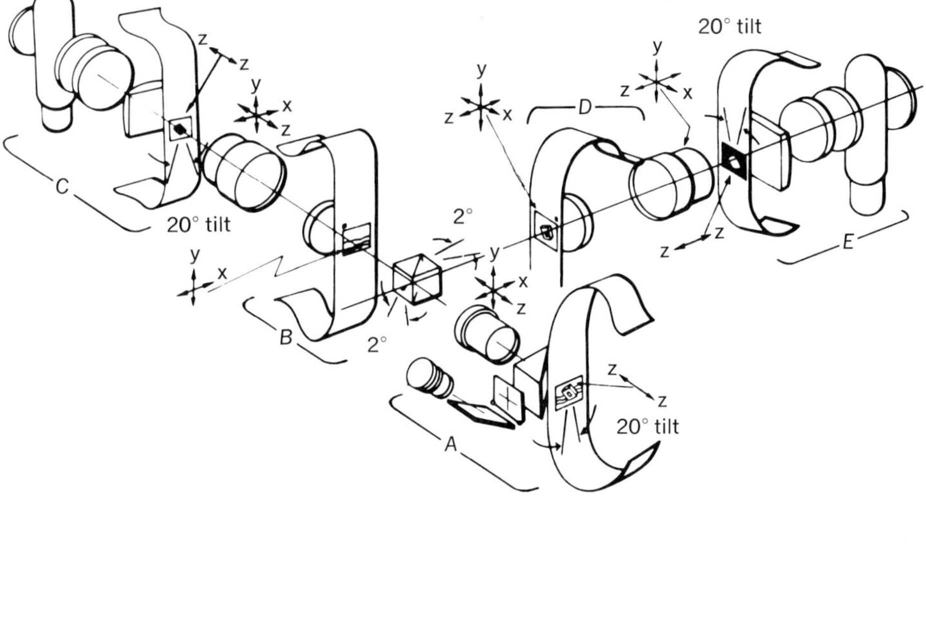

The printer can do blowup work from the 16mm size to 35mm or reduce the size of images from 35mm film to the 16mm size. The printer can also perform any type of skip-frame work such as adding frames or dropping them to slow down or speed up an action.

The 28 inches of working space between the camera and the projector permits the use of various specialized lenses. Counters indicate the positioning of the camera as well as the position of the lens mount. In addition to these counters, other counters keep track of the projected film, in frames, as well as the in-camera film. These frame counters operate in forward or reverse and are located between the projector and the camera where they are readily visible to the operator.

The variable-speed drive permits the printer to operate from a range of 60 to 240 revolutions per minute (rpm). An electric tachometer indicates the motor speed. Two high-speed rewind motors operate at approximately 650 rpm in forward and reverse.

The combination 35–16mm projector head is mounted on a finely controlled compound movement. There are two dial indicators to position the projector. These assure fine alignment and offset positioning when needed as in the case of off-center zooms. An air-cooled lighthouse attached to the projector head receives from 300- to 1,000-watt bulbs. A constant-voltage transformer in the base eliminates variations in line voltage. Instead of standard condensers in the lamp house, the 1000 printer has a built-in achromatic condensing system. This permits straight-through printing without a diffusion glass at 1 : 1. The condensing system can deliver 22,000 footcandles with a 500-watt lamp.

The shuttles on the Oxberry 35–16mm camera are designed with fixed pilot-pin registration. The spring-loaded pressure plate accommodates three films without adjustment. The threading of film is facilitated by a latched lift button.

The viewfinder permits the viewing of film in the projector for checking action and composing. Since the printer may be run at one-half live-action speed, the viewing would be comparable to screening film on an editor. The projector head also has a hand control for advancing film, thus permitting adjustment while viewing.

Traveling-matte devices, light meters, and automatic-dissolve mechanisms are available as accessories. Automatic follow-focus is optional. See Figure O-7*e*.

Research Products, Inc.: models 1000, 1001, and 1002. Research Products optical printers have been designed around commercially available camera and projector units. The printers feature belt-drive mechanisms, variable-speed-drive motors, electronic programmers for both camera and projector, an automatic fader, and an oversized optical bench. The modular construction permits rapid changes in format and convenient access to the replaceable mechanical-drive-system modules.

Model 1001 incorporates continuous automatic zoom from 4× enlargement to 5× reduction. The printers have convertible and interchangeable heads which make possible the printing of all format sizes from 70 to 8mm, in either black and white or color. The fixed-pin registration camera has a manually controlled variable shutter, as well as an automatic drive for fades and dissolves in lengths ranging from 16 to 128 frames. An integral reflex viewfinder incorporates a centering reticle on the ground glass for image placement.

The lamphouse is mounted so that it moves with the projector head, maintaining a flat, even field of illumination. Motor speeds, either continuous run or stop motion, range from 2.5 feet per minute (40 frames) to 20 feet per minute (320 frames); 40 feet per minute (640 frames) is provided with continuous run only. The printer has camera drive forward and reverse, with any combination of forward and reverse in relation to the projector head.

While skip-frame and multiple-frame printing can be accomplished manually, an optional accessory is the electronic programmer.

Fig. O-7e Oxberry 1000 optical camera.

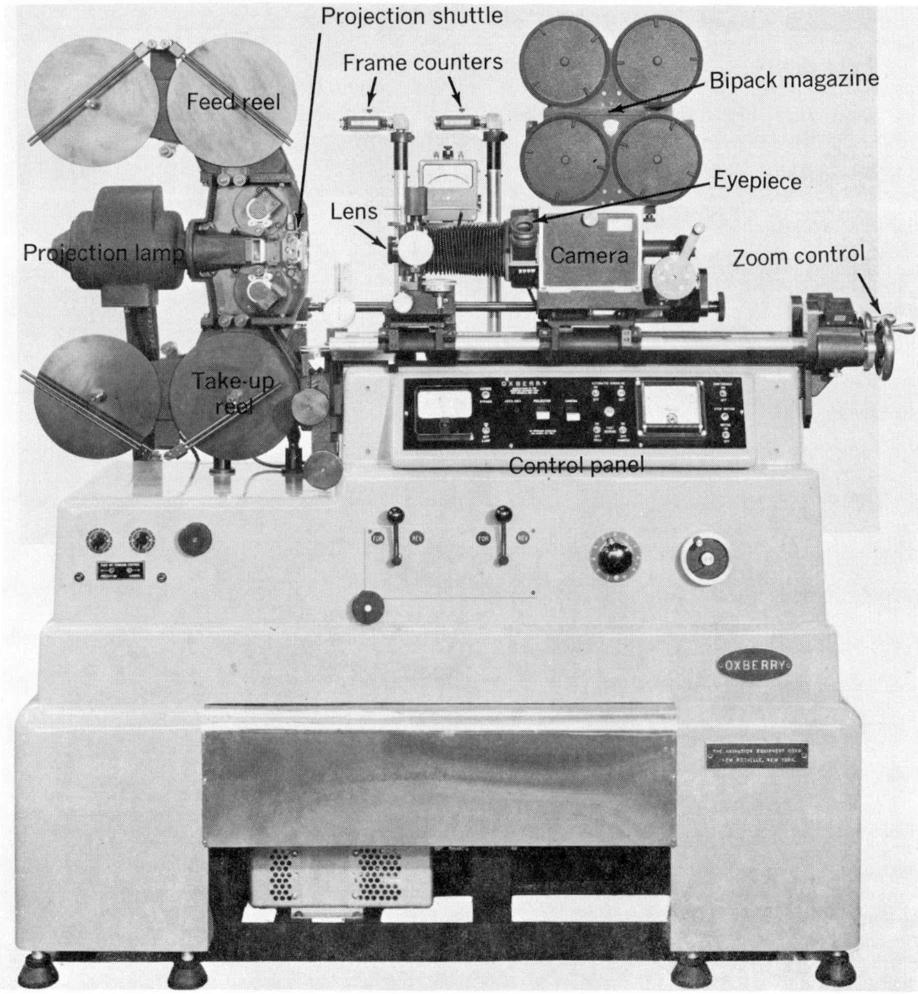

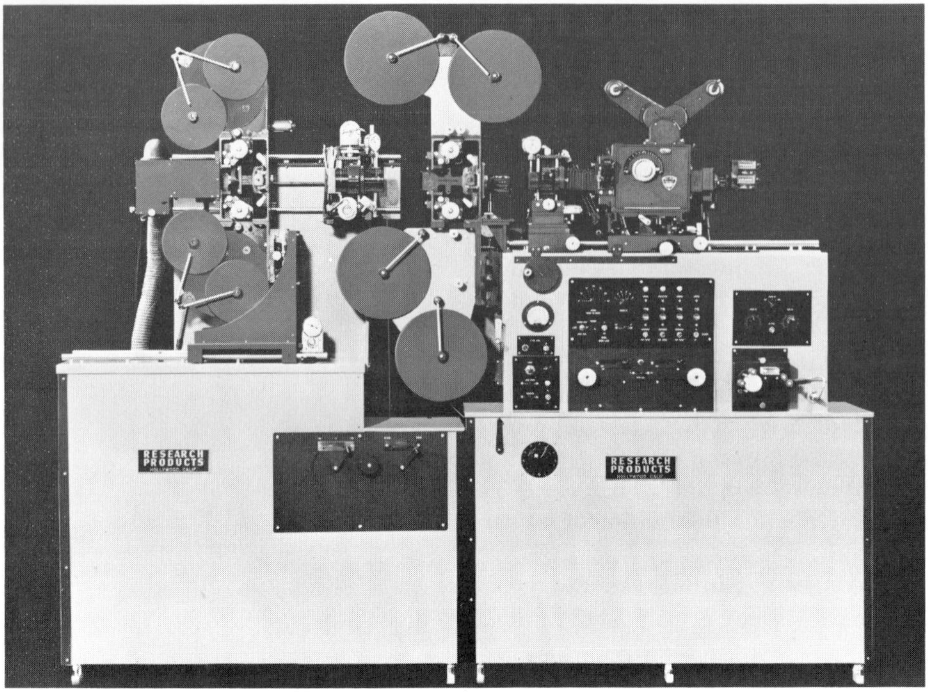

Fig. O-8*a* Research Products optical printer 1000. The aerial image unit with special effects and titling equipment setup. Film-to-film aerial image printing is accomplished by means of the second unit, on the left, in line with the basic optical printer and utilizing a second 35mm projector head.

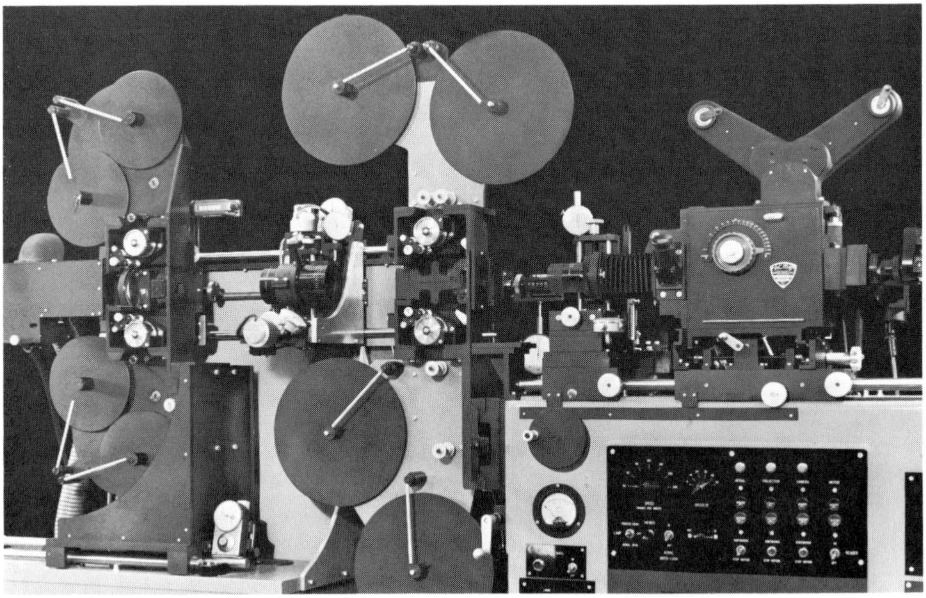

Fig. O-8*b* Research Products optical printer 1000, showing control panel for the modular design motor drive, the automated skip-frame programmer, camera, and projection head.

The programmer can also be used to actuate the automatic fader mechanism for automatic fades and dissolves from A & B rolls. See Figures O-8*a* and *b*.

Research Products, Inc., model 999. The model 999 optical printer is a low-cost laboratory type intended for printing operations not requiring the special effects capability represented by the 1000, 1001, and 1002 models. It is a basic unit and does not include such refinements as the modular drive unit or automatic zoom. It is, however, a precise unit, designed for continuous duty and high production.

optical recording The recording of photographic sound in the track area of the film. The modulations in the sound-track area are electrically converted from light impulses to audible sound during projection. See VARIABLE-AREA and VARIABLE-DENSITY SOUND TRACKS.

optical system The lens arrangement in a camera, optical printer, or projection device.

optical wedge See STEP WEDGE.

optics The science of light.

original picture negative The negative produced from the raw stock exposed in the camera during the actual filming process.

original sound negative A photographically obtained sound track which, after exposure in a film recorder, is processed to produce a negative sound image on film.

orthicon, image A television camera tube in which the optical image falls on a photoemissive cathode which emits electrons that are focused on a target at high velocity. The target is scanned from the rear by a low-velocity electron beam. Return-beam modulation is amplified by an electron multiplier to form an overall light-sensitive device.

orthochromatic Ortho for "right" and chromatic for "color." Actually, right color. See ORTHOCHROMATIC FILM.

orthochromatic film A type of film used during the industry's pioneer days. Its layer of emulsion was sensitive mainly to blues and violets. Other colors reproduced as dark grays or blacks.

orthopanchromatic film A panchromatic film which "sees" colors with the same degree of sensitivity that the human eye sees them.

Although sensitive to all the colors in the visible spectrum, the film's emulsion is less sensitive to red than the faster panchromatic emulsions.

oscillator An electronic device for generating "pure" tones or sine waves of varying frequencies.

oscillograph Instruments used for making graphically visible electric signals or flows of current.

output In sound recording or videotape, a connector such as a jack or terminal, that makes possible the transfer of signal output from one piece of equipment to a second piece of equipment.

out takes Alternate takes of the same scene, that are not used in the film.

overdevelopment As the term implies, periods of development in excess of those normally specified by the manufacturers. The usual result of overdevelopment is an extremely contrasty negative. Similar results are caused by developing solutions whose temperatures are higher than they should normally be.

overexposure An undesirable effect caused by any of several errors committed during the filming process. Overexposure, resulting in unusually dark prints, can be caused by periods of exposure that are too long, wrong aperture settings, or the use of extremely sensitive film that is unsuitable for the lighting being used.

overhead clusters Lighting units that provide soft, shadowless illumination. See Figure L-9*b*.

overhead strips Lighting units designed to produce a wide spread of even illumination. See Figure L-9*b*.

overlap The overlapping of the two ends of film that are to be joined by splicing. See SPLICER and EDITING.

overlay In animation, a part of the background that is designed to cover or overlay portions of the cels containing the animated action. These background overlays help create the illusion of depth in the scene.

overloading The recording and reproduction of sound at volumes in excess of the capacity of the equipment to record or reproduce those sounds without distortion.

P

paint pots A studio term for the controls that are used to adjust the red, blue, and green outputs of a live or film-chain camera to create the desired color balance.

pan The movement of the camera, in a horizontal plane, from one point of the set to another. Camera movement in a vertical plane is referred to as a *tilt.*

pan-and-tilt head A tripod mounting which makes it possible for the camera to be moved horizontally or vertically to achieve panning movement during the filming process. In many cases, calibrated scales on mountings indicate the amount of movement (from its normal setting) and the position of the camera in degrees.

Panaphonic reproducing system A sound-reproducing system in which a stereophonic effect is achieved without multiple sound tracks. The sound, on a monaural track, is directed to any of several loud-speakers in the theater or auditorium by means of a cueing sensor. Sound selection is based on the spacing of the darkened areas of the film which are located between the edge of the frame and the sprocket holes.

Panavision 35 See WIDE-SCREEN PROCESSES.

pan background A panoramic background that is positioned on the compound table of the animation stand and moved, via stop-motion processes, past the camera lens, each pan movement receiving one exposure.

panchromatic film Black-and-white film that is sensitive to all the colors in the visible spectrum. The term is also used to indicate broad spectral operation when used with filters, light sources, and the like. Compare ORTHOCHROMATIC and ORTHOPANCHROMATIC FILMS.

panchromatic master positive The black-and-white print made from a color negative. Printed on a panchromatic film, the pan master positive is subsequently used for making a black-and-white dupe negative.

pan glass A filter used by the director or cameraman to evaluate the tone values in the scene being filmed. The bluish green or brownish yellow filters approximate the tones and indicate how they will appear when photographed on panchromatic emulsions.

pantograph In animation photography, a device attached to the animation stand—actually, a pointer used in conjunction with a field guide. The pointer moves with the camera compound table and indicates the position of the camera in relation to the artwork during pan and zoom movements.

In lighting, the pantograph is a device with a scissor-type action to which light units are attached. Counterbalanced by springs or weights, the device is adjustable, thereby making it possible to vary the position of the light in relation to the subject matter on the set.

parabolic mirror A metal reflector or mirror surface in the shape of a parabola. A light source focused on the geometric center of such a shaped surface will reflect parallel rays of light. Parabolic reflectors or silvered mirrors may be used in spotlights in place of condenser lenses.

parallax The apparent difference or displacement between the area viewed through a camera viewfinder and the actual area photographed through the lens. It is especially troublesome with extreme close-ups. See Figure P-1.

parallel A platform, above the level of the studio floor, capable of supporting cameras, their operators, or overhead lights.

para-phenylenediamine A brownish powder used in fine-grain developers. It dissolves at normal room temperature when mixed with water.

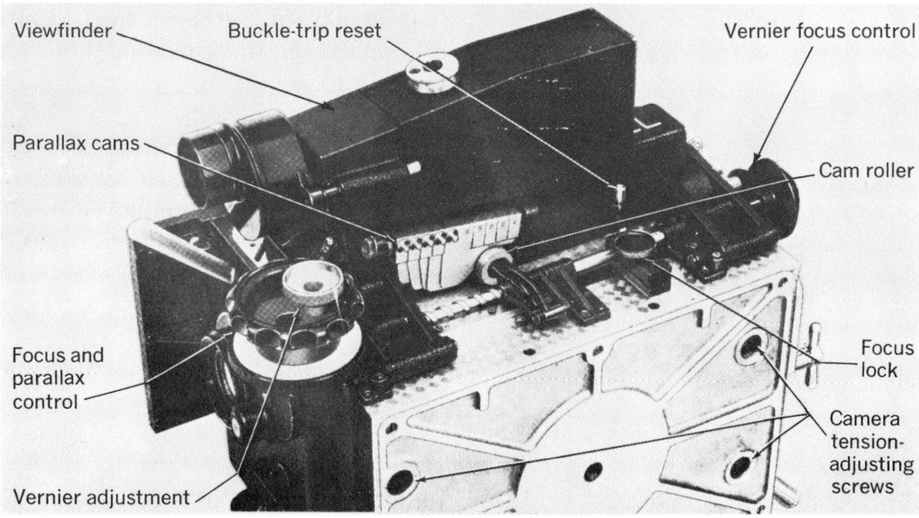

Fig. P-1 The Mitchell (BNC) camera controls that are used for parallax correction.

para-phenylenediamine hydrochloride A more stable fine-grain developer than *para*-phenylenediamine. The whitish powder dissolves at normal room temperature when mixed with water.

paraxial Those parallel rays of light that are nearest the optical axis of a lens. Compare ABAXIAL.

passing shot A photographic effect in which the camera remains in a stationary position while the subject moves past it. As an alternative, the subject matter may remain in a more or less stationary position while the camera moves. Compare RUNNING SHOT.

patch panel An electrical assembly with provision for plugs or jacks. Patch panels are used on the motion picture set or television stage to facilitate making temporary connections of electric equipment.

peak to peak The amplitude (voltage) difference between the most positive and the most negative peaks of an electric signal.

Also, the distance—with reference to phase and time—between successive peaks of the same polarity.

pedestal In lighting, a base or support on which light units are attached or mounted. The larger lighting pedestals are power-driven. See MOLEVATOR.

In videotape, the level of black in a picture. When raised, the picture is washed out.

pegs The fixed pins or pegs on an animation board or animation camera compound table that are used for registering drawings, cels, and backgrounds. These pegs can also be screwed into movable metal bars that fit into grooved tracks (on the compound table) so that cels or backgrounds registered to them can be moved from one position to another.

pellicle A sheet of transparent plastic material, approximately 0.0005 inch thick, that can be used as a substitute for the two-way mirrors that are often employed in the production of special-effects shots. Used as a beam combiner, it has an advantage over other reflective surfaces in that it is too thin to produce undesirable double images.

pencil Relatively narrow, parallel beams of light rays. Their degree of divergence from or convergence to a point is minimal.

pencil test A test in which the drawings made by the animator are photographed for the purpose of reviewing the animated action. If necessary, changes or corrections are made before the drawings are inked and opaqued. See INKER and OPAQUER.

penumbra A double-shadow effect produced by large unfocused light sources. The effect is that of a heavy shadow with a lighter shadow ringing it.

Pepper's ghost The achievement of the effect known as Pepper's ghost is simplicity itself. To produce the effect, a pane of clear glass,

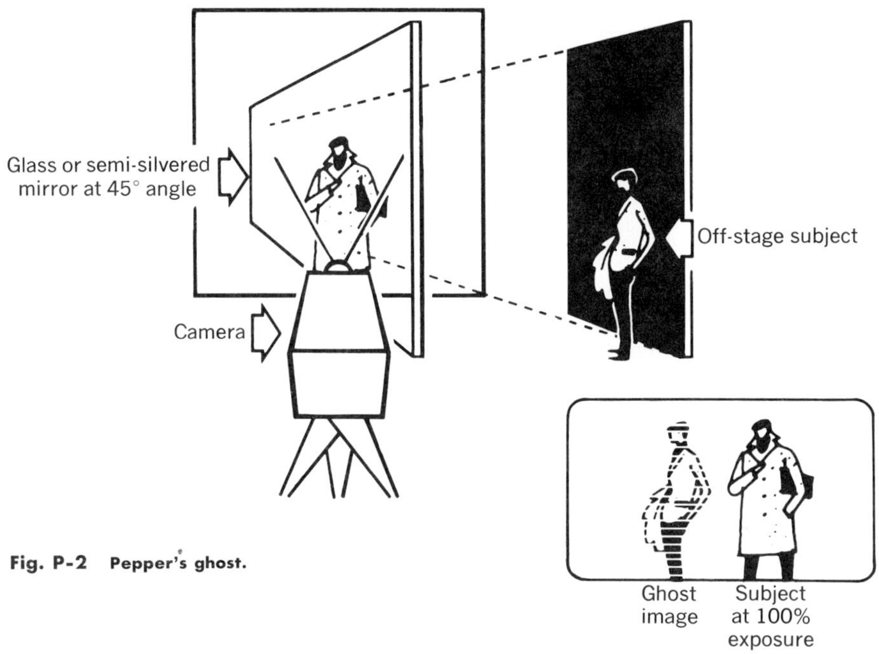

Fig. P-2 Pepper's ghost.

or half-silvered mirror, is positioned at an angle between the subject and the camera.

The subject is visible directly through the glass. The image of the off-stage subject, positioned in front of a black background, is reflected into the lens. The off-stage image can be made to appear as prominent or as subdued as is desirable by increasing or reducing the amount of light falling on the off-stage or "ghost" image.

Half-silvered mirrors have an advantage over plain glass in that the reflected images are brighter and more distinct. See Figure P-2.

percent sync The ratio, expressed as a percentage, of the amplitude of the synchronizing signal to the peak-to-peak amplitude of the picture signal between blanking and reference-white level.

percent transmission The measurable amount of light that is transmitted through a frame of processed film. Measurements are made on the basis of opacity. See OPACITY.

perforations, film The regularly spaced sprocket holes along the edges of a length of film that assure precise registration in photographic, projection, processing, optical, and editing equipment. Perforations are designated as either negative or positive and long or short pitch.

Negative perforations are designed for in-camera use and provide the needed precision during the interval of exposure. Positive perforations, by comparison, are shaped to withstand the stress and strain of repeated projection.

Both types of perforations are available in either short or long pitch sizes. *Pitch,* in motion picture terminology, refers to the distance between perforations. Measured from the leading edge of one sprocket hole to the leading edge of the next perforation, the short pitch is used during the photographic process while the long pitch is generally used for laboratory printing purposes. The distance from the leading edge of one sprocket hole to the leading edge of the next is 0.1866 inch for the short pitch and 0.1870 inch for the long pitch.

During the laboratory printing process, the in-camera negative and the raw stock on which the images are duplicated pass over the same sprocketed wheel, simultaneously, with the "print" film over the negative film. Measured from the center of the sprocketed wheel, the greater radius of the print film requires proportionate compensations in the spacing of the sprocket holes to prevent slippage and the resultant afterimages or "ride" in the subsequent print. The amount of compensation is 0.0004 inch—a minute amount that assures precise registration and eliminates the inherent danger of damaged sprocket holes.

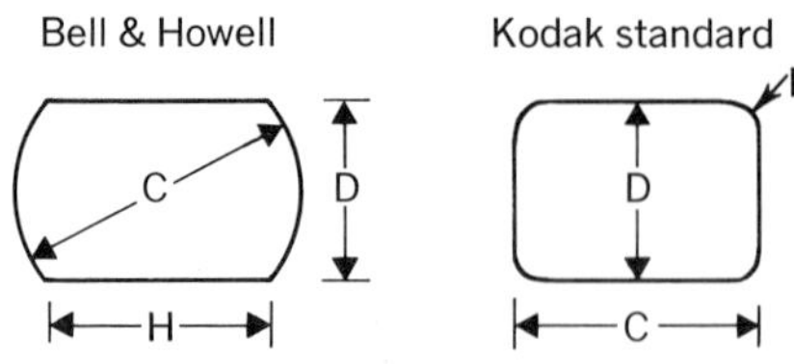

Types of perforations

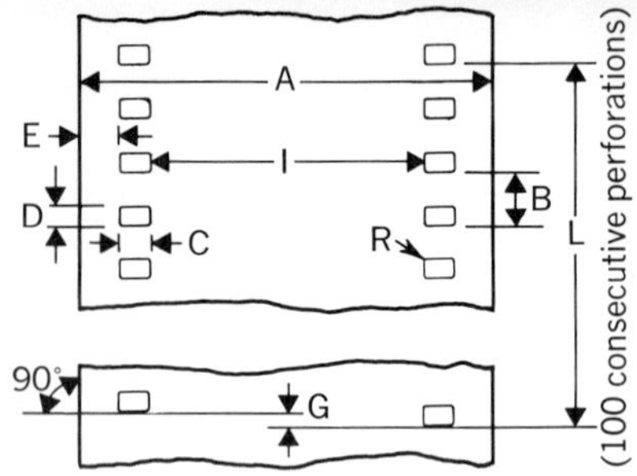

Dimensions

TABLE P-1 Film Perforations and Film Dimensions

Film	ASA Standard	ASA dimensions, inches									
		*A**	*B*†	*C*	*D*	*E*	*G* (max.)	*H*	*I*	*L*	*R*
16mm											
2R-1500 perforated 8mm	PH22.17	0.629	0.150	0.072	0.050	0.036	0.001	—	0.413	15.00	0.010
2R-3000	PH22.5	0.629	0.3000	0.0720	0.0500	0.036	0.001	—	0.413	30.00	0.010
1R-3000 perforated one side	PH22.12	0.629	0.3000	0.0720	0.0500	0.036	—	—	—	30.00	0.010
1R-2994 perforated one side	PH22.109	0.628	0.2994	0.0720	0.0500	0.0355	—	—	—	29.94	0.010
2R-2994	PH22.110	0.628	0.2994	0.0720	0.0500	0.0355	0.001	—	0.413	29.94	0.010
Tolerance allowed, ±	—	0.001	0.0005	0.0004	0.0004	0.002	—	—	0.001	0.03	0.001
35mm											
BH-1870‡	PH22.34	1.377	0.1870	0.1100	0.0730	0.079	<0.001	0.082	0.999	18.70	
BH-1866§	PH22.93	1.377	0.1866	0.1100	0.0730	0.079	<0.001	0.082	0.999	18.66	
KS-1870¶	PH22.36	1.377	0.1870	0.1100	0.0780	0.079	<0.001	—	0.999	18.70	0.020
Tolerance allowed, ±	—	0.001	0.0005	0.0004	0.0004	0.002	—	—	0.002	0.015	
70mm											
Type I	PH1.20—1956	2.754	0.234	0.1300	0.0800	0.099	<0.002	—	2.296	23.400	0.016
Tolerance allowed, ±	—	0.002	0.001	0.0008	0.0008	0.004	—	—	0.004	0.030	0.005
Type II	PH1.20—1956	2.754	0.1870	0.1100	0.0780	0.079	<0.003	—	2.376	18.70	0.020
Tolerance allowed, ±	—	0.002	0.001	0.0004	0.0004	0.004	—	—	0.004	0.05	0.005

Courtesy of DuPont.

*For low-shrink film, as defined in ASA Standards, dimension *A* shall be 0.628 ± 0.001, and *E* shall be 0.0355 ± 0.00200.

†In any group of four consecutive perforations, maximum difference in dimension *B* shall not exceed 0.001, and should be as much smaller as possible.

‡Long-pitch negative.

§Short-pitch standard negative.

¶Standard positive.

The perforations used in high-speed cameras approximate the size and shape of 35mm positive stock. The difference between the two lies in the extremely critical tolerances of the high-speed perforations.

Some rolls of 16mm film are perforated for in-camera use while others are perforated for duplicating or printing purposes. The raw stock used in "silent" cameras has two rows of perforations, one row along each edge of the film. Single-system cameras, by comparison, record sound and picture simultaneously and use film with only one row of perforations. The sound track, in the standard emulsion position, appears on the right side of the frame and replaces the second row of perforations. See Table P-1 and Figure P-3.

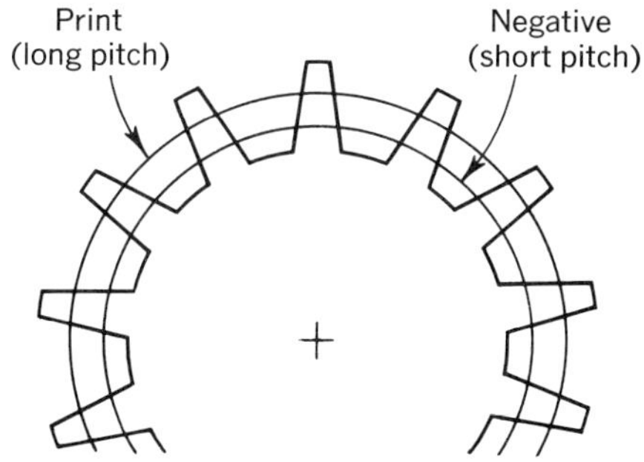

Fig. P-3 **Long- and short-pitch perforations.**

perforation pitch See PERFORATIONS, FILM.

persistence of vision The first serious article on the subject of image retention was submitted by Peter Mark Roget of England in 1824. It was called "The Persistence of Vision with Regard to Moving Objects," and was directly responsible for the development of the Zoetrope and, subsequently, other devices for viewing and projecting pictures.

The eye needs but a mere quarter of a second to record an image and transmit it to the brain for registration and interpretation. Thus, if two images are seen within the same $1/4$ second, the information will overlap and the images will be superimposed, thereby creating the illusion of motion.

If a length of motion picture film traveled through a projector without some means for stopping the action long enough to allow the picture information in each frame of film to be recorded by the viewer, the eye would see a continuously changing, unintelligible, vertical pattern made up of dark and light areas. The "means" for stopping the action is the shutter. This device cuts off the light falling on each frame of film after it has been projected for approximately $1/60$ second, thereby separating one picture from the next. The viewer is thus able to see 24 still pictures during each second of projection. The ability of the eye (and brain) to retain impressions, or the persistence of vision, causes a blending of picture information and the resultant illusion of movement.

In television, the action is stopped and each frame is separated from the one following it by what is referred to as *blanking*. Blanking takes place after the scanning cycle is completed and the electron beam returns to the upper left-hand corner of the screen to begin scanning the frame that follows. The process is repeated 30 times each second. See Figure P-4 and PROJECTION and MULTIPLEXED TELECINE OPERATION.

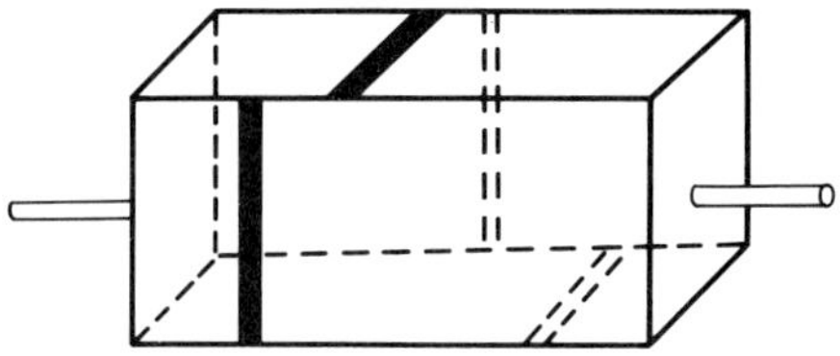

Fig. P-4 **Persistence of vision.** The persistence-of-vision theory can be proved quite simply by painting lines on each of four sides of a block of wood, with each line positioned a little further along the wood than the previous one. When the block of wood is made to spin slowly, the four lines seem to become one line which moves continually from left to right. If the block of wood is made to spin rapidly, each of the four lines, from each plane, spins in place and is visible at the same time.

perspective The representation of solid, three-dimensional subject matter on a flat plane in such a manner as to create the impression of depth and distance. In photographic processes other than 3-D, the camera lens is the "eye" that judges spatial relations and translates the information in terms of an image on a flat plane—the film.

Perspective can be changed by moving the flat plane in either a horizontal or a vertical direction in relation to the subject matter. Moving the plane closer or farther away from its original position will only make the subject increase or decrease in size without affecting its perspective. Objects that are closer to the camera will be proportionately larger than similar objects at greater distances from the lens.

Parallel lines (as in railroad tracks, for example), directly in front of a camera, will appear to meet at the horizon or "vanishing" point. When the lines are parallel in relation to the horizon, the vanishing point may be completely outside the limits of the film frame. The converging lines will, nevertheless, appear to meet at the same point on the horizon.

Basically, a scene is more acceptable to the eyes of the viewer when it is photographed normally. Forced perspective, even when held to a minimum, can be disturbing. As a general rule, the deliberate distortion of subject matter by extreme perspective, should be used only to create dramatic effects.

Perspective depends upon the camera's point of view. The subject matter can be filmed with the same magnification from different positions or working distances, by using different focal-length lenses. Even though the image size is the same in each instance, the scenes will not be alike. Each will have a different perspective—varying with the filming distance.

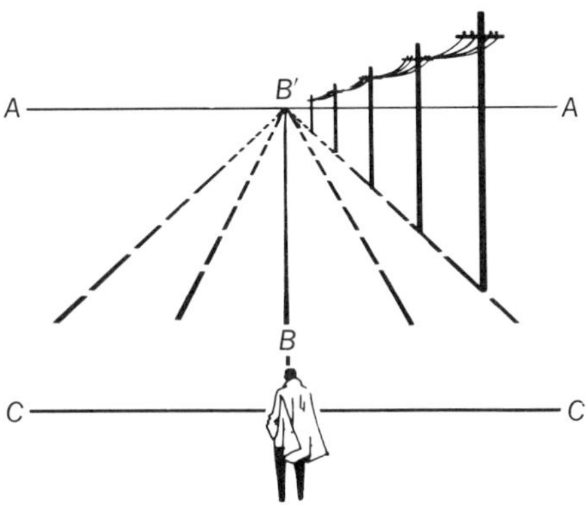

Fig. P-5 Perspective. (*AA*) Horizon line. (*BB*′) Line of sight. (*CC*) Ground line. (*B*′) Vanishing point.

From short distances and with short-focal-length lenses, objects in the background appear small and far away. From long distances and with long-focal-length lenses, background objects are larger and appear to be closer since the perspective is "flattened."

The long-focal-length lenses, because of their narrow angle of view, include a small background area and are ideal when it is necessary to eliminate disturbing background elements. By simply moving the camera to a slightly different position, the Telephoto lens can include a completely different background. Short-focal-length lenses include large background areas. Changing the camera position has little effect on the size of the area that is covered by the camera lens.

Linear perspective refers to the representation of distance through converging lines. In aerial perspective, by comparison, the distance between objects is represented by gradations of tonal values and colors. See Figure P-5.

Petzval lens A lens with two sets of widely spaced, color-corrected elements. Covering a relatively small, curved field, this type of lens was originally designed for portrait work. Modifications of these lenses are used for projection purposes and for photographic applications where losses of definition at the edges of the picture are tolerable.

pH The symbol for expressing the numerical composition of a processing solution with reference to its degree of acidity or alkalinity. Solutions that are strongly acid have a pH value of 0 to 2. Strongly alkaline solutions, by comparison, have a pH value of 12 to 14.

phase A term indicating the synchronous operation of the shutter and the transport mechanism in the camera. In a broad sense, the term is applicable to a sound track that is in editorial synchronism with the corresponding picture on another reel. The term is also used in connection with the operation of a projector, as in rear-screen processes, in synchronization with the motion picture camera that is filming the projected background and the action in front of it.

In television, the relationship, expressed in terms of time, between specific, maximum points in recurring electric impulses either in equipment or functions.

phase modulation AM (amplitude modulation) is used for transmitting radio signals as well as black-and-white television images. FM (frequency modulation) is used for transmitting radio as well as television sound signals. In both cases, either the amplitude or frequency of the carrier is changed.

Phase modulation, for conserving frequencies and providing a means of color coding, is the method used for changing a carrier's phase from its unmodulated characteristic at the time of transmission to the modulated phase at the receiving point. The difference between the unmodulated (or reference phase) and the received phase is the degree of modulation.

In color television, the hue values of the transmitted colors are separated—the brightness or Y level is transmitted as a black-and-white signal while the colors are phase-modulated. The color signals are converted into the RGB signals at the receiver. See RGB CHANNELS.

photoelasticity A photographic process, dependent on changes in the indices of refracted light, that is used for measuring and evaluating the characteristics or physical changes that occur in materials subjected to stresses and strains.

In operation, polarized light is transmitted through a scaled-down transparent model of the material or body that is to be analyzed. Stress, in the panchromatic film on which the distribution patterns become visible, is seen as a series of white and black lines, the former showing the contours and the latter showing the direction.

Studies based on the patterns derived through photoelastic stress-analysis processes are invaluable for improving the design and durability of materials while reducing their weight, size, and—of greater importance from a technological viewpoint—cost.

photoelectric cell A light-sensitive tube that is capable of either producing or changing the value of electric current. In exposure meters, the indicating pointer responds in proportion to the intensity of the light striking the photocell. In sound projectors, light passing through the sound-track area of the film is directed to the photocell. The corresponding variations in current produced in the cell by the modulations in the sound track are subsequently converted into audible sound.

photofabrication The manufacture of parts by photographic processes. Chemical rather than physical or mechanical action is used to produce as well as change the shape of component parts that are used for a wide variety of purposes—printed circuits, for example.

photofloods Incandescent tungsten lamps capable of providing a high-intensity light.

photogrammetry Also called **phototopography**. Photogrammetry is a filming technique that enables scientists to measure and study the planet inhabited by them. Developed by Col. Aimé Laussedat, a Frenchman who combined photographic principles with the art of surveying to create a new science.

Phototopography is used to measure relatively small land areas. Photographed by two cameras positioned at opposite ends of a designated area, specific details framed in each of the camera's view-finders are subsequently used as reference points and the basis for mathematically precise measurements. Correlated and oriented, the information from the phototheodolite pictures (photographs taken with cameras mounted on tripods at ground level) is transposed and given dimension in the form of relief features or surface configurations.

Photogrammetry, a related film technique developed by the same man in 1851, was used to produce the first mathematically accurate maps. The technique, in its original form, depended on information obtained with ground-level cameras and provided adequate results for more than a half-century.

With the invention of the airplane, the phototheodolite or terrestrial pictures were replaced by aerial images that showed the contours and elevations of relatively large areas. Superimposed to create a composite picture known as a *mosaic,* the overlapping photographs obtained with photogrammetric processes are analyzed and used for a wide variety of purposes other than map making.

For industrial as well as commercial purposes, photogrammetry has played an extremely important part in geological surveys, oil conservation programs, and construction projects. During World War II, stereophotogrammetric processes provided the Allied Forces with a completely reliable source of dimensional information unobtainable with other photographic techniques.

Far less spectacular than other motion picture techniques, photogrammetry and phototopography have already been used to help map the cratered surface of the moon. Someday these processes will be as useful to man when he explores other planets as they have been in helping him to measure and become more familiar with his own earthly surroundings.

photographic emulsion Suspensions of light-sensitive materials in a colloidal medium, usually gelatin, supported by an acetate film base. Negative emulsions are, generally, more sensitive than positive emulsions. See FILM.

photographic speed The sensitivity of an emulsion in relation to the way it is exposed and processed. See FILM SPEED and EXPOSURE-INDEX RATINGS.

photographic track An optical sound track that is produced photographically—the variable-area and variable-density sound tracks, for example. See VARIABLE-AREA and VARIABLE-DENSITY SOUND TRACKS.

photography The term is derived from the Greek: photo for "light," and grapho for "write" or "describe."

The process of forming an optical image on sensitized surfaces by the action of light or other forms of radiant energy. The invisible or latent image formed on the sensitized plate or film is subsequently made visible by the chemical action of the materials used for developing purposes.

The tonal values of the images produced on the negatives are reversed, or exactly opposite to the values in the original subject matter. These light-and-shadow patterns are restored to their original values by exposing the developed negative on a second sensitized surface. See Figure P-6 and FILM and PROCESSING.

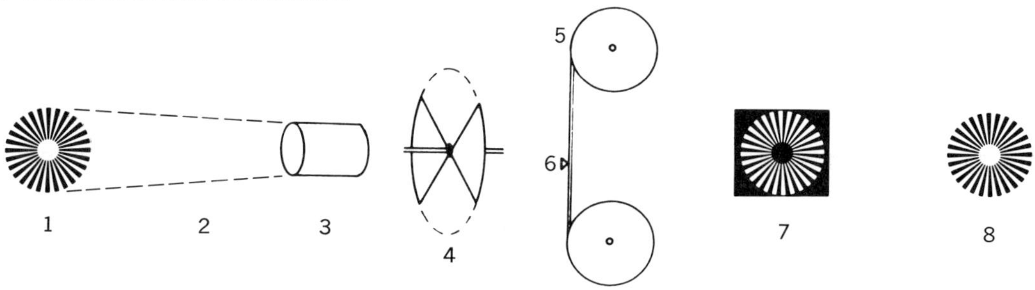

Fig. P-6 Photography. The light (2), reflected from the subject matter being photographed (1), passes through a lens (3). The shutter (4) allows a predetermined amount of light to expose the individual frames of film (5) as they are brought into position by a claw (6), or similar device, that engages the sprocket holes and holds the film in place during the interval of exposure. The negative image, after the film has been developed (7) has tonal values that are exactly opposite to those seen in the original subject matter. The true tonal values are reproduced in the subsequent print (8).

photoinstrumentation The techniques used for studying, analyzing, and correlating the information obtained with high-speed and other photographic processes.

Photoinstrumentation, as a scientific system for transposing photographic data into mathematically precise information, has been used for analyzing the composition as well as the stress and strain characteristics of newly designed industrial materials, tracing the routes of orbiting satellites, studying explosive phenomena, evaluating assembly operations, chemical studies, and for a wide variety of purposes serving government, industry, and scientific needs.

photomacrography The process of photographing small objects without using a compound microscope. Unlike the high-magnification photographs obtained through photomicrography, the degree of magnification is usually no more than approximately 20× diameters enlargement. Compare PHOTOMICROGRAPHY.

photometer An instrument, not unlike an exposure meter, that is used to measure the intensity of light. For positive lighting analysis, photometers provide the professional photographer with accurate, complete preexposure information. Photometers measure overall scene-brightness range and the relationship of any scene element to this range.

These meters simultaneously compute exposure on the exact constant-density principle. The high-definition spot photometer (shown under LUMINANCE ANALYZER) features ½-degree acceptance angle (1/4,000 of a typical scene) and adjustable spectral characteristics for film or eye. See LUMINANCE ANALYZER.

photometric filter Lamp filters designed to alter the color temperature of a specific light source by a definite amount. Produced for use with a photometer, this type of filter reduces the differences in color between lights that are operating at different color temperatures. See Table P-2.

photometry An important adjunct to all cinematographic techniques, photometry is used to measure light and the intensity of radiation. It provides a precise system for comparing the densities of images in processed film.

Visual methods of measuring light, in the past, depended on physical observation. Eye fatigue, caused by prolonged viewing under difficult conditions, made even the most accurate calculations unreliable. Readings from duplicate strips of processed film under similar viewing conditions usually resulted in opinions and estimates rather than in positive statistical analysis.

TABLE P-2 Wratten Photometric Filters

Filter number	Color temperature, °K	Mired shift value
78	2360–5500	−242
78AA	2360–4400	−196
78A	2360–3200	−111
78B	2360–2800	−67
78C	2360–2500	−24
86	5500–2360	242
86A	3200–2360	111
86B	2800–2360	67
86C	2500–2360	24
Dummy	For setting 0°	

Courtesy of Eastman Kodak.

In operation, light rays are directed at the sensitive surface of a photoelectric cell. The emitted electrons are transmitted to an indicating instrument such as a galvanometer or microammeter. This information is, in turn, translated into lumens, a standard of light measurement.

photomicrography Photomicrography is the process of photographing subject matter through the eyepiece of a microscope. *Microphotography,* by comparison, refers to the production of microscopic-size photographs.

The highly magnified picture information achieved through photomicrography provides the means for studying and analyzing minute images on large screens without the tiring effects that accompany direct microscope viewing. The additional advantages of a permanent record of the viewing are, of course, quite obvious.

Although the microscope is, unquestionably, one of the most important instruments available for the scientific observation of images that can only be seen by magnification, it has one definite disadvantage. Obviously, direct viewing can be done by only one person at a time. Others must wait their turn or make use of other instruments. Photomicrography has, of course, made such viewing methods impractical and a thing of the past. Motion pictures and transparencies can now be made by photographing the subject matter directly through the microscopes and then projecting the images onto large screens. With this method large groups can study and analyze the picture information simultaneously and discuss their findings practically.

For low-magnification results short-focus lenses are preferable to the high-magnification lenses that are used for obtaining extreme enlargements. In some cases, the photo eyepiece on the microscope tube is used as the camera lens. In either case, the subject can be illuminated by either reflected or transmitted light. For still photographs, electronic flash provides an ideal light source. This type of illumination also makes available a bright, constant light that can be used for time-lapse studies.

To avoid vibration during the filming process, and to simplify focusing and composition of the minute images, both the camera and light sources require rigid support. Where contrast is more important than detail in the processed print, complementary color filters are used. When detail is the primary consideration, filters that are similar in color to the values in the subject matter give the desired results. In some instances, infrared is used for filming images that are opaque to visible light.

Photomicrography is extremely valuable to the field of medical science. The study and observation of stained specimens and microscopic formations on large screens offers innumerable advantages over direct viewing methods through microscopes. In recent years advertisers have recognized the value of this filming technique and applied its principles to commercial television. The high-magnification demonstrations provide the viewer with a close-up look at a product's properties and make the salesman's task comparatively simple. See Figure P-7.

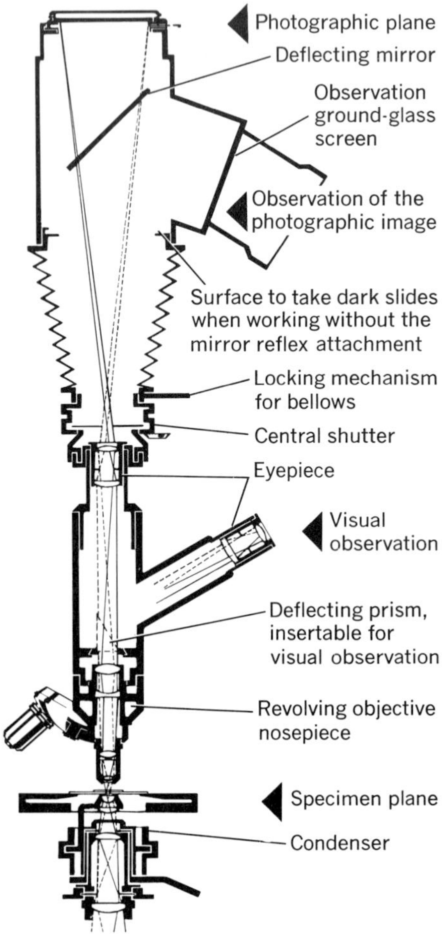

Fig. P-7 Beam path when taking photomicrographs, demonstrated on the Aristophot, in combination with the bellows camera 9 by 12 cm (3¼ by 4¼ inches) and the Ortholux microscope. (*Ernst Leitz, Inc.*)

Interchangeable Phototubes

For Leitz research microscope Ortholux

Orfotsine	Straight phototube 0
Oyeedsine	Photo tube FP with lateral inclined monocular obervation tube
Oiyeesine	Phototube FS with lateral inclined binocular observation tube
Orfax	Phototube FSA* with lateral inclined binocular observation tube, automatic focusing compensation, beam-splitting prism (80:20) interchangeable with 100 percent reflecting prism for visual observation
Opwim	Straight phototube of large diameter and light-screening collar for taking general-survey photographs (photographs without eyepiece)

For Leitz microscopes Laborlux and SM

Nacel	Straight phototube 0
Naseb	Phototube FS with lateral inclined binocular observation tube
Orfob	Phototube FSA* with lateral inclined binocular observation tube, automatic focusing compensation, beam-splitting prism (80:20) interchangeable with 100 percent reflecting prism for visual observation (only for use with Laborlux)

Design subject to alterations without notice.

*Especially recommended if acquisition of automatic microscope camera attachment Orthomat is intended.

photomultiplier Vacuum tubes in which electrons are caused to be emitted from a solid substance by exposure to electromagnetic radiation. These emissions are used to amplify currents.

In conventional flying-spot scanners, the photomultiplier produces an electric signal by reconverting the modulated light generated in the cathode-ray tube. See ELECTRON-BEAM RECORDER.

photoplastic recording A recording and instant developing process invented by Joseph Gaynor and announced by the General Electric Company in 1963. With this process, the image is recorded on a plastic film, either clear or opaque. The illumination used for exposing the image is an ordinary light beam which strikes the electrostatically charged photoconductive film.

The film after being exposed is heated, and microscopic depressions conforming to the pattern of the exposed image are formed on the surface of the film. The cooling process that follows seals the depressions. The heating and cooling actions both take place within a fraction of a second.

There are several differences between this process and the thermoplastic recording process invented by W. E. Glenn. Electric signals, for example, are used for recording with the thermoplastic process, whereas ordinary light in a normal atmosphere is used to record images with the photoplastic system. And, unlike the electrostatically charged photoconductive film on which the photoplastic images are recorded, the thermoplastic image is recorded by an electron beam on a low-melting thermoplastic film. The thermoplastic image can be projected in color or, if desirable, converted to an electric signal. See Figure P-8.

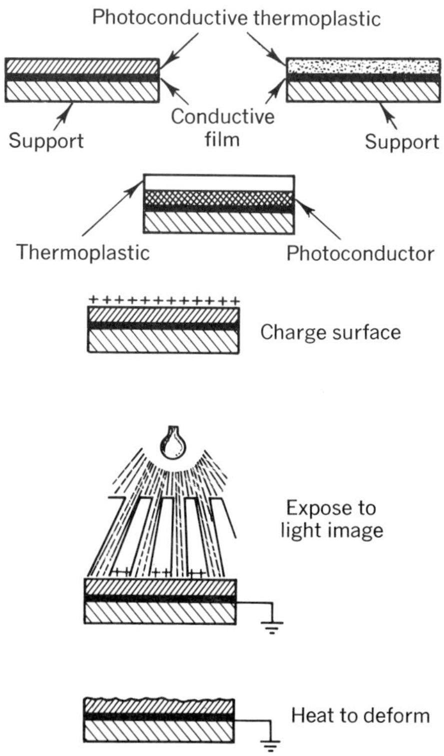

Fig. P-8 **Photoplastic recording.**

photostat An inexpensive type of reproduction on photographic paper. Exposed with copy cameras designed for that specific purpose, the original subject matter may be enlarged or reduced to produce either a positive or negative image.

Photostats are much used in the graphic arts to give a concrete indication of how a printing piece will appear in its final size. Their drawbacks are their relative conciseness, which makes them unsuitable for duplicating photographs or other halftone images, and the fact that they are subject to blemishes unless carefully produced.

phototopography See PHOTOGRAMMETRY.

photronic cell A photoelectric cell that generates electric current when activated by light; it is the active agent in exposure meters.

Pickley process See MATTES and SELF-MATTEING PROCESSES.

pickup tube An electron-beam tube used in a television camera in which an electron current or a charge-density image is formed from an optical image and scanned in a predetermined sequence to provide an electric signal.

pic-sync system The synchronizing pulse, in the Fairchild pic-sync system, is recorded at the same time as the audio signal. The two are subsequently separated by a pass filter. See SYNC GENERATORS.

picture duplicate negative Also called **dupe negative**. The negative made from color, black-and-white, or separation master positives. It may be edited and cut in with the picture release negative, or it may be used for obtaining additional prints. See PROCESSING and flow charts under PROCESSING.

picture element The most minute component part of any frame or picture reproduced either on film or by television transmission.

picture image A photographic likeness of subject matter on a photographic plate or motion picture film.

picture negative A negative made by laboratory processes from the raw stock exposed in the camera. See PROCESSING and flow charts under PROCESSING.

Picture-Phone A process developed by the Visual Research Group at Bell Telephone Laboratories which uses television system principles so that persons on a two-way telephone hookup can see each other while carrying on a conversation.

picture print A photographic plate or length of motion picture film that contains a positive image of previously photographed subject matter after processing. See PROCESSING and flow charts under PROCESSING.

picture release negative The final picture negative (either original negative, dupe negative, or a combination of the two) from which release prints are made. See PROCESSING and flow charts under PROCESSING.

picture signal That portion of the composite video signal which lies above the blanking level and contains the picture information.

picture tube A tube in receiving equipment in which the transmitted image is reproduced by means of variations in the intensity of a scanning beam.

piggyback A term used in connection with television commercial production to describe two commercials that have been spliced together to fill a one-minute time spot. The two commercials (a 40- and a 20-second spot, or two 30-second spots) usually feature products that are sold by the same manufacturer.

pillow distortion A linear distortion effect caused by the relationship between the diaphragm and the lens. With the diaphragm positioned behind the lens, lines forming a square will lean toward the center of the square. The distortion can be corrected by adding lens elements with complementary errors. See BARREL DISTORTION.

Pilote sync system A method of recording a sychronizing pulse during the filming process at the same time that the audio signal is recorded.

The Pilote synchronizing head has a slit opening of approximately 10 mils and a slit length of about $\frac{1}{16}$ inch for a tape speed of $7\frac{1}{2}$ inches per second. It is placed at right angles to the audio slit and records a well-balanced push-pull track down the center of the $\frac{1}{4}$-inch tape. It is claimed that this system, when recorded with bias, has less crosstalk than the U.S. systems. Compare MAGNETIC HEADS, ECHELON SYNC SIGNAL, NEO-PILOTTONE SYNC SYSTEM, and RANGER SYNC SIGNAL. See Figure P-9.

pilot film The name used to describe the first completed film or videotape production for a potential series. If successful, the pilot becomes the pattern for future production. Subsequent films have the same format and cast of characters and are seen at the same time each day or week.

pilot pins Also called **register pins**. The part of the camera transport mechanism that engages the sprocket holes and precisely positions each frame of film in the camera gate during the interval of exposure.

Fig. P-9 Pilote system. ***(Sketch courtesy of K. E. Gondesen.)***

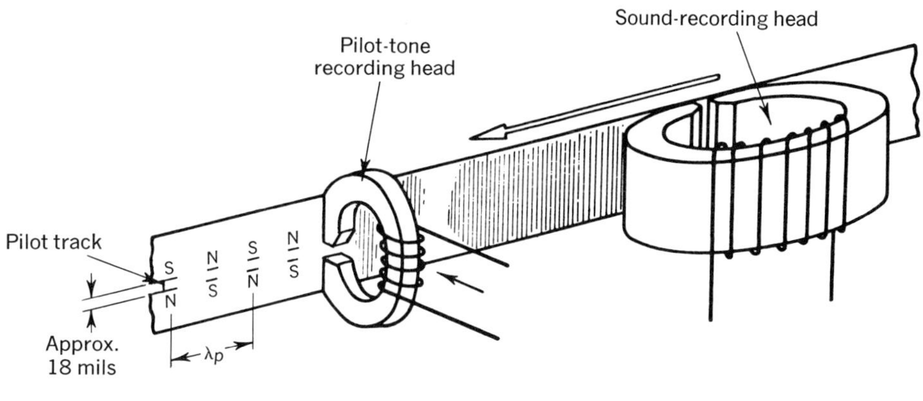

pinch roller Also called **pressure roller**. The rubber roller in recording mechanisms that engages the capstan and keeps the tape from slipping as it is being transported across the record or playback head.

pincushion distortion A type of distortion produced by an uncorrected lens. The defect causes straight lines at the outer edges of the film frame to curve inward toward the center of the field.

pinhole camera A camera that uses a tiny opening (rather than a lens) through which light from the subject matter enters and exposes the film. A sufficiently small opening will form a properly focused image without benefit of a lens.

pinhole optics A photographic process in which an image is obtained through a relatively small hole in the front of a camera instead of through a conventional lens system. It is employed mainly for creating the displays needed for simulating conditions used for scientific studies, where any degree of image distortion could be expected to have an influencing effect on those studies. The distance between the pinhole and the focal plane is a factor that determines the size of the pinhole (made by perforating a sheet of soft metal with a needle). See DISPLAY.

pip See FIDUCIAL MARKING and TIMING LIGHT GENERATORS.

pixlock In videotape recording, pixlock is a mode of operation that must be used when color program material is required. The pixlock mode provides phasing of the off-tape horizontal and vertical sync with that of an external reference. Pixlock synchronizes switching between tapes, studio signals, and other sources; and permits fades, lap dissolves, superimposures, and other special effects. See LINE-LOCK.

plain lighting Lighting in which the principal light source approximates the position of the morning or afternoon sun.

plane The distance from the rear node of the camera lens to the film (plane) where the image is focused.

Also, a relatively flat surface lacking in depth, or the impression of apparent depth in a picture. Sharp delineation between areas having distinct tonal values, through differential focusing, helps create the impression of depth.

platen A flat surface of optically clear glass that holds the cels in position over the background during the filming of animation. See Figure P-10.

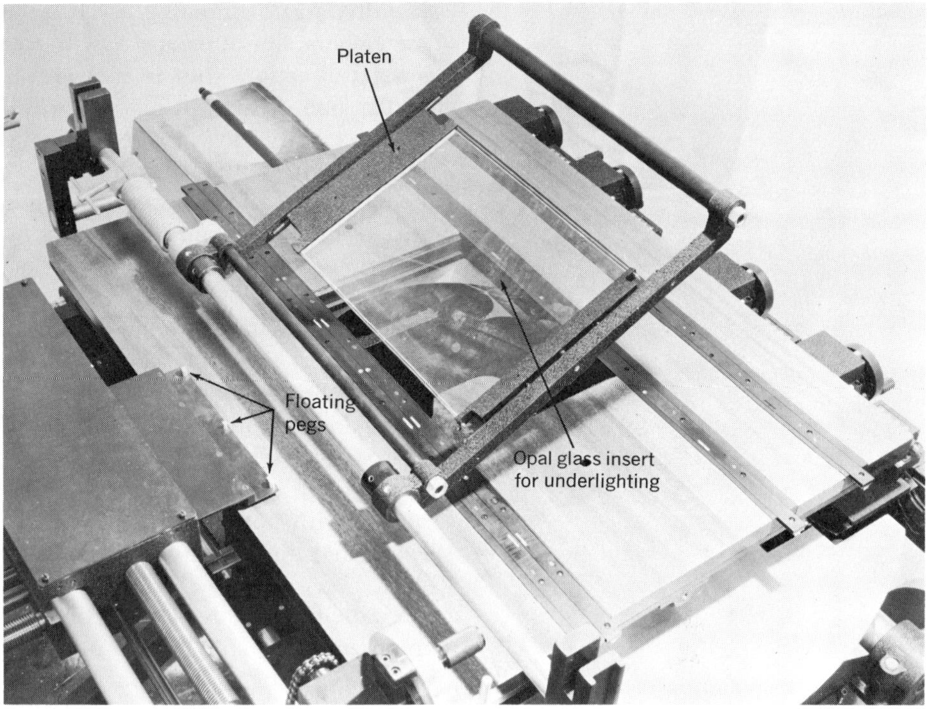

Fig. P-10 **The compound table showing the insert for underlighting, the platen, and the floating pegs. (*Animation Equipment Corp.*)**

playback The reproduction of videotape picture and sound at the studio, for monitoring purposes, any time after recording.

plumbicon A color television pickup tube patented by N. A. Phillips. It supercedes the image-orthicon and vidicon tubes. A plumbicon camera is a color or black-and-white television camera featuring a plumbicon pickup tube.

polar curve A graph showing the intensity, distribution, and emission characteristics of a light source.

polarity control A television camera control that selects positive or negative video polarity signal output. This control makes it possible to record a negative (rather than a positive) television signal on film, with kinescope processes, in order to obtain a direct positive print when the film is processed. See KINESCOPE RECORDING.

polarized light Although light travels in a straight line, it resembles a wave, with its accompanying crests and troughs. Unlike their watery counterparts, however, light waves travel vertically, horizontally, and diagonally. A light wave combines all these different movements simultaneously, with each wave moving in the same general direction—but in its own plane.

Any one of these transverse waves can be isolated or separated from the general flow to create a polarized light wave. Special plates or lenses, with millions of needle-shaped crystals, can be used to block the passage of all light waves with the exception of those that vibrate in the same direction. When used in pairs, the plates or lenses can be arranged to form a parallel opening through which waves in the same plane can pass. The two openings or slits can also be placed at right angles to each other to prevent the passage of all light waves, regardless of their direction of travel.

The reflection of sunlight and its subsequent refraction by atmospheric dust and moisture are an everyday example of polarized light. Mirror images from reflective, nonmetallic surfaces are another form of polarized light. Although objects may appear the same to the viewer, with polarized or nonpolarized light, the use of polarizing lenses in photography can cause distinguishable differences—on film.

Undesirable glare or reflections that are strong enough to cause an observer to shade his eyes or squint can be eliminated completely with the aid of a polarizing lens. In outdoor locations, where the brightness of sky or water makes filming conditions difficult, Pola-screen filters play a most effective part by reducing or controlling the strong accompanying glare. Contrasty subject matter, similarly, can be given proper exposure by creating a favorable balance between dark and light areas.

Polarization, as a result of recent experiments, is currently being used in frame-by-frame stop-motion photography to produce movement in inanimate subject matter. Celluloid patterns, applied to the surface of transparencies, can create the illusion of movement in various parts of the subject matter. Spinning patterns cause wheels on stationary vehicles to turn magically, while other patterns on Celluloid, cut out and affixed to other parts of the transparency, produce billowing clouds of smoke, or create wave effects within specific outlined areas. These Technimation effects and many others are produced by placing a polarizing disk between the camera lens and the subject matter and revolving the disk to a new position for each exposure.

Light does, indeed, travel in a straight line. Like a football, however, it can be made to take funny bounces.

polarizing color filters Polarizing color filters, produced by Harrison and Harrison in Hollywood, are available in all the colors listed in the Spectro-Wedge system, including neutral density, with the polarizing feature built into each.

By combining polarization with color these filters can be used wherever standard color filters would normally be used. The polarizing feature, however, makes it possible to minimize reflections and control color saturation, both quite difficult to manage with ordinary color filters.

Each of the polarizing filters has a permanently placed dot at the edge of the filter to indicate the axis of polarization. By pointing this dot at the light source a maximum amount of reflection is eliminated. When the dot is positioned so that it points upward, average reflections are minimized, and the filters produce results that are quite similar to the effects seen by the wearer of Polaroid glasses.

The normal exposure difference, due to the addition of the neutral polarizing feature, is a 40 percent or M transmission loss as indicated on the Spectro-Wedge scale. The separation of the exposure loss due to color from the exposure loss due to polarization is very important where the rated speed of a film is stated only in connection with the use of conversion or correction filters. Some films, for example, are rated for either daylight or tungsten use with compensating filters. The basic speed of the film, however, is not stated. Consequently, when polarization is added to the filter, its separate factor must be known in order to determine the new film speed based on the combination of color and polarization in one filter.

When determining the basic exposure of reflected light, major reflecting surfaces should be avoided since the illumination from those areas is usually greatly reduced by the polarizing filter. However, by taking readings from areas that are not shiny or highly reflective, a truer overall average of the basic exposure can be obtained since these areas are not greatly affected by the polarizing filter.

Many photographers use the incident-light method of measuring exposure since it is based on the mean average of reflectance. Shiny or highly reflective areas are not especially influenced by this method of basic exposure determination.

Since a color polarizing filter, in addition to color correction, reduces many of the surface reflections, the resultant image has more detail and color saturation. Prints, in turn, tend to appear slightly darker than those made without any correction. This apparent difference should be carefully analyzed before increasing exposure. The cameraman must determine that there actually is an exposure difference and not just an apparent difference caused by surface reflections. Overexposure, even without a polarizing color filter, tends to wash out detail and color saturation just as surface reflections do.

polarizing prisms Especially applicable for ultraviolet and high-power laser work, polarizing prisms are made of calcite or quartz (uniaxial birefringent crystals), and have two different indices of refraction along all axes but one—the optical axis. A light beam entering such a crystal along any axis but the optical axis is separated into two beams having different propagation directions and different propagation velocities. Each beam is completely polarized, and the polarization forms of the two beams are orthogonal. The prisms are available through Barnett Associates in Hauppauge, New York.

Polaroid filter Also called **Pola screen**. A type of filter used for controlling glare and brightness. Like the neutral-density filters, Polaroid filters are not used for the purpose of altering color values but for eliminating glare and reflections.

Pola screen See POLAROID FILTER.

polyester backing A relatively strong base for the ferromagnetic coatings on which tape recordings are made. Their resistance to changes in humidity make them extremely useful as tape supports. Compare ACETATE, CRONAR, ESTAR, and MYLAR BASES.

Pomeroy process See MATTES and SELF-MATTEING PROCESSES.

portrait lens A lens with a relatively narrow angular field of view that is especially applicable for portrait work. This type of lens is designed to minimize distortion in close-up filming.

positive A photographic reproduction in which the colors and tonal values in the original subject matter are directly reproduced. In a negative, lights and darks are reversed in black-and-white photography, and colors are represented by their complements.

positive lens Thicker in the middle than it is around the edges, a positive lens converges light rays.

postemphasis That portion of the equalization which is applied in audio and video reproducing systems. See EQUALIZER. Compare PREEMPHASIS.

postsynchronization The recording and synchronization of the sound track with the corresponding picture information after filming has been completed. Compare PRESYNCHRONIZATION.

potash alum Also called **potassium aluminum sulfate**. Many of the chemical compounds containing potassium are used in several areas of laboratory processing. For example:

POTASSIUM ALUMINUM SULFATE. Used as a hardening agent in laboratory fixing baths.

POTASSIUM BICHROMATE (orange-red crystals). Used as an intensifier.
POTASSIUM BROMIDE (colorless crystals). Used as a restrainer and a bleach.
POTASSIUM CARBONATE (white powder). Used as an alkali in developing solutions.
POTASSIUM CHLORIDE (white powder). Used as a bleacher and sensitizer.
POTASSIUM CHROMATE (yellow crystals). Used as a toner.
POTASSIUM CHROMIUM SULFITE, also known as **chrome alum** (violet crystals). A hardening agent used in fixing baths.
POTASSIUM CYANIDE (white, poisonous). Used in reducers and as a silver solvent.
POTASSIUM FERRICYANIDE (red, poisonous). Used in reducers as well as a bleach and toner.
POTASSIUM HYDROXIDE (white pellets). Used as an alkali in developing solutions.
POTASSIUM IODIDE (white crystals). Used as a bleacher, intensifier, and reducer.
POTASSIUM METABISULFITE (white crystals). Doubles as a preservative in developing solutions and as an acidifier in fixing and stop baths.
POTASSIUM PERMANGANATE (purplish black crystals). A powerful oxidizer used by laboratories as an intensifier and reducer of negatives, as a silver solvent during reversal processing, and as a hypo eliminator.
POTASSIUM SULFIDE (white powder). Used as a toner.
POTASSIUM SULFOCYANIDE (white crystals). Used in fine-grain developing solutions.

power belts, packs, and chargers Rechargeable sealed battery cells (Cine 60) built into a belt that can be worn by the cameraman. Automatic overload switches are built into the belt system to protect the battery and camera motor from damage in case of a short circuit. A built-in charger can be plugged into any 115-volt a-c outlet for recharging.

Dynamax power packs (Gordon Enterprises) provide a source of portable power for cameras, lights, and accessory equipment. Doubling the footage available with standard lead-acid batteries, the power packs contain a sintered-plate nickel-cadmium storage battery with a capacity of 8 ampere-hours. (16-ampere-hour capacity is obtained from the 8-16 volt Dynamax power pack for the Arriflex 16mm cameras.)

The Dynamax is considered a lifetime battery, with no known limit of service. The unit charges to 100 percent overnight and retains the charge to within 70 percent after one year in storage. Its cells are designed to withstand extreme shock and vibration. The power packs are available in various voltages both with or without built-in chargers.

Battery cells are contained in a stainless steel housing and cased in either leather or metal cases, with shoulder strap. The batteries are leakproof and may be carried in camera or accessory cases.

preamplifier An amplifier capable of boosting weak signals to a usable level by the power amplifier.

preemphasis Also called **preequalization**. That portion of the equalization which is applied in audio and video recorders. See EQUALIZER. Compare POSTEMPHASIS.

premiere The first showing of a new motion picture at a theater (rather than in a studio projection room). The audiences invited to attend the opening night festivities are usually made up of members of the cast, their escorts, and celebrities. Compare FIRST RUN and PREVIEW.

premix The process of combining several sound tracks into a composite track that will subsequently be combined with other sound tracks. Premixing provides the sound technician with a greater degree of flexibility during the final mixing process. During the production of the final composite track, the premixed sound effects and musical background can be additionally controlled while the dialogue is added at the desired level.

preservative Any of several chemicals that are added to developing solutions to keep the developing agent from oxidizing—sodium sulfite, for example.

pressure plate The plate in the camera gate that holds the film in the focal plane during the interval of exposure.

presynchronization The recording of a sound track before filming takes place. Usually associated with an animated production, the animator fits actions and lip movements to the frame counts supplied by the film editor after the sound track has been analyzed. See BAR SHEETS. Compare POSTSYNCHRONIZATION.

preview The showing of a new motion picture (generally in a studio theater or projection room). The audience is usually made up of studio personnel along with those guests who may have a financial interest in the production or are potential distributors.

Also, a special showing of a new motion picture or the "pilot" for a new television series to an audience selected from average moviegoers or television viewers. The purpose of the showing is to get a general reaction to the film and, perhaps, some honest, constructive criticism. In some cases, a new series for television has been made or dropped—based on the comments from the distributed questionnaires. Compare FIRST RUN and PREMIERE.

primary colors The three basic colors in the spectrum. Although artists, who must use pigments, have popularized the idea that they are red, yellow, and blue; optically and photographically the primary colors are red-orange, green, and blue-violet.

principal axis An imaginary line passing through the center of each of the lens elements in an optical system. See Figure P-11.

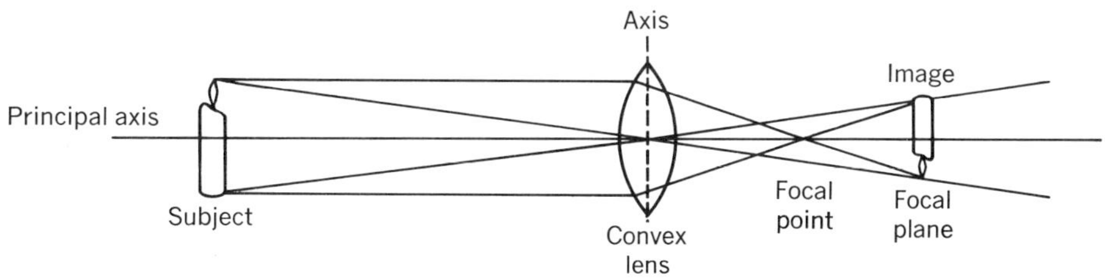

Fig. P-11 **Principal axis.**

print The positive picture made from an original or dupe negative in which the tonal values and colors in the original subject matter are duplicated.

printing A laboratory process in which a negative or positive image on film is used as the light modulator and reexposed on raw stock to produce either a negative or positive image, depending on the film used for copying.

printing light A controlled light source employed to expose the film stock used for duplicating purposes during the laboratory printing process.

print-through Magnetism that is transmitted from one segment of the tape to those segments physically above or below it in the roll. Overrecording and high temperatures during storage are usually responsible for this undesirable effect.

print-through gamma A laboratory term describing a gamma value measured by plotting the density of a gray scale on the horizontal axis of a graph, and the density of the print, from the same gray scale, on a vertical axis.

The print-through method takes into consideration variations in printing caused by inconsistencies in color temperature and specularity of the light used for printing as well as differences in the intensity level.

print-through ratio Usually the ratio, measured in decibels, between a 500-cycle signal at 2 percent distortion level and the output level from a signal induced into the preceding unrecorded layer of tape by the 2 percent distortion signal. Tape undergoing such a test is usually rewound and stored before comparisons are made.

print-up (print-down) The process of changing the transmission level for optical sound tracks. The transmission level is usually raised or lowered at certain points in the film for dramatic purposes.

Increases in output or volume are accomplished by printing the track lighter then normal, thereby increasing the transmission level and the dynamic range. Volume decreases when the track area is printed darker than normal.

prism A transparent, wedge-shaped piece of glass which because of its shape will cause differential diffraction of the constituent colors in a ray of "white" light, thus separating it into a rainbow-like spectrum. See Figure P-12.

prismatic lens A lens capable of changing the path of light rays being reflected from the subject matter while forming an image. In practice, it is a lens attachment with a number of facets, used to produce multiple-image effects, there being as many images in the composite as there are facets in the lens. The image may be rotated by turning the lens during the filming process.

prism beam splitter An arrangement of prisms that is used either to divide a ray of light and direct it to two different paths or to bring together two separate beams. A cube is formed of two right-angle prisms, with reflective or aluminum coatings, joined or cemented together. A part of the light entering the face of the prism passes through to expose the film in the camera. The remainder of the light is reflected at right angles and directed to a second length

Fig. P-12 **Prism.**

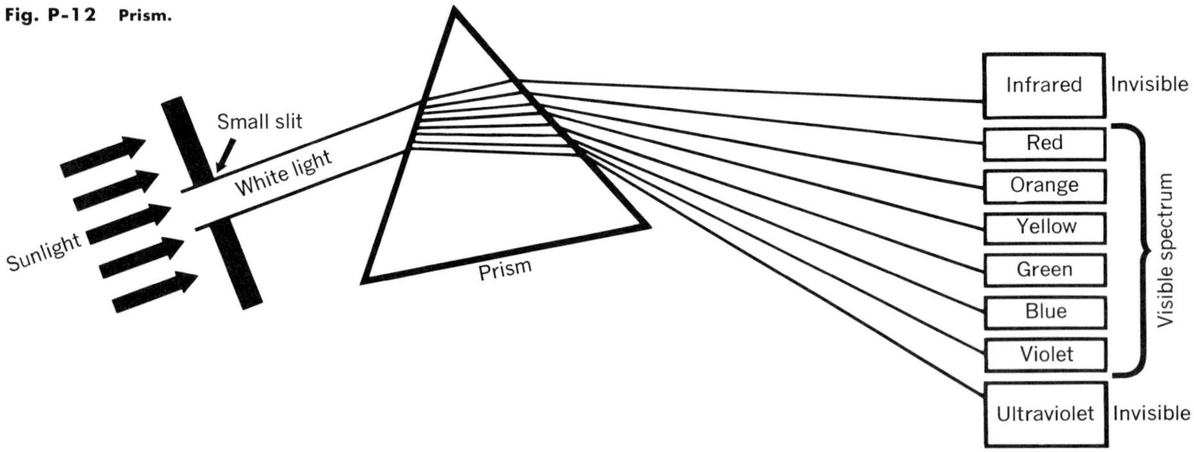

of film. In other applications, light striking two faces of the prism at right angles to each other will pass through and exit at the same face. See Figure P-13 and SELF-MATTEING PROCESSES.

probe A type of light-metering device that is used for checking the intensity of the lighting source in optical and contact printers.

process cameras Cameras designed specifically for optical printing, animation, and filmstrip work.

Acme 35mm and 16mm reflex. Pressure plates in process cameras are usually spring-loaded and are designed to take as many as three films without the necessity of adjusting the cam. The viewing systems in these cameras make it possible to "see" a composite or matte shot by inserting a piece of positive film in the viewer and lining up with the objective scene. The viewer has two registration pins located in the same position in relation to the aperture as the movement. By reversing the procedure, it is possible to project the image to paint mattes. There is no need to rack over as the viewer is positive reflex. A switch on the viewer assures that the motor will not start while the reflex mirror is in the viewing position.

An antibuckle switch is incorporated in the camera so that if for any reason the film does not take up, the motor stops. The switch works in either forward or reverse operating mode.

The camera has a full 170-degree dissolving shutter which is controlled from a large knob on the door. Automatic dissolves (in or out) can be made in 1- to 8-foot lengths, or in ½-, 1½-, 2-, 3-, or 4-foot lengths. The camera has a forward and reverse footage and frame counter that is resettable.

There are no pulleys or belts within the camera. The entire drive is gear-driven with overriding clutches on both sides of the magazine, so that the film is automatically taken up when the camera is running in either direction. The camera is machined to take Mitchell NC magazines.

With the Acme stop-motion motor, the film can be made to run forward or in reverse—either single-frame or continuous exposure. A selector switch makes it possible for the cameraman to expose single frames for black-and-white filming, or three successive frames for color work. Exposure times range from ¼ to 4 seconds, with provision for 1/16-second exposure for title work or fast rewinding. The latter speed cannot be used for stop-motion work.

The variable-speed motor allows the cameraman to select a convenient filming speed by turning a selector switch. At sound speeds of 24 frames per second (fps), the motor operates at 1,440 revolutions per minute (rpm); at 16 fps, 960 rpm; at 12 fps, 720 rpm; and so on to 1½ fps at 90 rpm. These motors deliver full power regardless of the filming speed.

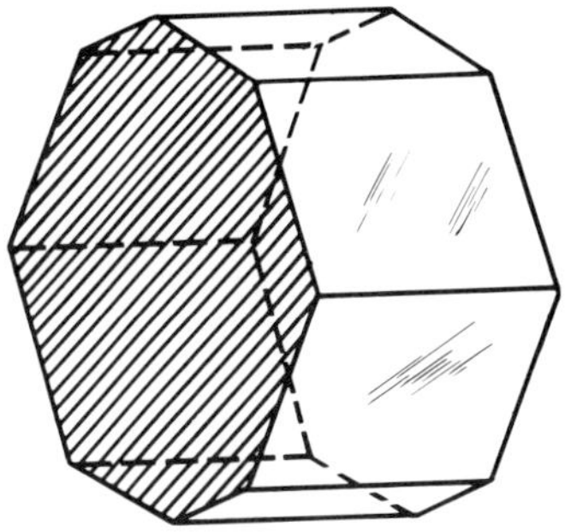

Rotating or intermittent prism

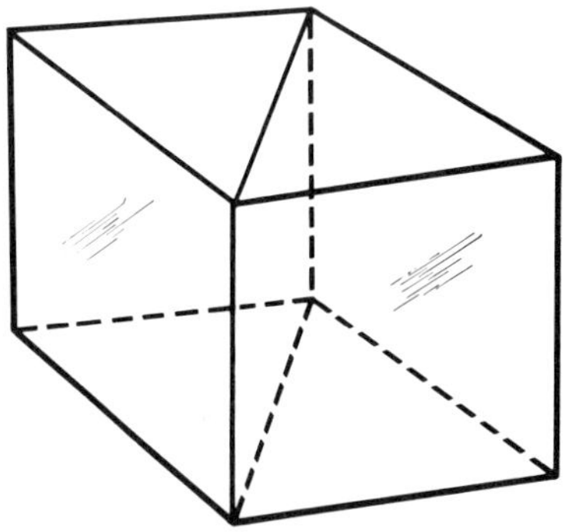

Prism beam splitter

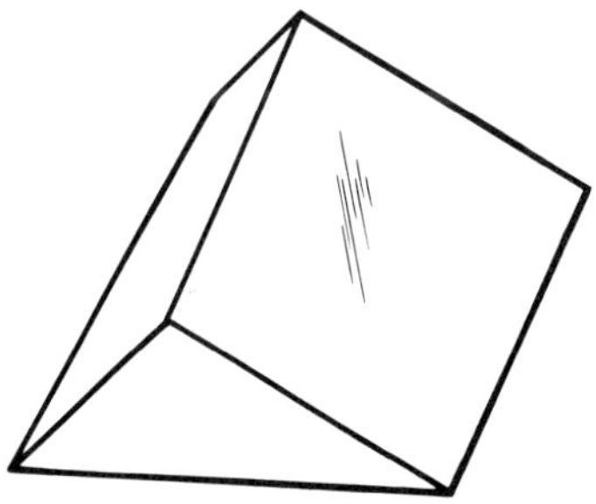

Right-angle prism

Fig. P-13 Various types of prisms.

Oxberry process camera 35mm–16mm. The Oxberry camera can be converted from 35 to 16mm filming by changing shuttles, lenses, transport mechanisms, and magazines. The camera remains in alignment for both 35 and 16mm filming.

Capable of accommodating two thicknesses of film, including splices, the camera's fixed-pin movement provides the precise registration needed for multiple runs, special effects, and matte shots. Apertures for full frame, CinemaScope, and filmstrips can be interchanged by removing four screws in the shuttle.

The 170-degree shutter can be used to make fades and dissolves, manually or automatically. The length of the desired fade or dissolve, from 8 frames to 8 feet, is controlled by presetting a pointer to the corresponding scribe mark on a round metal plate. These transitional effects can be made by running the camera continuously or with frame-by-frame, stop-motion tripping.

The camera on the animation stand can be racked over for viewing and composing. The optical printers have reflex viewing systems. Fixed locating pins in both viewing systems correspond with the position of the pins in the camera.

The animation stand is equipped with a 55mm Micro-Nikkor lens for 35mm filming and a 28mm Micro-Nikkor lens for 16mm work. The optical camera has a 90mm Ektar lens but also accepts anamorphic lenses for wide-screen work. Automatic focusing cams (optional) are also available.

The stop-motion motors provide filming speeds ranging from 30 to 240 frames per minute, and a 360-frame-per-minute speed for continuous shooting and fast rewind. The resettable frame counters are of the Veeder-Root type.

The cameras accept Mitchell-type double-compartment magazines with 400-, 1,000-, and 1,200-foot capacities, as well as 400-foot-capacity bipack magazines. The tension in the magazine's automatic take-up motors is adjustable.

processing An all-inclusive term indicating the various laboratory procedures needed to produce a visible image on film that can be viewed or projected.

To produce a visible image on the film exposed during the photographic process, the in-camera film is treated chemically. In the first step of this continuous process, the unwound film passes over rollers that guide it through a developing solution. Continuing on its mechanical journey, the developed film is washed; then carried along over the guiding rollers to a fixing bath, through another wash, and finally to a drying chamber. The developed negative is then rewound and used for printing the positive image. For this printing process, the unexposed emulsion side of the print film is placed in contact with the emulsion side of the developed negative. Light passing through this negative exposes the print film. The laboratory procedure for developing the newly exposed film is quite similar to the methods used for processing the original in-camera film.

In normal developing, black-and-white film produces a negative image, or one in which the tonal values are exactly opposite to those in the subject matter at the time of exposure. In the negative, the lighter areas in the subject matter have proportionately greater silver deposits than the darker areas. Whites, or lighter areas, therefore appear as the denser portions of the film frame, and blacks as clear areas in the negative.

For the low-contrast results desired in the processing of original black-and-white negative film, graphol or graphidone and hydroquinone are used as developing agents in order to produce the desired density and contrast. A sizable amount of sodium sulfite is used as the preservative. Borax or sodium metaborate acts as the low-contrast accelerator, and sodium or potassium bromide in small amounts is used as a restrainer.

For the higher contrasts needed for processing black-and-white release prints generally, the developing agents are proportioned to produce a higher reactivity. Sodium hydroxide or sodium carbonate or possibly both are used as the accelerator. Dichroic or chemical fog is avoided by the use of restrainers.

In reversal processing generally, the in-camera raw stock is designed to produce a direct positive image. While the silver halide deposits duplicate the formations in negative film, with the buildup greatest in those areas where the density is lowest, it is the unused silver halide in the emulsion that produces the tonal values of the positive image.

During the actual processing, a negative image is converted into a positive image by bleaching out the developed silver on the film exposed in the camera. Before the fixing process, the same film is reexposed and redeveloped to produce the positive image. The basic steps needed to develop color film are similar to those used for developing film generally. In place of the graphol, graphidone, or hydroquinone, color-developing agents are used in the developer baths.

For positive color films, the bleach baths contain potassium bichromate. The bleach baths for negative films contain potassium ferricyanide. Water rinses or washes between each of the processing stages remove excess chemicals from previous solutions. The print material's emulsion after the film has been dried contains only the dyes in the gelatin, the silver salts having been removed earlier.

The reproduction of color values is made possible by a chemical reaction described as *coupler development.* During the developing process, the exposed silver halide, in the first stage of the reaction, forms metallic silver. As the developing agent oxidizes, it combines with the chemical coupler to produce dyes in direct proportion to the amount of metallic silver formed during the initial reaction. The insoluble dyes duplicate the color values in the original subject matter. The negative images are approximately complementary to the color values in the original.

In color-reversal processing, the positive images are produced by developing each of the three layers of emulsion separately if the coupler components must be extracted from the developing solutions, or by developing the original film in a single solution if the couplers were added to the layers of emulsion by the manufacturer. Either way the basic process is the same.

Developed in a black-and-white bath to produce a negative image, the remaining silver halide in each of the emulsion layers is reexposed. The resultant positive dye images, produced by the chemical action of the couplers, duplicate the color values in the original subject matter. Including the developing, coupling, bleaching, fixing, and washing processes, the exposed film passes through as many as 10 (and more) separate baths before it is dried and readied for projection. Unlike color negative which can be corrected during the printing process, adjustments in the color balance of reversal film are, at best, limited.

The choice of chemicals and developing agents, generally, differs with the various film processes and manufacturer's specifications. Not mentioned previously are the chemicals used for lesser-known steps in processing—reducers, for example, for decreasing the density of a photographic image; intensifiers for increasing the density of a silver image; preservatives for stabilizing the dyes in color prints; hardeners that minimize the possibility of abrasions or surface scratches in print material; the chemicals used in prebaths that prepare the film for processing; and the application of wax (actually paraffin dissolved in a solution containing carbon tetrachloride) over the film's sprocket area to eliminate the undesirable clicking sound normally present during projection.

recommended laboratory practices—black and white. The normal gamma for the development of *original negative* is 0.65 to 0.70.

Fine grains or *master positives* should be developed to a normal print gamma of 1.4 (print-through method). The minimum highlight density in the print should be above 0.90.

Dupe negatives should be developed to a control gamma that will assure a duplication of the overall contrast and values in the original negative. The approximate gamma on *contact dupes* is 0.65 as measured by the print-through method. The approximate gamma on *optical dupes* is 0.55.

Prints should be developed to a contrast, on positive stock, so that the values in the negative are exactly complementary. If the contrast in the negative is low or flat, it should be printed and developed at approximately 2.35 gamma as read by the print-through method. Where the negative is contrasty, it should be printed and developed to a lower print contrast or a gamma of 1.85 (print-through method).

Prints intended *for telecasting* should, in general, be timed so that the density of the highlights is lighter. *Prints for* projection in *theaters* should have more contrast than those made for television use.

The densities of normal middle-gray tones in prints should be timed to read 0.75 ± 0.15. The minimum density reading of any

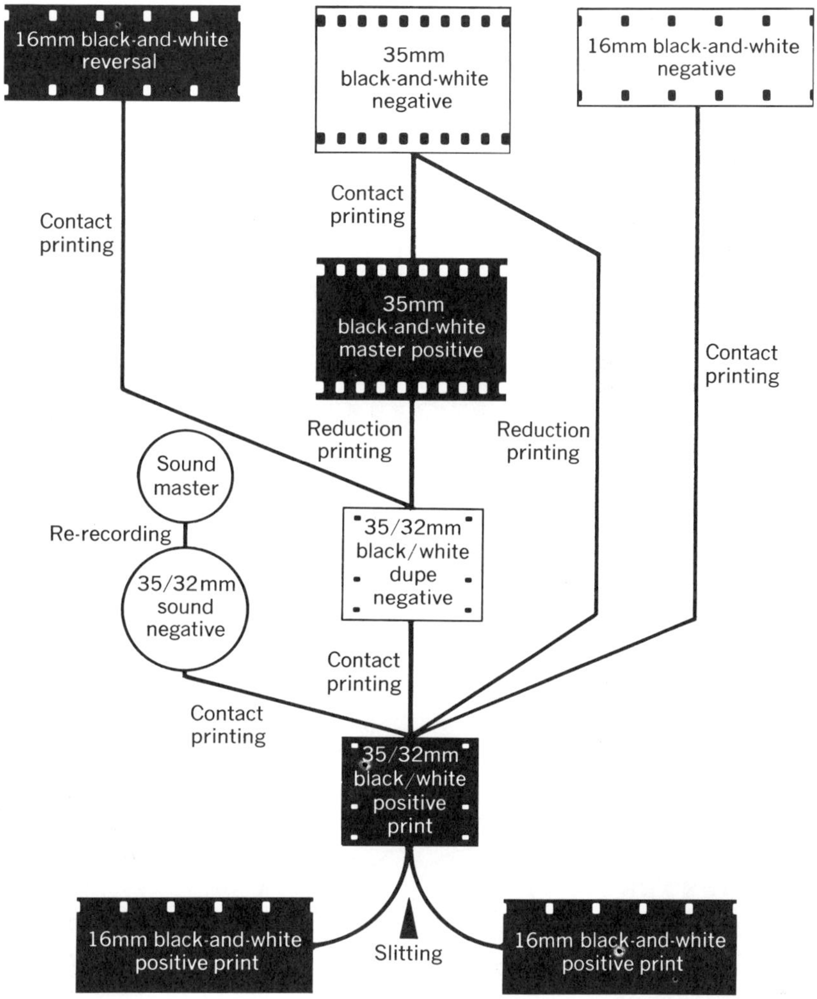

Fig. P-14*a* **The 35/32mm method for producing 16mm positive prints from black-and-white films. (*General Film Laboratories.*)**

area should be at least 0.40. The maximum reading for black areas should be 2.0 or less. The overall density of a print and the scene-to-scene balance should be as consistent as possible.

Sound negatives should be processed to negative gammas and densities in order to minimize distortion on variable-area tracks and provide maximum cancellation on variable-density tracks. Figures P-14*a* to *d* show printing flow charts for black-and-white films.

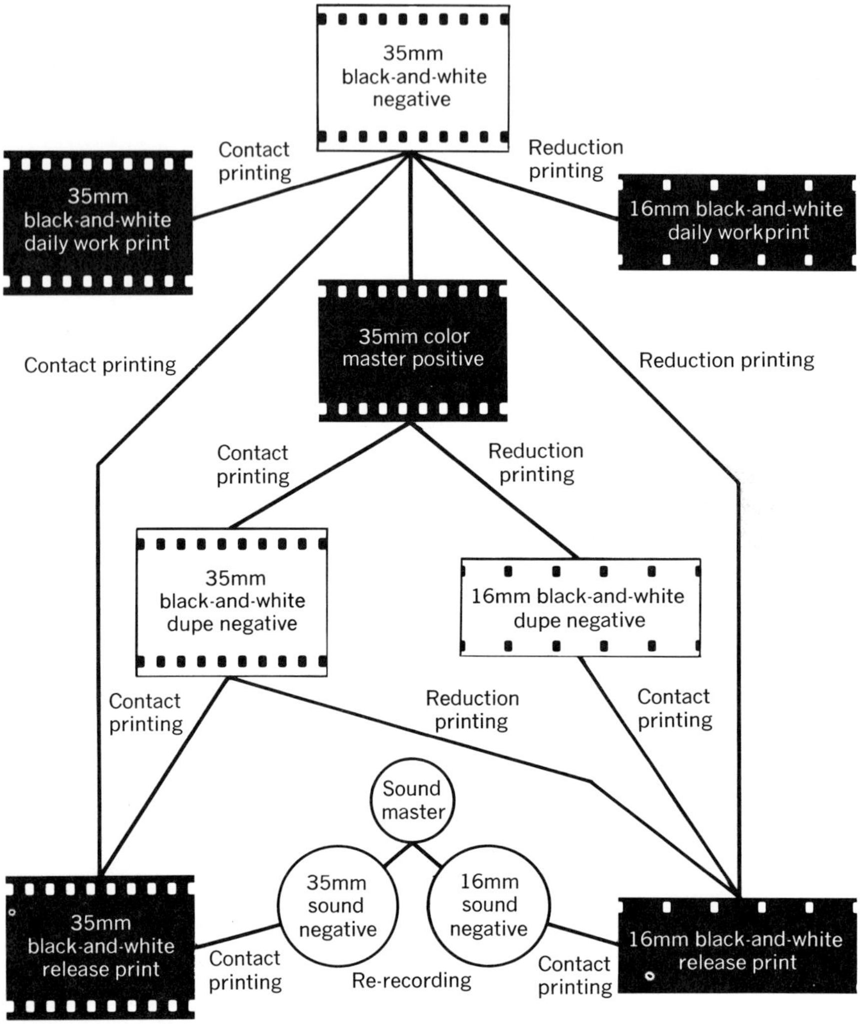

Fig. P-14*b* Printing flow chart for 35mm black-and-white negative (*General Film Laboratories.*)

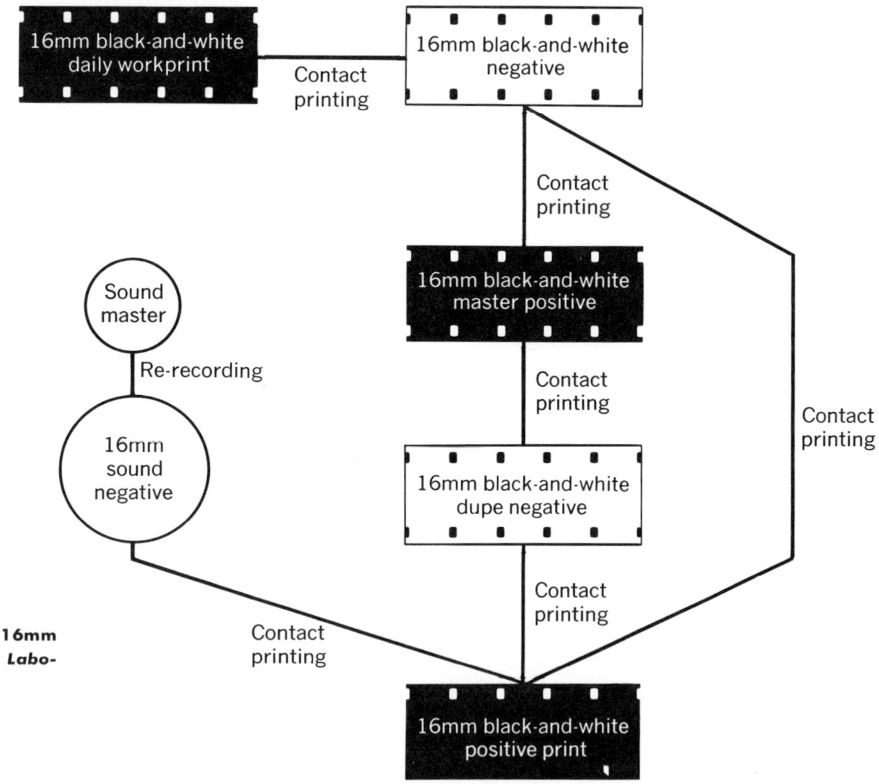

Fig. P-14c **Printing flow chart for 16mm black-and-white negative. (*General Film Laboratories.*)**

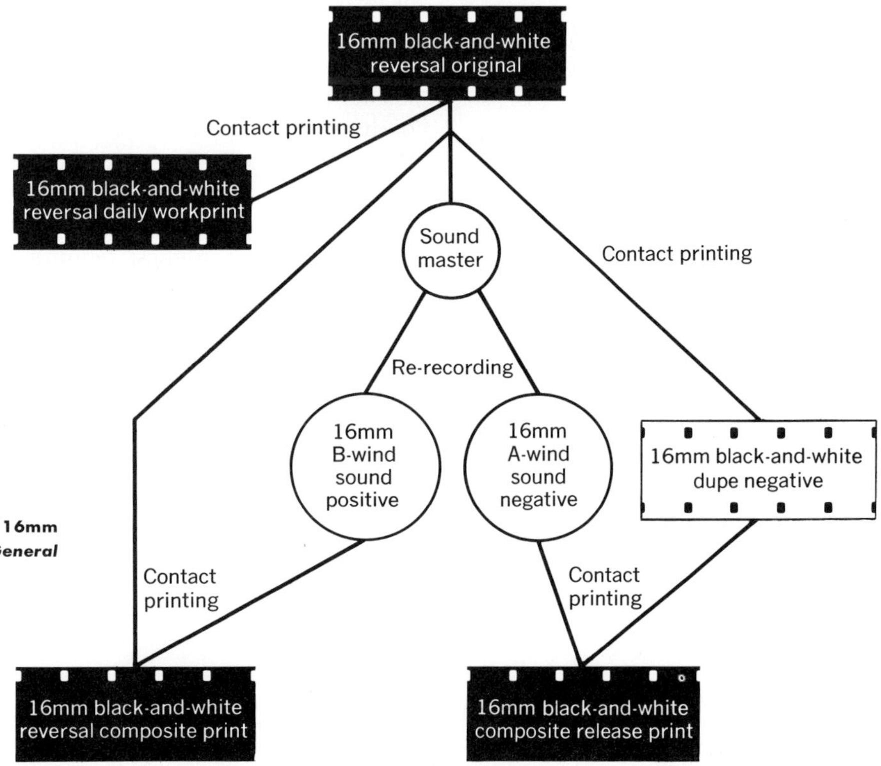

Fig. P-14d **Printing flow chart for 16mm black-and-white reversal original. (*General Film Laboratories.*)**

recommended laboratory practices—color. *Negatives* should be developed according to the manufacturer's specifications as soon after the film has been exposed as possible.

Color master positives and *separation master positives* should be exposed so that the tonal values in the gray scale will reproduce properly. *Separation positives* must be developed to gammas that will provide good color balance in the *internegative.*

Color dupe negatives and *internegatives* should be made on the film stock best suited for reproduction from Kodachrome, Ektachrome, color master positives (interpositives), or separation master positives.

Good tonal values in the gray scale and optimum color saturation are a must since the values and colors in these films will be used for the production of answer and release prints.

Prints should be timed to give good color saturation without muddying dark areas. The values of the gray areas should duplicate those in a black-and-white print.

A & B printing: In order to avoid the mixing of original material with second-generation material, and the possibility of a mismatch of colors, A & B printing is advisable for the production of prints for television from original negative or for making color master positives. Transitional effects such as cross-dissolves and fades can be added during the printing process.

Black-and-white prints are not usually made from original color negatives but from dupe negatives. First, a panchromatic master positive is made from the color negative. It should be printed to the same gammas that are used for producing normal black-and-white master positives. The dupe negative, made as with a normal master positive, is then used for producing the black-and-white prints. Figures P-15 *a* to *e* show charts for various color films.

In Figure P-15*e*, an image size change indicates optical reduction printing. Where reduction stages are called for, it is best in the interest of obtaining the highest definition in the final print to postpone reduction until the latest practicable stage. The dotted lines therefore indicate the less preferable method. (Figure P-15*e* is in color insert.)

The edited negative generally consists of both original camera negative and duplicate negative in which special effects have been incorporated. The original and duplicate negative footage may also be edited, in some instances, in A & B rolls in order to introduce special effects.

The choice of printing system depends on a number of factors, including the types of printing and processing equipment available, the physical and chemical processing requirements for a given type of film, and certain economic considerations. On this account, certain compromises may have to be accepted. See Figure P-15*f*.

Bell & Howell Color-additive Printing System. A dichroic mirror system separates the single 1,000-watt light source into primary color beams. Each beam can be automatically varied by a light valve through 50 steps of 0.025 log E. The simultaneous control of these light valves provides for color and light variations at the printing aperture.

With this Bell & Howell color-additive printing system, color corrections and fades, as well as "zero close" (for printing from A & B rolls) can be made to occur on cue, during high-speed printing from original negatives.

Dichroic mirrors and precision light valves provide over 125,000 variations of color control and light intensity. See Figure A-7*b* under ADDITIVE COLOR-PRINTING SYSTEM.

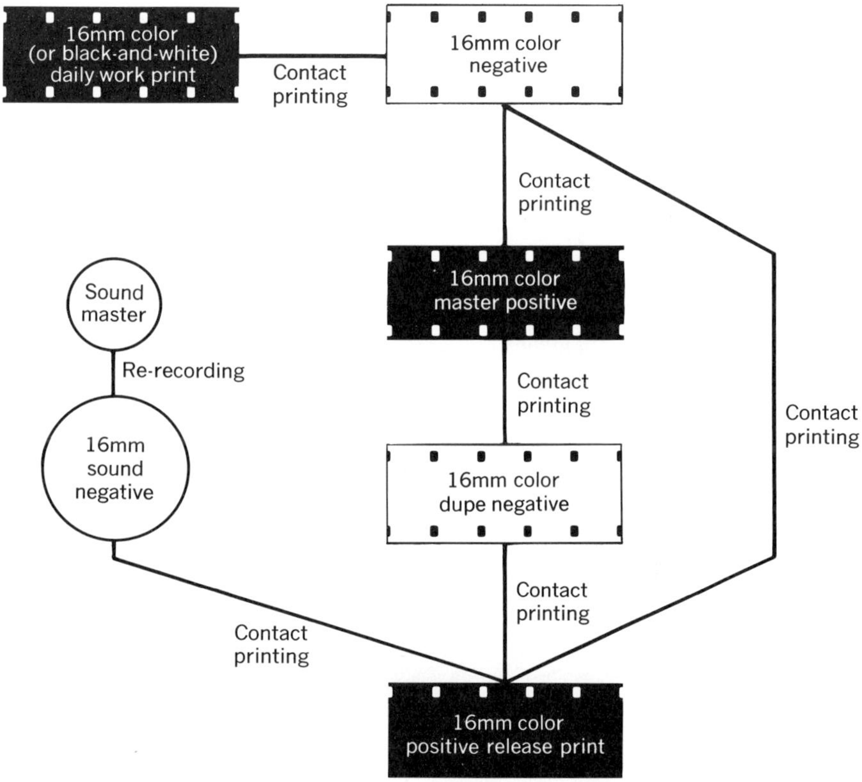

Fig. P-15*a* Printing flow chart for 16mm color negative. (*General Film Laboratories.*)

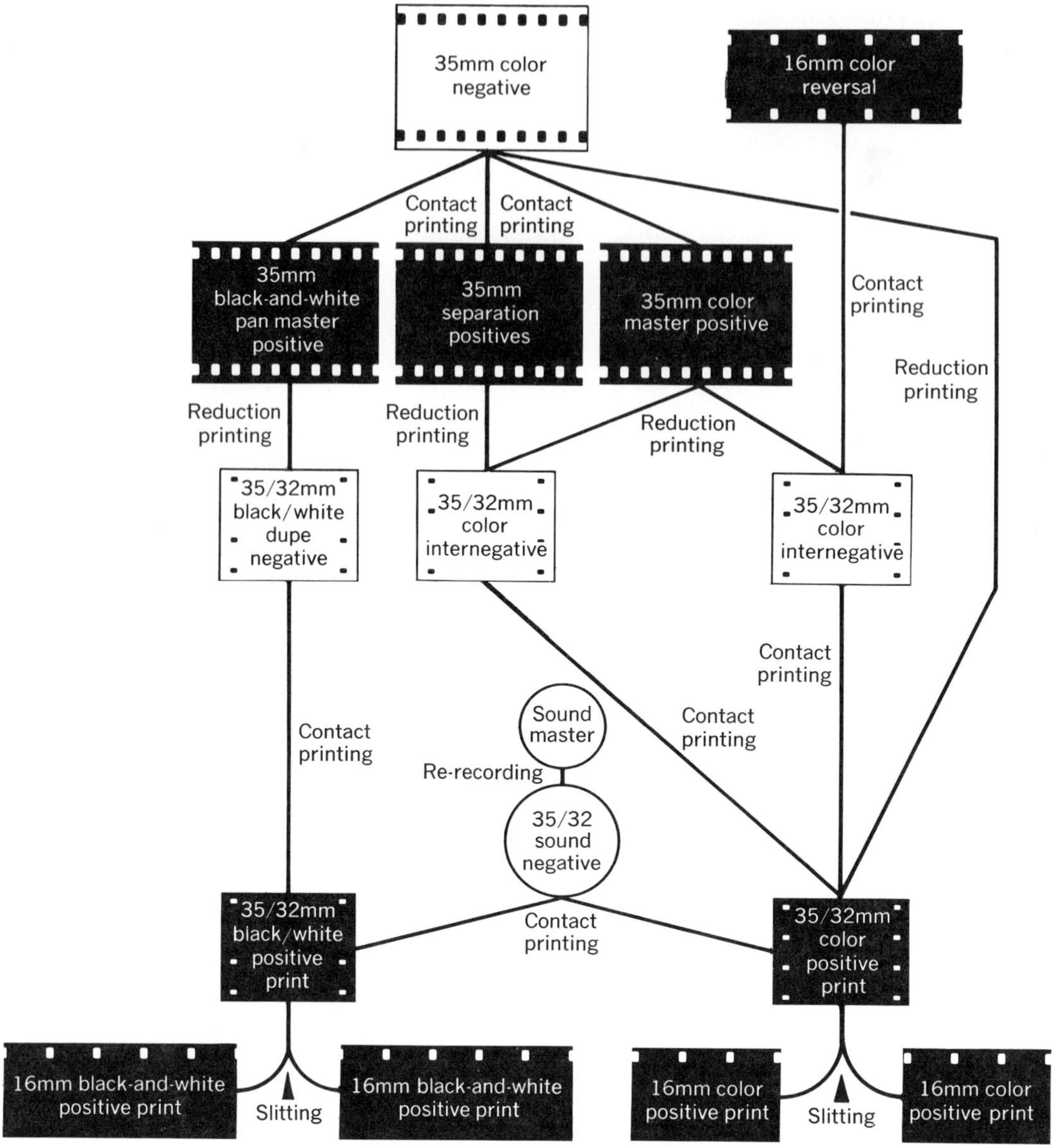

Fig. P-15*b* **The 35/32mm method for producing 16mm positive prints from color films. (*General Film Laboratories.*)**

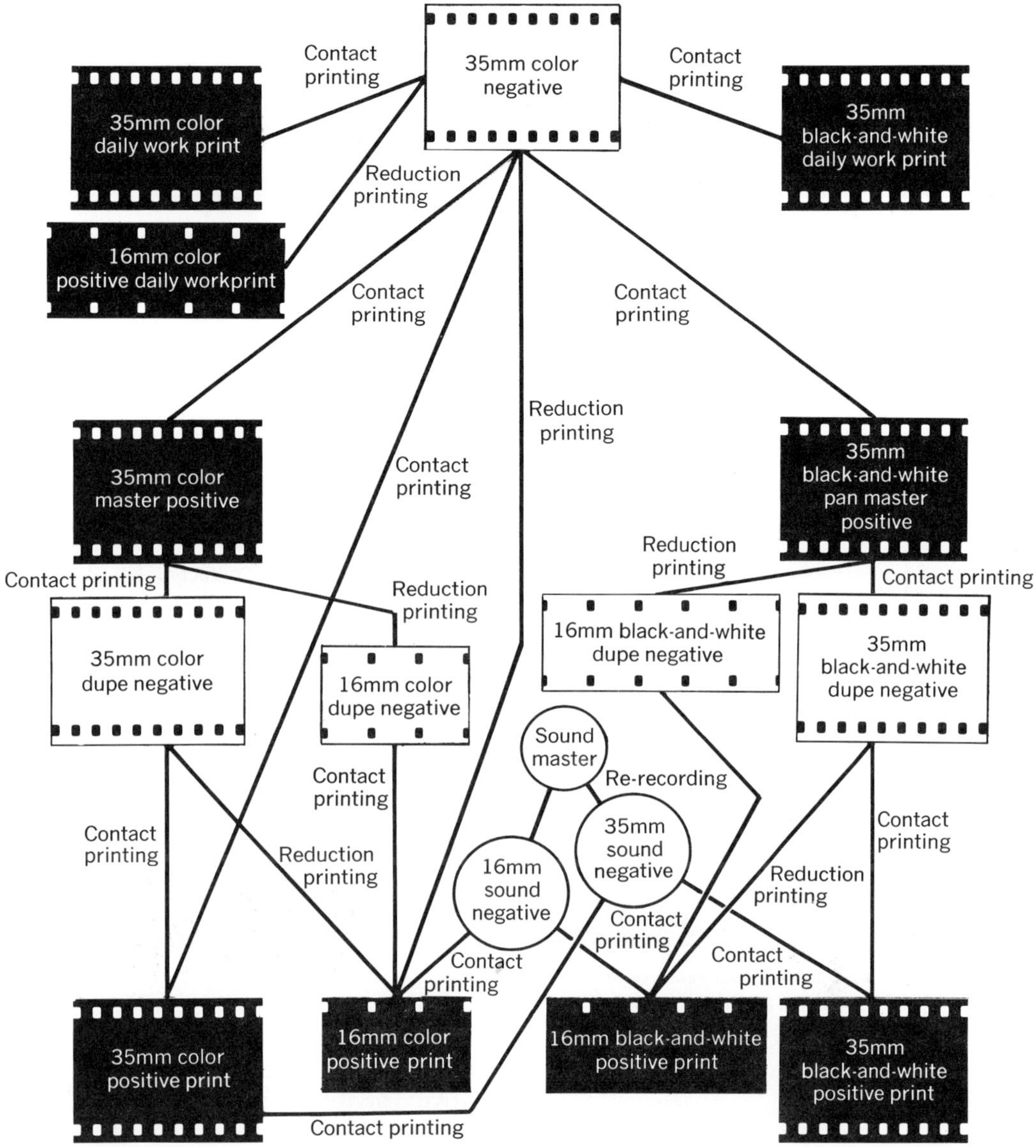

Fig. P-15c Printing flow chart for 35mm color negative. (*General Film Laboratories.*)

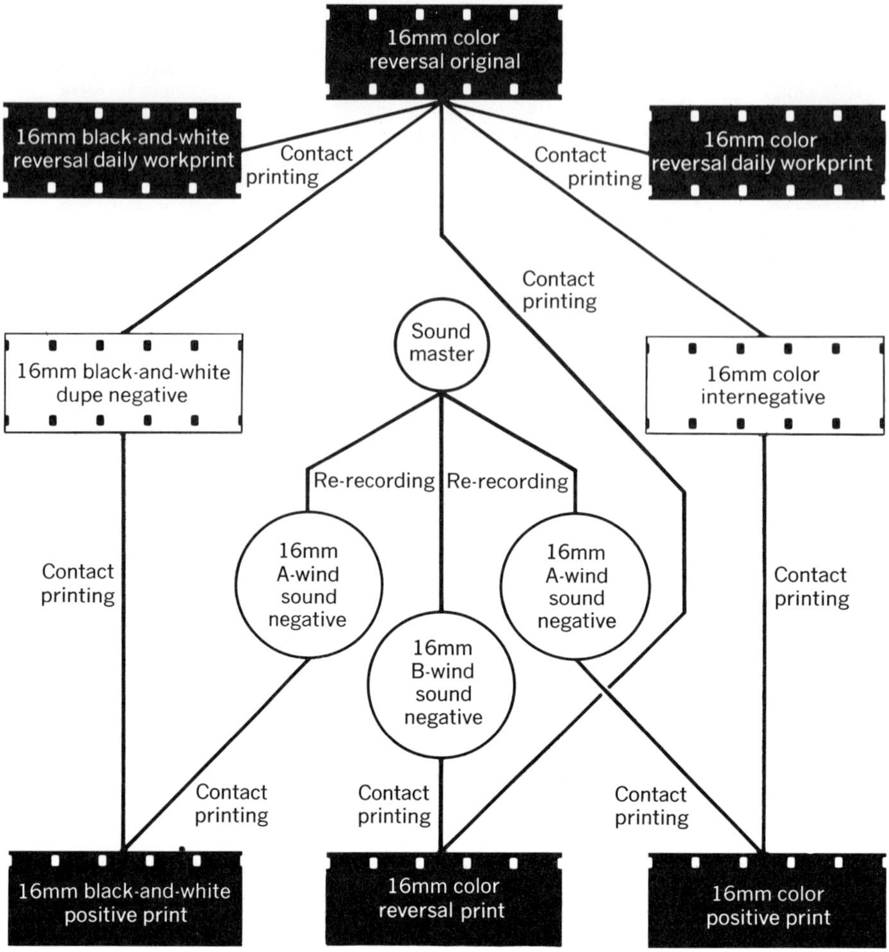

Fig. P-15*d* **Printing flow chart for 16mm color reversal original. (*General Film Laboratories.*)**

General Film Laboratories 35–32mm Printing Process. The 16mm color internegative (or black-and-white dupe negative) is made on 35mm-width stock, perforated for two 16mm prints. The sound track is re-recorded, or printed, on a separate 35–32mm printing roll. A single negative is made unless the initial print order is a large one. The release prints are contact-printed on 35–32mm positive stock, two to a roll, each with the same emulsion position. See Figure P-16*a* (in color insert).

Movielab Proposed Calendar Revision. See Figure P-16*b*.

processing amplifier—Ampex AC 911 The primary purpose of the processing amplifier is to provide a noise-free composite video output

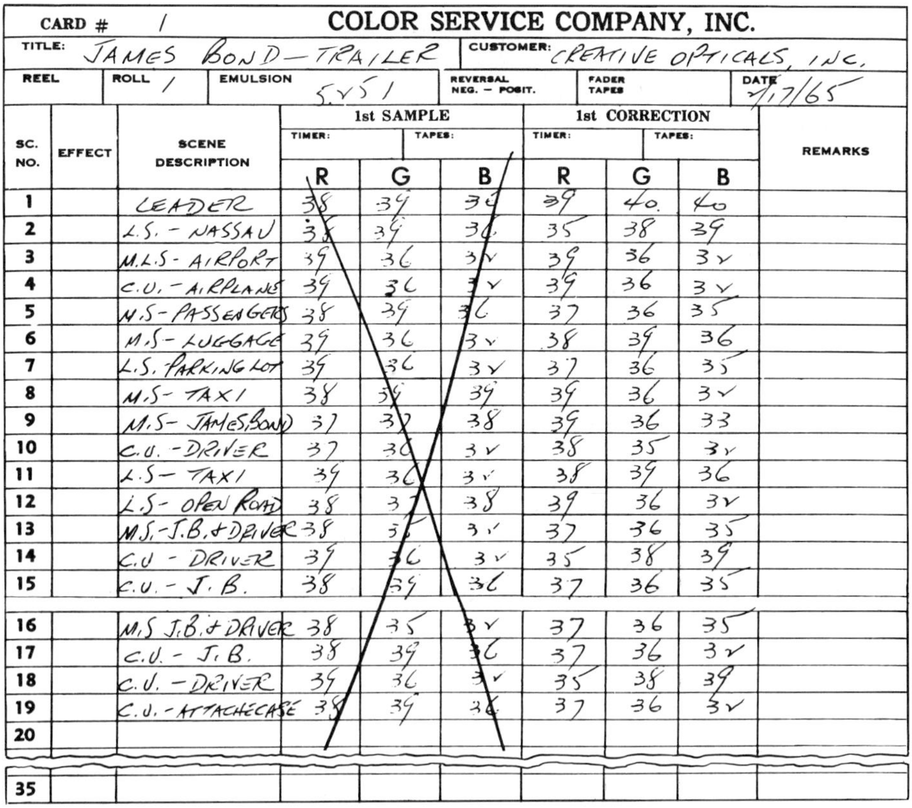

CARD # 1 — COLOR SERVICE COMPANY, INC.

TITLE: JAMES BOND—TRAILER | CUSTOMER: CREATIVE OPTICALS, INC.

REEL | ROLL 1 | EMULSION 5251 | REVERSAL NEG. — POSIT. | FADER TAPES | DATE 7/17/65

SC. NO.	EFFECT	SCENE DESCRIPTION	1st SAMPLE (TIMER: / TAPES:) R	G	B	1st CORRECTION (TIMER: / TAPES:) R	G	B	REMARKS
1		LEADER	38	39	36	39	40.	40	
2		L.S. - NASSAU	38	39	36	35	38	39	
3		M.L.S - AIRPORT	39	36	3✓	39	36	3✓	
4		C.U. - AIRPLANE	39	36	3✓	39	36	3✓	
5		M.S - PASSENGERS	38	39	36	37	36	35	
6		M.S - LUGGAGE	39	36	3✓	38	39	36	
7		L.S. PARKING LOT	39	36	3✓	37	36	35	
8		M.S - TAXI	38	39	39	39	36	3✓	
9		M.S - JAMES BOND	37	37	38	39	36	33	
10		C.U. - DRIVER	37	36	3✓	38	35	3✓	
11		L.S - TAXI	39	36	3✓	38	39	36	
12		L.S - OPEN ROAD	38	37	38	39	36	3✓	
13		M.S. - J.B. + DRIVER	38	35	3✓	37	36	35	
14		C.U - DRIVER	39	36	3✓	35	38	39	
15		C.U. - J.B.	38	39	36	37	36	35	
16		M.S J.B. + DRIVER	38	35	3✓	37	36	35	
17		C.U. - J.B.	38	39	36	37	36	3✓	
18		C.U. - DRIVER	39	36	3✓	35	38	39	
19		C.U. - ATTACHECASE	38	39	36	37	36	3✓	
20									
35									

Fig. P-15f Laboratory timing and color correction chart.

that meets recognized FCC specifications. The AC 911 has two groups of functions. The first group includes those common to color and monochrome operation; the second group includes those specifically required for color operation.

The functions common to color and monochrome operation are: The processing amplifier clips noise from incoming sync, reshapes the pulse form, and recombines it with video to produce a composite video signal. During monochrome operation, the processing amplifier limits noise to a fixed level in the video portion of the signal by preventing negative noise peaks from extending below the black level. The amplifier clamps to black and inserts phase-locked horizontal sync pulses into the dropout. The video and sync controlled by the processing amplifier is adjustable at the unit.

The function specifically required for color operation is: The processing amplifier removes the chroma portion of the video signal before the noise clipping section and reinserts the chroma after the noise clippers, thus allowing the chroma portion to pass without change. See VIDEOTAPE RECORDING.

In the EAST it's... MOVIELAB

MOVIELAB BUILDING 619 W. 54th St., New York 19, N.Y. • JUdson 6-0360

Neg.	Fri.	Fri.	Thu.	Wed.	Tue.	Mon.
8	7	6	5	4	3	2
16	15	14	13	12	11	9
23	22	21	20	19	18	17
31	30	29	28	27	26	24
38	37	36	35	34	33	32

MOVIELAB believes our present day calendar should be revised for present day use. This is what the proposed calendar will do for MOVIELAB and its customers:

1. Every job is a rush. Everyone wants his film yesterday. With this MOVIELAB calendar, a customer can place his order on the 7th and have it delivered on the 3rd.

2. All customers want their jobs on Friday — so there are two Fridays in every week.

3. There are seven extra days at the end of the month for those end-of-the-month rushers.

4. There will be no "first-of-the-month" bills to be paid as there isn't any "first." The "tenth" and "twenty-fifth" have also been omitted in case you have been asked to pay them on one of those days.

5. There are no bothersome non-productive Saturdays and Sundays. No time and a half or double time to pay.

6. There's a new day each week called Negotiationday. Contracts may then be reopened each week.

Fig. P-16*b* **Movielab proposed calendar revision.**

processing equipment Machines designed to process rolls of exposed film in a continuous, automatic operation. During the processing procedure, the exposed film is spliced to the film that is already traveling through the processor and is guided through developing, rinsing, fixing, and washing solutions before reaching the drying chamber.

processor A module in vidicon film cameras that inserts gamma correction into the video signal. The module contains the circuitry for manual and automatic black and automatic sensitivity functions.

process shot Any of several methods used for photographing foreground subject matter in front of a background that was previously filmed. The backgrounds or "plates," in the form of running film footage or a "still" transparency, are projected during the filming process, on translucent screens. See FRONT-SCREEN PROJECTION and REAR-SCREEN PROJECTION.

producer The person responsible for managing the production of a film. He is accountable for the success or failure of the enterprise, and all supervisory personnel assigned to a motion picture are under his jurisdiction.

production An all-inclusive term for the various operations involved in making motion pictures.

projection The term *moving pictures,* used frequently and loosely since the industry's magic lantern days, reflects a collective opinion rather than actual fact. The illusion of motion is actually created by projecting 24 still pictures, each properly composed within its own frame, each second. Projected in rapid succession, the information in each frame of film creates an impression that is retained by the brain for the brief interval during which the screen is darkened or blanked out by the shutter ($1/60$ to $1/80$ second) and the next picture is brought into position for viewing. If the viewer saw these 24 pictures move through the projector without this blanking action, he would experience an unintelligible blur of vertical light and dark patterns. The persistence of vision, and the constant repetition of the projection cycle, thus provides a continuous flow of picture information or "moving pictures." See PERSISTENCE OF VISION.

To motion picture audiences, the word *projection* is a relatively familiar term that is simply defined and easily understood. The devices that are used to illuminate the theater screen with images that have been exposed on lengths of film are equally familiar.

Each 1,000-foot reel of processed film has a running time of about 10 minutes. Standard 35mm projectors that accommodate 2,000-foot reels of print material can run continuously for approximately 20 minutes. Since this amount represents only about one-sixth the running time of a full-length feature film, the projectionist would have no alternative but to stop his machine long enough to change reels, an objectionable practice that distracts audiences and disrupts continuity. Two projectors, sharing a common booth, provide a practical solution to the stop-and-go viewing process. While one projector is lighting up the screen with the film's images, the next reel, already threaded in the second projector, is ready to roll. Punches or scribe marks, on four successive frames of film exactly 12 feet from the end of the reel, are repeated at the one-foot mark and alert the operator. On cue, the second projector throws its beam out over the audience, the reel from the first machine is removed and rewound, and the projection process continues uninterrupted.

The equipment used for projecting 35mm film is usually housed in booths in theaters and auditoriums. And, although most of the flammable rolls of nitrate stock are now carefully stored in film vaults and libraries, booths are expected to have sprinkler systems and fire-fighting equipment.

The projectors used for narrower-gauge film (16mm, 8mm, and Super 8, for example) are portable and require no special safeguards. Some 16mm projectors can accommodate 2,000-foot reels of film and project images continuously for almost a full hour.

During the projection process, the print passes from the feed reel to a take-up reel. Sprocket wheels that engage the film's perforations during the circuitous trip through the projector, and pins or claws that bring each frame into position in the projector gate, assure precise registration at all points. The stop-and-go intermittent movement of the film on its way to the aperture allows the projector's optical system to focus its light beam through each frame of film, in turn. After the individual frame has been projected, the shutter moves to its closed position and shuts out the beam of light until after the next frame of film has been advanced to the projection position.

Motion picture sound film is projected at the standard rate of 24 frames per second, or 16 frames per second at silent speed. The sound track precedes the corresponding picture by 20 frames on 35mm film and by 26 frames on the narrower 16mm film size. With the exception of a few types of film that are projected on wide screens, the viewer sees exactly 1,440 individual pictures each minute. The table shows the number of frames in a foot of film and the number of feet of 35, 16, and 8mm film projected in one minute.

Film	35mm	16mm	8mm
Number of frames per foot	16	40	80
Footage projected in 1 minute	90	36	18
Number of frames projected in 1 minute	1,440	1,440	1,440

The accompanying sound track, after the film leaves the projection aperture, must move constantly rather than intermittently at the point where the sound is projected. This change in speed, from intermittent to continuous, is effected when the film passes over the sound drum that is attached to a flywheel. Moving into position in front of an exciter lamp, a narrow beam of light is focused on the sound-track area of the film. As the objective light passes through the film, the constantly changing patterns in the variable-area track and the dark and light modulations in the variable-density track produce proportionate variations in the amount of light that subsequently reaches a photoelectric cell. The resultant electric impulses pass through an amplifier where they are converted into sound waves and sent to the theater's loudspeakers.

Including 3-D's superimposed images, and film with aspect ratios that are wider than the 1.33:1 standard, the picture information in each film frame is focused on opaque white matte, aluminum, or beaded screens. See PROJECTION SCREENS. The term *opaque,* as applied to screen materials, disregards the minute perforations through which the film's accompanying sound emerges from speakers positioned behind the screen. These perforations have no effect on the quality of the projected pictures.

Matte white screens, with their magnesium oxide surfaces, provide the desirable brightness without exaggerating the graininess of the projected film. They are capable of distributing light over a wide area, and color values remain relatively undisturbed by them. Aluminum screens have higher reflective characteristics than matte white screens and tend to increase contrast and graininess. They are basically designed for 3-D projection, and color values are influenced by the screen's metallic surface. Beaded screens, like aluminum screens, reflect in a smaller angle, about 30 degrees, and are best suited for narrow theaters and auditoriums. Beaded screens have millions of tiny glass beads, less than $1/100$ inch in diameter, imbedded in their white plastic surfaces, to lend sparkle to the projected images.

Translucent screens, by comparison, are used mainly for displays, demonstrations, and classroom study. And, unlike the images that are focused directly on the surface of the opaque screens in darkened theaters, the picture information that appears on the rear surface of translucent screens is reflected from a mirror and can be viewed in well-lighted areas.

The electromechanical devices used for projecting motion picture film do not exhaust the types of viewing equipment. Projectors for viewing filmstrips and slides are also relatively familiar items. A less familiar piece of equipment, to those not directly involved in motion picture production, is the moviola used by the film editor during the assembly and postproduction processes.

Other types of projectors are used for educational as well as commercial purposes. Overhead projectors, for black-and-white or color transparencies, and opaque projectors, for screening photographs, maps, drawings, and other graphic material, allow the operator to double as a lecturer and face the audience throughout the projection process. Extremely useful for meetings as well as for classroom work, the newer overhead and opaque projectors combine the best features of both models in a single unit. For transient audiences, continuous projectors that repeat lengths of film can be left unattended since they do not have to be rethreaded. This type of projector is widely used for demonstration purposes at exhibitions, conventions, and trade shows.

Usually taken for granted, the less familiar devices and the standard equipment used for projecting motion picture film are equally important to the producer. Without them, the industry has no way of exhibiting or displaying its collective product. See Figures P-17*a* to *c*, and Table P-3. Compare MULTIPLEXED TELECINE OPERATION.

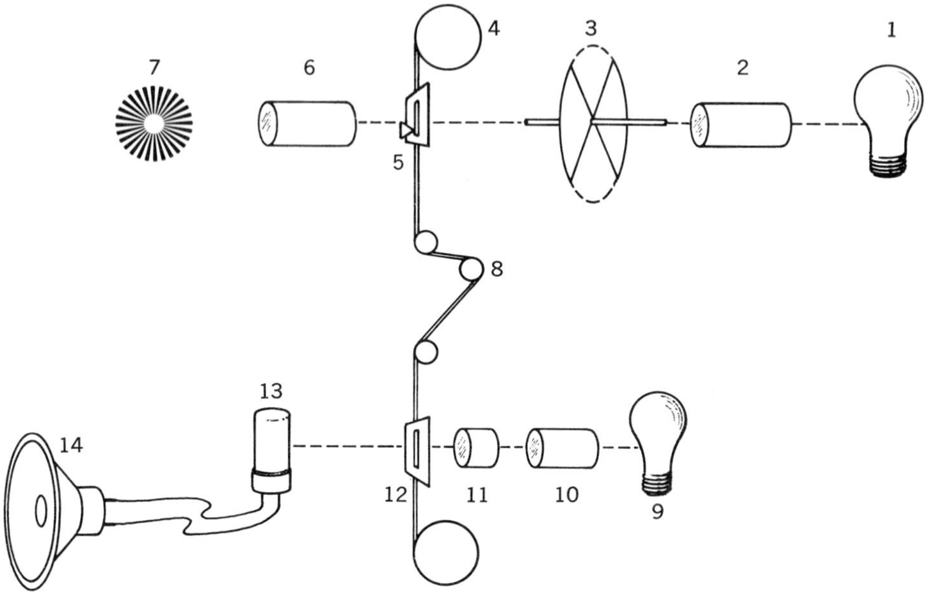

Fig. P-17*a* Projection. The same basic components that are needed to expose the film in the camera are also used to project the processed print onto a screen for exhibition purposes. The beam from a light source (1), directed through a condensing lens (2), is broken by the blanking action of the shutter (3)—*flickers*—as the individual frames of film (4) are advanced by a claw (5) into position in the projector gate. The lens (6) can be adjusted manually so that the image seen on the screen (7) is in sharp focus. Passing over rollers (8), the film continues to move to a second gate. Another light source (9) sends its beam through a second condensing lens (10) and, in turn, through an objective lens (11) focused on the film's sound-track area as it passes through the second gate (12). The electric output falling on the photoelectric cell (13), from the sound track, in the form of light waves is converted into sound waves and transmitted to the theater's speaker system (14) where it becomes audible sound.

TABLE P-3 Maximum Projectable Areas of Films

Film	Aspect ratio	Inches (width × height)	Millimeters (width × height)
8mm	—	0.173 × 0.130	4.39 × 3.30
Super 8mm	—	0.278 × 0.158	7.06 × 4.01
16mm	—	0.380 × 0.284	9.65 × 7.21
35mm			
Anamorphic (Squeeze ratio 2:1)	2.35:1	0.839 × 0.715	21.31 × 18.16
Nonanamorphic	1.33:1	0.825 × 0.602	20.96 × 15.29
	1.66:1	0.825 × 0.499	20.96 × 12.67
	1.75:1	0.825 × 0.474	20.96 × 12.04
	1.85:1	0.825 × 0.448	20.96 × 11.38
70 mm	—	1.913 × 0.870	48.59 × 22.10

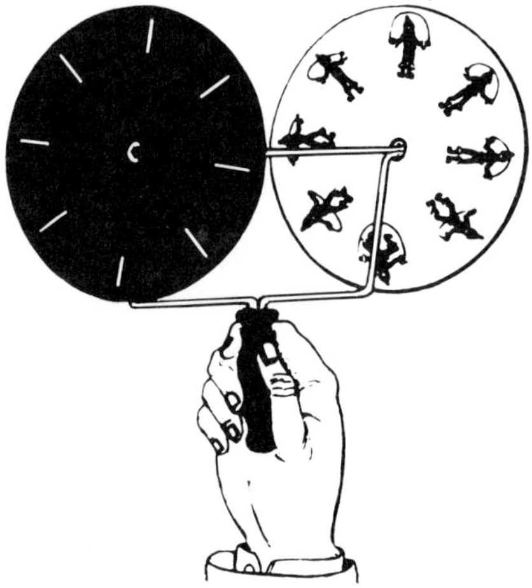

Fig. P-17*b* **Projection.** Two disks mounted on a common rotating shaft. The graphic material is mounted on one disk, and the viewer looks through slits in the second disk.

Fig. P-17*c* **The test pattern used for checking focus and picture steadiness. (*Motion Picture Research Council, Inc.*)**

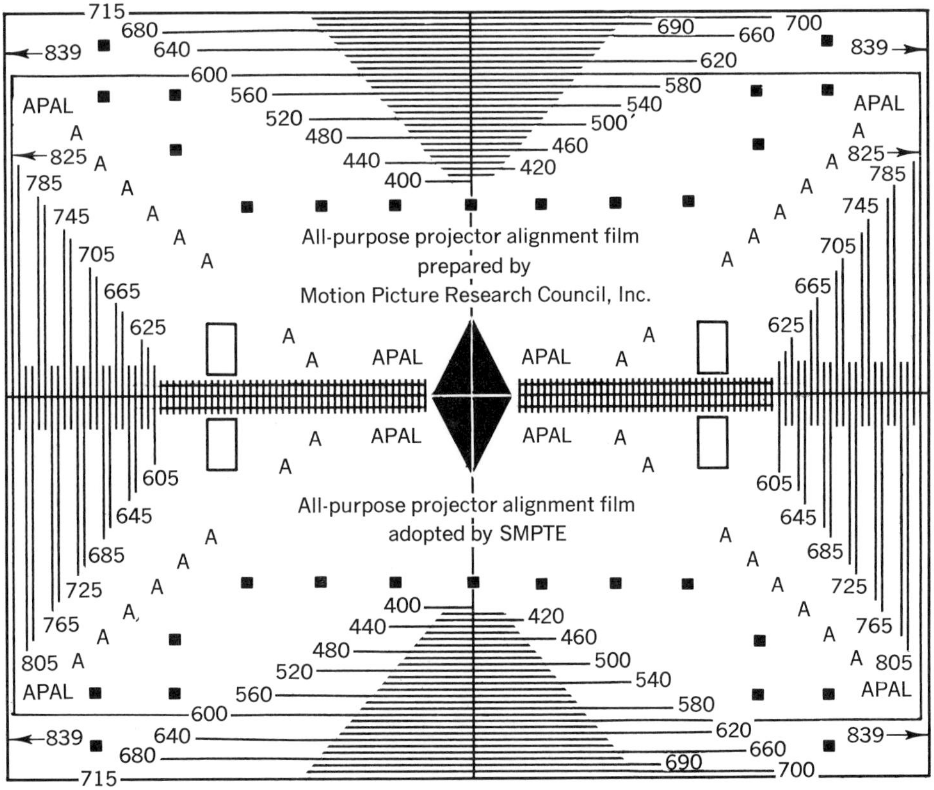

projection cue See PROJECTION and LEADERS.

projection formulas

$$\text{Picture size} = \frac{\text{length of throw} \times \text{width of aperture}}{\text{focal length of lens}}$$

$$\text{Throw} = \frac{\text{focal length of lens} \times \text{width of screen}}{\text{width of projector aperture}}$$

See PROJECTION LENS.

projection lens Projection lenses are used to focus images existing on motion picture film onto a screen that is used for viewing purposes. As such, they are designed to throw a well-defined image over the entire screen area.

Unlike camera lenses which transmit light that is reflected from the subject matter to expose the raw stock, projection lenses are corrected for color aberration and selected to conform with the type of light emitted by the projector's illuminating system. Since a projector lamp produces heat along with the emitted light, projection lenses are designed with a minimum of cemented surfaces. Except for those lenses that are used for projecting wide-screen pictures (anamorphic images), normal projection lenses are not intended for use over wide angular fields.

The formula for determining the required focal length of a projection lens in relation to the screen size, film size, and distance between the lens and the screen is expressed in this manner:

$$\text{Focal length of lens} = \frac{\text{distance between lens and projection screen}}{\text{magnification} + 1}$$

Magnification is obtained by dividing the width of the screen by the width of the film.

$$\text{Magnification} = \frac{\text{width of screen}}{\text{width of film}}$$

projection printing Also called **optical printing**. A process in which the film image that is to be duplicated is projected through an optical system for reexposure on the raw stock in the camera. By changing the relationship between the optical camera and the film being copied, it is possible to reposition a scene as well as to enlarge or reduce it.

projection screens

beaded screens. These screens are excellent for frame or roller mount where the audience is grouped within 25 to 30 degrees on each side of the projector. Beaded screens are unperforated.

Cordoscreens. Vinyl-coated fiber-glass screens with matte white surfaces. The screens are washable and rugged and are available in perforated or unperforated types. The material can be used for all roll screens.

Da-Lite projection screens. Highly reflective screens supplied in White Magic V-4 glass-beaded surfaces for sharp, bright pictures in color and black and white, or V-3 matte white for widest-angle viewing. The screen's surfaces are flame- and mildew-resistant.

Lenticulite screens. Matte white lenticular screens with heavy embossing for maximum diffusion. Available as perforated or unperforated, they are not intended for use as roll screens.

Mirro Claric screens. Seamless, white, tearproof, antistatic screens whose surfaces resist static charges normally created in theaters. Available in solid or perforated heavyweight plastic or metallic finishes, the screens have excellent brightness characteristics.

Superglo screens. Plain pearl-coated screens with a symmetrical distribution pattern. These are available perforated or unperforated and are not intended for use as roll screens.

Superoptics screens. Pearl-coated lenticular surfaces with head-on or tilt patterns. The vertical distribution of light is excellent for large balconied auditoriums. The screens are available either perforated or unperforated.

Vicralite screens. Aluminum-coated lenticulated screens, with good horizontal distribution. Suitable for all types of projection (including 3-D) with high ambient light conditions, they are not intended for use as roller screens.

projection-screen brightness formulas

$$\text{Footcandles} = \frac{\text{lumens}}{\text{size of screen area}}$$

$$\text{Footcandles} = \frac{\text{footlamberts}}{\text{reflectivity of surface}}$$

$$\text{Footlamberts} = \frac{\text{reflectivity of surface} \times \text{lumens}}{\text{size of screen area}}$$

$$\text{Footlamberts} = \text{footcandles} \times \text{reflectivity of surface}$$

$$\text{Lumens} = \text{footcandles} \times \text{size of screen area}$$

$$\text{Lumens} = \frac{\text{size of screen area} \times \text{footlamberts}}{\text{reflectivity of surface}}$$

projection shuttle The shuttle used for projecting previously processed film in the optical printer. If necessary, the shuttle can accommodate several thicknesses of film simultaneously. See OPTICAL PRINTER.

projection synchronism The spacing, in terms of frames, between any specific point in the sound track and the corresponding picture during projection. When the sound track is printed on the same length of film that contains the picture information, the track precedes the corresponding picture by 20 frames in the 35mm print and 26 frames in the 16mm print.

projector A device for screening motion pictures and their accompanying sound tracks. See PROJECTION.

projector lamps See Figures P-18*a* and *b* and Tables P-4 and P-5.

prop Also called **property**. A studio term used loosely to describe the miscellaneous items that are needed to decorate a set—lamps, a telephone, or some framed pictures, for example. See PROPERTY.

property A property or prop, as defined by the Propertymen's Guild, includes "anything that is not structurally a part of the set."

In the average film or television presentation, the prop man is concerned mainly with the selection and placement of furniture and decorations appropriate to the setting designed by the art director. Since most of the major studios store many varieties of furniture and fittings, ranging from contemporary through period and ancient, the needed props are taken from storage, dusted, and made available in a relatively short time. These assignments are routine and present few problems.

Assignments that call for the construction of special props, breakaway, trick, or mechanized, for example, are far from routine for the members of the property department. In many cases, the specialized demand requires imagination, individual treatment, and unprecedented departures from accepted methods.

TABLE P-4 Replacement of Old-style with New High-performance Sylvania Projector Lamps*

Old style		*High performance*		
Lamp	Wattage	Adapter +	HGTF† lamp	Wattage
CZX	500	350	DAY	500
DAB	500	350	DAY	500
DAG	500	390	DAY	500
DDB	750	350	CWA	750
DDW	750	350	CWA	750
DEJ	750	390	CWA	750
DFD	1,000	350	CTS	1,000
DFK	1,000	390	CTS	1,000
DFT	1,000	350	CTS	1,000
DFY	1,000	390	CTS	1,000

*See Figure P-18*b*.
†Hard glass Tru-Focus.

TABLE P-5 Sylvania Projector-lamp Interchangeability Guide

Lamp specified	Watts	Possible replacement	Watts
BEC	150	BEJ*	200
BEH	150	CAR†	150
BEJ	200	BEC	150
BFK	750	DGH	750
CAL	300	CWD CXP	300 300
CAR	150	BEH	150
CAS/CAV	50	CBS/CBC*	75
CAX	50	CBX/CBS*	75
CBJ/CBC	75	CAS/CAV	50
CBX/CBS	75	CAX	50
CCM/CHD	200	CGP/CFK	150
CEL	100	CBX/CBS	75
CEM	100	CBJ/CBC	75
CEW/CFC	150	CEM	100
CGP/CFK	150	CEL	100
CGW/CGT	200	CEW/CFC	150
CHK	150	CEW/CFC‡	150
CYK	400	DAK*‡	500
CYM/CYF	300	CYC‡	300
CYS	1,200	DBH‡ CTS	1,200 1,200
CZA/CZB	500	DAK CAL	500 300
CZX/DAB	500	CXK	300
DAG	500	DES*	750
DAH	500	DEK‡	500
DAK	500	DAY§	500
DAX	1,000	CTS§ DBM CWA	1,000 1,000 750
DAY	500	DAK CWD	500 300
DBH	1,200	CYS	1,200
DBJ	500	DFR	500
DBM	1,000	CTS‡	1,000
DCA	150	DEF	150
DCF	150	DCR DEF	150 150
DCH	150	DCL DJA	150 150
DCL	150	DCH DSA	150 150
DCR	150	DCF	150
DCS	500	DES*	750
DDB	750	DDW‡ DDY CZX	750 750 500
DDW	750	DDB§ DDY CZX	750 750 500

Lamp specified	Watts	Possible replacement	Watts
DDY	750	DGH§ DDB‡ DDW‡	750 750 750
DEC	750	DGS*	1,000
DEF	150	DCA DCF	150 150
DEJ	750	DAG	500
DEK	500	DAH	500
DEP	750	DGR§ CWA*	750 750
DES	750	DCS	500
DFA	150	DFG DCH	150 150
DFC	150	DJL	150
DFD	1,000	DFT§ DDW‡	1,000 750
DFF	150	CAR‡	150
DFG	150	DFA DSA	150 150
DFK	1,000	DFY§	1,000
DFR	500	DBS	500
DFT	1,000	DFD‡	1,000
DFW	500	DEK DAH	500 500
DFY	1,000	DEJ DFK‡	750 1,000
DGA	300	CRA	300
DGF	500	DFW† DGA	500 300
DGH	750	DDY	750
DGL	1,000	DGS‡	1,000
DGR	750	DEP	750
DGS	1,000	DGL§	1,000
DHB	1,000	DES	1,000
DHT	1,200	DFD DFT	1,000 1,000
DJA	150	DCH DCL	150 150
DJL	150	DFC DFG	150 150
DKM	250	DLR	250
DKR	150	DLS DEF	150 150
DLG	150	DCA DEF DKR	150 150 150
DLH	250	DJL	150
DLR	250	DKM	250
DLS	150	DEF DKR	150 150
DRC	1,000	DRS	1,000
DRS	1,000	DRC	1,000

Standard-voltage foreign lamp type					
6280C	6281C	6282C	6283C	7066W	6212N
Sylvania substitute					
CWD	DGA	DAK	DGF	CYC	CLS

*May be used if wattage does not exceed projector limitations.

†Proximity reflector may not operate in some projectors.

‡Replacement lamp will give much shorter life than lamp specified.

§Recommended replacement. This lamp should be used in place of specified lamp for longest life.

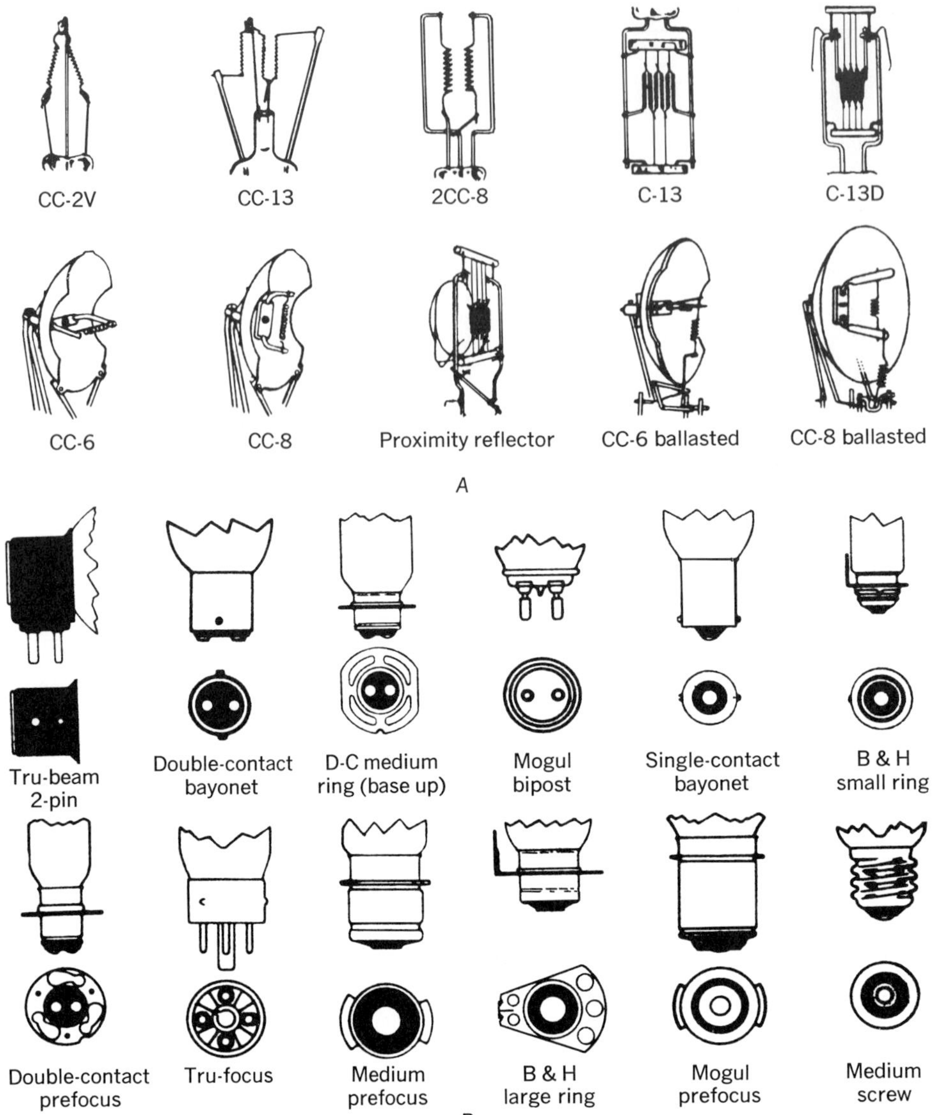

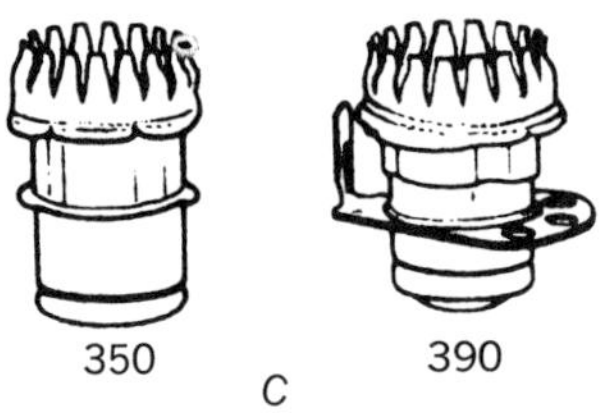

Fig. P-18a Sylvania projector lamps. (*A*) Filaments. (*B*) Bases. (*C*) Adapters: 350 Converts projectors from old-style medium-prefocus to high-performance hard-glass Tru-Focus lamps. 390 Converts Bell & Howell 16mm projectors to use high-performance proximity reflector hard-glass Tru-Focus lamps.

Refer to Table P-4 for replacing old-style lamps with new high-performance Sylvania projector lamps.

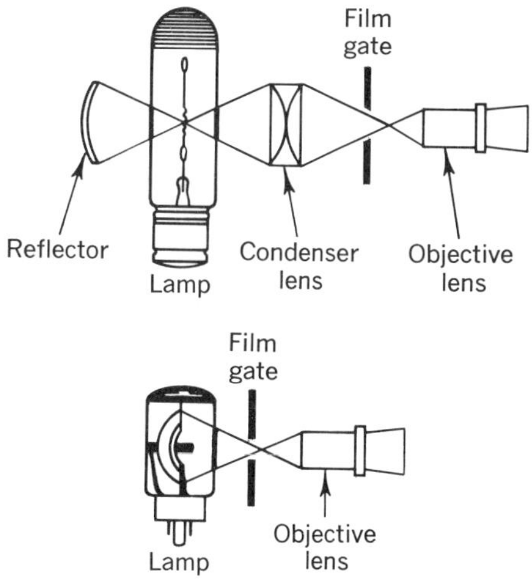

Fig. P-18*b* The Sylvania designed Tru-Flector projector lamp. Compact and powerful, and with a sealed-in silver condensing reflector, this lamp simplifies the optical systems of movie projectors by eliminating the need for external reflectors and condensing lenses.

In the standard projection system, a reflector behind the lamp and a condenser in front collect only a small portion of the light and direct it onto the film. An objective lens collects the light that passes through the film and directs it at the screen. The efficiency of this system is quite limited.

In the Tru-Flector or Tru-Beam system, an internal reflector acts as a condenser, collects almost a full hemisphere of light, and focuses it directly on the film. Instead of the light loss with standard optical systems, the objective lens collects almost the full light output of the lamp and directs the brighter image onto the screen.

The most frequently used weapons in cowboy thrillers and mystery chillers are bottles, glasses, and miscellaneous wooden items such as bats and bar stools. To preserve the sanity of directors and performers, the motion picture industry takes pains to devise methods for rendering such weapons harmless. Needless to say, the collective genius of the prop men is quite equal to the assignment.

Early glass weapons were made from a mixture of cooking sugar and cream of tartar. The two ingredients were blended and later poured into molds. When the liquid had jelled to the proper consistency, the prop had to be stored carefully. Exposed to the hot studio lights, the almost transparent glass representations sagged and stuck to an actor's hands. The mixture was a huge success since an injured actor could lick his wounds with relish. To add a touch of realism to the effect, the wound was sometimes covered with a bandage and contained one or two splashes of catsup. Over larger areas, however, more startling results were achieved with splotches of red oil-base paint.

Wooden breakaway props, made of balsa, were not quite as tasteful. Papier-mâché, quite useful for the construction of many types of props, is not employed too often because of the length of the drying process.

In recent years, several new types of plastic have been used for experimental purposes with excellent results. "Glass" breakaway props and other fabrications made from these new materials offer two distinct advantages over their prototypes. Not only are the new props more transparent, but also the cameras can remain focused on them for much longer periods—without having them wilt.

In spite of these advantages, however, new methods and materials are still being sought. The cutting remarks from fallen gladiators, after being struck by the make-believe bottles, are as sharp as the broken edges of the plastic props.

The production of mechanized props, or trick props with movable parts, requires special skills and individual treatment. The materials used for the construction of these props varies with each assignment. For budgetary reasons, each new prop, after use, is catalogued and stored—"in case." Quite often, these props are rebuilt and repainted in order to fill specific script requirements in subsequent productions.

Although a large number of these custom-made props play an important part in the production and appear "on camera," an equally large number are never seen on motion picture screens or television receivers. These invisible, off-stage props, or gimmicks, help the technician achieve visible on-screen illusory effects. The daily production schedule may call for the construction of a monster-type figure with movable parts, or for a covered wagon with jointed sections that will separate at a touch. At other times, the property department will be asked to construct a full-scale set showing the interior of a submarine (correct to the most minute detail), devise a mechanism that will cause a rifle to fall apart when the trigger is pulled, produce an invisible harness that will make it possible to suspend an actor in midair, or concoct a liquid that will change color on cue. To accomplish these "impossible" tasks, the technician may have to experiment with a variety of materials or make use of power sources ranging from simple battery-driven motors to giant rocket engines.

In rarer instances, these props are actually a means to an end. After the prop has been completed and photographed, animation and sometimes mattes are combined with the processed film in the optical printer to create an unusual composite effect.

For the property department technician, the familiar quotation credited to the Army Engineer Corps might serve as an appropriate slogan. "The difficult we do first. The impossible takes a little longer."

protection master positive A positive film that duplicates the picture information in the black-and-white or color-release prints. As the name implies, it can be used to make a new duplicating negative if and when the release negative is in any way damaged. If in color, the master positive could be either a color master positive or a set of three black-and-white separation master positives.

proxar lenses Supplementary lenses designed to shorten the focal length of a camera lens.

proximity pickup A device that senses the location of special aluminum foil that is applied to the film (at cut points) by the videotape editor. When this cueing device "senses" the foil, a signal is auto-

TABLE P-6 Time and Footage

16mm × 100 feet	= 4,000 pictures			
35mm × 100 feet	= 1,600 pictures			
70mm × 100 feet	= 400 pictures			
2 pictures per second	= 120 per minute	or	7,200 per hour	
1 picture per second	= 60 per minute	or	3,600 per hour	
1 picture every 2 seconds	= 30 per minute	or	1,800 per hour	
1 picture every 5 seconds	= 12 per minute	or	720 per hour	
1 picture every 30 seconds	= 2 per minute	or	120 per hour	
1 picture every minute	= 60 per hour	or	1,440 per day	
1 picture every 5 minutes	= 12 per hour	or	288 per day	
1 picture every 10 minutes	= 6 per hour	or	144 per day	

matically imparted onto the videotape during the mixing process. See FOIL.

pulse An electronic signal or system of modulation that is used to supply a means of reference for setting controls, or to activate a circuit. The term is also used to describe the module in vidicon film cameras that generates the black-level pulse, horizontal-drive stop pulse, and gating pulses. The module also generates test pulses for use while the camera is being set up.

pulse cameras Time-lapse photography is accomplished through the use of pulse cameras* in conjunction with intervalometers or timers. The intervalometer, or timer, is preset at the desired interval, and issues the power pulse which activates the camera for a single exposure each time the pulse is received. Some standard motion picture cameras are capable of taking time-lapse photographs through the use of a single-frame motor.

Table P-6 shows the frequency of exposure and the number of frames that will be obtained during a specific time period.

pup A relatively small, focused light source.

puppets Practically all the processes used in the production of motion pictures are also employed when the make-believe people referred to as puppets have the spotlight focused on them. After being assembled, the puppets are painted, positioned in front of three-dimensional backgrounds built to scale, and photographed with motion picture cameras driven by stop-motion motors. The actions of the puppets are subsequently synchronized with the music and

*16mm pulse cameras: Bell & Howell KD-7, Bell & Howell 200 P, Flight Research 111B, Siemens model B.

35mm pulse cameras: Aremac 1013, Automax G-1 and G-2, Beattie Varitron DR-2 and model E, DeHavilland MK-7, Flight Research IV C, Gordent W-B 35, Pacific Labs V-10.

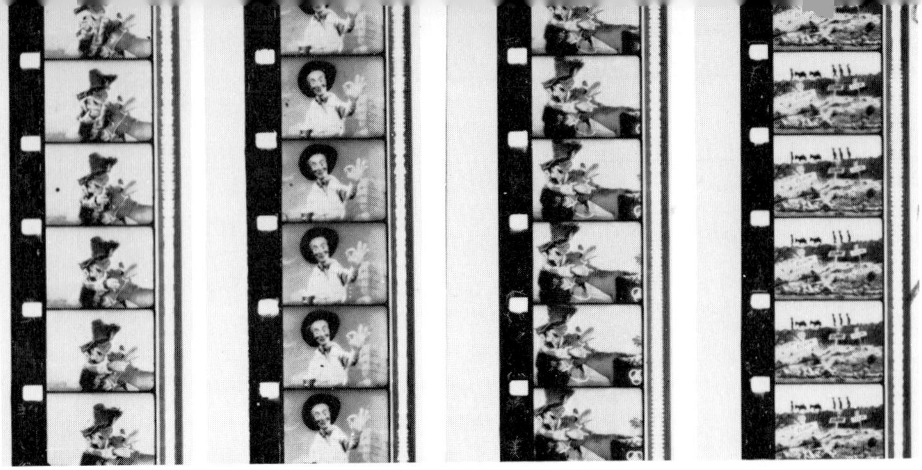

Fig. P-19 **Puppets—every move a picture!**

dialogue on the prerecorded sound track. Between exposures, technicians move arms, legs, mouths, and props to predetermined positions corresponding to the directions indicated on the exposure sheets. See Figure P-19.

pushing exposure Any of several methods used for achieving a higher exposure index than the normal recommendations listed by the ASA with respect to the emulsion speeds of a film. Prefogging of the film by light and acetic acid latensification are two of the methods used for pushing exposure. Extended developing, for color films, is accomplished by exposure increases of approximately one-half to one stop during developing. Beyond that point, there is an increase in graininess that could be undesirable. See LATENSIFICATION.

push-pull In electronics, a system or circuit in which two similar components or subsystems are arranged so that both operate simultaneously in opposite phase.

In amplifiers, for example, two tubes designed for push-pull operation are arranged so that one of the tubes is 180 degrees out of phase with the other. The arrangement, in this case, increases the power output and reduces distortion.

When push-pull operation is used in connection with optical sound tracks, a single track is divided into two parts—the modulations in each half being 180 degrees out of phase with the other. Push-pull tracks are not used for reproduction in conventional equipment since the sound would be quite distorted. They are, however, used for studio re-recording purposes, in conjunction with push-pull photocells, to reduce extraneous noise caused by splices or bloops.

pyrogallic acid Also known as **pyro**, the whitish powder is used as a developing agent. It oxidizes rapidly in alkaline solutions and dissolves readily in water.

Pyroxylin A substance used in the manufacture of Celluloid film bases. The white cottony mixture of cellulose nitrates dissolves in water or alcohol.

Q

quadlite An efficient floodlighting unit accommodating four 500-watt reflector-type lamps, each individually controlled by separate switches. Several of these units can be stacked on a pedestal and swiveled or tilted so that each quadlite throws light in a different direction.

Quadraplex recorder A videotape recorder with a revolving headwheel assembly which contains four record/playback heads over which the tape is transported. The 2-inch-wide tape used for recording travels at the rate of 15 inches per second over the headwheel which revolves 240 times each second. The equally spaced heads take turns recording. As the first head completes its sweep over the width of the tape, the second one begins its sweep. The first head begins a new cycle after the fourth head has completed its phase of the recording process. Two additional heads are used to record an audio track and a control track along the length of the tape.

This method of recording provides a higher degree of image resolution and color quality control than that obtained with single-recording heads in helical-scan equipment.

Transfers from tape to film or from film to tape are possible in the following combinations.

1. Quadraplex 2-inch videotape to Quadraplex 2-inch tape for duplication purposes, in either black and white or color.
2. Quadraplex 2-inch tape to helical-scan 1-inch tape, in black and white or color.

3. Quadraplex 2-inch tape to 35mm black and white, or to 16mm color or black and white. 35 and 16mm film may also be transferred to tape.

4. Helical-scan 1-inch tape to 16mm film, in black and white only, or 35mm or 16mm film may be transferred to tape.

quadrature error Distortions in a picture caused by misalignment of the video heads in Quadraplex recorders. See QUADRAPLEX RECORDING.

quantum efficiency The very minimum amount of exposure that is needed to produce an image that can be developed. The unit of radiant energy indicating such minimal amounts of exposure are expressed in quantums. The quantum efficiency for current emulsions is approximately 40.

quarter track The quarter-track tape serves a purpose similar to that of the dual-track tape in that monaural or stereophonic sound can be recorded on a small part of the tape's total width. With four tracks on $\frac{1}{4}$-inch rolls of magnetic tape, tracks 1 and 3 can be paired to record in one direction, and tracks 2 and 4 in the opposite direction. Actually a dual-track dual track.

quartz-iodine lamps Lighting sources that use iodine as the gaseous element. These lamps have a life rating of several thousand hours during which the rate of depreciation, with respect to light output, is relatively slow. See QUARTZ LAMPS.

quartz lamps Relatively new, these lamps are encased in quartz instead of glass. They are comparatively small in size and have a relatively long life-span. Operating at normal voltages, quartz lamps are light in weight, produce unusually brilliant illumination, and maintain constant color temperature. Their small size and light weight make them particularly useful for location work. Their output is rich in ultraviolet radiation.

When used outdoors, quartz as well as tungsten and incandescent lamps require the use of blue filters in order to convert 3200 or 3400°K illumination to the average daylight color temperature of 4000 to 8000°K. See Tables Q-1 to Q-4.

quartz lenses Lenses made of fused quartz instead of glass. Ultraviolet rays pass through this type of lens unobstructed.

quick cut A series of still shots or short scenes, spliced together in sequence, to create a composite impression. The term *still shots,* as used here, refers to graphic material that is photographed on motion picture film.

quinol A commonly used contrasty developer.

Qwart lights A light unit with provision for two different types of fixtures in a single housing. The unit features a soft light (tungsten-halogen lamp) as well as a focusing Fresnel spot (tungsten incandescent lamp). These units can be suspended over the acting area of the set and the lights used individually or in combination.

TABLE Q-1 Double-ended Quartzline Lamps with Single-contact Recessed Base

ASA lamp code	Watts	Volts	Life, hours	Approximate source size	Approximate lumens	Color temperature, °K	Bulb
				Lamps with 2⅝-inch Overall Length			
DWW	460	120	10	0.437 × 0.160	15,000	3400	T-4
FAL	420	120	75	0.437 × 0.160	11,000	3200	T-4
FFJ	600	120	75	0.600 × 0.170	17,000	3250	T-4
				Lamps with 3⅛-inch Overall Length			
DWY	650	120	25	0.625 × 0.160	20,000	3400	T-4*
DWZ	375	30	1,000	0.375 × 0.160	7,500	2900	T-4
DXV	800	230	16	0.875 × 0.160	22,000	3400	T-4*
DXX	800	230,240	75	1.00 × 0.160	20,500	3200	T-4*
DYX	800	230,240	75	Frosted DXX	20,500	3200	T-4*
FAB	650	120	16	0.625 × 0.170	20,000	3400	T-4*
FAD	650	120	100	0.625 × 0.170	16,500	3200	T-4
FBX	650	120	100	Frosted FAD	16,500	3200	T-4
FCA	650	120	25	Frosted DWY	20,000	3400	T-4*
FFM	420	120	75	0.437 × 0.160	11,000	3200	T-4
FGA	600	220,240	60	0.875 × 0.160	14,000	3200	T-4
				Lamps with 3¾-inch Overall Length			
BRH	1,000	120	75	0.750 × 0.220	30,000	3300	T-5
DWN	800	230	75	1.00 × 0.160	20,500	3200	T-4*
DXN	1,000	120	50	0.750 × 0.220	33,000	3400	T-5*
DXW	1,000	120	150	0.875 × 0.250	28,000	3200	T-5
FBY	1,000	120	150	Frosted DXW	28,000	3200	T-5
FBZ	1,000	120	50	Frosted DXN	33,000	3400	T-5*
FCB	600	120,130	75	0.650 × 0.170	17,000	3250	T-4
FEA	600	240	60	0.875 × 0.160	14,000	3200	T-4
FEB	600	220	60	0.875 × 0.160	14,000	3200	T-4
FGB	750	120	125	0.630 × 0.200	19,500	3200	T-5*
FGC	750	120	40	0.630 × 0.160	23,000	3400	T-5*
FGD	500	120	100	0.500 × 0.160	13,000	3200	T-4*
FGE	500	120	30	0.500 × 0.180	15,000	3400	T-4*
				Lamps with 4⅜-inch Overall Length			
DYA	1,000	120	150	0.875 × 0.250	28,000	3200	T-5
DYN	1,000	120	150	Frosted DYA	28,000	3200	T-5

Courtesy of GE Photo Lamp Department.
*Ultraviolet-absorbing bulb.

TABLE Q-2 Single-ended and Other Quartzline Lamps with Two-contact Base

ASA lamp code	Watts	Volts	Life, hours	Approximate source size	Coil	Approximate lumens	Color temperature, °K	Bulb	Base
				Single-ended Lamps with $2\frac{3}{8}$-inch Overall Length					
DVY	650	120	25	0.41 × 0.19	CC-6	20,000	3400	G-7*	Minimum 2-pin
DYH	600	120	75	0.50 × 0.22	CC-6	17,000	3200	G-7*	Minimum 2-pin
DYJ	650	230	20	0.44 × 0.38	CC-2P	20,000	3400	G-$7\frac{1}{2}$*	Minimum 2-pin
DYP	600	120	75	0.50 × 0.22	CC-6	17,000	3200	G-7*	2-button
				Other Two-pin Lamps					
EJM	150	21	40	Mounting distance $1\frac{1}{2}$ inches	CC-6	—	3200	Quartz integral 2-inch-diameter external dichroic reflector	2-pin
FCS	150	24	50	0.23 × 0.11	C-6 oval	—	3200	T-3	2-pin

Courtesy of GE Photo Lamp Department.
*Ultraviolet-absorbing bulb.

TABLE Q-3 PAR 36 Quartzline Uniflood Lamps*

ASA lamp code	Description, beam spread	Life, hours	Candle-power	Color temperature, °K
DXK	Medium beam, 30° × 40° spread	16	30,000	3400
FAY	Medium beam, daylight dichroic filter	16	30,000	5000
FBJ	Spot beam, 15° × 22° spread	16	60,000	3400
FCW	Wide beam, 70° spread	100	7,000	3200
FCX	Medium beam, 30° × 40° spread	100	20,000	3200

Courtesy of GE Photo Lamp Department.
*6,500 watts, ferrule-contact base, 120 volts, filament CC-6. Maximum overall length $2\frac{7}{16}$ inches.

TABLE Q-4 Quartzline Studio and Reprographic Lamps

GE code (ASA code)	Watts	Volts	GE code (ASA code)	Watts	Volts
Q300T3/2CL	300	120	Q1000T3/4	1,000	120
Q425T3/2CL	425	120	Q1000T3/3CL	1,000	185
Q500T3/4CL	500	120	Q1000T12/4CL	1,000	120
Q500T3/CL(FCL)	500	120	Q1000T14/4CL	1,000	120
Q750T3/4CL	750	120	Q1500T4/4CL	1,500	120
Q750T3/4	750	120	Q1500T$2\frac{1}{2}$/CL	1,500	230
Q750T20/4CL	750	120	Q1000PAR64/1	1,000	120
Q800T3/4CL	800	230	Q1000PAR64/2	1,000	120
Q800T3/4	800	230	Q1000PAR64/5	1,000	120
Q1000T3/4CL	1,000	100	Q1000PAR64/6	1,000	120
Q1000T3/4CL(FCM)	1,000	120			

Courtesy of GE Large Lamp Department.

R

rack (1) The process of threading film in a viewing or projection device. (2) The process of aligning or adjusting the framing of the picture in the moviola or projector gate. (3) A framelike device that is used for winding film during processing. (4) The movement of a lens in relation to the film plane during focusing.

radiation The emission, absorption, and transmission of radiant energy, especially those rays generated by radioactive substances. Also, any of the forms of electromagnetic energy: Hertzian (radio) radiation, infrared, ultraviolet, gamma rays, and x-rays.

radiography A photographic technique in which x-rays are used as the "light" source. X-rays cannot be reflected or refracted. They are capable of penetrating almost any substance, including metal, but can be stopped by enough thickness of any material.

In practice, an opaque body is positioned between a sensitized plate and the source of the x-rays that are used to form the image. In addition to its use in medicine, radiography makes possible the examination of the internal structure of many types of materials used for industrial purposes.

radio microflash A high-speed, single-flash light source* with a duration of 2 microseconds designed by Dr. Harold Edgerton. The

*General type 1530-A.

microflash differs from a true stroboscope in that it cannot be used for continuous flashing, to permit visual observation. Duration between flashes is 10 seconds. It is used for blurless photography of engineering or industrial objects moving at high speeds, such as projectiles, the disintegration of high-speed rotors, and the propagation of fractures in various materials.

The microflash may be used with any conventional camera. The flash can be tripped manually, by a contactor or by use of a sound- or pressure-sensitive microphone pickup. The power supply and trigger circuits are assembled in one case. The lamp is housed separately, but the cases lock together.

rails Pipes or scaffolding over a set to which lamps are secured.

rain From the gentlest of drizzles to the most torrential downpour, rain falls, looks wet, and causes varying degrees of discomfort.

Unlike snow effects, in which the artificial, plastic material looks as real as the actual flakes, studio technicians have found no adequate substitute for rain. With water readily available and as easily controlled, it is pointless to search for other "wet" materials.

On the studio set or on location, the falling "rain" is easily controlled by directing the flow of water through well-aimed hoses or perforated metal pipes installed over the heads of the performers. Good drainage facilities are a necessity.

Glycerin or varnish applied to floors and walls produces realistic reflections and supplies the necessary sense of wetness to the scene. If stock shots of rain effects are to be superimposed optically over the scene, presoaked clothing is worn by the performers to lend credibility to the composite effect.

range finder An instrument for measuring the distance to a relatively remote point. When viewed from two different positions, the lines of sight between the viewing positions and the subject matter converge. The converging angle increases or decreases in direct proportion to the distance between the viewing positions and the subject. As the distance between them increases, the angle decreases. Conversely, as the distance decreases, the angle increases. In either case, the distance is determined by the converging angle which is converted into feet and inches.

Ranger sync signal A method of recording a synchronizing pulse, during the filming process, at the same time the audio signal is recorded.

In the Ranger system, the synchronizing head is located after the recording head. It makes a recording 25 to 30 mils wide down the center of the tape, which is also down the center of the audio signal.

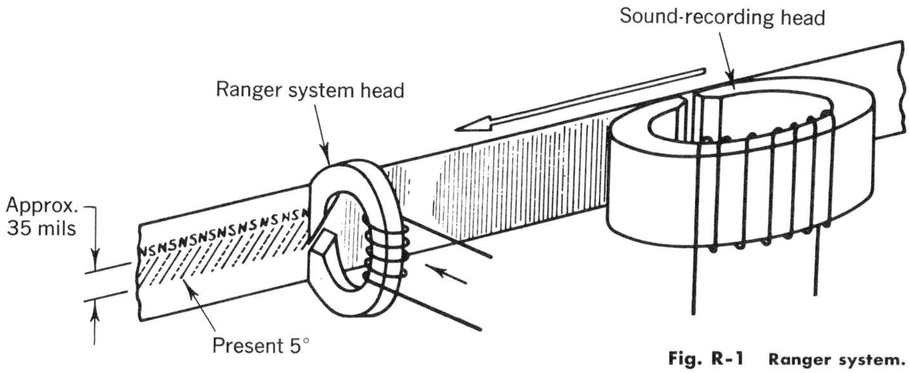

Fig. R-1 Ranger system.

The narrow slit of the Ranger head is about $\frac{1}{4}$ inch long and is placed almost at right angles to the audio signal. The recording of the synchronizing signal does not conflict with the audio signal.

Instead of being at a full 90 degrees to the audio, the Ranger head is rotated 5 degrees to provide a wider signal of sync recording and thus minimize trouble due to tape weave. The audio signal is, in effect, erased in the area used by the synchronizing signals. Thus, a 200-mil studio sound track is reduced to about 170 mils (this does not affect the signal-to-noise ratio enough to be of consequence). See Figure R-1. Compare ECHELON SYNC SYSTEM, NEO-PILOTTONE SYNC SYSTEM, and PILOTE SYNC SYSTEM.

raw stock Unexposed film.

reader, sound A sound-reproducing device containing an amplifier and a small loudspeaker. The magnetic recording (or optical sound track) is positioned over a drum where the recorded track makes contact with the reproducing head. The reader is used for editing and analyzing sound tracks. See Figure R-2.

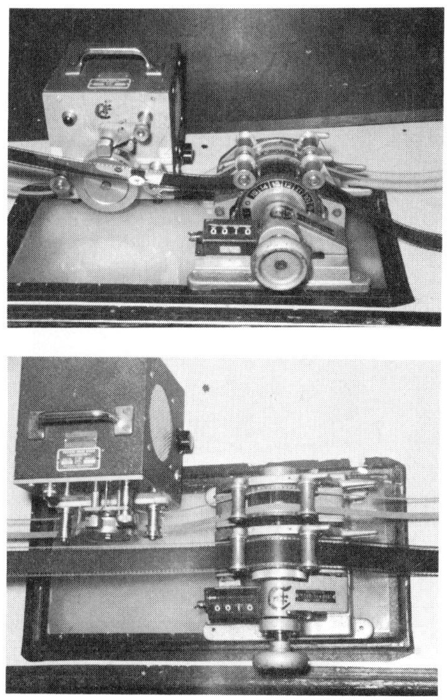

Fig. R-2 The sound reader and synchronizer. (*Above*) Front view. (*Below*) Top view.

readings, animation The frame counts made available to the animator by the film editor. Taken from the analyzed sound track, the frame counts are used by the animator to plot the length of an action or as a guide for the animation of mouth movements corresponding with prerecorded dialogue. The animation, after being filmed, is subsequently synchronized with the same previously recorded and edited sound track from which the frame counts were obtained.

readout Basically, the visual display of information. The term is generally used in connection with the picture that is produced in television camera pickup tubes, or the picture that is reproduced on a monitor.

rear projection screens Double-sided screens used for viewing only. (They are not used for rear-screen projection filming.)

The matte surface is forward when the screen has front-surface light interference, and the gloss surface is used when lighting interference is no problem, giving the picture better contrast and quality. The size of the screen varies depending on the size of the area needed for projection. Designed for use with all types of projection equipment, these screens present no hot-spot problem and provide unusually brilliant images that are viewable in sunlight.

rear-screen projection A background plate or a print of a previously photographed scene projected on a translucent screen between the projector and foreground stage, where it serves as a background for the action being filmed. Such backgrounds may be either stationary or traveling.

Stationary background plates contain subject matter that remains in a fixed position throughout the filming of a foreground action. Traveling background plates, by comparison, are used to create the illusion of movement. A stationary automobile, in the foreground action, would appear to be traveling at great speeds in relation to the movement in the projected background. Many similar effects are made possible with this photographic technique.

Fig. R-3a **Rear-screen projection.**

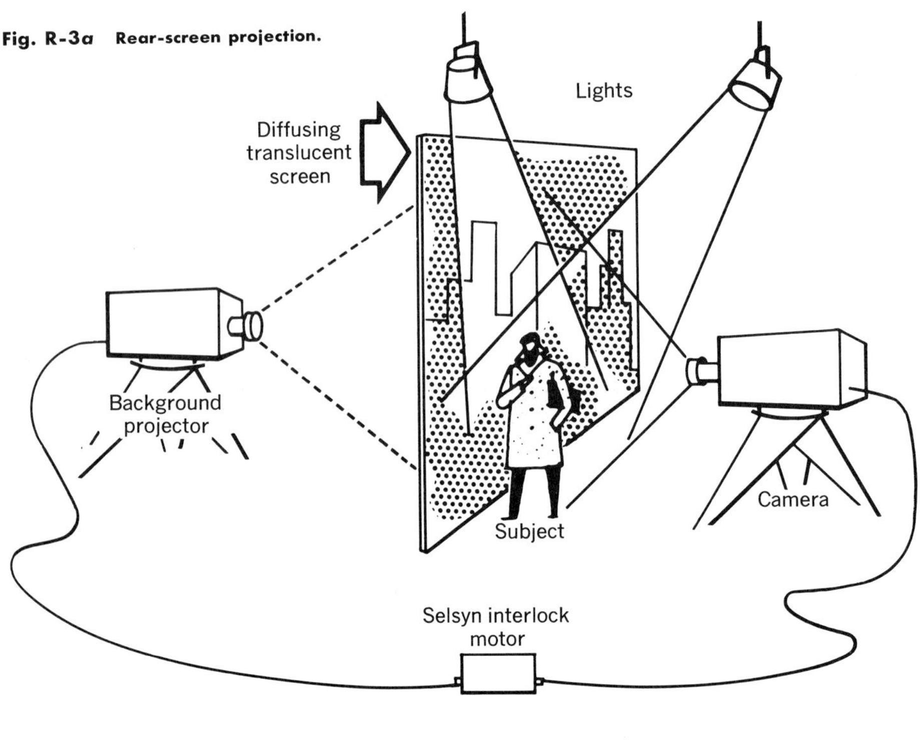

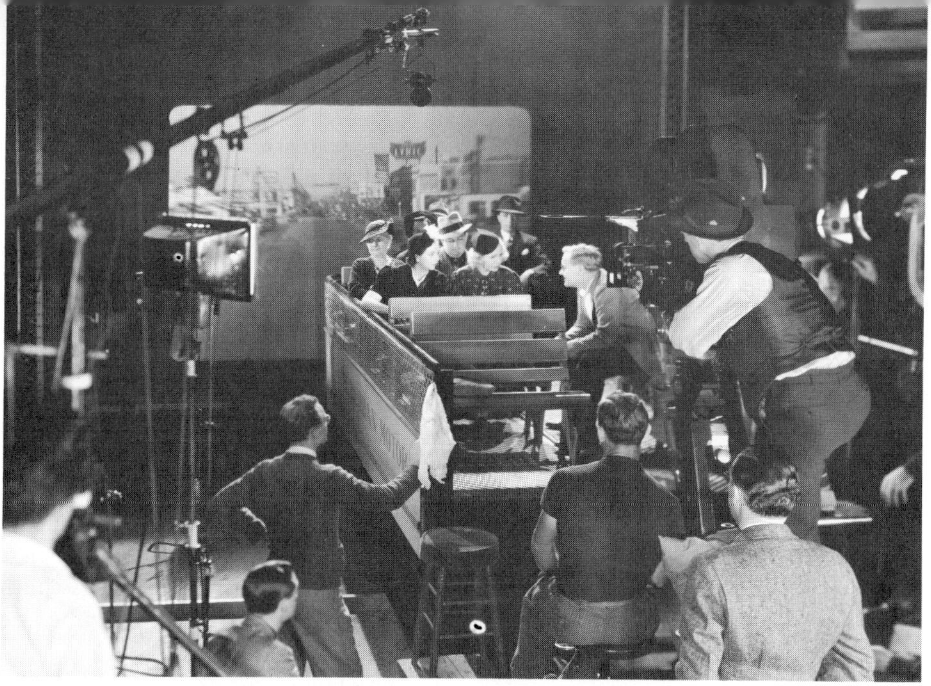

Fig. R-3*b* To the movie audience, the actors will appear to be riding atop an open bus on New York's Fifth Avenue. The ride, including bumps, is shown being filmed on a studio set with rear-screen projection processes.

Overexposed backgrounds, with good blacks and pronounced highlights, are preferable for this type of process shot. The relative speeds of traveling backgrounds, as well as their angle, perspective, and horizon position, are all equally important and add to the realism of the composite scene.

The foreground action is composed and staged to create a favorable relationship between both planes of action. Because this relationship can be immediately established, the director and cameraman can easily make changes in their basic setups for the purpose of compensating for differences in size and perspective between foreground and background.

Studio illumination used for foreground action should not be allowed to fall on any portion of the screen used for projection. The lighting should also be arranged so that the shadows, in the foreground action, will coincide as closely as possible, with the shadow pattern in the projected background. Obviously, projector and camera must be in perfect synchronization.

Optical processes for achieving composite effects are unnecessary with rear-screen filming techniques. The results of this one-step in-camera photography can be viewed in its final state immediately after the film has been processed. When time and tight deadlines are the important considerations, rear-screen photography has a distinct advantage over traveling-matte processes that require optical combination.

The use of matteing processes will sometimes give better results and prove more flexible. In other instances, where a foreground action is directly related to the action of the background, the rear-screen projection process is preferable. See Figures R-3*a* and *b*.

rear-screen projection screens Seamless, translucent screens that are either bound and grommeted and laced to frames or in other ways supported during the filming process. Available in a variety of sizes, running footage or transparencies containing appropriate backgrounds are projected onto their translucent surfaces during rear-screen projection shots. The actors, positioned in front of these backgrounds, create the illusion that they are being photographed on location. See REAR-SCREEN PROJECTION.

reciprocity law and effect The reciprocity law states that proportionate increases in exposure time will compensate for insufficient illumination.

When filming conditions dictate the need for longer exposure times than those normally used for a particular film type, or when exposure meters do not provide readings that compensate for the conditions that make longer exposure times necessary, the reciprocity effect becomes a factor.

The law of reciprocity, with respect to exposure, does not apply when lighting levels are too low. In actual practice, much greater increases in exposure are necessary. The required exposure compensations vary with the characteristics of the film being used. Light-balancing or color-compensating filters that change the color quality of the light source or the overall balance of a scene are used to make necessary adjustments in exposure. See Figure R-4.

Sensitivity equals the reciprocal of exposure required to produce some unique density value above gross fog when processed as recommended by the film manufacturer.

If the correct exposure for a particular photographic material under certain conditions is known, the photographer can estimate

Fig. R-4 Reciprocity characteristics. (*Eastman Kodak.*) Curves are for average product and for a density of 0.3 and 1.0 above gross fog when developed as noted.

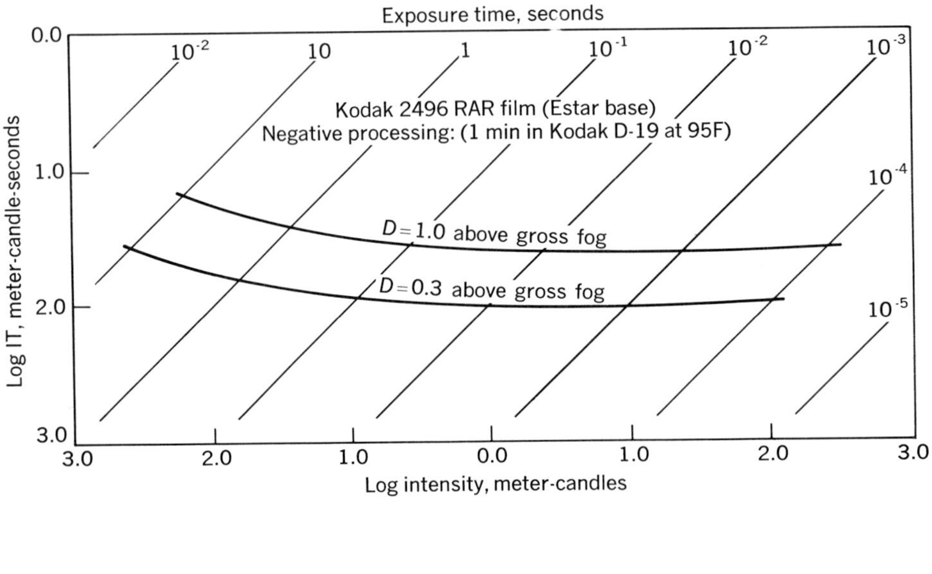

the required exposure for other materials under the same conditions. Because higher-speed photographic materials are generally associated with higher-speed numbers, speed is usually expressed as the reciprocal of some indicator of the quantity of energy required to make a correct exposure.

Every sensitometric speed system is based on a speed point—that particular point on the characteristic curve that corresponds to the exposure required to produce a specific photographic response. A characteristic curve is a plot of the logarithm of the exposure versus the optical density of the image. Exposure is usually expressed in meter-candle-seconds, and density is equal to the logarithm of the opacity, or the logarithm of the reciprocal of the transmission of the image.

Some photorecording situations require only minimum densities for detection. For such speed systems, a low-density speed point above gross fog is chosen.

When the subject consists of many (or continuous) tones, the speed is chosen to give the best correlation between sensitometric speed and effective camera speed and is not always a matter of achieving some unique density value above gross fog. The "above gross fog" is important since the photographer is interested only in density resulting from exposure to the subject, not from chemical fog and base density.

recording head The device in the tape-transport mechanism that makes it possible to convert the electric output of the microphone into magnetic patterns on coated tape.

rectifier An electric device that conducts current in one direction only, thus converting alternating current into pulsating direct current. Associated circuit elements normally smooth out the output.

reducer The activating agent in developing solutions. In the presence of light, it breaks down halide compounds such as silver bromide. Cutting reducers add contrast while reducing the intensity of a negative. Flattening reducers decrease both the contrast and density of an image. Proportional reducers decrease the density of a negative but have little or no effect on the contrast. Subtractive reducers decrease density without affecting contrast. Superproportional reducers are used to correct underexposed and overdeveloped negatives. These reducers have little effect on density but increase the degree of contrast.

reduction That part of the developing process in which silver bromide is changed to metallic silver. Also, the process of decreasing the density of photographic images.

reduction print The optical or laboratory process in which the picture information on 35mm film is duplicated on 16mm film. Unlike blowups (16 to 35mm) in which the grain is magnified proportionately, the reduction does not affect the quality of the narrower-gauge print.

reel A metal or plastic spool on which film is wound.

reference blacks A method used for controlling the "low" contrast during television transmission. Unlike reference whites—the use of the brightest portion of the scene to determine the standard for the control settings—the darkest or blackest areas in the scene become the reference points for the picture that is to be transmitted. Compare REFERENCE WHITES.

reference generator The reference generator in videotape recorders processes local sync to produce horizontal-rate reference, field-rate reference, and frame-rate reference.

reference whites A television camera control setting calibrated so that the brightest portion of a scene, referred to as reference white, is transmitted at 100 percent signal voltage. This setting produces a picture with maximum transmission contrast. Compare REFERENCE BLACKS.

reflectance A measure of how effectively a surface will reflect light. The degree of reflectance is determined by comparisons with the brightness of a magnesium oxide deposit on a silver surface (the standard of such measurements). Basically, the ratio of the reflected lumens to the incident lumens on the surface.

reflected-light meter Reflected-light meters are used in a number of ways to determine the lens setting needed for proper exposure. These meters are, for example, used to read both the lightest and darkest portions of a scene. The average of the two readings is then evaluated to determine the proper *f*-stop setting.

As an alternative procedure, exposure can be determined by using the meter in conjunction with a Kodak neutral gray test card. With this method, the card (18 percent reflectance value on the gray side and 90 percent reflectance value on the white side) is positioned in front of the subject matter when the reading is made. With the third method, the meter is aimed at the overall scene to produce a reading that indicates the general brightness. More sophisticated reflected-light meters measure the foreground subject matter as well as the background and provide readings that take both into account.

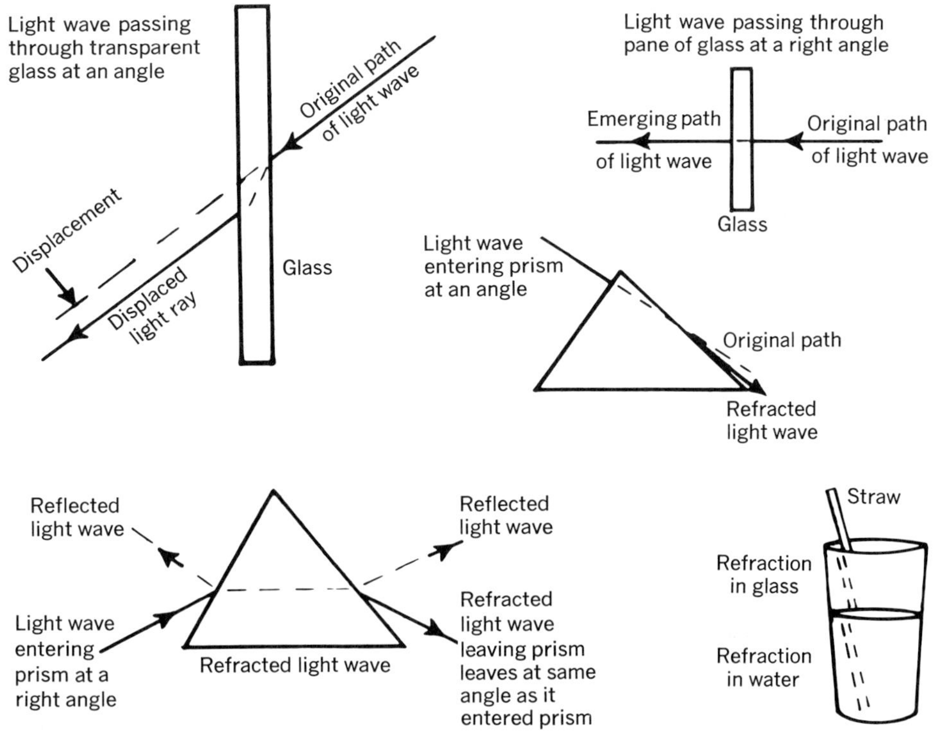

Fig. R-5 Reflection and refraction.

reflection Indirect light that is visible from a secondary surface. The angle at which a ray of light strikes a reflective surface is known as the *angle of incidence.* The angle at which the light ray is reflected from that surface is the angle of reflection. The angle of incidence is equal to the angle of reflection. A familiar example of this phenomenon is the image that is seen by reflection in a mirror. The reflected image is visible in the same plane and at a similar angle as far behind the surface of the mirror as the subject is in front of it. See Figure R-5 and REFLECTION EFFECTS.

reflection, sound The passage of sound waves through the air and their subsequent blockage by another medium. As with light waves, a portion of the sound is absorbed by the medium—the remainder of the sound being reflected from it. Reflected sound adds a degree of reverberation which, if excessive, distorts the sound considerably.

reflection effects Reflected light can be exploited to achieve a variety of interesting visual effects.

Distortion, for example, with an aquatic background can be produced by aiming the camera at any watery surface in which the subject matter can be reflected. The ever-present ripples or waves that disturb the tranquillity of the reflective surface create a fluid effect and varying degrees of distortion in the exposed subject matter. When the processed film is projected, the reflected images will be upside down, but the images can easily be righted optically. As a transitional effect, the camera can be panned up from the reflected image, on the surface of the water, to the actual subject matter. Similarly, zooms from a close-up of the reflection to a long shot showing both the reflected and real images in their relative positions in the film frame make interesting and effective establishing shots.

Similar distortion effects can be produced, on a smaller scale, without leaving the studio stage. To the shallow tanks that are substituted for the natural aqueous locations, ink or dye is added to conceal the sides and bottom of the container. The reflected images on the surface of the water can be changed from clear to rippled by the slightest disturbance. In time, the wavelets and accompanying distortion both subside, leaving the image relatively clear. The scene may be reversed optically to achieve the opposite effect. Fans or wind machines, running at low speeds, will keep the surface of the water in a constant state of turbulence during the entire filming process. The light sources can be used to create highlights on the edges of the ripples, or positioned to produce a variety of interesting effects.

The proximity of water can also be suggested by reflection without actually showing a single drop of it. These reflected ripple or wave effects are achieved with direct, in-camera photography by focusing a light source on a glass tray filled with water. Reflected to any part of the background by a mirror positioned at a 45-degree angle beneath the tray, the effect is quite convincing. Similar effects can be produced by placing pieces of mirror at the bottom of a shallow pan or tank. The light source, striking the water and reflective surface of the mirrors at a 45-degree angle, redirects the reflections to any part of the studio set.

The cameraman's ingenuity, the viewer's imagination, and several short lengths of linen can be combined to produce realistic reflections without water. These background reflections are created by tacking several lengths of the material to a wood slat that is positioned in front of the light source. The slightest movement of the suspended linen ribbons during the filming process will reflect flickering light and dark patterns on the studio background. (Oddly enough, the same lengths of linen can also be used to produce amazingly realistic flame reflections.)

Reflections are equally useful for titling work that lends itself to this type of treatment. The lettering can be rendered on cards to produce a variety of interesting effects. Since the title is reflected as a mirror image, the lettering should be upside down and reversed with respect to right and left.

Each title, after the required footage has been exposed, can be distorted beyond recognition by agitating the surface of the water in which the title card has been submerged. Replaced by a second title, the filming continues until the waves subside and the new lettering becomes legible. For these effects, the camera should be tilted at a 45-degree angle to the surface of the water. The title cards should be positioned at approximately the same angle. Exposure-meter readings can only be used as a guide. In practice, the lens should be opened to almost three full stops over the indicated meter reading.

reflector In lighting, a reflective surface (usually metal) that is capable of blocking light rays traveling away from a source and redirecting them to a specific area. Flat reflectors spread the light over relatively large areas. Curved reflectors concentrate the rays and reflect them to comparatively small areas.

In television, elements added to a TV antenna system. They are used to block signals coming from one direction while boosting the signal coming from the opposite direction.

reflex focusing The process of focusing in which the image is projected on a small screen in the camera by a mirror positioned behind the lens. The mirror swings clear of the lens during the interval of exposure.

In cameras having twin-reflex focusing, the viewing and camera systems work independently through coupled lenses—one for viewing, the other for focusing.

reflex shutter, mirror The mirror reflex shutter* rotates between the lens and film plane at a 45-degree angle to the axis of the lens. In its closed position, it reflects the image to the camera's viewfinder; in its open position, the image passes directly onto the film.

It is provided with a fixed shutter opening equivalent to 180 degrees. In each phase of its operating cycle, 100 percent of the transmitted light reaches the film and the eye intermittently. The bright ground-glass image permits focusing even at small *f*-stops. Through-the-lens focusing and viewing eliminate parallax problems and the need for multiple finders, masks, and rack-over devices. Irrespective of focal length, viewing through the taking lens permits accurate framing and critical follow focus. See Figures R-6*a* and *b*.

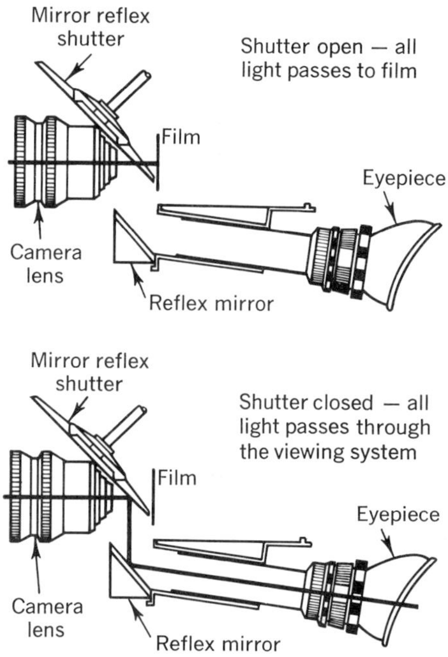

Fig. R-6*a* **Schematic explanation of the mirror reflex shutter. (*Arriflex Corp. of America.*)**

*Arriflex Corp. of America.

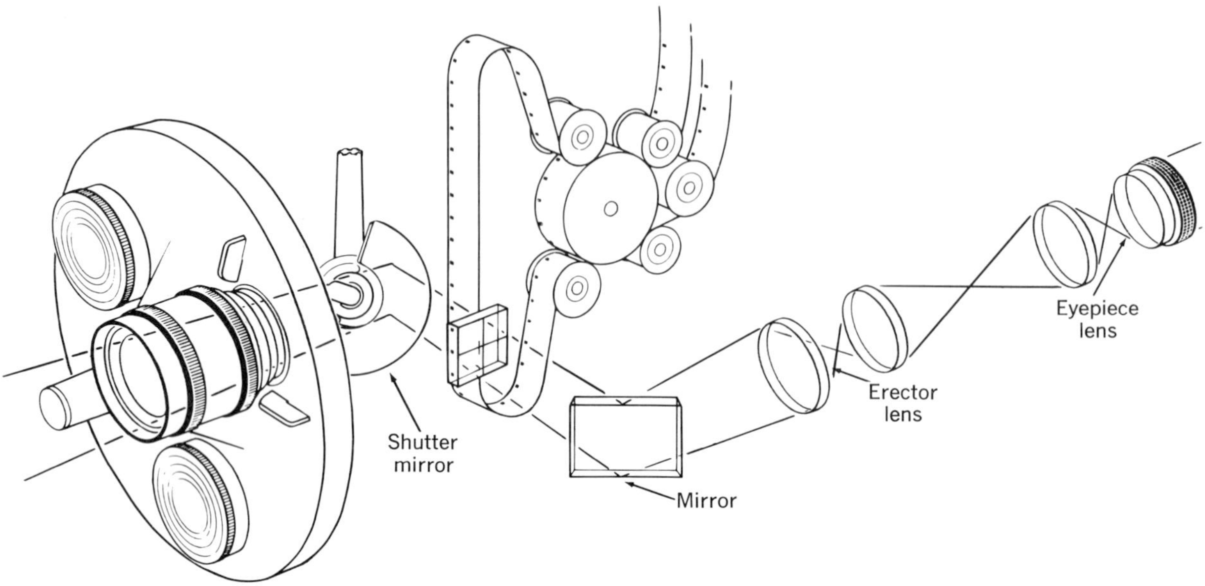

Fig. R-6*b* The optical system in the Kodak reflex special camera.

refraction The change in the direction of a light ray passing obliquely through mediums of different densities. The changes are caused by the differing densities of the two mediums and the resultant increases or decreases in the velocity of the light rays. See Figure R-5 and REFRACTIVE INDEX.

refractive index The index of refraction is the degree of change in the direction of light rays passing from one transparent medium to another. The angle of refraction varies according to the density of the medium through which the light ray passes.

The refractive index is the ratio between the sine of the angle of incidence and the sine of the angle of refraction.

regenerator An auxiliary module in vidicon film cameras that regenerates sync, blanking, and clamp pulses. The module contains the circuitry for horizontal-drive advance to compensate for camera cable delays.

register In animation, the positioning of a drawing, cel, background, or other graphic material, on the pegs of the animation drawing board or on the pegs in the camera compound table, so that it will align correctly with other graphic material.

register pins Also called **pilot pins**. The camera mechanism that engages the sprocket holes and precisely positions each frame of film in the camera gate during the interval of exposure.

relative aperture The relationship between the aperture of the diaphragm and the focal length of a lens. The measure of that relationship is used to determine the effective aperture for other lenses—regardless of their size or focal length. In practice, different lenses set at relative apertures produce images that are similar in brightness and quality. See EFFECTIVE APERTURE.

release An industry term that is used to describe a completed motion picture ready for distribution.

release negative The final negative picture and sound track from which the release prints are produced.

release print The final composite print, containing both the picture and sound track, that is distributed to theaters for exhibition.

remote broadcast A broadcast produced away from the studio.

repeat In animation, a term indicating that the same drawings or cels are to be reexposed. A repeat cycle in which a cartoon character is shown walking or running, for example, may be used over and over again until the needed film footage has been exposed.

replenishment The replacement of exhausted chemicals in processing solutions with fresher and more active chemicals. See EXHAUSTION.

reproduction (1) The duplication on the processed print of the tonal values and colors as they appear in the subject matter. (2) The process wherein previously recorded sound signals, on tape or film, are transformed into audible sound.

re-recording Also called **mix**. The process of combining the various sound elements (such as dialogue, musical background, and sound effects) into a composite sound track that will subsequently appear on the film in synchronization with the corresponding picture.

residual A secondary fee paid to a television commercial performer for repeat showings. See Figure R-7.

resolution Also called **resolving power**. The ability of a lens to form images with fine detail. Also, the sharpness of the image that is reproduced in the processed print.

In a television picture, horizontal resolution is the amount of resolvable detail in the horizontal direction. It is usually expressed as the number of distinct vertical lines, alternately black and white, that can be seen in three-quarters of the width of the picture. This

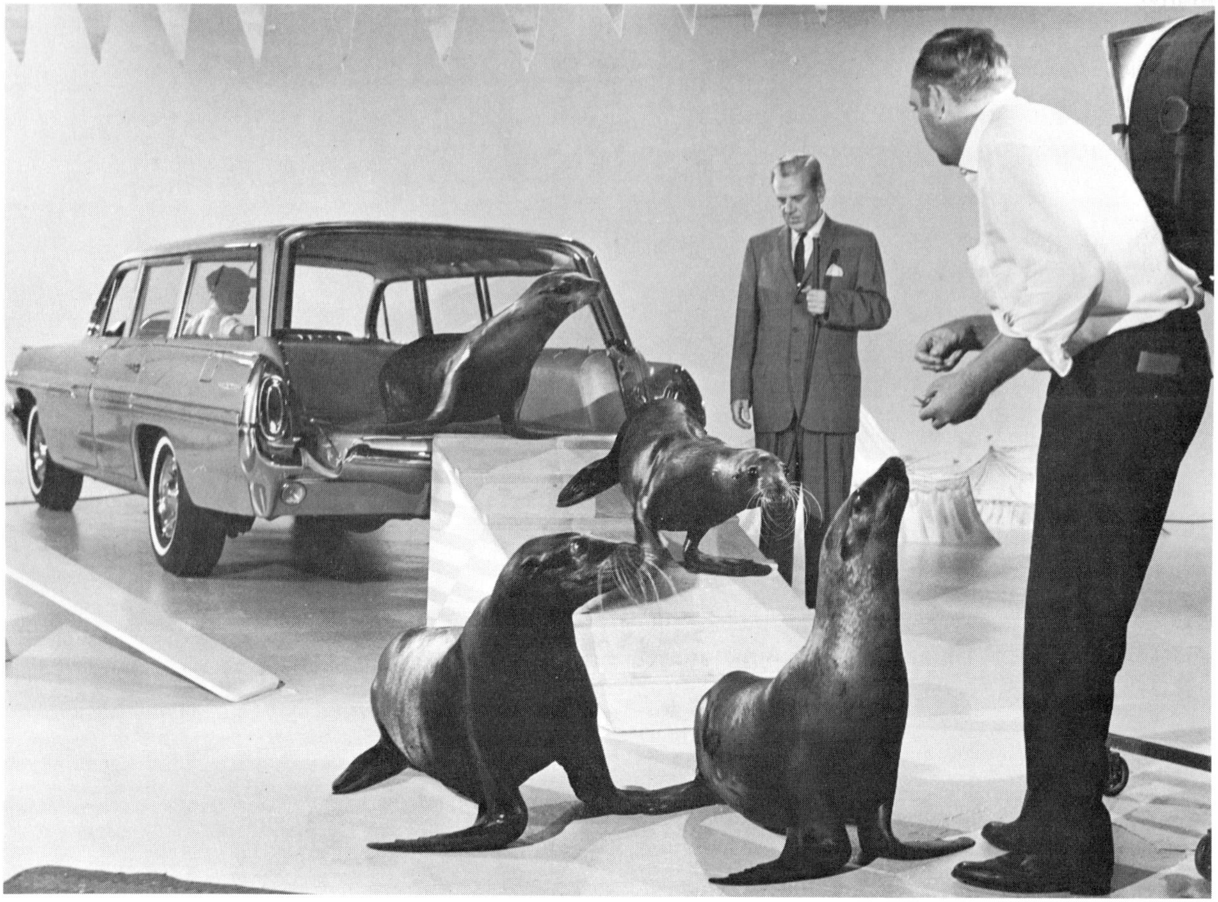

Fig. R-7 No residuals here. The performers in the Delco commercial are paid off in fish and chips.

information usually is derived by observation of the vertical wedge of a test pattern. A picture that is sharp and clear and shows small details has a good, or high, resolution. If the picture is soft and blurred and small details are indistinct, it has a poor, or low, resolution. Horizontal resolution depends on the high-frequency amplitude and phase response of the pickup equipment, the transmission medium, and the picture monitor, as well as the size of the scanning spots.

Vertical resolution is the amount of resolvable detail in the vertical direction. It is usually expressed as the number of distinct horizontal lines, alternately black and white, that can be seen in a test pattern. Vertical resolution is primarily fixed by the number of horizontal scanning lines per frame. Beyond this, vertical resolution depends on the size and shape of the scanning spots of the pickup equipment and picture monitor. It does not depend on the high-frequency response or bandwidth of the transmission medium or picture monitor. See Figure R-8.

resolver amplifier The resolver amplifier produced by the Ryder Company provides for automatic synchronization during the transfer of $\frac{1}{4}$-inch magnetic tape to sprocket-driven film. The sync signal, during the transfer process, is reproduced by the sync head and directed to the amplifier. The amplified signal is used as the direct drive of the synchronous motor on the sprocket-driven magnetic recorder, thus assuring synchronization between the camera film and the sprocketed magnetic film. See Figure R-9.

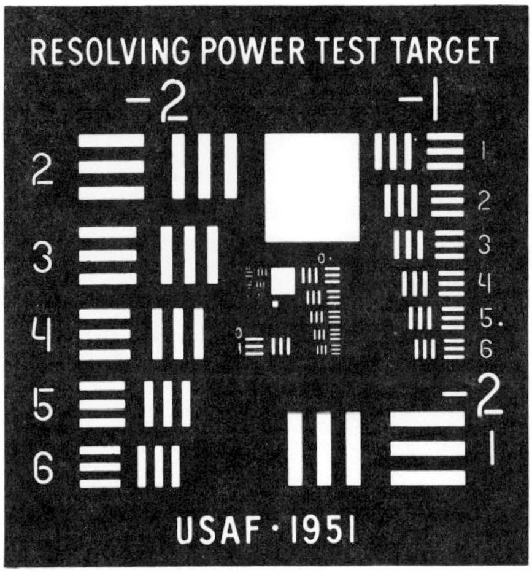

Fig. R-8 Resolving power test chart.

response, acoustic The measurable frequency and inherent characteristics of the studio, stage, or auditorium used for the recording and reproduction of sound.

restrainers Chemicals that are used in processing solutions to slow the rate of development. These chemicals also act as fog inhibitors in some developing solutions.

Some operate mechanically; others chemically. Water and chrome alum, for example, are mechanical. Among the chemicals are bromides and chlorides, used for restraining highlights, and potash and citrates of soda, which increase density.

retake A studio term for reshooting.

reticulation The wrinkling of the film's emulsion layer, caused by excessively warm developing solutions, an abundance of alkali in the developer, or other undesirable conditions in processing.

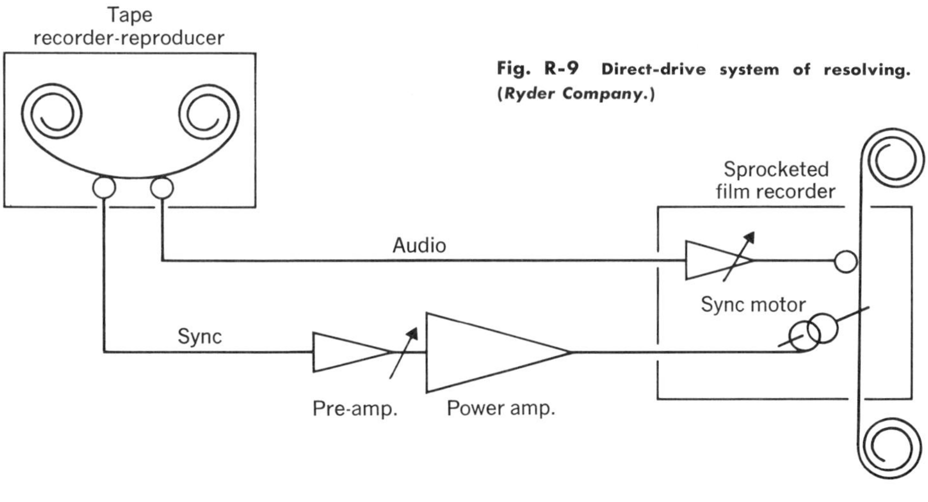

Fig. R-9 Direct-drive system of resolving. (Ryder Company.)

reticule Also called **graticule**. The etched fine lines on the surface of the ground glass in camera viewfinders that are used for framing and composing. See Figure R-10.

revamp The process of altering background "flats" or set props for reuse at a later date in another production.

reverberation Echo effects produced by reflection of sound from any of several surfaces in a recording area. See ECHO CHAMBER.

reversal film Film with an emulsion layer which, after exposure and processing, becomes a direct positive. See PROCESSING.

reversal original A positive image produced by reversal processing from the original in-camera stock. During projection, the emulsion side of a reversal original must face the lens in order to read correctly.

reversal print A reversal-type film, usually a reversal original, that has been used to expose a positive image and subsequently processed to produce a direct positive image. See REVERSAL PROCESSING.

reversal processing A process wherein the in-camera film, after special processing, produces a positive image. The film is first developed so that the silver in the emulsion is exposed. Light areas in the original subject matter show as pure metallic silver while the darker areas show as unexposed silver.

The exposed developed silver is then bleached away by chemical action, leaving the unexposed developed silver. The silver that had not been previously exposed is now reexposed, redeveloped, and fixed to produce the positive image on the original in-camera film. The bleaching action, redevelopment, and fixing process, during which

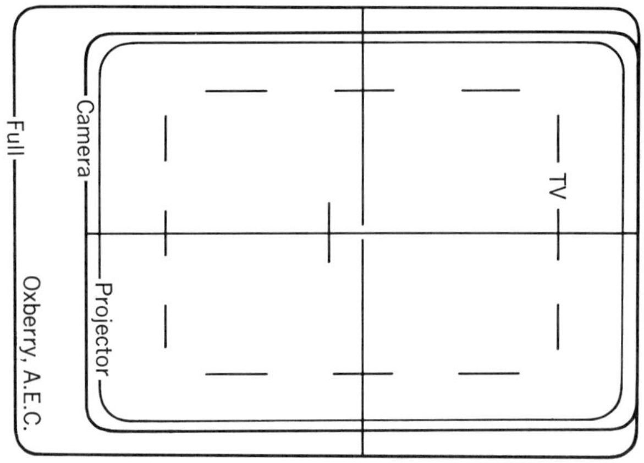

Fig. R-10 Reticule markings. The reticule markings shown in the diagram indicate the full aperture, camera aperture, projection aperture, and the safe title area for television commercial production. The solid horizontal and vertical lines meet at the exact center of the 35mm film frame. The short vertical line, at the point where it meets the horizontal line running through the center of the frame, indicates the center of the 16mm frame (in those cameras where the transport mechanisms are interchangeable.)

as many silver granules as possible are removed, reduce the grain effect considerably.

reverse scene The process wherein the negative is "flipped" during printing so that the picture information is reversed as in a mirror image. The editor must exercise judgment when reversing scenes since signs and other information in the background will also read backward. The resultant print, because the negative is flipped, will not be as sharp as it would be if it were printed in the conventional manner.

reverse shot The arrangement of cameras in scenes where two actors face each other. The film from each camera, when intercut by the editor, will show each of the actors facing the camera, in turn.

reverse Telephoto objective This type of lens, in addition to providing the cameraman with the possibility of a high relative aperture, has a wide-angular field and a long back-focal distance. Larger than conventional lenses, it can be used on both motion picture and single-lens reflex cameras.

reversing action Since the industry's pioneer days, technicians have taken lengths of film and reversed them so that the last frame on a reel subsequently became the first frame projected. Although many of the unusual illusory effects have been produced for entertainment purposes, reverse projection and photographic techniques have also been used to achieve a wide variety of seemingly impossible and implausible effects. Regardless of the usage, reversed actions are certain to hold the amazed viewer's wide-eyed attention with a minimum of effort and with maximum effect.

The most familiar examples of the technique are the newsreel shots that have shown shapely females or muscled males diving into a pool. While the narrator explained the difficulties involved in executing a perfect one-and-one-half forward somersault, the diver emerged from the water, with hardly a splash, and defied the laws of gravity by returning to the original position on the diving board.

Actions can be reversed by any of several methods. Many professional motion picture producers prefer to photograph an action normally and achieve the desired effect with optical processes. With this method, a fine-grain print of the original action is made to run backward on the projection side of the optical printer while the raw stock in the camera runs in a forward direction. In the processed film, the last frame in the original scene will be at the beginning of the roll exposed in the optical camera and thus the first frame that will be visible when the film is projected. Since most camera motors can be reversed and the film made to travel backward, cameramen can also use this more direct method for photographing actions in reverse. With this filming method, the cameraman must first close the camera shutter and run the film for the length of the action—assuming that the length of the action is known before filming begins. Similar effects can also be achieved with cameras that are not designed to run backward. With that type of equipment, the cameras must be mounted or held in an inverted position during the photographic process. The filmed sequence, however, after processing, must be cut out of the roll and spliced to adjoining scenes so that the section containing the reverse action will travel through the projector in the desired direction.

Although each of these methods will produce similar results, the producer must decide which of the processes is the most practical for a particular action. Decisions must take into consideration the difficulties involved in staging along with the print quality of the processed film. Though differences in the former may be negligible, it must be remembered that prints made from the original negative are superior to the second-generation characteristics of an optical print.

No less spectacular than the scenes that show a diver returning to the board are the sequences in which performers, in rapidly moving vehicles, collide or engage in equally dangerous daredeviltry. The attendant risk can be removed by filming such scenes in reverse. Before filming begins, the vehicles are placed in their postcollision positions and made to back away from each other when the cameras roll. Intercut with close-up shots of the same action, and accompanied by appropriate "stock" sound effects selected by the film editor, the action will be completely believable when the film is projected.

Scenes in which a pedestrian steps off a curb into the path of a speeding vehicle can also be filmed in reverse. To produce the desired effect, the pedestrian assumes a prone position in front of the car and later, while the cameras roll, the 'struck' jaywalker rises quickly and walks backward toward the curb while the car backs away. Here again, the processed film is projected so that the last frame photographed becomes the first frame visible on the screen. The accompanying sound effects (screeching brakes and crowd noises) provide the touch of realism needed to lend authenticity to the action.

In recent years, art directors have used reverse-filming techniques, vivid imaginations, and stop-motion processes to perform similar feats of magic with titles. Simpler effects that show individual letters appearing gradually or moving into position without visible assistance are achieved by photographing the titles in reverse with stop-motion processes. Titles in which portions of the letters appear in successive frames of film until the entire title is revealed are painted on the flat surfaces of sheets of glass or Celluloid. Placed over a suitable background, a predetermined portion of the lettering is removed between exposures during the reverse, stop-motion filming. When the photographic process has been completed, the entire title will have been removed from the surface of the glass. Since the last frame of film to be photographed will again be the first frame projected, the individual letters in the title appear on successive frames of film. Obviously, since the order in which the letters were removed during the photographic process is reversed during projection, none of the letters are visible in the first frame of film. They do, however, appear in their proper sequence during projection, with the last frame showing the complete title.

More unusual titling effects can be achieved with similar stop-motion techniques and reverse photography. Arranged on a flat surface, cutout or three-dimensional letters can also be moved (between exposures) according to a predetermined pattern to form a composite title. With the camera lens focused on the lettering and background from a vertical position, the letters can be arranged in their final position and then moved out of the shooting field, in gradual stages, on successive frames of film. When the processed film is projected, each of the letters will be seen moving into its final reading position from off-screen.

Variations of this effect are equally popular. With similar photographic processes, all the letters in the title are arranged in their final position and photographed for the required footage and reading time. With the film running backward through the camera, each of the letters can be moved away from its reading position, frame by frame, until an unintelligible, alphabetical pile of letters is formed in the center of the shooting field. The effect, when the processed film is projected, will show the letters unsnarling themselves and regrouping in the final reading position. Wipe effects, in which white titles needed for superimposures are painted on black cards and then revealed by gradually uncovering the title with a second black card can be shot normally with the film running forward in the camera. Here again, stop-motion techniques are used to achieve the desired effect.

In television commercial production, reverse actions have been used to achieve equally spectacular results. That "greasy kid stuff," for example, can be squeezed from a tube and then, by reversing the action, be made to return from whence it came. Similarly, liquid can be poured from a glass or bottle and the film reversed optically to show the same liquid reentering the empty container. Other illusory effects, involving products that lend themselves to this type of treatment, have shown cigarette smoke being exhaled to form a rather large ring and then inhaled as a continuous action. In other startling demonstrations, dishes have been dropped, broken, and then, by reversing the action, restored to their original form.

The possibilities are, indeed, unlimited. And, whether accomplished optically or with direct photographic processes, the viewer's reaction to the projected effect is completely predictable—wide-eyed amazement and undivided attention. See Figure T-12 under THREAD AND WIRE.

rewinds Devices attached to the editing table to support the feed and take-up reels that contain the picture and sound track. When two, three, or four films are being synchronized, the reels will usually revolve at different speeds. The difference in speeds results from the variation in diameter of film buildup on each reel.

Conventional rewinds, with friction-driven reels on a single shaft, require frequent manual adjustment of reels to take up film slack or to halt runaway reels. Differential gear rewinds, because of their torque-equalizing characteristics, provide for the transmission of positive, nonslip driving or braking torque equally to each reel regardless of its speed of rotation.

r-f The abbreviation for **r**adio **f**requency, the waves used to transmit sound or video signals.

r-f pattern A term used to describe a fine herringbone pattern in televised images. The effect may cause a slight horizontal displacement of scanning lines resulting in rough or ragged vertical edges of the picture. The pattern is caused by high-frequency interference.

RGB Channels The **r**ed, **g**reen, and **b**lue channels in the television or videotape recorder's color system.

rheostat A resistance coil that may be tapped at any point to adjust the amount of current flowing through a circuit.

ride The process of adjusting controls normally performed during a broadcast or telecast. Engineers speak of "riding" the video while on the air, for example.

rigs and riggers Rigs, also know as **rails** or **scaffolds**, are supports over the studio set to which lights are secured. While the gaffers handle the electric equipment, and the grips (stagehands) are responsible to the chief cameraman, the riggers handle the construction of scaffolding and the moving of heavy equipment.

right-angle prism Prisms used to invert images in photography, as light rays reflect through at a 90-degree angle. These prisms are also found in astronomical telescopes, projectors, photoelectric systems, range finders, and other optical instruments. They may be coated or silvered.

rim light A halo or rim of light that outlines the general shape of the subject matter. The effect is produced by positioning the light source behind the subject.

ring head A magnetic recording head in which the magnetic core material forms an enclosure with one or more gaps. The magnetic tape bridges one of these gaps and is contacted by the pole pieces.

R mount A lens mount designed by the Eastman Kodak Company for use with their reflex cameras. With these mounts, clamps wedge the lens tightly into the seat where it cannot become unscrewed or vibrate loose.

robot photography In industry and science, the inadequacy of the human senses, due to visual delay, individual variation, or limits to concentration, may dictate the use of recording and measuring methods that overcome such shortcomings.

Photographic registration is often the only means to solve recording problems, as only the photographic picture gives an irrefutable visual documentation. When the 35mm camera was introduced, it soon became obvious that it was a perfect medium for systematic record keeping. Its economy in multishot application, its ready adaptability to a variety of tasks, its minimum space requirements, the superior range of high-quality optics, and the practically unlimited possibilities of the construction-unit system—all have contributed to this success.

Photographic automation and instrumentation was introduced more than twenty-five years ago. Built around the basic Robot cameras,* automation, in this system, means power-driven film transport and self-cocking shutter, for rapid one-by-one shots or kinographic sequence, and exposure counting. In photographic registration, the picture records subjects that may be extremely small and almost imperceptible to the human eye, or even completely invisible because of their transitory occurrence. Unattended, the camera registers counters, scales, dials, curves, and similar industrial data, giving a totally objective, permanent documentation. Recordings can be made simultaneously from several locations, or two recordings can be combined on one negative. See Figure R-11.

rock, camera A vibration or off-register effect achieved by moving the camera slightly for each frame of the action. These effects are especially valuable for violent actions such as explosions, crashes, and falls.

*Karl Heitz, Inc., New York.

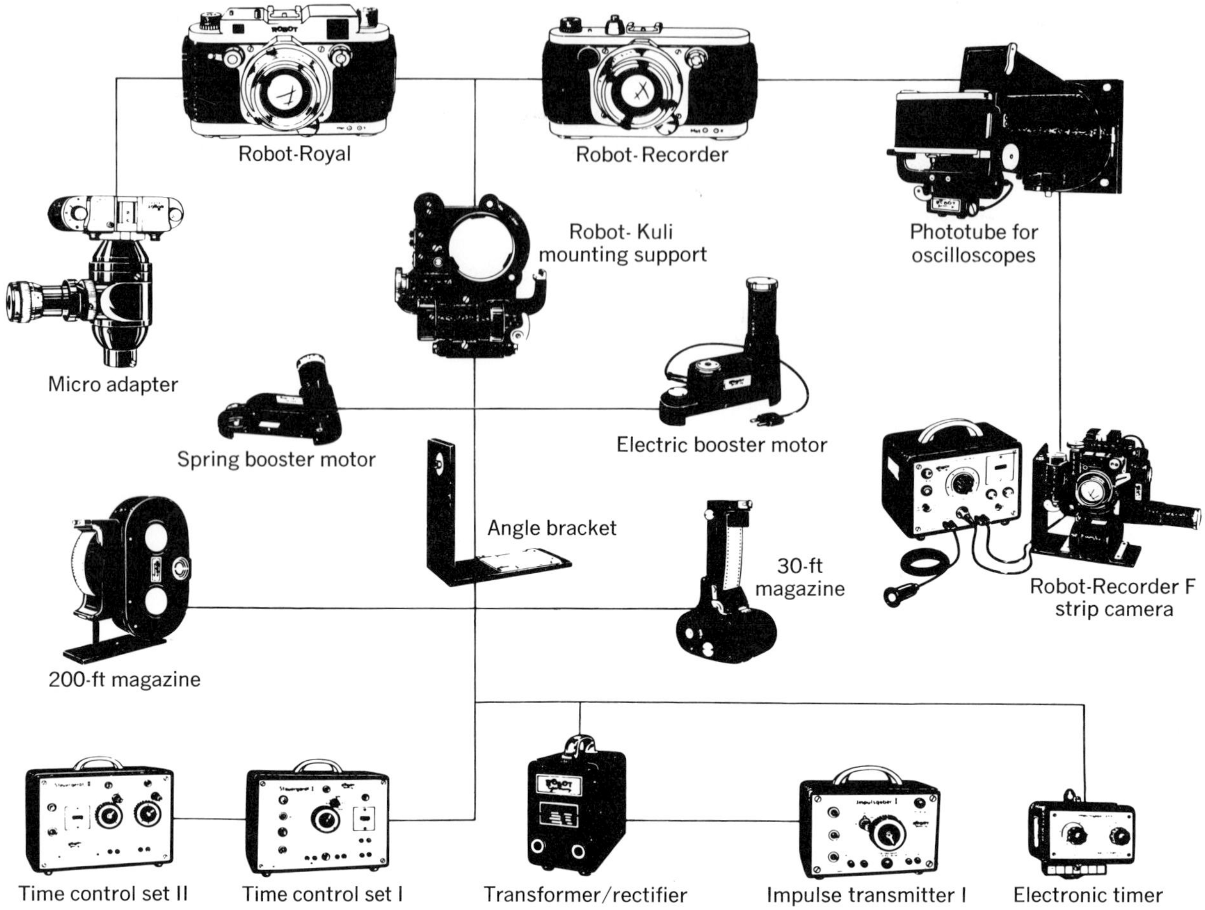

Fig. R-11 The Robot constuction unit system. ***(Karl Heitz, Inc., New York.)***

rocking The process of moving a section of recorded videotape back and forth, by hand, across an audio-head assembly in order to locate the beginning of a sound, or a specific frame.

roll (1) Lengths of film wound on cores. (2) The signal to start cameras. "Roll 'em!" (3) A lack of vertical synchronization which causes a televised picture to move upward or downward.

rolling loop A relatively new method for transporting film through the camera. With no reciprocating, oscillating, or intermittently moving parts, a continuously moving rotor forms a series of loops in the film which is fed, through a confined path, to the aperture for the interval of exposure. In operation, the film's perforations are engaged by a set of undersized pilot pins ahead of the standard register pins. The loop formed by the rotor reduces progressively after leaving the aperture while the incoming loop is being formed. The movement is applicable for television film recording. See Figure R-12.

rotating-mirror camera See OPTICAL COMPENSATING CAMERA.

rotating prism Also called **intermittent prism**. Lens arrangements that are used in some high-speed cameras to take the place of the

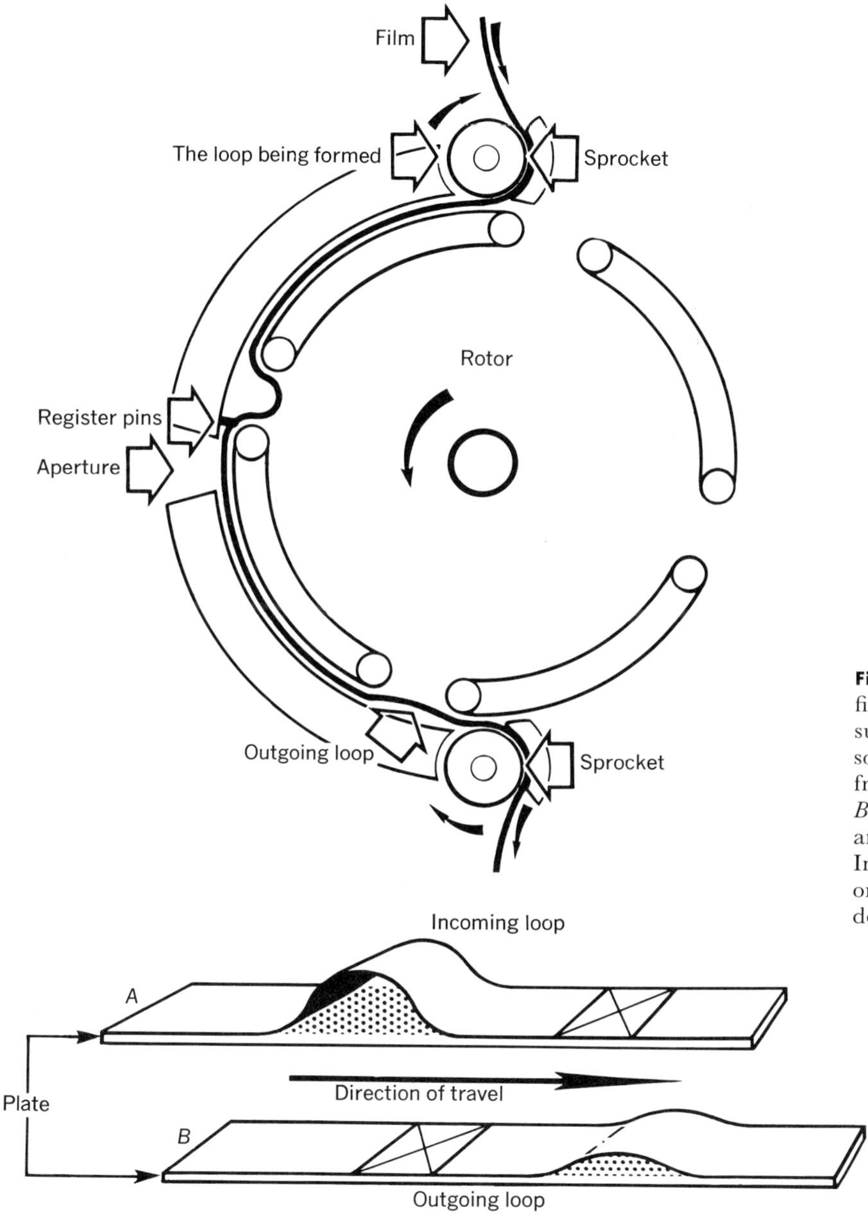

Fig. R-12 The rolling loop. The length of film in *A* is one frame longer than the supporting plate. The extra frame is absorbed by the loop which precedes the frame marked with a cross. The loop in *B* has, theoretically, passed the aperture and is progressively being reduced in size. In this way, the film is advanced, from one point to another, without lineal pull-down.

shutter. These lenses, in projection equipment that is used with certain types of telecine systems, make it possible to project a flickerless picture. See HIGH-SPEED PHOTOGRAPHY and MULTIPLEXED TELECINE SYSTEMS.

rotoscope A process during which each frame of film in a scene or a portion thereof is projected to achieve a composite special effect. Projection takes place, frame by frame, either on the animation stand or in the optical printer.

During the projection process, tracings of specific areas serve as guides for making the mattes that are used to insert new picture information in specific areas of the projected scene. In other cases, the tracings are used for positioning new material, usually from a second length of film, in relation to the subject matter in the original scene.

rough cut The first arrangement of scenes in their approximate length and order.

run The exposure of film during the photographic process. A double run or second run indicates that the same length of film is to receive a second exposure. Compare FIRST RUN.

runners The overhead scaffolding or supports for lights, backgrounds, and large props.

running shot A photographic effect in which both the camera and the subject are in motion, but maintain a constant relationship. See Figures R-13*a* and *b* and compare PASSING SHOT.

running title Also called **crawl title**. Credits or acknowledgments that pan from the bottom to the top of the screen. During the panning movement, the lines of the title appear at the bottom of the film frame, move up gradually, and finally disappear as they pan beyond the top limits of the film frame. The effect is like a scroll unwinding. The speed at which the lines move should be paced to what is considered a comfortable reading speed for most viewers—between 120 and 130 words per minute.

run-through A prefilming rehearsal during which lines are spoken and the performers go through their actions. For television programming, where time is the all-important factor, the rehearsal is timed with a stopwatch, and scenes are lengthened or shortened—as required.

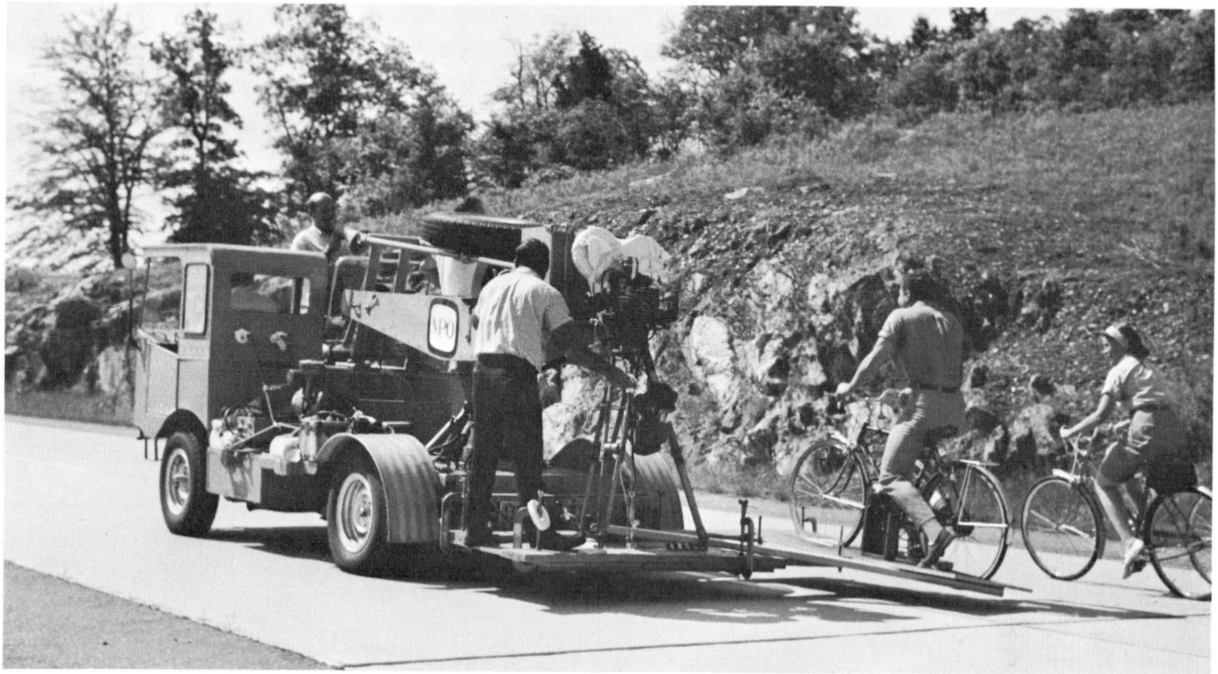

Fig. R-13*a* Running shot. *(Peter Paul, Almond Joy, Westchester, N.Y.)* Pedaling to nowhere. Six-day bicycle racers never had it this good. The stars, after this running shot has been filmed, will be fresher than the camera crew and the technicians who arranged it.

Fig. R-13*b* Running shot. "Get a horse!" Sound effects will make the audience think that the Pepperidge Farm wagon is actually being pulled by a horse.

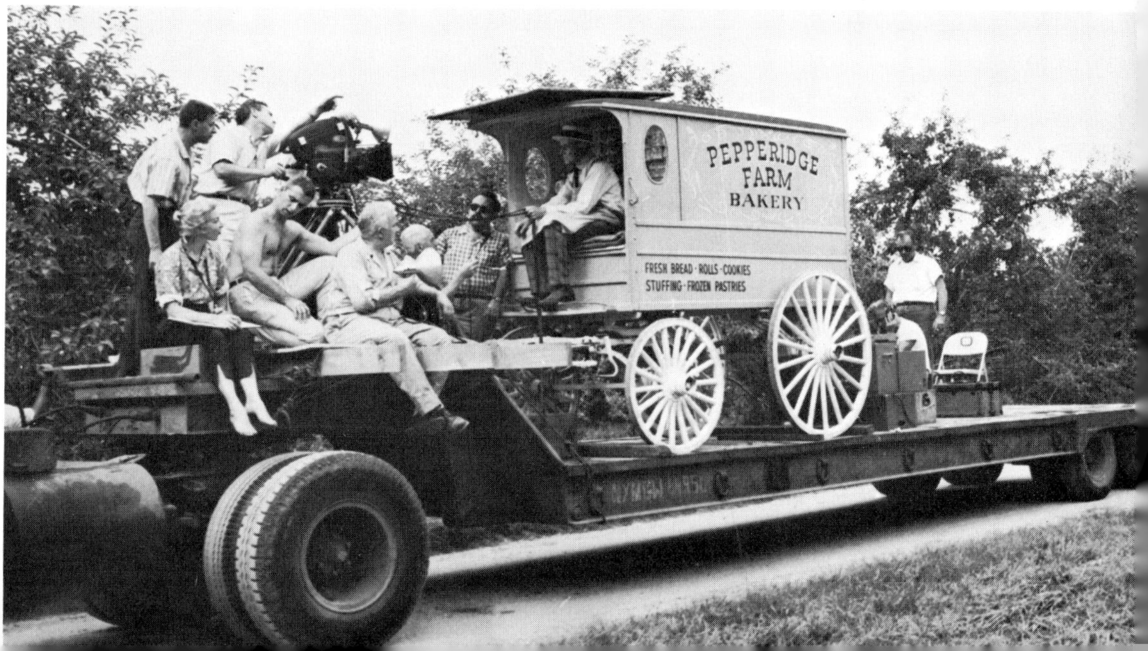

run-up That part of the changeover procedure, when two projectors are being used, in which the reel of film being projected is replaced by the reel that follows. The projectionist is alerted by punch or scribe marks that appear in the upper right-hand corner of four consecutive frames of film. The second projector is started when the operator sees this motor cue. Exactly 168 frames later, similar punch marks are placed on 4 more frames of film to let the operator know that there are only 24 frames of picture remaining on the reel. This is the changeover cue, and the second projector is at operating speed. The picture information continues uninterrupted from one reel to the next, and the audience is not aware of the changeover procedure.

rushes Also called **dailies**. The processed results of the previous day's photographic efforts.

S

safelight A light used in darkroom work. Safelights produce a high intensity monochromatic light beam in a wavelength region of the visible spectrum where many color films and print stocks are insensitive.

Yellow lights may be used with films or plates that are relatively slow and not color-sensitive—materials used in contact printing, for example. Fast films that are not color-sensitive require the use of brown or light-red safelights. Red is used for orthochromatic films and plates. Bluish green or deep green safelights are recommended for panchromatic films. These lights should not be allowed to expose the film for more than 1 or 2 seconds. Yellow-green lights are used for infrared work.

Screens for darkroom use may be made of colored fabric, Celluloid, plastic, or glass (tinted). These materials are positioned in front of the light source. See Figure S-1.

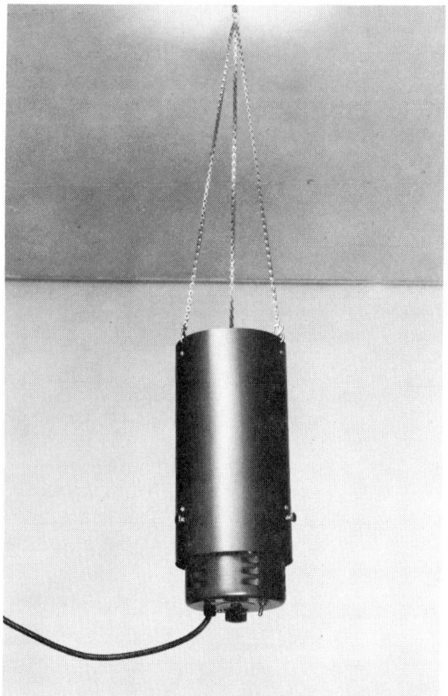

Fig. S-1 ColorTran Spectroluxe safelight, model SL-1—a blue safelight with a very narrow bandpass for use with color films.

safety area Also called **cutoff**. The outer areas of the frame that are not visible during television transmission. Titles and other important picture information, exposed on film for TV transmission, should be kept within the area outlined in ASA specifications. See Figure S-2.

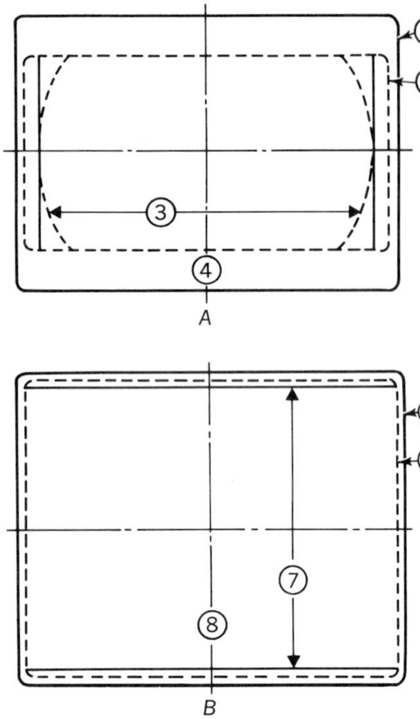

Fig. S-2 Safety area.

(*A*) Wide screen. (1) Camera aperture: 0.868 × 0.631 inches. (2) 1.85:1 aspect ratio: 0.825 × 0.446 inches. (3) Safe title area: 0.757 × 0.446 inches. (4) Center of scanning area.

(*B*) Anamorphic release prints. (5) Camera aperture: 0.870 × 0.735 inches. (6) 2.35:1 aspect ratio: 0.839 × 0.690 inches. (7) Safe title area: 0.839 × 0.678 inches. (8) Center of scanning area.

safety film Light-sensitive layers of emulsion supported by a cellulose or acetate noncombustible film base.

SAG The abbreviation for the Screen Actors Guild.

Samenco controls A specially designed motor equipped with timing devices for time-lapse photography. The timers can be regulated to take single exposures, at predetermined intervals, ranging from 1 second to approximately 10 hours. If necessary, several cameras can be set up at different points in relation to the subject matter, each in synchronization with the others, and activated by a common control unit.

saturation The degree of intensity of a color. Vivid colors are those that are highly saturated. Pastel colors, by comparison, have a low degree of saturation. See MUNSELL SYSTEM.

scale A term indicating the degree of contrast or the proportionate differences between light and dark areas in a print. *Long scale* describes prints of low contrast in which there are many halftone values or intermediate shades of gray. *Short scale* describes contrasty prints, usually used for matteing and title work, in which there are relatively few (if any) intermediate shades of gray.

scalloping In television, picture distortion caused by improper vertical alignment. Compare SKEWING.

scanning The process of breaking down an image into a series of electronic elements representing light values and transmitting this information in time sequence. Also the reverse process, in which the tonal values of the transmitted television picture are reproduced by the scanning action of the cathode-ray tube.

scanning line One of 525 narrow, horizontal lines (U.S. standard), that fill the television screen. These lines are created by the electron spot in a cathode-ray tube. Variations in the intensity of the spot re-create the highlights, shadows, and halftones in the original subject matter.

scanning spot The cross section of an electron beam at the point of incidence (the fluorescent screen) in a camera or picture tube.

ScanoScope See WIDE-SCREEN PROCESSES.

scenario Also called **script** or **treatment**. The plot and basic dialogue from which the shooting script is subsequently developed.

scene A part of a motion picture in which a single situation or action is presented in a specific location. Also, an individual action or related incident forming a part of a sequence.

Scheiner system A European system for rating the speed of an emulsion. The method of rating is based on the minimum amount of exposure needed to produce an image on film. The average ratings in the Scheiner system for similar film types are approximately two to six points faster than those in the Weston ratings. See EXPOSURE-INDEX RATINGS.

Schufftan process An optical process for in-camera combination of prepared background material with the subject to be filmed.

Developed by the German film technician, Eugen Schufftan, during the 1920s, it depends on the use of a front-silvered mirror, positioned at a 45-degree angle to the axis of the lens, from which a portion of the silver surface has been removed. The removed silver portion of the mirror must match the shape of the material, either artwork or miniature representations, that is to be used as the background for the scene being filmed.

The camera is positioned so that the performer is reflected off the mirror surface, in relation to the background material, to form a composite effect directly on the in-camera film.

Although mirrors will always have a place in photography, professional producers generally find that matteing processes are more practical in that they provide a greater degree of flexibility in the production of most composite effects.

scoop Also called **basher**. An incandescent lamp used for floodlighting.

score An original theme or musical accompaniment written for a specific motion picture or television commercial.

Scotchlite process See FRONT-SCREEN PROJECTION.

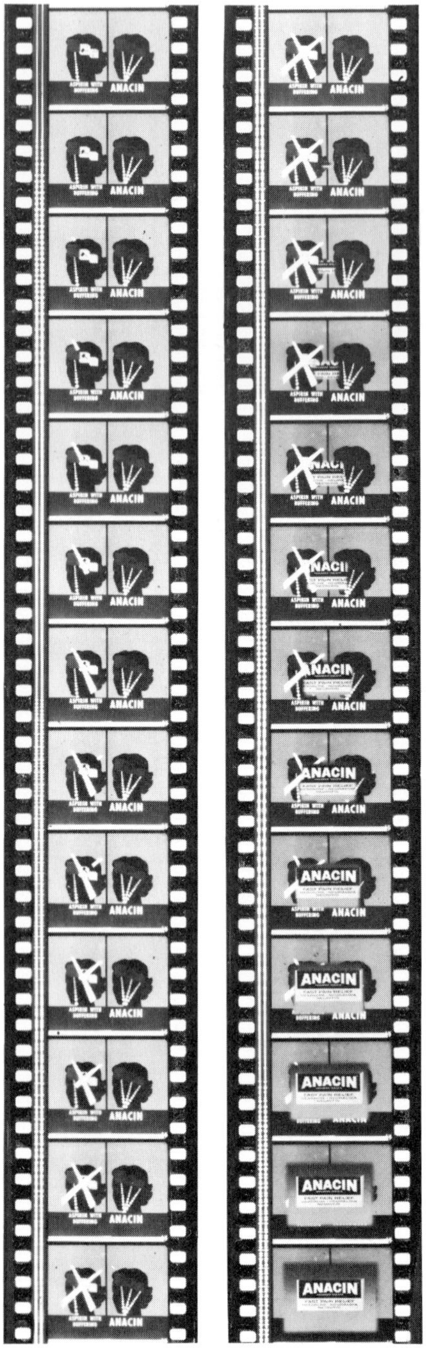

Fig. S-3 The scratch-off effect.

scraper That part of a splicer used for scraping away the emulsion from the portion of the film frame that is to be spliced. Film cement is applied to the scraped area which is then joined to the second length of film.

scratch-off effects A process wherein portions of artwork on inked and opaqued cels are scratched away or in some way concealed during the filming process.

The animation cameraman removes a portion of the artwork or graphic material on successive frames of film, usually photographed in reverse, as stipulated by the animator in his exposure sheets. When the processed film is projected, the last frame photographed becomes the first frame that is visible to the viewer. The scratched-off areas appear on successive frames of film, exactly opposite to the order in which the areas were removed during the filming process.

As an alternative to the scratch-off, titles that seem to appear and disappear are photographed on a high-contrast film. In the preliminary phase of production, the title is prepared in its entirety by rendering it in white on an acetate cel or on a black card. During the filming process, a second black card is used to either reveal or conceal portions of the title according to the instructions given to the cameraman in the exposure sheets. The title, after it has been filmed and processed, can be burned in or superimposed optically over live-action scenes.

The latter method is used whenever possible since the title or artwork remains intact after the filming process has been completed. If a retake is necessary, the title does not have to be redone. See Figures S-3 and S-4.

scratch print Stock or library footage used by the film editor for inserts. The scenes are deliberately scratched to prevent unauthorized duplication. Usable prints are made from the negative which remains in the hands of the supplier.

scratch track In the absence of the actual sound track, a temporary, nonprofessional recording is made and used by the film editor for timing purposes. Synchronization, when scratch tracks are used, can only approximate the actual timing.

screen Opaque white matte, aluminum, beaded, or translucent materials on whose surfaces motion pictures are projected.

Also, a term generally used to describe the viewing of print material as on moviolas or with studio projection equipment.

scrim A gauze screen that is positioned in front of a light source in order to diffuse or soften the light originating from it.

script Also called **scenario** or **treatment**. The voluminous pages of dialogue from which the shooting script is subsequently developed.

Although critics and reviewers have openly questioned their existence on many occasions, motion pictures are produced from scripts. Each contains a plot or story; it has been held that producers have successfully managed to repeat approximately 30 basic plots since the industry's magic lantern days, with only minor variations from the original themes.

Each script, through a series of transitory processes, is subsequently translated into its final, visual form—film. Often referred to as the blueprint for production, motion picture scripts are usually adaptations of well-publicized, best-selling novels or hit shows originally produced for theatrical or television purposes. Rewritten by the playright or scenarist, in collaboration with the film producer or director, these voluminous scenarios include pages of dialogue; descriptions of the central and secondary characters in the story, their attitudes, and moods; suggestions for staging and filming; word sketches of the backgrounds, costumes, and props; and, finally, detailed information outlining transitional or special effects. After approval, and long before actual filming begins, copies of the script are distributed to each of the studio production departments as well as to the performers who have been selected to portray the various characters. For every member of the cast who memorizes a page of dialogue, a small army of talented craftsmen and technicians play their behind-the-scenes roles, quietly and efficiently, before one frame of film is finally exposed. Some design costumes; others build backgrounds and prepare props. Those not directly involved in the preproduction processes select remote locations, prepare shooting schedules, attend to hundreds of other miscellaneous details, and keep a continuing record of costs and expenditures. With a specific number of days allotted for filming each of the sequences, the complex studio production machinery moves with the precision of a fine watch.

Responsible for translating the written script into its final, visual form, the director is the dominant figure on and away from the studio set. With as many ways of telling a story as there are listeners or viewers, the director's imaginative interpretation and treatment is usually influenced by the contents of the script. As the absolute monarch of all that the camera surveys, each director, in his own way, attempts to blend all the production elements into a dramatically effective, well-paced film presentation that takes full advantage of the medium's intimacy.

In the theater, the emphasis is placed on broad, dramatic representation—and the spoken word. Separated from the audience by

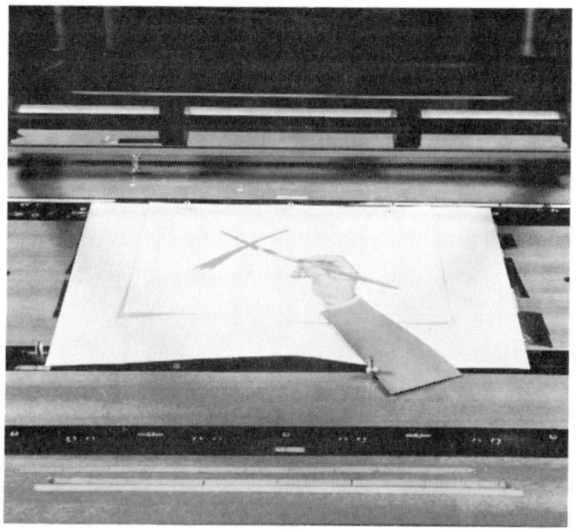

Fig. S-4 The same principles that were used in the "Out of the Inkwell" animated cartoon series produced by Max Fleischer during the 1920s for Paramount Pictures are as applicable for scratch-off effects today as they were then.

footlights, subtleties too often go unnoticed. Similarly, the performer's hushed whisper must still be loud enough to reach those seated in the very last row of the theater.

The motion picture, by comparison, because of the all-seeing camera lens and more intimate approach, is less dependent on dialogue. In that medium, the gesture, expression, and reaction serve as adequate substitutes. Indeed, pantomime, one of the oldest forms of histrionic endeavor, was the generally accepted form of expression long before the industry discovered its voice.

But directors also have their personal styles. Some—who may have theatrical backgrounds and are occasionally accused of being frustrated actors—bring their performers to dramatic heights and mechanical perfection through long rehearsals. Actors and actresses, influenced by these tireless thespians, speak their lines and repeat stage business with monotonous efficiency before a single frame of film is exposed. For other directors, the effectiveness of long, tiring rehearsals is of questionable value. The latter group, collectively, is more interested in achieving a feeling of spontaneity than it is in flawless delivery.

Imbued with a sense of the spectacular, some directors naturally "think big." The thousands of extras and elaborate settings used in the production of their prodigious efforts are items not usually associated with limited budgets. Generally reserved for wide-screen processes, these extravaganzas require months of planning and an expenditure of millions of dollars. Other directors, with a special talent for keeping an audience perched on the edges of their collective seats, are assigned to off-beat horror chillers or suspense-filled mystery thrillers with a twist or surprise ending.

Some directors, especially those who served their apprenticeships behind the camera, are as familiar with lighting effects, photographic processes, and editing procedures as they are with staging and dramatic techniques. For them, cameras are all-seeing, mobile pieces of equipment that help establish a feeling of intimacy and presence between the performers and the audience.

Since multiple-camera setups are the rule, the director and film editor can select and intercut alternate shots of the same action in order to add interest and dramatic impact to the production. Extreme close-ups that focus attention on detail (eyes or hands, for example) are quite effective when combined with close-ups of the performer's face and medium shots that include the full figure. Similarly, establishing shots, or long shots, that show the general area in which an action occurs, and angle shots, in which the camera is above, below, or to one side of the performer, all play an important part in the overall pattern. Projected on large theater screens and backed up by an appropriate sound track, the camera is allowed to tell its story in a most convincingly effective manner to an appreciative audience.

Because one background often serves as the setting for a number of scenes, shooting schedules are planned accordingly. In spite of the apparent lack of filming continuity, the director must visualize each day's production in relation to the entire script and maintain a scene-to-scene consistency. With little more than judgment and experience to guide his efforts, the director must sense the eventual reactions of the audience. Faced with tight schedules and influenced by budget limitations, his decisions and evaluations must be accurate. Retakes are too costly.

Obviously, good scripts and award-winning film presentations do not necessarily go hand in hand. For the motion picture director, however, the better script can be the first long step down the aisle leading to the stage where the awards are distributed.

script girl Although she is mainly responsible for checking dialogue during filming, the script girl also keeps a record of the filmed scenes; checks continuity with respect to costumes, props, and action; and, in general, acts as an assistant to the director while filming is in progress.

seamless backgrounds Photographic backgrounds produced by spraying a light-sensitive emulsion on a large sheet (up to 60 feet wide in some cases) attached to a wooden frame. The material has a self-diffusing film base that is similar to the screens on which backgrounds are projected during the filming of process shots. This material may be rolled up and stored without damaging the background.

Exposed by a projected negative of the scene that will subsequently be used as the background, the sheet is developed to produce a sharp enlargement. Oil paints matching the colors in the background are applied by artists to the rear surface of the background to eliminate the feeling normally associated with projected transparencies. The translucent material eliminates the possibility of the hot spots that appear in backgrounds used for process shots. During the filming process, the backgrounds are front-lit to reduce contrast. Backlighting brings out detail and creates the desired feeling of depth. The process was developed by M. B. Paul, a director and expert in the field of special effects, who received an Academy Award for his efforts in 1950.

Seaquartz underwater lights Lighting units for underwater filming. Designed to operate at varying depths, the lights are battery-operated and have a working capacity of approximately 45 minutes or, at half power, about 90 minutes. Some lights (the Birns & Sawyer Mark IX, for example) can be used at depths of 9,000 feet.

segue A musical transition.

selective filters Also called **separation filters**. A group of filters sensitive only to the colors within their own spectral range. This type of filter rejects or absorbs colors complementary to the filter color. These filters are also used for making black-and-white separation negatives from material originally photographed on color film.

selectivity The ability of a receiver to reproduce desirable signals while filtering out undesirable signals from neighboring wavelengths.

self-matteing processes The continuing search for new methods of combining picture information from different sources for the purpose of creating composite effects has paralleled the growth of the motion picture industry. Of the many production techniques that have been developed through the years, the self-matteing process is, undoubtedly, the most important.

Self-matteing, or traveling mattes, have had a great influence on motion picture production. They have made possible the achievement of unusual, eye-arresting effects that would be either difficult or impossible with direct in-camera photographic techniques. Additionally, the advantages of filming actions conveniently, for subsequent superimposition over a choice of background material that exists on other lengths of film, is quite obvious and needs no qualifying explanation.

For the producer with a limited budget, the expenses normally associated with location filming and the added costs arising out of unpredictable weather conditions can be avoided. Money set aside for costly set construction can, likewise, be channeled into other aspects of production.

The prephotographed backgrounds that are used with this type of matteing process provide the director with almost unlimited possibilities for planning foreground action. If necessary, the backgrounds can be repositioned optically, thus giving the producer with limited staging facilities additional flexibility. Miniature props and flats, used in conjunction with the previously filmed backgrounds, offer added staging possibilities to the imaginative producer. Pinpoint sharpness or soft-focus effects, in either foreground or background, are possible and easily accomplished during the optical combinations of the various film elements.

With self-matteing processes, dangerous filming situations can be completely neutralized. Actors need not leave the confines of comfortable studios to face the hazards that normally accompany the filming of relatively dangerous sequences. The staging of such actions can be preplanned in relation to the angle and perspective of the selected film background, and the actor is seldom required to do anything more hazardous than remember lines of dialogue.

With self-matteing processes, normal-sized actors can be made to look like giants or dwarfs by optically enlarging or reducing the film frames that serve as backgrounds. For illusory effects, actors can be rephotographed and made to appear in a scene several times. If necessary, an actor can also be made to appear in front of or behind himself in the same frame of film. This illusion is easily created by the optical cameraman who merely changes the position of the mattes in relation to the film that contains the image. Similarly, an actor can engage himself in conversation, go through complicated dance routines with himself as the partner, and, if need be, play all nine positions on a baseball field.

Some self-matteing processes depend on the use of color film or on an application of color principles for creating the film components needed for optical combinations. Other self-matteing processes, such as the infrared, ultraviolet, and sodium systems, depend on the absence of visible light and color for their success.

Because the traveling mattes must match the size, shape, and position of the original image (on another length of film), the methods used for the filming of the necessary film elements are of primary importance. To ensure precise registration, the image and matte can be photographed simultaneously (as in the Spectrumatte process) through a common lens.

Specialized photographic equipment, similar to the three-strip Technicolor cameras, are used for this type of dual-film self-matteing process. The beam splitter, or prism, positioned behind the camera's lens, is capable of transmitting identical images to different lengths of film—one image to a panchromatic film stock in one of the magazine compartments while the matte image is exposed on a high-contrast film in a second compartment.

Because photographic emulsions are sensitive to certain colors, and not to others, the repetition of similar colors in foreground and background is not advisable. If a blue background is used (as in the blue matteing process, for example), the use of that color in clothing and accessories should be avoided since a portion of the background would be visible in the hole created by the matte. Actresses, also, must be careful in their choice of colors in both costume and makeup. The combination of blue eyes and blue eyeshadow, for example, could produce effects that would be quite unusual, at best.

As an alternative to dual-film, self-matteing processes, the single-strip or color-difference system depends on color sensitivity. As the name implies, only one film is used during the photographic process.

Eastman color negative film, for example, has three layers of emulsion, each sensitive to one of the photographic primaries. A blue background will, therefore, expose the blue-sensitive emulsion layer but will have no effect on either the red or green layers. Similarly, any color that can be used to expose one of the layers of emulsion without affecting the others may be used for the background. Because flesh tones are usually the most important in a scene (in terms of exposure), and can be readily appraised in relation to a blue background, blue is preferable to red, green, or yellow for background purposes. The blues generally recommended are M-107-1 mixed with a plastic binder Q-50-1 available through the Krieger Color and Chemical Company in Hollywood and Idding's number 159 ultramarine blue.

Lighting, an important adjunct to the photographic process generally is especially important in the filming of infrared, ultraviolet, and sodium self-matteing processes. When these processes are used, the foreground and background must be lit separately and with complementary colors. Foreground lighting may not be allowed to interfere with the lighting of the background. Neither may the illumination that is used for lighting the background be allowed to reflect on the foreground action.

The infrared, ultraviolet, and sodium matteing processes are all dual-film systems that depend on the use of light rays that are invisible to the eye. This invisible light is used to illuminate the background only and, although it provides the means for properly

Fig. S-5a Self-matteing with + and − infrared lighting.

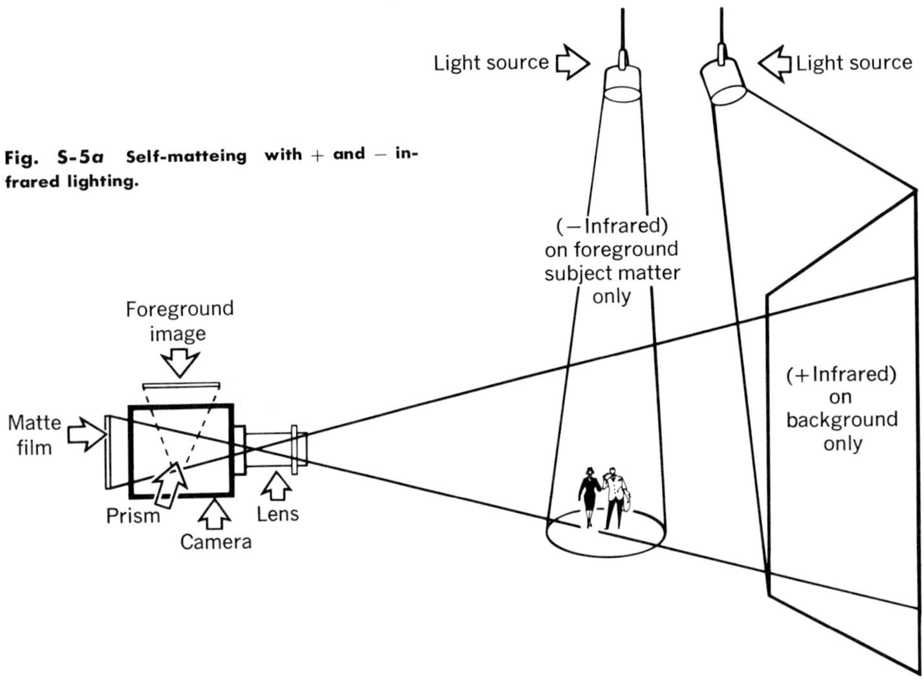

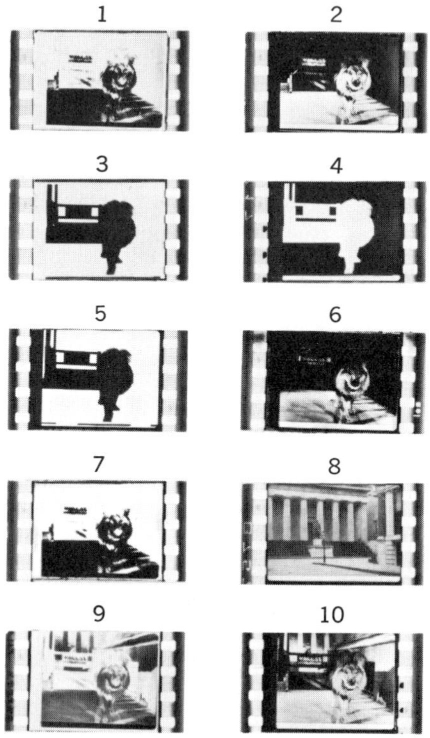

Fig. S-5*b* **Self-matteing processes.** The following elements are needed for the optical combination: (1) Eastman color negative. (2) Pan master—fine-grain print. (3) High-contrast print made from color negative. (4) High-contrast negative. (5) High-contrast positive made from high-contrast negative. (6) Separation print from original color negative. (7) High-contrast negative from separation print. (8) Background scene—fine grain. (9) Composite negative. (10) Composite print.

In the first step of the optical combination, the positive matte of the lion (5) was photographed with the background showing the Wall Street area (8). That film was then run back to the starting point and combined with the subject (2) and the negative matte (4) to produce the composite (10) after the film had been processed.

exposing the matte film, it exerts little, if any, influence on the panchromatic film that receives the foreground image.

The special black nylon velvet material that is used as the background for the infrared matteing process is highly reflective and produces the best results for that system. Foreground lighting is simplified since extraneous light will have no influence on the black velvet background

For the ultraviolet system, colorless backgrounds, similar to the translucent screens used for rear projection, are lit with fluorescent tubes that provide ultraviolet light. Filters must be placed over the standard studio lamps that are used to light the foreground action to block the ultraviolet rays that they also emit.

In the sodium system, the prism in the camera reflects the light from the sodium-vapor lamps to the film that will subsequently contain the matte. The background, for the sodium matteing system, may be either white or yellow fabric. Translucent screens, illuminated from the rear by groups of sodium tubes, may also be used for backgrounds. Since the sodium light source is not intense enough to expose the color negative, standard studio lights, without filters, can be used for the foreground action.

The first American patent for a monochrome traveling-matte process was issued to F. D. Williams in 1918. Nine years later, a patent for a color process was granted to J. A. Ball. The principles that served as the basis for the original processes have been used as stepping-stones for further development. Improvements in optical equipment have kept pace with these new developments thus making possible the achievement of unusual, eye-arresting effects that truly deserve the name cinemagic.

Of all the techniques and processes used in motion picture production, self-matteing is, perhaps, the most versatile. Its limitations are unknown. Its potential has not yet been fully realized. See Figures S-5*a* to *d*.

Fig. S-5c The Dreyfus Fund commercial. To the delight of everyone in New York's Wall Street area, the lion was photographed in a studio under the watchful eyes of its trainer. Filmed against a treated background, the lion climbed the "prop" subway steps behind his ever-present trainer. The Wall Street area which was used as the background was photographed separately. The mattes, achieved with the MPO Spectrumatte process, were combined optically with the background and the film of the lion to create the memorable effect.

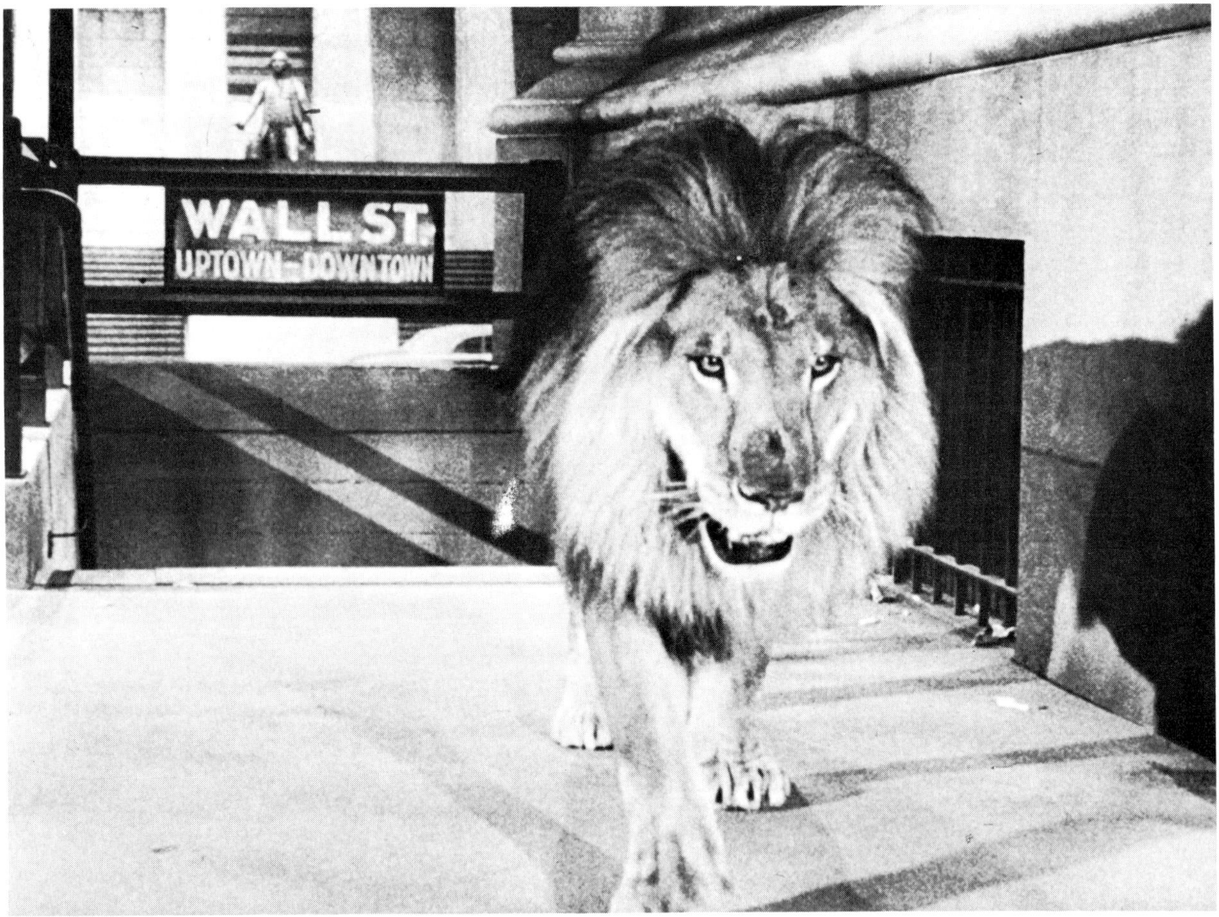

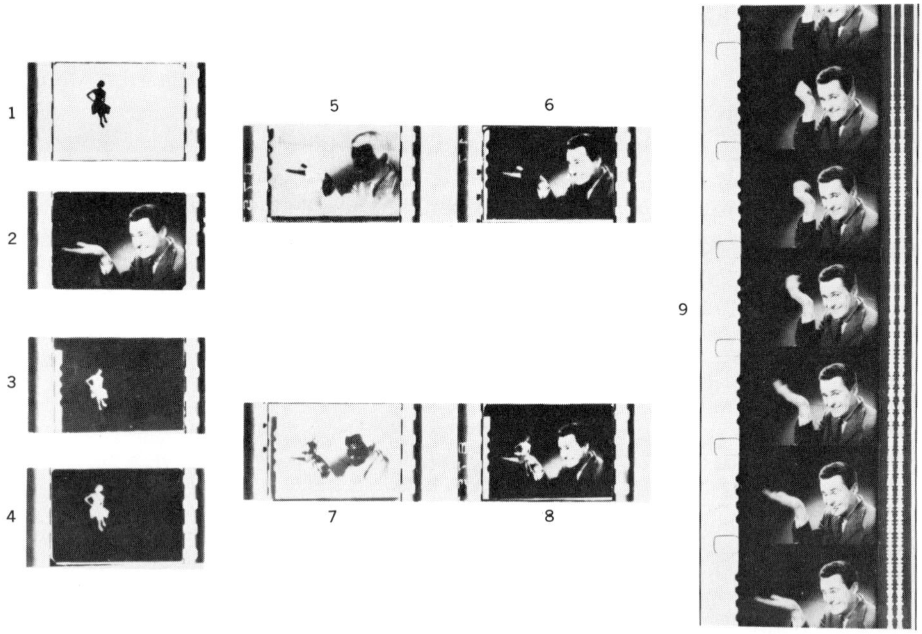

Fig. S-5*d* The commercial produced for Command, a hair-grooming product, assures the viewer that "it's a snap to have girls in the palm of your hand." The composite effect was anything but a snap from a production point of view. Achieved with matteing, the man and woman were each filmed at different times. The cameraman, when photographing the woman, used a dual-film camera equipped with a prism behind the lens. This enabled him to record, simultaneously, the image of the woman on one length of film and a silhouette image (matte) on a second length of film.

In the first step of the optical combination, the positive matte (1) was photographed with the fine-grain film that served as the background (2). During the second optical run, the fine-grain film of the woman (3) was photographed with the negative matte (4).

If the film had been developed and subsequently printed after the first run, the positive matte would be the white area as shown in (5) and, in the print, as a black silhouetted area (6). Frame 7 is the composite negative, and frame 8 the composite print.

The series of frames showing the finger snap that preceded the sudden appearance of the woman in the palm of the man's hand is shown at 9.

black-backing traveling mattes. The first traveling mattes were produced many years before color film and color-separation processes were made available to the film technician. Considering their limitations, the general approach to their self-matteing processes, between 1915 and 1920, was not too different from the practices used in today's more sophisticated dual-film multiemulsion processes.

Instead of the blue flats or dye-impregnated fluorescent screens that are now being used as backgrounds, the performers were positioned in front of backings that had been painted dead black. Overexposed during the filming process, the image in the developed negative was denser than normal. The background area, by comparison, was reasonably clear. The negative, in the next stages of production, was printed, reprinted, duped, and reduped on high-contrast film until the subject matter was opaque black and the background seemed to be completely clear.

Where the subject matter was photographed in front of a white background, the film was underexposed during the filming process. The steps leading to the production of the mattes were, however, basically similar to those used for the production of the mattes after the performers had been exposed in front of a black background. Successive stages of printing and duping eventually produced the elements that could be used for the optical combination of the subject matter with the background existing on another length of film.

The composite effect left much to be desired. The many printing and duplicating steps needed for the preparation of the mattes, along with film shrinkage and the subsequent loss of registration, produced matte lines and fringing around the inserted image. The effort did, however, lay the groundwork for many of the self-matteing processes in use today.

blue-backing self-matteing process. Also called **color-separation self-matteing system**. The end products of the filming, optical, and laboratory processes provide the components that are needed for the optical combination of the insert material or foreground image with a background existing on another length of film. This is true of most matteing systems.

The success of the blue-backing matteing system, however, is based on the ability of a color film's several layers of emulsion to absorb certain colors while transmitting others. This degree of selective sensitivity in an emulsion makes it possible to record certain colors on the in-camera film during the photographic process and negate other colors by laboratory processes.

During the filming process, the subject matter is illuminated with conventional white light in front of a blue-painted flat or dyed translucent screen. The color negative, after development, is used to produce the matteing elements as well as the master positives that are needed for the optical combination and the achievement of the composite effect.

In the first of the basic intermediate printing steps, by either laboratory or optical processes, the original in-camera color negative is step-printed with a black-and-white master positive. Since only the blue is recorded, the blue background area in the resulting separation positive is rendered clear.

To record the subject matter's red components, the process is repeated, and the original color negative is again printed in contact with a black-and-white master positive. The blue backing, in this color-separation positive, is rendered black.

The high-contrast film produced from this color-separation positive has the subject matter as a negative image and the background as a clear area. The female, or clear core matte, results from the optical combination of the blue positive and the red duplicate negative. The high-contrast film, after processing, shows the subject matter as a clear area against a black, opaque background.

During the optical combination, the master positive containing the background picture is printed with the high-contrast male matte, the one in which the subject matter is represented by a black area surrounded by a clear area. On a second run, the master positive that has the subject matter is printed with the female high-contrast matte. The subject matter is thus inserted in the previously unexposed area of the film frame while the matte prevents the background, exposed during the first optical run, from being contaminated.

The film exposed in the optical camera, after processing, has the subject matter precisely positioned over a background that may have been photographed years earlier and thousands of miles away from the area in which the performers were exposed in front of a blue background.

The second of the color-separation blue-backing systems also uses a color negative film to produce the elements that are needed for achieving the composite effect. Unlike the conventional white lighting that is used to illuminate the foreground subject matter in other blue-backing processes, yellow-filtered or orange light is used.

After the original color negative has been developed, color-separation filters are used to absorb the blue in the background while transmitting the yellow or orange light in the foreground material. The result is a self-matteing black-and-white master positive.

To produce the traveling matte, the process used previously for making the master positive is reversed. This time, the color-separation filters transmit the blue in the background but absorb the orange or yellow light from the foreground material. The result is a clear core matte on high-contrast film with the foreground subject matter as the clear area and the blue backing as the black. This high-contrast print is used to produce the shadow side, the matte in which the subject matter is opaque black and the background area is clear.

During the optical combination, the background is exposed with the male matte, the one in which the subject matter is black and the surrounding area is clear. In the second run, the film that has the subject matter is printed with the female high-contrast matte, the one in which the subject-matter area is clear and the surrounding area is black. The film exposed in the optical camera, after processing, has the composite effect.

brightness-separation overlay. Also called **combined-montage amplifier** and **moving electronic matte**. An electronic matteing process that makes possible the combination of picture information from two separate sources for the purpose of forming a composite picture. The success of the process depends on the selective sensitivity of the two cameras in use, as well as on the degree of contrast in the background and subject matter.

In this automatic electronic switching system, the subject matter is positioned in front of either a white or black background, depending on the dominant color in the subject matter. If the tonal values are relatively light, the background is black. If the tonal values are generally dark, the background color must be white.

The camera focused on the background "sees," evaluates, and electronically selects and exposes those portions of the background in which the background color is the dominant one. The second camera, the one focused on the subject matter, makes the same electronic evaluations and exposes accordingly. In the composite picture, only those portions of the background in which the tonal values are the dominant ones are seen. The subject matter or insert material, selected by the second camera, is seen in its position over the background.

The process has its shortcomings in that portions of the background may seep through parts of the subject matter. This happens when the values selected by the two television cameras are almost similar.

chroma-key. A matteing system in which complementary colors are used to mask specific areas of the picture in order to insert other material in the previously matted areas. This matteing process is used in television and videotape production and is based on the electronic selectivity of complementary colors in both background and subject matter.

color-difference self-matteing process. Developed by Peter Vlahos, all the elements needed to produce the composite effect in this self-matteing system are derived from the original in-camera negative. In this, as in other self-matteing systems using blue backings, the subject matter or foreground material is illuminated with white light and photographed in front of a blue-painted flat or translucent screen impregnated with dyes of the same color.

Black-and-white separation positives are made from the original negative. In the blue separation positive, the blue background around the subject matter is a clear area. This clear area is subsequently rendered black by exposing the original color negative and green separation positive to blue light. The backgrounds, in both the red and green separation positives, are also black. Each of the three separation positives, with its own black background, is at this stage self-matteing. The negative and positive mattes that are needed for the optical combination are prepared by the laboratory, on high-contrast film, after a series of printing and duping processes.

This system has several advantages over other self-matteing systems along with some minor disadvantages. With each of the separation positives having its own black backgrounds, many of the problems connected with fringing around foreground objects are minimized. Similarly, ghost images resulting in semitransparent items such as smoke and glass, are practically nonexistent. Another plus. The big minus lies in the system's inability to reproduce all the values in the blue and violet range.

Improvements in the system are credited to Walter Beyer of Universal Studios and W. E. Pohl of the Technicolor Corporation.

color-separation electronic matteing system. With this electronic separation system, the subject matter and background are each lit separately and photographed by two cameras through a common mirror system.

The subject is illuminated by yellow light and is photographed by the first camera through a yellow filter. This produces a normally lit figure against black. The second, or keying camera, focused on the blue background, shoots through a blue filter and "sees" a silhouetted figure against a relatively bright background.

The subject seen by camera one is electronically matted into another scene by the keying signal from camera two which actuates special circuits. Switching, between the picture information from camera one and the background scene, is automatic. See Figure S-6.

Fig. S-6 Color-separation electronic matte.

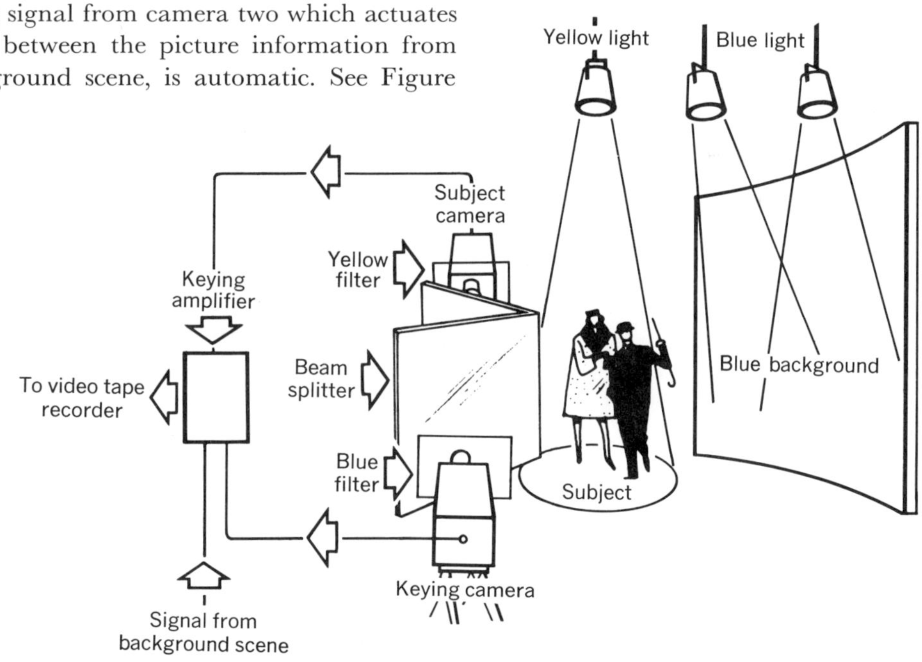

Dunning process. During the 1920s, C. Dodge Dunning developed a process for insetting the performers on a set into previously photographed backgrounds to produce a direct, in-camera composite negative.

The performers were positioned in front of a blank screen illuminated by blue light. Orange-filtered light was used for illuminating the performers. In the camera, the background scene on the master positive ran bipack with the raw stock or negative film. The master positive had previously been bleached and dyed orange.

During exposure, the orange-filtered light reflected from the performer passed through the master positive to expose the panchromatic stock. The dyed portions of the background scene had little effect on these images. The blue light that was being reflected back from the screen on the set served as a printing light for the background on the master positive. Wherever the performer's body blocked the passage of blue light from the screen, the background could not print through to the raw stock in the camera, thus making the performer a matte for himself.

The painstaking dyeing and bleaching of the master positive and the complicated preparations that preceded the actual filming process did not discourage other pioneering technicians who used Mr. Dunning's process as a basis for the development of more efficient matteing systems.

Pickley process. Also called **infrared self-matteing system**. The Pickley process is a self-matteing system that uses plus-infrared lighting for the background and minus-infrared lighting to illuminate the subject matter.

A beam splitter, within the dual-film Technicolor camera that is used during the photographic process, exposes the background and foreground subject matter on separate lengths of film—the matte record on high-contrast film and the subject on color stock. Both films are exposed simultaneously through a common lens. This patented process was invented by Leonard Pickley and is licensed by MPO Videotronics, Inc., New York.

This method of self-matteing is also referred to as the SPECTRUMATTE SYSTEM.

Pomeroy process. A process developed during the late 1920s by Roy J. Pomeroy for insetting the performers on the set into previously photographed backgrounds to produce composite effects by direct, in-camera filming.

An improvement over the Dunning system developed several years earlier, the Pomeroy process combined some of Dunning's principles with the "glass" shot. During the filming process, the performer was sandwiched between a large transparency, positioned before the camera, and a blank screen. The transparency, bleached and orange-dyed, contained the background for the composite scene.

Orange-filtered light was used to illuminate the performers and blue light for the blank screen behind them. Short-focal-length lenses and relatively small apertures were needed to ensure a composite in which both the performer and the transparency would be in acceptable focus.

This system had several advantages over the Dunning process in that bipacking was eliminated, thus making it possible for a producer to use conventional cameras during the filming process. And, since it was possible to "see" the composite effect in the viewfinder, composition became a matter of choice rather than a hit-or-miss, hope-for-the-best effect.

While the Pomeroy and Dunning processes left much to be desired, they did lay the groundwork and establish some of the outlines for future matteing processes. For this, the industry is indebted to these pioneer technicians. See GLASS SHOTS.

sodium self-matteing system. A multifilm, self-matteing process that is similar in operation to the infrared and ultraviolet systems. The yellow background used for this matteing process is front-lit by sodium-vapor lamps which produce monochromatic yellow light. Incandescent lamps with didymium-coated filters that subtract monochromatic yellow are used to light the subject matter in the foreground.

The beam splitter, in the multifilm Technicolor camera that is used during the photographic process, transmits monochromatic yellow light to the high-contrast film (type 5362). The foreground image is exposed on Eastman color negative (type 5251).

The sodium system has one distinct advantage over both the infrared and ultraviolet systems. The matte image produced on the high-contrast film duplicates the size of the foreground image on the color negative stock and does not have to be enlarged or reduced in the optical printer during the combining process.

ultraviolet self-matteing process. Similar in operation to the infrared self-matteing system, the subject matter and background are each lit separately and exposed through a common lens and beam-splitting prism in the dual-film Technicolor camera on separate lengths of film.

Incandescent lamps, with filters for subtracting the ultraviolet radiation, are used to illuminate the subject matter in the foreground. The background is a clear translucent screen which is lit from behind by ultraviolet light. Fluorescent lamps are used as the light source.

The coated beam-splitting prism in the camera reflects ultraviolet radiation to the high-contrast stock that is used for recording the matte image. The color negative for the subject matter receives only the visible wavelengths, the ultraviolet being absorbed by the coating on the prism.

The patents covering this self-matteing system are owned by Warner Brothers Studios.

selsyn motor Also called **interlock motor**. A **sel**f-**syn**chronous combination (from which the term is derived) of motor and generator. This type of motor is used to drive related equipment (the camera and projector in process shots, for example), keeping both in synchronization during the filming process.

semisilvered mirror A front-silvered mirror from which portions of the silver surface have been removed to produce a clear area in the glass. The removal of a portion of the silver makes it possible for the camera to photograph a subject through that particular area while simultaneously exposing anything that is being reflected in the mirror. The mirror is thus capable of reflecting as well as transmitting light. The Pepper's ghost shot and the Schufftan process owe their success to this type of mirror.

senior Solarspot An all-purpose heavy-duty spotlight that is especially suitable for floodlighting. The senior has a 14-inch Fresnel lens and takes a 5,000-watt bulb. See Figure S-7.

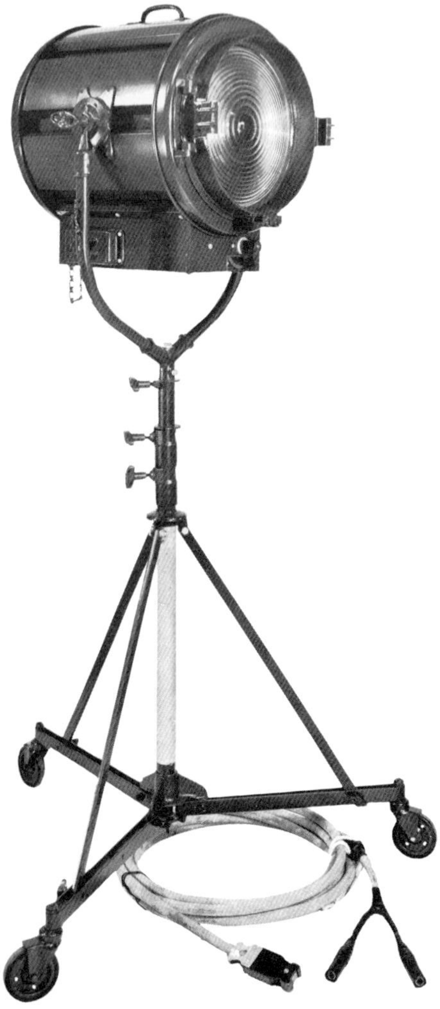

Fig. S-7 Senior Solarspot. (*Mole-Richardson.*)

sensitivity An emulsion's sensitivity to light and other radiations. The degree of sensitivity or the speed of the emulsion is indicated by *speed* numbers. See EXPOSURE-INDEX RATINGS. Spectral sensitivity is the sensitivity of an emulsion to specific colors. As applied to television, sensitivity is a receiver's ability to reproduce comparatively weak signals with acceptable quality.

sensitivity control A television camera setting that controls the operation of the iris servo amplifier for correct exposure.

sensitized material An all-inclusive term for motion picture film as well as photographic paper and plates that have one or more light-sensitive layers of emulsion on a supporting base.

sensitizer A solution capable of making the surface to which it is applied sensitive to light.

sensitometer A measuring device used for determining the sensitivity of a film's emulsion. See SENSITOMETRY.

sensitometry Sensitometry is a mathematically precise method of measuring the image-recording properties of an emulsion. These measurements, based on the response of an emulsion to light, are subsequently used to control film exposure and processing.

A sensitometer creates a series of precise exposures on a length of film by means of a step wedge so that the developed result is a number of bands of uniform density. Each exposure in the series is

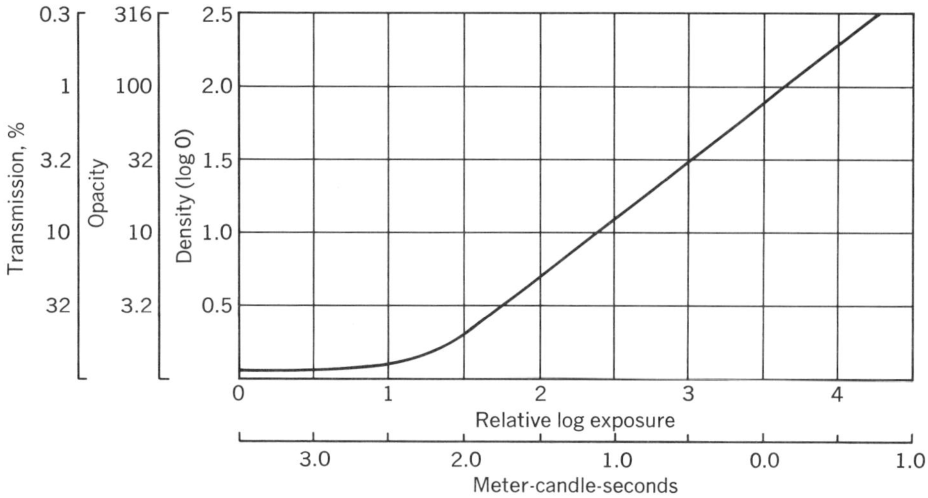

Fig. S-8 Sensitometry. (*Ilford, Ltd.*) The vertical, or ordinate, scale shows how the developed image may be measured in terms of the percentage transmission of light, or the opacity which is the reciprocal of the transmission, or density in terms of log opacity. It is this last measurement which is commonly used and which provides the units of density that are normally quoted. The horizontal scale relates the actual exposures in meter-candle-seconds to the steps created in the sensitometer and shows how these relate to the log exposure scale normally used. The straight-line portion of the curve shows that in this region density increases in proportion to exposure. It can also be seen that the curve has a minimum density value (since it does not touch the base line) which is the density of the film base plus fog; this density may increase with development increases but is unaffected by exposure variations.

greater by a calculated amount than the preceding step, and the exposure range covers all practical conditions.

A densitometer is used to measure the densities of the developed film in terms of logarithmic units. See Figure S-8. Compare DENSITOMETER.

separation The basis of successful color photography. All the colors visible in the subject matter must be separated into three individual components: blue-violet, green, and red-orange. See COUPLER DEVELOPMENT, DYE-TRANSFER PROCESS, SEPARATION POSITIVE, and TECHNICOLOR THREE-COLOR SYSTEM.

separation filter See SELECTIVE FILTER.

separation negatives A set of negatives made from a positive color transparency (as in the Kodak dye-transfer process). These negatives are processed to produce relief images in the gelatin. The thickness of the relief images is proportional to the density of the image. This process makes it possible to control each of the component colors that make up the composite image. See DYE-TRANSFER PROCESS.

separation positive A black-and-white film in which the red, green, and blue image component of a color negative appears as a positive image. The separation positive is produced from a color negative by printing through filters on panchromatic black-and-white film. See PROCESSING and flow charts under that entry.

sequence A related group of scenes in a film with each sharing a common location or period of time.

serial photographs A succession of still pictures exposed by a mechanical tripping device. This method of exposure is made possible by using a number of still cameras that are positioned at different points within which the action will occur. In an alternative method, the action is stopped by using a single camera and repeated shuttering to obtain a multiple-exposure effect on a single film or plate.

servo systems Electric mechanisms such as the selsyn motors that are designed to control the operating speed of one or more motors driving interrelated equipment. These mechanisms are used, for example, for synchronizing and controlling the speeds of several camera shutters, operating in matching positions, during the filming process. They are also used to link a television camera's output with a remote-controlled console. See SELSYN MOTOR.

set, the Backgrounds, as the name implies, should be exactly that and nothing more. Like lighting, they should not compete with, detract from, or attempt to share the spotlight with the on-stage performers. Paradoxically, poor representation, rendering, perspective, and coloring can "draw" the audience's attention away from the spotlighted thespians. Backgrounds must, therefore, be accurately represented, subtly rendered, and remain where they belong—in the background.

Chronologically, the history of scenic art is as old as the dramatic presentation itself and can be traced back to the days when individual performers and itinerant troupes sang, danced, and acted their way from arena to amphitheater. Early stages were usually barren, and audiences were required to use their collective imaginations for visualizing nonexistent exits, curtains, and other miscellaneous items usually associated with a theater stage. Props were used sparingly and only when their presence was absolutely essential for conveying thoughts that needed visual support.

Moved indoors, early backgrounds crudely suggested the general setting with painted backdrops of various interior and exterior locations. Artists made no attempt at realism but contented themselves with reasonably adequate representations that satisfied the story and mood requirements of the play being enacted.

During the pioneer days of the motion picture industry, these crude scenic impressions were replaced by rendered "flats." The critical eye of the camera, however, focused attention on the inadequacies of the representations. The studio theater thus became the classroom, and the small area set aside for set construction on the back lot became the laboratory. In this setting, scenic designers and art directors fashioned their trades. Because studio facilities were available and location filming was so costly, the pioneer producers naturally favored studio production. The economic advantages of shooting on the studio set were immediately obvious. But the studio set, although convenient, still had to be transformed into an appropriate setting, either interior or exterior, that duplicated the specific locale referred to in the script.

The critical demands on art directors and scenic artists paralleled the higher standards required in motion picture production. Earlier efforts and methods were no longer acceptable and representations were no longer adequate substitutes for realism.

Today the art director and scenic designer are responsible for transforming the barren studio stage into an accurate setting. The description, or word picture, must first be visualized and then faithfully translated into layouts, drawings, and miniature models that serve as guides for the construction of the flats and the renderings brushed on their canvas surfaces.

In the initial stages of production, discussions among the producer, director, writer, cameraman, and art director result in a common understanding of the overall scenic requirements. These requirements are diagrammed and later given dimension. Since the details of construction, painting, and decoration are done by a small army of artists and technicians, the ability to communicate visual concepts to others, through diagrams and layouts, is an extremely important part of the art director's job.

Besides being familiar with each of the recognized arts, the art director should have a working knowledge of photography and a general understanding of the basic production processes. In addition to this, he must be able to evaluate the advantages of matte paintings, glass shots, rear-screen projection, and photographic prints and transparencies as alternatives or additions to the standard rendered backgrounds.

After the preliminary layouts have been completed and approved, they are redrawn to scale and traced onto hundreds of square feet of canvas. The camera position becomes the focal point and is used to establish the horizon line and determine perspective. Individual styles and techniques have no place in the giant pattern and the actual rendering follows an established routine. The sky area is painted first, distant objects next, and the foreground last. Detail and shadows are, in turn, painted over the previously rendered areas. Telephone poles, ship's masts, trees, and similar objects, extending through one or more planes, add depth to the background and are painted last.

Foreign settings, and the architecture and foliage native to the particular area, must be researched in order to achieve technical accuracy in subsequent renderings. Similarly, the ornamentations and trappings of different periods must be duplicated in minute detail if the set is to look authentic. Inaccuracies in rendered detail will, almost always, be detected and not by the expert alone. Even the casual observer becomes a vituperative critic, and the accompanying flood of letters, pointing out the discrepancy or inconsistency, serves as a reminder that there is no substitute for realism.

In place of the normally accepted artists' tools, the scenic artist uses large brushes, paint rollers, and spraying devices to cover the huge canvases that are stretched over the wooden framework of the giant flats. On the scaffolds that carry the artists high over the set, brushes are dipped into gallon cans, and the primer and paint is applied with bold strokes. Backgrounds, up to 40 feet high and 100 feet long, absorb considerable quantities of both.

Wide-screen productions have affected the size of the background flats proportionately and added to the woes of art directors, scenic artists, and construction crews. Since each of the camera positions requires its own picture plane, and the horizon line must, of necessity, remain constant, each of the planes must have its own vanishing point and corresponding perspective. This creates unprecedented problems commensurate with the size of the screen itself.

Unlike the painted flats that are prepared for use as backgrounds, the art director must be capable of designing "architecturally." He must evaluate the shooting script and, when necessary, turn the designing of the background over to draftsmen, who then prepare the plans needed for the construction of buildings and other relatively large props that serve as backgrounds. These dimensional structures must, of course, be built to scale and include all the recognizable details normally associated with the particular type of structure.

With the exception of those assignments that specify "the interior of a submarine," similar to the one constructed for the TV series, "Voyage to the Bottom of the Sea," or for an *Enterprise* type spaceship ("Star Trek"), most architectural assignments are generally regarded as routine.

Many major studios have built-up areas (on studio grounds) with houses of every type and period lining both sides of paved streets. Their false fronts are, indeed, misleading since well-kept lawns, shrubs, and trees (along with curtained windows) give these houses a lived-in look.

Actors walking through the front doors of these dwellings, however, can look back at a braced framework that supports the front of the structure. Obviously, filming and editing are dependent on each other for the achievement of realism. As the actor walks through the front door, the editor cuts, fades out, wipes, or cross-dissolves to the scene that follows. In the next shot, the actor may be photographed from a reverse angle and be shown (on another set) closing the door through which he apparently entered. In subsequent scenes, the same actor might be shown climbing a long flight of stairs (on a third set) that lead to—a solid studio wall.

Although each of these sets, both exterior as well as interior, may be some distance from one another, a relationship will be apparent and a sense of continuity will be created when the individual scenes are cut together by the editor. In all cases, however, the lighting must be consistent and natural to the scene.

Productions photographed in color magnify some of the art director's problems and minimize others. Generally, the advantages outnumber the disadvantages. In the black-and-white production, for example, the designer is limited to black, white, and approximately eight shades of gray. For the color production, however, the art director can choose from an infinite number of tints and shades in the entire visible range of the spectrum. The wide choice makes it easier to achieve a more favorable balance between subject matter and background, and virtually eliminates the problems normally associated with contrast between the two elements. Obviously, the greater freedom and latitude must be used wisely and with a measure of restraint. The effects of difficult colors (yellow, for example) on the sensitive layers of the film's emulsion must be taken into consideration, and compensations must be made to ensure their accurate reproduction. Where a doubt exists, the color in question can be applied to a flat and photographed under conditions similar to those that will exist during the actual filming. The processed film will serve as a guide.

In addition to the responsibilities connected with the production of the motion picture to which he is assigned, the art director must also be aware of the requirements and limitations of the neighboring medium—television. The choice of colors, in black-and-white production, is limited to black, white, and only five shades of gray compared with the eight shades available in motion picture production. Thus, if the subject matter is rendered in the middle shades of gray, no. 3 or 4, for example, the background and other objects in the scene can only be rendered three shades lighter or darker than

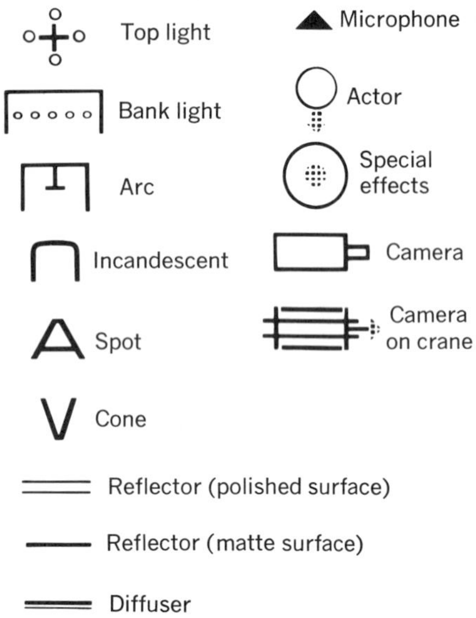

Fig. S-9 Symbols for planning lighting, action, and camera movement.

the shade selected for the subject matter. Obviously, the art director and scenic artist must plan carefully and confine their choices to the limited range of tonal values.

The use of stark whites and blacks over large areas is not recommended. To avoid the monotony of large, flat-toned backgrounds, soft-focus and shadow patterns are created through the use of kukaloris effects. The *cookie*, a flat piece of metal with irregular-shaped cutout areas can be positioned in front of a light source that is focused on the background to provide interesting designs that help relieve the otherwise monotonous effect.

As in motion picture production, the art director has at his disposal an infinite number of tints and shades from which to select colors that will create the desired separation between background and subject matter. Since the degree of contrast is a matter of choice, the art director must make his decisions wisely—and with a measure of restraint.

Black, in costumes, decorations, and props, loses detail and should never be used with dark backgrounds. Bold, evenly spaced horizontal and vertical stripes should also be used with discretion to avoid the accompanying moiré interference in the TV receiver's scanning lines. Small patterns, usually wasted, and designs that are capable of causing distortion, of one kind or another, should also be avoided.

In television production, the art director is not only responsible for the design, construction, and decoration of the set but is also usually held accountable for the lighting. Fortunately, many of the rules that guide the lighting of a motion picture set also apply to the TV studio stage. See Figure S-9.

set lights Illumination focused on the background rather than on the performers or other subject matter on the set.

setup The term used to describe the arrangement of cameras, lighting units, and sound-recording equipment in relation to the performers and the background.

In videotape recording, the level of black in a picture. When the level is raised, the picture is washed out proportionately.

shading A monitoring process during which necessary light adjustments are made so that the tones and values in the image produced by the television camera's pickup tube will approximate those in the original subject matter.

shadows Unless they serve a specific purpose, shadows are generally undesirable and may even be objectionable. Paradoxically, when they are used for dramatic effect or for decorative purposes, shadows play a silent but important role in motion picture production.

Complementing the eerie atmosphere in mystery or horror pictures, shadows are quite often more frightening than the strange characters from whom they originate. As a visual device for suggesting the time of day or as a barometer that reflects the prevailing weather conditions, shadows provide immediate clues by their very size and direction.

For decorative purposes, shadow effects are invaluable for relieving the monotony of solid-toned backgrounds. These kukaloris effects are produced by placing a flat piece of cardboard or metal with irregular-shaped cutout areas in front of a light source. When projected on studio backgrounds, the cutout areas produce a wide variety of interesting dark and light shadow patterns.

The shadows cast by a performer can be made larger or smaller by changing the relationship between the performer and the light source. Projected on studio backings with subtle softness or focused to crystal clarity, shadows can also be distorted beyond recognition by angling either the light source or the camera in relation to the performer. If desirable, several light sources can be used simultaneously to create interesting multiple-shadow effects.

The illusion of movement, in stationary objects on the studio set, can be achieved by rotating circular pieces of cardboard in front of a light source. See CELLO GLASS. As the irregular, cutout areas in the cardboard pass in front of the light source, corresponding light and dark shadow patterns are alternately focused on the faces of performers or flashed across stationary objects on the set. Especially effective when focused on foreground subject matter during the filming of process shots, these passing shadows that suggest moving trains, speeding vehicles, or overhanging branches are amazingly realistic.

sharp A term used to indicate the amount of detail that can be exposed on film and subsequently reproduced in print material. A comparative term, image sharpness or acutance relies on visual observation and assessment of a number of images photographed under similar conditions with different types of equipment. See CIRCLES OF CONFUSION, DEFINITION, DEPTH OF FOCUS, and RESOLUTION.

sheet film Also called **cut film**. Light-sensitive layers of emulsion (negative) supported by a film base and cut in a number of sizes to suit individual needs.

shoot The process of exposing film.

shooting schedule The order of filming indicated by convenience and economics rather than by plot continuity. Scenes using the same background or location, whether at the beginning or end of a motion picture, are photographed together.

shooting script The blueprint for production. The shooting script, in addition to the dialogue, contains information on the location, camera movement, stage business, and other details.

short Also called **featurette**. A relatively short motion picture with a running time from about 5 to 20 minutes. These films cover a wide variety of subjects and usually precede the showing of the featured motion picture. The popular travelogue series is included in the general category.

short-focus lens A lens with a short focal length. Lenses of this type do not cover the diagonal of the negative. They produce a smaller image, at the same distance, than those yielded by normal or long-focus lenses but cover a greater part of that image. Compare LONG-FOCUS LENSES.

shot (1) A general term that is used to indicate the setup for a scene. (2) A description of the scene itself.

shoulder pad A double-shoulder tripod * that can be used with all conventional cameras having $\frac{1}{4}$- or $\frac{3}{8}$-inch base sockets. The tripod affords mobility, leaving both hands free for handling the camera and accessories.

shutter The mechanical device that prevents light passing through the lens from exposing the film while the individual frame is being brought into position for exposure.

The rotary dissolving shutter in the Mitchell NC described here, is designed for hand-operated dissolves and fades. It has a maximum opening of 175 degrees, calibrated in increments of 10 degrees, and can be closed down to any smaller opening and locked and operated in any desired position. The shutter-opening scale and hand-dissolve lever are mounted on the rear of the camera. A miniature shutter at the rear of the camera shows visually the relation of shutter opening to aperture. The shutter is synchronized with the movement so that when its open segment is in front of the aperture, the film is at rest. The film moves while the opaque segment of the shutter covers the aperture.

The shutter in the Mitchell BNC model is the automatic-dissolving type. Two buttons at the rear of the camera operate the shutter automatically to make fades, either in or out. A hand-dissolve lever is also provided. Shutter position is indicated by a pointer which moves in a calibrated segment. This segment is graduated from 0 degrees to the maximum shutter opening of 175 degrees. A monitor-

*Cine 60 Prenzel.

ing miniature shutter is visible through a window at the rear of the camera.

Since the industry's infancy, three basic designs (the diaphragm, focal plane, and sector shutter) have set the pattern for most of the shutters in use today. These, along with several other basic types, and their method of operation are described in the following paragraphs.

A between-lens shutter makes use of a thin blade between front and rear lens elements. This type of shutter is especially popular in cameras that have no provision for the interchanging of lenses. The well-known Compur shutter is of this type. The term *between-lens* shutter is also used to describe shutters that are positioned behind the lens.

The diaphragm shutter has joined, overlapping metal blades that open outward to form a circle through which light passes to expose the film. The size of the opening controls the amount of light transmitted through the lens. The efficiency of this type of shutter is high since a minimum amount of time is required for the opening and closing action of the metal blades. See SHUTTER EFFICIENCY.

The Everset shutter has a spring-loaded trip which activates two metal plates—a shutter plate and a capping plate. When the shutter plate is activated, the capping plate swings clear of the lens and allows light to be transmitted to the film. The capping plate automatically returns to its original position after the exposure has been made. This type of shutter mechanism is usually found in relatively inexpensive cameras. The fact that it does not require tensioning before being activated makes it convenient for average use.

The focal-plane shutter is designed to operate at relatively high speeds in a plane close to the film's light-sensitive layer of emulsion. This curtainlike device, either cloth or metal, has a slit that varies in width. The speed with which the shutter moves and the width of the slit are the factors that determine the length of the interval of exposure.

Similar in operation to the focal-plane shutter, the guillotine shutter has two metal blades, spring-actuated, that move in a flat plane over the film's surface during the interval of exposure. Also known as a *drop* shutter, its original design made it impractical for use with today's fast emulsions. Subsequent modifications have improved its efficiency, and instantaneous exposure is now possible with this type of shutter.

Designed for high speeds, the Kerr cell shutter has two crossed polarizing screens that have liquid mounted between them. A high-voltage current changes the polarization of the liquid and allows light to be transmitted through it. This shutter makes possible exposure times of 1/100,000,000 second.

A mirror shutter is a reflective camera shutter that makes it possible for the cameraman to view the subject matter as it is being exposed. The images are reflected into a prism and, in turn, viewed through a lighttight finder.

The original sector shutter had a round hole cut into a circular plate and was mounted either in front of or behind the lens. In some cases, the plate was mounted between the lens elements. As the circular plate turned, light passing through the hole in the plate was transmitted through the lens and exposed the film. The amount of exposure was determined by the speed of the rotating plate and the length of time needed for the hole to pass in front of the lens.

The variable shutter has an adjustable blade (or secondary shutter) which can be locked in different positions. These shutter positions are usually indicated by a pointer which moves in a calibrated segment. This segment is graduated from 0 degrees to the maximum shutter opening in 10-degree steps. This type of shutter permits the cinematographer to control exposure without changing *f*-stops. Transitional effects such as fades and cross-dissolves can be made with this type of shutter.

shutter, lighting A slatted metal frame used for controlling the intensity of a light source.

TABLE S-1 Shutter Exposure Time Obtained with Various Camera Speeds and Shutter Openings

In Fractions of a Second

	Camera speeds			
Shutter opening degrees	8 frames per second	12 frames per second	16 frames per second	24 frames per second
170	1/17	1/25	1/34	1/51
160	1/18	1/27	1/36	1/54
150	1/19	1/28	1/38	1/57
140	1/20	1/30	1/41	1/60
130	1/22	1/33	1/44	1/66
120	1/24	1/36	1/48	1/72
110	1/26	1/39	1/52	1/78
100	1/29	1/43	1/58	1/87
90	1/32	1/48	1/64	1/96
80	1/36	1/54	1/72	1/108
70	1/41	1/62	1/82	1/123
60	1/48	1/77	1/96	1/144
50	1/57	1/91	1/115	1/182
40	1/72	1/108	1/144	1/216
30	1/96	1/144	1/192	1/288
20	1/144	1/216	1/288	1/432
10	1/288	1/432	1/567	1/864
5	1/576	1/864	1/1152	1/1738

Courtesy of Mitchell Camera Corp.

shutter efficiency A measure of the true amount of light that is being transmitted through a lens during the exposure cycle. The exposure cycle is the interval between the time the shutter begins to open and the time when it is in its completely closed position. The amount of light passing through the lens during that interval is measured against the amount of light that would have entered the lens if no time had been consumed by the shutter's opening and closing action. The efficiency of the shutter is determined either by sensitometric measurements or by oscillograph.

shutter speed, minimum The minimum shutter speed that may be used efficiently in order to prevent image blur in moving objects. See BLURRING.

shutter time The specific amount of time allotted for exposure. See Table S-1.

shuttle The camera mechanism in which each frame of film is held in precise registration during the interval of exposure.

sideband Frequency bands immediately above or below the frequency of the carrier wave.

signal Electric impulses that are subsequently converted into visible pictures or audible sound.

signal-to-noise ratio A measure of the ratio between (1) the maximum undistorted signal output and (2) the noise level output from material that does not carry a signal. Signal-to-noise ratios depend much more on the equipment in use rather than on the recording tape. Because of this, an alternative figure of great interest is the ratio of maximum undistorted output to saturation noise. It is important to check which type of signal-to-noise ratio is meant when any figures are being compared. Signal-to-noise ratios are expressed in decibels. The average signal-to-noise ratio for recording on 35mm film is 40 to 50 decibels. For 16mm film, the average is 35 to 40 decibels.

signal-to-noise ratio, unweighted standards (NAB) Unweighted noise is measured over the frequency range of 20 cycles per second to 20 kilocycles. The response of the measuring system should be uniform ($\pm$0.3 decibels from 30 to 15,000 cycles per second).

Tape speed	Full track	Two track	Four track
15 ips*	50 db	45 db	Not used
7½ ips	50 db	45 db	45 db
3¾ ips	46 db	46 db	45 db

*Inches per second.

signal-to-noise ratio, weighted standards (NAB) Weighted noise is measured using the weighting curve based on the ASA "A" curve (ASA Standard S1.4). The noise measurement is made using a tape previously recorded with bias but with no signal. Calibration is made (with the weighting network inserted) at 1,000 cycles per second, using the 1,000-cps standard level which is included for this purpose on the NAB standard test tape.

Tape speed	Full track	Two track	Four track
15 ips*	58 db	53 db	Not used
7½ ips	60 db	55 db	52 db
3¾ ips	57 db	54 db	52 db

*Inches per second.

silent track An unmodulated area of track that is inserted between two adjoining sections which contain music or dialogue. Used to eliminate dead spots in the sound track, the variable-area silent track shows only the bias lines. In the variable-density silent track, the entire width of the track area is exposed to a light or medium shade of gray. See VARIABLE-AREA and VARIABLE-DENSITY SOUND TRACKS.

silhouette An effect in which the subject matter is illuminated by backlighting so that, in the print, it is almost completely black with little or no distinguishable detail.

silk Translucent material, usually silk, positioned in front of light sources and used for diffusion.

silver bromide One of the halogens, yellow in color, silver bromide is a light-sensitive salt and the principal constituent in the manufacture of negative emulsions. Silver bromide is used alone or in combination with chloride or iodide. The compound turns dark when exposed to light.

silver chloride This compound is a light-sensitive salt in positive photographic emulsions. Unlike the relatively fast silver bromide emulsions, this emulsion is quite slow. White in its pure form, silver chloride turns blue quickly when it is exposed to light.

silver halide This compound is found in film emulsions that contain any of the light-sensitive halogens such as bromine, chlorine, or iodine. After exposure, the halides form latent images which are subsequently developed to produce visible silver images.

silver iodide A light-sensitive salt that is added to and used in combination with other silver compounds in fast negative emulsions. This pale yellow compound darkens when it is exposed to light.

silver nitrate An ingredient found in all emulsions. It is used in intensifiers and is responsible for the formation of bromide and chloride in the emulsion during the development process. The colorless crystals dissolve readily in water.

simulcast The simultaneous broadcast of program material, over radio and television facilities. The programs originate from the same radio station or point of transmission, in both AM and FM.

single broad An all-purpose floodlight using 500- to 750-watt bulbs. Excellent as a fill light, the single broad produces a soft light that minimizes shadows.

single-system recording Also called **single negative sound**. The process of recording picture and sound simultaneously on the same film. Used mainly for newsreel work, or in those cases where the quality of the sound track is of secondary importance, the sound and picture are developed together. Synchronization is, of course, automatic.

In double-system recording, the picture information is exposed normally in one camera while the sound is recorded on another length of film (sound positive stock) in a sound camera, or on ¼-inch tape in a magnetic recorder. The sound positive stock used for recording is exposed as a negative and produces a black image when the film is developed.

The motion picture camera and recorder are both driven by a synchronous motor. This system provides a greater degree of control with regard to the quality of both sound and picture than is possible with single-system recording. Additionally, the editor has more freedom and flexibility in synchronizing picture and sound.

skewing In television, picture distortion caused by improper horizontal alignment. The resultant zigzag pattern is corrected by Amtec. See AMTEC. Compare SCALLOPING.

skip frame The process of eliminating frames from the film being duplicated and exposing the selected frames on the raw stock in the optical camera. Skip framing is used to speed actions or to shorten scenes. See OPTICALS.

TABLE S-2 Sky-brightness Rendering

Negative material	Kodak Wratten filter	Monochromatic rendering of a clear blue sky
Non-color-sensitized	None	Lighter than correct
Orthochromatic	None	Lighter than correct
Panchromatic	None	Lighter than correct
Orthochromatic	K2 (no. 8)	Practically correct
Panchromatic	K2 (no. 8)	Practically correct
Panchromatic	G (no. 15)	Darker than correct
Panchromatic	A (no. 25)	Very dark
Panchromatic	F (no. 29)	Almost black
Infrared-sensitive	A (no. 25)	Black

sky-control filters Graduated sky-control filters produced by Harrison help prevent overexposure and the washed-out effect of sky and clouds in location photography. Produced in a number of combinations, one-half clear and one-half color, half one color and half another color, or in combinations ranging from neutral density to clear, sky filters have been used to create beautiful sky effects in black and white as well as in color where only drab gray skies were in evidence during the actual filming process.

Although the standard combinations, color to clear, color to color, and neutral to clear, have been used for more than a quarter of a century with excellent results, new filters have been developed that complement new film types. The 85-to-clear filter, for example, produces beautiful blue sky effects with foreground correction. Similarly, the 85-to-neutral filter enhances the blue in sky colorings when negative color film is being used, or half-blue-to-clear to produce blue skies with daylight color film.

The line of demarcation between the clear area and the color portion of the filter creates no serious problem during the filming process since the filter is positioned close to the lens and is, therefore, out of focus. See Table S-2.

sky light A lighting source used for illuminating relatively large areas. It is especially suitable for lighting backgrounds.

sky pan A powerful light source that can be used in combination with other sky pans to produce a wide expanse of shadowless illumination. Sky pans can be suspended from scaffolding or mounted on stands. The term is also used to describe small reflectors that are positioned in front of light sources.

slant track See HELICAL SCAN.

slate A card or blackboard that contains information pertinent to the scene or scenes being photographed. Exposed for several frames immediately before each take of a scene, the slate provides the film editor with a means for identifying the miles of footage that will subsequently pass through his moviola.

sled A slang term for a lighting accessory that is used to support lamps that must be positioned over a set. Like the "trombone" devices that serve almost similar purposes, these supports are suspended over a stage and not attached to walls.

slidefilm Unlike any of the other motion picture techniques, each frame in a slidefilm tells its own story. Although standard motion picture film is used during the photographic process, each frame of the slidefilm is projected in sequence on specially designed equipment for the length of time needed for the accompanying narration.

Usually the sound track is recorded on tape or disks, and the operator advances each frame of film manually to synchronize the picture with the recorded narration. Photographs, artwork, cartoons, and other types of graphic material can be combined and used successfully in the production of slidefilms. See Figures S-10*a* to *c* and Table S-3.

TABLE S-3 Slidefilm Camera-aperture Formats

Inches*	Ratio	Format
0.590 × 0.890	2:3	Single frame for 2 × 2s only.
0.630 × 0.840	3:4	Single-frame Salesmate.
0.670 × 0.890	3:4	Single-frame standard.
1.340 × 0.500	1:2.66	Double-frame wide screen.
1.420 × 0.940	2:3	Double-frame standard 2 × 2.
0.630 × 0.840	1:2.54	Anamorphic on single frame, B aperture.
0.735 × 0.980	3:4	Single frame, full size, CinemaScope, very thin frame line.
1.490 × 0.634	2.35:1	Vista-Strip aperture. For filmstrips so arranged that no frame line appears between frames. The film moves at a constant rate from right to left, giving the appearance of a continuous strip.
0.885 × 0.294	3:1	Half of a single frame with two-sprocket hole advance.
0.906 × 0.421	2.2:1	Wide-screen format with four-sprocket hole advance.

*All dimensions in thousandths.

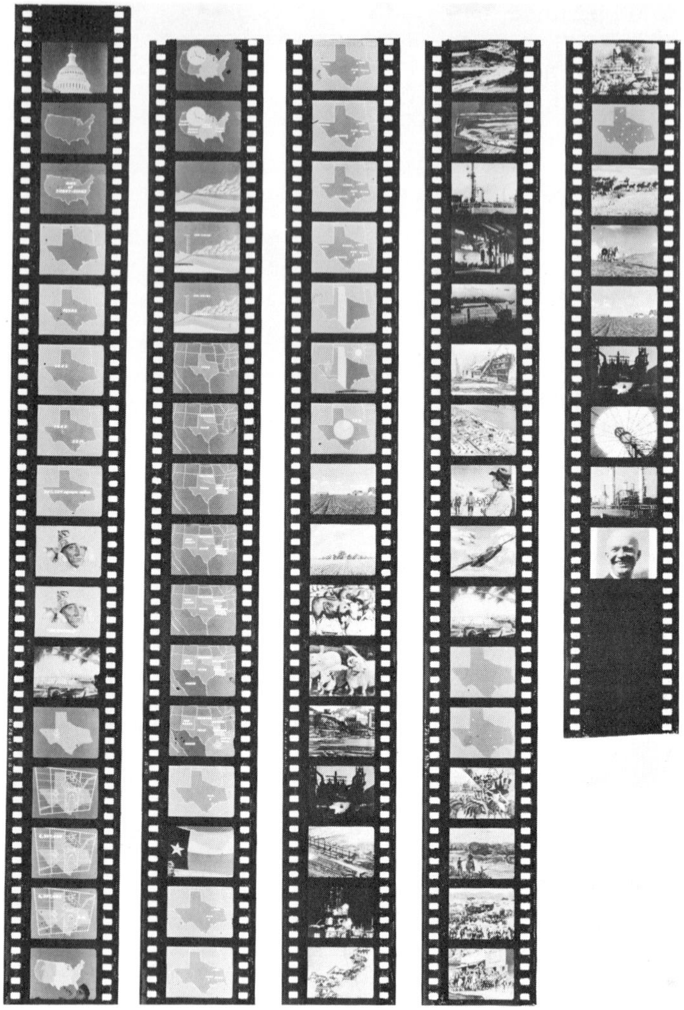

Fig. S-10*a* A typical slidefilm.

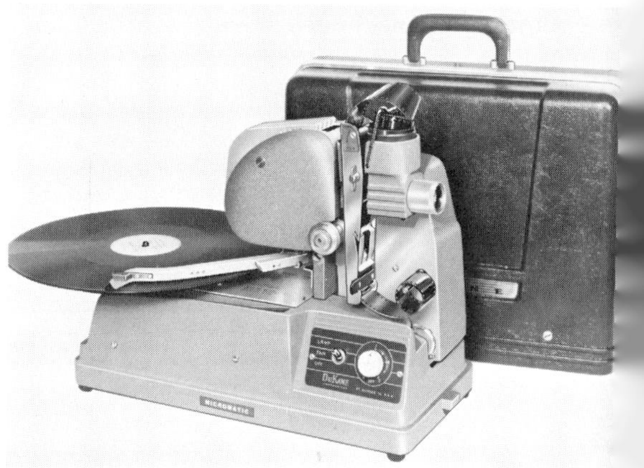

Fig. S-10*b* A slidefilm projector and the accompanying sound track.

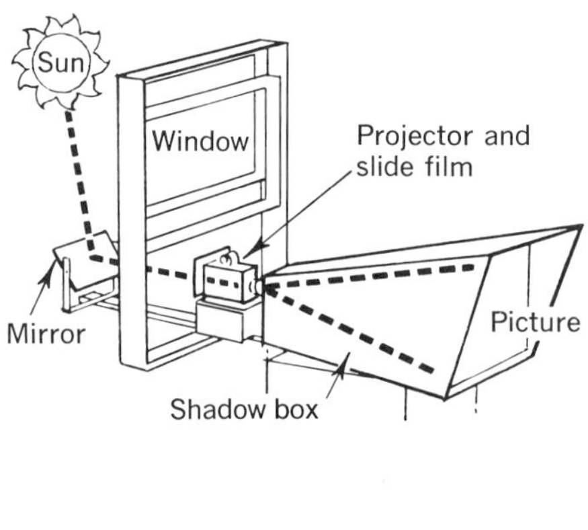

Fig. S-10*c* Slidefilm projector developed by United States foreign service employees in the Far East. It does not require electricity, but uses reflected sunlight.

standard specifications for 35mm single-frame filmstrips. A filmstrip made according to these specifications will conform to a standard now widely accepted in the slidefilm industry. Producers who follow this procedure will find that their films can be used on all major lines of equipment and by both skilled and unskilled projectionists with little or no difficulty. The problems that result from lack of uniformity in frame size, length of leader, and method of identification will be virtually eliminated. See Figure S-11.

A The trimming of the leading edge of the film will be a straight cut located between the sprocket holes at 90 degrees (±3 degrees) with the edge of the film.

B The first three frames of the film will have the word START in large white block letters appearing in each frame.

C A white dot no less than ⅛ inch in diameter will appear in the upper left-hand corner.

D A filmstrip produced in color will show green as the background of the three START frames. A black-and-white production will use black as the background color.

Fig. S-11 Standard specifications for 35mm single-frame filmstrips. See text for explanation.

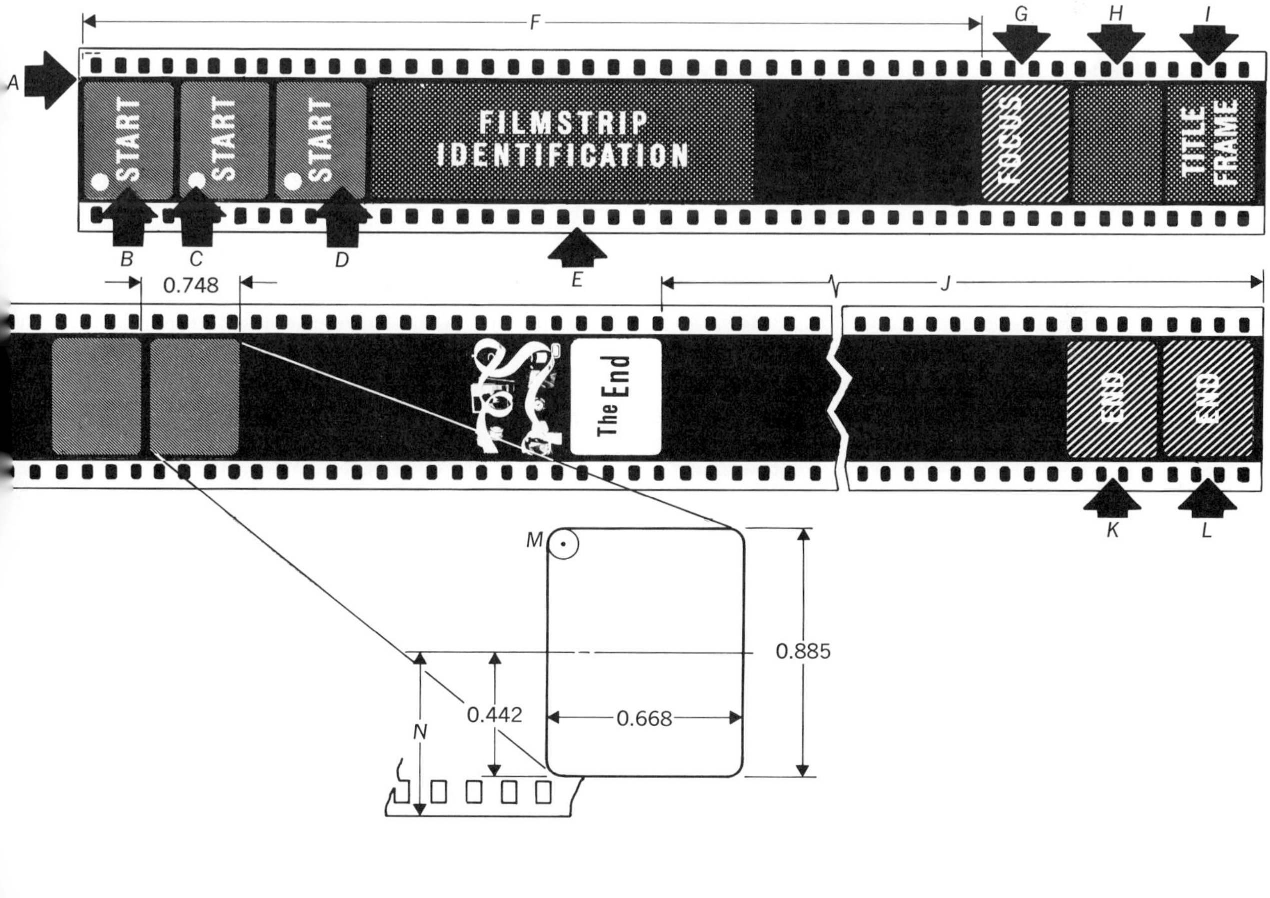

E Information identifying the filmstrip should appear parallel to the film edge in the four frames following the three START frames.

F The distance from the cut leading edge of the film to the focus frame is to be no less than nine frames or approximately $6\frac{3}{4}$ inches.

G The focus frame should contain a single design so that the projected picture can be accurately located and sharply focused. Instructions about starting the record should also appear in this frame (i.e., ADVANCE TO BLACK FRAME AND START RECORD).

H It is recommended that a black frame appear between the focus frame and the title frame so that the focus frame is not kept in view and the title frame is not revealed until the presentation begins.

I The title frame (or first information frame) will be no less than the 11th frame from the leading edge of the film.

J The distance from the end title or last frame of the filmstrip should be equal in length to the circumference of the can in which it is stored. Damage to the picture portion of the film will thus be avoided.

K The last two frames of the film before the cut edge will have the word END in large white block letters appearing in each frame.

L A filmstrip produced in color will show red as the background color of the two END frames. A black-and-white production will use black as the background color.

M The corner is created by taking the indicated portion of a circle having a radius of 0.16 inch.

N The position of the frame in relation to the center of the film is critical. The center line of the frame must fall along the center line of the film. If all measurements of frame size and position are made from this center line, then proper framing will be maintained regardless of variations in total film width.

Allowable tolerance on all dimensions is ± 0.003 inch.

slide-motion film

Slidefilm + camera movement = slide-motion film

Unlike the exposure sheets used for photographing cel animation, cue sheets provide the information needed by the animation cameraman to make the still pictures move.

Movement, in slide-motion films, generally refers to zooms, pans, the superimposition of titles, and the use of fades and cross-dissolves as transitional effects. Photographed on motion picture film and combined with a sound track that is in synchronization with the picture information, the slide-motion film uses standard motion picture projection equipment.

slo-syn motor A pulse-type, stop-motion motor that is used in conjunction with a translator. It is especially valuable where precise positioning of shutters, lens mounts, and interrelated camera equipment must be remotely controlled. See TRANSLATOR.

slow scan A process for transmitting still pictures or a fixed video frame, when a loss in quality is tolerable, over audio lines (telephone cable, for example). Slow-scan transmission offers two advantages. First, it is a less expensive method of transmission, and, second, it makes possible the conservation of the frequency spectrum for other uses.

smoke, fog, mist, and haze effects Long before ready-to-use kits were made available to the motion picture technician, smoke, fog, mist, and haze effects were created in a number of ways by using a wide variety of materials. In many cases, the professional results did not reflect the crude methods used in obtaining them.

On a small scale, these effects, including vapor and smog, were achieved by placing Dry Ice in a metal container and then pouring water over it. A cardboard or newspaper, waved by an assistant, directed the "effect" to the part of the studio set where it was most needed. For large-scale effects, petroleum products were pumped through pipes to provide the required results. The kerosine and diesel fuel, when ignited, created clouds of smoke that either billowed skyward naturally or clung to the ground because of the well-directed downward gusts from the giant studio fans.

Today portable commercial devices produce amazingly realistic, odorless smoke effects easily and inexpensively. The units can be used outdoors as well as on the studio stage and can provide considerable quantities of smoke, fog, mist, or haze in varying densities. The compound used in the generating unit, when cooled, forces the vapor to hug the ground, thus simulating the effect of low-lying fog or mist. When necessary, the device is capable of producing sufficient vapor to blanket extremely large areas.

The effect of a dense ocean fog or of low-hanging smoke from a forest fire can be quickly and easily created without the attendant discomfort of real smoke. The mechanism not only has a capacity of 1½ hours of continuous operation, but can even be used for dissipating the smoke previously produced.

Heat haze, a valuable distortion effect that can be used for many purposes, is easily achieved with any of several types of heating units.

These are positioned just below the camera lens and provide a shimmering or fluid effect that makes even the freshly made-up hero look as though he were suffering the effects of exposure to a blistering desert sun. See Figures S-12*a* to *c*.

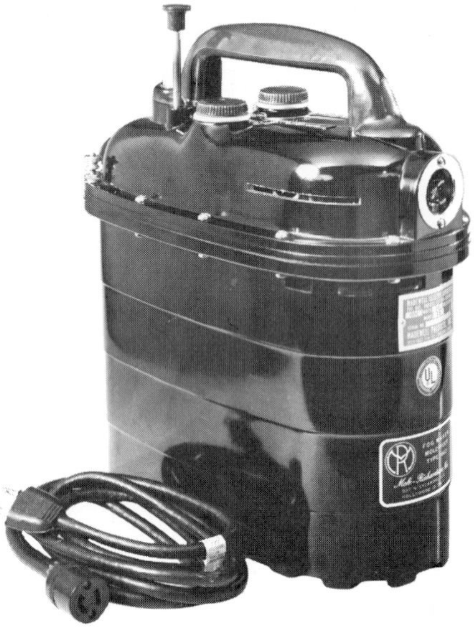

Fig. S-12*a* Mole-Richardson type 1963 Fogmaker Moleffect.

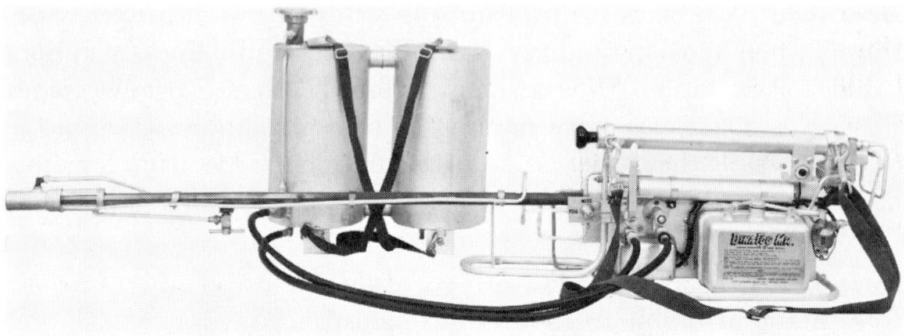

Fig. S-12*b* Mole-Richardson type 1983 Mister Fogmaker Moleffect. Lightweight, portable generator powered by a small pulse-jet engine produces realistic fog, mist or smoke effects of substantial duration by vaporizing liquid compounds fed from a reservoir carried by the operator.

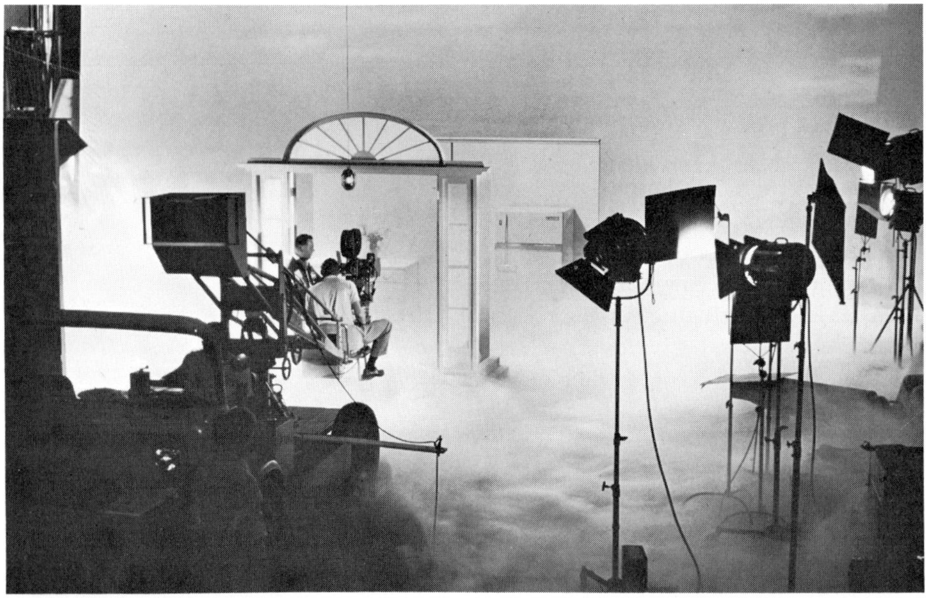

Fig. S-12*c* Fog effect produced on motion picture stage by Mole-Richardson type 1963 Fogmaker Moleffect.

smoke pot Also called **smoke kettle**. When plugged into an ordinary electric outlet, the smoke pot produces clouds of smoke in approximately two minutes. The smoke pot (and smoke powder) is available through S.O.S. Photo-Cine-Optics.

SMPTE The abbreviation for the Society of Motion Picture and Television Engineers.

snap-on mount A camera housing in which the lenses are mounted and held in position by means of spring clips.

snoot A metal, conelike attachment that fits over the front of a light source in order to direct the beam to a specific part of the set.

snow In television, an undesirable video effect (irregular white patches) caused by heavy random noise in the transmitted signal.

snow effects For many years, the illusion of falling snow was achieved with mechanical hoppers; large quantities of crispy, crunchy white cornflakes; and large fans. The painted cornflakes were dispensed by the swaying hoppers and caught in the wind streams from giant fans positioned high above the studio stage.

Although light plastic flakes are currently being used for the same purpose, the method of distribution has not changed. Moreover, the plastic flakes do not have to be vermin-proofed or treated to prevent deterioration. The flakes, with today's more efficient production methods, are simply swept into a neat pile, stored, and reused when necessary.

For covering large outdoor areas, crushed ice, salt, and limerock provide the white ground cover wherever required. Obviously, this covering does not have to be removed when the set is struck. If footprints, or other similar impressions, play an important part in the scene, a variety of powdered materials can be added to the ground cover with excellent results.

Frost on windows, in both interior and location settings, adds the touch of authenticity to the overall wintry effect. The unlikely combination of stale beer and epsom salts, when applied to the corners of a glass pane, dries quickly and can only be distinguished from the real thing under close examination.

sodium salts The salts of this metallic element are used in practically all phases of laboratory processing. For example:

SODIUM BICARBONATE is used in toning baths.

SODIUM CARBONATE is an accelerator used during the development process. In combination with other developing agents, this salt alkalinizes the solution and speeds development.

SODIUM CHLORIDE is a relatively weak restrainer used during the development process.

SODIUM HYDROXIDE is an accelerator used in strong developers.

SODIUM PHOSPHATE is an accelerator.

SODIUM SULFITE is a preservative used in fixing and developing formulas.

SODIUM THIOSULFATE is a reducing agent used after development to dissolve residual silver.

sodium self-matteing system See SELF-MATTEING PROCESSES.

sodium-vapor lamp A light source that emits an intense yellow light. Used as a source of illumination in the sodium self-matteing process, these lamps are also adaptable for darkroom work where the materials being handled are not affected by yellow light. See SELF-MATTEING PROCESSES.

soft contrast Describing prints in which the halftones or intermediate shades of gray are well balanced in relation to one another. Contrasty prints, by comparison, have extremes of light and dark values.

soft focus Describing prints in which lines forming the image are lacking in sharpness and not well defined.

soft gradation The gradual rather than abrupt change between intermediate shades of gray in print material.

soft light Light sources producing soft or diffused illumination that minimizes or eliminates shadows.

solarization A reversal effect produced by exposing a photographic image to intense light for a comparatively long period of time. The sun, for example, would appear black in a photograph exposed under these conditions instead of white.

solenoid An electric device capable of activating camera shutters and film transports from remote locations.

sound and recording Sound is defined as waves or vibrations that produce audible sensations in the organs of hearing.

In 1927, the year the motion picture industry discovered it had a voice and uttered its first syllables, no one would have dared predict that the voice within twenty-five years would be making commercial announcements advising millions of people to "put a tiger in their tanks" and "leave the driving to us."

Historically, two of Thomas Alva Edison's early inventions provided the means (the projector and phonograph), and showed the industry how they could be combined to produce a sound background for the silver screen's flickering images. The Wizard of Menlo Park demonstrated his idea and met with what might be graciously described as limited success. Intrigued by the very novelty of the projected images, the small audiences cared little about Edison's attempts to synchronize picture and sound. The lack of interest left the industry speechless for more than a quarter of a century.

"Don Juan" and "The Jazz Singer" which starred Al Jolson are generally recognized as the first talkies. The music and dialogue for the latter picture were recorded separately on wax disks. Synchronization with the visual portion was left in the hands of the theater's projectionist.

The enthusiasm for the two film productions spurred experimenters, who had already made some successful attempts to record sound directly on film, to even greater efforts. By 1931, a mere four years later, their efforts produced not one but two separate systems for recording sound optically—the variable-area and variable-density systems.

Recorded photographically on motion picture film at the projection rate of 24 frames per second, the variable-area track, with its irregular light patterns along one side of the length of film, is produced by converting sound waves into light rays by means of a galvanometer, an instrument used for measuring the strength and direction of an electric current.

In the variable-density sound track, the sound waves are transformed into a narrow band of horizontal light gradations ranging from black through gray and white. The sound from the variable-density track depends on the degree of opacity in the recorded dark and light patterns. In both cases, the patterns and tonal values in the sound-track area, along the edge of the length of film, are converted into sound waves by means of an objective light that impresses the constantly changing patterns and tonal gradations on a photoelectric cell. The resulting electric impulses are amplified and transmitted to the loudspeakers, which convert them back into sound waves.

In the single system, or single-negative method of recording, the picture and sound are recorded simultaneously on the same length of film during the photographic process. The developed film contains the picture image as well as the sound track and can be projected immediately after processing. In newsreel work, or where time is the important factor, the single-system method provides adequate results—if the quality of the sound track is considered of secondary importance.

Although improved methods for recording variable-area and

variable-density tracks were subsequently developed, the results were still inferior to the quality of the sound produced with high-fidelity recording equipment. An alternate sound-recording system was needed to keep pace with the ever widening theater screens and the quality of the motion pictures that were being projected on them.

The new process turned up quite unexpectedly. Successful experiments with tape, begun by German scientists during World War II, were continued by American engineers after the war ended. The best features of high-fidelity wire recorders, perfected previously, were used as a guide for further experimentation. By 1950, motion picture producers had virtually abandoned direct optical or photographic recording processes in favor of magnetic recording.

With the new technique, tape coated with iron oxide was passed through a magnetic field in the recording equipment and "exposed" by the sound waves that were fed into the microphone. Because processing was unnecessary, the tape could be played back immediately. Changes and corrections were made by re-recording over the original tape surface, and, if necessary, the entire contents of the previously recorded tape could be erased and reused, either immediately or at some later date.

The excellent quality of the original recordings and the negligible loss due to transfers and mixes in subsequent re-recordings made magnetic tape immediately acceptable to professional sound technicians. Since the original tape could also be transferred to film, 35mm magnastripe and 35mm full coat, for example, and used in existing editing and projection equipment, the new recording process was greeted with enthusiasm by the motion picture industry.

The relationship between sound and picture is of particular importance in the theatrical film as well as the commercial production. Separate sound tracks of the dialogue, music, and sound effects give the film editor complete control over synchronization. After each sound track has been aligned with the corresponding picture, they are all re-recorded or "mixed" into a single composite track. During the mix, the sound engineer can fade out musical passages or sound effects from one track while increasing the volume of the dialogue track in accordance with the film editor's instructions. The change in levels creates the proper balance and mood for the overall scene which is projected during the mixing process. The combined track and the edited film are sent to the laboratory for processing. The resultant composite print has the picture and sound track on the same length of film and is ready for projection.

Musical backgrounds, in both live and animated productions, help establish the mood for entire scenes or actions. Several bars of music can accomplish as much as several pages of dialogue. In animated films, the music and sound effects are especially important since many visual effects need "sound" support to augment the picture.

In cartoon animation, sound tracks are sometimes recorded after the film has been photographed and processed. This is known as *post-sync recording.* For commercial films, however, the recording is usually made before a single frame of film is exposed. In animated commercials, the animator times the scene's action to the analyzed sound track and the frame counts supplied to him by the film editor. This process is known as *pre-sync recording.* In many studios, sound effects and musical excerpts from previous productions are catalogued and filed for reuse as needed. Under "train effects," for example, the editor might have at his disposal the clatter of an elevated train, the high-pitched shriek of a diesel streamliner, the chugging of a locomotive, the roar of a subway train, the rattle of an ancient trolley, as well as a wide variety of sounds related to that specific subject.

During projection, sound precedes the corresponding picture in order to provide proper synchronization between both. In the 16mm film, the sound track precedes the corresponding picture by 26 frames. In the 35mm film, the track is advanced 20 frames ahead of the picture. These compensations are necessary because of the intermittent movement of the picture, through the gate of the projector, and the constant advance of sound track over the drum in front of the projector's objective light. The spacing between picture and sound in the composite print automatically places each element in its proper synchronous relationship. See Figures S-13*a* to *c* and MAGNETIC SOUND and MAGNETIC RECORDING.

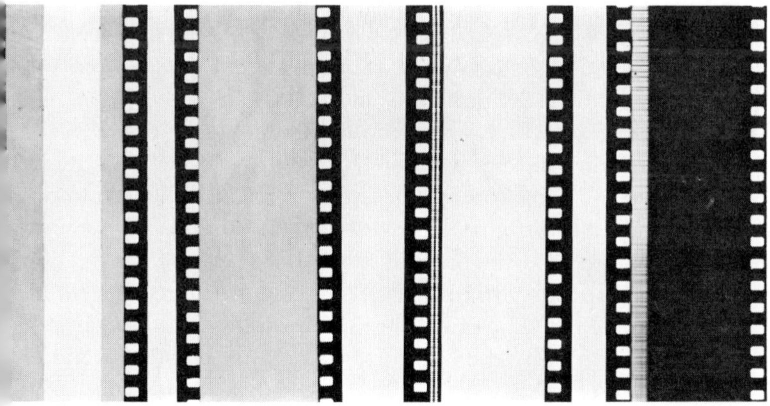

Fig. S-13*a* From left to right, 35mm MagnaStripe, 35mm magnetic full base, the variable-area sound track, and the variable-density sound track.

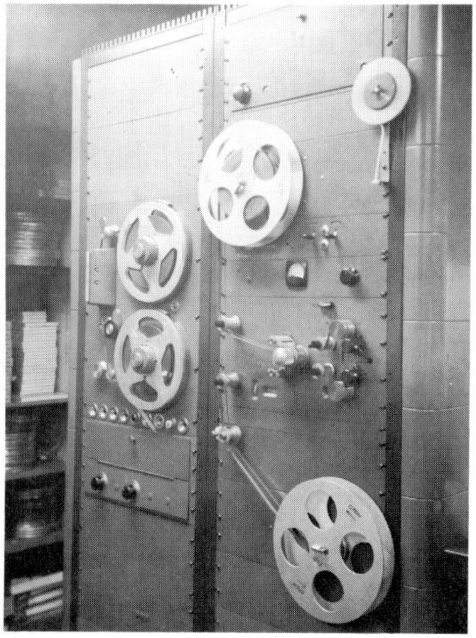

Fig. S-13*b* The original sound track, on ¼-inch tape, being transferred to magnetic film.

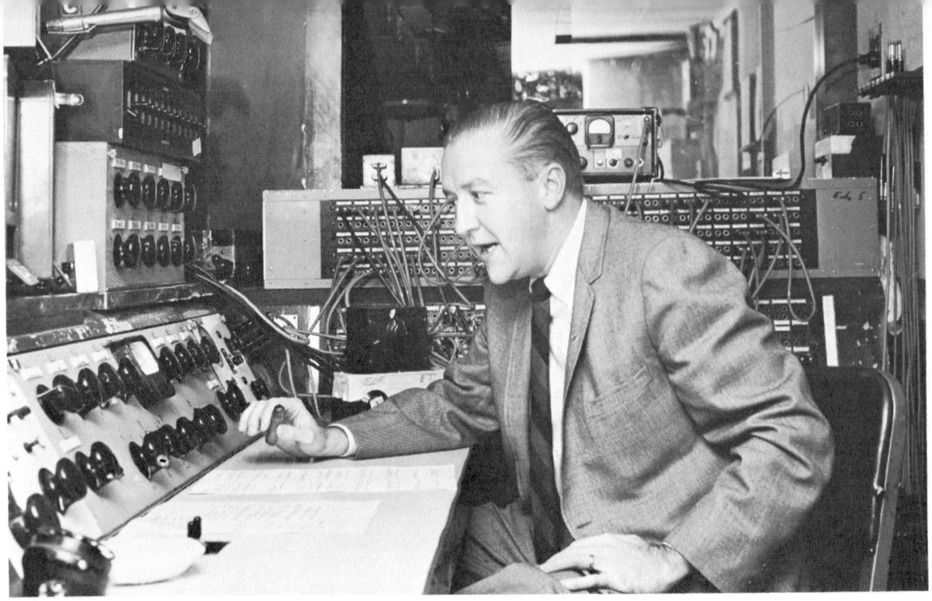

Fig. S-13c All microphone levels are adjusted before videotape time on "The Tonight Show" starring Johnny Carson. Audio engineer Bobby Bugg is shown at the audio board which mixes all of the program's sound.

sound effect A special effect on the sound track such as the sound of glass breaking or of a door closing.

sound-effects console The sound-effects console looks like an electronic organ and has 26 main sound groups, including industrial, traffic, animal, weather, and war sounds. Any of 1,260 different sound effects can be located within 20 seconds and injected directly into motion picture films, television, and sound broadcasting at either the dubbing or initial recording stages or in direct transmission.

Push-button switches above the keyboard select the range of effects, and the keys on the piano keyboard then control the individual sounds or related sequence. Sound effects can be played simultaneously, with individual spot effects synchronized.

The sound effects, which are compiled from the British Broadcasting Corporation's library, are stored on seventy $\frac{3}{8}$-inch magnetic tapes, each carrying three tracks. The tracks are each divided into six sections of recorded information, providing 1,260 sections. The sound effect can continue for 8 seconds at a normal speed of $7\frac{1}{2}$ inches per second, with speeds variable at ± 20 percent to vary the pitch and speed of the effect. The unit contains amplifiers and loudspeakers. It is manufactured by the Mellotronics organization in England and is distributed exclusively in the United States by Todd-AO Corporation in Hollywood.

sound-effects print A print containing the sound effects to be used in the composite track that is made from the sound-effects negative, or from a print by reversal processing.

sound head That part of a tape recorder that converts the recorded signal into electric impulses.

sound image A sound track that is produced photographically and one in which the modulations are visible as in the variable-area or variable-density sound tracks.

sound negative A negative sound track produced by direct recording or by exposing a positive sound image.

sound on sound The recording of dialogue, music, or sound effects over a prerecorded track. The new sound is added by re-recording the existing material on a second track—while simultaneously adding the new material.

sound print A positive sound track, on film, produced from a sound negative or direct positive recording.

sound projection rate The rate of projection for 35mm film is 90 feet per minute, or a linear speed of 18 inches per second. The 16mm print is projected at 36 feet per minute or 7.2 inches per second. 8mm prints are projected at 18 feet per minute or 3.6 inches per second. The projection rate for Super 8mm film is 20 feet per minute, or 4 inches per second. In all cases, the number of frames projected in one minute is 1,440. See FILM CONVERSION.

sound reader A device used by the film editor for analyzing optical and magnetic sound tracks for synchronization purposes. See Figure S-14 and READINGS.

sound track The recorded patterns or modulations in a specific area of the tape or film that represents the dialogue, sound effects, and background music complementing the picture information in the film. On 35mm film, the sound track occupies a narrow area on the left side of the film frame when the emulsion faces the viewer. In 16mm A-wind tracks, the track is toward the editor when viewed through the emulsion surface with the head start to the editor's right. The B-wind track is away from the editor when viewed through the emulsion surface with the head start to the editor's right. See Figure S-15.

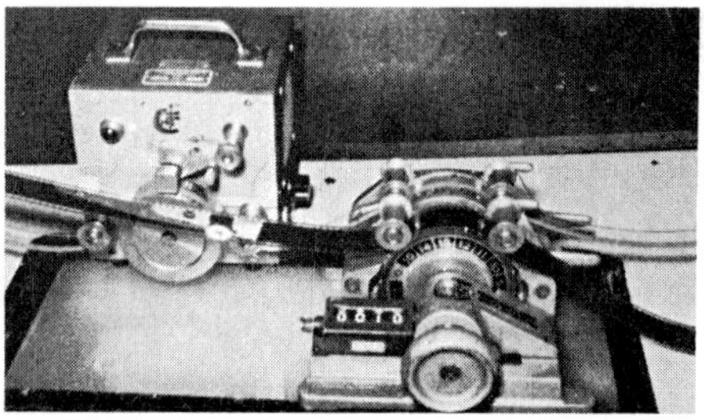

Fig. S-14 **The sound reader and synchronizer.**

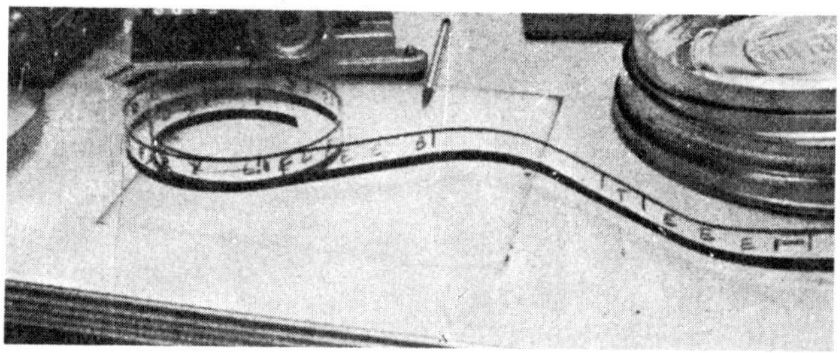

Fig. S-15 **The analyzed sound track.**

sound-track advance In 35mm film, the sound track precedes the corresponding picture by 20 frames. In 16mm film, the advance is 26 frames for optical sound tracks and 28 frames for magnetic tracks. In 8mm film, the magnetic* track is advanced by 52 frames. The 30-mil track, on Super 8mm film, precedes the corresponding picture by 18 frames (±½ frame).

The advance, in all cases, is a compensation made necessary by the physical displacement of the sound head in projection equipment in relation to the picture projection head.

sound work print An intercut sound track containing sections of dialogue, music, and effects prepared by the editor during the completion process. Each individual portion of the track is synchronized with the corresponding picture in the film.

soup A slang term for developing solution.

spark photography A method of illuminating the subject matter in high-speed photographic processes. The spark, or light source, is activated by a triggering condenser connected to a pair of leads within which the spark is produced. Exposures shorter than 1/10,000,000 second in duration are possible with this type of light source.

sparks A slang term for a studio electrician.

spatial relations See PERSPECTIVE.

special effects An all-inclusive term for illusory effects. The term includes fades, cross-dissolves, and wipes, as well as trick shots, montages, split screens, and multiple-image shots.

Only on rare occasions are special effects written into a motion picture script because of their eye-arresting appeal. In practically all cases, the unusual effect, either audio or visual, fills a specific need and serves a useful purpose. In some instances it is used to suggest or augment an action. At other times it can quickly establish a mood or theme for an entire sequence. In commercial film production, however, special effects are used freely in order to capture the viewer's attention and hold his interest.

*Although an optical track can be recorded on 8mm film and Super 8mm film, the quality of the reproduction is considerably poorer than that of the tracks recorded on magnetic recorders.

Today's special-effects technician is expected to be talented, resourceful, ingenious, inventive, imaginative, and original. He must also have a working knowledge and practical experience as a photographer, set designer, architect, painter, engineer, researcher, chemist, artist, and magician. While the dreamer enjoys certain advantages, the modern da Vinci must be able to dream practically in black and white and color, and in wide screen as well as 3-D.

On short notice, the effects man should be equal to any and all assignments, from creating rain and snow to producing waterfalls and cobwebs. At other times, he will be asked to simulate a full-scale naval battle that will be fought with miniatures, construct a scale model of a London waterfront, or create a mechanized monster with movable eyes, arms, and legs.

In recent years, motion picture technicians have not only succeeded in their attempts to duplicate some of Mother Nature's best efforts but have, in fact, surpassed them. Their earnest desire to improve the quality of production generally, prompted them to reveal their technical secrets to others in the industry. As a result, some of the more frequently used effects are now available in kit form: the wind machines, fogmakers and cobweb spinners, for example.

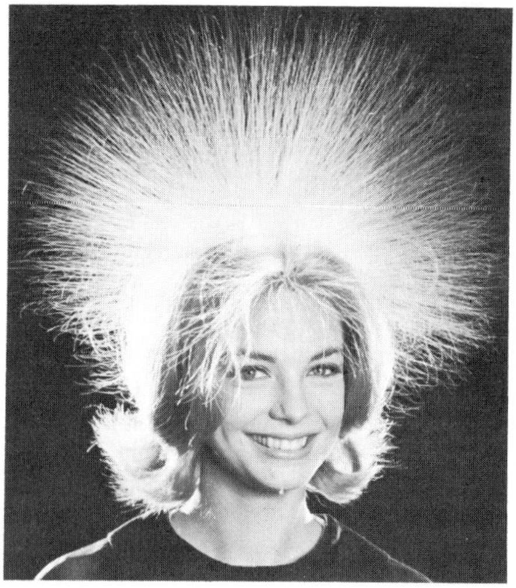

Fig. S-16*a* Hair-raising! Produced with the aid of a static electricity machine that had been placed on top of alternate layers of rubber mats and glass blocks. Similar layers of insulation raised the bare-footed model 2 feet above the level of the studio floor. The model's right hand rested on the dome-shaped top of the machine.

At the end of each "take," the director brought the coin he had been holding in contact with the coin held by the model. The current absorbed by the two coins reduced the size of the spark that jumped between the machine and the model when her hand was withdrawn.

Wind machines, capable of producing the gentlest of air flows as well as the most violent of hurricanes, make such effects readily available to the film producer. A companion item, the wind howler, may be used "live" or as a means for recording the desired sound for later use. Here, too, the producer has at his disposal a variety of sounds ranging from the whisper of a gentle breeze to the shrill screech of a whistling winter wind. The latter effect is especially useful for drowning out the anguished cries of the critics who complain about the loudness of sound tracks in commercial film productions.

The fogmaker unit is another versatile device that can be used to create smoke, haze, and mist effects. It also doubles as a dispenser for deodorants and insecticides. On location in tropical climates, the latter function may be of greater value than the uses for which the device was originally intended.

The cobweb spinner, also amazingly simple to operate, can make even the most proficient spider look to his laurels. With this unit, the effect is produced by simply aiming the device at the area where the webbing is needed. In operation, the rubber cement compound used to produce the effect is forced through a tiny valve. A multibladed fan attached to the unit transforms the compound into thin threads of fluid that fool the most discerning spider. See Figures S-16*a* to *c*.

The wind machine, wind howler, fogmaker, and cobweb spinner are available through Mole-Richardson, Hollywood, California.

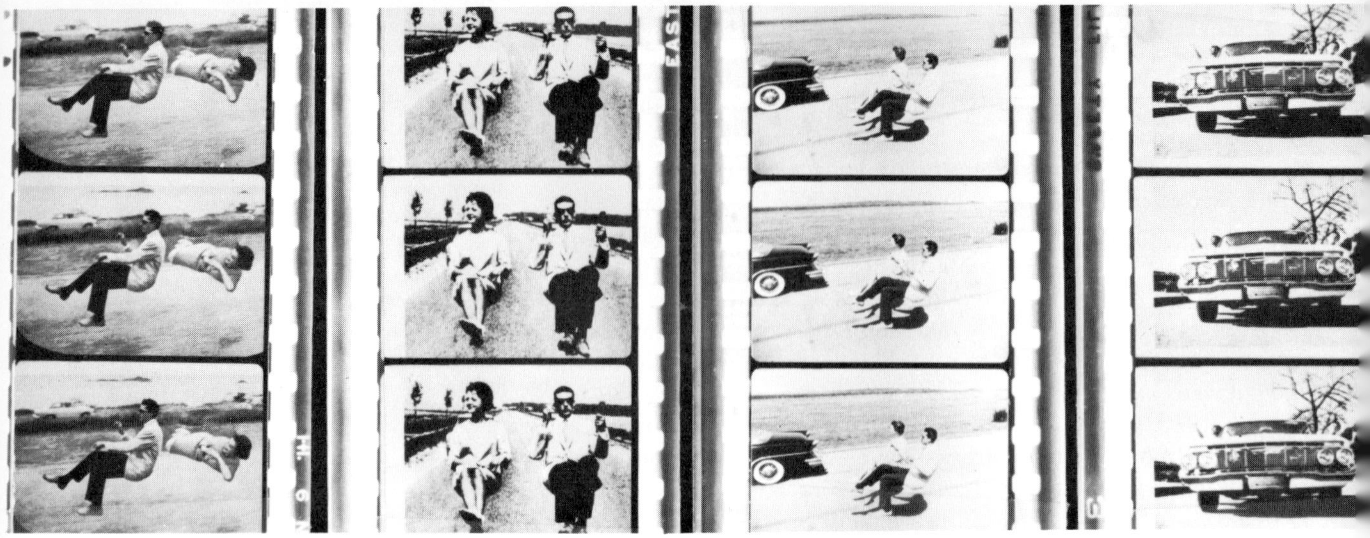

Fig. S-16*b* **Magic ride. (*Arco Productions, New York.*)** Photographed in France, Chevrolet's magic ride commercial showed a young couple driving along a country road in an "invisible car."

During most of the ride, the couple was seated on the end of a steel beam attached to the camera car. A specially built prism, positioned in front of the camera lens, concealed the beam and showed the couple suspended in midair at exactly the right driving height in relation to the road. The shadows cast by the couple gave the entire effect an added touch of realism.

For some scenes, mirrors and specially designed Rube Goldberg devices, built to fit inside the camera, were used to achieve the desired illusory effect. The entire commercial, filmed as a one-step, in-camera process with no optical work needed at any stage of production, was photographed on 35mm Eastman color negative film.

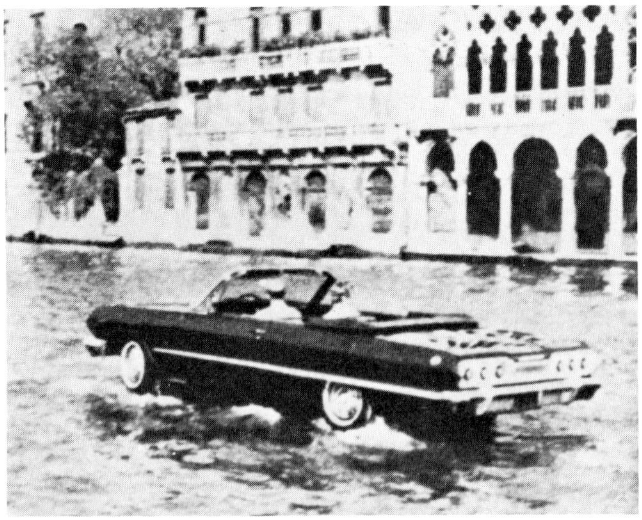

Fig. S-16*c* **Floating car? (*Arco Productions, New York.*)** Citizens of Venice who lined both sides of the Canal to witness the filming of this Chevrolet commercial, were as puzzled as American TV viewers who saw the eye-arresting effect on their home screens.

The more ambitious do-it-yourselfer can float Junior's toy car in the bathtub by building a miniature raft and weighting it so that its top is just below the water line. To make the car's wheels turn, simply add rollers and anchor the car to the raft.

special-effects generator Also called **matteing amplifier**. The special-effects generator is one of the most versatile tools ever placed in the hands of the videotape technician.

The equipment makes it possible to create and record special effects on the spot during production. Performers or products on one length of recorded videotape, for example, can be inserted into another scene; split-screen conversations can be recorded between two people from different scenes or by one person positioned on both sides of the frame; giant or midget-sized props can be handled by normal-sized people; titles or logos can be made to move in sync with a prerecorded jingle; and so on.

CBS Super-Wipe, another type of effects amplifier, is capable of generating approximately 300,000 wipe and transitional effects between two or more picture signals.

spectra filters A series of corrective filters for use in color photography. The S series (pinkish in color) and the T series (turquoise) are used to change the color temperature of a light source. The M and G series (magenta and green) are used for color correction. The mounting rings, for each of the filters in the four series, state the filter factor and the required amount of exposure increase in *f*-stops.

spectrograms Wedge spectrograms, for measuring the color sensitivity of film, are prepared by exposing a length of film to a light spectrum through a neutral-density wedge. The wedge is almost completely opaque at the top and transparent at the bottom. The high points of the image, on the scale of the spectrograph, are in direct proportion to the log sensitivity of the emulsion. See Figure S-17.

Fig. S-17 Color sensitivity—panchromatic. This spectrogram demonstrates the response of type 140 to light of approximately equal energy over the visible range.

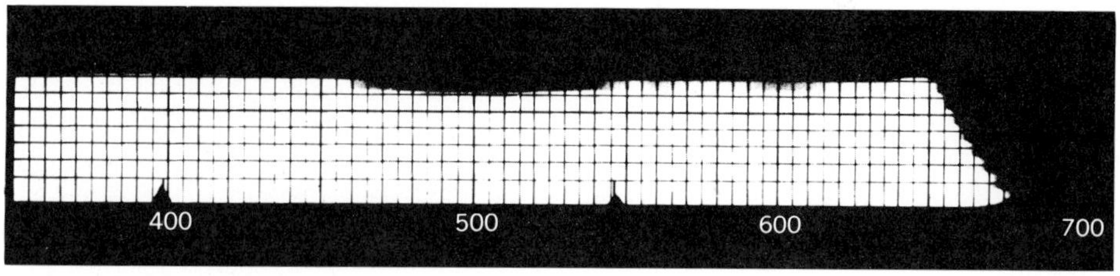

spectrography The photography of colors in the visible spectrum as well as colors in portions of the invisible spectrum for the purpose of analyzing the composition of materials.

During the photographic process, light reflected from the material is admitted through a narrow slit in a spectrograph and is dispersed by a prism. The light is focused by a lens and is transmitted to the panchromatic film where the image is recorded. The image is visible on the developed plate as a series of lines of different intensities. These lines, measured by microdensitometers, provide the only practical method of analyzing and determining the composition of materials used by physicists, chemists, and other scientists generally engaged in research work. The facility with which minute quantities of elements can be detected makes this photographic measuring system far superior to the less sensitive chemical methods used previously.

In combination with ultraviolet and infrared photography, spectrography has been used quite successfully in criminal investigations to identify materials and determine the presence of elements that would go unnoticed with other analytical methods. Practically applied to the current space-exploration program, spectrography offers astronomers and scientists a method of acquiring information about the structure and composition of heavenly bodies.

spectrometry The analysis of the spectral values of a color in terms of frequency and wavelengths.

spectrophotometric absorption curve Graphs showing the density of a color plotted against wavelength. Expressed in millimicrons, one wavelength is equal to 0.000001 millimeter. Density is the common logarithm of 1/transmittance.

A 1 density indicates a 10 percent transmittance. A density of 2 is equal to a 1 percent transmittance. A 3 density is equal to 0.1 percent transmittance. Similarly, a 0.10 density would indicate an 80 percent transmittance.

spectrophotometry The measurement of the amounts and intensities of the spectral lines in color.

spectro-wedge color system A guide designed to aid the cameraman in the selection of filters. With the spectro-wedge color system, each color is assigned a definite number, and the intensity of each color is identified by a letter of the alphabet.

spectrum The division of white light into its component parts and the resulting bands of color. The visible colors, from one end of the band to the other, range from violet to deep red.

The great spectrum includes all electromagnetic radiation, from the longest radio waves known to the shortest cosmic rays at the other extreme. It encompasses the visible spectrum. Since wavelengths vary inversely with frequency, the long waves have low frequencies, and the short waves very high frequencies. See Figure S-18.

spectrumatte self-matteing system See Pickley process under SELF-MATTEING SYSTEMS.

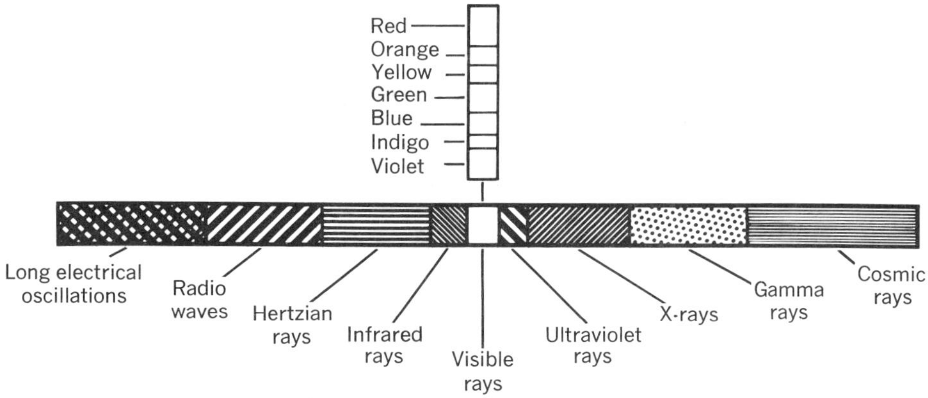

Fig. S-18 Wavelength scale.

specular In optics, the term describes either a metal or metal-coated mirror—used in telescopes or reflectors. When used in connection with lighting, the term describes a controlled beam of light that is reflected from this type of surface—a spotlight, for example.

speed The cue indicating that camera and sound-recording equipment are in sync and running at the proper operational speed—24 frames per second, sound speed.

speed numbers Basically, the degree of sensitivity, or the speed of an emulsion. The photographic speed of an emulsion is related to the sensitivity of the emulsion, the developer formula, the technique of the processing laboratory, the nature of the subject, and the illumination. The exposure indices quoted for each film are expressed in terms of speed values. These values should not be taken as indicating the absolute speed of the film and should be adjusted to the requirements of equipment, the method of use, and the development.

The average exposure rating gives an indication of the setting that will ensure a satisfactory image in a wide range of exposure and processing conditions. The minimum exposure rating is recommended when it is necessary to achieve the highest film speed and best grain characteristics of each emulsion. See FILM SPEED and EXPOSURE-INDEX RATINGS.

spherical aberrations Disturbances on the face of a lens that prevent all light rays from converging at the exact focal point of the lens. See ABERRATIONS.

spherical lens A lens designed to give wide-angle coverage without the accompanying spherical image distortion normally associated with fisheye lenses. With this type of lens, the image remains undistorted throughout the entire film frame.

spider A flat, triangular support for the tripod on which the camera is mounted.

spider box Also called **junction box**. A conveniently located unit that contains receptacles for the various sources of lighting used on a set.

spill light Excess or undesirable illumination falling on the motion picture set.

spin, camera A spinning effect produced by rotating the camera compound table of the animation stand and the artwork positioned on it. Similar effects can be produced with the optical printer.

spindles, supply and take-up Spindles that hold the reels of tape in position on the recorder during the recording and reproducing process.

splicer A device used for joining lengths of film. In splicing lengths of film, one length is positioned on the pins on the left side of the splicer (when the plate hinges are away from the editor). The emulsion or dull side of the film faces up.

The left cover plate is closed over the film, and the emulsion is scraped from the exposed strip that extends beyond the plate. The right-hand side of the splicer has two plates—a cutting plate and a pressure plate. Registered to the pins, the second piece of film is trimmed to the proper length when the cutting plate is lowered. Film cement (acetic acid) is brushed on the previously scraped strip, and the pressure plate is brought down to its closed position. A small amount of pressure should be applied until the splice is set. The excess film cement should be wiped away before the film is wound or projected.

splicing block A device for holding the tape in place while splices are being made. Scored lines on the block act as guides for the diagonal cuts that are made on each end of the tape before they are joined.

When making splices, the editor brushes a solution of carbonyl iron particles on the surface of the recorded tape to bring out the magnetic pulses that indicate the fields at which cuts are to be made. The lengths of tape that are to be joined are then placed on the splicing block and cut. The aluminized tape used for joining the lengths of tape does not affect subsequent transmission.

split-field diopter lens Plus-diopter lenses are cut in half and arranged so that one-half of each lens is used to photograph the foreground area while the background is being photographed through the open or clear portion of the lens. These lenses are similar to bifocal lenses in principle.

split reel A two-sectioned spool used by the film editor. A roll of film on this type of reel may be removed without rewinding by simply separating the cover section of the reel from the supporting section.

split-screen shot An optical effect in which the frame of film is divided into a number of areas with different picture information in each. Achieved by multiple exposures and matteing processes, each part of the frame is masked in turn, and the insert picture information is exposed in the previously matted areas. See Figures S-19*a* and *b*. Compare MULTIPLE-IMAGE SHOT.

Fig. S-19*b* **A hidden split made this effect possible. A soft-edge vertical wipe successfully hid the line of demarcation in each of the two camera "runs."**

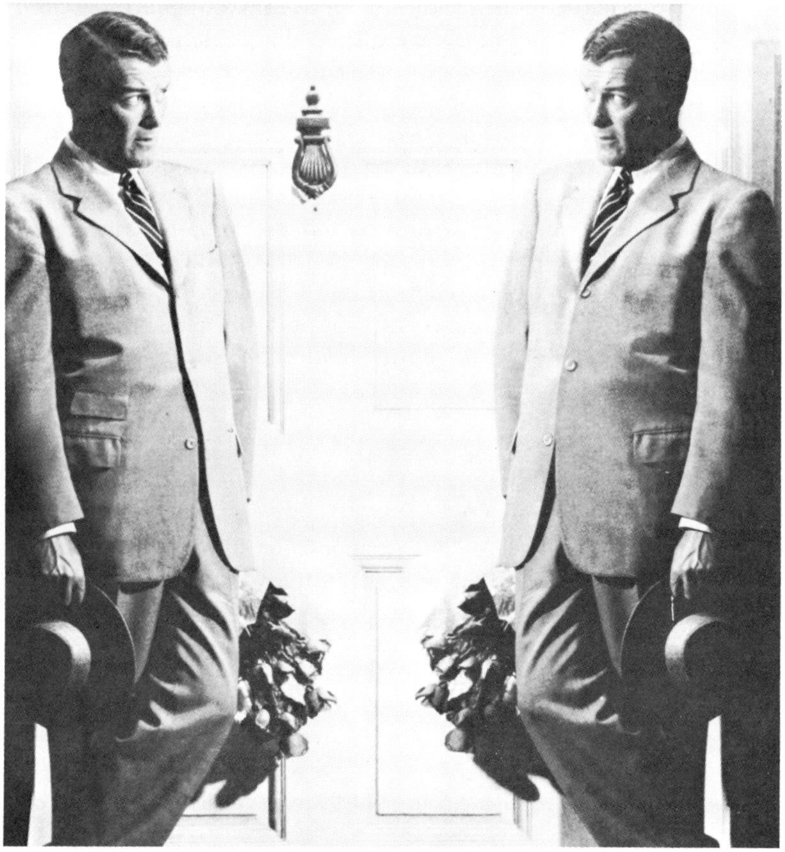

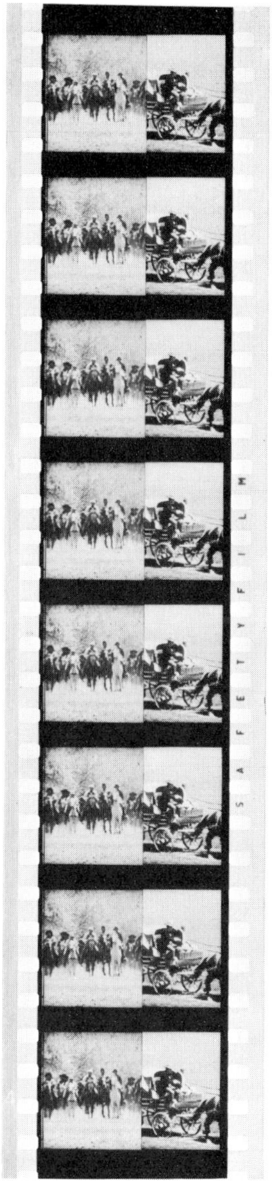

Fig. S-19*a* **Split-screen shot.**

spot The scanning point of light at any specific part of the 525 lines in the target area of a TV pick-up tube. The spot moves continuously to create 60 fields or 30 complete frames each second.

spot-brightness meter A meter that measures the reflected light from the background as well as the foreground subject matter from the camera location. Unlike the method used for taking light readings with the reflected-light meter (separate readings of the brightest and darkest points of a scene), the spot-brightness meter provides an indication of both, as a guide to lens settings, from one vantage point.

spotlight A lighting source capable of projecting a focused beam of light to a specific part of the set or stage.

spray bath A laboratory developing process in which the developing solution and also the solutions used for fixing and washing are applied to the film by means of jet sprays instead of by immersion in tanks. The solutions are fed into pipes positioned in the developing machine and, subsequently, forced out through regularly spaced holes. Spray baths make possible high-speed and high-volume negative and positive processing under accurately controlled temperatures.

sprocket Toothed projections in the film-transport mechanism (as in cameras, moviolas, or projectors) that engage the film's perforations or sprocket holes as the film travels through the equipment.

sprocket holes Also called **perforations**. The holes along the edges of film that fit over the sprockets in transport mechanisms and keep the film precisely aligned as it travels through cameras, moviolas, and projectors.

SPSE The abbreviation for the Society of Photographic Scientists and Engineers.

Spyder An eight-wheel dolly* designed for filming in relatively small or confined areas. This lightweight dolly is easy to maneuver and can be steered and locked into almost any position. The dolly features a hydraulic lift, swivel seats for the cameraman and his assistant, and rubber wheels that can be converted for track operation.

The Jonathan jib-arm assembly is an accessory that is mounted on the Spyder dolly to convert it into a dolly crane capable of revolving 360 degrees and raising the camera to a lens height of 7

*Elemack.

feet. It is equipped with a pantograph indicator which enables the "grip" to follow the shot precisely as directed.

squash and stretch Exaggerated changes in the proportions of an animated cartoon character. Like the flattening or compressing action of a bouncing ball when it strikes the ground, cartoon characters and objects are sometimes drawn to show the same type of compression (when falling) or elongation (when traveling at great speeds). See Figure S-20.

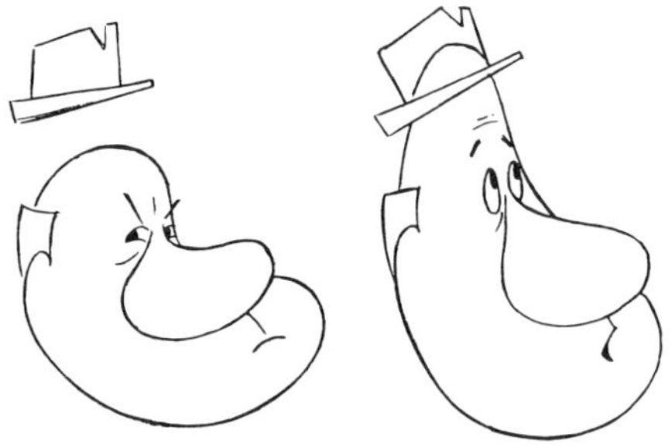

Fig. S-20 **Squash and stretch.**

squeeze lens Also called **anamorphic lens**. A type of lens used for compressing images, on standard-width films, for subsequent wide-screen projection. During projection, the images are restored to their proper proportions by complementary anamorphic lenses.

stabilization The process of fixing prints quickly (with special solutions intended for that purpose) where speed is of primary importance. Sodium thiosulfate (hypo) is a stabilizing agent that will perform this task in approximately one-half minute. The print is not permanently fixed, however, and does not have to be washed after the fixing process has been completed.

stage The general area in which the scene's action takes place.

stagehand A studio employee who helps prepare the set for the filming process.

standard field In animation, "standard" fields are actually misnamed since they vary from studio to studio. Generally the term standard field is assigned to the one selected and used most frequently for aligning the compound table, filming, projecting frames of film for matteing purposes and composition generally, and for photographing titles and other material for subsequent optical superimposition.

standard film 35mm motion picture film. Any film having a narrower gage than the 35mm width is referred to as *substandard film.*

standard junior A light source similar to the baby junior with flood-to-spot control. Equipped with a 10-inch Fresnel lens, the lamps used with this unit range from 1,000 to 2,000 watts.

standard senior A 5,000-watt light source for relatively large sets. At the flood setting, the intense light produces a spread of approximately 40 degrees.

stand-in A minor performer or extra whose features or physical characteristics are similar to those of the actor playing the leading role. The stand-in is used as a guide while the set lighting is being arranged or, during the actual filming, as a substitute for the star in potentially hazardous situations. The term *double* is frequently used in place of stand-in.

star Also called **lead**. The person playing the leading role in a motion picture is the *star,* but any well-known actor or actress will at times be called a star.

starlet An aspiring actress presumably on her way up the long ladder leading to stardom. Not all complete the ascent.

start mark Also called **sync mark**. An X mark placed on both the sound track and corresponding picture leaders to indicate a common synchronization point.

static Undesirable markings on film. The faint, irregular patterns are the results of static electricity produced either by friction on the film or, in cold weather, by actual contact between the film and the cameraman. Compare INTERFERENCE.

static electronic matte See INLAY.

steam and vapor Dry Ice, technically carbon dioxide in the frozen solid state, is a versatile material that is used for many purposes in motion picture production. Although its low temperature and peculiar properties can cause blistering if it is mishandled, Dry Ice can also release clouds of harmless white vapor when warmed or be dissolved when placed in water or other aqueous liquids.

In the studio laboratories that serve as backgrounds for horror chillers or mystery thrillers, small pieces of Dry Ice can be dropped into retorts or tubes filled with colored liquids to produce bubbling and foaming effects. Added to warmer solutions, the Dry Ice instantly creates billowing clouds of vapor. For television commercial production, where its uses are as many as they are varied, a relatively small piece of the icelike material at the bottom of an on-camera cup will keep the cold liquid steaming for long periods of time.

When heated, the Dry Ice can be used to produce mist effects on a miniature set, provide ground-hugging fog or vapor that makes performers seem to be walking on clouds, or duplicate the smoke-filled haze that surrounds the participants in a barroom card game. Since gaseous CO^2 released by Dry Ice is heavier than air, it can be "poured" (from an off-camera container) over a relatively small flame to extinguish it mysteriously. Similar effects, using the same principle, can be achieved just as easily.

step-contact printing A process in which a negative or positive length of film is placed in direct contact with the raw stock and advanced intermittently to produce either a negative or positive image, depending on the film used for printing. Each frame of film is exposed to the printing light, in turn. The intermittent movement of the transport mechanism is similar to the movement of film traveling through the camera. See CONTACT PRINTING.

step wedge Also called **step tablet**. A length of negative film in which the light tones at the beginning of the strip grow progressively darker. The wedge is used for making exposures for subsequent sensitometric analysis.

stereophonic recording A method of recording in which the sounds picked up by several microphones are played back on separate channels through speakers placed to simulate the positions of the microphones. The process gives depth and direction to the reproduced sound.

stereoscopic photography Basically, the illusion of depth and dimension produced with photographic processes.

In 1838, an Englishman named Charles Wheatstone outlined his theories on spatial relations and the perception of depth to a group of interested scientists. Subsequently knighted for this and other scientific findings, Mr. Wheatstone explained that while depth, as an illusion, could be realized and assumed by inference and association, the physical sensation of actual dimension would still be missing. Converging lines, as well as changes in perspective and light patterns in two-dimensional pictures, were used as obvious examples to help support the scientist's well-founded theories.

Two important developments in 1891 paved the way for further experimentation with depth perception and dimensional photography. The first, a suggestion by L. D. Duhauron, advanced the theory that two photographs of the same object, taken with lenses spaced to approximate the distance between a viewer's eyes, could simulate the natural impression of depth. Printed in complementary colors, the pictures could then be superimposed and viewed through complementary lenses. The second, a prediction by J. Anderton, was based on similar photographic processes, but favored polarized light as a less fatiguing method for viewing the twin images.

By the time Thomas Edison introduced his first crude motion picture projector in 1896, stereoscopic viewing devices could be found in many homes. With each viewer turning knobs to adjust focus, two almost identical photographs, positioned side by side on a single card, were being blended into a composite, dimensional picture.

Three years after the development of practical synthetic polarizing filters by E. H. Land, audiences attending the New York World's Fair were treated to dimensional motion pictures, in color. The year was 1939. Several months later, the designer of the Polaroid camera also succeeded in exposing two oppositely polarized images on the same length of film. By 1950, the spectacular, stereoscopic effects that had made audiences blink in amazement were also responsible for many cases of eyestrain and nausea. In order to make viewing a more comfortable process, producers in various parts of the world experimented with many different filming and projection techniques, all designed to create the impression of depth by adding another dimension to the standard flat screen.

On some lengths of film, the twin images appeared side by side. On others, the dual images were positioned vertically in the same film frame. In some of these experiments beam-splitting devices were used during the filming process, and in others the images for each eye were projected alternately.

Most of the proposed processes, however, incorporated one or more of the features suggested by the original developers of the technique. In its basic form, the anaglyph or superposed images are exposed on separate lengths of film and projected simultaneously, each of the images passing through complementary-colored or light-polarizing filters positioned in front of each projector lens. The filters

Fig. S-21 **Stereoscopic projection.**

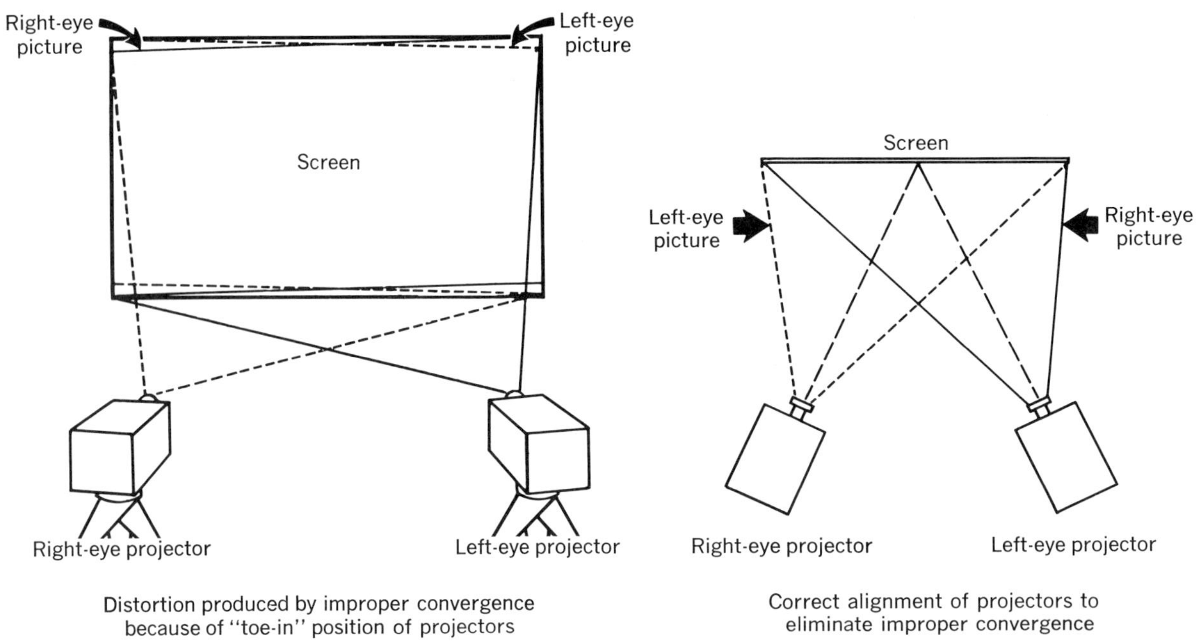

attached to the synchronized projectors are arranged so that their polarizing axes slant in opposite directions—one filter set at a 45-degree angle to the right and the other slanted at the same angle to the left. If the processed images are printed in complementary colors, they can be projected without filters.

In the spectacles worn by the viewer, each filter blocks out or rejects one of the images while allowing the correct image to pass through—relatively undisturbed. The right eye, for example, sees its appropriate image through red-orange filters while the left eye looks through a complementary blue-green filter.

Unlike live filming processes in which the picture information for each eye is photographed simultaneously through matching lenses on separate lengths of film, the photographic process for animation and titling work is repeated with the camera lens in a different position for each of the two runs. After the first run, in which the subject matter is photographed normally on one length of film, the images for the second eye are exposed on a second length of film.

For the second run, the camera and its lens are shifted, horizontally, to a new position that approximates the distance between a viewer's eyes. The interaxial spacing, or the distance between the optical centers of the two lenses, varies with the subject matter and the amount of scene depth desired. Scene depth becomes proportionately greater as the distance between lenses is increased. In most cases, the amount of lens shift needed to achieve average scene depth ranges from $\frac{1}{2}$ to approximately 3 inches. An integral part of the viewing process, the same spectacles, with their complementary-colored filters for each eye, must be worn when the two lengths of film are projected.

Photographed on separate lengths of film or combined on one reel, the aluminum screens on which the superposed images are focused have little, if any, effect on the polarization of the light. When the same images are projected on glass-beaded or white matte screens, however, polarization disappears, and the picture information blends into a composite and undesirable double-exposure effect. The images beamed at plastic screen surfaces with rear projection processes may be either brilliant or disturbing—depending on the position of the viewer in relation to the screen.

Obviously, the success of dimensional motion pictures will depend largely on the industry's ability to film and project the dual images properly. And although the problems associated with image synchronization and alignment cannot be classified as minor, they are not insurmountable. Technicians will in time make it possible for theater audiences and home viewers to watch dimensional films and television pictures in complete comfort—and without spectacles. See Figures S-21 and S-22.

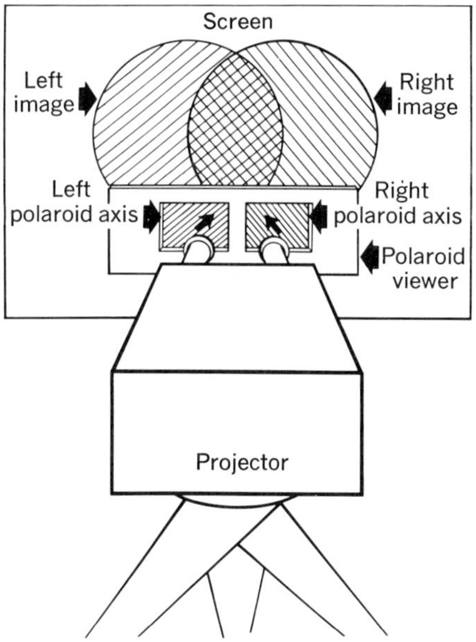

Fig. S-22 Polarized light and stereoscopic projection.

stilb A unit of brightness equal to one candela per square centimeter. This photometric unit of measurement is used to indicate the intensity of a light source falling on a specific area.

still A single photograph taken with an ordinary camera, as opposed to the sequence of pictures taken with motion picture equipment.

still background In animation, a background that remains in a stationary position on the camera compound table during the filming of an animated scene.

stock Also called **raw stock**. Unexposed film.

stock shot Also called **library footage**. Scenes or lengths of film from earlier productions that can be used as insert material by the film editor in subsequent motion pictures. Classified and cataloged so that specific shots can be made available quickly, library footage includes almost any subject from animals in Alaska to zinnias at the Zuyder Zee. See SCRATCH PRINT.

stock solution Developing solutions in concentrated form which, because of their characteristics, are kept in separate containers and combined only before being used. In many cases, these solutions will remain effective for long periods when stored separately. Combined, they either spoil or deteriorate rapidly.

stop Lens diaphragm settings, calibrated by *F*-numbers. See *f*-STOPS.

stop bath A developing solution (up to 5 percent acetic acid, citric acid, or potassium metabisulfite, for example) which stops functioning immediately after the alkalinity has been neutralized. These solutions are particularly useful after high-speed developing, or for processing in warm or tropical climates.

stop motion A process in which objects are photographed frame by frame, and moved from one position to another between exposures. When projected at normal speeds, the objects appear to move without visible assistance to create unusual and eye-arresting effects.

Unlike time-lapse photographic processes, in which an action that cannot be seen as a continuous movement must be photographed at regular intervals over long periods of time, stop-motion techniques are used to create the illusion of motion with inanimate subject matter. Stop-motion actions are produced by manually changing the position or altering the shape of the subject matter between exposures. These gradual changes, in each of the successive frames of film, create the illusion of continuous movement when the processed film is projected.

Stop-motion photographic processes were developed during the industry's infancy. The first crude methods for exposing individual picture information on successive frames of motion picture film have been replaced by more sophisticated equipment and modern production techniques. Precision and accuracy, the essential requirements for all stop-motion production, are made possible by registration devices in the camera, viewfinder and tabletop controls. Solenoid-controlled stop-motion clutch units are driven by single-phase synchronous motors that are attached to the camera, either directly or through adapters. These variable high-speed motors provide a range from 60 revolutions per minute (rpm) and $\frac{1}{2}$-second exposure to 240 rpm and $\frac{1}{8}$-second exposure.

The motors, designed to handle the tremendous shocks that the shutter mechanisms must withstand during filming, have four forward speeds that permit a wide choice of lens stops. The higher speeds, 240 frames per second, save valuable time on long continuous runs for titles and other work where hold footage is required. Fast rewind speeds, 720 rpm or the equivalent of $\frac{1}{24}$-second exposure, speed the filming process for multiple-exposure work. Suitable for practically every type of stop-motion production, the range makes provision for slow or fast film stock and for variations in lighting systems. Frame and footage counters attached to the camera provide an accurate and constant measure of the amount of film exposed during the photographic process.

Aside from the drawings or cels used in the production of animated films, the inanimate objects that may be given movement with this photographic technique include photographs, diagrams, three-dimensional scale models of mechanisms, puppets, commercial products, or indeed any subject matter that must be moved manually between exposures. For the stop-and-go photographic process, technicians with experience in both live-action filming and animation arrange the lighting for the backgrounds or miniature three-dimensional sets and move or manipulate the subject matter for each frame of the action with the precision of a skilled surgeon.

The comparatively slow and painstaking filming process requires the same meticulous attention to detail as that given to live-action photography. The resultant action or illusion of motion in the projected film makes the great amount of planning and preproduction effort completely worthwhile. Puppets, for example, can be made to duplicate their real-life counterparts without the temperamental outbursts or on-stage histrionics occasionally displayed by professional performers.

Cigarettes, beer bottles, and other products can be arranged in a variety of ways and paraded, without visible assistance, before the wide-eyed viewer. Similarly, motors and other mechanisms can be shown in operation and combined with titles and graphic material for educational purposes. More recently, wrappers have been shown folding themselves neatly around products while bottles and glasses have been filled or emptied, without visible assistance, on successive frames of film.

For entertainment, educational, or commercial purposes, stop-motion photographic techniques have provided the means for magically transforming the inanimate into the unbelievable. Regardless of production costs, the unusual amount of preplanning that is necessary before a single frame of film is exposed, and the time-consuming photographic process, the results justify the efforts. See Figures S-23*a* to *e*.

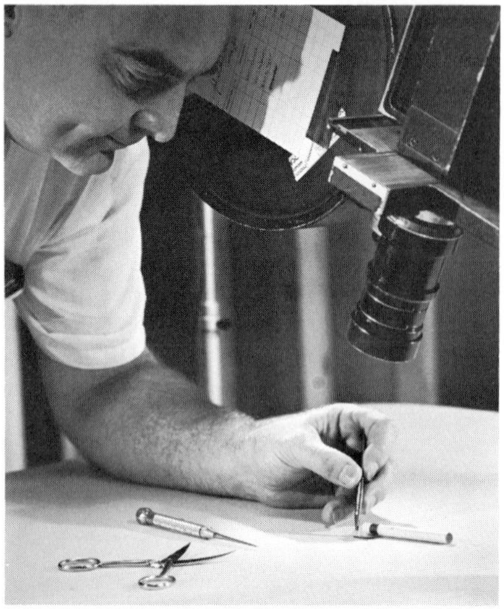

Fig. S-23*a* **Stop motion—with all the patience and dedication of a skilled surgeon.**

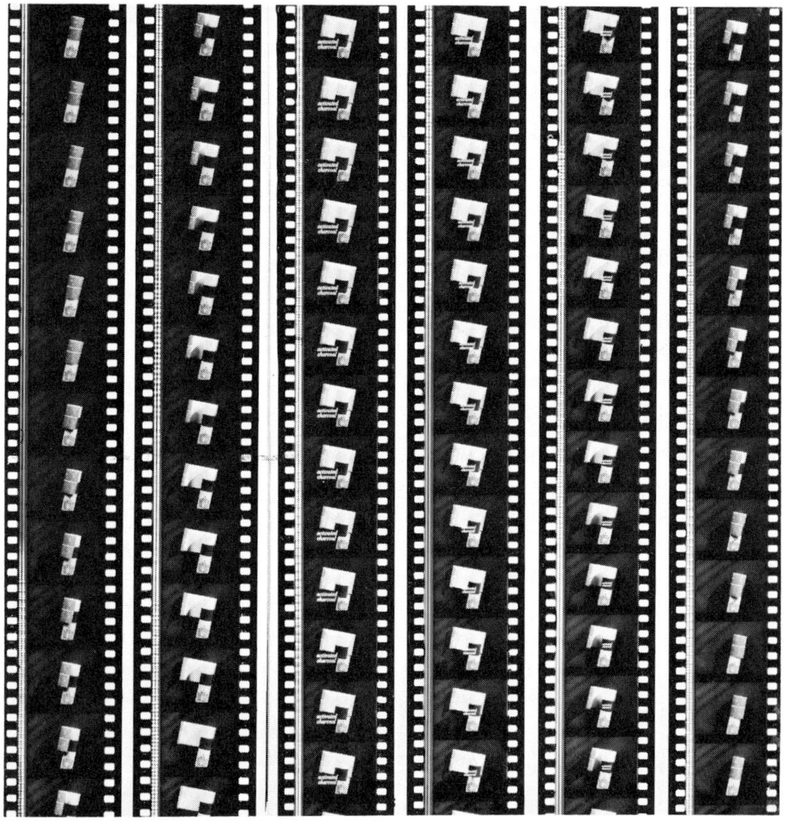

Fig. S-23*b* **The stop-motion sequence subsequently seen by the home viewer. The title was superimposed optically.**

Fig. S-23c Dance sequence. These pats of margarine didn't have to take dancing lessons. Highly skilled technicians "twisted" and moved each of the pats in synchronization with the beat on the sound track recorded for the Kraft commercial.

For each frame of film in the dance sequence, each pat was carefully moved to a new position and attitude with the patience, precision, and skill of a surgeon performing a delicate operation. The marching cigarettes of yesteryear were also put through their paces with stop-motion techniques. In many instances, these sequences have higher production budgets than some of the more complicated live-action shots.

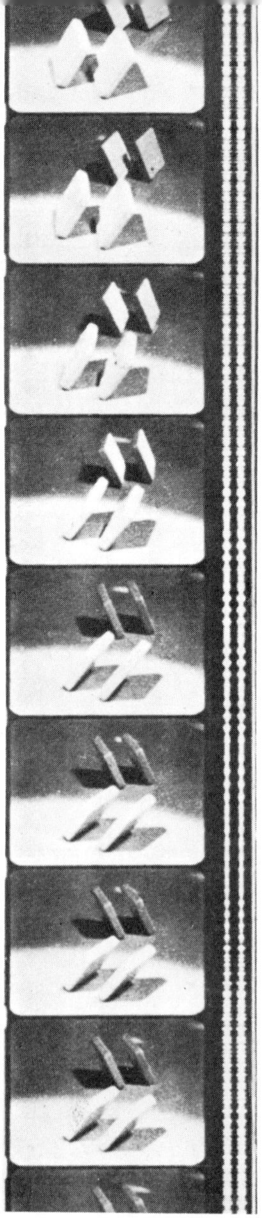

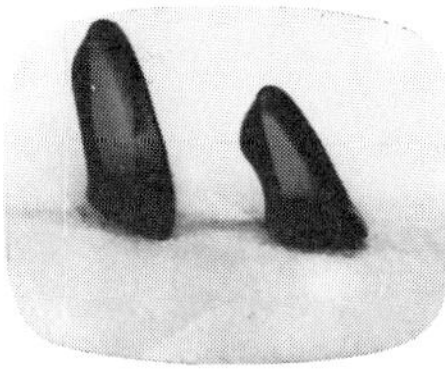

1. WOMAN: It's unbelievable. Now you can change your shoe color . . .

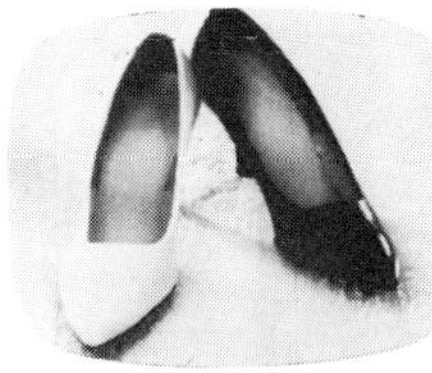

2. to any other color you like. Even black to white.

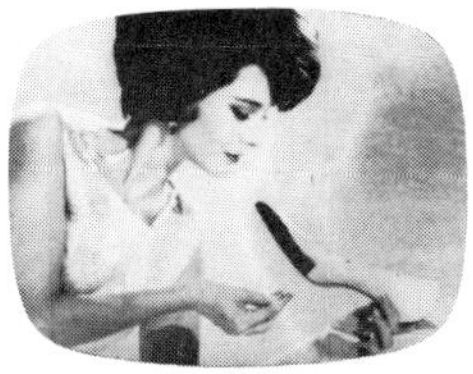

3. Yes, at last you can change your shoe color as easily . . .

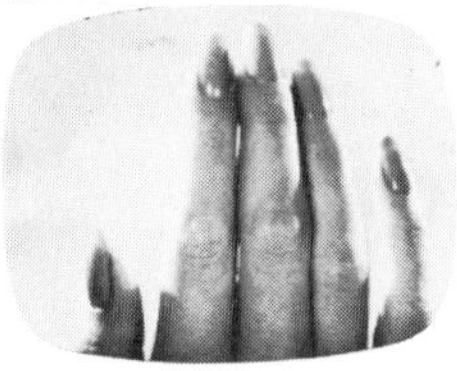

4. as your nail color.

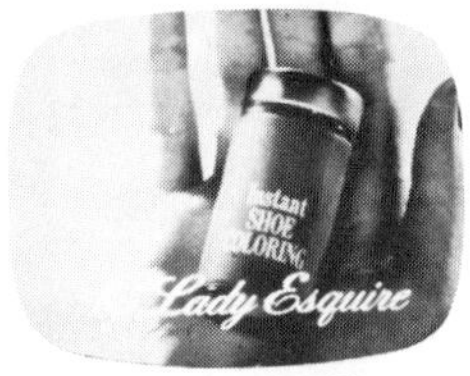

5. With new Instant Shoe Coloring by Lady Esquire.

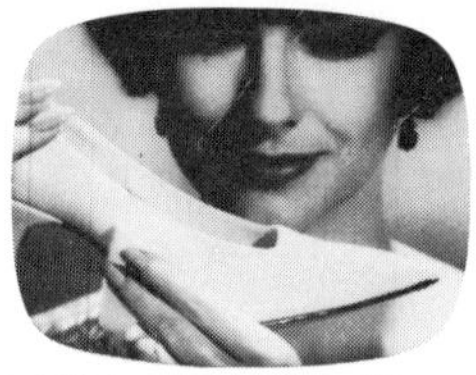

6. Wisk off old color with conditioner cleaner.

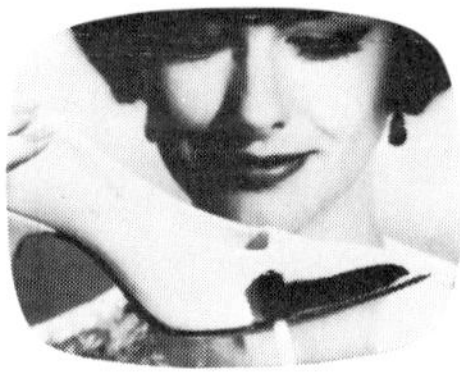

7. Smooth on new color in seconds

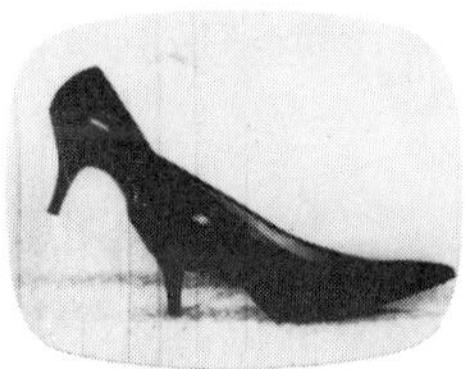

8. Incredible! They look new!

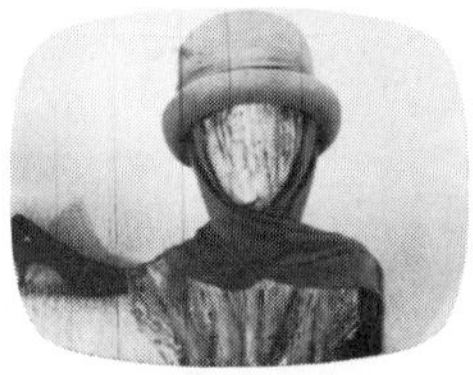

9. Change your blue shoes . . .

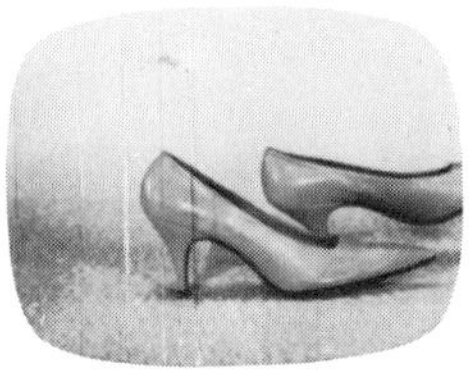

10. to lemon yellow. Change your yellow shoes to teak brown.

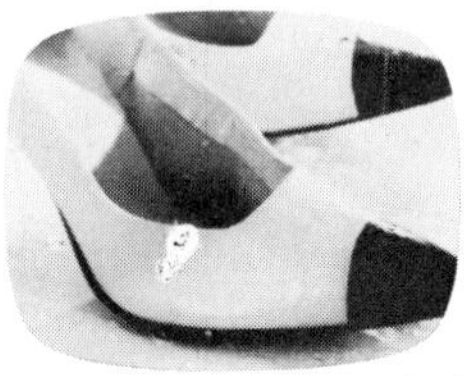

11. Two-tone the same pair of shoes. And Instant Shoe Coloring won't chip, smudge or rain away.

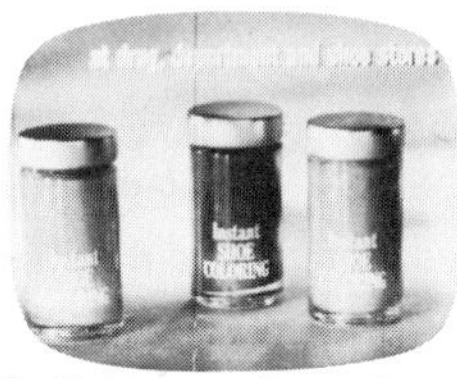

12. Eighteen custom colors only $1.00 each, all made by Esquire. World famous for shoe care products.

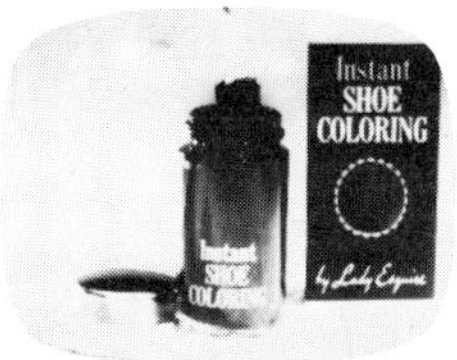

13. So start coloring with exciting new Instant Shoe Coloring by Lady Esquire.

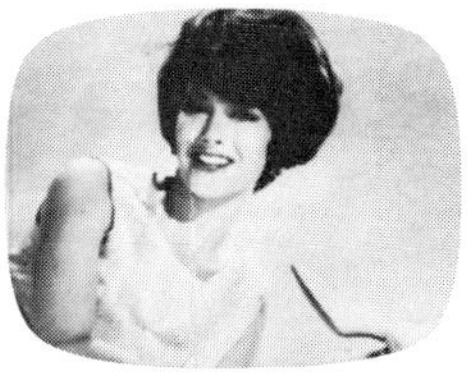

14. It's as easy as coloring your nails.

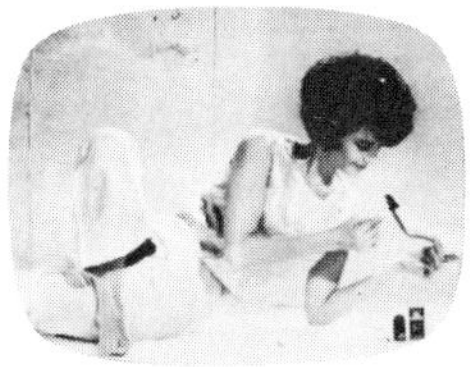

15. (MUSIC)

Fig. S-23*d* This story board for a television commercial contains the sequence illustrated under stop motion.

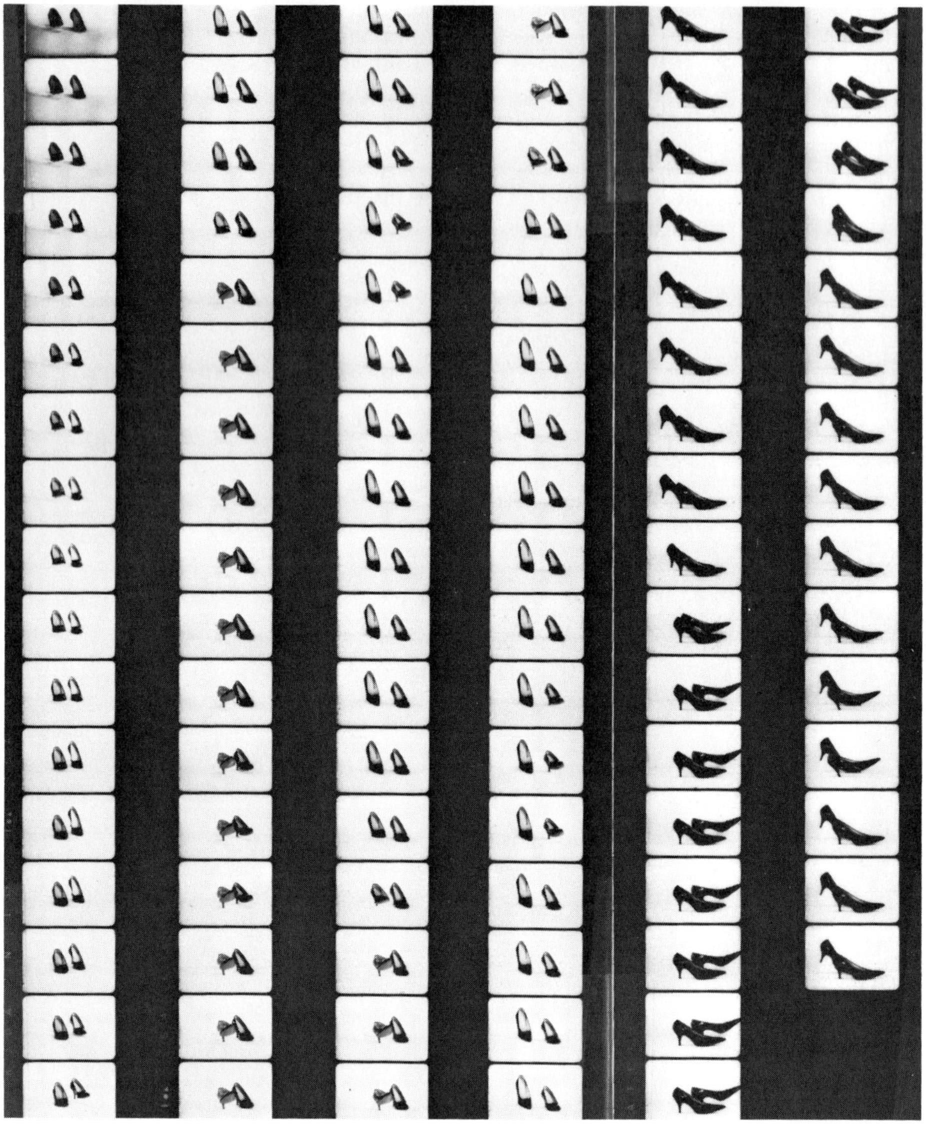

Fig. S-23e Lady Esquire's magic shoes. The commercial produced for Esquire shoe-coloring products includes a stop-motion sequence that shows that technique at its best. In the commercial, a pair of woman's shoes walk from one frame of film to those that follow—without visible help—in sync with the musical background on the accompanying sound track.

Each of the shoes was moved to a new position, according to a predetermined pattern, for each frame of film. Extremely flexible, the specially made shoes used for filming were suspended by wires. For some of the more difficult movements, the shoes were actually nailed to the studio floor.

stop-motion motor To make animated films, one frame is exposed at a time. Also, when refilming silent movies, four frames must be made for each three in the original, as projector speeds have increased from 16 to 24 frames per second. This is done by driving the camera and optical equipment with a stop-motion motor. See Figures S-24*a* to *c*.

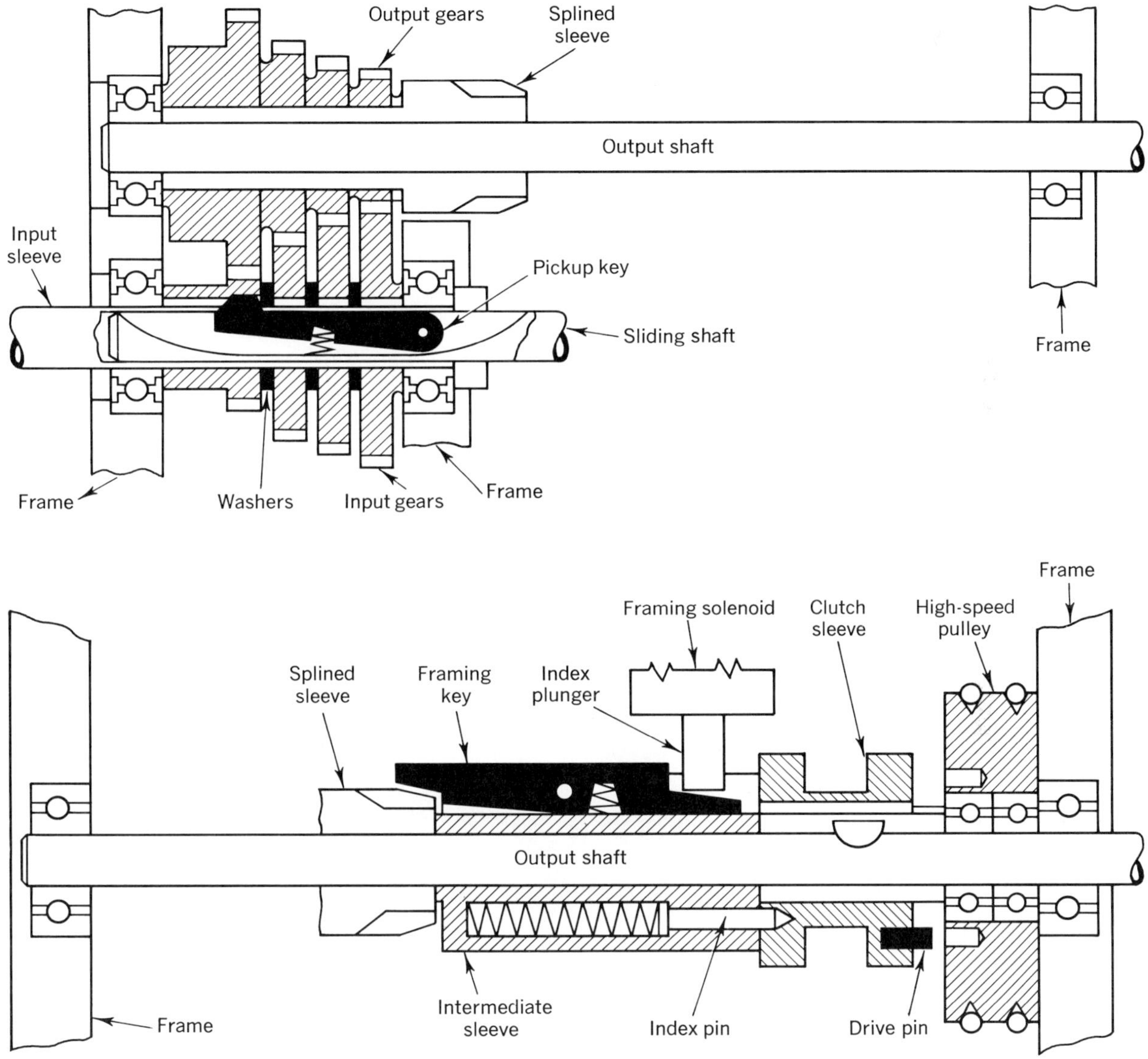

Fig. S-24*a* **The stop-motion motor designed by John Oxberry. (*Animation Equipment Corp.*)**

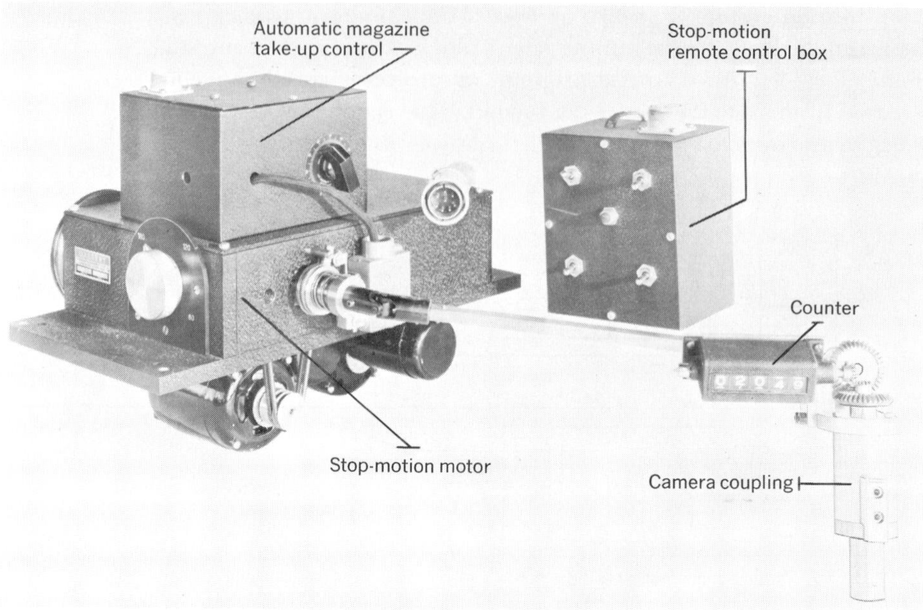

Fig. S-24*b* **The stop-motion motor and controls.**

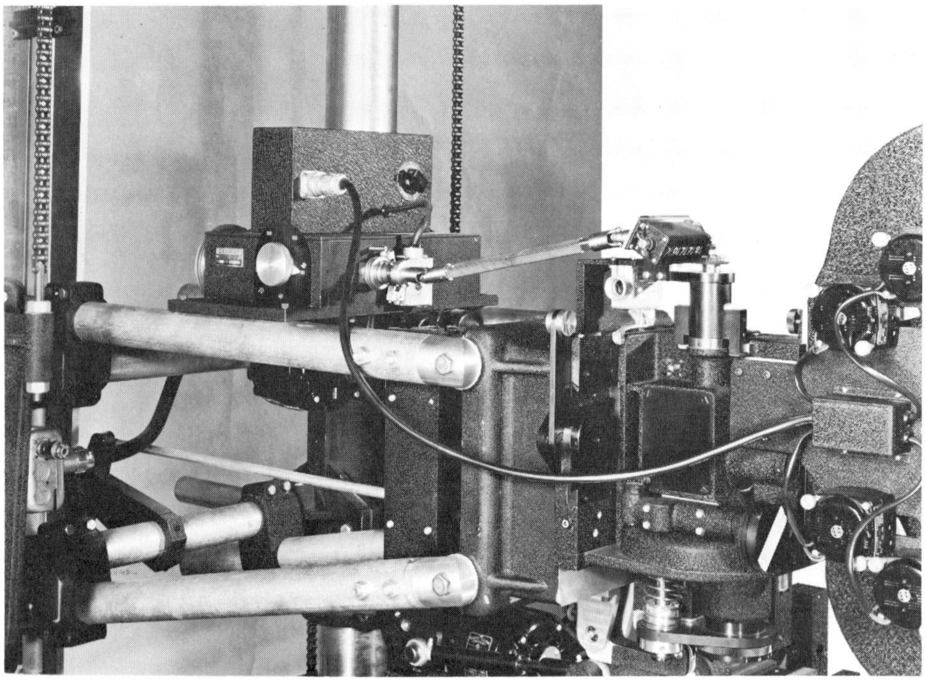

Fig. S-24c **The stop-motion motor mounted on the animation camera stand.**

The framing motor drives the input sleeve of the four-speed transmission. The gears are not keyed to this sleeve and are separated by hardened spacers. The sleeve drives the gears through the pickup key carried by the sliding shaft. The shaft is positioned by the speed selector. Spacers prevent the key from engaging two gears. The output gears are keyed to the driven sleeve which is splined to receive the framing key. This sleeve rotates on the output shaft but does not drive it.

The framing action starts when the solenoid pulls the index plunger from the slot in the intermediate sleeve. The framing key then engages the splines in the transmission output sleeve. When released, the plunger reenters the slot, depresses the key, and stops motion. The output shaft is driven by the sliding clutch sleeve which is positioned by a second solenoid. The clutch is shown engaged with the framing drive. A single-pin connection couples the shaft only in the indexed position. When moved to the right, the pin on the clutch engages the pulley of the high-speed winding drive.

stop-motion projector A projector for the study of picture information on film. This type of projector has forward or reverse variable speeds from 8 to 24 frames per second, frame counters, and provision for single-frame and manual operation. A heat-absorbing glass prevents film damage during single-frame operation.

story board The story board used as a guide for the production of television commercials is not unlike a comic strip in appearance. Arranged in sequence, each panel illustrates a part of the action that is to be duplicated during production. The accompanying narration or dialogue is usually written or typed under each panel.

straight part Unlike the role played by a comedian, heavy, or specialty performer, a straight part requires an actor to play a specific character (as described in the script) in a straightforward and convincing manner.

streaking (television) A term used to describe a picture condition in which objects appear to be extended horizontally beyond their normal boundaries. This is more apparent at the vertical edges of objects when there is a large transition from black to white or from white to black. The change in luminance is carried beyond the transition and may be either negative or positive.

Streaking is usually expressed as short, medium, or long. Long streaking may extend to the right edge of the picture, and in extreme cases of low-frequency distortion, can extend over a whole line interval.

stress markings Undesirable streaks that appear in the print after the film has been developed. These markings are usually caused by improper handling (unnecessary friction) between the time of exposure and the time the film is developed.

stretch Exaggerated changes in the proportions of an animated cartoon character. Unlike the compressing or flattening action of a squeezed or squashed character, the cartoon figure is elongated in animated actions to indicate that the figures are moving at great speeds.

striations The modulations in a variable-density sound track. See VARIABLE-DENSITY SOUND TRACK.

strike The process of dismantling a set after filming has been completed.

strip In lighting, a row or line of lights hung or fastened to a single support.

stripe The band of magnetic coating on motion picture film that is used for recording sound.

striplight A rectangular-shaped light source that produces an even spread of light over a relatively large area. The unit accommodates five 1,000-watt bulbs.

strobe Strobe may be defined as an illusion, or impression, created by a regular relationship between the speed of the subject matter being photographed and the interval between exposures. Wheels that appear to be turning backward, in spite of the obvious forward motion of the vehicle to which they are attached, offer excellent examples of this type of illusory effect. See Figure S-25*a*.

Strobe is usually unpredictable and, almost always, undesirable. An understanding of its causes can lead to its subsequent elimination.

If a wheel moves an amount equal in distance to the spacing of the spokes between exposures, the projected wheel will appear stationary since each spoke will occupy the same position in successive frames of film as the one before it (example 1 in Figure S-25*a*). When the wheel moves an amount equal to half the distance between spokes for each frame of film the projected wheel will also appear to be standing still, but it will look as though there are twice as many spokes in the wheel (example 2).

When the spokes of a wheel move an amount that is slightly less than the distance between spokes, or to a position that places the spoke slightly behind the original position, the eye will accept the lesser movement rather than the "true" or actual movement, and the wheel will appear to be turning slowly backward (example 3).

A spoke moving at any speed that is equal to or slightly less than the thickness or width of the spoke will provide the proper illusion and make the wheel appear to be turning correctly (example 4).

The illusion of strobe can be minimized by changing the number of spokes in the wheel, thereby creating a new and more favorable relationship between spacing and the exposure or projection time, $\frac{1}{24}$ second per frame of film. An increase in the speed of the turning wheel will cause blurring and make the projected effect negligible and relatively unimportant. Simple solutions for the elimination or reduction of the undesirable illusory effect are also possible by changing the angle or perspective of the moving spokes or objects in relation to the position of the camera. See Figures S-25*a* and *b*.

Fig. S-25*a* Examples of strobe.

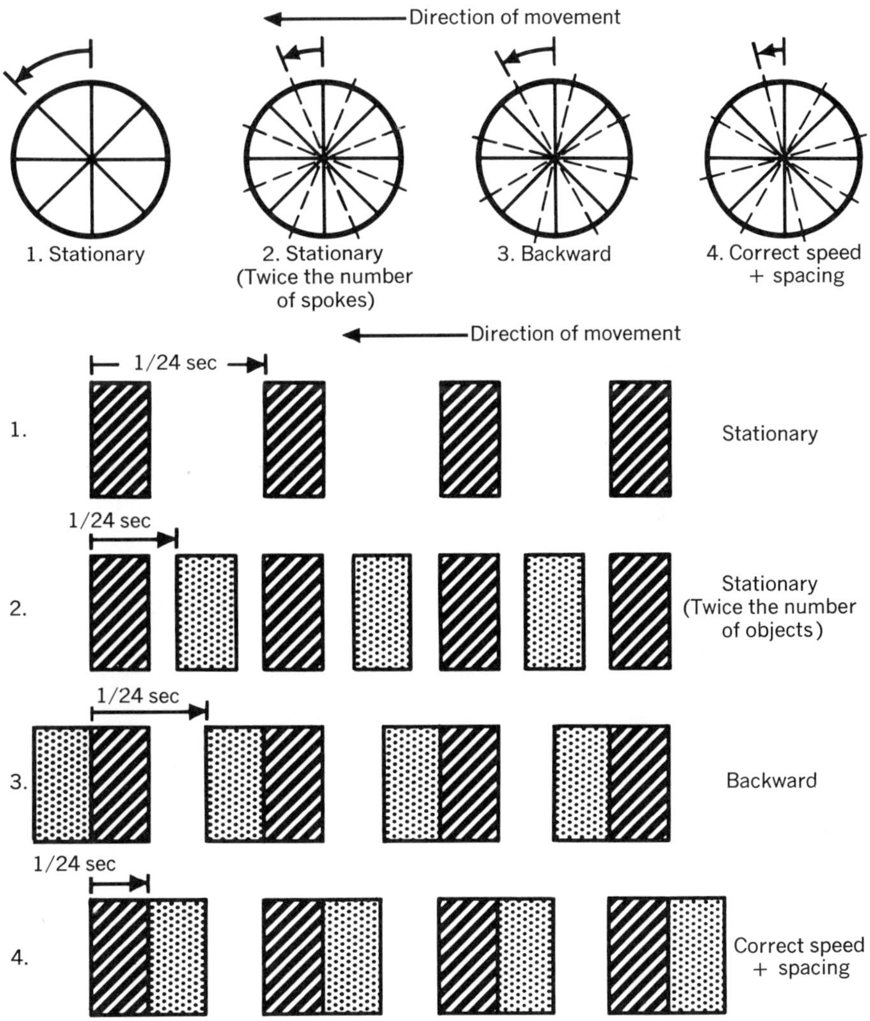

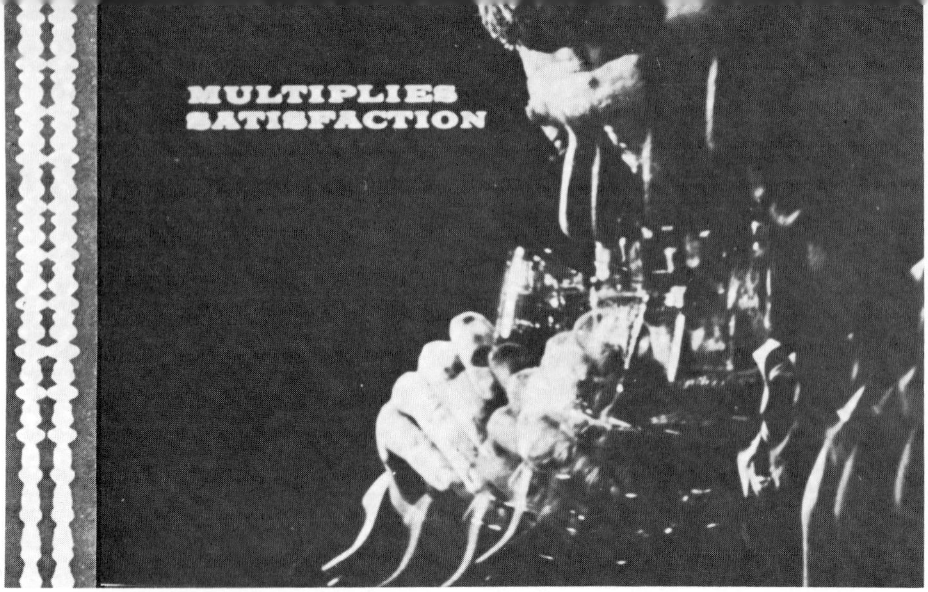

Fig. S-25*b* Multiple-exposure effect similar to strobe. The multiple-exposure effect produced for the Canada Dry Ginger Ale commercial is, in some ways, similar to the illusion created by a strobe effect. To achieve this, the original scene was duplicated optically over the same length of film. For each of the several runs that were required, the start frame of the action was advanced by several frames and reexposed on the raw stock in the optical camera.

strobe light resolver A process that permits synchronous sound recorded on ¼-inch tape to be transferred to a sprocket-driven recorder. During the transfer process, a synchronizing signal on the tape activates a stroboscopic disk on the sprocket-driven recorder. With speeds controlled manually, the lines on the disk appear to be stationary when both recorders are running at the identical speed. See Figure S-26.

Fig. S-26 Strobe light resolver.

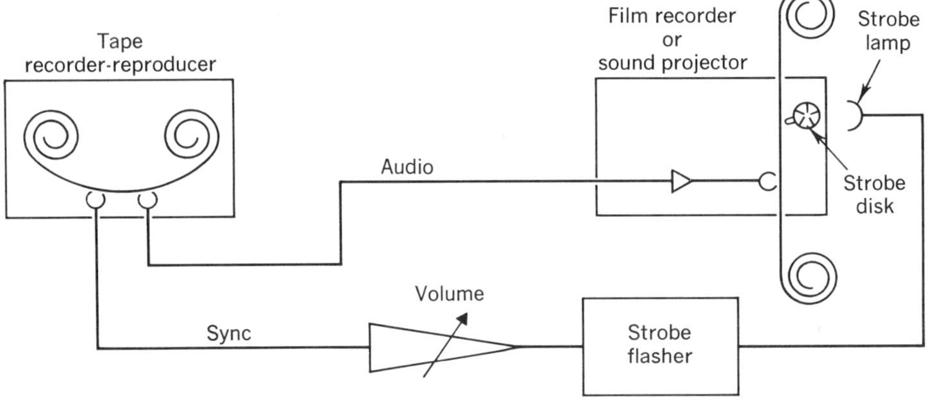

stroboscopic disk A disk containing radial lines used to check the speed of recording or reproducing equipment. If the lines appear to be stationary under the 60-cycle-per-second illumination of fluorescent light, the operator is assured that the equipment is functioning properly.

stroboscopic light A flashing light source, about 1/1,000,000 second in duration, used mainly as a source of illumination in high-speed photography. The light is produced by electric sparks that are set off inside a gas-filled lamp.

studio An all-inclusive term indicating the premises owned or leased by a production organization. In some cases, the administrative offices and production personnel are housed under one roof. In larger studios, each department has its own buildings, and areas are set aside for filming. Studio back lots are often used for location filming.

stunt man The performer who stands in for the star in a motion picture and performs the risky or hazardous actions called for in the script.

substandard film Narrow-gage film less than the standard 35mm width—16mm, Super 8mm, and 8mm film, for example.

subtractive correction A laboratory process in which the printing light passes through filters that subtract or negate undesirable colors.

subtractive filter A group of filters sensitive to practically all colors of the spectrum with the exception of the filter color itself.

subtractive system A method of producing color prints by exposing films having three layers of emulsion, each of which is sensitive to one of the photographic primaries. The term subtractive is used to describe this type of printing system because the final color results from the removal or subtraction of certain components from each of the three emulsion layers.

As a result of the developing process (in color negative film), both a dye image and a silver image are produced in each of the layers of emulsion. A stopbath ends the development process, after which the silver image is removed by bleaching. The result is a color image in which the tonal values are similar to those found in the original subject.

In color-reversal films that do not have the color formers in the emulsion, a black-and-white developing solution is used in the first stage of the processing procedure. In subsequent stages (bleaching and developing) the silver is bleached out, and the dye images in each of the layers of emulsion are produced to form the color print.

In color-reversal films where the color formers or couplers are added to the emulsion by the manufacturer, a negative silver image is produced after development. The remaining or undeveloped silver salts are reexposed, and the dyes are produced in each layer of emulsion (in proportionate amounts) to form the color print. Compare ADDITIVE COLOR-PRINTING SYSTEM. See COUPLER DEVELOPMENT, EKTACHROME, and EKTACOLOR.

substratum The layer or coating on the Celluloid base to which the emulsion adheres.

sun arc An extremely intense arc light.

sun gun A relatively small, portable light source with reflecting surfaces which provide an even field of illumination over the entire picture area. Thumb controls allow several degrees of flood-to-spot adjustments.

sunk mount A lens mount in which only a small part of the lens protrudes in front of the camera's lens panel. The remainder of the lens, actually the larger portion, is positioned behind the lens panel.

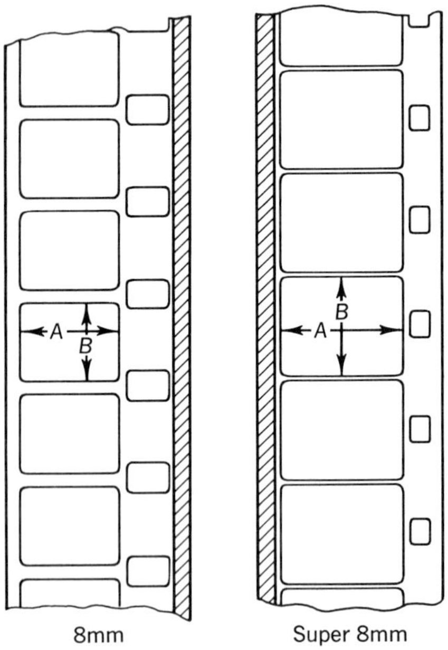

Fig. S-27 A comparison of 8mm and Super 8mm camera apertures.

	Inches	*Millimeters*
8mm:		
A	0.173	4.39
B	0.130	3.30
Super 8mm:		
A	0.211	5.39
B	0.158	4.03

sunlight The difference in color temperature between sunlight and daylight can be as great as 20,000°K or more, an important consideration in terms of exposure and cinematography generally. Sunlight is that amount of light reaching us directly from the sun only, whereas daylight reaches us from all directions and includes light from the sun and sky as well as light that is visible by reflection.

Sunlight encompasses the entire visible spectrum (violet to red) and has a range of approximately 4000 to more than 7000 angstroms. Sunlight has a color-temperature range of 2000 to approximately 5000°K. When sunlight passes through atmospheric haze or moisture, the light is diffused in proportion to the density of the atmosphere. Compare DAYLIGHT. See COLOR TEMPERATURE.

sunshade See MATTE BOX.

Super 8mm film A format designed to provide a picture area on 8mm film that is 50 percent larger than the conventional 8mm frame. This format yields pictures considerably brighter and sharper than those obtained in the standard 8mm frame. See Figure S-27 and Table S-4.

TABLE S-4 Super 8mm Conversion Table

Feet of film*	18 frames per second Minutes:Seconds	24 frames per second Minutes:Seconds
1	0:04	0:03
2	0:08	0:06
3	0:12	0:09
4	0:16	0:12
5	0:20	0:15
10	0:40	0:30
15	1:00	0:45
20	1:20	1:00
25	1:40	1:15
50	3:20	2:30
100	6:40	5:00
150	10:00	7:30
300	20:00	15:00
500	33:20	25:00
1,000	66:40	50:00

*There are 72 frames in 1 foot of Super 8mm film.

Super-80 A powerful fill or flood unit for illuminating large areas. The unit takes 1,300-watt R-80 quartz-iodine lamps.

Super-Beam 1000 A lighting unit that uses a single-ended 1,000-watt tungsten-halogen quartz frosted lamp (3200°K), operating directly from 120 volts a-c or d-c. The Super-Beam has an intensity ratio of 11:1, producing 50 to 560 footcandles at 20 feet from the flood to the spot position. The light provides a well-defined beam with a minimum of "spill" and is ideal for color work in motion picture and television studios requiring a constant color temperature of 3200°K. See Figure S-28.

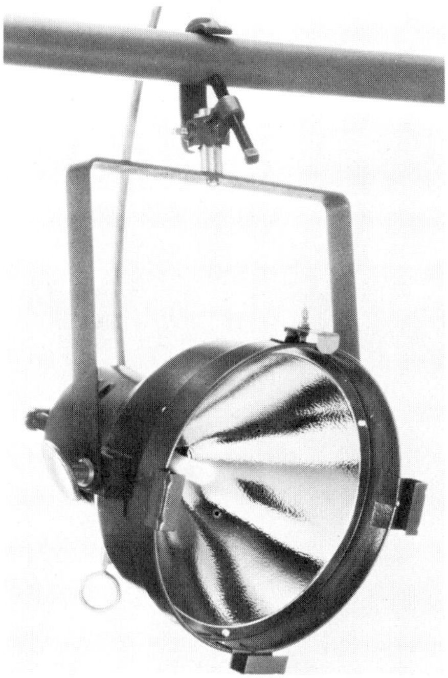

Fig. S-28 Super-Beam 1000—television model.

superimposure Also called **burn in**. The exposure of a title or picture information over previously exposed or processed film. Undesirable double-exposure effects are eliminated by using mattes.

Super Panavision 70 See WIDE-SCREEN PROCESSES.

SuperScope See WIDE-SCREEN PROCESSES.

Super Technirama 70 See WIDE-SCREEN PROCESSES.

supplementary diopter lenses A set of supplementary lenses, including a special range finder that is used for critical focusing. These lenses are extremely useful for close-up and insert shooting.

swan cube A cubical double prism, made up of two right-angled prisms, both prisms being cemented to form a single unit. Also known as a **beam splitter**, the double prism is capable of transmitting a portion of the light from the subject matter in the original direction while reflecting some of the light rays at right angles. Beam splitters are especially useful in dual-film, self-matteing processes.

sweep The repeated horizontal and vertical movement of the electron beam across the chemical coating or fluorescent screen of the cathode-ray tube.

swish pan Also called **zip pan**. A transitional effect used for indicating a change in both time and location. The camera, during the transition, seems to pan very quickly from the scene being photographed to the one immediately following it. The effect is a blur during which specific objects or detail are not distinguishable.

switcher, FM In videotape recorders, the FM switcher is used to activate either the record or playback head. In the playback mode, the head scanning the tape is connected to the output. See PLAYBACK and RECORDING HEAD.

switching In television or videotape, the process of selecting a single picture from the output of two or more cameras.

symmetrical lenses Lenses in which the front half is the mirror image of the rear half, so that the front element is matched by an identical element at the back, and each following element is likewise matched by a counterpart one step further forward in the rear half. See ASYMMETRICAL LENS.

sync Abbreviation for the words *synchronization, synchronizing,* or *synchronism.*

Sync, as used in television, for example, applies to the synchronization signals or timing pulses that lock the electron beam of the picture monitors in step, both horizontally and vertically, with the electron beam of the camera pickup tube.

sync generators, sound Sync generators produce a synchronizing signal that is recorded on the audio tape along with the sound signal. This provides a synchronizing index between picture and sound. During reproduction, the synchronizing signal controls the position of the tape relative to the sprocket-driven film, thus effecting synchronization.

The $\frac{1}{4}$-inch tape may be transferred to either 16 or 35mm magnetic-striped film for side-by-side double-system editing. The edited magnetic sound track may be transferred to 16mm sound-striped film or re-recorded to optical track for composite release prints.

Sync generators are available for custom fitting to cameras or battery-powered d-c camera motors. Permanent magnets are used in the generators, eliminating the need for batteries or amplification between generator and the synchronizing record head. These generators will supply the sync pulse to the Perfectone recorder, equipped with sync-pulse recording head, or to any recorder equipped with a Rangertone sync-pulse recording head. They will also provide the pulse to a Pilotone-equipped recorder.

Sync generators are available in various models that provide a 48-, 60-, or 72-cycle sine-wave sync signal. The various sync signals may be resolved on a resolver with a similar cycle. If other than the 60-cycle sync signal is resolved on a 60-cycle resolver, it is necessary to go through a frequency converter. Thus a 48–60 or 72–60 frequency converter is required to resolve these respective signals through a 60-cycle resolver.

These sync generators (the Ryder is described here) are designed to work with 24 frames-per-second sound speed. They can also be used at 16 frames per second, in which case frequency doublers may be required. See Figure S-29.

sync generator, television Also called **sync gen**. The generator at the transmission station that provides the synchronizing pulses that precede the scanning of each horizontal line and vertical field, and control the brightness of the picture. If improperly synchronized, the picture may roll, or two narrow pictures may be visible (one over the other), or the picture may appear in the form of diagonal strips. These pulses or signals are also used to synchronize the scanning operation in the receiver.

synchronism The process of aligning the sound track with the picture information so that both elements will be seen and heard in their proper relationship when projected from separate reels as in an interlock.

synchronization The correct coordination and relationship between component parts of a motion picture or videotape production (sound and picture information, for example) or in circuits (the horizontal and vertical scanning pulses in television transmitting and receiving equipment).

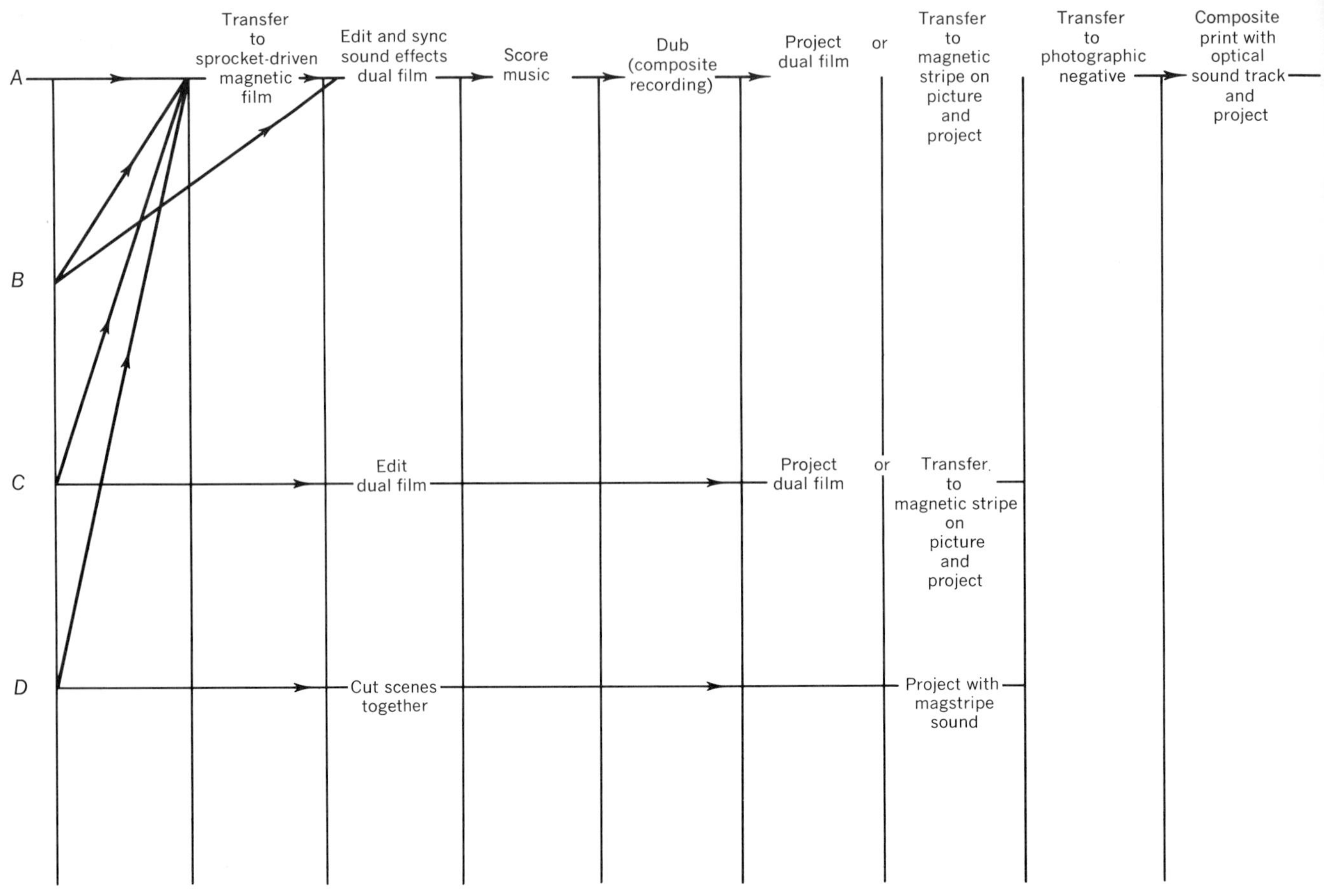

Fig. S-29 Sound-synchronizing methods. (*A*) ¼-inch sync tape: 80% of world production. Simplest. Lightest for location and studio. Most reliable. Least expensive. Requires clapsticks or equivalent. (*B*) Sync drives of sprocketed magnetic film, 16 or 35mm: Now obsolete. Requires too much drive equipment for location. Too heavy. Expensive magnetic film. Requires clapsticks. (*C*) Interlock drive of sprocketed magnetic film ¼ inch wide: Erlson-type drive. Sync from start, no clapstick. For amateurs only. No protection without transfer. (*D*) Single-system cameras: Most compact. No clapsticks. Makes camera too heavy. Questionable sound quality. Good for news and documentary. No protection.

synchronizer In editing, the sprocketed measuring devices used by the film editor. Also, a module in vidicon film cameras that takes the sync fed to the camera head and horizontal-drive stop pulse from the noise module, and generates and distributes vertical gating, vertical drive, timing pulse, and horizontal drive. The module contains the circuitry for automatic time-delay compensation and also clips and amplifies sync for use in other modules.

synchronous motor Camera drives in which the speed of the camera fluctuates to maintain a direct relationship to frequency changes originating at the power source. Changes in line frequency, during the filming process, produce equivalent changes in the camera's running speed. Synchronous motors are quite efficient, even when the source of power is nothing more sophisticated than an ordinary a-c wall outlet. (Electric utility companies generally exercise careful control of the frequency of their a-c power, as the accuracy of any electric clock can attest.)

sync logic In videotape recorders, sync logic generates horizontal and vertical blanking and combines them into composite blanking. Sync logic also combines tape sync and regenerated horizontal sync into composite regenerated sync. It also generates a start pulse which phases the counting of the vertical advance circuitry. See VERTICAL ADVANCE.

sync mark Also called **start mark**. An X mark placed on both the sound track and corresponding picture leaders to indicate a common reference point for synchronization.

sync-pulse generation $\frac{1}{4}$-inch tapes are used extensively in conjunction with motion picture cameras and in the production of feature films and commercials. This is only possible because the motion of the tape can be precisely synchronized to the picture film speed. The technique for achieving this is known as sync-pulse generation. With this technique, synchronizing pulses at frequencies much higher than the audible range are generated in the motion picture camera and fed to the tape-recording equipment. During reproduction, these pulses are fed to a control system which automatically governs the tape speed to maintain synchronism with the motion picture film.

sync separation See VIDEO FILTER.

synopsis A condensed version of a motion picture scenario. Stories for full-length features have been told in a single typewritten page—and sold in less time than it took to read its contents. In other cases, a 10-page synopsis has failed to impress a producer. There are no specific rules that govern the length of a synopsis.

system noise Unwanted sound, usually a high-frequency low-volume hiss, originating in random variations in amplifying and other circuits. It should not be confused with ground noise, which usually has a much lower frequency and is due to imperfections in the recording material.

T

T-stop A system of lens calibration based on the true amount of light that a lens is capable of transmitting. The T stands for transmission in terms of light.

The *f*-stop system of diaphragm markings, by comparison, is a mathematical system that does not make provision for losses of light resulting from absorption or reflection.

tabletop photography Also called **insert stage**. The photography of products or miniature models of subject matter on a relatively small table or platform, or on the compound table of the animation stand. The subject matter can be positioned over any appropriate background (large cards, fabrics, rendered paintings, or dimensional reproductions of a specific area, for example) and illuminated to provide the desired effect.

Because zoom and pan movements can be made on the compound table of the animation stand during the filming process, the cameraman has a greater degree of control and flexibility than he does when filming on ordinary tables or platforms. See COMPOUND TABLE.

tachometer An instrument used for measuring the speed of rotation of a motor. When used in conjunction with a camera motor, for example, a tachometer indicates the operating speed and the possible need for adjustment during the filming process.

tag A trademark or product shot at the end of a television commercial. Also, a term used to describe a visual sequence that is repeated during the closing segment of a regularly scheduled television series.

tail leader See LEADERS.

tails Also called **afterimage**. A secondary or ghost image that trails the main or more distinct image. The undesirable effect is caused by cameras or other photographic equipment in which the shutter and transport mechanism are not in phase or are improperly synchronized.

take (1) Any one of several near-identical shots made during the filming process. The selected take is the one that is cut into the work print. (2) In acting, a reaction indicating surprise.

take-up reel The reel on which the film is wound after it has been screened either by projection or on a moviola, for example.

talking clock In videotape recording, the audible numbers on the videotape at one-second intervals. The cue numbers occur every two seconds—"forty-eight, thirty-one, beep, forty-eight, thirty-three, beep," and so on. The numbers, forty-eight and thirty-one, indicate minutes and seconds, respectively. The *beep* represents the missing number—thirty-two seconds.

tally light The light on a television camera that indicates to the performer that the output of that particular camera is the one whose picture is being transmitted. The director in the control room sees the output of all the cameras, on as many monitor screens as there are cameras, and selects (switches to) the desired picture.

tape An all-inclusive studio term for magnetic tape generally. More specifically, unperforated lengths of ferromagnetic coatings supported on polyester or mylar bases, wound in roll form. Magnetic tape is used for recording sound or picture.

tape cartridge Unrecorded or prerecorded lengths of magnetic tape, usually wound or arranged to form continuous loops, enclosed in containers or magazines that can be positioned in a cartridge recorder. The tape can be used for recording, or if prerecorded, can be played back without the tape having to be threaded through the recorder's transport mechanism.

tape loop Lengths of magnetic tape in which both ends are joined to form a continuous loop. This type of arrangement makes it possible to repeat previously recorded sound or picture without having to rewind the tape.

tape speed The speed at which magnetic tape moves past the recording or playback heads. Speeds used most frequently are $1\frac{7}{8}$, $3\frac{3}{4}$, $7\frac{1}{2}$, and 15 inches per second.

Professional recordings are made at the faster speeds because of the improved high-frequency response. Tape recorded at $3\frac{3}{4}$ inches per second and played back at $7\frac{1}{2}$ ips will sound one octave higher than normal because the frequency of the sound is doubled as well as the rapidity with which it is heard. Similarly, reducing the speed by half during the playback lowers the tone by one octave. See VIDEOTAPE RECORDERS and MAGNETIC RECORDING.

tape-sync processor The tape-sync processor in videotape recorders processes tape sync to produce horizontal-rate reference, field-rate reference and frame-rate reference. See HORIZONTAL AFC and PROCESSING AMPLIFIER.

tape-to-film transfer process See TECHNICOLOR TAPE-TO-FILM TRANSFER PROCESS.

target The chemically coated screen at the end of the cathode-ray tube—also known as **mosaic**. The electron beam scans the target area to expose the image being photographed by the television camera. See CATHODE-RAY TUBE and TELEVISION.

T-bone A metal bracket to which light units are attached.

technical animation An animation technique in which diagrams, charts, symbols, and similar graphics are substituted for the cartoon characters that are featured in animated films produced for entertainment purposes. The technical animation production is generally used for instructional and training purposes.

Although the nature of the subject matter may not keep an audience on the edge of its collective seat throughout the entire showing, the presentation need not be dull. The graphic symbols, for example, can be superimposed over other picture information and be made to zoom or pan to different parts of the film frame. In other cases, arrows and directional lines are used to show the movement of materials in an assembly-line operation, the most efficient method of wiring component parts into an electronic circuit or to show how gasoline flows through an automobile motor. Charts or graphs can also be superimposed over live scenes to show a relationship between the two. Time, distance, weight, and other pertinent information may also be shown as lines that appear and then increase in size (via the scratch-off process) or animate into some appropriate shape. Technical animation can be as interesting as the film's director allows it to be. See Figure T-1.

Technicolor

three-component system. In the Technicolor three-component system, light reflected from the subject matter is transmitted through the single lens of a special camera where it strikes a prism. One part of the light is passed through the prism and a green filter to a green-sensitive negative. The remainder of the light is reflected from the prism and absorbed by negatives sensitive to red and blue light. Each of the negatives is developed to produce new negatives which resemble black-and-white negatives.

These three-color-separation negatives are used to produce positives with a relief image, actually varying thicknesses of hardened

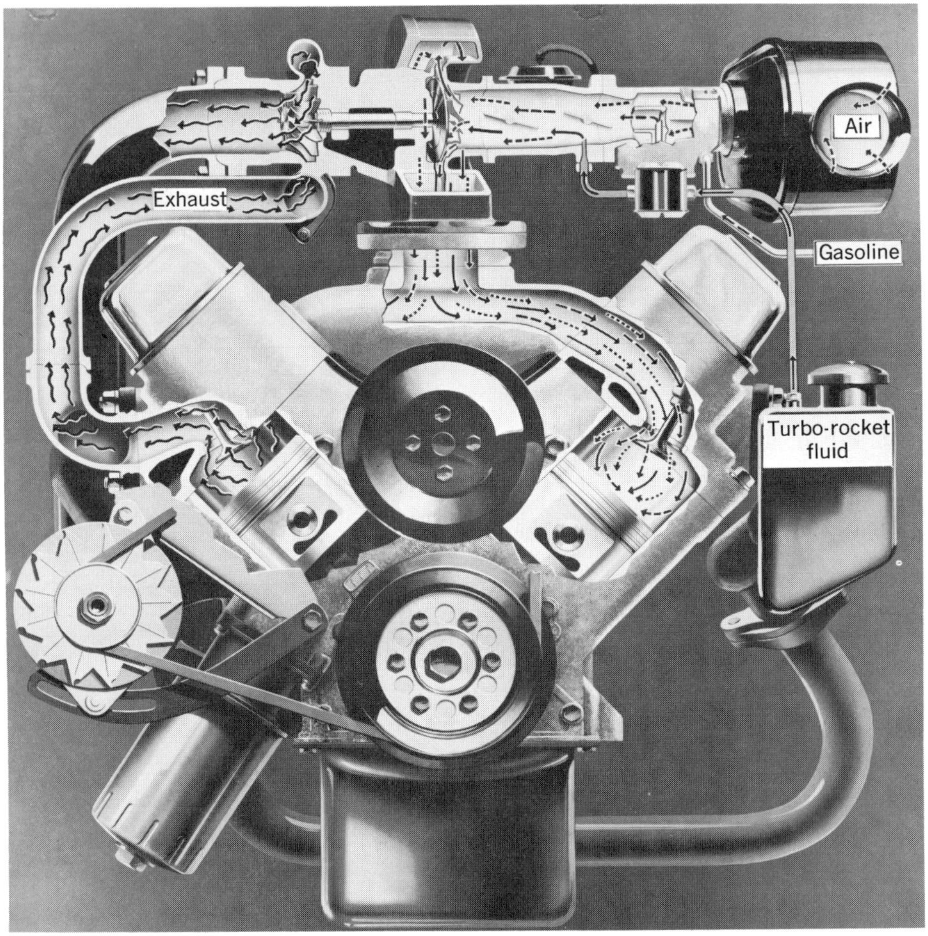

Fig. T-1 Technical animation. Arrows, along with titles, are superimposed over a cutaway motor in order to show the directional flow of gasoline and air in a commercial produced for Oldsmobile.

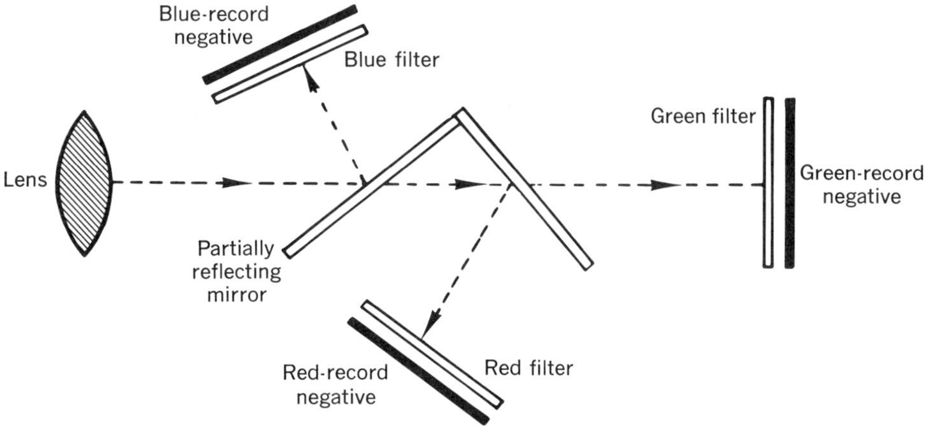

Fig. T-2 **Technicolor three-component system.**

gelatin that represent the gradations in the filmed subject matter. These matrices are used for printing. The colored dyes absorbed by the three matrices are transferred, each in turn, to the final print.

The actual printing process is similar to that used for making lithographs. See Figure T-2.

monopack. Unlike the Technicolor three-component system in which each of the primary colors is exposed on different lengths of film, the monopack is a single film with three layers of emulsion, each sensitive to one of the primary colors. It is developed as a negative and reversed in processing, thus becoming a positive color print.

The surface emulsion is blue-sensitive, the middle emulsion layer is sensitive to green, the third emulsion layer to red. After the film has been developed as a negative, the images are bleached out, reexposed, and developed in coupler developers.

Prints are made from separation negatives. These negatives are produced on an optical printer where different filters are used to reproduce each of the primary colors. The subsequent print is made with the Technicolor imbibition process.

desaturation process. A process in which the color components in the original negative are separated and then recomposed by the superimposition of selected matrices. Actually, an effect in which the colors in a scene are desaturated to produce a monochrome effect.

In order to maintain the values in a scene while the colors themselves are subdued or muted, a standard color negative and a precisely registered black-and-white negative* of the same scene are used. By double-printing this set of matrices, a dye-transfer image is produced in which the colors reproduced from the original negative are altered by the dye components derived from the black-and-white negative.

*The black-and-white negative is made by contact printing into panchromatic stock from a color intermediate master positive.

The amount of exposure given to each matrix, during the two printing operations, affects the degree of desaturation. It is thus possible to control the printing and to produce a full-color reproduction in which the values in the original subject matter are faithfully duplicated, or a print in which the colors are a monochrome representation.

In place of the black-and-white component, any of the three basic dye components (yellow, cyan, and magenta) or any hue derived from a combination of those colors may be used as the desaturator.

The effect can be produced throughout an entire scene or for a portion of it without cutting the negative. In fact, one color can be cross-dissolved to a second color, within the same scene, in order to achieve unusual mood effects. See Figure T-3 and IMBIBITION and DYE-TRANSFER PROCESS.

Fig. T-3 Technicolor desaturation process.

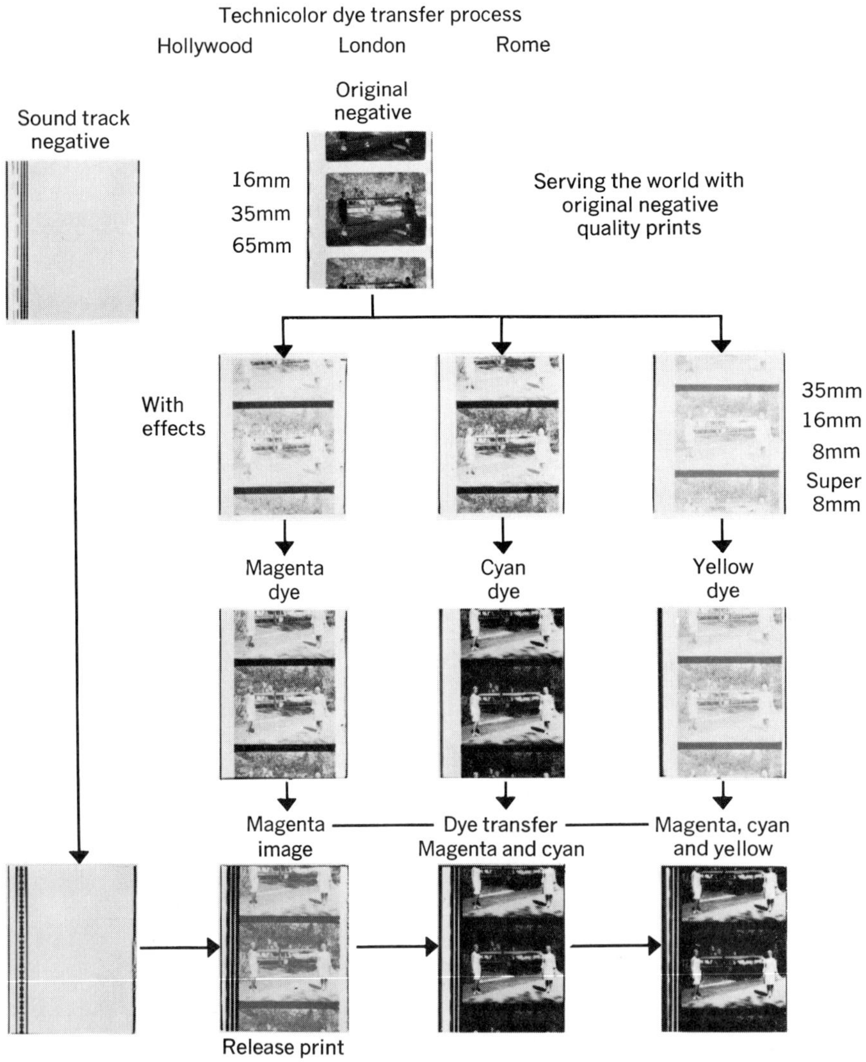

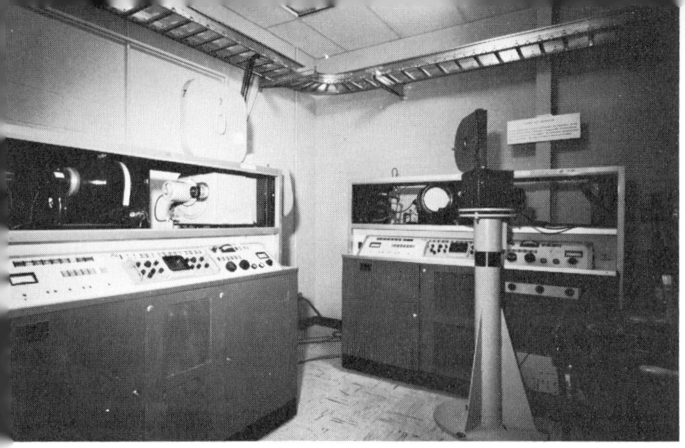

Fig. T-4*a* **Two electronic film recorders in the equipment array of the Technicolor Vidtronics Division.**

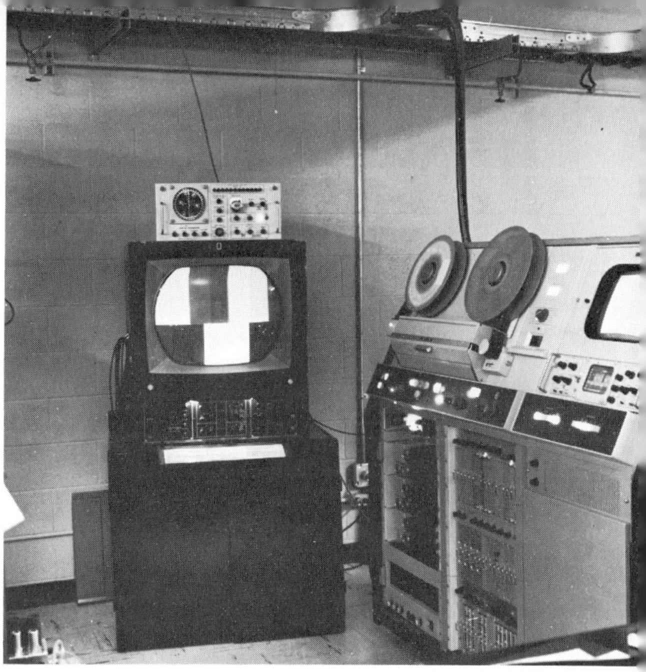

Fig. T-4*b* **Color videotape machine and monitor are key elements of the Technicolor Vidtronics tape-to-color film process.**

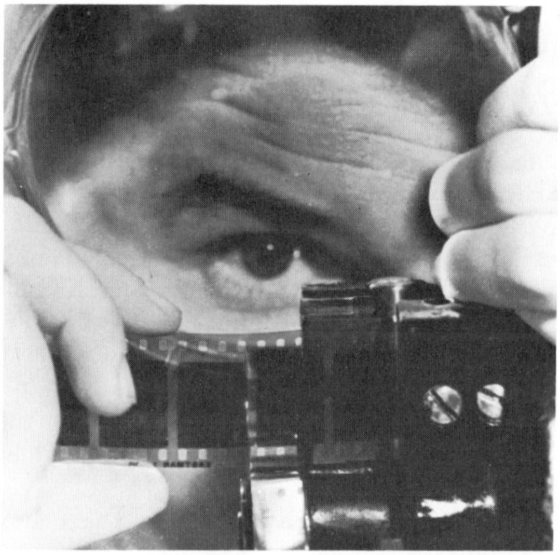

Fig. T-4*c* **Constant checking assures correct tolerance in splices of Technicolor television prints.**

tape-to-film transfer process. A process for producing color film transfers from videotape. The process opens a whole field of color film usage for networks, producers, and syndication companies who can place their color tape product on film and syndicate it throughout the world.

The transfer process can be described as a marriage of an electronic process to the Technicolor print-processing techniques. The transfer system involves breaking a videotape down into its red, green, and blue images and recording them separately.

These images are specially enhanced electronically before processing. The separations are then registered to produce a quality composite print by integrating the components into the final release print. See Figures T-4*a* to *c*.

Technirama See WIDE-SCREEN PROCESSES.

Techniscope See WIDE-SCREEN PROCESSES.

Telco Line The transmission or reception of audio and video signals over telephone company lines.

telecast A television broadcast.

telecine process The process of projecting motion picture film for television transmission.

telecine systems Since a large part of all network as well as local television broadcast schedules consist of film programs, the faithful reproduction and transmission of film picture information is of prime importance.

One of the major problems regarding film transmission, generally, is the difference between European and American standards. European standards, for example, call for a transmission rate of 25 pictures per second. In the United States, by comparison, the rate of projection is 24 frames per second, the same rate as in normal sound photography.

Several systems have been designed for the purpose of making both standards compatible with respect to the number of frames of film projected per second. One of the most frequently used methods is referred to as the 50- and 60-fields-per-second system.

In the European 50-fields-per-second system, each frame of film remains in the projector gate for two full scans. Frame 1, in each series of two successive frames of film, is replaced by frame 2 which also remains in a stationary position in the projector for two field scans. The total amount of time for the scanning and film advance, for each frame, is $1/25$ second, or a projection rate of 50 fields per second.

In the American 60-fields-per-second system, the first of two successive frames of film is scanned for two fields. The second film frame in the series is then brought into projection position where it is scanned for three fields. This sequence, 2:3, is then repeated. The scanning time for the first frame of film, two fields, is $1/30$ second. The scanning time for the second frame of film, three fields, is $1/20$ second. Scanning and film advance time for both frames totals $1/12$ second, thus duplicating the standard sound-speed projection rate of 24 frames per second.

The several telecine systems available for the transmission of film fall into two basic categories. The first of these is the *nonstorage* system in which film is scanned by the flying spot.* In this system, light

*Disks or mirror drums for scanning, used during television's infancy, have been replaced by cathode-ray tubes which now provide the flying spot.

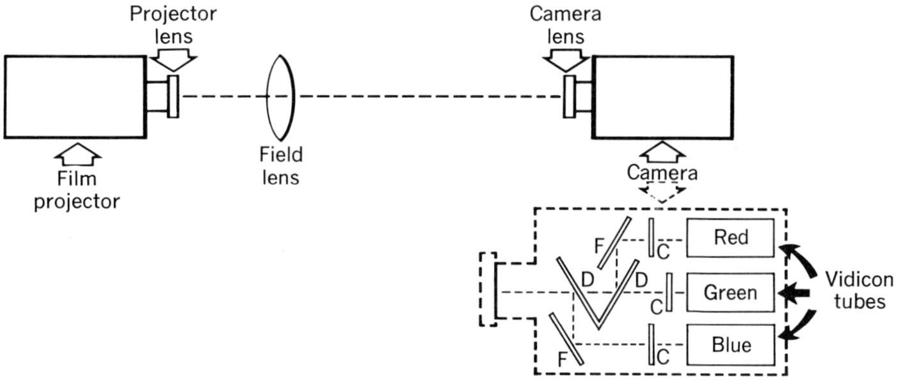

Fig. T-5 **Telecine system.** (F) Front-silvered mirrors. (D) Dichroic mirrors. (C) Filters.

modulated from each film frame is directed into a photoelectric cell. The output, after being amplified, is the required signal needed for transmission. In *storage* systems, the electric output signal is produced by projecting the film onto the photosensitive surface of a camera tube. Each frame of film, with this type of projection system, is stored by the camera tube before it is scanned.

Conventional intermittent as well as continuous-motion projectors in which the film moves continuously with constant velocity may be used for both nonstorage and storage telecine systems.

The rapid pulldown, nonstorage system, for example, utilizes an intermittent-motion projector to transport the film. Similarly, intermittent projectors are used in camera-tube storage systems where the light source is either incandescent or pulsed. In flying-spot nonstorage systems, continuous-motion projectors are utilized for the twin lens and hopping patch, as well as the optically compensated systems which include the mechau mirror drum, polygonal prism, and twin-mirror compensator systems. Camera-tube storage systems using continuous-motion projection processes are the mechau mirror drum and twin-mirror compensator systems. Optically compensated, continuous-motion projectors can be used in camera-tube telecines as well as for flying-spot scanning.

In either case, the differences between the various systems mentioned are based on the methods used for generating the signal needed for transmission, as described in an earlier paragraph. See Figure T-5.

Telek lens Lenses used for Telephoto effects. With these lenses focusing is accomplished by means of a ground glass. Their basic use is with cameras that have bellows or other devices for extending the lens.

telenegative lens A supplementary lens attached to the camera lens for the purpose of converting it into a Telephoto lens.

telephone pickup Induction-coil devices used in conjunction with any ordinary telephone that make it possible to record both ends of a conversation on magnetic tape. Intercut with the normally recorded sound in the sound track, the recorded telephone conversation provides the touch of authenticity and can be regarded as an effect.

Telephoto lens A long-focus lens that enables a photographer to take close-up shots of a subject from relatively great distances. Technically an image must be magnified to a size at least 50 percent larger than the size of the image produced with the normal camera lens, in order to be classified as a Telephoto lens.

Telephoto lenses actually have two lens groupings. The elements used in the front part of the lens form a converging system. The elements in the rear portion of the lens are the diverging units. See Figure T-6.

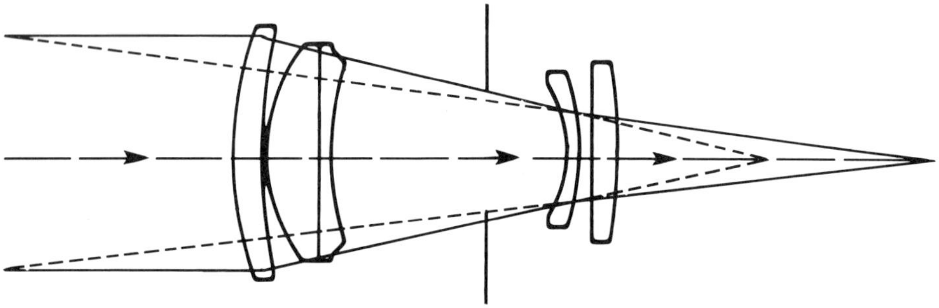

Fig. T-6 **Telephoto lens.**

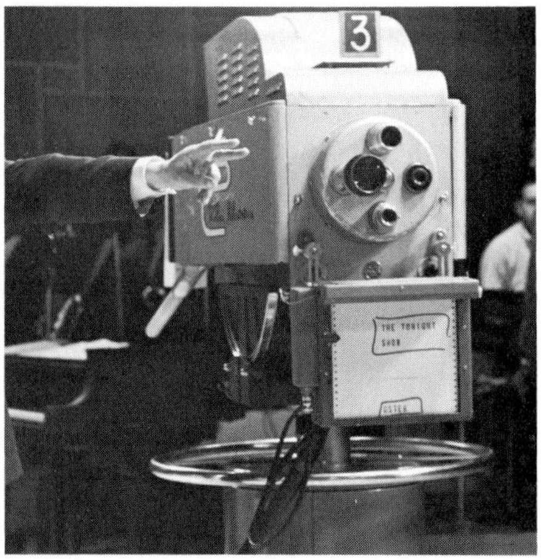

Fig. T-7 **The teleprompter mounted on camera 3.**

teleprompter A cueing device that allows a performer to read rather than memorize lines for the filming process. Positioned near the camera, the lines of dialogue move continuously on a roller. See Figure T-7.

telerecording Also called **kinescope** and **teletranscription**. The process for transferring a televised picture to motion picture film for subsequent transmission.

teletranscription See KINESCOPE.

television Television is a system of transmitting and receiving visual images, over great distances, by electric means. Generally, light waves carrying the picture information are converted into electric signals and carried over radio relay and coaxial cable circuits to the receiving set. There, the process is reversed, and the electric waves are changed back into light waves or visible images, and the original picture information is received by the viewer.

The initial concept was advanced in 1877 by a Frenchman named Senlecq. The first crude but practical method devised for the electric transmission of an image was developed seven years later. Named for its inventor, the Nipkow disk scanner revolved in front of a light source. Each of the small holes in the circular disk scanned a circular segment of the picture during the disk's rotation. See Figure T-8.

By 1907 a process was proposed in which an electron beam would scan the latent image electrically. Differences in light and shade, along each of the parallel scanning lines in the picture area, would thus be in direct ratio to the tonal values in the original image. This arrangement temporarily satisfied transmission requirements. For the reception and reproduction of the picture information, the first cathode-ray tube, an equally important part of the electronic system, was constructed during the same period. Almost a quarter of a century later, theory was turned into actuality with the granting of the first patent for the Iconoscope, an early version of the pickup tube in electronic TV cameras. During the same year, 1925, transmission over wire lines was first demonstrated before a wide-eyed audience.

Subsequent improvements, in the Iconoscope and the cathode-ray tube, led to large-scale field experiments. The tests conducted by the Bell Telephone Laboratories and by the Radio Corporation of America, between 1927 and 1936, played a major part in the development of the medium. Although the first public telecasts were begun in 1939, the large-scale production of TV receivers did not begin until after World War II. By 1947 commercial television had arrived in the home.

Color television, first demonstrated by the Bell Laboratories in 1929, became a reality during 1953 with the development of a compatible system for reproducing color as well as black-and-white images in the same receiver.

Today's television cameras, the image orthicon especially, are extremely sensitive to light. The flame from an ordinary candle, within a reasonable distance from the subject matter, provides sufficient illumination for the transmission of identifiable picture information. This amazing degree of sensitivity, by comparative standards, makes even fast motion picture film seem slow. The vidicon, a more recent development, differs from the image orthicon in principle as well as in sensitivity. The image orthicon uses photoelectric principles for transmitting images whereas the vidicon records picture information by photoconductivity. Because it is more sensitive to light, the image orthicon is used for transmitting live telecasts. The vidicon is used mainly for film and closed-circuit transmission.

The image, or picture information, is focused on the photosensitive surface of the camera tube by means of a lens. The electrons on the sensitized surface of the plate within the tube, are released by ex-

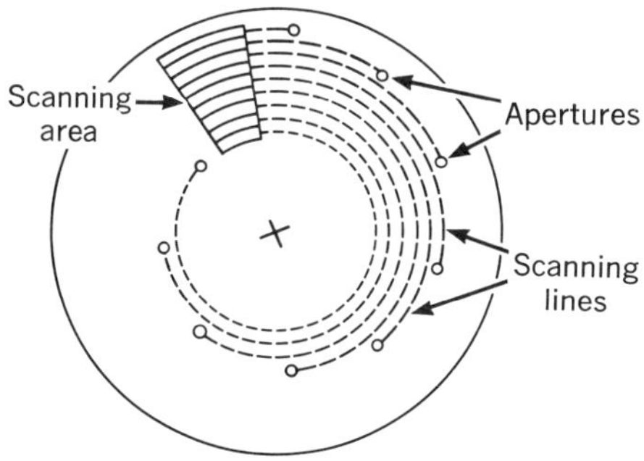

Fig. T-8 The principle of the Nipkow disk—1884. With this scanning method, there must be as many lines as there are apertures. Each aperture exposes its own line, in turn.

posure to the picture source. The regrouping and arrangement of these released electric signals simulates the dark and light tonal values in the original picture, thus creating the image seen by the viewer. The electron beam, or scanning device, causes this process to be repeated 30 times in every second.

Standard television screens contain 525 horizontal scanning lines and are divided into two interlaced fields. The electron beam, or moving light spot, traces each picture during its electronic trip across the screen. The actual picture is re-created in approximately 490 lines. The other 35 lines remain blank while the spot returns to its starting position ready to repeat the tracing process.

Each of the 525 scanning lines contains approximately 450 dots, or picture elements. These dots can be likened to the halftone screens or dot patterns in newspaper and magazine photographs. If the number of lines actually used for the reproduction of the image (490) is multiplied by the number of dots on each line (450), the total number of picture elements on the screen for every $\frac{1}{30}$ second of viewing time is 220,500. A staggering figure indeed. A frame of 16mm motion picture film by comparison contains approximately 250,000 individual grains. The rate of transmission, determined by multiplying the 220,500 picture elements by 30, gives a total of 6,615,000 picture elements per second.

The mathematics are further complicated by the addition of the three electron guns used for reproducing the pictures in color receivers. The screens, in these sets, contain tiny fluorescent dots. One-third of the several hundred-thousand dots, fluoresce in red light. The remaining two-thirds fluoresce in equal amounts of blue and green. Each of the three guns is able to expose only one set of dots. The three superimposed images combine to produce a composite picture in "glorious, living color."

television—the principles Television is an all-electronic system that picks up an original scene, transmits it through the air, and then re-creates the original scene at receiving points. The programs are broadcast directly from television stations which, for network operation, are connected by microwave radio relay circuits or coaxial cables.

But a complete picture cannot be sent at one time. It must be broken up into exceedingly small units. Each unit is then transmitted as an electric signal which represents the relative brightness of that unit. By arranging for the orderly transmission of these signal units, the one following the other, it is then possible at the receiving end to reassemble the picture with each unit in its exact position in its original shading, ranging from black to white. By transmitting a number of consecutive scenes rapidly (30 per second for television) the viewer is given a sense of motion in much the same way that the motion picture film gives the impression of lifelike movement.

In television, as in photography, a camera is essential for picking up an image of a scene. The heart of a television camera is the image orthicon which performs a function similar to the film in a motion picture camera.

This camera tube, in appearance, resembles a large-size flashlight. At one end is a flat photosensitive glass surface on which a lens focuses a light image of the scene to be televised. The appearance of the visible image instantly creates on a target inside the tube an identical image formed of electrons. The number of electrons at each point of this invisible image corresponds to the intensity of the light at that point in the original scene.

The next step is to "read" this electron pattern so that each unit area in its composition can be evaluated and then transmitted to receivers. This is done by scanning this charged image by a beam of electrons shot from a gun at the opposite end of the image orthicon. This beam, pinpoint small, scans the electron image from side to side and top to bottom just as the human eye scans a printed page, but a thousand times faster. When this beam reaches the image, it looses different amounts of electrons, depending on the charges it meets, and then returns to a spot near the gun, where the electrons are amplified millions of times and led off to the transmitter as the television signal, which is radiated into space by the transmitting antenna.

At the receiving end of the television system, an antenna intercepts a bit of the electromagnetic energy sent from the transmitter. This energy causes a current to flow which is fed to the receiver. After the signal is amplified and processed by an array of tubes and components in the receiver, it is impressed on a picture tube, called the *kinescope.* An electron beam in the kinescope paints the received picture on the face of the kinescope. The screen of the picture tube is coated with fluorescent material, a phosphor, which glows at any point when stimulated by the electron beam. The image of the scene is, in effect, divided into 525 horizontal strips, a small fraction of an inch in width. The scanning beam at both camera and receiver completes the scan of each picture 30 times a second. As the beam moves along its directed path in the tube, its energy varies constantly to conform to the differences in illumination of the original scene. As a consequence, the impact of the beam on the fluorescent surface of the picture tube rapidly constructs a pattern made up of tens of thousands of these dots of light which, in $\frac{1}{30}$ second, reproduce for the eye an entire scene in black and white.

By electronic means, the scanning beams at transmitter and receiver are locked together, synchronized, throughout the telecast so that under normal conditions the resulting picture is a good reproduction of the original scene, free from distortion and unnatural movement.

The scanning rates and other operating specifications of black-

and-white television have been standardized by the Federal Communications Commission; hence the signal from any television station can be received by any television receiver within range.

To prevent the stations from interfering with one another, the FCC has allotted channels or "highways" in the radio spectrum for the transmission of television programs, as it has for other types of radio transmission. The television channel, because of the mass of information (the number of details per picture per second) that it must carry, is 6 million cycles (6 megacycles) in width compared to 10,000 cycles for AM radio broadcasts and 20,000 cycles for FM programs.

The TV transmitter is, in effect, two separate units. One sends out the picture, and the other the sound. Visual transmission is by amplitude modulation (AM). The sound portion employs frequency modulation (FM) and operates much like regular FM radio broadcasting.

Like other forms of modern communication, such as radio and the telephone, television consists of taking something that is evident to our senses in one place, converting it into a series of electric impulses that match the original information in strength or duration, pumping these impulses into a wire or into the air, capturing them at a distant location, and then converting them into a reproduction of the original information.

The diaphragm in a telephone mouthpiece, for example, changes sound waves into electric signals which will travel over wires either across the street or across the nation. The diaphragm in the earpiece at the receiving end reconverts these signals into sound waves. Radio uses a microphone in place of the telephone mouthpiece, electromagnetic waves instead of wires, and a loudspeaker in place of the earpiece. Although the process is much more complicated, television does the same thing with light that the telephone and radio do with sound.

The TV camera employs a lens to focus light rays reflected from a scene or an object in the same way that a camera lens is used. But, instead of focusing these rays on a piece of film, it directs them to a light-sensitive surface. The action of these light rays is to release from the surface electric charges corresponding exactly to the amount of light reaching that point.

recording the image. Initially, light reflected from the subject matter positioned in front of the camera is transmitted through an objective lens to the camera pickup tube where it is focused on a light-sensitized plate that is referred to as the *target* or photocathode.

The electron gun located in the neck of the glass vacuum tube directs an electron beam in the direction of the 525 scanning lines on the light-sensitive plate. Every part of the image focused on this plate or target area releases an amount of electricity that is in direct proportion with the light and dark areas in the original image.

The beam, when creating the image on the target, moves across every odd-numbered line, from left to right, beginning with the first scanning line at the top of the target. After reaching the end of the last odd-numbered line in the lower right-hand corner of the target area, the beam from the electron gun swings back to the upper left-hand corner to begin scanning each of the even-numbered lines, in turn. Each scan of all the odd- or even-numbered lines completes one field, or half of the total picture. Thus, two complete scans, from left to right and top to bottom, are needed to form one frame of the entire picture. The scanning beam, when it completes its sweep of any one line or when returning from the lower right-hand corner to the upper left-hand corner of the screen, is blanked out. This blanking action duplicates the function of the shutter in a motion picture camera and is repeated 525 times horizontally and twice vertically during the scanning of one frame.

By scanning or "reading" these charges, left to right and top to bottom, with an invisible, pencil-like beam of electrons, the TV camera is able to put out a video signal in which each individual charge or picture element is represented in the proper order. In the sensitive image orthicon tube, the "eye" of the TV camera, the strength of this signal is multiplied many times before the signal leaves the camera.

Since the scanning beam travels completely over the target 30 times in every second, the signal continuously leaving the camera represents a series of many pictures in rapid succession, varying as movement occurs in the scene. This provides the illusion of motion when the scene is reproduced, in the same way that this illusion is achieved in motion pictures.

The electric impulses that are produced in the conductive coating in back of the photocathode during the scanning process are directed through a wire to the outside of the tube and on to an amplifier which builds up the impulses so that they can be used for transmission purposes.

These amplified impulses are fed through a cable (coaxial) that leads to the control room. Pulses from a generator designed for that specific purpose synchronize the scanning speed of the electron gun with the scanning speed of the receiving equipment. The timing between both is controlled to 1/10,000,000 second. Speeds are interlocked to ensure proper transmission. The sync generator also controls the blanking signal.

Each picture, during the scanning process, is visible on an electronic oscillograph. The original subject matter is seen on a black field as video waveforms or narrow green lines, indications to the technicians in terms equivalent to the volume controls in sound equipment.

The amplified impulses, during the transmission process, are fed into a second electron gun in another cathode-ray vacuum tube known as a *monitor*. In this studio monitor or studio receiver, the fluorescent target area is scanned by the electron beam in a manner identical with the scanning action that took place at the time the image was received and exposed originally. As the electric impulses strike the fluorescent material that coats the screen area, each scanned spot glows in direct proportion to the dark and light areas in the original subject matter. Adjustments with respect to shading are made electronically, and the monitored signal is sent to the master control room from which it is transmitted, along with the accompanying sound signal.

transmission. In the master control room, the producer or director may switch the station output instantaneously from one camera to another to change scenes, move from a long shot to a close-up, or vary the angle from which the scene is viewed. Or, he may switch to the output of a TV film pickup unit and to televised outdoor or other remote-location sequences from film previously made or obtained from a library.

The heart of any television station is its transmitter plant which contains complete picture and sound-transmitting equipment, the antenna and related apparatus needed to broadcast the programs originating in the station studios. In effect, the cameras and microphones in the studios are the eyes and ears of television; the transmitter may be regarded as its beating heart.

A television transmitter is actually two transmitters, one unit being used for sound broadcasting, and the other for picture transmission. Each unit serves a threefold purpose. It picks up the picture or sound signal from the control room and greatly amplifies or strengthens it. But, even after amplification, these impulses have neither the properties nor the power to travel alone from the studio to home receiving antennas. The electronic delivery system that is required to carry them to their destination, is termed the *carrier wave*. The creation of this wave is the transmitter's second function. Its final task is to impress the amplified sound and picture impulses on the

Fig. T-9a Television pickup tube.

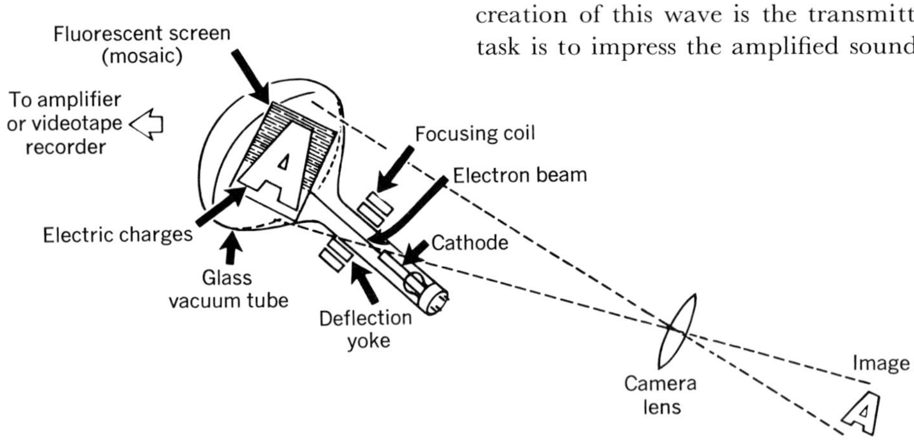

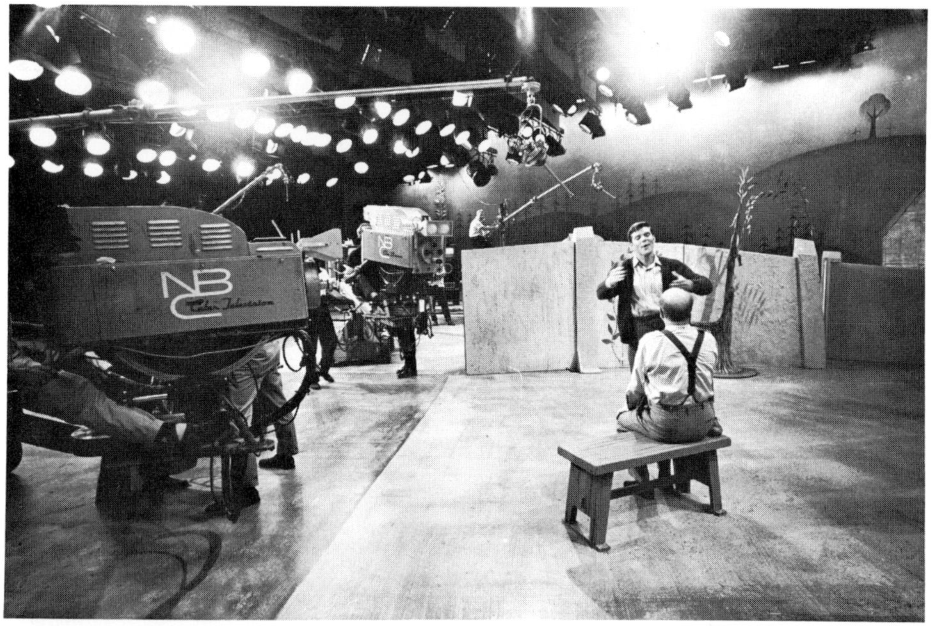

Fig. T-9*b* The rehearsal or "walk-through."

carrier wave, for delivery to home receivers. This process is called *modulation.*

The amplification function of the transmitter is performed by tubes and circuits that have been designed to impress on stronger currents signals obtained from the TV studio. While this is being done in one section of the transmitter, other circuits are creating the carrier wave from 60-cycle electric current, the same type used to light homes and run electric appliances. Since this type of electric current is used to run both transmitters and receivers, it provides the time basis for the entire broadcasting system.

The modulator stages of both the sound and picture transmitter now perform their third function—combining the amplified program signals with the carrier wave. The two signals are carried through separate transmission lines to the diplexer, a coupling unit which combines them so they can be radiated from the same antenna at the same time.

When these signals reach the antenna, electromagnetic waves are set up in space, spreading out from the antenna in the same way that ripples spread over the surface of a pond from the point at which a disturbance is caused by tossing a stone into the water. Carrying in their variations all the detailed information impressed on the original signal, these waves spread throughout the coverage area of the TV station. When they encounter a TV receiving antenna, corresponding electric impulses are created in this antenna and carried by the transmission line to the television receiver.

In the receiver, the complex radiated signals are again amplified and separated into their original three parts. The picture signal is fed to the scanning beam that "paints" the picture on the TV screen. The sync signal is fed to the deflection circuit that controls the horizontal and vertical movements of the beam as it deposits its information on the screen. The audio or sound signal is fed to the loudspeaker.

The scanning beam, varied in intensity by the picture signal, reproduces the picture by exciting to varying degrees of luminosity, according to its strength at each instant, the particles of phosphor with which the back of the viewing screen is coated. See Figures T-9*a* to *e*.

And, now, a word from the sponsor.

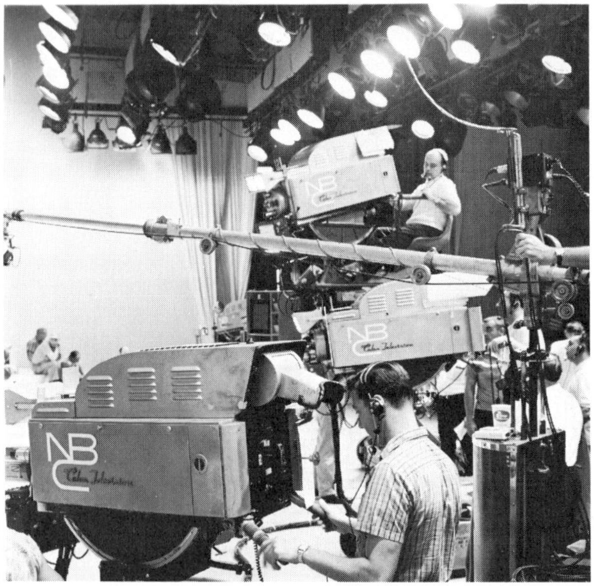

Fig. T-9*c* "On the air!"

Fig. T-9*d* The control booth.

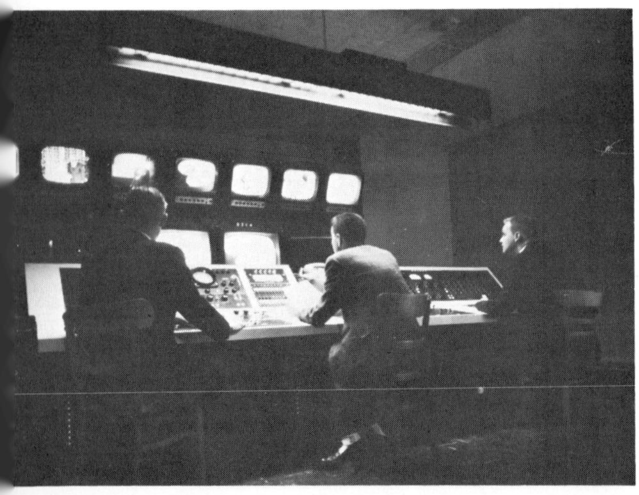

Fig. T-9*e* Cut to camera 2.

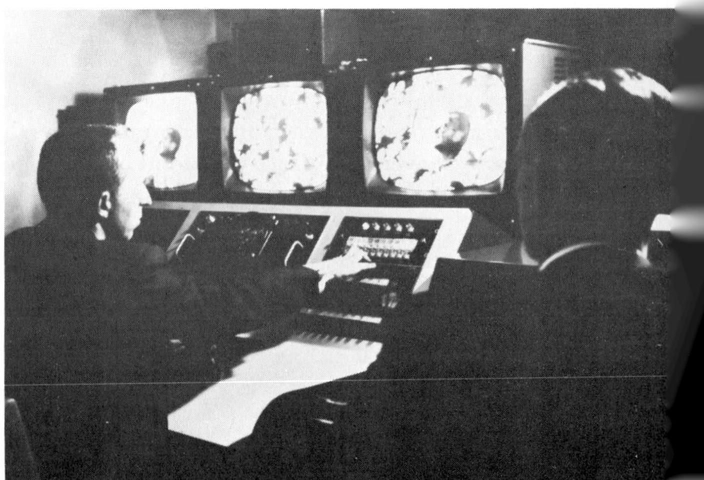

television—color The color television picture tube is probably the most complex device ever to be mass-produced for use in the home. It is a precision electronic device built to exacting tolerances measured in thousandths of an inch. Yet the color tube is produced with assembly-line techniques by the millions each year.

The color television picture tube employs three electron beams to excite the fluorescent screen and produce the picture by principles similar to those employed in the far simpler conventional black-and-white picture tube. It is a high-speed information-handling device that displays several different types of synchronized image data simultaneously to create 30 complete full-color pictures every second.

Common to all television picture-tube display systems, color or black and white, are these elements:

1. An electron source at the rear of the tube, to emit a beam of electrons varying in density in accordance with the picture information of the received TV signal.

2. A phosphor screen applied on the interior of the tube face, consisting of tiny crystals of phosphor material that emit visible light under bombardment by the high-energy electron beam.

3. A scanning process that causes the electron beam to sweep at a high uniform speed from side to side over the phosphor screen, forming a field of 525 parallel lines.

These basic elements appear in far more complex form in the color television picture-tube system. The color tube requires not one but three electron beams, emitted by three separate electron sources at the rear of the tube. Each of these electron beams is varied in intensity in accordance with the TV picture information corresponding to the three primary colors of the television system—red, green, and blue.

In contrast to the black-and-white tube, whose phosphor screen is a continuous surface of white light-emitting material, the phosphor screen of the color tube consists of approximately 1.3 million phosphor dots arranged precisely in triangular groups. Each of these triads is formed by one red, one green, and one blue light-emitting dot of phosphor.

Interposed between this intricate phosphor pattern and the electron guns at the rear of the tube is an element unique to the color tube—a *shadow mask.* This is a metal perforated sheet containing over 400,000 tiny holes, each of which is precisely aligned with one of the triangular groups of phosphor dots to a design tolerance of less than the diameter of a human hair.

The function of the shadow mask is evident from its name. It masks the three electron beams so that the electrons of each beam can excite only the phosphor dots of a single color. Thus the beam varied by the red signal information will excite only the red light-emitting dots; the green beam will excite only the green dots; and the blue beam will excite only the blue dots. The combination of the emitted light in varying intensities produces a picture reproduction on the screen in "living color."

The light emitted from the vast number of individual phosphor dots cannot be seen individually. The eye blends the numerous bits of picture together, resulting in a full-color picture detail display 30 times per second—the speed at which the process occurs. The picture dot structure can be discerned with the aid of a magnifying glass while the tube is operating.

When a program is broadcast in black and white instead of color, the incoming signal controls all three of the electron beams simultaneously. The red, green, and blue phosphor dots are excited in proper proportion to reproduce the monochrome picture. It is recognized that the three primary colors can be added in proper proportion to produce white light. The overall light output is then varied to produce black-and-white pictures.

RCA TK-42 color TV camera. The TK-42 color television camera utilizes four pickup tubes, one 4½-inch image orthicon to develop a separate wideband luminance signal, and three electrostatically focused 1-inch vidicons to develop the chrominance signals. This is a new concept to produce both color and monochrome video signals of highest quality. The separate luminance channel provides two very important advantages over earlier three-tube cameras. First, the output signal, when viewed on a monochrome receiver or monitor, is equivalent in resolution, gray scale, and overall quality to that obtained from the finest black-and-white cameras. Second, the high-resolution luminance component of the colorplexed video signal enhances the gray scale and resolution of the color picture.

Transistors and other solid-state components are utilized throughout the camera system to perform all circuit functions. The only exceptions are the four pickup tubes, viewfinder kinescope, high-voltage rectifier, and the tubes used in the picture and waveform monitors.

Built into the camera is an extremely versatile high-quality zoom-lens system. The basic lens and an accessory wide-angle adapter cover the entire range of focal length from 1.6 inches (40mm) to 40 inches. The superior optical quality and wide range of focal length of the system do away with the need for a conventional lens turret and assortment of lenses. In addition, the single-lens system eliminates the problem of color matching between lenses of various focal lengths on a single turret.

The built-in lens and wide-angle adapter will handle virtually any requirement for either studio or field operation—with the added advantage of an infinitely variable range and focal length to frame any scene precisely within the desired field of view. See Figures T-10*a* to *c*.

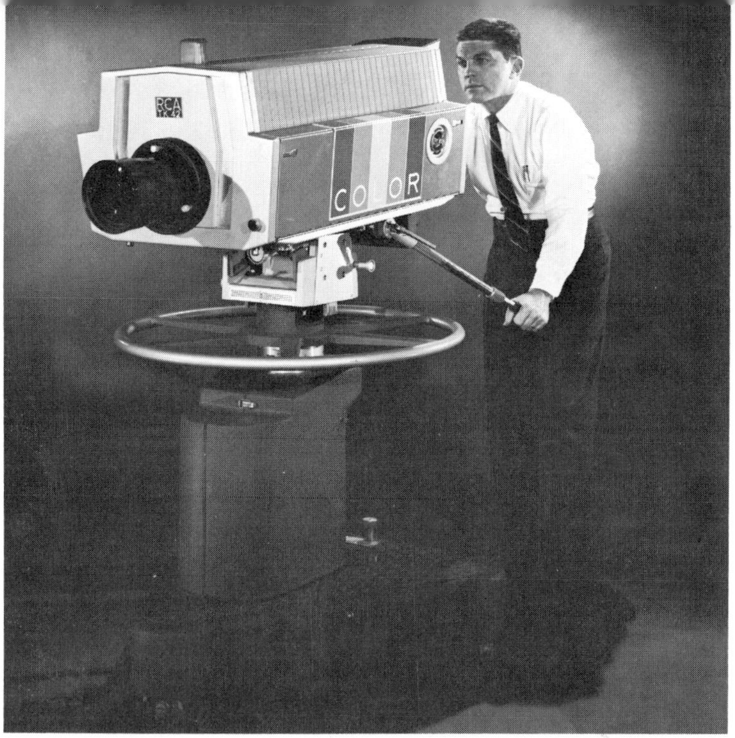

Fig. T-10*a* RCA TK-42 color television camera.

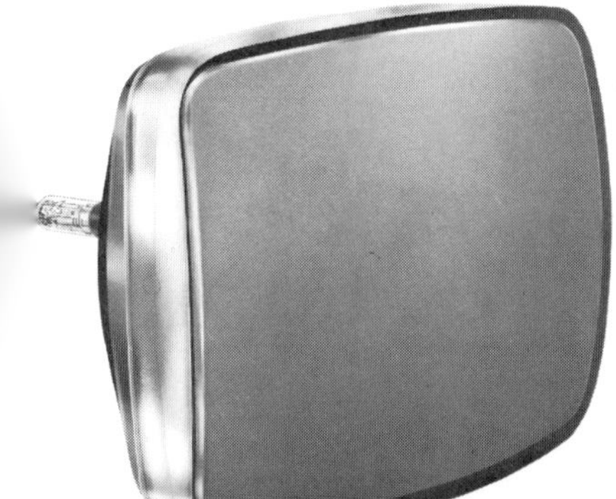

Fig. T-10*b* RCA 25 EP 22 Permachrome 90-degree rectangular High-Lite color picture tube.

Fig. T-10*c* Each of the RCA color camera chain's tubes are checked, before taping begins for "The Tonight Show" starring Johnny Carson, by NBC TV network engineer Joe Kolb.

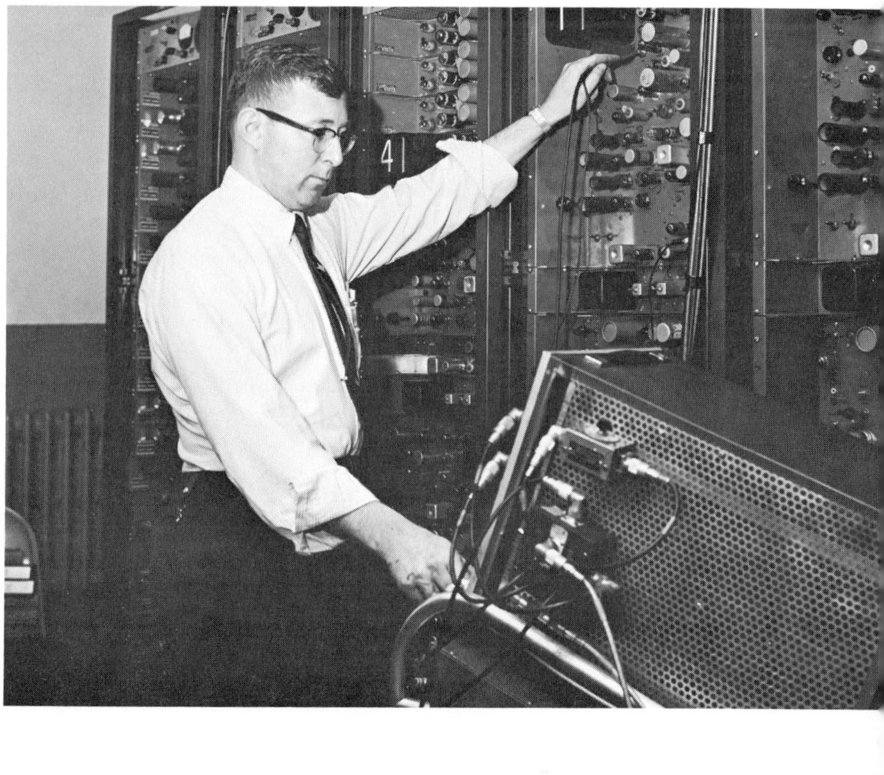

temperature coefficient The factor used to determine the required compensation in development time made necessary by reductions in the temperature of the developing solutions.

tener A heavy-duty spot that is especially useful as a key light. Its Fresnel lens is 20 inches in diameter. The tener, as the name implies, takes a 10,000-watt lamp.

Tessar lens A triplet-type lens designed to correct for curvature of field while providing wider relative apertures. The angular field of these lenses ranges from approximately 50 to 60 degrees. Modifications by the various manufacturers have been based on the original arrangement of lens elements in which the collective and dispersive units are separated by air spaces.

test chart A chart containing lines of various thicknesses that are arranged to form an overall design or pattern. The chart is photographed through the lens, and the resulting print is used for determining the general performance characteristics of the lens.

test pattern A geometric linear pattern that is used for lining up transmitting and receiving equipment.

test strip An exposed length of film at the end of a roll that is sent to the laboratory for processing. The exposed length of film is developed first and used as a sample to determine how the remainder of the roll of film should be developed.

test tape A test tape is a recording of various known frequencies at known amplitudes, usually for the purpose of testing and measuring reproducing equipment.

theodolite and triangulation cameras These cameras are used in pairs to determine the trajectory and velocity of aircraft and missiles, as well as for plotting contour maps to show the relative elevations of an area, and for survey work, generally.

Each filming unit consists of a 35mm motion picture camera* that operates in conjunction with a telescope. The camera measures the horizontal and vertical angles of a point and records a picture of that point along with measurements of its angles. The angles are indicated on mechanical counters and photographed together with a time base on each frame of film. A target may thus be pinpointed for azimuth, elevation, and distance. Triangulation data obtained from the two different positions permit determination and subsequent

*Mitchell 35mm.

plotting, in all dimensions, of the absolute position of the target or subject matter, its speed in fractions of a second, and the direction of movement.

A more ambitious program, in which the earth itself will be measured to ascertain its true shape, is already in progress. The measurements needed for this survey of surveys are being made with this type of photographic equipment.

Equipped with synchronous motors, the cameras can be operated at a number of speeds and use 35mm film in 400-foot magazines.

thermography A photographic process, actually a form of radiography, used for recording body-surface temperatures, in either black and white or color. The photographic reproductions are produced by radiant energy, within the infrared region, which is emitted by an object whose temperature is above absolute zero.

Images are formed as a result of emitted infrared radiation which is converted into electric signals and, subsequently, to visible light which is used for exposure. Bright and dark tones are in direct proportion with the degree of emitted heat. Greater emissions of heat produce lighter areas (in the print) than the tones produced by lesser emissions.

Thermography is used mainly for scientific investigations, medical studies and diagnoses, and for analyzing the composition of materials that are used for industrial purposes.

thermoplastic recording

Glenn process. One of the newer systems for recording visual and sound images is known as thermoplastic recording. The inventor of the electrophysical process is W. E. Glenn of the General Electric Research Laboratory in Schenectady, New York.

The film material used for recording has a base with a high melting point and a thermoplastic layer that has a relatively low melting point. Sandwiched between these two layers is an electrically conductive layer. In some ways, the film can be likened to videotape since it can be restored to its original condition, after recording, and reused at a later date.

In operation, the subject matter is recorded by an electron gun that scans the film area as it is played off a reel at a constant speed. This electrostatic image is heated by radio-frequency current. The physical characteristics of the film are altered as a result of this heating process. The resulting deformations, such as dimpling and rippling, with the almost instantaneous dissipation of the heat, become fixed on the film.

The subject matter recorded on the thermoplastic material can be viewed optically when refracted light is used, by reflection, or by projection on a screen. As an alternative, the picture information can be viewed electronically by scanning with a flying-spot scanner similar to those used for reproducing television images.

This type of recording process suggests a possible future compromise between videotape and film. Videotape's images, although invisible, can be played back immediately after the recording process has been completed without processing and with reasonably good results. The latent film image, however, must be developed to render it visible. The subsequent color values in the print may or may not contain a faithful reproduction of the colors in the original subject matter. The film images do give the editor a greater degree of latitude, with respect to cutting and individual color correction, than that offered by videotape. Thermoplastic recording promises the advantages offered by both—videotape and film.

Kalvar photographic process. A dry photographic system that does not use conventional camera stock but a thermoplastic resin coating on a transparent polyester base. Comparatively slow with respect to film speeds and slightly less than 0.0005 inch thick, the coating contains dispersions of an ultraviolet compound that are sensitive to light in the near ultraviolet range.

Unlike the incident light that is absorbed by the silver grains within other silver halide materials, the light, in Kalvar emulsions, is reflected and refracted by the scattering centers within the resin coating.

Upon exposure to ultraviolet radiation, the photosensitive dispersions in the film's coating decomposes and releases nitrogen and other volatile products. The created pressures, within the emulsion, create a latent image.

The application of heat softens the resin crystallites and produces microscopic vesicles by reorientation and recrystallization of the polymer. With a different index of refraction from that of the medium surrounding them, the vesicles scatter light incident upon them, thereby creating an extremely stable image. See Figure T-11.

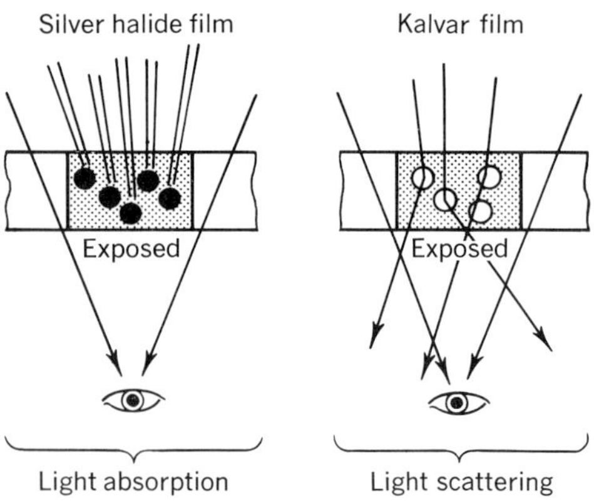

Fig. T-11 Kalvar film's process.

thin print A processed print lacking in density.

thread and wire Doors and windows that open by themselves, typewriter keys that spell out messages without visible assistance, steering wheels that turn while stick shifts move from one position to another, and hats that fly off a performer's head on cue are all eye-arresting, illusory effects that become conversation pieces when the projected images fade and the silver screen darkens.

Magic? Illusion?

Not to the special-effects department. For those talented technicians, the production of this type of effect requires little more than planning, preparation—and some thread or wire.

Obviously, these fine lengths of fibrous or metal material must be concealed or camouflaged to preserve the behind-the-scenes secret process. Since concealment is not always easy or practical, the threads or wires are usually made inconspicuous by painting or covering the strands with colors or materials that duplicate those in the background. And, while many of these effects are achieved with normal camera speeds, others require frame-by-frame positioning and stop-motion photography.

Of the several effects mentioned previously, the invisible typist is, perhaps, the most representative as well as the most complicated to produce. The description of the preproduction preparations that help make this effect possible can be used as the basis for the production of less complicated but equally puzzling "look, mom, no hands" actions.

Actually, the power of suggestion and imaginative editing play as important a part in the production of these effects as the rigging process itself. If, for example, a black typewriter is being used for the filming, the color of the thread must also be black. With one strand controlling the lever that advances the typewriter's platen, other threads are tied to the undersides of those keys that will be moving during the filming—with an invisible assist from an off-camera technician.

Close-ups of portions of the typewriter, showing the "controlled" keys being depressed, are the "key" shots around which the composite effect is built. Establishing shots, or long shots of the entire typewriter, help convince the audience that they are not being tricked. Extreme close-ups, showing the individual letters impressing themselves on the paper in the typewriter carriage, are intercut during the editing process to add credibility to the effect. For the latter shots, the keys are struck manually with the operator's hands just outside of camera range. Thread and wire, indeed. Take a letter, Mr. Technician. See Figure T-12.

Fig. T-12 Thread and wire. Don't bother looking for wires. They're painted to blend with the background. For this effect, the driver is lifted out of the car rather than lowered into it. The film footage is reversed optically when the driver and car are matted in over the background—on another length of film.

This type of scene can be filmed as a process shot, in its composite form on location or via the optical printer and matteing processes. If the car remains in the same relative position on the screen, even when the road is unusually rough and bumpy, the reader can be sure the vehicle never left the studio.

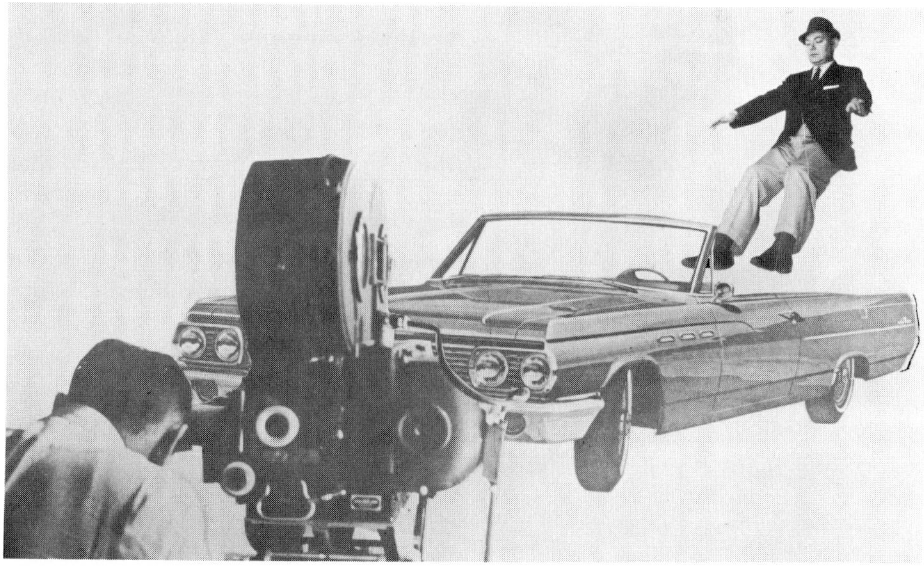

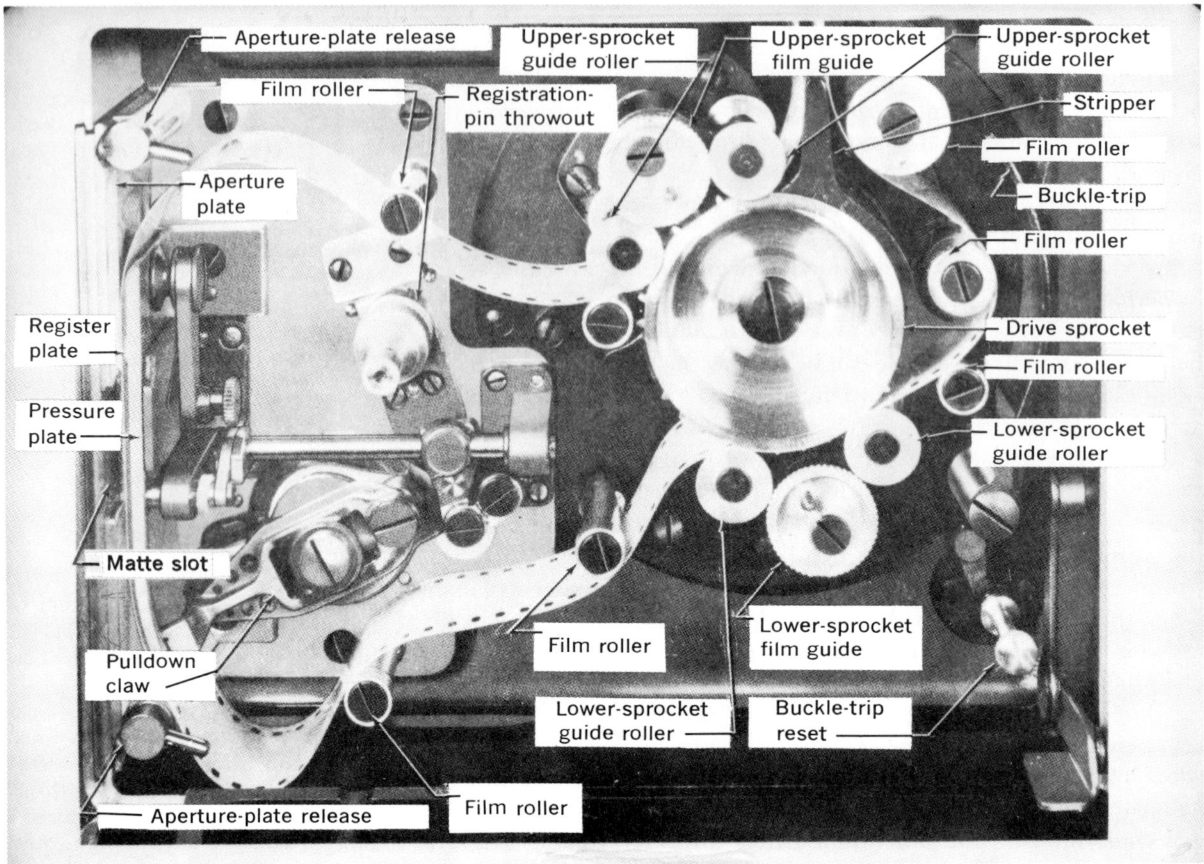

Fig. T-13 Threading camera.

threading The placement of film in a camera, projector, moviola, or other photographic equipment so that the film's perforations fit precisely over the sprocket wheels that transport the film through the equipment. See Figure T-13.

threshold exposure The very minimum amount of exposure that is needed to produce a visible image on the film negative.

throw The distance between the projector and the screen upon which the film is shown. Also, the maximum distance that a light beam can illuminate objects effectively.

thunder-effects sheet A suspended thin sheet of metal which, when vibrated, produces low-pitched sounds simulating thunder or heavy guns.

tightwinder Basically, a guide roller positioned at the end of a hinged metal spring-tension arm attached to rewinders. The tight-winder is used by film editors for winding film quickly and easily on cores or hubs rather than on reels or flanges.

tilt A term used to indicate that the camera is positioned or panning in a vertical or diagonal plane rather than horizontally.

time base Any type of time-reference scale that is photographed with the subject matter. These scales, usually positioned in the lower portion of the film frame, show the relationship between an action and the interval of time in which it occurs.

time exposures Relatively long periods of exposure. The film is exposed by tripping a device which opens the shutter for a predetermined amount of time. Tripping the device a second time closes the shutter and ends the exposure period. Technically, time exposures must be of greater duration than one second in order to be considered such.

time-gamma curve A graph designed to show the relationship between gamma and development time. These graphs indicate the amount of time that is needed to develop a particular film in a specific solution.

time-lapse photography In time-lapse photography each frame of film is exposed at predetermined time intervals, thereby enabling technicians to photograph continuous actions which move too slowly to be detected by the eye. Some actions may require days or weeks of intermittent exposing. The comparatively short length of processed film, when projected, contains a continuous visual record of the subject matter showing the progressive stages of growth, development, or evolution.

Depending on the subject matter, single exposures are made at one-minute to one-hour intervals, both day and night, for long periods of time. Spring-driven cameras, operated by timing devices, automatically turn the studio lights on several seconds before an exposure is to be made and shut them off immediately after the picture has been taken. The set lights, prolonged by the on/off timing device, are controlled by voltage stabilizers. Especially important in color production, these stabilizers keep the illumination at a constant level and prevent changes in color temperature, or the degree of brightness of the light source. Where necessary, the timers can also be coupled with other devices, thus providing a means for synchronizing any auxiliary equipment that may be utilized during the photographic process.

Extremely valuable for scientific observation, time-lapse photography can be combined with photomicrographic techniques. In such cases, the subject matter is photographed through the eyepiece of a microscope. Equally practical for industrial and commercial purposes, time-lapse photographic techniques have been used to show, in minutes, chemical processes that would require long periods of constant observation. Similarly, flowers have bloomed magically from buds in a matter of seconds and lumpy pieces of dough have been transformed into mouth-watering delicacies in the same amount of projection time it takes the viewer to blink in amazement. See Figure T-14.

timing and color correction The laboratory terms timing and color correction are used interchangeably, and in many cases, incorrectly. Although one term is used frequently as a synonym for the other, the only basis for comparison lies in the fact that both terms are associated with laboratory printing processes.

Timing for correct printing exposure is determined by measuring the density of each scene against a standard, for the purpose of achieving a consistent balance throughout the entire production. Light scenes are darkened and dark scenes are lightened by controlling the volume of the printing light that is used to expose the film stock employed for printing purposes. Resultant changes in tonal

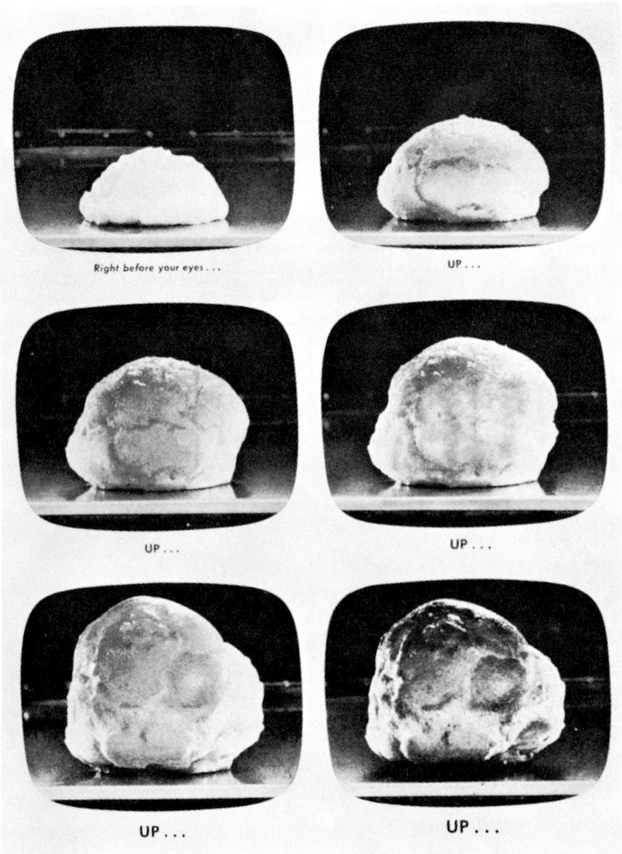

Fig. T-14 Time-lapse photography. In order to achieve this mouth-watering effect, the General Foods Corporation used time-lapse techniques. The baking process, in which mounds of dough are transformed magically into golden, luscious shells, is shown in a matter of seconds with this technique.

Page 1 of / Title TV – CLAIROL SHAMPOO – 60 SEC.

REEL NO. / Description	SCENE NO.	1ST CORRECTION				2ND CORRECTION				3RD CORRECTION				4TH CORRECTION				5TH CORRECTION			
		LIGHT	C	M	Y	LIGHT	C	M	Y	LIGHT	C	M	Y	LIGHT	C	M	Y	LIGHT	C	M	Y
MOTHER + CHILD	1		8	21	24		7	22	25												
C.U. MOTHER	2		8	22	23		8	22	24												
M.S. CHILD	3		6	21	25		7	22	25												
2-SHOT – MOTHER + CHILD			7	22	23		6	21	24												
M.S – PRODUCT ON SHELF			6	21	24		6	21	23												
M.S – MOTHER + PRODUCT			7	22	25		7	22	24												
C.U. PRODUCT	7		8	22	24		7	21	24												
	8																				

Fig. T-15 **Timing card.** (***Pathé Laboratories, Inc.***)

values are coincidental rather than intentional. Experienced timers use a subjective approach and rely on years of experience and a trained eye. Since color is generally regarded as a matter of individual taste, even the most highly regarded timers do not always see eye to eye with respect to timing standards.

Color correction, as the name implies, refers to the deliberate alteration of specific color values within a scene through laboratory printing processes. Bordering on the ironic is the fact that automation, in an industry that was built on a foundation of individual and collective creativity, is now a reality. In many laboratories, prepunched tapes automatically control timing and color-correction processes. With this automatic system, exposure information, predetermined by the laboratory timer, is transferred to control tapes, thus ensuring consistent print quality—electronically.

Obviously, there is more latitude in the printing of black-and-white material than in the production of color prints. During the printing process, the exposure and developing time may be varied to produce more desirable results. Similarly, the composition of the developing solutions used for black-and-white film may be altered in order to effect necessary changes in print quality.

With color, however, it is almost impossible to make similar changes without affecting the overall density, color balance, etc. The laboratory, for example, is able to regulate the maximum or the minimum density in a print by controlling the exposure. It cannot, however, control both and hope to achieve predictable results. Overall differences in density and color imbalances of minor proportions, on the other hand, are everyday modifications for the laboratory technician.

Generally, color film for television use should be processed normally according to the film manufacturer's specifications. The density of the highlights should be as low as possible without sacrificing detail or compromising flesh tones. The contrast ratio should be kept within acceptable limits—approximately 40 or 50 to 1. See Figure T-15.

timing circuits, automatic In videotape recorders, these circuits eliminate the requirement for front panel controls and the need to adjust. Picture geometry and burst correction are automatically achieved—freeing the operator from constant touch-up of controls.

timing-light generators* A timing light is provided in most photo-instrumentation cameras to place marks, commonly called *pips,* on the edge of the film and so give an accurate time base when data are evaluated and analyzed. The lamp, generally a miniature bayonet-based neon glow lamp, is positioned in the camera at a place where the film is constantly moving. The lamp extends into an opening in a film guide which is fitted with a slotted mask to form more clearly defined pips.

The lights have their own connectors and use a separate power supply from that which powers the camera motors. Timing-light generators are available which provide signals of 10, 100, and 1,000 cycles per second. Any other suitable source may also be used provided the end result is acceptable. The timing light, however, should be driven by a square-wave pulse of approximately 450 volts peak to peak at 4 milliamperes with a duration of approximately 30 microseconds. This driving voltage is available from the listed timing generators—Fairchild HS 10001, Red Lake Milli-Mite, Wollensak WF 311.

tiny Mac A McAllister spotlight used for fill lighting. The compact unit accepts lamps ranging from 50 to 200 watts.

titles During the motion picture industry's pioneer days, printed and hand-lettered titles spelled out the credits, advised ladies to kindly remove their hats, and silently called attention to the offerings of local merchants during the intermission. An integral part of the silent film presentation, the titles complemented the projected images by replacing long lines of dialogue with short printed phrases and captions. Given a voice in the form of a sound track, the annotations that had provided a running commentary were no longer needed. Except for the musical background, the voice remained respectfully silent while the main and credit titles were being projected.

Now, as then, titles offer the viewer pertinent information regarding the projected images. Specific rulings relative to their production are nonexistent. Regardless of style or technique, however, titles must be legible, the only obvious requirement. And today, as in the industry's infancy, specially designed titles that establish a mood or theme for the film footage that follows have always found favor with studio art directors.

*Timing-light generators are available through Gordon Enterprises, North Hollywood, Calif.

In some cases, these highly original efforts are optically superimposed over the opening scenes of the picture. In others, suitable graphics are used as the background for the main title and subsequent credits. In more imaginative applications, titles suggest or reflect the film's subject matter and act as transitional effects between introductory credits and the picture itself. Titles inscribed on a sandy beach, for example, immediately suggest an aquatic theme. Washed away by incoming tides, the main title is subsequently replaced by subtitles and secondary credits.

Similarly, feature films with higher production budgets and aerial backgrounds have used sky-writing planes to spell out the opening title against a heavenly blue background. Obviously, the variety of title treatments is limited only by the collective imaginations of art directors—and production budgets.

In many of the more recent film presentations, the titles are preceded by several minutes of story. The titles and credits follow several of the picture's opening scenes and are not introduced until after the audience's complete attention has been gained. Used sparingly at first, this subtle approach is rapidly gaining popularity in both mediums, motion pictures and television.

In the latter medium, the sponsor's message often receives as much attention and consideration as the show itself. Sandwiched between portions of the story, the supersalesman known as "the commercial" has replaced yesterday's bell-ringing, foot-in-the-door hawker.

Since scheduled appearances are limited to one minute, the message must be delivered quickly and effectively. And, because each tick of the clock is an expensive sound, the aural and visual information are specially designed to complement each other. Sharing the spotlight with the accompanying picture information, titles have become an integral part of these abbreviated film presentations. As such, they are no less important than the sound-track information they usually echo. Newly developed devices for turning off the sound in television receivers while the commercial is "on" have no effect on the titles.

The same psychological reasoning that influenced pioneer producers in their choice of colors, light-colored clothing, and white horses for the heroes in western movies, as opposed to dark clothing and black horses for villains, applies similarly to titles. This type of association and identification is evident in television commercials where white titles are synonymous with the "good guys" and dark or black titles are used to create negative impressions.

Since identification, in these 10- to 60-second presentations must be immediate, mere readability is no longer a standard by which titles are judged with respect to effectiveness. Titles, in today's general pattern, are expected to establish and maintain the commercial theme or, in some other way, reflect and define the accompanying sound track. The word "big," for example, should either grow in size or be proportionately larger than the word "small," if both appear in the same film frame. "Flavor," by comparison, should stir the taste buds or whet the viewer's appetite for the advertised product. The letters in the word "go" or "speed" can and should be designed to suggest movement.

Other words or phrases that lend themselves to similar treatment can be used as visual suggestions or as a substitute for adjectives. Since the choice of lettering and the method of presentation is a matter of individual taste, specific rules and standards are of negligible value. Regardless of content, technique, or method of introduction (animation, pop-on, or fade-in), titles must appear on cue and in synchronization with the narration in the accompanying sound track.

Most titles are hand-lettered on the surface of optically clear glass sheets in any combination of colors, placed over suitable backgrounds, and photographed on animation or titling stands. Where production budgets permit, individual letters cut out of paper, or three-dimensional letters made of wood or plastic, are moved from one position to another, within the photographic field, until the final reading position is reached. The infinite variety of effects made possible with this frame-by-frame, stop-motion titling technique is limited only by the cameraman's patience and the director's imagination.

For superimposure effects, white titles on black cards are exposed on high-contrast film. After developing, the resultant high-contrast negative and positive film lengths are combined, optically, with previously processed scenes or other graphic material selected for background purposes.

Double-exposure effects are produced by variations in the amount of exposure during the optical printing process. With the shutter completely open, the high-contrast title will "burn" through the picture in the background scene and be completely opaque. With the shutter half-open, the background scene would be partially visible through the white title. Further reductions in exposure cause propor-

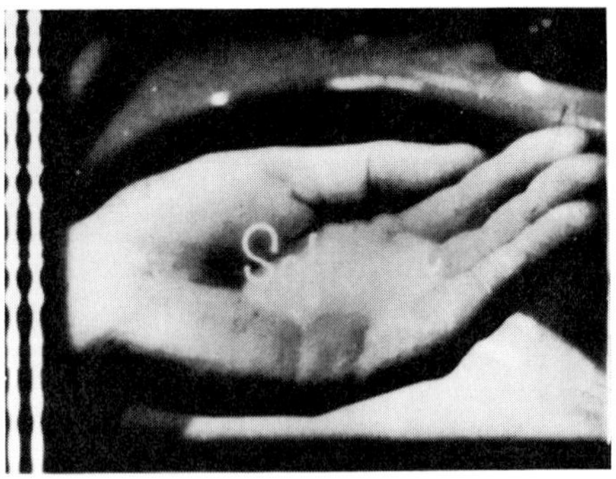

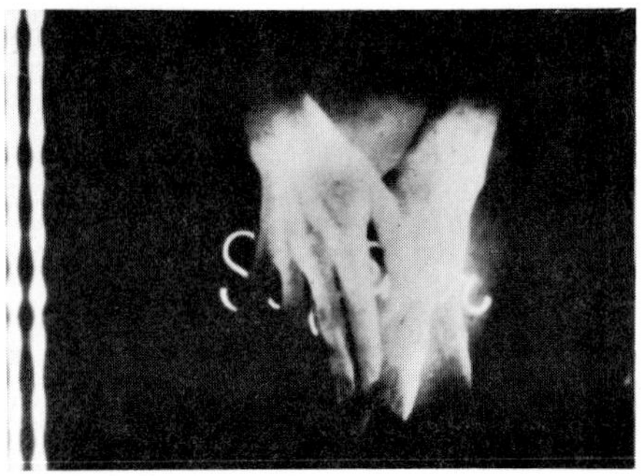

Fig. T-16a Titles. The use of an animating title for the purpose of visually suggesting the qualities of a product is not new in commercial production. A simple but effective transition, at the beginning of the Proctor & Gamble Thrill commercial, shows the animating word "softener."

In the sequence that follows, the title is sandwiched between a woman's hands. Separate mattes were needed for each of the 290 frames of film in which any part of the woman's hands covered any part of the animating word that was used to describe the qualities of the dishwashing aid.

tionate decreases in opacity and give the double-exposed title a transparent or ghostlike quality.

With the emphasis on unusual, eye-arresting illusory effects, zooming and panning titles are the rule rather than the exception. Wipes and scratch-off effects, in which parts of the title gradually become visible on successive frames of film, are also classified as routine. The latter effects are achieved by removing portions of the lettering from the complete title according to previously prepared guides.

Photographed in reverse, the entire title will have been scratched away by the time the last frame of film has been exposed. After processing, however, the last frame of the scene to be photographed becomes the first frame to travel through the projector. With the scratched-off letters appearing in reverse sequence on successive frames of film, the title is not visible in its entirety until the last frame of the effect has been projected.

All titles, regardless of type style or presentation technique, can be made more readable by adding a dropped shadow. No optical illusion, the effect helps separate the title from the background and is achieved optically by exposing both the high-contrast positive title and its negative over the background scene. The negative title (black lettering within the clear area of the film frame) acts as a matte and is exposed during the first run along with the background scene. The positive title (white lettering in the opaque black frame) is offset slightly during the second optical run. The offset shadow may be higher, lower, or to the right or left of the burned-in title.

The suggestion "if one is good, then two must be better" is especially applicable with respect to titles and their importance in complementing the corresponding track information in the motion picture as well as in the television commercial. The simultaneous appeal to both senses, aural as well as visual, is a most effective combination. See BURN-IN, DROPPED SHADOW, and SCRATCH-OFF EFFECTS. See Figures T-16*a* to *c*.

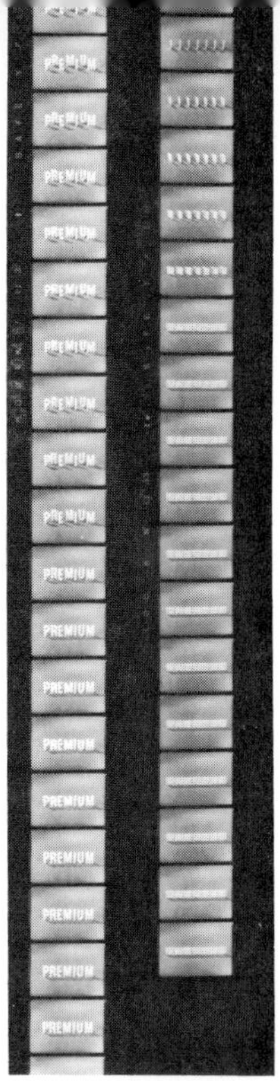

Fig. T-16*b* Good taste. In this stop-motion sequence from a Nabisco commercial, the lettering (actually the name of the cracker) animates into a neat package.

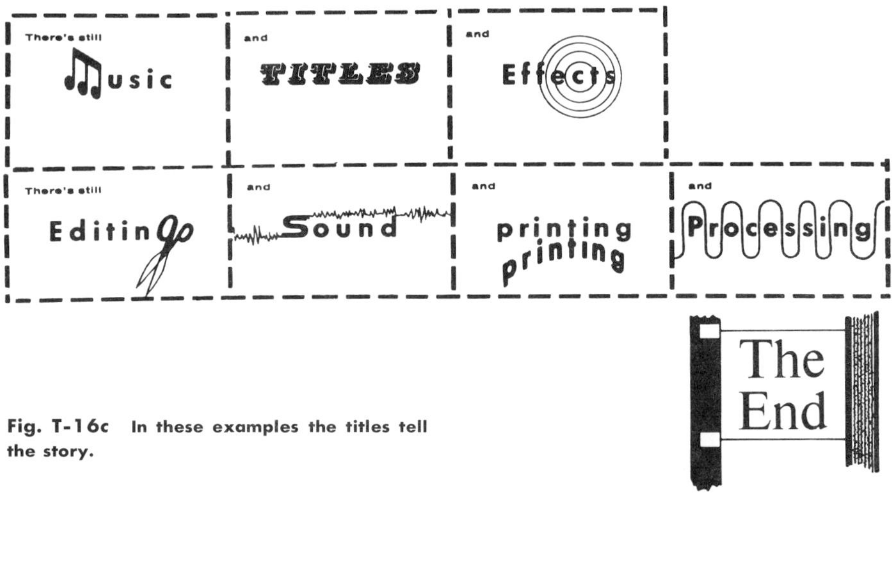

Fig. T-16c In these examples the titles tell the story.

titration A method of chemical analysis by which the degree of acidity or alkalinity in photographic solutions can be determined.

Todd-AO See WIDE-SCREEN PROCESSES.

tonal key (lighting) The prevailing tones on a set or in a photograph. A relatively light set or print would be termed *high key. Low key* would indicate a set or lighting arrangement in which there are relatively few light areas.

tone Halftones and/or intermediate shades of gray in a print. Also, a term used to indicate the overall effect with respect to light and dark areas in the print.

toning The process of chemically altering the color or increasing the brilliance of a print.

top hat A type of camera mount especially useful for low-angle shots.

top lights Lighting units positioned over a set or stage. These light sources are intended to provide fill light as well as to minimize contrasts and shadows. The intensity of the top lights, in relation to the fill light, should be between 30 and 40 percent.

trace back An animator's instructions to an inker, indicating that some portions of the drawing must be traced back from another drawing in the same scene.

tracking The process of adjusting the three pickup television camera color tubes so that their electrical characteristics are practically identical. Once adjusted, with respect to overall balance, the controls should remain undisturbed. Compensations for scene-to-scene balance, during transmission, are not considered technically advisable.

tracking mount Basically a gun mount, this type of mount is designed to provide smooth operation for use with cameras in tracking aerial or ground targets. The mount* accepts a pair of 16, 35, or 70mm cameras, with eye-level binoculars for sighting. Movement of the mount and the "firing" of cameras are controlled by pistol-grip handles. Vibration-free, it has a tracking rate of 60 degrees per second in azimuth and elevation and is equipped with a self-contained power supply.

*Model 249GE, Gordon Enterprises.

tracking shot A shot in which the camera moves forward, backward, or sideways in order to follow similar movements by the performers during the filming process.

tracks Rails or similar devices that are used as guides for moving the camera, its support, and the cameraman from one position of the set to another.

trade associations

Academy of Motion Picture Arts & Sciences	Hollywood
Academy of Television Arts & Sciences	Hollywood
	New York
	Chicago
Actor's Equity	New York
Affiliated Property Craftsmen	Hollywood
American Association of Advertising Agencies	New York
American Federation of Musicians	New York
American Federation of Television-Radio Artists (AFTRA)	Hollywood
American Guild Music Arts	New York
American Guild of Variety Artists (AGVA)	Hollywood
	New York
American Society of Cinematographers (ASC)	Hollywood
American Standards Association (ASA)	New York
Association of Cinema Labs (ASC)	Washington, D.C.
Association of Film Craftsmen	New York
Association of Motion Picture & TV Producers	Hollywood
Association of National Advertisers	New York
Association of Theatrical Agents & Managers	New York
Broadcast Ad Producers Society of America	New York
Broadcast TV Recording Engineers	Hollywood
Chicago Federation Advertising Club	Chicago
Directors Guild of America	Hollywood
	New York
	Chicago
Dramatists Guild	New York
Film Exchange Employees Union	New York
Film Producers Association	New York
Film Technicians (Laboratory)	Hollywood
IBEW (Electricians) International Brotherhood Electrical Workers	Hollywood
Independent Motion Picture Producers Association	Hollywood
International Alliance of Theatrical Stage Employees (IATSE)	Hollywood
	New York
International Sound Technicians	Hollywood
International Television and Radio Society	New York

IPMPI (Cameramen) International Photographers of the Motion Picture Industries	Hollywood New York Chicago
League of Broadway Theatres	New York
League of New York Theatres	New York
Legitimate Theatre Employees	New York
Make-Up Artists & Hair Stylists	New York Hollywood
Motion Picture Assistant Directors	New York
Motion Picture Association of America	Hollywood New York
Motion Picture Costumers	Hollywood
Motion Picture Electricians	Hollywood
Motion Picture Film Editors	Hollywood New York
Motion Picture Illustrators	Hollywood
Motion Picture Lab Technicians	New York Chicago
Motion Picture Screen Cartoonists	Chicago
Motion Picture Script Supervisors	New York
Motion Picture Set Painters	Hollywood
Motion Picture Studio CineTechnicians	Hollywood
Motion Picture Studio Grips	Hollywood
Motion Picture Studio Mechanics	Chicago
Motion Picture Studio Projectionists	Hollywood
Moving Picture Machine Operators	New York Chicago
National Association of Broadcast Employees & Technicians (NABET)	New York Hollywood
National Association of Radio & Television Broadcasters	New York Washington, D.C.
National Visual Presentation Association	New York
Operating Engineers	New York
Publicists Guild	New York Hollywood
Radio & Television Broadcast Engineers	New York
Scenic & Title Artists	Hollywood
Screen Actors Guild (SAG)	Hollywood New York Chicago
Screen Cartoonists Guild	Hollywood New York
Screen Producers Guild	Hollywood
Script Supervisors Guild	Hollywood
Service Employees—Theatrical	New York
Set Designers & Model Makers	Hollywood
Society of Motion Picture Art Directors	Hollywood
Society of Motion Picture & Television Engineers (SMPTE)	New York
Society of Photographic Scientists & Engineers (SPSE)	Washington, D.C.

Sound Construction	Hollywood
Stage Directors & Choreographers	New York
Station Representatives Association	New York
Studio Transportation Drivers	Hollywood
Television Bureau of Advertising	New York
Television Information Office—NAB	New York
Theatre Wardrobe Attendants	Chicago
Theatrical Stage Employees	Hollywood
Theatrical Wardrobe Attendants	New York Hollywood
Treasurers & Ticket Sellers	New York
United Scenic Artists	New York
Writers Guild West	Hollywood

trailer Also called **teaser**. A group of selected scenes from a motion picture. The trailer is shown as a part of the regular movie program in order to advertise the next feature attraction. Appropriate titles are usually superimposed over the scenes along with the star credits.

training films Motion pictures produced for the purpose of familiarizing personnel with equipment, processes, or techniques. See TECHNICAL ANIMATION.

transfer The re-recording of a magnetic original to magnetic sound film.

Transflex screen A special, self-collimating reflex-type screen used for front projection filming. In operation, the camera lens looks at the screen through a two-way mirror set up at a 45-degree angle to the optical axis. The projector casts the background illusion, via the reflective portion of the same mirror, onto the Transflex screen.

The screen itself is a brilliant reflector returning a high percentage of projected light directly along its axis and into the camera. Persons or objects in front of the screen block corresponding portions of the projected background. The front projection process is patented in the United States, and a license is issued with the purchase of the screen.

transformer An electric device that changes the voltage of alternating current.

transients Electronic signals that generally endure for a brief time before the attainment of a steady-state condition.

transistor A miniature electronic device capable of controlling and amplifying an electron current.

transitional effects Visual effects that help blend a number of independent scenes into a related sequence and, in turn, into a complete, full-length film presentation are known as transitions. Collectively, they offer the viewer a smoother flow of picture information, improve the general continuity of the presentation, and add the subtle touches that distinguish the professional effort from the cut-after-cut choppiness of the amateur production.

Some between-scenes effects are custom-made to suit the individual and imaginative requirements of the producer. Other film makers prefer to link scenes with the more traditional, transitional effects, such as fades, cross-dissolves, and wipes.

Although each of these effects serves a specific purpose, the most frequently used transition is, by coincidence, also the easiest to produce. As the name implies, fades are effects in which the picture is darkened, in successive frames of film, until the detail is lost and the film is completely black. Indicating a feeling of finality, a fade-out may be compared with the theater curtain that heralds an intermission or signals the end of the play.

By comparison, the first frame of a fade-in is completely black. Details in subsequent frames of film become progressively more distinguishable until finally the dark and light portions of the original images duplicate the values in the normally exposed picture.

Achieved during the filming process, or added optically after the film has been processed, the fade effect is produced by opening or closing the shutter, by degrees, thus varying the amount of light reaching the film. Reductions in light cause proportionate decreases in exposure and a darkening or fading effect. Increases in the amount of light that passes through the lens produces the opposite effect—the fade-in.

For animation or stop-motion, frame-by-frame photography, the fade is produced by closing or opening the shutter, in fractional amounts, for each exposure in the effect. If the camera has a 170-degree shutter, for example, the shutter indicator for the first frame of a one-foot fade-out would point to the 160-degree mark. The second frame would be exposed with the shutter at the 150-degree opening. For each of the frames in the 16-frame effect, the shutter is closed down by an additional 10 degrees until the indicator points to 0—the fully closed shutter position. For the one-foot fade-in, the shutter is opened 10 degrees, for each of the 16 frames, until the 170-degree mark, or fully open shutter position, is reached.

Fade effects can also be produced by reducing the lens aperture, by rotating polarizing filters in front of the lens until the filters shut out all light, and by dimming the studio lights gradually until the scene and film are both completely dark. Passed in front of the lens, fading glasses and strips of film, with gradations from clear to black, can also be used in place of the variable shutter for producing similar fade effects.

The cross-dissolve, another frequently used effect, helps indicate lapses in time and changes in location. Achieved by combining the fade-out of one scene with the fade-in of another scene, over the same length of film, picture information is literally blended during the transition.

Because of the overlapping picture information, the effect is also referred to as a *lap dissolve.* Whether achieved with in-camera photography or in the optical printer, the production process is the same. The first or outgoing scene is faded out normally. This length of exposed film is then rewound, with the shutter in its closed position, so that the first frame of the fade-out is again in shooting position in the camera shuttle. To complete the cross-dissolve, the incoming scene is merely faded in over the previously exposed film.

Wipes, where portions of the outgoing scene are replaced by portions of the incoming scene on successive frames of film, add visual interest, and depending on the wipe pattern, can be used to suggest movement, time lapses, or changes in location. The wipe patterns that serve as mattes during the transition are animated and photographed on high-contrast film. The resultant negative and positive lengths of processed film are carefully cataloged for quick and easy identification by the film editor or optical cameraman.

A hard-edge wipe, in which one scene is replaced by a second scene, with no separation line visible between the two, can be used to follow an action from one location to another. In a soft-edge wipe, by comparison, the line of demarcation between the incoming and outgoing scenes, appears as a narrow, gray, out-of-focus area consistent with the leading edge of the wipe pattern.

A horizontal-wipe effect, from a scene in which an automobile is shown speeding along a highway to the same car refueling at a gas station, immediately suggests a change in time. A second horizontal wipe, in the same direction in which the car is facing, can show the car arriving at its destination.

If the camera position and background remain the same, in scenes that follow one another, a cross-dissolve to indicate a lapse of time would be meaningless since the picture itself would remain unchanged. A soft-edge clock wipe could, in such cases, make the time lapse immediately obvious. These every-day examples illustrate but a few of the more practical ways in which wipe effects can be used. Other wipe patterns can be used to speed up long sequences and, by suggestion, eliminate the need for entire pages of explanatory dialogue.

During the optical process, the positive wipe is exposed for the length of the effect with the film of the outgoing scene. Acting as a matte, the wipe patterns mask out portions of the picture, in successive frames of film, until the entire scene is wiped out completely. After the film has been rewound, with the shutter closed, the negative wipe is photographed with the incoming scene. On this second optical camera run, the new scene is exposed in the previously masked areas, and, finally, when the last frame of the wipe effect has been photographed, the incoming scene is visible in its entirety.

Paradoxically, the negative and positive wipe patterns that act as mattes are never visible. Theoretically, if the negative and positive mattes were placed over one another in the projection shuttle of the optical printer, the leading edges of the wipe pattern, on the two lengths of film, would match each other precisely. It would be completely opaque, and no light could pass through the two mattes to expose the film in the optical camera.

Although the fade, cross-dissolve, and wipe are the most frequently used transitional effects, the imaginative film producer can easily create effective visual links that are consistent with the subject matter in the scenes to be joined. Out-of-focus effects, for example, can be added during the actual filming, if practical, or optically, anytime after both scenes have been processed. Especially valuable for dream sequences, the scenes are photographed normally for the required footage and then blurred beyond recognition for half the length of the effect. The second scene, out of focus when filming begins, gradually sharpens to complete the transition. If the effect is to be achieved optically, both scenes are filmed in sharp focus, and the blurring is left to the optical cameraman.

Distortion and diffusion effects for transition purposes are produced in similar fashion. The materials used for achieving the effect are introduced during filming or added optically when needed. In out-of-focus as well as in distortion and diffusion effects, the last few frames of the outgoing scene and the first few frames of the incoming scene are either so blurred or so distorted that they can be joined together as a cut.

Another effective transition that requires little planning or preparation is the *zip pan.* Achieved during the actual filming process, the camera is panned away from the subject matter quickly. The resulting unintelligible blur of horizontal dark and light patterns is spliced to a similar length of film that introduces the scene that follows.

Scenes that contain smoke, fire, and/or water can also be used to create interesting transitions. A smoke-filled screen, following an explosion, can serve as a lead-in for the scene that follows—after the smoke has cleared sufficiently. Similarly, the entire screen can be darkened by having a performer walk toward the camera until the lens is completely covered. At that point filming stops, and the director introduces a new scene by having the same performer, or another actor, walk away from the covered camera lens. Temporarily blackened by objects thrown at the camera, or by performers passing in front of it, the screen lightens on a new setting to provide other interesting transitions.

In many instances, transitional effects require more thought than preparation. And, as is often the case, the most effective visual transitions are also the least expensive and the easiest to produce. The possibilities are, indeed, limited only by the producer's imagination. See Figure T-17 in the color insert and also CROSS-DISSOLVE, FADE, and WIPE.

translator A switching and pulsing device which converts information fed into it by a computer-type system into electronically controlled movement (of equipment, for example) that eliminates the need for manual operation. See SLO-SYN MOTOR.

translucent Describing a material that transmits light but breaks up the rays so that a sharp image cannot be seen through the material.

translucent screen A type of screen used generally for rear projection.

transmission (1) The process of sending picture and sound information from the point of origin to the receiving units. (2) In optics, the amount of light passing through a filter. The percentage of a specific color being transmitted is known as *specific* transmission. The percentage of the light falling on a filter and being transmitted through it is known as *total* transmission. See TRANSMITTANCE.

transmission factor A measure of how effectively a material will transmit light—the ratio of the transmitted lumens to the incident lumens on the material, for example.

transmittance Also called **transmission**. The amount of light passing through a lens to expose the film in the camera.

transmitter (1) The sound-sensitive portion of the microphone. (2) The electric equipment that provides the needed power for generating and transmitting television signals.

transparency A photographic print, either color or black and white, on Celluloid or glass, which is viewed by transmitted light. Each frame in a length of processed motion picture film can be considered a transparency.

transparent Describing a material that transmits light without breaking up the rays of light. Compare TRANSLUCENT.

trapeze A device that makes it possible to suspend lighting units from overhead ropes or chains. The trapeze can be guyed to prevent swinging.

trapezoidal distortion The distortion of a projected image which results when the axis of the projector lens is not properly aligned with the screen axis. The difference in distance between the two axes is measured in degrees of arc. The maximum acceptable limit of distortion is approximately 2 percent from normal.

traveling matte Unlike mattes that are used to mask one specific area of the film frame for the subsequent insertion of picture information existing on another length of film, traveling mattes are used to mask portions of a number of film frames corresponding with the constantly changing position of subject matter within those areas. See MATTES and SELF-MATTEING PROCESSES.

traveling pegs In animation, the traveling peg bars on the camera compound table that are used for panning long cels or backgrounds. The long cels are known as sliding cels and usually contain cycle animation. Long backgrounds are generally referred to as pan backgrounds. See ANIMATION.

treatment Also called **script**. The general theme or pattern for a motion picture. Similar scripts, in the hands of different directors, would receive completely different treatments.

tricolor filters Also called **color-separation filters**. A set of three filters that duplicate the transmission values of the primary colors—red-orange, green, and blue-violet. These filters are used for making the color-separation negatives from which the color prints are produced. These filters allow only light of their own color to be transmitted through them and, in that respect, are similar to contrast filters with the same coloring. See CONTRAST FILTER and SEPARATION NEGATIVES.

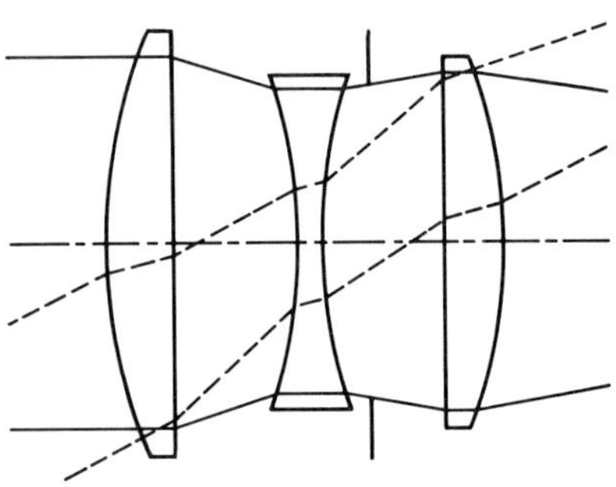

Fig. T-18 A simple triplet lens designed by a program-controlled electronic computer. (*Carl Zeiss, Oberkochen.*)

trims Unused short lengths of film cut from a scene or action by the film editor.

tripack Three emulsions on as many separate film bases. The three emulsions, in this pioneer photographic process, were placed in contact with one another and exposed simultaneously to produce separation negatives. Compare INTEGRAL TRIPACK.

triplet lens A rather popular type of lens developed in 1893 and originally known as the Cooke triplet. The lens has three air-spaced elements—a converging lens made of crown glass, a diverging lens of flint glass, and a second converging crown-glass lens. This type of lens is capable of exposing an area with a diagonal that is equal to the lens' focal length. See Figure T-18.

tripods Some tripods are sold as complete units, consisting of legs and pan-and-tilt head. Others are sold separately so that the cameraman may choose whatever set of legs and head he desires. The availability of friction, fluid, gear, gyro, and balanced (spring-loaded) pan-and-tilt heads in various sizes permits fitting the tripod head to the filming job. Tripod legs are available in wood or metal and in various lengths. Standard-length legs will suffice for most setups. "Shorty" or "baby" legs should be used for low camera angles. Hi-hat (low-mount) adapters allow a tripod head to be mounted to title boards, airplanes, helicopters, boats, and other places where a tripod cannot be positioned because of space or other limitations.

A tripod combination should be heavy enough to handle the camera and accessories in use. This is particularly important with heavy blimped sound cameras. A recent innovation is the ball-socket adapter, placed between the pan-and-tilt head and the tripod top casting, which permits leveling the tripod head without adjusting the legs. Some tripods possess this as an integral part; others can make use of a separate ball adapter (with the properly mated top surface), allowing almost any head to be leveled in this manner. See Figure T-19.

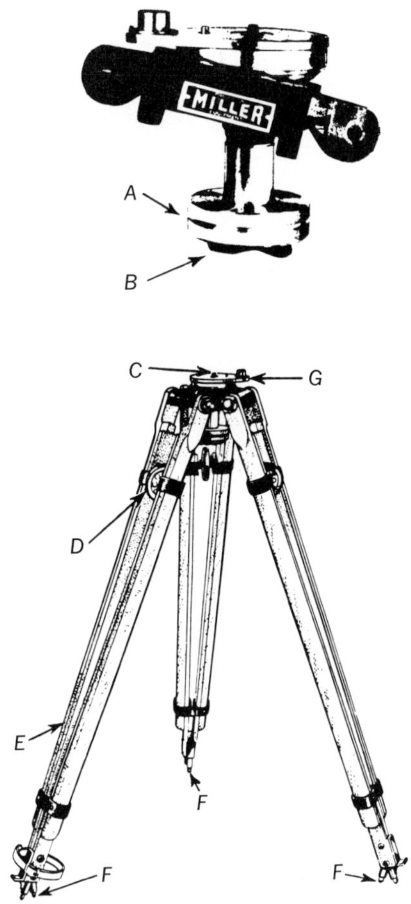

Fig. T-19 The Miller professional tripod. (*A*) Leveling lock wheel. (*B*) Mounting screw, spring-loaded. (*C*) Mounting screw, spring-loaded, $\frac{1}{4}$-inch standard thread. (*D*) Positive automatic equalization with Ezi-Grip lock wheel on leg sections—ensures rigidity and prevents leg warp. (*E*) Vee section. Sliding members of selected kiln-dried white mountain ash. Dressed with silicone oils and lacquered for smooth telescoping action and long life. (*F*) Double-replacement heavy-duty leg tips on each leg—with hold-down spur. (*G*) Ball leveling top with spirit level. 25-degree leveling movement for Miller fluid-action senior heads and all others with $\frac{1}{4}$-inch mounting thread. Flat top with spirit level for Miller fluid-action senior heads and all others with $\frac{1}{4}$-inch mounting thread. Other mountings to fit Miller fluid-action professional heads, also pro junior and SOS heads.

tripod head That part of the tripod on which the camera is mounted.

trombone A support, positioned over a motion picture set or television stage, to which light units are attached.

Tru-Broad A focusing flood light capable of providing smooth coverage over relatively large surfaces. Particularly suitable for confined areas, the center of the beam is very flat and does not peak. Full-focusing control permits both the coverage and intensity of the light to be varied. A large choice of single-ended tungsten-halogen quartz lamps, operating at 120 volts a-c or d-c may be used to obtain the desired coverage. The Tru-Broad light unit is available through ColorTran.

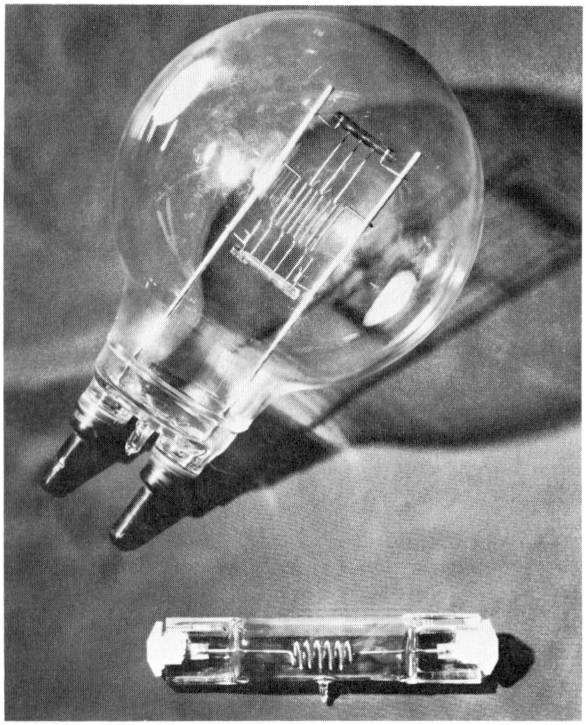

Fig. T-20*a* Tungsten-halogen lamp. One of the new Sylvania lamps, a double-ended 2,000-watt unit, is shown with a larger conventional incandescent studio lamp. The effective use of tungsten-halogen lamps in studio and theatrical lighting equipment will result in a dramatic increase in optical efficiency and could reduce production costs.

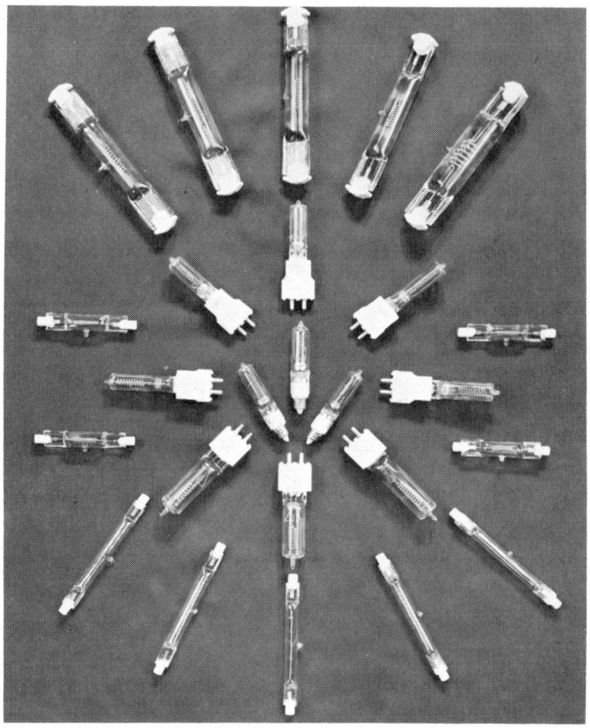

Fig. T-20*b* The development of the 25 compact tungsten-halogen lamps shown here represents a major advance in studio and theatrical lighting. (*Sylvania Electric Products, Inc.*) The tungsten-halogen units offer the same illumination at lower wattages, or greater illumination at the same wattages, throughout the life of the lamp. This offers possibilities of lower costs in applications where maximum lighting efficiency must be maintained over long periods of time.

truck Also called **dolly** and **zoom**. A photographic effect in which the camera is moved from one position of the set to another in order to change the spatial relationship between the camera and the subject matter.

tungsten-halogen lamps Lamps in which the light output and color temperature remain at a reasonably constant level during the entire life of the lamp. With self-cleaning properties, filament particles mix with the gaseous element within the lamp and are redeposited on the filament, thereby preventing a blackening of the lamp housing and the subsequent reduction of light output. The lamp housings, usually quartz or high-silica glasslike material, have high melting points, thus making possible the use of more powerful filaments than those normally found in other light sources used for motion picture illumination. See Figures T-20*a* to *c* and Tables T-1 and T-2.

turret A circular lens mount. Any of the several lenses mounted in the turret can be brought into position for filming by merely rotating the mount. See Figure T-21.

TVola (videotape) A device that converts videotape's magnetic impulses to frozen pictures, thereby providing editors with a means of studying taped television frames one by one.

The portable device which plugs into any videotape recorder couples a 200-transistor timing circuit with four memory tubes which

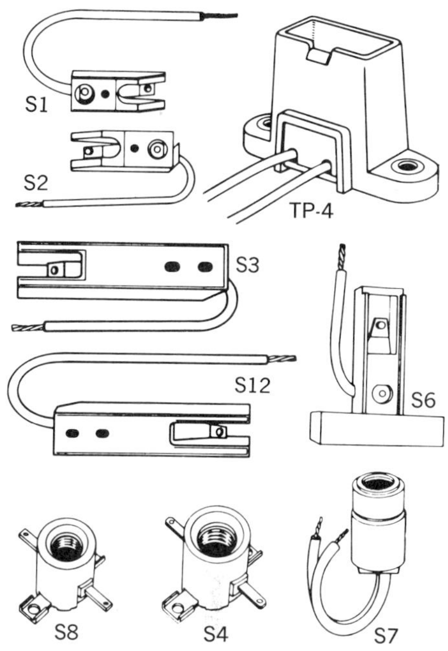

Fig. T-20c Tungsten-halogen lamp sockets—not to scale. (*Sylvania Electric Products, Inc.*) For explanation see Table T-1.

TABLE T-1 Tungsten-Halogen Lamp Sockets*

Ordering abbreviation	Base	Maximum watts	Maximum current, amperes	Maximum temperature, °C
		Double-ended Lamps		
S1	RSC only	650	6	250*
S2	RSC only	1,500	10†	350*
S3	RSC & rect. RSC	1,500	10†	350*
S6	RSC & rect. RSC	1,500	10†	350*
S12	RSC & rect. RSC	2,000	20	350*
		Single-ended Lamps		
S4	Mc	1,000	8.5	300‡
S7	Mc	1,000	8.5	300‡
S8	MS	75§	8.5	300‡
TP4	Med. 2P	1,000	10	200¶

*At rear of contact button.
†Higher current rating can be attained with forced cooling.
‡At rear of central contact button.
§With special applications up to 350 watts.
¶At lead-wire junction.

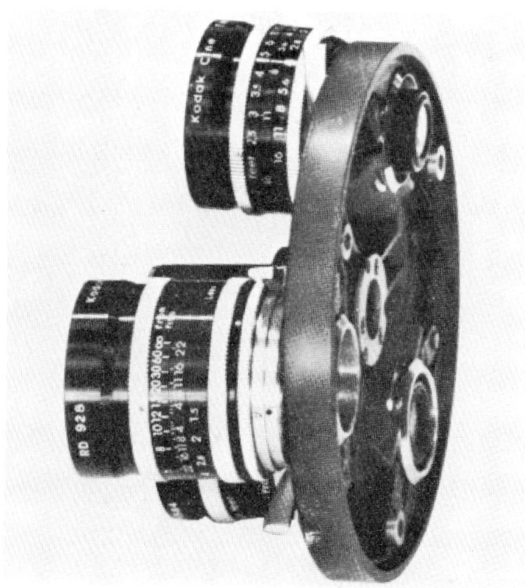

Fig. T-21 The lens turret.

TABLE T-2 Sylvania Tungsten-Halogen Lamps for Studio, Theater, Motion Picture, and Television Lighting

ASA code	Watts	Life, hours	Lumens	Color temperature, °K
Double-ended Universal T3				
EHM	300	2,000	6,000	2950
EHN	300	400	7,000	3100
FCL	500	2,000	10,500	3000
FDF	500	400	13,000	3200
FCM	1,000	300	27,000	3200
Double-ended Intermediate Wattage				
EHP	300	2,000	5,000	3000
EHR	400	2,000	7,500	3000
FDA	400	500	8,500	3100
FAD	650	100	15,500	3200
Double-ended High Wattage				
DWT	1,000	2,000	21,000	3000
EHS	1,000	500	24,000	3100
FER	1,000	150	26,000	3200
DWS	1,000	30	33,000	3400
FEY	2,000	150	52,000	3200
Single-ended Low Wattage				
EHT	250	2,000	5,000	3000
EJX	250	500	6,000	3100
EHV	325	500	7,800	3100
Two-pin Intermediate Wattage				
EHD	500	2,000	10,000	3000
EHB	500	500	12,000	3100
EHC	500	150	13,000	3200
EHG	750	2,000	15,000	3000
EHF	750	500	19,000	3100
EHH	1,000	2,000	21,000	3000
EHK	1,000	500	24,000	3100
FEL	1,000	150	26,000	3200

can hold a single picture for as long as 20 minutes. The editor first selects the approximate spot at which he wants to make his cut on the tape. Running the tape through the recorder in normal playback fashion, he engages the TVola to view on its four monitors up to 90 frames contained in the three-second segment of tape on each side of the edit point.

Isolating a single frame on the monitors is accomplished in several sequences of elimination, each merely requiring the push of a button on the TVola. The entire operation of refining the segment down to the specifically marked single frame takes less than one minute.

Four images are frozen on the monitors in one-second intervals: at the start beep, start plus one second, start plus two seconds, and start plus three seconds. The editor then selects the two frozen images which bracket the desired editing point.

twin-track recorder A tape-recorder head designed to record on less than half the width of the total tape area. This arrangement makes it possible to flip the roll, after completing a recording, so that a second recording can be made on the same tape, using the second recording area.

two-color meters Color-temperature meters designed to provide artificial light readings in degrees Kelvin. The Spectra meters indicate the corrective filters that are needed to ensure consistent lighting and proper color balance. See COLOR-TEMPERATURE METER.

two-shot A scene in which two performers are featured with more or less equal prominence.

U

ultrahigh frequency Frequencies of 300 or more megacycles—those above the very high frequency range. A megacycle is a unit of measurement equal to wave frequencies of 1 million cycles per second. See ULTRASONIC.

Ultra Panavision 70 See WIDE-SCREEN PROCESSES.

Ultra Semi-Scope See WIDE-SCREEN PROCESSES.

ultrasonic Describing high-frequency sounds—above approximately 15,000 cycles per second—that are inaudible to the human ear. The original term *supersonic* has been replaced by ultrasonic.

Frequencies above 20,000 cycles per second (macrosonics) are used for a number of purposes including soldering, cleaning small metal parts, and aging beverages containing alcohol. See ULTRASONIC CLEANER.

ultrasonic cleaner A device used for cleaning film. Ultrasonic energy is the most effective and economical way to clean motion picture film thoroughly and rapidly without mechanical scrubbing and wiping. The boiling effect (cavitation) of ultrasonic energy performs the entire operation. Only a solvent (methyl chloroform technical by Dow) touches the film and a forced-air, flash dry-off removes all solvent and residue. The solvent must be handled carefully since its vapor may be harmful. See Figure U-1.

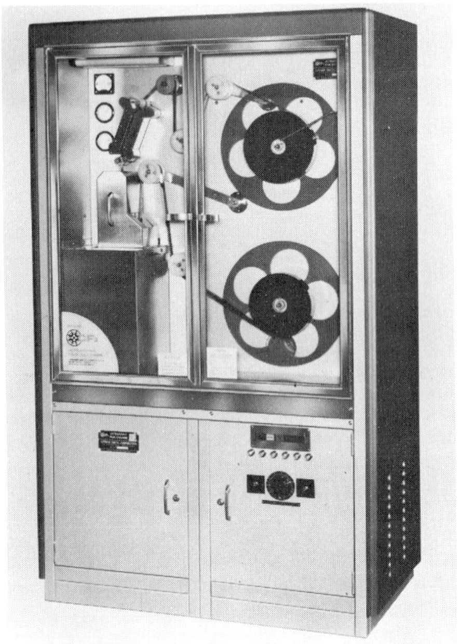

Fig. U-1 The ultrasonic cleaner. ***(Lipsner-Smith Corp.)***

ultraviolet The band of radiation above the highest visible color (violet) in the spectrum. Ranging between wavelengths of approximately 4000 to about 140 angstroms, ultraviolet is also referred to as *black light.* Although invisible to the human eye, ultraviolet radiation will expose many types of film. See ULTRAVIOLET PHOTOGRAPHY and SELF-MATTEING PROCESSES.

ultraviolet condenser lenses Condenser lenses for use in the near-infrared and ultraviolet ranges. These lenses are suitable for a wide variety of applications where high temperatures are a factor. Fused-quartz condenser lenses have an extremely low temperature coefficient of expansion, and melt at approximately 1800°C. Since these lenses are used in condensers only, photographic imaging quality is not important. These condenser lenses are available through Barrett Associates in Hauppauge, New York.

ultraviolet filter A type of filter for ultraviolet "light." Like an infrared filter, its ability to absorb visible light while transmitting a very narrow band of ultraviolet makes it especially important to dual-film, self-matteing processes. Ultraviolet filters are also extremely useful for reducing the effects of atmospheric haze during location filming, since haze is especially strong in the blue and ultraviolet regions of the spectrum. This type of filter is positioned over the camera lens. See ULTRAVIOLET PHOTOGRAPHY and SELF-MATTEING PROCESSES.

ultraviolet photography Ultraviolet photography, like infrared, is especially valuable because of its sensitivity to colors in the invisible region of the spectrum. Although ultraviolet radiations range between 4000 and 140 angstroms, wavelengths beyond the near ultraviolet (less than 2000 angstroms) are unsuitable for photographic purposes since the radiation is completely absorbed by the gelatin in a film's emulsion. For all practical photographic applications, the area between approximately 3500 to 4000 angstroms is usable without specialized equipment. Below about 1800 angstroms the radiation is absorbed by the air.

In the first of two photographic methods used for gathering picture information, filters allow only a very narrow band of ultraviolet "light" to expose the film in the camera. In the second or fluorescence method, the ultraviolet radiation is absorbed, and only the visible fluorescent light reaches the film.

Because of the differences in the fluorescent qualities of various materials, ultraviolet processes are used for many unusual purposes. As an unofficial but respected member of law-enforcement agencies, ultraviolet has been used not only for detecting invisible writing and fingerprints but also for determining the authenticity of documents suspected of forgery or alteration. In more commercial applications, ultraviolet is used in self-matteing processes for achieving unusual, illusory effects.

Because optical glass absorbs wavelengths below 3500 angstroms, quartz lenses must be used for radiation below that level. The visible light from mercury-vapor and carbon-arc lamps can be removed by placing ultraviolet filters in front of the light sources. This filtered light can then be used for self-matteing processes as well as for photomicrography. Filters that are used to absorb ultraviolet radiation are positioned over the camera lens. Protective goggles should be worn when working around sets where ultraviolet radiation is being used. Excessive exposure to these rays can result in severe cases of sunburn and other skin irritations.

ultraviolet self-matteing process See SELF-MATTEING PROCESSES.

umbra The sharpest and darkest portion of a shadow. Compare PENUMBRA.

umbrella As the name implies, an umbrella-shaped lighting accessory whose reflective surface is intended to provide "bounce" illumination.

undercrank A general term, handed down by motion picture pioneers, indicating photography at speeds slower than the accepted standard. Filming at a rate of 12 frames per second (half the normal projection rate) causes actions to appear twice as fast when projected at conventional speeds. See LOW-SPEED PHOTOGRAPHY.

underdevelopment Insufficient developing time or improper solution temperatures. Negatives, as a result of underdeveloping, are generally lacking in contrast and show some loss of detail in shadow areas. Prints made from such negatives have a washed-out look. Bleaching and redeveloping will usually make the negative more suitable for printing. See INTENSIFICATION.

underexposure An undesirable effect caused by any of several errors committed during the filming process. Underexposure, resulting in unusually light prints, is caused by improper lens settings, insufficient light, or filming at excessive camera speeds. Unfortunately, there are no corrective measures that can be taken that will put detail into a print where none exists in the negative. Obviously, the only effective measure that can be taken to prevent a recurrence of the undesirable effect is to make a series of tests that will subsequently indicate the cause.

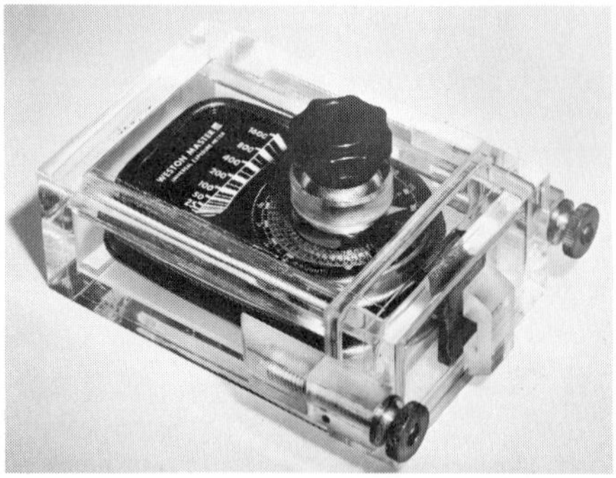

Fig. U-2a An underwater exposure meter.

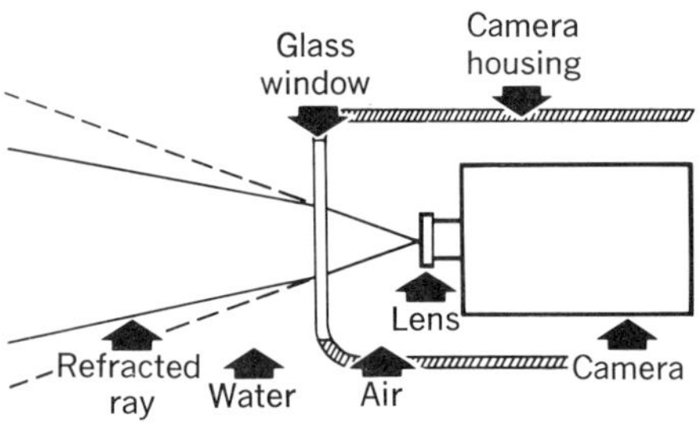

Fig. U-2b Underwater filming and refraction. Refraction, in underwater filming, causes objects to appear approximately 25 percent larger than their actual size and proportionately closer.

underwater photography Paradoxically, the first underwater photographs were made on "dry" plates in 1893. The crude equipment used for the subsurface filming consisted of a 5- by 7-inch camera that was encased in a waterproof housing.

With the development of commercially acceptable equipment, scientific exploration of the uncharted depths proceeded at an accelerated pace. Slowly, and reluctantly, underwater secrets revealed themselves—on film.

Unable to withstand the pressures at extreme depths, scientists soon devised new methods for continuing their studies of marine life and underwater formations. The lowering of cameras to unprecedented depths was only a temporary solution that was soon replaced by diving apparatus capable of carrying both explorer and camera to the deepest levels. The famed Swiss scientist and deep-sea explorer, Dr. Auguste Piccard, spent many years designing a gondola that could descend to the very bottom of the ocean floor. In 1960, his *bathyscaphe* made a record dive of approximately 38,000 feet off the coast of Guam. Mercury-vapor lights illuminated the area around and under the gondola and made it possible for motion picture and still cameras to take pictures through the 5-inch-thick plastic portholes in the bottom of the hull.

For salvage operations in relatively shallow water, divers equipped with aqualungs and other breathing devices are able to carry out their underwater assignments at 200-foot levels for relatively long periods of time. With the invention of the Rebikoff Torpedo and other lighting units, subsurface cameramen were able to use high-intensity lights for filming underwater subject matter.

Recent improvements, such as propeller-driven units that tow the cameraman, allow more freedom and flexibility in the handling of underwater photographic equipment.

Obviously, cameras must be protected in waterproof and pressure-proof housings. Some of these contain flat glass pressure plates, or windows, positioned in front of the camera lens. Because of the distance between window and lens, focusing is difficult for the underwater cameraman. Since objects appear to be nearer and larger than they actually are, accurate focusing becomes a problem of compensation and mathematics. Newer lens arrangements, the submarine type specifically, eliminate the need for the protective window by bringing the front of the lens in direct contact with the water. Magnification of underwater objects is thus reduced to a minimum and allows the cameraman to determine accurate focusing distances easily.

Underwater photographs, both still and motion pictures, can be taken without auxiliary lighting sources, at shallow depths. Light, penetrating from outside, is reflected to a lesser degree when the sun is directly overhead, thus making it possible for the cameraman to use near-normal exposure. Below the 20-foot level, however, high-intensity lights are used to provide the needed illumination.

Undistorted picture information, even under ideal conditions that include a minimum of turbulence and strong sun, is almost impossible at great depths. In black-and-white filming, picture clarity and image definition drop sharply when the subject matter is more than 10 to 15 feet from the camera. Because red and blue light are absorbed in great quantities by the water itself, the critical distance, in color photography, is even less. Maximum quality, for these reasons, is best at distances less than 10 feet from the camera. Color-corrected filters, at shallow depths, are effective to a limited degree.

In any event, proper "makeup" is not a serious problem in underwater photography. See Figures U-2*a* to *e*.

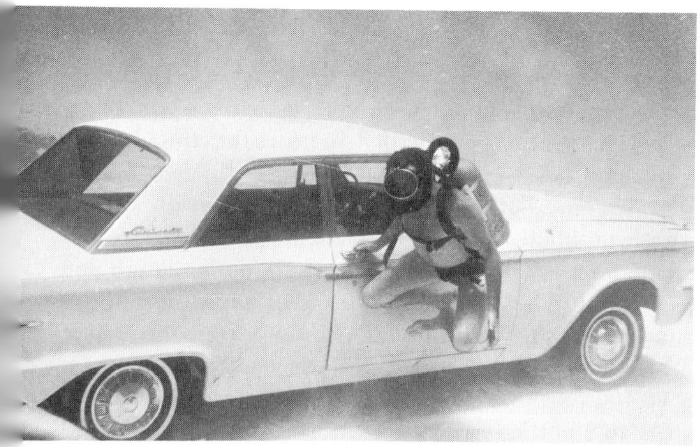

g. U-2c The big trick here was getting the Ford Fairlane ıder water. (*Photos by Jordan Klein, Miami, Florida.*) The arkling new, watertight automobile refused to allow ; bumpers to be dampened. The car was finally bmerged by holding its doors open and allowing ıture to take its course.

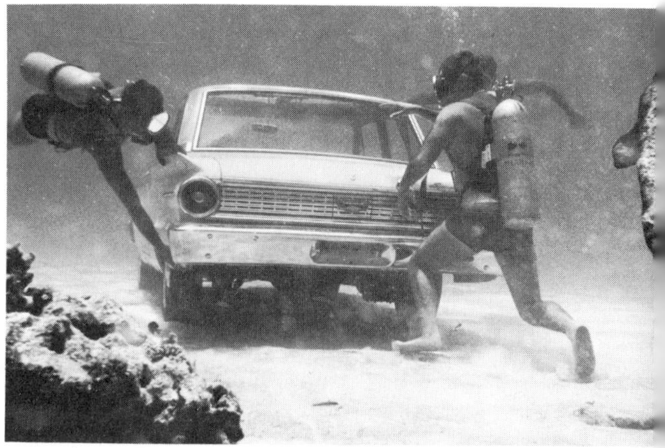

Fig. U-2*d* The coral formation, 25 feet below the surface off Nassau, provided the background for the commercial. The cameras, in waterproof housings, were lowered to the subsurface location along with air compressors, generators, and 15,000 watts of light.

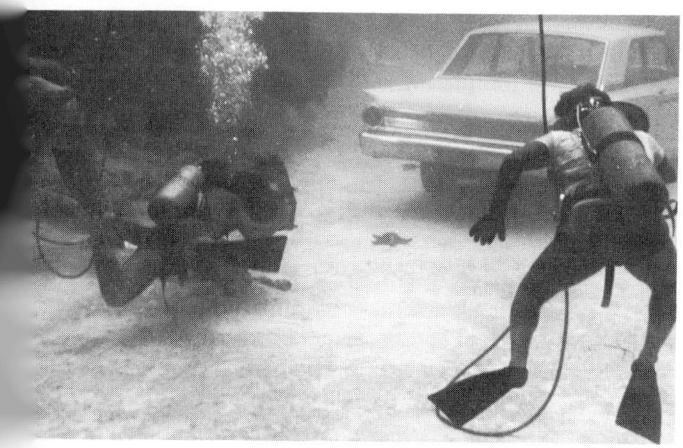

Fig. U-2e Sorry—but no one thought of testing the windshield wipers under water.

underwater cameras

Costeau Bathygraf underwater camera, 35mm. The camera is automatically pressurized at any depth to 330 feet to maintain one pound buoyancy. Compressed air from a built-in cylinder with automatic demand level ensures against water penetration and makes the camera weightless under water, allowing the unit to rise to the surface when released.

The chromium-plated brass housing is mounted on a shaft equipped with two pistol grips which allow the operator to aim the camera. The pistol grips rotate to control aperture and focus, provided on dial readings at the cameraman's face port. The finder sight automatically sets parallax when focus is adjusted. Both lens aperture and focus may be changed during a scene. The camera is a modified LeBlay newsreel camera. 25mm *f*/2 and 50mm *f*/1.3 Kinoptik lenses are interchangeable under water. The camera may be equipped with 200- or 400-foot film magazines. A 24-volt d-c silver-potassium battery drives a $\frac{1}{4}$-horsepower motor which operates the camera from 8 to 80 frames per second.

Eclair Aquaflex, 16mm–35mm. An underwater version of the Camerette with a modified elongated magazine that reduces the height of the camera and allows a slender silhouette design for the underwater housing. The camera has all the basic features of the Eclair 16–35mm camera. It is equipped with external controls for operating diaphragm, focusing, and motor drive. The camera is supplied with two 400-foot magazines, a special eyepiece for external reflex viewing, a 6- to 8-volt nonspillable battery, and a photoelectric exposure meter. The blimp has a compressed air bottle and a standard Costeau spirotechnique apparatus for automatic regulation of air pressure within the housing.

The following Kinoptik lenses are furnished: 25mm *f*/2, 35mm *f*/2, 50mm *f*/2, and 75mm *f*/2. Detachable wings and a vertical rudder aid greatly in transporting and stabilizing the camera under water. The camera can be guided with one hand while the operator is swimming. The complete unit weighs 107 pounds in air and can be adjusted to positive, negative, or neutral buoyancy under water.

Fenjohn underwater motion picture camera, 16mm. The camera is constructed of cast aluminum machined to tolerance. Aperture, focus, filter, and speed settings can be made under water. Focus calibrations, *f*-stops, and the filter in use can be seen through the viewfinder while filming. The right thumb operates the starting trigger. The housing is equipped with a 16mm Bell & Howell magazine, a 13mm *f*/1.5 Elgeet wide-angle lens, a large brilliant viewfinder, and an electric motor which operates on flashlight batteries and exposes approximately 1,000 feet of film. The camera has speeds of 12, 16, 24, 32, 48, and 64 frames per second. Four filters may be mounted and changed while filming. The housing, complete with camera, weighs 21 pounds in air and only $3\frac{3}{4}$ pounds under water.

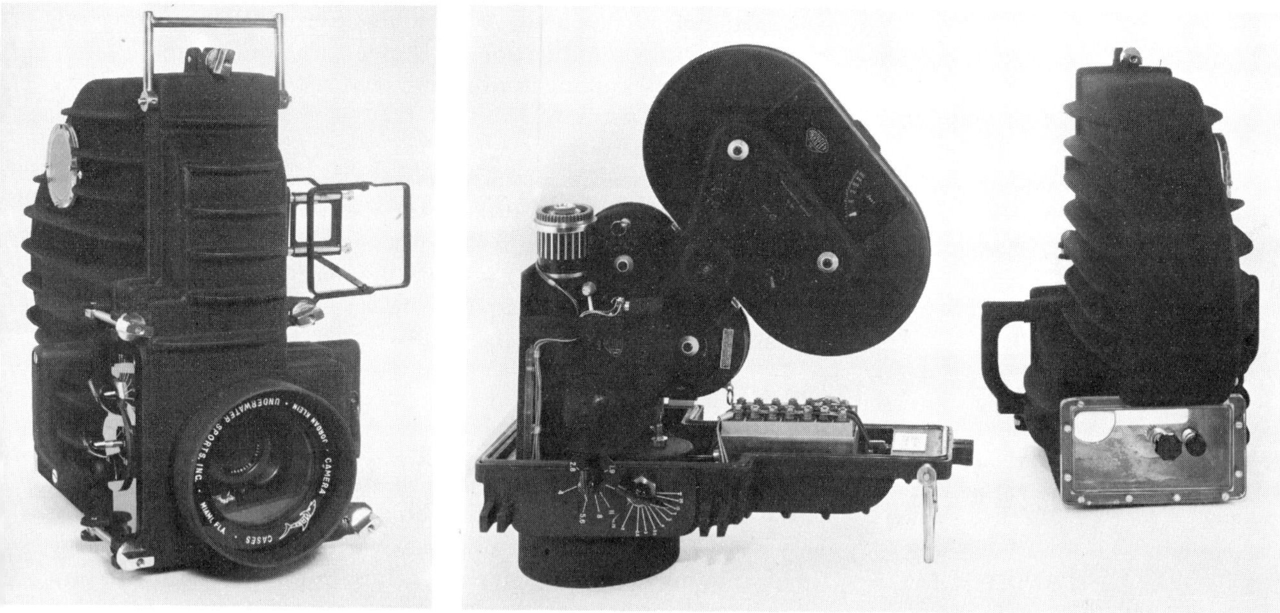

Figs. U-3*a* and *b* The Arriflex 16mm with 400-foot magazine. (*Mako.*) The Arriflex underwater housing is constructed of special impregnated and anodized aluminum castings and selected tempered aluminum plate. All fittings are stainless steel. The underwater housing is tested to pressures equal to a 300-foot depth. The lens-aperture control knob is operated from the rear. Offset pistol grip provides excellent stability and control.

underwater housings

Bolex. The Bolex underwater housing accepts Bolex Rex and reflex cameras with 10mm Switar lenses. It has two viewfinders—one full-view iconometer viewfinder with built-in parallax correction, and a sports finder for rapid tracking. The front port has a neutral filter to prevent reflections and increase picture contrast. The camera can be readily inserted and removed from the housing which has two side handles. The housing has been tested to depths of 300 feet.

Mako. The Mako housings accept 16 and 35mm Arriflex cameras with 400-foot magazines. The housing features cast aluminum O ring seals and stainless steel or brass shafts. The cameras mount on plates which slide into the housing and bolt in place. The housings have external controls for all camera functions except turret rotation. Tested to 150-foot depths, the housing is supplied with batteries (16 or 18 volts), a charger, and a carrying case. See Figures U-3*a* and *b*.

Sampson-Hall. The housings are designed to accept 16mm, 35mm, and high-speed cameras (Bell & Howell 70DR, Bell & Howell single-lens Eyemo, and Fairchild HS-100). The housing is constructed so that the light meter (photocell at the front and the scale at the rear), is visible through the rear port. Pistol-grip handles on each side of the housing ease the cameraman's task. The unit has a slight positive buoyancy in water.

unidirectional microphone A microphone capable of picking up sound from only one direction. It is useful in reducing the lévelof background noise coming from other directions. Compare OMNIDIRECTIONAL MICROPHONE.

uniformity A measure of the freedom from sensitivity variations along the entire length of the recording material.

uniform system Also, U.S. System. A system of marking lenses in which numbers are used in place of *f*-stops. The markings in this system are proportional to the *f*-values.

f-values	1.4	2	2.8	4	5.6	8	11	16	22	32	45
U.S. System	—	—	—	—	—	4	8	16	32	64	128

Unilux 500 A computer-operated lighting system. An adaptation of the strobe light used by still photographers, the Unilux system* provides a brilliant flash of illumination that is synchronized with the precise interval of exposure. Synchronization is coordinated by a computer to assure a predetermined number of flashes per second—the amount varying with the camera's preset rate of exposure. With this type of illumination, high-speed actions can be stopped completely (in the individual frame of film) so that numbers printed on the whirring blades of a fan, for example, can be read without difficulty.

When electronic flash lighting is used, energy from a power supply is stored in an energy storage system and is discharged at a rate determined by any type of synchronizing circuit or mechanism. See Figure U-4.

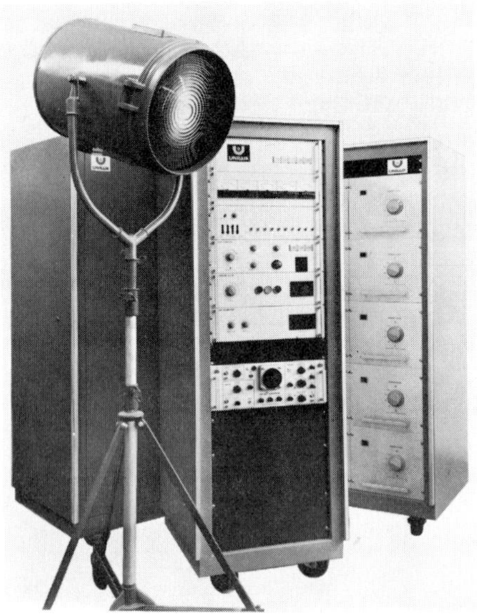

Fig. U-4 The Unilux energy storage rack.

universal finder A versatile viewfinder with adjustable masks and, in some cases, a series of scribe lines. This type of finder shows the field of view that will be covered when lenses of different focal lengths are used for filming.

uranium chloride A compound used in toning prints. The compound decomposes at relatively low temperatures and is poisonous.

uranium nitrate A yellow-green salt used as a toner and intensifier. It dissolves readily in water at normal temperatures.

USASI Now changed to ANSI, the abbreviation for the American National Standards Institute. This independent industry-wide association establishes standards to promote consistency and interchangeability among manufacturers. This organization was formerly known as the American Standards Association or ASA.

*Unilux, New York, N.Y.

V

vanishing point An imaginary point on the horizon at which converging parallel lines appear to meet. Parallel lines directly in front of a camera (as for example railroad tracks) appear to meet at the vanishing point on the horizon. See PERSPECTIVE.

vapor-discharge lamp An electric lamp that uses a vapor or gaseous element, instead of a filament arrangement for producing light.

variable-area sound track An optical sound track in which the modulations take the form of widening or narrowing black areas in the film's sound-track area.

variable-beam light controls Controls that make it possible to regulate the light beam emanating from a light source. In most cases, these controls permit adjustments from spot to full flood with provision for stops between the two extremes.

variable-density sound tracks An optical sound track in which the modulations take the form of light gradations ranging from black, through intermediate shades of gray, to clear.

variable diffuser The variable diffuser* is used for special diffusion effects and consists of two plates of optical glass, clear at one end and gradually becoming diffused toward the other. These plates are carried in two metal slides which are operated simultaneously—

*Mitchell Camera Corporation.

usually by a knurled knob. In one extreme position the two clear ends overlap. As the knob is turned, the overlap increases, and the degree of diffusion may be increased to its maximum capacity. A graduated dial with a pointer provides the means for reading the amount of diffusion. The diffuser mounts on the matte box and is held in place with a lock screw.

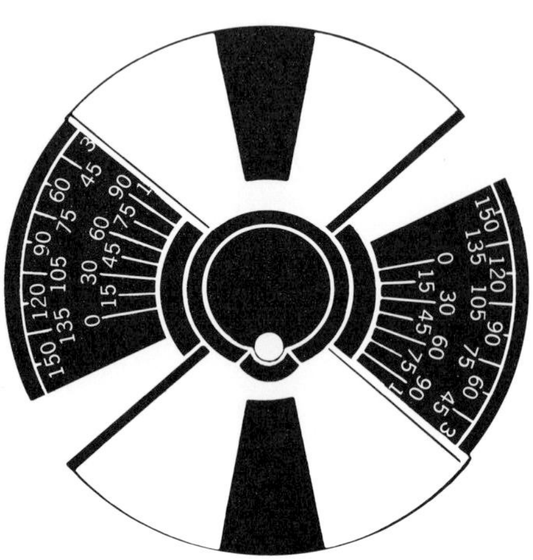

Fig. V-1 Variable shutter. (*Arriflex Corporation of America.*)

variable-focus lens Also called **zoom lens.** A lens whose focal length can be changed in order to increase or decrease the size of the film image. This type of lens makes it possible to achieve zoom effects without having to move the camera from its original position.

variable shutter This type of shutter has two metal plates—one fixed and one adjustable in relation to the other. During the interval of exposure, the plates are engaged (regardless of the shutter setting and degree of opening) and rotate simultaneously to allow light to be transmitted to the film.

This type of shutter permits the cinematographer to make fades and cross-dissolves; and to control exposure without changing *f*-stops, depth of field without changing selected exposure, and image sharpness of fast-moving objects.

During the production of fades or dissolves, the secondary shutter, or adjustable plate located behind the fixed shutter, closes or opens gradually, in successive frames, for the entire length of the effect. See Figures V-1 and V-2.

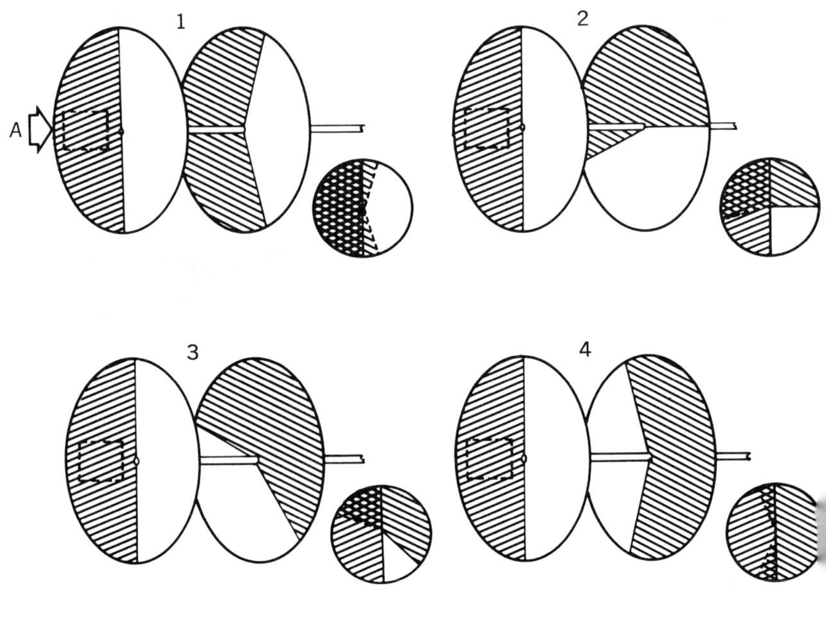

Fig. V-2 The variable shutter. The arrow at *A* indicates the position of the camera aperture. At the top left (1), the two shutter plates (shaded areas) are in the nonphotographing position even though the 170-degree shutter is in its completely open position. Each of the small circles shows the open segment of the shutter.

At the top right (2), the shutter is set at 85 degrees and is half closed in the nontaking position. In the diagram at the bottom left (3), the shutter is only 25 percent open—42.5 degrees. In diagram (4) at the lower right, the two plates are shown in the closed position.

variable-speed motor Used in conjunction with animation stands and optical printers, this type of motor allows the cameraman to select a convenient filming speed by turning a selector dial. At sound speeds of 24 frames per second (fps), a variable-speed synchronous motor will, for example, operate at 1,440 revolutions per minute (rpm); at 16 fps, 960 rpm; at 12 fps, 720 rpm; and so on. These motors deliver full power regardless of the filming speed.

Also known as **wild motors**, their use for sync sound filming is not recommended.

variable telephoto lens A lens capable of filming at a number of magnifications. See TELEPHOTO LENS.

varioscopic cinematography A system developed by the Scientific Cine and Photo Research Institute which makes it possible to change the aspect ratio of a motion picture frame from approximately 1:2 through 1:1 to 2:1. Proposed by S. M. Eisenstein in 1931, the system uses 70mm film that has 10 perforations per frame. The same results, with this system, can be achieved with 35mm films by using the full height of the frame, or with an optical system designed for this system (a squeezing factor that can be varied within the 0.5-to-2 range).

vault A room or an area, temperature- and humidity-controlled, that is used for storing film.

vectorscope An electronic control used for setting or adjusting the color balance in a videotape recording. At the proper setting the photographic primaries and their complements are visible against a black background (at specific points) on the round screen. If all the colors are not visible in their precise positions at the same time, the technician must make additional adjustments to be assured of a good color balance in the subsequent recording.

velocity compensator The accessory, in videotape recorders, that compensates for banding. See BANDING, COLOR and BANDING, NOISE.

vertex The point of intersection at which the optical axis and ocular surface of a lens meet. See Figure V-3.

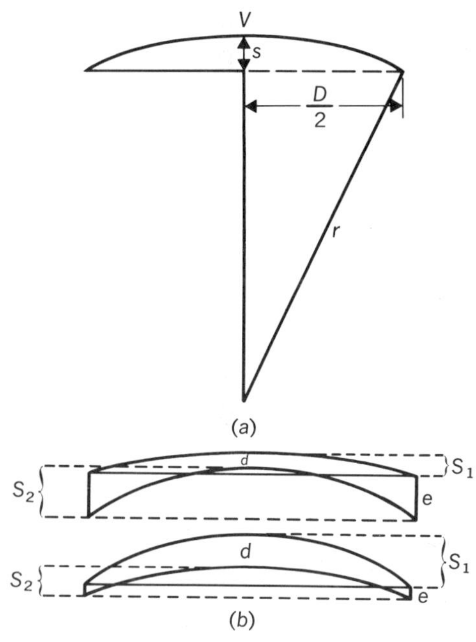

Fig. V-3 (a) Vertex depths of lenses. (b) Difference between center and edge thickness of a lens. (Bausch & Lomb Inc.) For the precise adaptation of deep-curved lenses, it is often necessary to take into account the vertex depth or sagitta value s of the lens: i.e., the perpendicular distance from the vertex V which is the center of the inner surface to the plane of the edge of the lens. This value may be computed by the following formula:

$$s = r - \sqrt{r^2 - \left(\frac{D}{2}\right)^2}$$

where s is the sagitta value, r the radius of curvature in millimeters, and $D/2$ the semidiameter of the lens.

vertical advance Special circuitry in videotape recorders which counts out the number of pulses in a field to determine the position for regenerated vertical blanking.

vertical sweep The vertical movement of the scanning beam as the picture image is re-created over the 525 lines in the television screen.

very high frequency Frequencies ranging from 30 to 300 megacycles, used for the transmission of television picture information and sound signals. Above this range are the ultrahigh frequencies. See ULTRAHIGH FREQUENCIES.

video From the Latin: "I see." A term pertaining to the bandwidth and spectrum position of the signal which results from television scanning and is used to reproduce a picture.

Also, a module in vidicon film cameras that amplifies the video signal received from the preamplifier and provides aperture compensation.

video amplifier The amplifier in television receivers where the approximate 1-volt signal received from the video detector is boosted to the required voltage (50 to 90 volts) needed for proper reception. These amplifiers also provide the correct video signal polarity.

video band Although the picture information and sound signal are transmitted simultaneously (and over the same channel), the sound is carried on frequencies that are 4.5 megacycles higher than those used for the video signal. The sound is frequency-modulated—the picture information is amplitude-modulated.

The spectrum is divided into transmission bands or channels and each channel is allotted a specific frequency by the Federal Communications Commission. Channel 2, the lowest channel number of the more than 80 in current use, is in the low VHF (Very High Frequency) band and has an assigned frequency of 54 to 60 megacycles (Mc). The picture is transmitted at 55.25 and the sound at 59.75 Mc.

Channels 3 to 6 are also in the low VHF band and have frequencies ranging from 60 to 88 Mc. Picture frequencies range from 61.25 to 83.25, and sound from 65.75 to 87.75 Mc.

Channels 7 to 13 are in the high VHF band and have assigned frequencies ranging from 174 to 216 Mc. Picture information for these channels is transmitted between 175.25 and 211.25 Mc. Sound is transmitted between 179.75 and 215.75 Mc.

Channels 14 through 83, the highest numbered channel, are in the UHF (ultrahigh frequency) band and have frequencies ranging between 470 and 890 Mc. For these frequencies, the picture is transmitted between 471.25 and 885.25, and the sound between 475.75 and 889.75 Mc.

video detector In television receiving equipment, the circuit that receives the incoming amplitude-modulated video signal and the frequency-modulated sound signal located 4.5 megacycles from it. In color sets, the color is received (along with the brightness signal) as a phase-modulated 3.58-megacycle signal. The fixed-time circuit in the detector "processes" both the sound and video signals, and relays them to the video amplifier where they are boosted to the voltage required for proper reception.

video disk recorder A recording device used to record picture information on a magnetic alloy plated aluminum disk. Resembling a phonograph record, the recorded pictures, or a fixed video frame, are selected for transmission by radially positioning the recorder's playback head. This type of recorder is used for stopping an action, or for providing instant playbacks during the telecasting of sports events. The recorder, for computer applications, uses several record heads and is capable of recording a number of signals and storing them until needed.

Videofilm color tape-to-film transfer The registered name of the Reeves Sound Studios process for transferring videotape to film. This direct-reversal transfer process provides a full-range 16mm color print with standard SMPTE leaders.

video filter Also called **sync separator**. Networks or circuits that clip the horizontal and vertical sync pulses from the composite video signal after the signal has been amplified. High-pass filters allow the horizontal pulses to be transmitted at the beginning of each horizontal scan. Low-pass filters allow vertical sync pulses to be transmitted at the beginning of each field scan. These sync pulses keep the scanning actions in synchronization. The process of clipping the pulse signals from the video signal and separating them according to frequency is referred to as sync separation.

Videograph display control A system* for generating alphanumeric video information. The unit produces titles, names, and other data by means of punched paper or magnetic tape for subsequent superimposition over a television picture.

Videola With the Videola system of editing, a kine machine first provides edge-numbered 16mm film of the videotape selected takes. The film work print is then edited.

*Developed by the A. B. Dick Company of Chicago and introduced by Visual Electronics.

After the rough-cut version is approved, the kine is put on the Videola where as each frame of film moves through the viewer, the charting tapes move a corresponding amount at videotape speed (15 inches per second). These charting tapes are synchronized and marked for A & B roll tape editing. Using the same charting tapes as a cutting guide, the editor then cuts the videotape itself. In the system's final phase, the Videola's tapes trigger a mixing automator to produce a composite videotape that conforms to the frame-by-frame editing done on the kine rough-cut.

VideoScene VideoScene is an electronic camera system which makes it possible to blend live action with miniature settings, still photos, or motion pictures in such a way that the performers seem to become part of the reproduced locale. It links two cameras with a servo system that synchronizes camera movements so that the background and the performer are always perfectly matched. As the master camera pans, tilts, or zooms when photographing the performer on a special

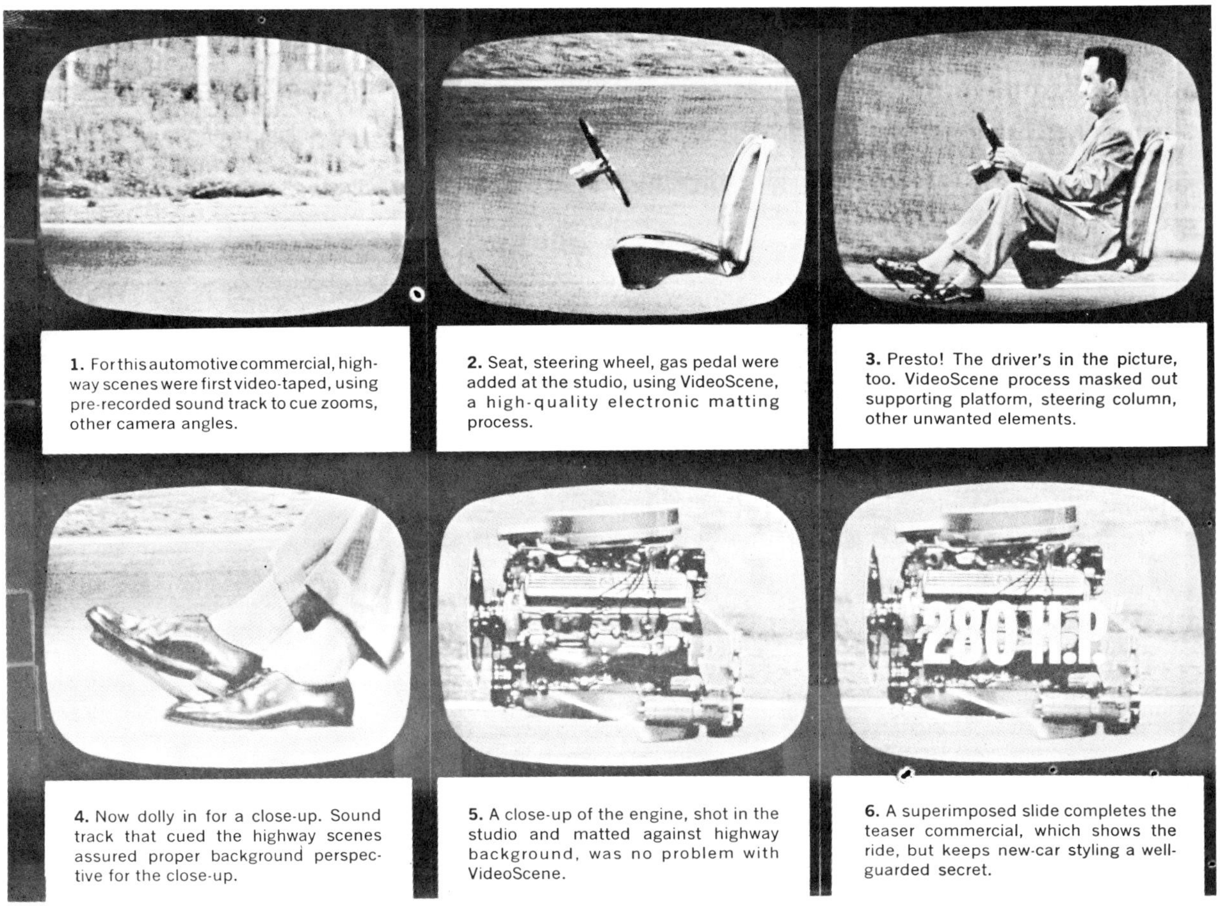

Fig. V-4 VideoScene—electronic matteing at its best.

1. For this automotive commercial, highway scenes were first video-taped, using pre-recorded sound track to cue zooms, other camera angles.

2. Seat, steering wheel, gas pedal were added at the studio, using VideoScene, a high-quality electronic matting process.

3. Presto! The driver's in the picture, too. VideoScene process masked out supporting platform, steering column, other unwanted elements.

4. Now dolly in for a close-up. Sound track that cued the highway scenes assured proper background perspective for the close-up.

5. A close-up of the engine, shot in the studio and matted against highway background, was no problem with VideoScene.

6. A superimposed slide completes the teaser commercial, which shows the ride, but keeps new-car styling a well-guarded secret.

matteing set, the linked or satellite camera, photographing the miniature background, also is made to pan, tilt, or zoom in synchronization.

During the production process, one camera photographs the miniature set which gives the impression of normal size on the frame. The second camera covers a blank set where the performers go through their movements against a background of special reflective sheeting developed by the Minnesota Mining & Manufacturing Company. Reflected light blacks out the background allowing the performers to be blended into the miniature set. See Figure V-4.

video signal The video signal, according to U.S. standards, should have a range between 30 cycles per second and approximately 4 megacycles. The range, it is felt, provides the degree of image resolution that would make it possible to see a dot that has a diameter equal in size to the height of a horizontal scanning line.

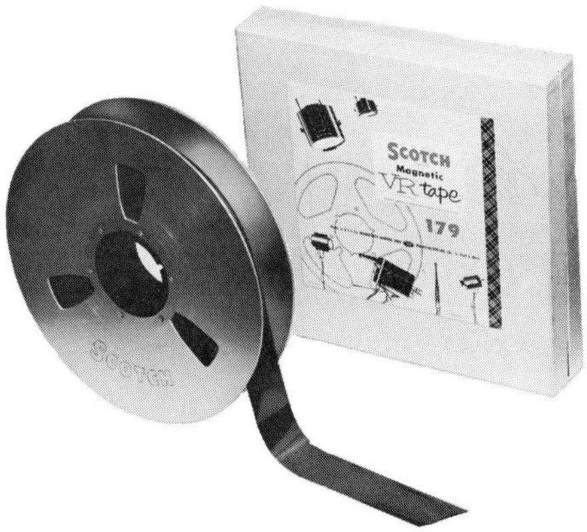

Fig. V-5 Videotape—before and after recording. (*Minnesota Mining and Manufacturing Company.*)

videotape The material used for recording picture and sound information on videotape recorders. Varying in width (½ to 2 inches), the tape's sensitive surface, comparable to the emulsion on motion picture film, consists of a layer of iron oxide particles approximately one-sixth the thickness of a human hair. The thickness of the tape is indicated in mils. See Figure V-5.

videotape recording Combining the versatility of film with the "presence" of live telecasts, videotape provides the means for recording synchronous sound with color or monochrome pictures—electronically.

The video signals or picture information, recorded on magnetic tape by a television camera system, can be played back immediately, or at a later date. Unlike motion picture film that must be processed before the latent image is rendered visible, videotape's sound and picture can be reproduced immediately after recording without the delays necessitated by developing and laboratory printing processes.

With improved production and editing techniques, videotape has become an extremely valuable means of communication for entertainment, educational, and commercial purposes. Instantaneous playback has not only helped speed the production and completion processes but has also completely eliminated the need for insurance takes. For the producer on location, its economic advantages are immediately obvious. Similarly, the tape technician's ability to add illusory effects and between-scene transitions during the recording process has virtually eliminated the need for time-consuming matteing and costly optical work.

During the recording process, the tape passes from the supply reel, on the tape-transport mechanism, to the take-up reel. Traveling at the rate of 15 inches per second, the videotape first passes over the master erase head, where it is degaussed to remove any residual picture or sound, and then over the recording headwheel. Rotating at 14,400 revolutions per minute (240 rps), the magnetic heads scan the tape a total of 960 times each second (64 scanning lines per inch). The sound signal, $9\frac{1}{4}$ inches ahead of the corresponding picture, is recorded as the tape passes over the audio head in the transport mechanism.

The brown iron oxide particles on the magnetic tape are rearranged, during the recording process, to form invisible images on the tape's sensitized surface. Following television transmission standards, the fields are arranged vertically across the width of the tape and are recorded at the rate of 60 per second. The intermeshed picture information from two adjoining fields form one composite frame, exactly $\frac{1}{2}$ inch wide. Each $\frac{1}{4}$-inch field contains a $262\frac{1}{2}$-line image, or half of the 525-line total of the completed picture transmitted each $\frac{1}{30}$ second in television.

The standard $12\frac{1}{2}$-inch videotape reel contains 4,800 feet of recording surface, almost enough to cover the diamond on a baseball field. During the reel's 64 minutes of projection time, at the tape speed of 15 inches per second, the viewer sees a total of 115,200 separate images. Since tape, unlike film, does not depend on sprocket holes for registration, the inherent danger of abrasions or torn sprocket holes during the recording and transmission processes is nonexistent.

Arranged from top to bottom along the 2-inch-wide strip of magnetic tape are four synchronous tracks, each serving a specific purpose during the recording process. Sound for the accompanying picture is recorded on a narrow track along the top edge of the tape. The wide video track is sandwiched between the sound track at the top of the tape and the control track. The latter contains an edit pulse that guides the technician during the editing and splicing processes. The fourth track, at the lower edge of the tape, is used for on-stage cueing purposes during production.

For the producer whose previous experience was gained in rival media, the relaxed on-stage atmosphere during a live production does not go unnoticed. For him, as well as for the director, cameraman, and performers, the waiting period that precedes the viewing of dailies is eliminated along with the potential ulcer, a common occupational hazard. Immediate playback and the opportunity to appraise the rushes before going on to the next shot take the guesswork out of production. In commercial taping, client approval can be as immediate as the subsequent playback.

Missed cues and flubbed speeches are no cause for panic. Bad takes are easily erased and the new scene is immediately re-recorded over the same length of tape. Flaws, unnoticed during the taping, are quickly detected during the playback and corrected before the set is struck and the performers leave the studio. If desirable, on-the-spot corrections may include new camera angles, lighting changes, and even script revisions. Performers, "schooled" during the playback, can make necessary changes in both timing and delivery. Troublesome sequences can be divided into shorter segments and accompanying long speeches can be broken up, thus eliminating the need for memorizing long monologues. When edited, all the actions are spliced together without affecting the continuity, the spontaneity of the performance, or the quality of the recording.

Program material from several remote locations can be recorded simultaneously, on a single length of videotape without optical camera work or sacrifice in quality. Newly developed electronic devices have also made it possible for the producer to combine on-camera performances with graphic material, animation, and/or sequences from existing film. The audio portion, for this blending of techniques, can be recorded immediately or added at a more convenient time. Synchronization, in either case, is achieved quickly and precisely.

In recent years, videotape mobile units have recorded complete hour-long programs as well as on-the-spot news events and commercials in places that can best be described as remote. Differences, in setup and shooting time, for the film and videotape productions, are negligible. Exposed film, however, must be developed, printed, synchronized with a sound track and, omitting optical camera work, reprinted again as a composite. By comparison, the sound and picture on the recorded tape can be played back and edited immediately after recording has been completed. With the last take approved, the production crew can pack its equipment and move to a new location or return to the studio secure in the knowledge that their collective effort is ready for transmission. Dupes made from the master recording reflect the on-mike, live quality of the original. And, unlike film which must be inspected, repaired, and cleaned periodically, tape will show only an inconsequential amount of deterioration after more than 100 playbacks. See Figures V-6*a* to *c*.

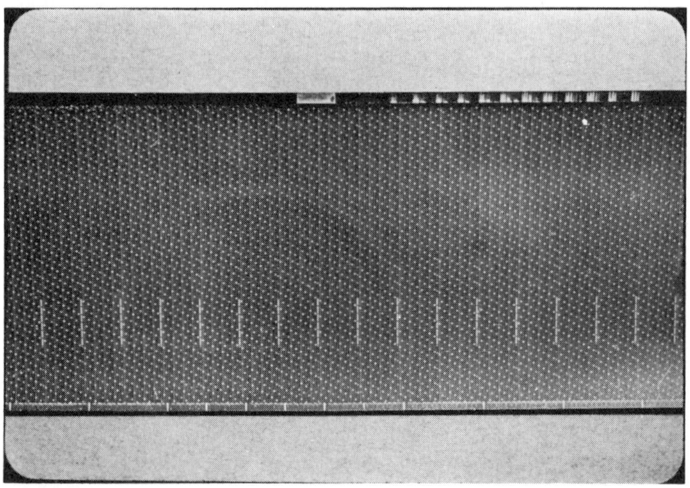

Fig. V-6a Videotape patterns. The narrow band along the top edge is the sound track. The wide area carries the picture portion. The control track and edit pulses are positioned below the picture portion. The cue track runs along the bottom edge of the tape.

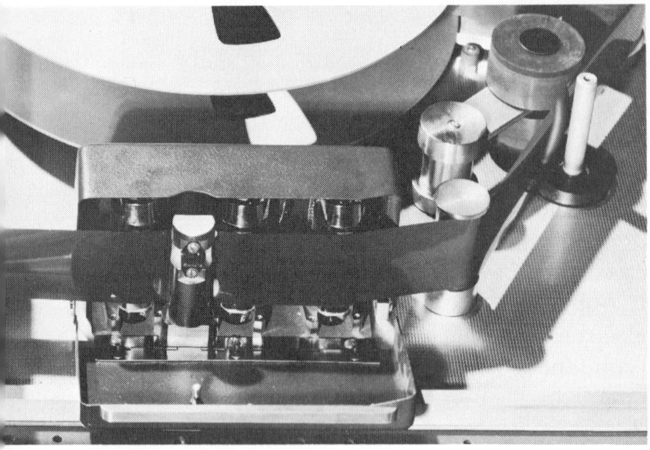

Fig. V-6b The videotape recorder's transport mechanism.

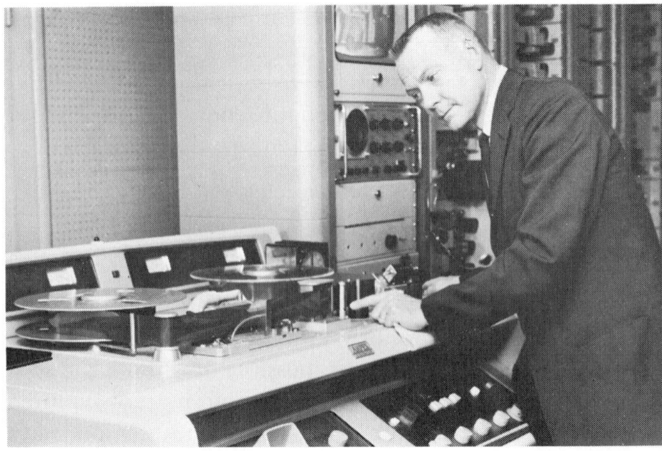

Fig. V-6c The playback.

videotape—editing and special effects Because standard motion picture processes could not be used for editing the invisible images on the recorded tape, technicians developed new methods and electronic devices that made frame-by-frame editing and splicing possible. Accurate splicing points, similar to the frame lines in motion picture film, are indicated by magnetized impulses along the lower edge of the tape. These field pulses, between each $\frac{1}{4}$-inch field, become visible when brushed with a carbonyl iron solution. Splices, with the exposed lines as a guide, produce jump-free cuts that equal the film editor's best efforts. More sophisticated electronic techniques, described under electronic editing, have simplified the editing process considerably.

The TVola, similar in principle and performance to the moviola used in motion picture editing, is a monitoring device that enables the editor to locate specific areas in the recorded tape. Sound is pinpointed by electronic readers resembling those used for analyzing sound tracks. Cutting points are marked with a grease pencil, and splices are made on a calibrated block with extreme accuracy. The aluminized tape that is used for joining lengths of tape does not affect subsequent transmission.

If sound and picture must be edited separately, or if sound from several previously recorded tapes must be combined, audible beeps on the cue track guide the synchronization and editing processes. Referred to as *edit sync,* these beeps (at one-second intervals) help the editor locate specific cut points. Individual frames, between signals or bursts, can be marked off with the aid of scaled rulers that divide each 15-inch length of tape into 24 sections, and are the equivalent in projection time of a frame of film.

Many special film effects that can be achieved only with mattes and optical camera work can be duplicated quickly, inexpensively—and electronically—with the special-effects generator, or matteing amplifier. Fades, cross-dissolves, superimposures, double exposures, split-screen shots, and wipes are all classified as routine effects by the videotape technician. The wipe charts that suggest a variety of transitional effects for the motion picture production are miserly compared with the more than 300,000 effects that can be produced with the effects generator.

More complicated illusory effects are achieved with other electronic recording devices, each designed to serve a specific purpose. VideoScene, for example, is a multiple-camera recording system that enables the technician to combine picture information from several sources into a single composite tape. The component subject matter, photographed from separate sets, is recorded simultaneously. Centrally controlled, each of the cameras is synchronized with the others. Complicated zoom and pan movements do not affect the predetermined relationship between the cameras and the subject matter. The resultant composite effects are similar to those achieved with mattes and optical camera work.

With either the Inter-Sync or A-B composite systems, picture information from previously recorded tapes can be combined to produce unbelievable superimposures, split screens, and montage effects. Similarly, playbacks from existing tapes can be used with on-the-spot studio recordings to form an equally wide variety of eye-arresting composites. Normal-sized performers can be enlarged or reduced, in controllable proportions, and superimposed over prephotographed backgrounds.

With imaginative planning and the zoom keyer, an electronic device for combining slides with other graphic material, performers and on-camera announcers can be positioned within a trademark or title. Several interlocked recording systems multiply the number of visual possibilities and place at the producer's disposal an unlimited variety of unusual illusory effects. See Figures V-7*a* to *d*.

Fig. V-7*a* **Monochrome videotape leaders.**

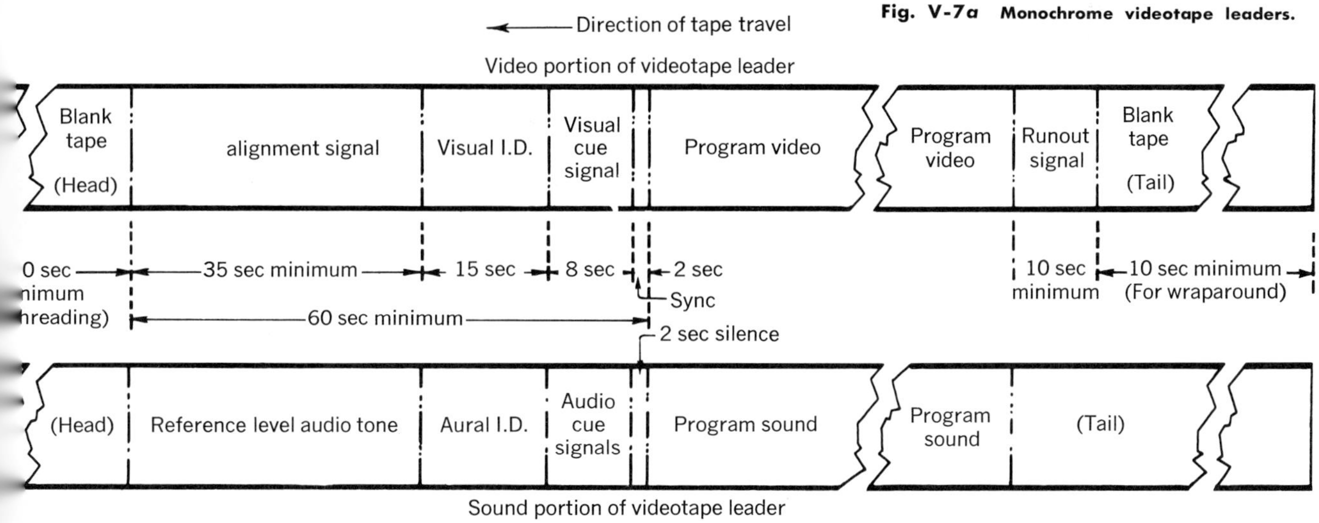

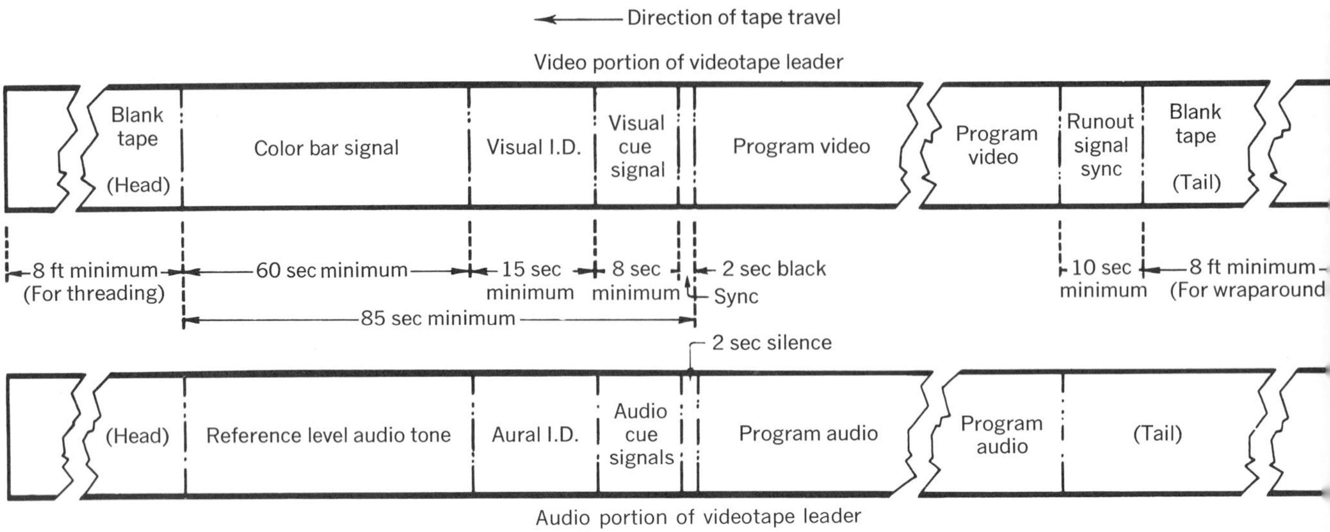

Fig. V-7*b* Color videotape leaders.

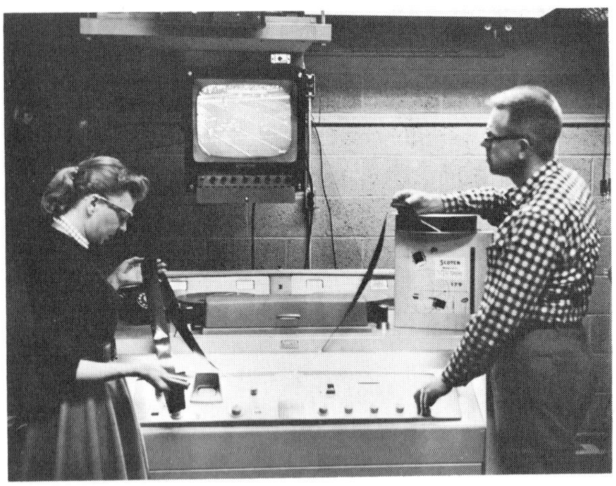

Fig. V-7*c* Editing videotape—the only picture is on the monitor.

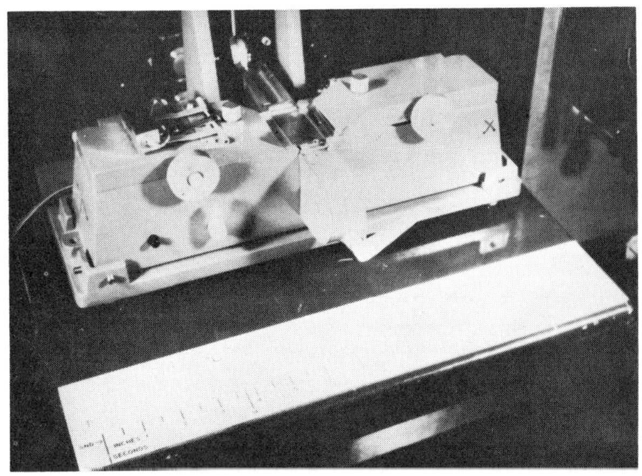

Fig. V-7*d* The videotape splicing block.

videotape in education Videotape's acceptance, for entertainment and commercial purposes, might have been generally anticipated. Its influence on traditional teaching methods, however, exceeded all expectations.

First used experimentally in mid-1958, the new medium of communication was greeted enthusiastically by professional educators, teachers, and students. Within three years, millions of students, in schools and in front of home receivers, were receiving taped instructions in a wide variety of subjects. Programs, recorded by outstanding instructors at their convenience, were being transmitted to the little red schoolhouse in the country as well as to larger schools in more populated urban areas. The classroom receiving sets were making every seat a first-row vantage point.

Called the greatest advance in the learning process since the invention of the printing press, recorded videotapes have done much to eliminate the strain on school budgets. Educational institutions with limited funds are able to offer courses they could not afford without tape facilities. Similarly, expanded curriculums included prerecorded programs by specialists in various fields.

The ever-present blackboard was replaced by field trips to libraries, laboratories, and museums—via tape. More remote locations, providing the students with a close look at the subject matter, were also made available through coaxial cable. Closed-circuit systems, capable of carrying an unlimited number of programs simultaneously, transmitted similar courses to entire schools rather than to individual classrooms. Linked by a system of microwave towers, recorded material was relayed over longer distances.

Unhampered by geographical boundaries, open-circuit transmission does not depend on coaxial cables or relay stations. Material is beamed in all directions simultaneously and competes with commercial programming on local channels. Continental Classroom, a network educational program that offers full academic credits, competed successfully when it made its spectacular debut in 1958. The transmitted programs have outlasted many of the more entertaining, high-budgeted efforts. The school bell, for these home sessions, rings at an hour when most viewers are studying nothing more important than the creases in their pillows—6:00 A.M.

Obviously, subject matter must be kept up to date. With videotape, outdated or undesirable portions of a program are spliced out of the recording and replaced with new material quickly and easily. On tapes made for transmission to foreign countries, the language spoken in each country is recorded on the audio channel that is normally used for cueing purposes. Visual information, for foreign consumption, can thus be translated and transmitted in more than one language—on the same length of tape.

Instructors and lecturers, unfamiliar with recording and dramatic techniques, can see themselves as others will see them—on studio monitors. Like professional performers who study their efforts during the playback of the recorded take, teachers can also school themselves on their pacing and delivery. They become more relaxed and confident as a result, and the second take will, almost always, show an amazing improvement in technique and poise. With justifiable respect for the capabilities of the new medium, the producers of educational programs, like the producers of programs designed for entertainment purposes, are continually looking for new ways of presenting their collective subject matter—electronically.

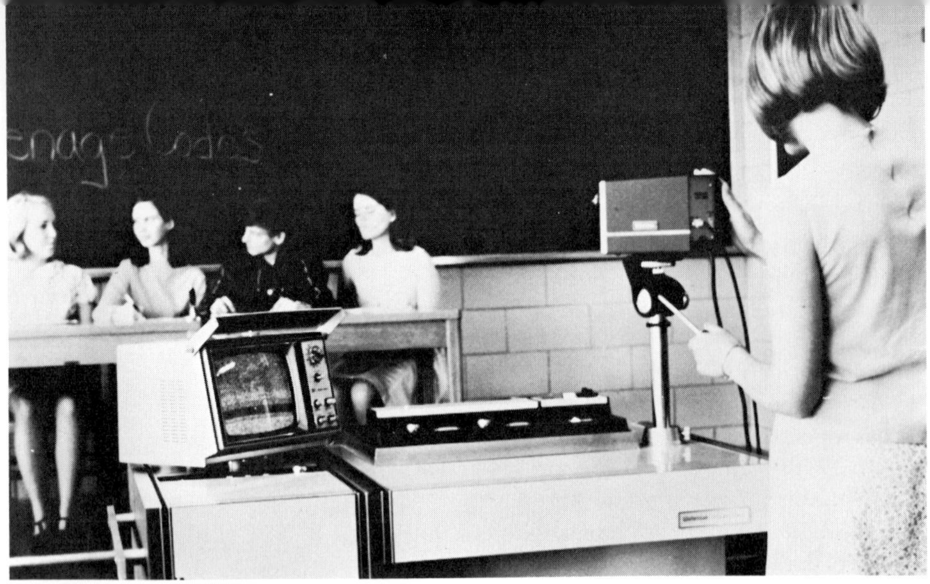

Fig. V-8 Videotape in education. (*Wollensak.*) The entire student body can become involved in panel discussions via videotape. The program can be presented to all students in small groups, without disruption of class schedules.

Combining the best features of motion picture production processes with the electronic wonders of television, the unusual or difficult videotape effect has become routine; the impossible effect merely requires additional planning. Certainly, the versatile medium has, quite adequately, demonstrated its effectiveness as an all-purpose means of communication by satisfying the needs of the most imaginative and demanding producer. Its future potential is even greater than its spectacular past. See Figure V-8.

videotape production In many cases, videotape recording techniques are basically similar to those used in motion picture production. In at least as many other cases, however, the processes and procedures are completely different. Possibly, the most difficult fact that must be accepted by the film producer is that there is no film in the camera that is used for exposing the picture information on videotape, and that the end product or recorded roll has no frame lines that separate one invisible image from the next.

Before taping begins, two small racks of electronic equipment—one containing a camera control unit, the other a sync generator, vectorscope, and a simple switcher—are moved into position on the studio floor, and the cables between the racks are interconnected. The camera itself is mounted on a conventional geared head. Power to the camera is switched on, and the registration pattern is illuminated. Focus and zoom controls are mounted on the side of the television camera, instead of at the rear in the usual television style. This allows a cameraman to concentrate on composition and movement, while an assistant cameraman operates the focus and zoom controls, just as on a motion picture camera.

After the equipment has warmed up for about a half-hour, the camera is electronically registered by the videotape technician. Each segment of the production is shot, scene by scene, and illuminated individually.

Lighting for tape is essentially the same as lighting for film. Color video cameras need only about 250 footcandles or approximately 80 percent of the lighting level required for film.

On film, each scene can be color-corrected to a certain degree during processing. On videotape, this must be done during actual production. Color correction on tape during postproduction is much more difficult than correcting film answer prints. With videotape, corrections are made after examining replays of the previous recording and making the necessary adjustments.

The producer also has a certain degree of color-correction capability by lighting a selected still frame and using calibrated painting controls in the video circuitry of the camera chain's red, blue, and green outputs.

All scenes are shot and slated exactly as they would be on film. The audio is handled in the same way, with the audio signal recorded directly on the videotape, on synchronous ¼-inch magnetic tape, or on magnetic film. After the filming of each take, the videotape is played back for the producer's approval. This procedure continues until all the scenes scheduled for shooting have been approved. A preproduction planning session can be used to tailor the scene filming sequence to the talent. If a principal appears in more than one scene, for example, the scenes can be shot together. The same type of scheduling is used when the backgrounds are similar.

On those occasions when chroma-key matteing techniques are needed, the producer should remember that the best quality is achieved if the matteing is done during the original shooting. If the desired result is impossible to achieve during the original shooting, chroma-keying in postproduction is possible if the scene to be inserted is shot on film against a cobalt-blue background.

As the video image of each scene is recorded, live audio is simultaneously recorded on the audio track of the videotape. At the same time, a separate audio track is recorded on the videotape's cue track. This is a talking cue track, called the *talking clock*. It is used to identify videotape picture frames.

Immediately after the completion of shooting, the selected takes of each scene are transferred to 35 or 16mm direct-positive black-and-white film. The talking clock is transferred to the audio track of the film. The live audio track from the videotape is transferred to 35mm magnetic sound-recording film.

The film editor cuts the film and the 35mm magnetic sound track exactly as he would for a film production. The audio track of the film print can be disregarded unless the scene is less than two seconds long. If it is, the editor must indicate the audio-track footage number that is closest to the in-position of the scene. He then counts the number of frames from that point to where the cut occurs. This is standard operating procedure because the talking clock counts at one-second intervals. The film editor must also record the sequence of scenes that he has used, as well as their corresponding take numbers. After the work print has been completed, it looks just like any other film work print, with all the cuts made on film and the transitional effects indicated with a marking pen or grease pencil.

If the work print is approved, the editor must foil the film for a certain number of frames before each cut. This is done with a small piece of special aluminum foil that serves as a guide during the transfer process that follows.

As the film is transferred to tape, a pickup device reads the location of the aluminum foil on the film and places an electronic cue on the videotape. This permits precise editing of the tape. Because the talking clock is on the original tape as well as on the film-to-tape transfer, the location of scenes is a relatively simple matter. Electronically edited A & B rolls are thus set up for the video mix. During the mixing process, transitional effects are added between scenes, and special effects and titles are inserted where they are needed.

The composite videotape is thus made possible through the use of the most efficient techniques from two different mediums. More important, however, is the fact that the producer is able to use those processes and techniques with which he is most familiar—and apply them intelligently and efficiently in either medium.

videotape versus film, comparative quality The production of a motion picture depends on a combination of optical, mechanical, and chemical processes. Television and videotape use optical and electronic processes to produce and transmit images. Obviously, the objective of both is the production on film or tape of an image that will duplicate, in definition and quality, the colors and tonal values that are visible (to the eye) in the original subject matter.

The first comparative tests were made in 1942 by Albert Rose of the Radio Corporation of America. These tests were conducted on the basis of sensitivity and line resolution. One of Mr. Rose's conclusions was that the eye is more versatile and sensitive than either of the two mediums.

During the filming process, the original picture information is exposed on a negative film that has three separate layers of emulsion. Each of these layers is sensitive to one of the photographic primaries, and each contains its own color dyes which are needed to restore the color balance and accurately reproduce the colors in the subject matter.

These colors are, in turn, transferred from the original negative to an intermediate color film which is used to make the interpositive, needed for optical printing. The internegative is used to produce a second interpositive that is needed to make a reduction internegative and, in turn, a 16mm contact print.

The films that are used for duping or printing are multilayer films that also have three sensitive layers of emulsion and their own dye components. The colors in the original print (35mm) and the colors in the contact print (16mm) usually have only little in common. The amazing fact is that second- and third-generation dupes are as good as they actually are.

The second- and third- (and even fourth-) generation tape dubbs, by comparison, are practically indistinguishable from the original videotape recording.

videotape recorders Like cameras, videotape recorders are produced in all sizes, with or without built-in accessories, each designed to fit the needs and budgets of the individual producer. The smallest recorders weigh under 50 pounds. The largest recorders weigh more than half a ton.

Recording speeds, measured in inches per second, vary considerably. Unlike the four basic speeds used in sound-recording equipment ($1\frac{7}{8}$, $3\frac{3}{4}$, $7\frac{1}{2}$, and 15 ips), videotape recorders are capable of speeds from $4\frac{1}{4}$ to 15.6 ips. Between these speeds, the producer has a wide choice and can record at 5.91, 7.5, 7.8, 8.57, 9, 9.6, 10.5, 12, 15, and 15.6 ips. The diameters of the tape reels are 7, 8, $8\frac{1}{2}$, 9, $9\frac{1}{2}$, $9\frac{3}{4}$, 12, $12\frac{1}{2}$, and 14 inches. Recording time varies, depending on the tape speed at the time the recording was made. The tape is available in $\frac{1}{2}$-, 1-, and 2-inch widths.

The more versatile color-compatible lenses available to the producer are equipped with electronic focus, electronic viewfinders, and devices that make it possible for the lens to adjust to light changes automatically. These lenses cost approximately $2,500. Other lenses call for expenditures up to $10,000.

Many of the newer recorders are capable of stopping an action, reversing it, and, if necessary, slowing it down or speeding it up. Built-in monitors with provision for still-frame monitoring are another feature found in newer models. Similarly, automatic recording head cleaners ensure the quality of the recording at the initial stage. Switching, too, is simplified with controls that enable the operator to use the equipment as a camera, camera/monitor, or as a monitor for playback. See Figure V-9.

Fig. V-9 Videotape recording—helical scan. *(Ampex Corporation.)*

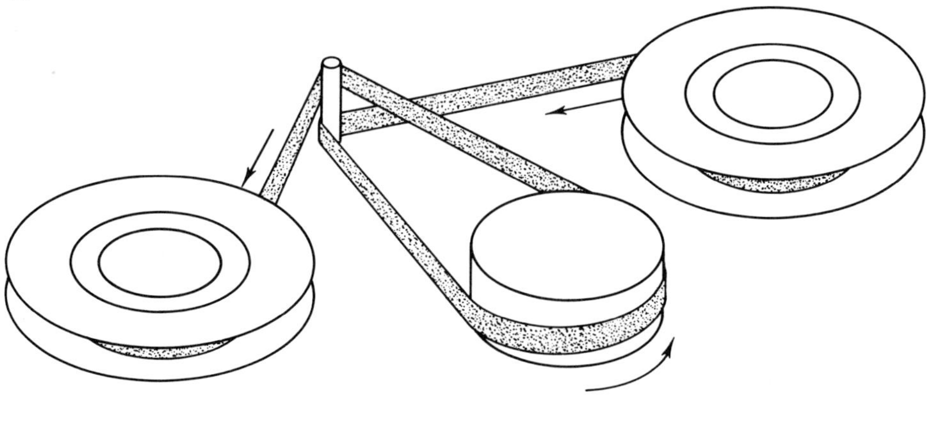

Ampex VR-7500C color. Similar in many ways to the Ampex VR-6000 and VR-7000 series recorders, the VR-7500C incorporates additional features that add to its versatility. The advancements include a full 4.2-megahertz bandwidth; variable slow motion—from 2 to 20 fields per second; rotary video-head transformer; automatic tension release around the scanning drum, permitting indefinite standby; a second audio (cue) track; a 4-minute fast forward time; a wider tension adjustment range; and improved differential gain and transient response.

The TR-7500C is equipped with ball-bearing motors, military-type epoxy glass circuit boards, and other heavy-duty components. The recorder uses a minimum number of controls, and thus little if any operator training is required.

Like the Ampex VR-6000 and VR-7000 series recorders, the VR-7500C uses the Ampex air film, single-head helical scan recording method. Black-and-white tapes made on VR-6000 and VR-7000 series recorders will play back on the VR-7500C, and low-band monochrome recordings made on the VR-7500C will play back on these earlier models, when these have been updated for color.

The VR-7000 is a portable videotape recorder incorporating features that meet all closed-circuit television requirements. It has a full 3.5-megahertz bandwidth with 350 lines horizontal resolution and a 42-decibel signal-to-noise ratio.

Ampex CC-327 studio camera. The Ampex CC-327 studio camera is designed for multicamera operation in closed-circuit television studios. The model's design and construction anticipates centralized control of two or more cameras from a program director's station. Remote control of various electrical functions, intercommunication, local and remote tally lights, single cable operation, and a number of other features are provided in the camera to make such centralized control efficient and convenient.

Ampex CC-326 studio camera. The CC-326's are electronic viewfinder cameras designed for use in closed-circuit television studios. The lower end of the series is an economy model random interlace camera. The top of the series is a broadcast-compatible plumbicon camera with the features needed for incorporation in the multicamera studio. The most important single feature of this series of cameras is that any of them can be upgraded to a higher model

Ampex CC-324 vidicon camera. The CC-324 is fully transistorized, requires only a 117-volt a-c power source, and accepts standard C-mount lenses in its three-lens turret. (A close-focus 25mm *f*/1.4 lens is supplied as standard equipment.) The only operating control is the on/off switch. When used in combination with Ampex series VR-7000 videotape recorders, the camera's internal crystal-controlled horizontal oscillator complements the stop-motion performance of the recorders.

Ampex HS-100 color Slow-Mo recorder. The HS-100 is capable of recording 30 seconds of picture information for subsequent playback at slow speeds. Playback speeds are preselected, forward or reverse operation, or in the alternative, for single-frame advance.

Built-in, electrically controlled cue markers allow the operator to locate a specific point on the tape quickly. The unit is especially useful for taping sports events where the instant replay has become an important part of the telecast.

Ampex HS-200. This model is a combination high-band, computer-controlled color disk recorder and editing system which can record picture information in a frame-by-frame operation. The HS-200 includes a push-button console which enables the editor to locate any previously recorded frame quickly.

An outgrowth of the HS-100 which is used during sport telecasts to slow motion and provide instant replays, the disk recorder is a self-contained unit housing the disk system, controls, and special-effects switches. See VIDEO DISK RECORDER.

Milliken videofilm recorder DBM-R1. This unit, introduced by the Milliken Company, features a 16mm camera with a film-transport mechanism that is compatible with the television scanning rate. In practice, the film is advanced through the camera in synchronization with the television vertical blanking interval, thereby eliminating the periodic appearance of the shutter bar.

Newell tape transport. Developed by Newell Associates, Inc., the high-speed transport provides tape speeds up to 1,000 inches per second with bandwidths over 10 megahertz. With only three rotating parts, videotape recorders using these transports are self-threading. Up to 50 tracks can be recorded on the ½-inch tape used with the transport.

Panasonic NV-8000, NV-8100, and NV-204. These Panasonic recorders operate at 117 volts a-c, 50 or 60 cycles, and have power consumptions of 100, 160, and 258 watts, respectively. Each of the three recorders has two rotary heads and operates at U.S. standards (525 lines, 60 fields per second). The NV-8000 and NV-8100 models record on ½-inch-width tape at 12 inches per second and accept 7-inch reels. The NV-204 records on 1-inch tape at 8.57 inches per second and uses 8.5-inch reels (NAB hub). The horizontal resolution of the NV-8000 and NV-8100 recorders is more than 220 lines; the NV-204 has more than 350 lines. Video signal-to-noise ratio is more than 40 decibels in the NV-8000 and NV-8100 recorders and more than 43 decibels in the industrial model. Audio signal-to-noise ratio is more than 40 decibels in the NV-8000 and NV-8100 recorders and more than 46 decibels in the NV-204.

RCA TR-70. The stabilized circuits in the TR-70 function to correct themselves, holding a high level of performance over long periods of operation. They compensate for changes that may occur with component aging. Operators are thus freed from constant attention and frequent touch-up of controls.

High band, a recording and playback mode that utilizes higher FM deviation frequencies for both color and monochrome, is incorporated in the TR-70. TV line standards and tape speed are accomplished on a push-button basis. Inhibit circuits, which are incorporated in the switching logic, will not permit an incorrect selection of standards.

All circuitry relating to the basic requirements of the system is built in. There are no extra modules required to operate on different standards.

The high degree of stability in the TR-70 automatic timing circuits eliminates the requirement for front panel controls and the need to adjust. Picture geometry and burst correction are automatically achieved. See Figure V-10*a*.

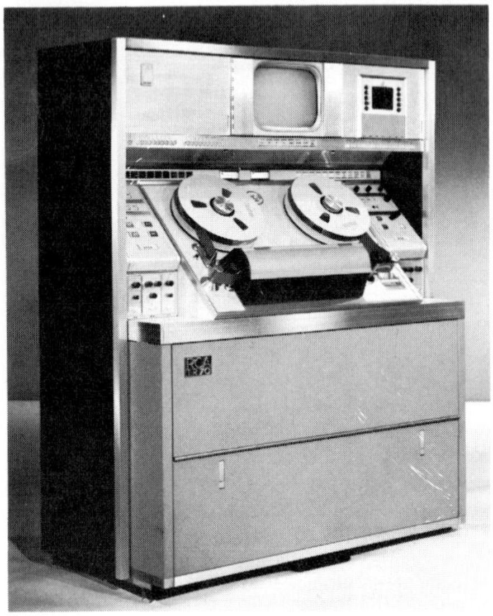

Fig. V-10*a* **The RCA TR-70 videotape recorder.**

RCA TR-50. The TR-50 is a compact, completely transistorized color videotape recorder. It is designed for high-band and low-band applications and produces tapes of uniform quality.

Circuits are provided in the TR-50 which allow for a choice of either the standard 15 inches per second operating speed or a half-speed operation of $7\frac{1}{2}$ inches per second. The use of a narrow-track head wheel in place of the head wheel normally supplied permits twice as much information to be recorded on the same length of tape.

High band, a recording standard that utilizes higher FM deviation frequencies for both color and monochrome, is basic to the TR-50. The selection of the various recording and playback standards is accomplished by a single switch. All necessary circuitry changes are automatically selected to place both the record and playback electronics in the proper mode.

The audio cue record-playback unit in this model permits the recording of cueing information along the edge of the tape. This information can be in the form of voice, tone, or digital cues. The audio cue unit permits independent recording of the audio signals without disturbing the prerecorded video. See Figure V-10*b*.

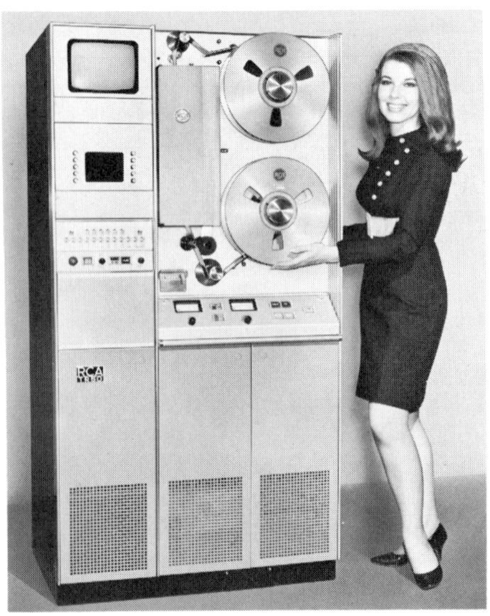

Fig. V-10*b* **The TV industry's newest high-band color television tape recorder—RCA's TR-50. The model (right) is Linda Boyce.**

RCA miniature recorder-reproducer. Developed for the National Aeronautics and Space Administration and designed for use in spacecraft, the miniature recorder is especially useful for recording information when ground stations are outside the range of the spacecraft. As many as 4 hours of slow-scan TV can be recorded on its 1-inch tape and subsequently transmitted to ground stations at a rate eight times faster than the rate used for recording.

Sony DVK-2400. The portable, solid-state DVK-2400 battery-operated videocorder uses $\frac{1}{2}$-inch-wide magnetic videotape, running at $7\frac{1}{2}$ inches per second. A dynamic microphone mounted on top of the camera picks up the audio while a built-in 1-inch television screen viewfinder helps frame the picture and shows exactly what is being taped. There is a remote-control trigger on the camera handle that starts the recording process. The lens on the camera is a 16–64mm *f*/2 C-mount zoom. The DVK 2400 features helican scan recording (playback on any Sony CV-2000 series videocorder), proportional frequency control, fully automatic video and audio level control, a rotary video head, and two-track audio control heads.

The videocorder accepts 5-inch reels, has a full-track erase head, and uses ½-inch-wide tape. The recorder's video signal-to-noise ratio is greater than 40 decibels. The audio signal-to-noise ratio is 38 decibels.

Sony DVC-2400. The DVC-2400 has a 525-line scanning system which provides the standard 60 fields per second with 2:1 interlace (sync is supplied by the DVK-2400 videocorder). It is equipped with a VCL-16 zoom-type lens, 16–64mm *f*/2. The video signal-to-noise ratio is greater than 40 decibels, and the automatic sensitivity control range is 600 to 100,000 lux. Power consumption is 8 watts.

Wollensak VTR-150. The Wollensak VTR-150 is available in two concepts: a portable recorder-playback device only, or mounted in a mobile console complete with camera, tripod monitor/receiver, microphone, patch cords, and one reel of Scotch brand helical videotape. Operation of the unit is similar to that of a sound recorder. The head design assures interchangeability so that all tapes produced on a Wollensak VTR can be played on another VTR 150. Operating at a speed of 7½ inches per second, one 7-inch reel of ½-inch-wide tape (2,400 feet) will record for one hour. See Figure V-11.

Fig. V-11 Designed as a training device for education, business, industry, and the military, this new Wollensak VTR-150 videotape recorder from 3M Company records and plays back television signals, "live" programs recorded with an accessory camera, or prerecorded tapes prepared on other VTR-150s.

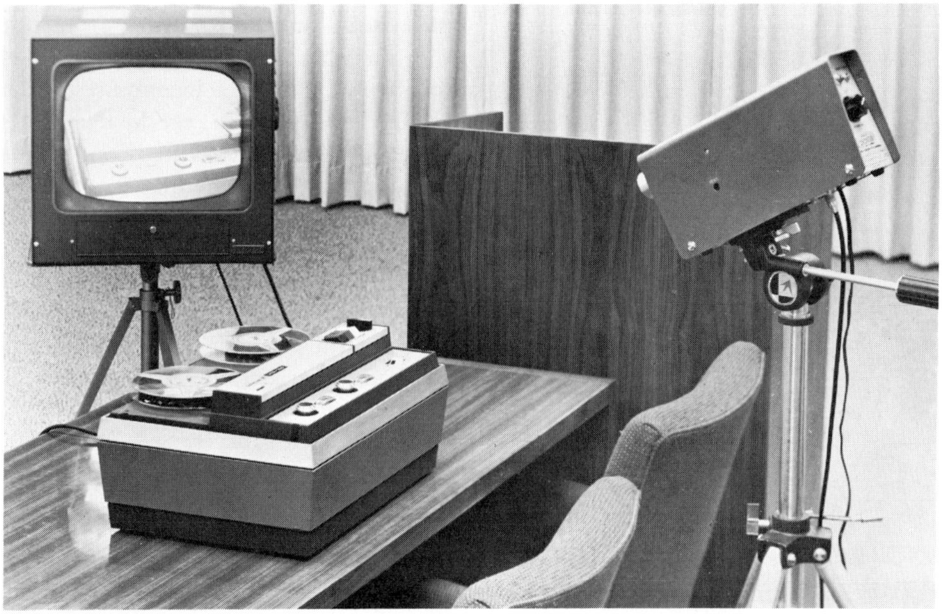

videotape sound readers Videotape sound readers* are basically similar to the sound readers used by film editors. Because of these devices, the larger and more expensive videotape recorders previously used for such editing functions as identifying, cueing, timing, and analyzing sound tracks are freed for other purposes. Portable, battery-powered, and transistorized, these program and cue-track sound readers have built-in timer counters. See READER, SOUND.

vidicon tube A low-velocity cylindrical glass scanning tube used in TV cameras. The 0.50- by 0.375-inch target, at one end of the tube, is a transparent signal electrode that is electrically conductive. Sandwiched between the electrode and a fine-mesh screen connected to a cylindrical focusing electrode (or wall anode) is the photoconductive layer of antimony trisulfide which forms a coating on the surface of the electrode. An electron gun, at the other end of the tube, produces the electron beams.

Focused at the surface of the target or photoconductive layer, the electrons are slowed down or decelerated by the mesh screen. The scanning beam thus reaches the photoconductive layer at near zero velocity and deposits the electrons on the positively charged area to form electric patterns composing the picture.

The electrons deposited on the target reduce the surface potential to equal that of the cathode. The variations in the current flow originating from the electrically conductive surface of the signal electrode develop the video signal voltage needed for subsequent transmission.

vidicon tube and intermittent projection Because of the different projection rates for film and television (24 and 30 frames per second, respectively), vidicon telecines use the 2:3 scanning sequence. With this nonstorage system, the first of two successive frames of film is scanned by two fields. The second film frame is then brought into projection position where it is scanned by three fields. This 2:3 sequence is then repeated.

The scanning time for the first frame of film, two fields, is $\frac{1}{30}$ second. The scanning time for the second frame of film, three fields, is $\frac{1}{20}$ second. Scanning and film advance time for both frames totals $\frac{1}{12}$ second, thus duplicating the standard sound-speed projection rate of 24 frames per second. This system requires the use of intermittent film projectors with normal film advance times and incandescent or continuous light sources. See Table V-1.

viewfinder Also called **finder**. A camera attachment that enables the photographer to see the area that will subsequently be exposed

*Magnasync/Moviola Corporation.

TABLE V-1 Videotape-to-Film Conversion

Frames		Frames		Frames	
Tape	Film	Tape	Film	Tape	Film
1	1	11	9	21	17
2	2	12	10	22	18
3	2	13	10	23	18
4	3	14	11	24	19
5	4	15	12	25	20
6	5	16	13	26	21
7	6	17	14	27	22
8	6	18	14	28	22
9	7	19	15	29	23
10	8	20	16	30*	24*

*One-second projection time.

on the film in the camera. The viewfinder is especially useful for composition purposes. (See FINDERS for different types.)

In the Mitchell NC viewfinder used here to describe their function generally, the magnifying prismatic viewfinder is attached to the left-hand side of the camera beside the focusing tube by means of a dovetail bracket. Adjustable mattes, controlled by two calibrated knobs, one on the top and one on the side of the finder, may be set to mark the fields of lenses of various focal lengths.

The image seen on the ground glass is upright and correct as to right and left. The finder has a picture area of about $1\frac{7}{8}$ by $3\frac{5}{8}$ inches. With the mattes fully open, the field corresponds approximately to that of a 35mm lens. With other lenses, adapters must be installed on the front of the finder to change the actual size of the image. This is necessary because the normal field of one lens might be greater than that of the viewfinder, whereas the normal field of another lens could be too small to be of practical value.

The focusing adjustment has a lock nut on top of the finder. A parallax adjustment bracket, at the rear of the finder, is fastened to the camera base. In order to open the camera door, it is necessary to release the parallax locking nut on top of the finder and to free the parallax adjustment bracket from the camera base. The finder can then be pivoted away from the camera door. In the BNC cameras, cams are provided for automatic parallax adjustment of the viewfinder when properly synchronized with the lens. See Figures V-12*a* to *c* and FINDERS.

viewpoint Basically, the relationship between the camera and the subject matter on a motion picture set or television stage. This all-inclusive term implies consideration of perspective, composition, and camera angles.

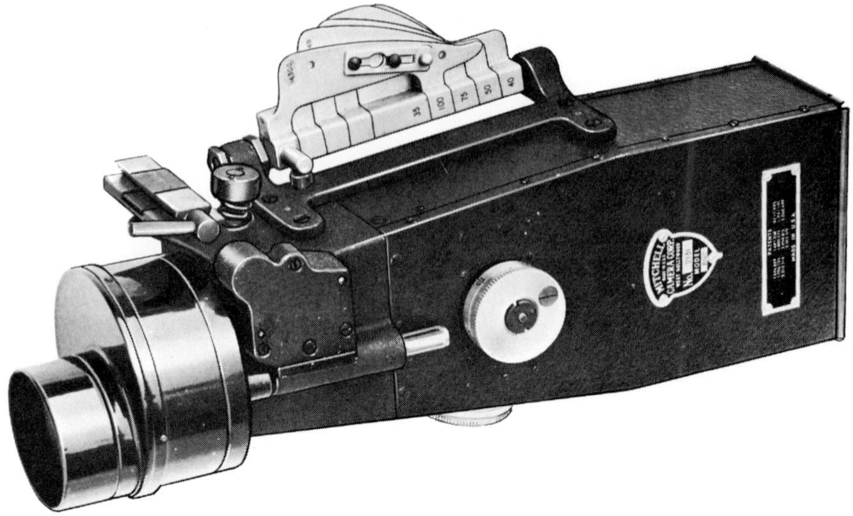

Fig. V-12*a* **The Mitchell BNC viewfinder.**

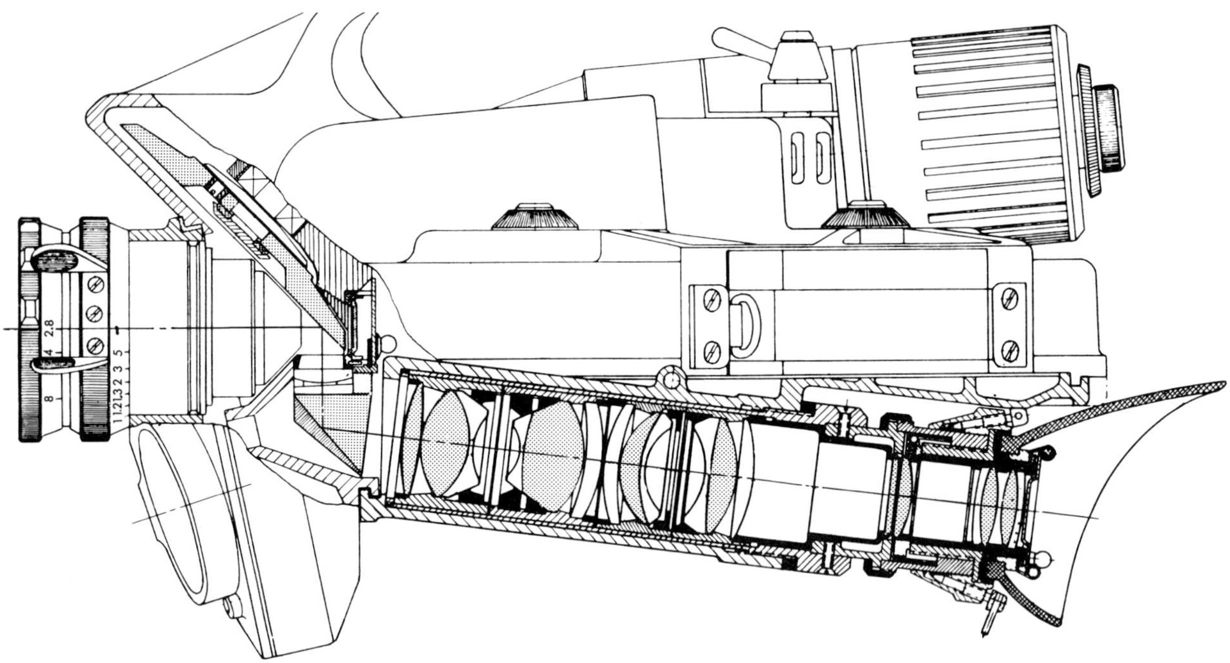

Fig. V-12*b* **Actual cross section of Arriflex 16 optical system.**

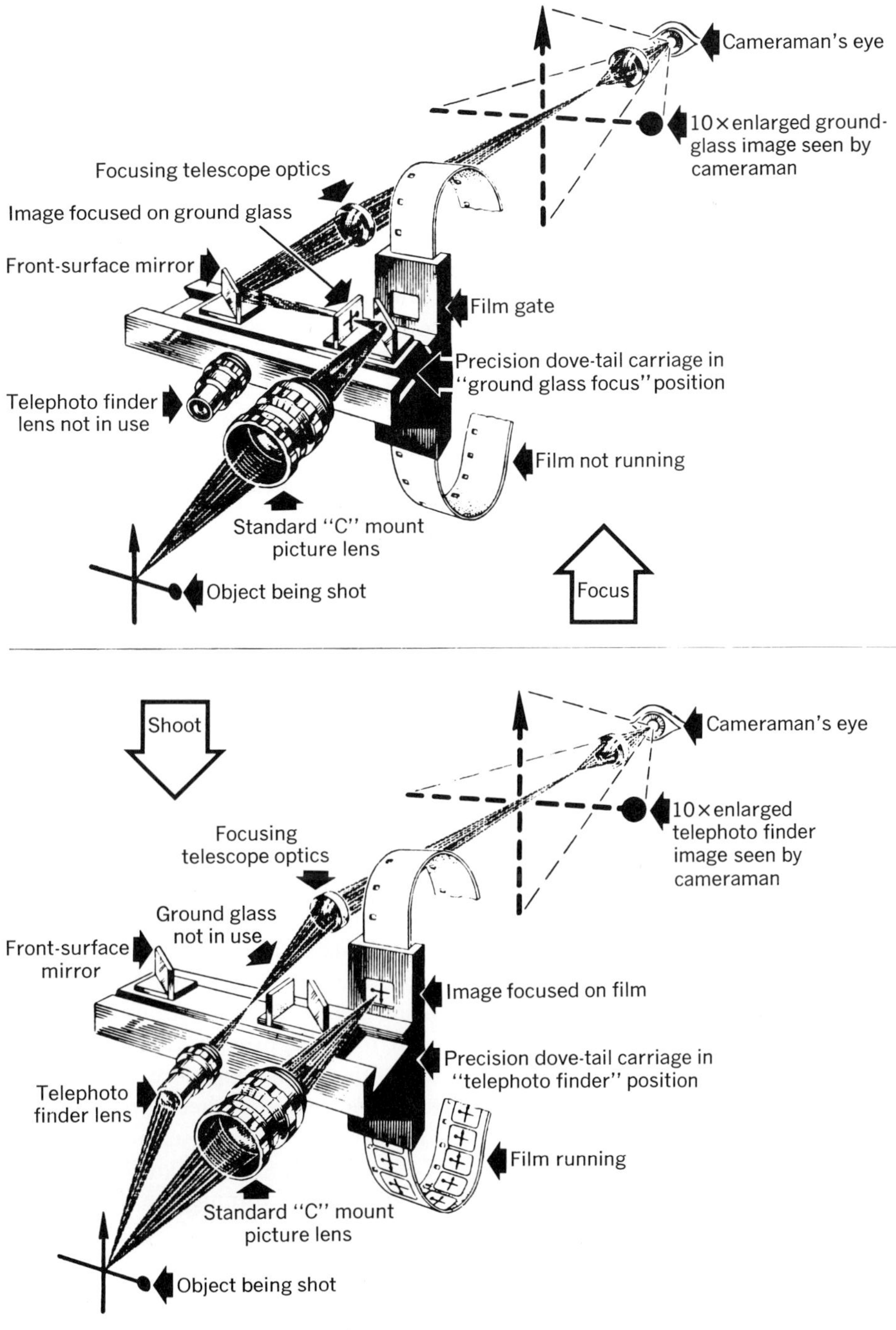

Fig. V-12c Auricon's full-frame reflex ground-glass focusing optical system.

(*Top*) In use for full-frame ground-glass focusing: Super 1200 reflex ground-glass focusing frame indicates the field covered by any focal-length lens at all distances. A 10× enlarged ground-glass image is seen by the cameraman, for needle-sharp critical focusing.

(*Bottom*) In use as telephoto finder during filming: No camera-weight shift on tripod legs when moving from focus to film-shooting position with the Super 1200. Camera body does not shift to focus, or shoot film, as optical-system dovetail carriage is only moving part.

viewing glasses Glasses designed to compress the visual range of light and dark areas in scenes into the photographic range. The number of discernible steps in a black-and-white scene can run as high as 2,000. The degree of contrast in a black-and-white print, by comparison, may show no more than about 20 steps or gradations. Viewing glasses, therefore, indicate to the cameraman the approximate results that will be obtained, under ideal photographic conditions, in the processed print.

Panchromatic viewing glasses for black-and-white film not only compress the visual contrast range to the photographic contrast range but also show the effect of various filters on the tonal values in the scene.

Placing a filter in front of the viewing glass and looking at the scene through both elements will change the color contrast of objects in the scene in accordance with the color of the filter. Through a viewing glass and red filter, all objects that are pure blue will appear to be black. Similarly, green objects will also appear to be black. If they are not pure colors, but are combined with white, they will appear darker in an amount equal to the density of their color. Red objects in the scene will, by comparison, appear lighter, depending on the purity of the color.

The monochromatic viewing glass for black-and-white film is deep yellow. The eye is more sensitive to yellow than to any of the other colors in the spectrum. Thus, the choice of yellow for the mono viewing glass seems quite natural since that color comes closer than any but blue to reducing highlights, shadows, and colors to a monotone value.

Since the monotone appearance of the scene creates the impression of overexposure, many cameramen prefer to interpret the lighting based on this visual method rather than on the possible photographic result, based on contrast, shown by the panchromatic viewing glass.

Viewing glasses for color film compress the contrast range from the visual to an approximation of the color contrast that can be expected with normal exposure. Since the color balance of the film determines the degree of filter correction, if any is needed, the viewing glass can only be used for determining the contrast range of the color film. It cannot be used with light-balancing or color-balancing filters as a means for determining the proper filter for a given scene.

vignette An image that gradually fades out near the edges of the picture frame, instead of being sharply defined by a rectangle (or other geometric shape), or by the shape of the subject itself, as in a silhouette. The effect can be achieved by positioning a specially prepared matte in front of the camera lens during the filming process. As an alternative, a soft-edge matte can be added optically to produce this effect.

Viscor filter The Viscor filter corrects the photronic cell response to match the standard luminosity curve adopted by the International Commission on Illumination. See Figure C-27 under COSINE CORRECTION FILTER.

VistaVision See WIDE-SCREEN PROCESSES.

visual squeeze A technique in which still photographs are exposed on the animation stand, on motion picture film, and the illusion of motion is created by zoom, pan, and tilt movements. During the filming process, the photographs are registered on the compound table of the animation stand (as in cel animation), and movement is generally achieved with stop-motion processes.

The visual squeeze is a relative newcomer to commercial television production compared with older and more established techniques. With this technique, more audio-visual information can be "squeezed" into a one-minute commercial than can be included in any other type of film presentation of similar length. The idea for the technique can be traced back to the animated flip-books where the illusion of motion was created by flipping a series of still pictures that were stapled together to provide a degree of registration. See Figures V-13*a* to *c*.

visual test film A film print containing a number of targets (or linear patterns) used for checking the focus and alignment of projection equipment. These targets, during projection, show picture unsteadiness (if present), ghost images, jumps, weaves, or a lack of sharpness in the projected film.

voice over Dialogue or narration, by off-stage or off-camera performers or announcers, which accompanies the picture information being projected or televised.

volume The level of sound during recording or reproduction.

VTC film-tape translator See ELECTRONIC EDITING.

VTR The abbreviation for videotape recorder.

VU In audio-frequency sine-wave measurements, a VU or volume unit indicates changes equal to one decibel in volume.

VU meters The meter in recording units that indicates, in volume units, the level of a recording or playback. The dynamic characteristics of these meters are similar to the response of the human ear. VU meters, calibrated in decibels, indicate power levels and are used mainly in test equipment and not for monitoring purposes.

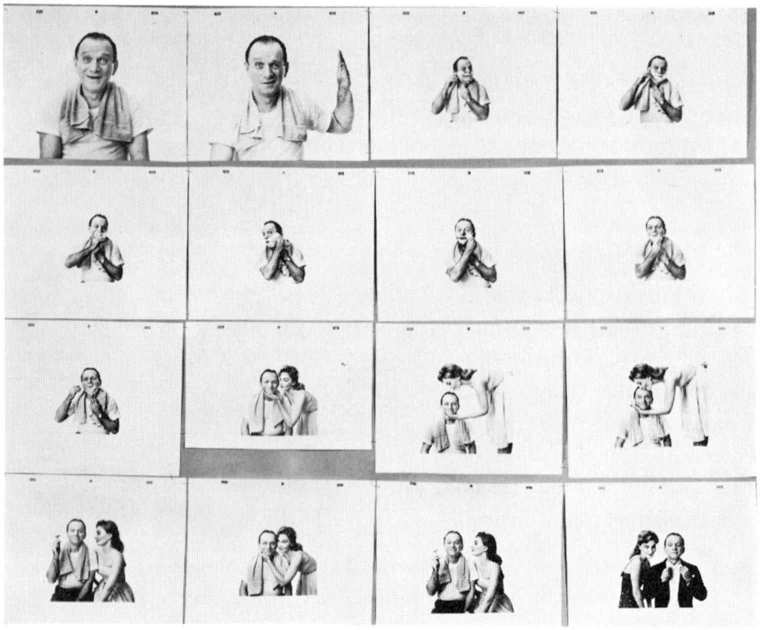

Fig. V-13*a* The visual squeeze technique. Instead of one still photograph (and camera movement), a series of still photographs depicting an action is often used to achieve additional movement. In the example here, a man is shown shaving. His body is cut out of a photograph and positioned on a cel. His head and arms are cut out of other photographs and mounted on other cels in registry with the cutout of the body. As the cels containing the heads and arms are positioned over the cutout of the body and exposed on successive frames of film, a flip-book type of movement is achieved. Many unusual effects can be achieved with this type of photographic technique.

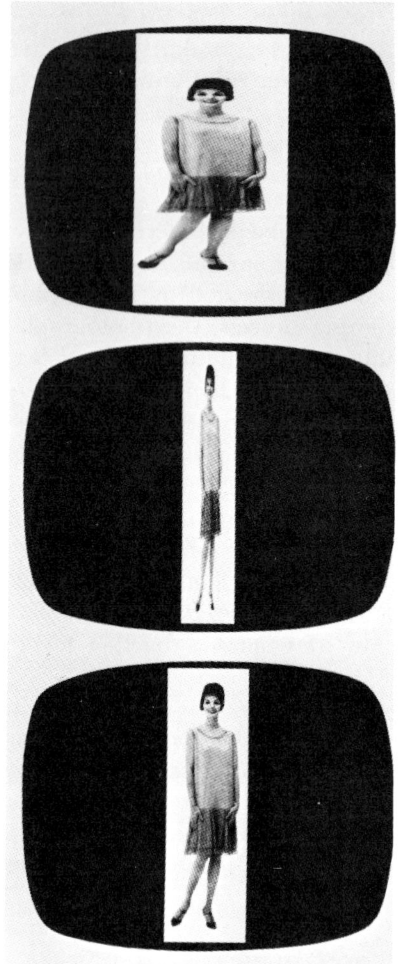

Fig. V-13*c* Won't shrink—won't stretch—always stays the same. The commercial that introduced the most imitated television commercial technique. Produced for the Chemstrand Corporation, this is one scene from that commercial—the visual squeeze technique, literally.

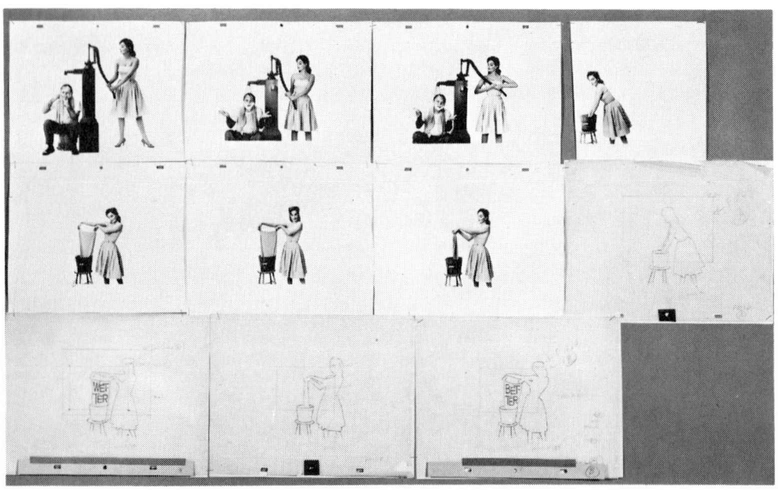

Fig. V-13*b* The illusion of motion with cutout still photographs.

W

washing The process of removing any of the developing or fixing chemicals remaining on the emulsion or film support after each of the several stages of the processing procedure. The washing process prevents residual compounds from subsequently damaging or discoloring the film image.

The efficiency of the process (actually the removal of the silver halide fixer complexes) is measured in terms of the ability of the system to remove these salts according to the residual hypo content of the film after it has been dried.

Normal washing techniques in processing laboratories ensure a residual hypo content not greater than 0.3 milligram per square inch of film. This ensures a film permanence of ten to fifty years. Any residual silver compounds will, in time, produce brown or black spots in the picture and, in some cases, cause fading. Unfortunately, time takes its toll, and even under ideal storage conditions, physical changes in the film (such as shrinkage) may make it impossible to project old prints. Periodic duping is, perhaps, the best method for extending the life of the film.

wave effects Waves, from lazy ripples to the abnormal crests and troughs that accompany hurricanes, can be produced in studio tanks without difficulty.

The subsurface displacement of water, by three-bladed propellors or by machine-driven paddles, produces a controllable agitating action that simulates the exact type of wave condition needed during the filming process. Especially useful for miniature sets and subscale photography, the results are quite adequate when intercut with other aquatic scenes.

waveform The shape of sound waves or electromagnetic radiation, either produced on an oscilloscope or drawn for purposes of analysis as on a graph.

wavelength The measurable distance between points of like phase in a light wave. Wavelengths, like frequencies, are variable. A wavelength is thus the distance between corresponding points of any one wave—regardless of the frequency.

Different colors have correspondingly different wavelengths ranging from approximately 0.4 micron for violet to 0.7 for red at the other extreme of the visible spectrum.

Since a micron is 1/1,000 millimeter, with approximately 25 millimeters to the inch, an estimate of the number of light waves that enter the eye each second is a task for a computer. With each color having an average of 50,000 wavelengths to the inch, the eye receives over 500 million million or 589 trillion, 248 billion impulses every second. See SPECTRUM, GREAT.

wave-retardation plates Also called **wave plates**, **retarders**, and **phase shifters**. These plates are used to change polarization forms of polarized monochromatic radiation, such as laser beams. A wave-retardation plate divides incident polarized radiation into two components, changes the relative phase between the two components by retarding one more than the other, and then combines the two components so that the polarization form of the emerging radiation is different from that of the incident radiation. These special plates are available through Barnett Associates in Hauppauge, New York.

weave An undesirable, lateral movement of the film caused by improper threading or faulty alignment of the transport mechanism in the projection equipment.

wedge spectrogram Wedge spectrograms, for measuring the color sensitivity of film, are prepared by exposing a length of film to a light spectrum through a neutral-density wedge. The wedge is completely black (opaque) at one end and transparent at the other. The high points of the image, on the scale of the spectrograph, are in direct proportion to the log sensitivity of the emulsion.

weighting characteristic The shaped response-frequency characteristic of a measuring device used to produce more realistic indications of the subjective effects than are obtained with unweighted (flat) measurements. See SIGNAL-TO-NOISE RATIO, UNWEIGHTED and WEIGHTED.

Weston speed A numerical system for rating the speed of an emulsion. Based on tests made with Weston exposure meters, the ratings can be related to other numerical systems used for this purpose. See EXPOSURE-INDEX RATINGS.

wet-gate printing A process used with either step-contact or continuous-contact printers wherein the negative and the raw stock are immersed in a liquid (tetrachlorethylene) in order to minimize the effects of surface scratches or abrasion marks. Modifications of the film path in the transport mechanisms place the negative and raw stock in close contact as immersion takes place immediately preceding and during the interval of exposure. The matching refractive indices of the liquid make scratches and abrasions invisible.

white balance A television camera control that varies the white level of color channels individually. See REFERENCE WHITES.

white level A television camera control that selects manual or automatic white-level control. In the manual position, the white-level reference pulse is controlled. In automatic, the control sets the level to which peak white is held. See REFERENCE WHITES.

wide-angle lenses Short-focus lenses that are especially valuable because of their ability to force perspective. In confined areas, these lenses help make a motion picture or television stage appear larger than it actually is. Distant objects, moving toward or away from the camera, appear to move faster than when long-focus lenses are used.

To prevent undesirable diffraction around the edges of the lens diaphragm, the cameraman should avoid stopping down wide-angle lenses of short focal length to very small apertures.

Wide-angle lenses can be either symmetrical or asymmetrical with respect to the component lens elements. In both types, the front elements curve broadly in order to gather the maximum amount of light from the widest possible area. Technically, a lens must have an angular field of view in excess of 60 degrees in order to be classified as a wide-angle lens. See Figure W-1.

Fig. W-1 An f/1.4 100-degree wide-angle lens designed by a program-controlled electronic computer. (*Carl Zeiss, Oberkochen.*)

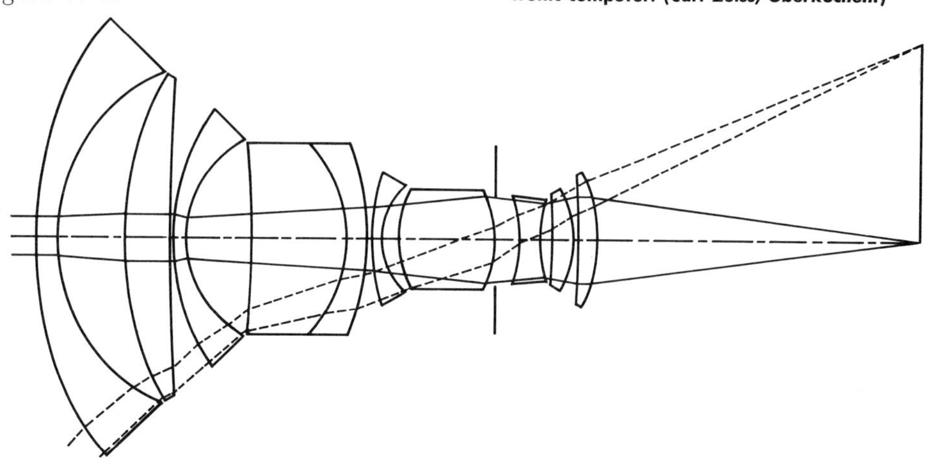

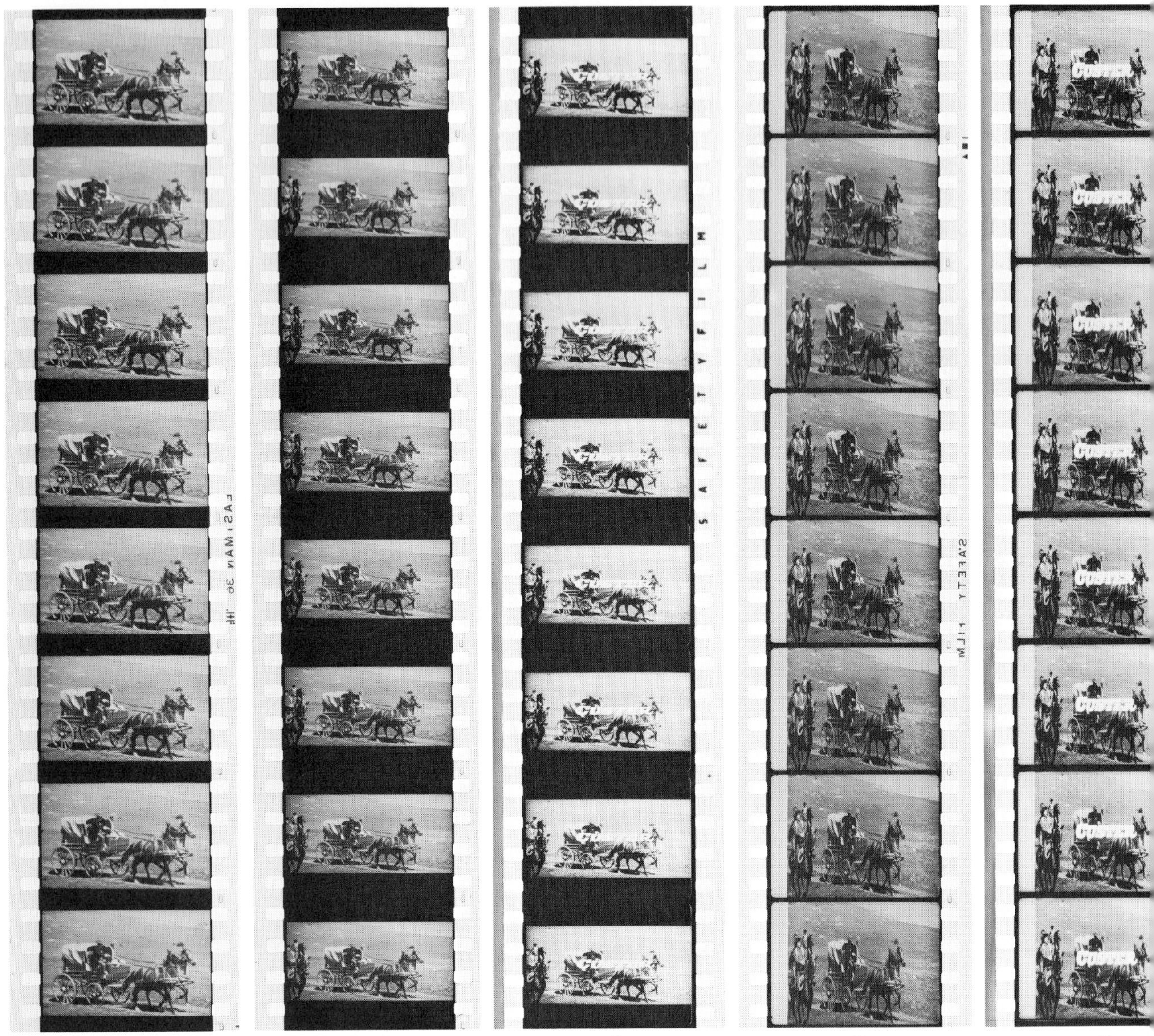

A B C D E

Fig. W-2 Wide-screen processes. (*A*) Unsqueezed to full screen. (*B*) Unsqueezed to wide screen. (*C*) Unsqueezed to wide screen with title. (*D*) CinemaScope. (*E*) CinemaScope with title.

wide screen Projection proportions larger than the standard 1.33 : 1 aspect ratio of the 35mm standard frame.

wide-screen print A print whose projection proportions are larger than the standard 1.33 : 1 aspect ratio. Wide-screen prints, unlike anamorphic prints, do not necessarily contain squeezed images.

wide-screen processes Any of the filming processes that are used to produce motion picture prints in which the projected images are larger than the 1.33 : 1 aspect ratio of the standard 35mm film frame. See Figure W-2.

CinemaScope. The CinemaScope wide-screen image is produced by using standard-width 35mm film in conjunction with an anamorphic lens designed by a French physicist, Dr. Henri Chrétien. The afocal or telescopic attachment, positioned in front of the regular camera lens, magnifies the image to twice the normal size in the horizontal meridian. The amount of distortion in the perpendicular meridian is inconsequential.*

The compressed images in the 35mm 4-perforation negative can either be printed on standard 35mm film and unsqueezed during projection or released as a 70mm unsqueezed print with 5 perforations per frame. The 35mm prints accommodate four magnetic or magoptical sound tracks.

The 35mm contact print has an aspect ratio of 2.35 : 1. The aspect ratio for the 70mm flat print is 2.2 : 1. See Figure W-3.

CinemaScope 55. CinemaScope 55, a variation of the basic CinemaScope system, uses 55mm film to produce a negative image that is twice the size of the 35mm CinemaScope image. The amount of anamorphic squeeze is the same for both film sizes, 2 : 1.

Although the 55mm negative has 8 perforations for each frame of film, the positive print is projected at the 24-frame-per-second rate, each frame being advanced by 6 perforations during projection. Prints are also made optically, on 35mm film, for subsequent projection on standard equipment. The aspect ratio for CinemaScope 55 is 2.35 : 1.

Cinemiracle. With an aspect ratio of 2.33 : 1, the Cinemiracle process uses 35mm film during the photographic process. The three synchronized cameras used for this wide-screen process are mounted on a single head in a position similar to the arrangement used for the Cinerama production.

The camera setup, with mirrors positioned in front of the outer cameras, is repeated during the projection process. Sharing a common booth, the three synchronized projectors direct the images to deeply curved, seamless screens.

*The camera aperture is 0.868 by 0.735 inch as compared with the ground-glass markings of 0.839 by 0.715 inch. The latter figures are the same as the dimensions for the projector aperture.

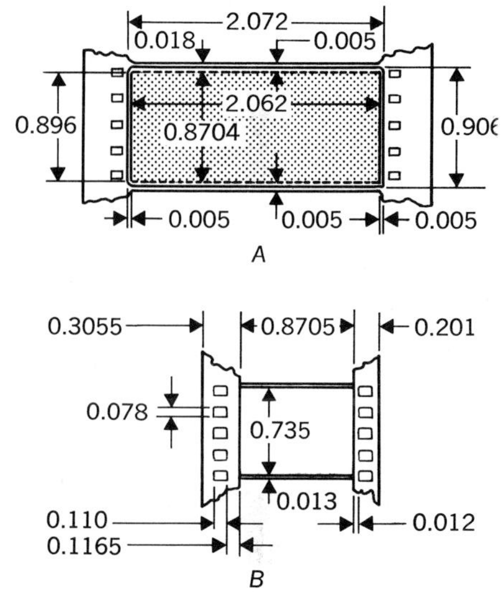

Fig. W-3 **CinemaScope.** (*A*) Usable area for CinemaScope—Magoptical reduction on 35mm film. (*B*) Head as seen from emulsion side. CinemaScope—Magoptical camera aperture. Print down from 65mm Todd-AO negative, 90 feet per minute.

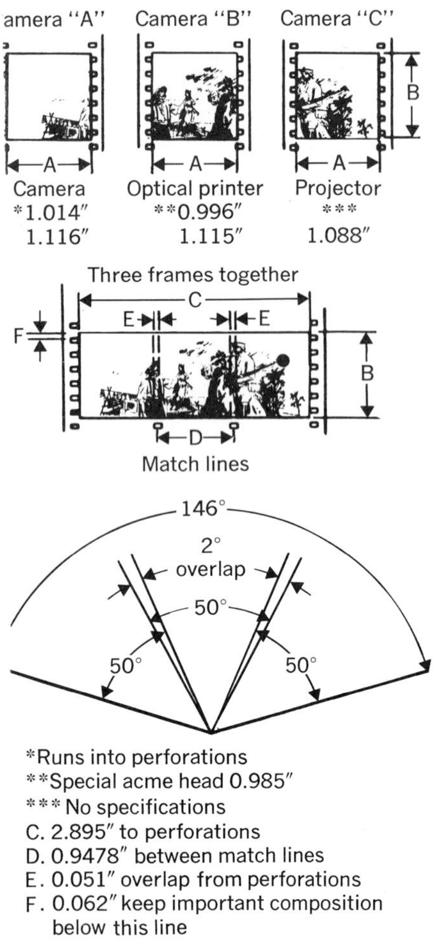

Fig. W-4 Cinerama. Camera aspect ratio 2.59:1. 27mm lens. Angular field coverage—camera aperture: horizontal: 146 degrees; vertical: 55½ degrees.

With 6 perforations for each frame of film, the three prints are projected simultaneously at the rate of 26 frames per second. Of the seven magnetic tracks that complement the picture information, five are directed to speakers behind the screen. The sound, on the remaining two tracks, is directed to side speakers in the theater. Compare CINERAMA.

Cinerama. This was first used during World War II for training purposes. Subject matter for the Cinerama production is photographed on standard-width 35mm film with 6 perforations per frame. The images are recorded by three cameras.

After processing, each of the three prints is projected simultaneously, by separate projectors, at the rate of 26 frames per second or 146.25 feet per minute. The three prints, focused on deeply curved screens, are blended into a single composite picture.

Unlike standard screens, the Cinerama screen consists of more than 1,200 vertical, overlapping strips of perforated material. As many as seven magnetic sound tracks, on separate lengths of 35mm full-coat magnetic film, are fed to the speakers positioned behind the screen and on both sides of the theater.

The aspect ratio for the Cinerama production is 2.59 : 1. The aspect ratio, as measured from the middle projector of the three in operation, is 2.06 : 1. See Figure W-4.

Circarama. A *wide* wide-screen process with a 360-degree picture, one that would completely encircle the viewer, is a source of amazement to Disneyland visitors who are surrounded by a picture with those unusual dimensions. The appropriately named Circarama process uses 11 cameras during the filming process and 11 rear-screen projectors that focus the images on as many screens. Viewers, however, including those with exceptional peripheral vision and a swivel chair, would have to see a picture twice in order to enjoy its contents fully. See Figures W-5*a* to *d*.

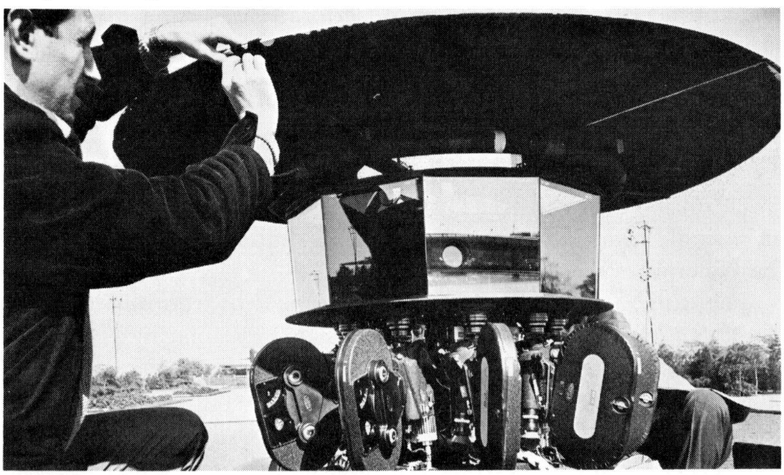

Fig. W-5*a* Circarama.

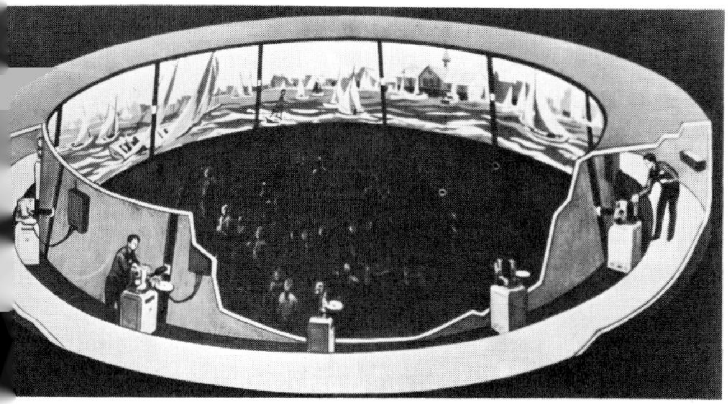

Fig. W-5*b* Walt Disney presents Circarama. (© *Walt Disney Productions.*) Artist's section drawing showing Circarama theater setup. The circular theater, 142 feet in circumference and with screen 9 feet high, is divided into 11 segments, opposite a similar number of projectors casting the separate but coordinated images through slight apertures, as seen in the drawing. The whole operation is integrated so that the viewer gets an impression of a 360-degree view extending to the horizon in every direction and in constant flow of scene and motion of people and objects as photographed by 11 synchronized cameras in battery. Spectators stand in the stationary central area.

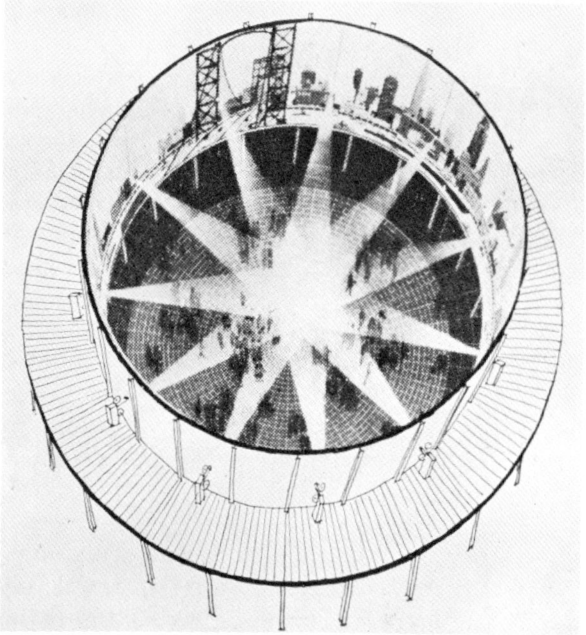

Fig. W-5*c* An artist's projection of the Circarama pavilion at the New York World's Fair. The panoramic scope of the 11 projectors cast their adjoining images on as many screens. Projectors are seen outside the rotunda casting their light beam against the opposite screens, with spectators in the center. One of the top attractions of the fair, the show inside is seen on a 360-degree screen.

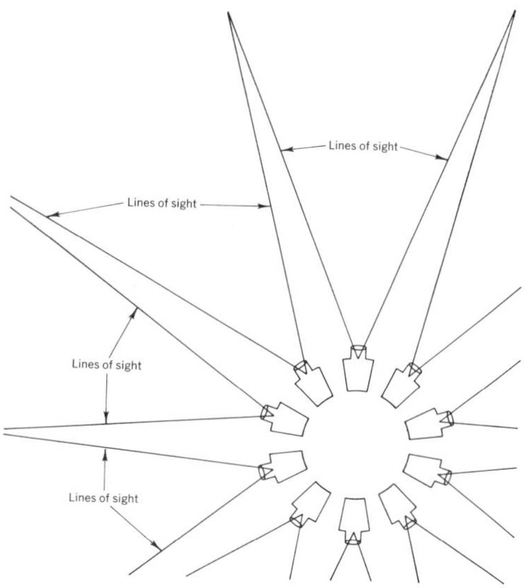

Fig. W-5*d* Lines of sight. This rough sketch shows lines of sight if mirrors were not used on the rig, and if the cameras pointed straight out. Note that if the lines of sight meet at infinity, they do *not* meet at closer distances. Therefore, an object passing close up from one field to another would not be continuous.

Dimension-150. Approximating the widest range of the human eye, Dimension-150, with its 150-degree arc sweep, is one of the newer wide-screen processes. Like the Todd-AO pictures, the images for Dimension-150 are recorded on 65mm film by wide-angle lenses; 50, 70, and 120 degrees, and the 150-degree bug-eye. The widest picture ever produced, on 70mm prints, can be projected on a deeply curved screen or reduced optically to the 35mm width and projected on flat screens with conventional equipment. Practically distortion-free, and with its entire width in perfect focus, the 150-degree picture is accompanied by six magnetic tracks similar to the arrangement used for the Todd-AO presentation. See ASPECT RATIO. Compare TODD-AO.

MGM Camera 65. MGM Camera 65, with an aspect ratio of 2.5 : 1, also uses a 65mm negative with 5 perforations for each frame of film. First used in 1957, the wide-screen pictures are produced by a Panavision lens that compresses the image. Unsqueezed during projection by a complementary 1.33 anamorphic lens, the prints, on either 65 or 70mm film, are projected on screens that have a slight curvature. While the 70mm release prints have the six magnetic tracks printed on the same length of film, the tracks for the 65mm positive are printed on a separate length of film. Prints, for CinemaScope projection, are produced from the 65mm negative by using anamorphic lenses with optical reduction printing processes. The tracks for the standard-width prints may be optical, four-track magnetic, or magoptical.

Naturama. Naturama, a wide-screen process with an aspect ratio of 2.55 : 1, uses an anamorphic lens and standard equipment for producing its images. As with other image-compressing lenses, the picture information is compressed, in the horizontal meridian, during the filming process and restored to its normal dimensions by complementary lenses during projection. Conventional equipment is used for filming and projection.

Panavision 35. Panavision 35 uses a 2-to-1 anamorphic lens to produce a squeezed negative on standard 35mm film. The camera aperture for the filming process is 0.868 by 0.735 inch. The ground-glass markings, by comparison, are 0.831 by 0.715 inch. The camera negative has a 2 : 1 squeeze ratio. The squeezed contact print, with a screen-aspect ratio of 2.35 : 1, requires unsqueezing during projection. The 70mm optical print, with 5 perforations per frame, has an aspect ratio of 2.2 : 1. See SUPER PANAVISION 70 and ULTRA PANAVISION 70.

ScanoScope. The ScanoScope wide-screen process, with an aspect ratio of 2.55 : 1, uses its own ScanoScope lens for producing squeezed images on 35mm film. The horizontally compressed images are restored to their normal proportions by anamorphic lenses during projection.

Super Panavision 70. The basic dimensions for the Super Panavision 65–70mm production, with respect to ground-glass and finder markings, are similar to those used in the Todd-AO wide-screen systems. The camera aperture for the 65mm unsqueezed picture with 5 perforations to the frame is 2.072 by 0.906 inches.

The 70mm print, with two of the six magnetic sound tracks positioned inside the perforations, is 1.913 inches wide and 0.868 inch high. The aspect ratio is 2.2:1.

The projector aperture for the 35mm print is 0.839 inch wide by 0.715 inch high. The aspect ratio for the 35mm squeezed picture is 2:1. See Figure W-6.

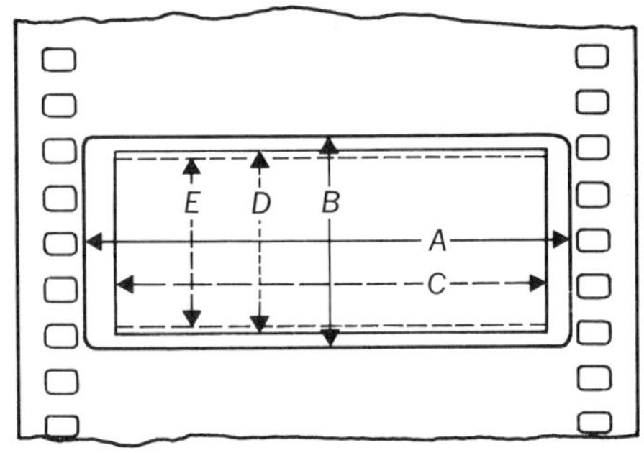

Fig. W-6 Super Panavision film dimensions.
A 2.072 inches } camera aperture
B 0.906 inch
C 1.913 inches } 70mm composition, 2.2:1
D 0.868 inch
E 0.814 inch
35mm, 4 perforations, 2:1 squeezed print in Panavision 35
Also CinemaScope from Todd-AO and Super Panavision 70

SuperScope. Although the images for the SuperScope process are recorded on standard 35mm film, the picture area extends from the sprocket holes on one side of the film to the sprocket holes on the other side. Unlike other wide-screen processes in which the squeezed image is produced during the photographic process or optically, in a dupe negative, the SuperScope image is compressed during processing (by Technicolor), and restored to its normal proportions by a deanamorphic lens during projection. Projected with conventional equipment on curved screens at the standard 24-frame-per-second rate, SuperScope has an aspect ratio of 2:1. In SuperScope-235, a variation of the system, the projected picture is somewhat larger and has an aspect ratio of 2.35:1.

Super Technirama 70. Technirama, the basic system, and Super Technirama 70 are similar in practically all respects—with the exception of the format used for release prints. The prints made from the Technirama negative, and those produced by filming with 65mm cameras and printing on 70mm film, are compatible. The Technirama negative, on 35mm film, has 8 perforations per frame (horizontal) and a squeezed image with a 1.5:1 ratio. The images on the 35mm print have a 2:1 squeeze and an aspect ratio of 2.35:1. The print for Super Technirama 70, on 70mm film, has 5 perforations per frame and an aspect ratio of 2.21:1. See TECHNIRAMA.

Technirama. The Technirama system uses a standard 35mm film that travels through the camera in a horizontal direction, right to left. The compressed or squeezed image, in the double-frame negative, contains approximately 50 percent more picture information than could normally be exposed in the standard-width 35mm frame. The negative image, rotated 90 degrees, is compressed additionally by an anamorphic lens in the optical printer. The 1.5:1 squeezed negative image along with the additional compression it receives in the printer (1.33) provides a total squeeze ratio of 2:1.

The 35mm squeezed release print has an aspect ratio of 2.35:1. Unsqueezed release prints, on 35mm film, have an aspect ratio of 1.85:1.

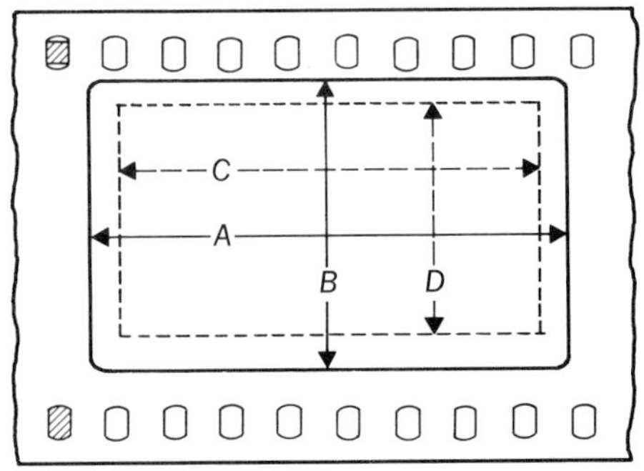

Fig. W-7 Technirama film dimensions.
A 1.486 inches } camera aperture
B 0.992 inch
C 1.312 inches } composition area
D 0.837 inch

The camera aperture for the Technirama process is 1.496 by 0.992 inches. The ground-glass markings, representing the field for the 2.35:1 aspect ratio, are 1.312 by 0.837 inches. See Figure W-7 and SUPER TECHNIRAMA 70.

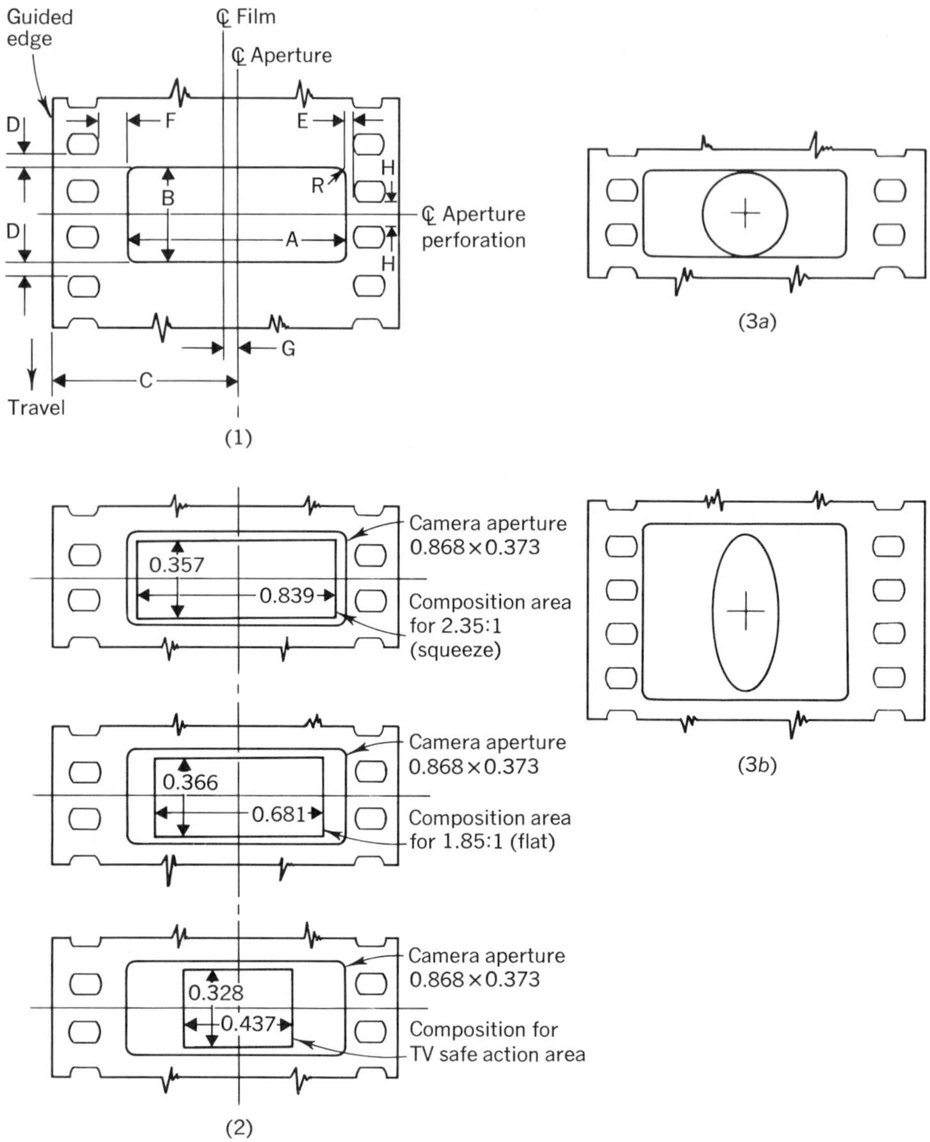

Fig. W-8a Techniscope system.

(1) Relative position of standard Techniscope camera aperture (as seen when looking through lens toward the film) and aperture dimensions. (*A*) 0.868 inch. (*B*) 0.373 inch. (*C*) 0.738 inch. (*D*) 0.0568 inch (refers to B & H perforations with 0.1866 pitch). (*E*) 0.026 inch. (*F*) 0.115 inch. (*G*) 0.0495 inch. (*R*) 0.03 inch approx. (*H*) = 0.113 inch.

(2) Techniscope camera aperture dimensions for three different picture formats, and the positions of same within the standard Techniscope picture area of 0.868 by 0.373 inches.

(3) Comparative picture areas of (*a*) camera negative and (*b*) release print of the Techniscope wide-screen process.

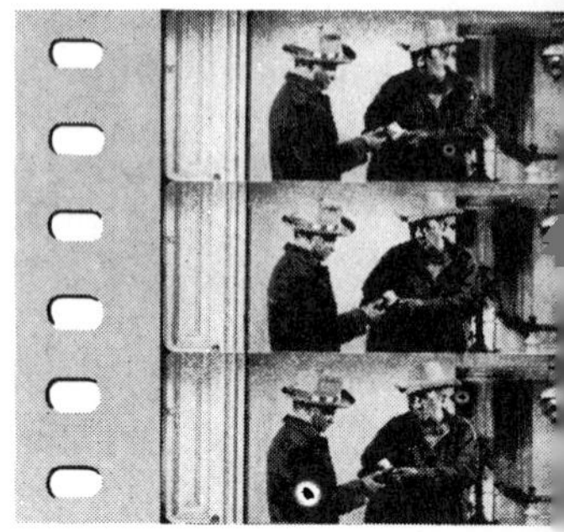

TECHNISCOPE FRAME as taken in car

A

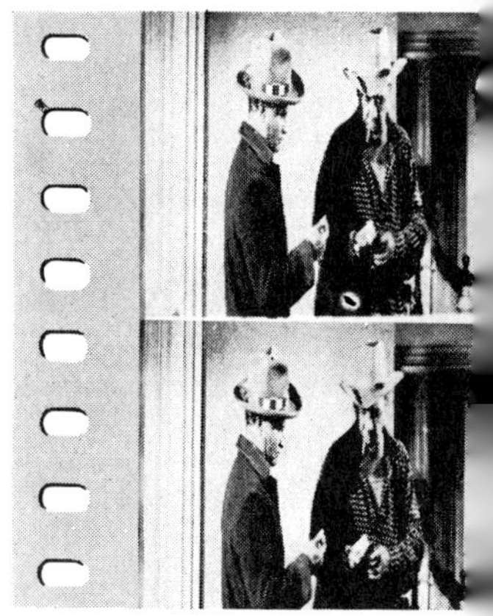

PROCESSED FRAME ready for projec

B

Fig. W-8b Techniscope frames. (*A*) As taken in camera. (*B*) Processed ready for projection.

Techniscope. A wide-screen system introduced by the Technicolor Corporation in 1963. Designed as an economy measure, it uses standard 35mm film during the photographic process.

Each frame in the print made from the original negative is only two perforations high and occupies an area half the height of a normal film frame. The prints match the aspect ratios of most 35mm anamorphic systems including CinemaScope and Panavision. The release prints made from the Techniscope negative are enlarged and printed in squeezed (anamorphic) format.

Mitchell and Arriflex cameras provide two-pin pulldown movement which is needed for the Techniscope process. See Figures W-8*a* and *b*.

Todd-AO. Most of the wide-screen processes, with the exception of CinemaScope 55, use standard-width 35mm negative in a single- or multiple-camera setup with the film traveling either horizontally or vertically through camera or projector. Todd-AO, the first of the wide-screen processes to expose its images on a negative film larger than the standard 35mm width, was first introduced in 1955. The single camera used for filming is complemented by a single projector, both specially designed to accommodate the wider 65mm negative and subsequent 70mm print. Lenses, specially designed for this wide-screen system, have a horizontal coverage of 37, 48, 64, and 128 degrees. The 128-degree lens is usually referred to as a *bug-eye*.

With five sprocket holes for each frame of film, and an aspect ratio of 2.2:1, the 70mm composite prints are projected on deeply curved screens at the standard rate of 24 frames per second, and a nonstandard 112.5 feet per minute.

Of the six magnetic sound tracks that accompany the picture, three tracks appear on each side of the film frame. One track is positioned on each side of the picture; the others are located outside the sprocket holes. The sound is directed to five loudspeakers behind the screen and to auxiliary speakers in other parts of the theater. See Figure W-9 and Table W-1.

Ultra Panavision 70. The 65mm cameras used for producing the unsqueezed, wide-screen pictures for Ultra Panavision 70 are similar to those used for the super Panavision and Todd-AO systems. The camera aperture permits exposure of a picture area that is 2.072 inches wide by 0.906 inch high. The camera lens has a squeeze ratio of 1.25:1.

The 70mm print (squeeze ratio 1.25:1), has to be unsqueezed during projection. In order to be consistent with Panavision and CinemaScope standards, 35mm prints require additional squeezing, (1.6:1), and have an aspect ratio of 2:1.

The Ultra Panavision frame, during the filming process, can be squeezed to provide an aspect ratio of 2.7:1. Camera apertures would then permit exposure of an area 2.050 inches wide by 0.870 inch

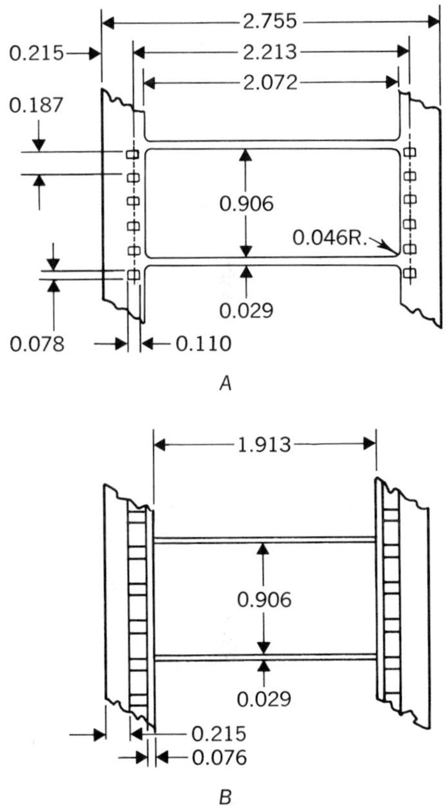

Fig. W-9 Todd-AO wide-screen process. (*A*) Camera aperture. (*B*) Composite usable area.

TABLE W-1 Todd-AO Camera Lens Speeds*

Lens focal length, millimeters	Lens speeds	Horizontal angle, degrees	Vertical angle, degrees
28	*f*/3.6 (*T*/4)	87	45
40	*f*/3.2 (*T*/3.5)	66	32
50	*f*/2.4 (*T*/2.6)	55	26
60	*f*/2.4 (*T*/2.6)	47	22
75	*f*/3.2 (*T*/3.5)	39	15
100	*f*/3.2	29	13
150	*f*/3.2	20	9
	Zoom Lenses		
60–150	*f*/3.5 (*T*/4)	47–20	22–9
100–300	*f*/4.5 (*T*/4.8)	29–10	13–4

*65mm: spherical (no squeeze).
70mm: release print (2.21/1).
Camera aperture: 2.072 by 0.906 inches.

high. This picture can be printed on three separate lengths of 35mm film and projected onto deeply curved screens in a manner similar to that used for the Cinerama production. The film is projected at the standard sound speed rate of 24 frames per second with a 5-perforation pulldown.

Ultra Semi-Scope. The Ultra Semi-Scope system is, in some ways, similar to the Techniscope system. Registered by the patent agency in Japan in 1965, the system projects a one-frame picture in CinemaScope format—2.35 : 1.

Modified camera and projection equipment make possible compatible use of either 4- or 2-perforation process films. Optical sound is projected at one-half the usual film speed.

VistaVision. VistaVision, like Technirama, uses 35mm film that travels horizontally through cameras and projectors designed for this type of wide-screen process.* The double-frame images can subsequently be reduced optically on standard 35mm film, for projection on conventional equipment. The optical sound track, at the left side of the frame in 35mm film, is positioned at the top of the double frame in the VistaVision print.

Projected horizontally on slightly curved screens, the film travels similar to the Technirama print, at a rate of 24 double frames per second, or a linear speed of 180 feet of film per minute. Single-frame prints obtained with optical processes are projected at the normal rate of 24 frames per second, or 90 feet per minute. Release prints, on standard 35mm film, can be projected with conventional equipment. The aspect ratios for these could be 1.66 : 1, 1.75 : 1, or 1.85 : 1, depending on the projection aperture. See Figure W-10.

*Paramount Pictures.

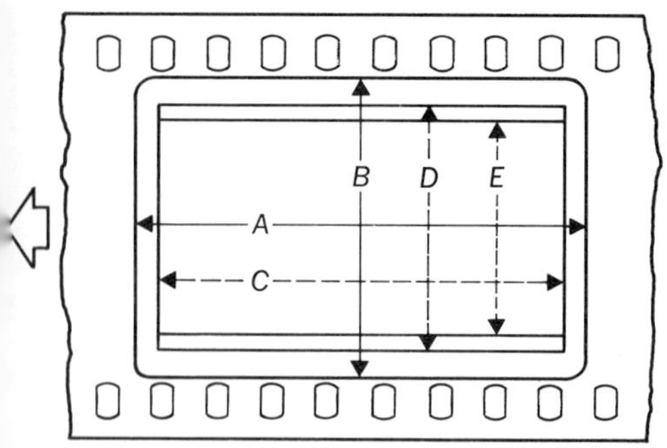

Fig. W-10 Vista Vision.
A 1.485 inches, *B* 0.991 inch } camera aperture 1.50:1
C 1.3365 inches, *D* 0.805 inch } composing area 1.66:1
C 1.3365 inches, *E* 0.723 inch } ground-glass markings 1.85:1
The outline arrow at the left indicates the direction of film travel.

wide-screen projection The screens used for projecting wide-screen prints are usually curved in order to shorten the distance of the throw on each side of the screen. The curvature assures a uniformly focused picture over the entire screen surface.

wild motor Also called **variable-speed motor**. An adjustable motor that can be regulated to operate at different speeds or predetermined settings. It is not recommended for sync-sound filming.

wild track In those cases where the sound track that is to be synchronized with the picture information is not available, the editor will sometimes have a track recorded (dialogue, music, or both) that approximates the content and timing of the missing or nonexistent track. This procedure is followed in post-sync productions, or where the editor (for any of several reasons) must give up the track for any length of time. Wild tracks are, basically, time-saving devices.

wind effects Wind effects, from a gentle zephyr to a howling hurricane, are created by fanlike devices just out of camera range. Commercially available, the fans can be as large as the giant-sized propellers used on airplanes.

The windhowler, a companion accessory, may be used live or as a means for recording a variety of sound effects from a gentle breeze to a whistling winter wind. The latter effect is especially useful for drowning out the anguished cries of the critics who complain about the loudness of television commercials.

During the filming process, studio microphones must be shielded to protect the sensitive instruments from mechanical noises. Known as wind screens, the wire-mesh strainer-type coverings use fine silk to reduce the level of ambient sound and protect the vibrating diaphragm in the sensitive microphones. See Figures W-11*a* and *b*.

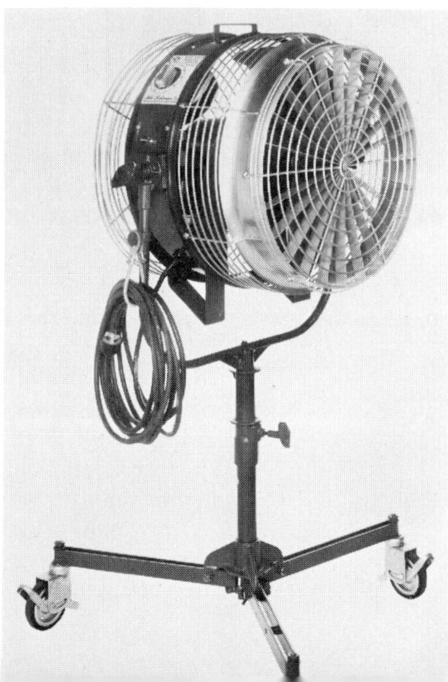

Fig. W-11*a* **Mole-Richardson type 1971 wind machine Moleffect.**

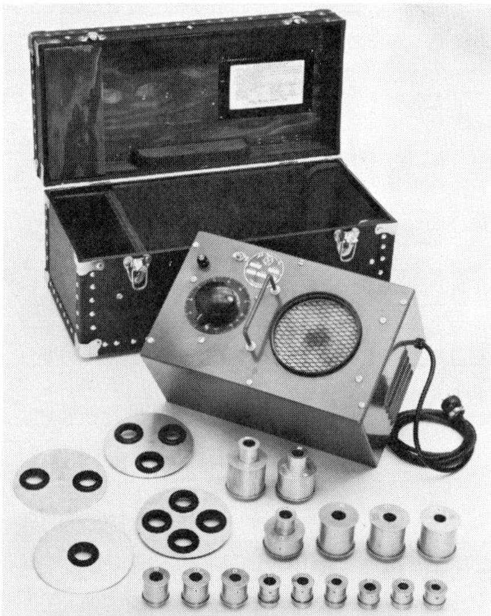

Fig. W-11*b* **Mole-Richardson type 2282 wind-howler Moleffect kit.**

window filters Window filters for controlling natural light coming through a window or changing daylight to tungsten balance are produced in the form of gelatin sheets or rolls of acetate. These sheets (MT-2) or rolls (Orange 85) can be cut to size and made to fit the individual panes of glass or simply hung over the outside of the window.

wipe An optical effect, made possible by matteing, in which one scene is gradually replaced by another scene over the same length of film. While many stock wipe patterns are available at every optical studio, special wipes to fit specific needs can be animated, photographed on the animation stand, and readied for optical combination in short periods of time. Unlike the changes in exposure that take place during a cross-dissolve (where one scene gradually fades out as the new scene fades in), both scenes in a wipe receive full exposure during the entire transition. See Figures W-12*a* and *b*.

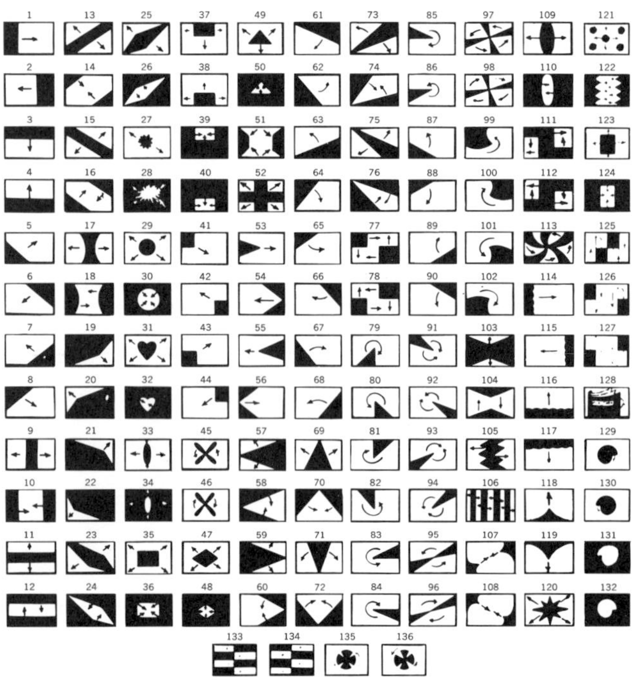

Fig. W-12*a* Typical wipe patterns. (*Creative Opticals, Inc.,* New York.)

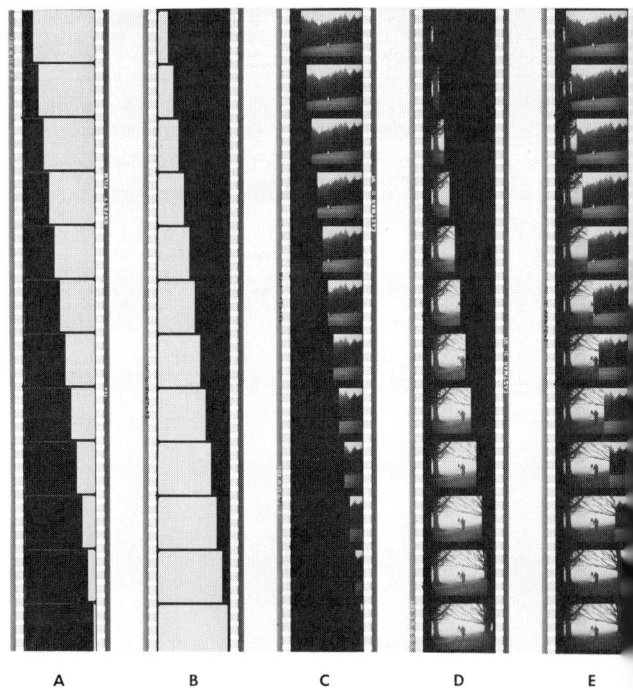

Fig. W-12*b* The wipe effect. Columns *A* and *B* show the mattes, on high-contrast film, that are needed to produce the composite effect in column *E*. In the first stage of the optical process, the mattes in column *A* are photographed with the outgoing scene in column *C*. The shutter is then closed and the film is run back to the frame where the effect begins. With the shutter again in its open position, the mattes in column *B* are exposed with the frames from the incoming scene in column *D* to complete the wipe effect.

Wood's lamp An ultraviolet lamp named after its developer. This source of black light has three basic components: a glass housing, vaporized mercury, and a nickel oxide filter layer. The nickel oxide filters out visible radiation, leaving the ultraviolet rays. Produced by the glowing arc within the housing, they are in the form of a violet glow.

These lamps must be allowed to cool sufficiently, after being turned off, before the arc can be reactivated.

working aperture The largest lens setting that can be used to produce a properly exposed image. Although larger apertures are available to the cameraman, the negative exposed at those apertures would show proportionate losses in definition. See EFFECTIVE APERTURE.

working distance Technically, the area within the camera between the film and the rear surface of the lens.

Less technically, working distance is also described as the distance from the camera to the subject. It is an important consideration in close-up photography because the cameraman will often find limitations as to how close he can approach a subject or how far away he can go from it.

In medical photography, for example, the cameraman must keep his equipment outside the working range of the surgeon and yet cover a relatively small field. In other situations, moving parts of a machine or high temperatures may make a close approach dangerous or impossible. In these cases, a long working distance is needed which also has the advantage of providing more space for the positioning of lights. Long working distances also minimize the problem of shadows from either the cameraman or the equipment. On the other hand, crowded quarters or smoke or fog in the air may require a short distance in order to obtain a good image on the film.

The same magnification can be obtained from completely different working distances by changing the focal length of the lens, or—expressing it in another way—the cameraman can cover the same area from different working distances by using different focal-length lenses.

work light A permanent lighting system installed on a motion picture set that provides the illumination needed while setting up lights, moving scenery and, generally, setting the stage before filming. It is usually not adequate for filming in itself.

work print The arrangement of filmed scenes in their proper sequence for continuity and review purposes. Used as a guide during the editing or completion process, the scenes are cut into the work print and synchronized with the accompanying sound track. Optical and transitional effects are indicated on the work print with a grease pencil. These markings are used as a guide for the optical process that follows.

wow Changes in pitch of recorded sound due to variations in the speed of the tape recorder's transport mechanism. See FLUTTER.

Wratten filters Gelatin cemented between glass disks or lacquered gelatin cut into squares that are used for controlling light in black-and-white as well as color photography. These filters are especially useful in outdoor photography where it is necessary to either increase or decrease contrast or alter the normal color balance of a scene. The filter is named after Frederick Charles Luther Wratten, an Englishman who developed several photographic processes and manufactured panchromatic plates and other photographic materials. See Table W-2.

TABLE W-2 Wratten Visual M Filters

Number	Color	Use
78AA	Blue	A photometric filter that can be used to convert the color quality of light from incandescent tungsten lamps of the common type to that which is approximately visually equivalent to daylight. Often employed for viewing colored specimens with their commonly accepted standard daylight appearance.
38A	Blue	A filter for increasing the apparent contrast in faintly stained yellow or orange preparations. Helps in the resolution of fine detail.
45A	Blue-green	Especially useful when the highest resolving power visually possible is required, as in the study of diatom structure. It has no red transmission and its dominant wavelength range is 470–480 $m\mu$.
66	Light green	A contrast filter for use with pink- and red-stained preparations. Preferred by some workers for general use in place of no. 78AA.
58	Green	A contrast filter for use with faintly stained pink or red preparations.
15 22	Deep yellow Yellow-orange	For increasing the contrast in blue preparations and for helping in the observation of detail in insect mounts by reducing the contrast between the preparation and the background.
25	Red	Contrast filter for use with preparations stained with methylene blue, methyl green, etc.
96	Neutral	A filter for reducing the intensity of the illumination. The density supplied (1.0) transmits 10% of the incident light.

Courtesy of Eastman Kodak.

X, Y, Z

X The abbreviation for exposure, or frame of film.

X-dissolve See CROSS-DISSOLVE.

xenon lamps Lamps with quartz glass housings filled with xenon gas having a life expectancy of 1,200 to 2,500 hours. They are characterized by constant color temperature (approximately 5500 to 6000°K), high illumination efficiency, and arc stability.

Xenon lamps (Wollensak) for high-speed motion picture photography consist of a control unit, a lamp housing containing a focusable paraboloid reflector, the xenon lamp, and interconnection cables. The quartz lamp bulb furnishes color temperatures of 6700°K. The small illumination source is designed to operate from a battery supply of 36 to 42 volts. A voltmeter and an ammeter indicate the power being dissipated in the lamp during operation.

Pulsed xenon arc lamps are designed to flash a specified number of times per second, each flash lasting approximately 1/1,000 second. These high-intensity lamps are used in displays as well as for graphic arts, medical, and some high-speed photographic processes. The wattages of these lamps range from 450 to about 8,000.

x-ray A form of electromagnetic energy capable of limited penetration of almost any substance. Although the rays can be used as a "light" source for the exposure of film, they cannot be reflected or refracted. X-rays are used in medical photography, to examine the interior of the body, and also have such industrial applications as the checking of welds, the soundness of castings, and the behavior of turbine blades and engine parts.

All types of film are sensitive to x-rays and can be used to produce visible images. In most cases, however, the radiographs are made on film which has an emulsion layer on each side of the film base. This type of film allows a greater portion of the rays to be absorbed.

X-sheets The abbreviation for the exposure sheets that are used for filming animation. See EXPOSURE SHEETS.

yoke A set of coils mounted around the neck of the cathode-ray tube that produces the electromagnetic deflection of the electron beam, both horizontally and vertically.

zero cut A method of hiding or making splices invisible in A & B roll editing. Each end of every scene that is cut into the checkerboard rolls prepared by the film editor is extended by at least two frames before being spliced to the blank leader that separates one scene from the others. During laboratory printing, the change from one roll to the next will not show splices because with this method there are none at the cut points. See Figure Z-1.

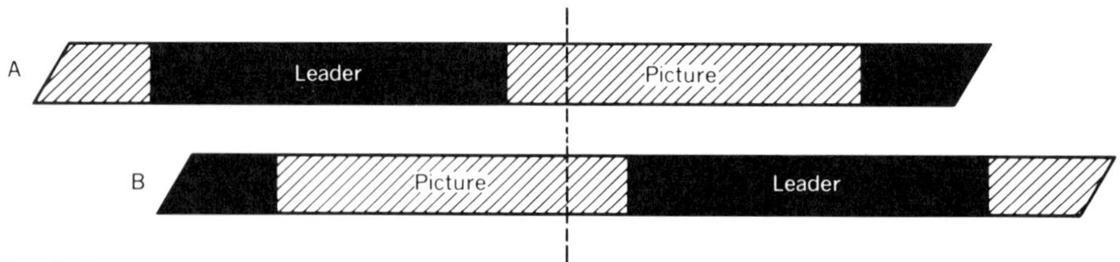

Fig. Z-1 Zero cut.

zip pan Also called **swish pan**. Actually, a transitional effect used for indicating a change in both time and location. The camera, during the transition, seems to pan very quickly from the scene being photographed to the one following it. The effect is a *blur,* during which specific objects or detail are not distinguishable.

zirconium lamp An arc lamp in which a relatively small tube of zirconium oxide is used to produce a concentrated, bright light. At sufficiently high temperatures, the color temperature of the arc produced between the ring-shaped electrodes within the glass housing is approximately 3000°K. The lamp is suitable for projection and enlarging equipment.

zone focusing The process of focusing on an area rather than on a specific plane. This method of focusing is used for photographing actions that may, at some time during the filming process, occur within a particular part of a larger area. Focusing, under these circumstances, is merely a matter of selecting a suitable lens setting that will provide the proper depth of field, and the correct shutter speed that will keep the subject matter in sharp focus at both the near and far limits in which the action is to occur.

zoom Also called **truck or dolly**. A photographic effect in which the camera moves from one position on the set to another in order to change the relationship between the camera and subject matter.

In animation, the effect is achieved by moving the camera from one field position to another, via stop-motion photography, in order to produce a close-up of the subject matter positioned on the compound table of the animation stand. The camera may also be moved away from the subject matter to produce a long shot. Both movements are referred to as zooms. Cams attached to the animation stand keep the subject matter in sharp focus at all points during the zoom movement.

An animated zoom, by comparison, is achieved by the positioning of artwork on the compound table of the animation stand. Produced with a series of drawings or photographs in which the subject matter changes gradually in both size and shape for each frame of the effect, the length of the effect governs the number of drawings or photographs that are required.

A zoom, in animation, is achieved with camera movement. An animated zoom is produced with artwork.

zoom keyer (videotape) A device used for inserting picture information from one camera into an area matted out by various matteing patterns existing on 35mm slides.

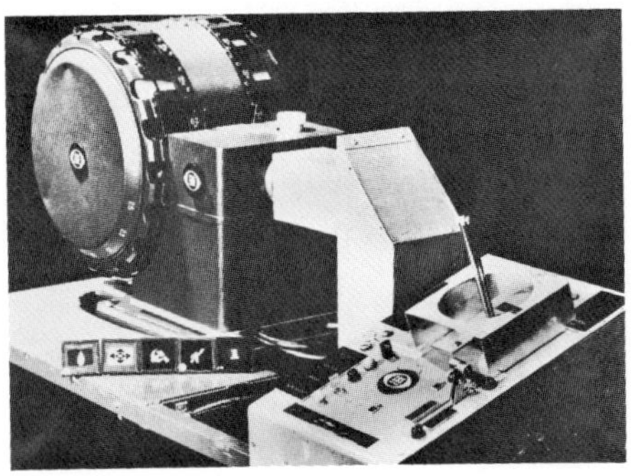

Fig. Z-2 The zoom keyer.

With the zoom keyer, from one to thirty-two 35mm slides may be inserted into the machine. Each slide is a white silhouette on a black field. The silhouette can be any shape, including trademarks, symbols, titles, etc.

Through a system of electronic controls, the operator makes the pattern any size that is desired over a range of 70 to 1. The operator can expand or contract the pattern over this entire range.

By means of a second electronic control the patterns are positioned any place in the picture. A pattern is fed to a standard matteing amplifier and can be matted over any desired background. By using the pattern as a key insert, it is possible to use one camera picture outside the pattern and a second camera picture inside the pattern. See Figure Z-2.

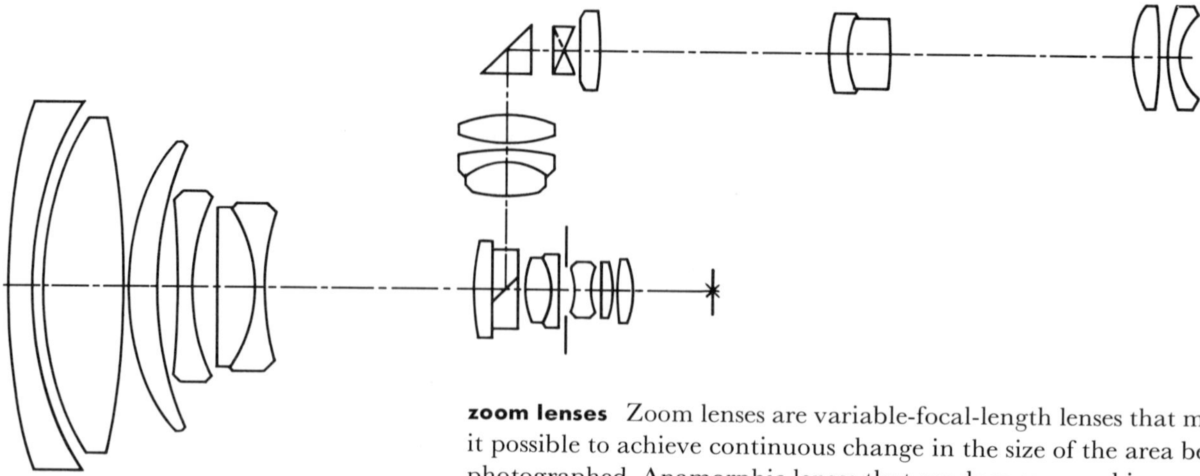

Fig. Z-3 The arrangement of lens elements in a typical zoom lens. Carl Zeiss 7.5-30mm Vario-Sonnar f/1.9 with 4:1 zoom ratio for Zeiss Ikon Movieflex super.

zoom lenses Zoom lenses are variable-focal-length lenses that make it possible to achieve continuous change in the size of the area being photographed. Anamorphic lenses that produce squeezed images for wide-screen motion pictures are used in the same manner as conventional zoom lenses.

Extension tubes cannot be used in conjunction with zoom lenses since accurate focus changes are impossible during zoom movements. With diopter lenses, however, zooms can be made to achieve extreme close-ups of the subject matter being filmed. Diopter lenses, unlike extension tubes, do not necessitate increases in exposure when used with zoom lenses.

Zoom lenses should be focused at the maximum focal length in order to ensure image sharpness at all points during zoom movements. Small motors are often attached to the lens in order to control and provide for even movement during zooms.

Zoom lenses have an advantage over other camera lenses in that they can magnify or change the relative size of the subject matter while the camera itself remains in a stationary position.

Although a wide variety of zoom lenses are available for motion picture filming, the zoom lenses used most frequently for commercial videotape recording are the Taylor-Hobson or the Angenieux-Evershed. The former is an *f*/2, 20–200mm lens. The latter is an *f*/2.2, 18–180mm lens. Each of these models can be used in conjunction with electric motors to achieve smooth zoom movement. See Figure Z-3.

In conclusion . . .

There can be no logical ending to any book that discusses motion picture production and photographic techniques. Necessity is the mother of invention and tomorrow, as in the industry's yesterdays, new processes, techniques, equipment, and terminology will be added to existing cinematographic glossaries. These in turn, will be used as stepping-stones for further experimentation and research.

Certainly, pioneer technicians, with their accumulated credits and experience, show no signs of fatigue or any inclination to rest smugly on their collective laurels. The younger technicians, hard on their heels in the continuing quest for new illusory effects, will not permit a slackening of the pace in the ever-accelerating chase.

Even now, while this book is being edited and the proofs read and checked, stories and releases describing new equipment, film types, processes, and techniques reach our desk daily. Some are still in the experimental stage. Others have already been tested on a limited basis. Until such time as they are practically applied and widely accepted by an industry as critical as ours, their description can only be sketchy and, in most cases, premature.

As one pioneer film producer pointed out recently: "The motion picture industry will always be in its infancy!"

Photo courtesy of Paillard Inc.

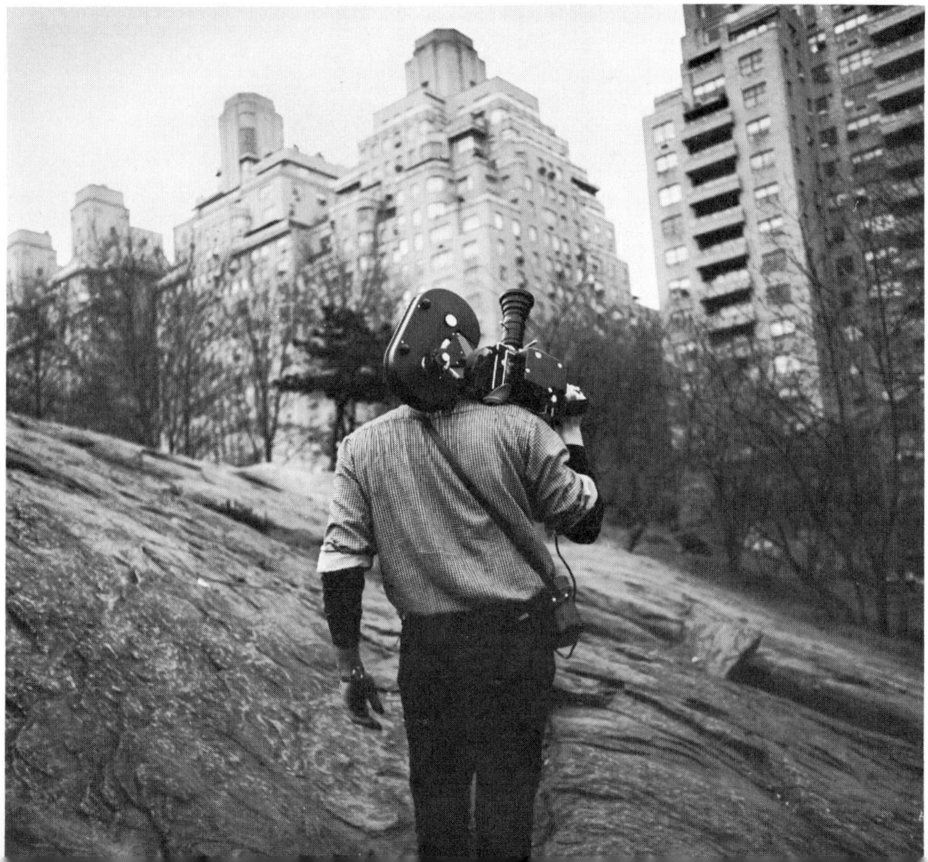

subject guide to entries

Readers interested in each of the following major subjects may wish to review the list of related entries that follow to determine which would be most appropriate for their purposes.

Animation

Aerial Image Photography
Animated Zoom
Animation
  Computerized
  Multiplane Effects
Animation Board
Animation Camera
Animation Stand
Animator
Anticipation
Assistant Animator
Background
Backlight
Bar Sheet
Breakdown
Calibration
Cam
Cel
Cel Level
Character
Checker
Collage
Color Model
Complete Drawing
Compound Table
Continuing Zoom
Control Panel
Crawl Title
Cutout
Cycle
Director
Exposure Sheet
Extremes
Field
Filmstrip
Floating Pegs
Gimmick
Graphics
Guide
Hold
Hook-up
In-betweens
Inker
Layout
Lead Sheet
Lily
Limited Animation
Lip Synchronization
Match
Mechanical Animation
Model Sheet
Newton Rings
Off Register
Ones, Twos, Threes
Opaque
Opaquer
Overlay
Pan Background
Pantograph
Pegs
Pencil Test
Platen
Post synchronization
Presynchronization
Process Cameras
  Acme 35-16mm Reflex
  Oxberry 35-16mm
Reading
Register
Register Pins
Repeat
Rock, Camera
Rotoscope
Scratch-off Effects
Slidefilm
Slidefilm Formats
Slide-motion Film
Spin, Camera
Squash and Stretch
Standard Field
Still Background
Stretch

Aspect Ratio

Camera (See also Camera Types)

Camera Types (See also Camera)

Motion picture cameras

Specialized types

Color

Editing

Exposure

Film

Filters

Lenses

Lighting

Mattes and Matte Processes

Opticals

Photography

Processing

Production

Projection

Sound and Recording

Special Effects

Television

Videotape